CALCIUM ORTHOPHOSPHATES

SERGEY DOROZHKIN

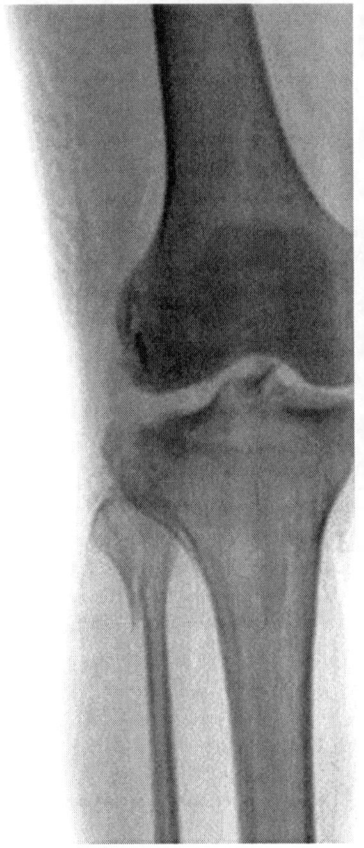

CALCIUM ORTHOPHOSPHATES

Applications in Nature, Biology, and Medicine

PAN STANFORD PUBLISHING

Published by

Pan Stanford Publishing Pte. Ltd.
Penthouse Level, Suntec Tower 3
8 Temasek Boulevard
Singapore 038988

Email: editorial@panstanford.com
Web: www.panstanford.com

British Library Cataloguing-in-Publication Data
A catalogue record for this book is available from the British Library.

Calcium Orthophosphates:
Applications in Nature, Biology, and Medicine

ISBN 978-981-4316-62-0 (Hardcover)
ISBN 978-981-4364-17-1 (eBook)

Printed in the USA

Contents

Preface

Calcium apatites and other calcium orthophosphates have been of considerable interest to mineralogists, chemists, material researchers, biologists, and clinicians for many decades. The reasons for this are clear: Calcium apatites form the mineral component of bones and teeth, while some other calcium orthophosphates are involved in biomineralization process in mammals. Furthermore, calcium orthophosphates are found in pathological calcifications. Therefore, calcium apatites and other calcium orthophosphates appear to be biologically friendly inorganics and, thus, they are increasingly used as implantable biomaterials for various types of bone fillers and bone substitutes. As a final point, calcium apatites and other calcium orthophosphates are widely distributed minerals in Nature, providing the worlds supply of phosphorus, particularly phosphates for the production of fertilizers.

Due to the aforementioned, the scientific databases reveal that the research on calcium orthophosphates has a very long history. However, it exploded in the 1960s, and since then the number of publications permanently increased. Simultaneously, the variety of both investigations and biomedical applications of calcium orthophosphates are greatly expanding. Namely, calcium orthophosphate-based bioceramics specific to the *in vivo* applications have been designed, synthesized, investigated, and applied. Furthermore, new synthetic processes for the fabrication of calcium orthophosphates with the desired properties (such as, the Ca/P ratio, crystallinity, phase composition, particle shape and dimensions, ion-substitutions, etc.) have been also developed. Methods of the structural and surface analysis have also greatly progressed. For example, in early studies, the biological responses of living tissues to implanted materials were evaluated by optical microscopy. Nowadays, the biological analysis is performed at the molecular level in combination with the high-end physical techniques. In addition, long-term clinical data are now available. All these findings give the important suggestions for designing new types of calcium orthophosphate-based formulations for biomedical applications.

Therefore, the aim in writing this monograph has been to provide an integrated account of the present knowledge on preparation, chemical composition, structure, properties, and applications of all available calcium orthophosphates, particularly in the biomedical context. Since the entire subject appears to be very broad (over 30,000 publications on calcium apatites and other calcium orthophosphates have been already published), a great number of references to the related publications detailing various specific aspects of the matter have been collected.

The monograph consists of eight chapters. The division arrangement of the chapters is generally based on the subject, with subdivisions on the major aspects, such as introduction, basic definitions and knowledge, structure, preparation, properties, biomedical application, and future directions, ended up by conclusions. This overall scheme is used to emphasize the mutual interrelationships among various calcium orthophosphates. The main purpose has been to group the material in the most natural way and, if appropriate, to provide cross-references from other sections. This is sometimes done explicitly and sometimes by giving section references, where other aspects of the specific subject are discussed. Namely, Chapter 1 contains the general information on all available calcium orthophosphates, including their geological and biological occurrence, chemical composition, structure, solubility, and a brief information on their location in calcified tissues of mammals (bones, teeth, and antlers), including the unwanted (pathological) calcifications. Furthermore, Chapter 1 also encloses an important section on biomimetic crystallization, including artificial simulating solutions. Chapter 2 is devoted to the structure, chemical composition, properties, and biomedical application of amorphous calcium orthophosphates (ACPs), which erroneously are considered as individual compound, whereas in reality, they are just an amorphous (or a very poorly crystalline) state of other calcium orthophosphates. Chapter 3 focuses on the similar topics of nanodimensional and nanocrystalline calcium orthophosphates. The following three chapters are mainly devoted to various aspects of the biomedical applications of calcium orthophosphates. Namely, Chapter 4 describes the available knowledge on bioceramics, Chapter 5 concentrates on the self-setting formulations, and Chapter 6 is devoted to biocomposites and hybrid biomaterials. Chapter 7 appears to be more specific because it is devoted to a narrower subject of the dissolution mechanism of both hydroxyapatite (HA) and fluorapatite (FA) in acids. This subject provides the chemical fundamentals for both dental caries and osteoporosis. Finally, Chapter 8 gives a brief historical overview on the calcium orthophosphate matter, which is just interesting to read.

To conclude, this monograph represents the author's vision on the topic, which by no means is ideal. Furthermore, since not each and every possible aspect of calcium apatites and other calcium orthophosphates has been described, various imperfections are possible. Thus, any criticism, opinions, or suggestions are always welcome. However, it is worth mentioning that the main goal is not only further development of the subject itself but also making a possible contribution to the welfare of human beings, in particular those with diseases potentially treatable by calcium orthophosphates.

Finally, I would like to acknowledge the continuous encouragement of my mother, Tamara, my wife, Elena, and my son, Denis. Hopefully, publication of this book will help the author find a suitable position in science or industry to provide a better financial support to the beloved members of his family.

April 2012

Sergey V. Dorozhkin
Moscow, Russia
sedorozhkin@yandex.ru

Chapter 1

Calcium Apatites and Other Calcium Orthophosphates

1.1 Introduction

Due to abundance in nature and presence in living organisms, calcium apatites [1] and other calcium orthophosphates remain the chemical compounds of a special interest in many fields of science, including geology, chemistry, biology, and medicine. Because of big problems with accessing to the scientific literature published in the nineteenth century and before, a historical description of the subject appears to be both fragmental and incomplete. Namely, according to the accessible literature [2], as early as in the end of the eighteenth century, a French chemist Joseph-Louis Proust (1754–1826) and a German chemist Martin Klaproth (1743–1817) proposed that calcium apatite was the major inorganic component of bones. In the middle of the nineteenth century, attempts to establish the chemical composition of calcium apatites and other calcium orthophosphates were performed by J. Berzelius [3], R. Warington Jr. [4] and R. Fresenius [5]. The chemical formula of perfectly transparent crystals of natural fluorapatite (FA) as $Ca_5(PO_4)_3F$ was established in 1873 [6], while the crystallographic faces of a natural calcium apatite were described in 1883 [7]. Furthermore, a paper on a behavior of an undisclosed calcium orthophosphate in organisms of carnivores was published in 1883 [8]. Further, the quantitative analysis of a calcium orthophosphate was performed in 1884 [9], followed by remarks by

Calcium Orthophosphates: Applications in Nature, Biology, and Medicine
Sergey Dorozhkin

ISBN 978-981-4316-62-0 (Hardcover), 978-981-4364-17-1 (eBook)
www.panstanford.com

C. Glaser in 1885 [10]. In 1880s, occurrence of a calcium apatite [11] and calcium orthophosphate [12–14] in a metallurgical slag was discovered. Chemical reactions between calcium orthophosphates and other chemicals were investigated as early as in 1891 [15]. Research papers on bone repairing are known since, at least, 1892 [16], while the earliest well-documented systematic studies of calcium orthophosphates were performed at the beginning of the twentieth century by F. K. Cameron with co-workers [17–21] and H. Bassett [22–25]. The majority of the aforementioned researchers already operated with individual chemical compounds. Further, historical details might be found in Chapter 8.

By definition, all calcium orthophosphates consist of three major chemical elements: calcium (oxidation state +2), phosphorus (oxidation state +5) and oxygen (reduction state −2), as a part of orthophosphate anions. These three chemical elements are present in abundance on the surface of our planet: oxygen is the most widespread chemical element of the earth's surface (~47 mass%), calcium occupies the fifth place (~3.3–3.4 mass%) and phosphorus (~0.08–0.12 mass%) is among the first twenty of the chemical elements most widespread on our planet [26]. In addition, the chemical composition of many calcium orthophosphates includes hydrogen, as an acidic orthophosphate anion (for example, HPO_4^{2-} or $H_2PO_4^-$), hydroxide (for example, $Ca_{10}(PO_4)_6(OH)_2$) and/or incorporated water (for example, $CaHPO_4{\cdot}2H_2O$). Diverse combinations of CaO and P_2O_5 (both in the presence of water and without it) provide a large variety of calcium phosphates, which are distinguished by the type of the phosphate anion: ortho-(PO_4^{3-}), meta-(PO_3^-), pyro-($P_2O_7^{4-}$), and poly-($(PO_3)_n^{n-}$). In the case of multi-charged anions (orthophosphates and pyrophosphates), calcium phosphates are also differentiated by the number of hydrogen ions attached to the anion. Examples include mono-($Ca(H_2PO_4)_2$), di-($CaHPO_4$), tri-($Ca_3(PO_4)_2$) and tetra-($Ca_2P_2O_7$) calcium phosphates [27–29] (here, prefixes “mono”, “di”, “tri” and “tetra” are related to the amount of hydrogen ions replaced by calcium). However, only calcium orthophosphates are considered and discussed in this book. They are listed in Table 1.1 [30, 31]. Since all of them belong to calcium orthophosphates, strictly speaking, all abbreviations in Table 1.1 are incorrect; however, they are extensively used in literature and there is no need to modify them.

The atomic arrangement of calcium orthophosphates is built-up around a network of orthophosphate (PO_4) groups, which gives stability to the entire structure. The majority of calcium orthophosphates are sparingly soluble in water; however, all of them are easily soluble in acids but insoluble in alkaline solutions. All chemically pure calcium orthophosphates are crystals of white color and moderate hardness. However, natural minerals of calcium orthophosphates are always colored due to impurities, the most widespread of which are ions of Fe, Mn and rare earth elements [32, 33]. Biologically formed calcium orthophosphates are the major component of all mammalian calcified tissues [34], while the natural ones are the major raw material to produce phosphorus-containing fertilizers [35–38].

1.2 Geological and Biological Occurrences

Geologically, natural calcium orthophosphates are found in different regions mostly as deposits of apatites (belong to igneous rocks), mainly as natural FA or phosphorites (a sedimentary rock) [36–39]. Some types of sedimentary rocks can be formed by weathering of igneous rocks into smaller particles [40]. Other types of sedimentary rocks can be composed of minerals precipitated from the dissolution products of igneous rocks or minerals produced by biomineralization (Fig. 1.1) [41]. Thus, due to a sedimentary origin, both a general appearance and a chemical composition of natural phosphorites vary a lot [42, 43]. It is a common practice to consider francolite (or carbonate-hydroxyfluorapatite regarded as its synonym) as the basic phosphorite mineral [39, 44–48]. A cryptocrystalline (almost amorphous) variety of francolite (partly of a biological origin) is called collophane (synonyms: collophanit, collophanita, collophanite, grodnolite, kollophan) [49–51]. It occurs in natural phosphorites predominantly as fossil bones and phosphatized microbial pseudomorphs: phosphatic crusts of chasmolithic biofilms (or microstromatolites) and globular clusters with intra-particular porosities [52–55]. Natural phosphorites (therefore, francolite and collophane as well) occur in various forms, such as nodules, crystals or masses. Occasionally, other types of natural calcium orthophosphates are found as minerals, for example clinohydroxylapatite [56], staffelite (synonyms: staffelit,

Table 1.1 Existing calcium orthophosphates and their major properties [30, 31]

Ca/P molar ratio	Compound	FORMULA	Solubility at 25°C, $-\log(K_s)$	Solubility at 25°C, g/L	pH stability range in aqueous solutions at 25°C
0.5	Monocalcium phosphate monohydrate (MCPM)	$Ca(H_2PO_4)_2 \cdot H_2O$	1.14	~18	0.0–2.0
0.5	Monocalcium phosphate anhydrous (MCPA or MCP)	$Ca(H_2PO_4)_2$	1.14	~17	[c]
1.0	Dicalcium phosphate dihydrate (DCPD), mineral brushite	$CaHPO_4 \cdot 2H_2O$	6.59	~0.088	2.0–6.0
1.0	Dicalcium phosphate anhydrous (DCPA or DCP), mineral monetite	$CaHPO_4$	6.90	~0.048	[c]
1.33	Octacalcium phosphate (OCP)	$Ca_8(HPO_4)_2(PO_4)_4 \cdot 5H_2O$	96.6	~0.0081	5.5–7.0
1.5	α-Tricalcium phosphate (α-TCP)	α-$Ca_3(PO_4)_2$	25.5	~0.0025	[a]
1.5	β-Tricalcium phosphate (β-TCP)	β-$Ca_3(PO_4)_2$	28.9	~0.0005	[a]
1.2–2.2	Amorphous calcium phosphates (ACP)	$Ca_XH_Y(PO_4)_Z \cdot NH_2O$, N = 3–4.5; 15–20% H_2O	[b]	[b]	~5–12 [d]

1.5–1.67	Calcium-deficient hydroxyapatite (CDHA or Ca-def HA)[e]	$Ca_{10-X}(HPO_4)_X(PO_4)_{6-X}(OH)_{2-X}$ $(0<X<1)$	~85	~0.0094	6.5–9.5
1.67	Hydroxyapatite (HA, HAp or OHAp)	$Ca_{10}(PO_4)_6(OH)_2$	116.8	~0.0003	9.5–12
1.67	Fluorapatite (FA or FAp)	$Ca_{10}(PO_4)_6F_2$	120.0	~0.0002	7–12
1.67	Oxyapatite (OA, OAp or OXA)[f]	$Ca_{10}(PO_4)_6O$	~69	~0.087	[a]
2.0	Tetracalcium phosphate (TTCP or TetCP), mineral hilgenstockite	$Ca_4(PO_4)_2O$	38 – 44	~0.0007	[a]

[a] These compounds cannot be precipitated from aqueous solutions.

[b] Cannot be measured precisely. However, the following values were found: 25.7±0.1 (pH = 7.40), 29.9±0.1 (pH = 6.00), 32.7±0.1 (pH = 5.28) [236]. The comparative extent of dissolution in acidic buffer is ACP >> α-TCP >> β-TCP > CDHA >> HA > FA [107].

[c] Stable at temperatures above 100°C.

[d] Always metastable.

[e] Occasionally, it is called "precipitated HA (PHA)".

[f] Existence of OA remains questionable.

staffelita) belonging to carbonate-rich fluorapatites (chemical formula: $Ca_5[(F,O)(PO_4,CO_3)_3]$) [5, 57] and DCPD [58]. Furthermore, calcium orthophosphates were found in meteoric stones [59]. The world deposits of natural calcium orthophosphates are estimated to exceed 150 billion tons; from which approximately 85% belong to phosphorites and the remaining ~15% belong to apatites [39].

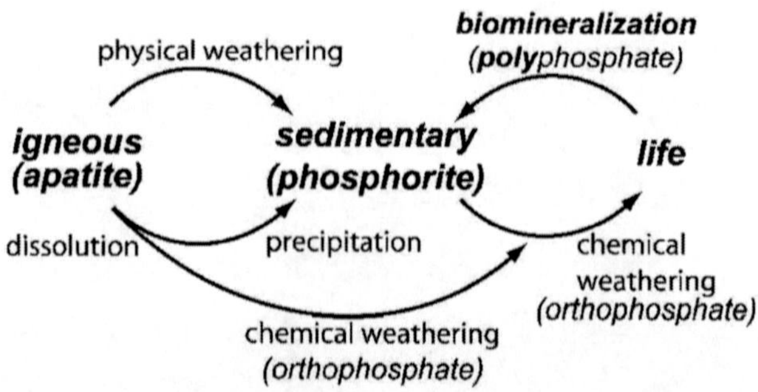

Figure 1.1 Simplified schematic of the phosphorus cycle from apatitic igneous rock to phosphorite sedimentary rock through chemical or physical weathering. Life forms accumulate soluble phosphorus species and can produce apatite through biomineralization. Reprinted from Ref. [41] with permission.

Natural calcium orthophosphates occur in most geological environments usually as accessory minerals (<5%). Concentrations sufficient for economic use (>15%) are also available. The largest world deposits of natural apatites are located in Russia (the Khibiny and Kovdor massifs, Kola peninsula [60, 61]), Brazil and Zambia, while the largest world deposits of natural phosphorites are located in Morocco, Russia, Kazakhstan, USA (Florida, Tennessee), China and Australia, as well as in the oceans [35–39]. Most of natural calcium orthophosphates occur as small polycrystalline structures (spherulitic clusters). Larger crystals are rare [62]. They usually have the crystal structure of apatites (hexagonal system, space group $P6_3/m$). Giant crystals including "a solid but irregular mass of green crystalline apatite, 15 ft. long and 9 ft. wide [63] and a single euhedral crystal from the Aetna mine measuring 2.1 × 1.2 m with an estimated weight of 6 tons [64] were found. None of them is a pure compound; they always contain admixtures of other elements. For example, ions of calcium might be partially replaced by Sr, Ba, Mg, Mn, K, Na, Fe; ions of orthophosphate may be partly replaced by

AsO_4^{3-}, CO_3^{2-} and VO_4^{3-} [65]; ions of hydroxide, chloride, bromide, carbonate and oxide may to a certain extent substitute fluoride in the crystal lattice of natural apatites [47]. Furthermore, various organic radicals have been found in natural apatites [66, 67]. In principle, the crystal structure of apatites can incorporate half the Periodic Chart in its atomic arrangement. In medicine, this property might be used as an antidote for heavy metal intoxication [68]. Ease of atomic substitution for apatite leaves this mineral open to a wide array of compositions. This might be related to the fact that the apatite structure type displays porous properties [69]. The substitutions in apatites are usually in trace concentrations but large concentrations and even complete solid solutions exist for certain substituents (e.g., F^- and OH^-). To make things even more complicated, some ions in the crystal structure may be missing, leaving the crystallographic defects, which leads to formation of non-stoichiometric compounds. Figure 1.2 shows examples of polycrystalline and single-crystalline samples of natural FA.

Figure 1.2 Polycrystalline (left) and single-crystalline (right) FA of a geological origin. The single crystal has a grey-green color due to incorporated ions of transition metals.

Manufacturing of elementary phosphorus (white and red) [70, 71], phosphoric acids [36, 72–75], various phosphorus-containing chemicals and, especially, agricultural fertilizers (namely, superphosphate [76–78], ammonium orthophosphates [79] is the major industrial application of natural calcium orthophosphates.

The annual consumption of a phosphate rock has approached ~150 million tons and about 95 percent of this production is utilized in fertilizer industry [80, 81].

In biological systems, many organisms, ranging from bacteria and isolated cells to invertebrates and vertebrates, synthesize calcium orthophosphates [41]. Formation of calcium orthophosphates in primitive organisms is believed to enable the storage and regulation of essential elements such as calcium, phosphorus and, possibly, magnesium. The morphology of precipitates in these organisms (small intracellular nodules of ACP often located in mitochondria) complies with the necessities for rapid mobilization and intracellular control of the concentration of these elements [82]. In vertebrates calcium orthophosphates occur as the principal inorganic constituent of normal (bones, teeth, fish enameloid, deer antlers and some species of shells) and pathological (dental and urinary calculus and stones, atherosclerotic lesions, etc.) calcifications [27, 83–88]. Except for small portions of the inner ear, all hard tissues of the human body are formed of calcium orthophosphates. Structurally, they occur mainly in the form of poorly crystalline, non-stoichiometric, calcium-deficient, Na-, Mg- and carbonate-containing HA (often called "biological apatite" [89–93] (which might be abbreviated as BAp [94, 95]), bioapatite [96–99] or dahllite [100]). The main constituents of human bones are calcium orthophosphates (~60–70 wt.%), collagen [104] (~20–30 wt.%) and water (up to 10 wt.%) [87, 96–98, 105–107]. Detailed information on the chemical composition of the most important human normal calcified tissues is comprised in Table 1.2. One should note that the values mentioned in Table 1.2 are approximate; the main constituents can vary by a percent or more [108].

1.3 The Members of Calcium Orthophosphate Family

In the ternary aqueous system $Ca(OH)_2$–H_3PO_4–H_2O (or CaO–P_2O_5–H_2O) [109–111] there are twelve known non-ion-substituted calcium orthophosphates with the Ca/P molar ratio within 0.5 and 2.0 (Table 1.1). An anhydrous phase diagram of CaO–P_2O_5 at temperatures within 200–2200°C is shown in Fig. 1.3 [112, 113]. Table 1.3 comprises crystallographic data of the existing calcium orthophosphates [28, 114–116]. The most important parameters of calcium orthophosphates are the ionic Ca/P ratio, basicity/acidity and solubility. All these parameters strongly correlate with the solution pH. The lower the Ca/P molar ratio is, the more acidic and water-soluble the calcium orthophosphate is [27–29]. One can see that the solubility ranges from high values for acidic compounds, such as MCPM, to very low values for basic compounds, such as apatites, which allow calcium orthophosphates to be dissolved, transported from one place to another, and precipitated, when necessary. Crystallization, dissolution, and phase transformation processes of different calcium orthophosphates under various experimental conditions have been reviewed recently [117].

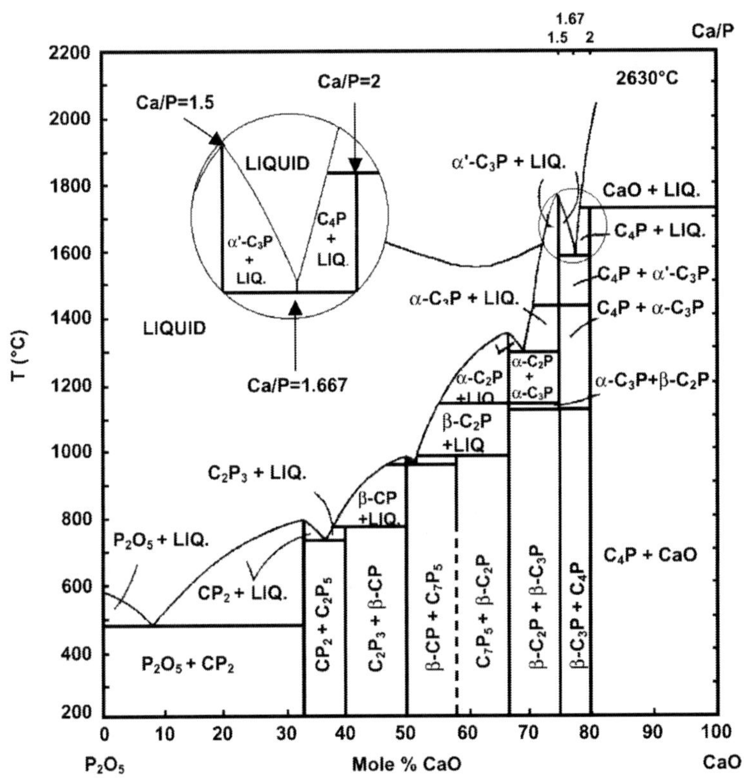

Figure 1.3 Phase diagram of the system $CaO–P_2O_5$ (C = CaO, P = P_2O_5) at elevated temperatures. Here: C_7P_5 means $7CaO·5P_2O_5$; other abbreviations should be written out in the same manner. Reprinted from Refs. [112, 113] with permission.

Table 1.2 Comparative composition and structural parameters of inorganic phases of adult human calcified tissues. Due to the considerable variation found in biological samples, typical values are given in these cases [27, 107]

Composition, wt%	Enamel	Dentine	Cementum	Bone	HA
Calcium[a]	36.5	35.1	~35	34.8	39.6
Phosphorus (as P)[a]	17.7	16.9	~16	15.2	18.5
Ca/P (molar ratio)[a]	1.63	1.61	~1.65	1.71	1.67
Sodium[a]	0.5	0.6	[c]	0.9	—
Magnesium[a]	0.44	1.23	0.5–0.9	0.72	—
Potassium[a]	0.08	0.05	[c]	0.03	—
Carbonate (as CO_3^{2-})[b]	3.5	5.6	[c]	7.4	—
Fluoride[a]	0.01	0.06	up to 0.9	0.03	—
Chloride[a]	0.30	0.01	[c]	0.13	—
Pyrophosphate (as $P_2O_7^{4-}$)[b]	0.022	0.10	[c]	0.07	—
Total inorganic[b]	97	70	60	65	100

Total organic[b]	1.5	20	25	25	—
Water[b]	1.5	10	15	10	—
Crystallographic properties: Lattice parameters (± 0.003 Å)					
a-axis, Å	9.441	9.421	[c]	9.41	9.430
c-axis, Å	6.880	6.887	[c]	6.89	6.891
Crystallinity index (HA = 100)	70–75	33–37	~30	33–37	100
Typical crystal sizes (nm) [454, 544, 546]	100 μm × 50 × 50	35 × 25 × 4	[c]	50 × 25 × 4	200–600
Ignition products (800°C)	β-TCP + HA	β-TCP + HA	β-TCP + HA	HA + CaO	HA
Elastic modulus (GPa)	80	23.8 ± 3.7	15.0 ± 3.6	0.34–13.8	10
Tensile strength (MPa)	10	100	[c]	150	100

[a] Ashed samples.

[b] Unashed samples.

[c] Numerical values were not found in the literature but they should be similar to those for dentine.

Table 1.3 Crystallographic data of calcium orthophosphates [28, 114, 115]

Compound	Space group	Unit cell parameters	Z[a]	Density, g cm^{-3}
MCPM	triclinic $P\overline{1}$	a = 5.6261(5), b = 11.889(2), c = 6.4731(8) Å, α = 98.633(6)°, β = 118.262(6)°, γ = 83.344(6)°	2	2.23
MCPA	triclinic $P\overline{1}$	a = 7.5577(5), b = 8.2531(6), c = 5.5504(3) Å, α = 109.87(1)°, β = 93.68(1)°, γ = 109.15(1)°	2	2.58
DCPD	monoclinic Ia	a = 5.812(2), b = 15.180(3), c = 6.239(2) Å, β = 116.42(3)°	4	2.32
DCPA	triclinic $P\overline{1}$	a = 6.910(1), b = 6.627(2), c = 6.998(2) Å, α = 96.34(2)°, β = 103.82(2)°, γ = 88.33(2)°	4	2.89
OCP	triclinic $P\overline{1}$	a = 19.692(4), b = 9.523(2), c = 6.835(2) Å, α = 90.15(2)°, β = 92.54(2)°, γ = 108.65(1)°	1	2.61
α-TCP	monoclinic $P2_1$/a	a = 12.887(2), b = 27.280(4), c = 15.219(2) Å, β = 126.20(1)°	24	2.86

β-TCP	rhombohedral $R3cH$	$a = b = 10.4183(5)$, $c = 37.3464(23)$ Å, $\gamma = 120°$	21[b]	3.08
HA	monoclinic $P2_1/b$ or hexagonal $P6_3/m$	$a = 9.84214(8)$, $b = 2a$, $c = 6.8814(7)$ Å, $\gamma = 120°$ (monoclinic) $a = b = 9.4302(5)$, $c = 6.8911(2)$ Å, $\gamma = 120°$ (hexagonal)	4 2	3.16
FA	hexagonal $P6_3/m$	$a = b = 9.367$, $c = 6.884$ Å, $\gamma = 120°$	2	3.20
OA	hexagonal $P\overline{6}$	$a = b = 9.432$, $c = 6.881$ Å, $\alpha = 90.3°$, $\beta = 90.0°$, $\gamma = 119.9°$	1	~3.2
TTCP	monoclinic $P2_1$	$a = 7.023(1)$, $b = 11.986(4)$, $c = 9.473(2)$ Å, $\beta = 90.90(1)°$	4	3.05

[a] Number of formula units per unit cell.

[b] Per the hexagonal unit cell.

Due to the triprotic equilibrium that exists within orthophosphate-containing solutions, variations in pH alter the relative concentrations of the four polymorphs of orthophosphoric acid (Fig. 1.4) [118] and thus both the chemical composition (Fig. 1.5) [119] and the amount of the calcium orthophosphates that are formed by a direct precipitation. The solubility isotherms of different calcium orthophosphates are shown in Fig. 1.6 [28, 29, 110, 111, 120–123]. However, recently, the classic solubility data of calcium orthophosphates [28, 29, 110, 111, 120–123] were mentioned to be inappropriate [124]. According to the authors of the latter study, all previous solubility calculations were based on simplifications, which are only crudely approximate. The problem lies in incongruent dissolution, leading to phase transformations and lack of the detailed solution equilibria. Using an absolute solid-titration approach, the true solubility isotherm of HA was found to lie substantially lower than previously reported. In addition, contrary to a wide belief, DCPD appeared not to be the most stable phase below pH ~4.2, where CDHA is less soluble [124].

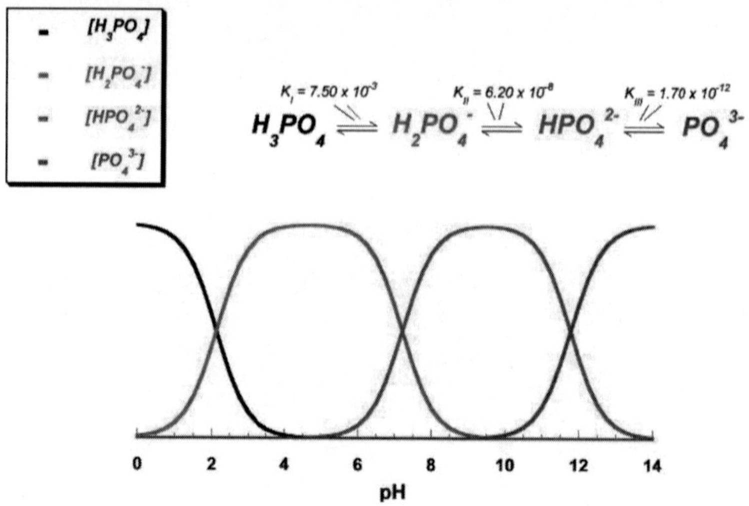

Figure 1.4 pH variation of ionic concentrations in triprotic equilibrium for phosphoric acid solutions. Reprinted from Ref. [118] with permission.

A brief description of all known calcium orthophosphates is given in Table 1.1.

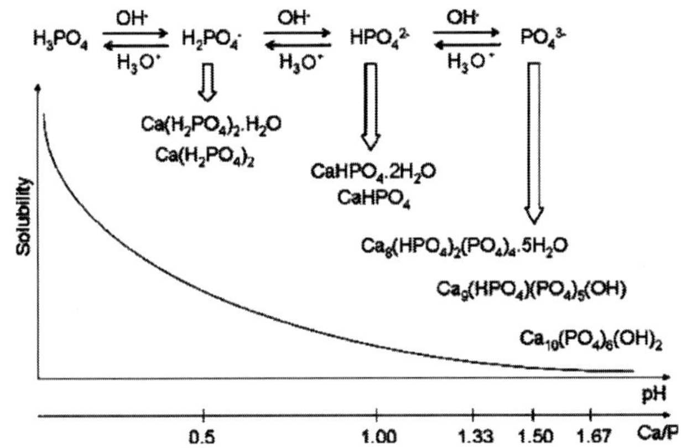

Figure 1.5 Various calcium orthophosphates obtained by neutralizing of orthophosphoric acid. Ca/P are reported in the figure. The solubility of calcium orthophosphates in water decreases drastically from left to right, HA being the most insoluble and stable phase. Reprinted from Ref. [119] with permission.

1.3.1 MCPM

Monocalcium phosphate monohydrate ($Ca(H_2PO_4)_2 \cdot H_2O$); the chemically correct name is calcium dihydrogen orthophosphate monohydrate) is both the most acidic and water-soluble compound. It precipitates from highly acidic solutions that are normally used in industry of phosphorus-containing fertilizer production ("triple superphosphate") [36]. Besides, MCPM might be fabricated by a simple precipitation method using $CaCO_3$ and H_3PO_4 in aqueous and acetone media at ambient temperature [125]. At temperatures above ~100°C, it releases a molecule of water and transforms into MCPA. Due to high acidity and solubility, MCPM is never found in biological calcifications. Moreover, pure MCPM is not biocompatible [126] with bones [129]. However, in medicine MCPM is used as a component of several self-hardening calcium orthophosphate cements [130–133]. In addition, MCPM is used as a nutrient, acidulant and mineral supplement for dry baking powders, food, feed and some beverages [134, 135]. Coupled with $NaHCO_3$, MCPM is used as a leavening agent for both dry baking powders and bakery dough. MCPM might be added to salt-curing preserves, pickled and marinated foods. According to the European classification of food additives, MCPM is marked as E341 additive. Occasionally, MCPM is added to tooth

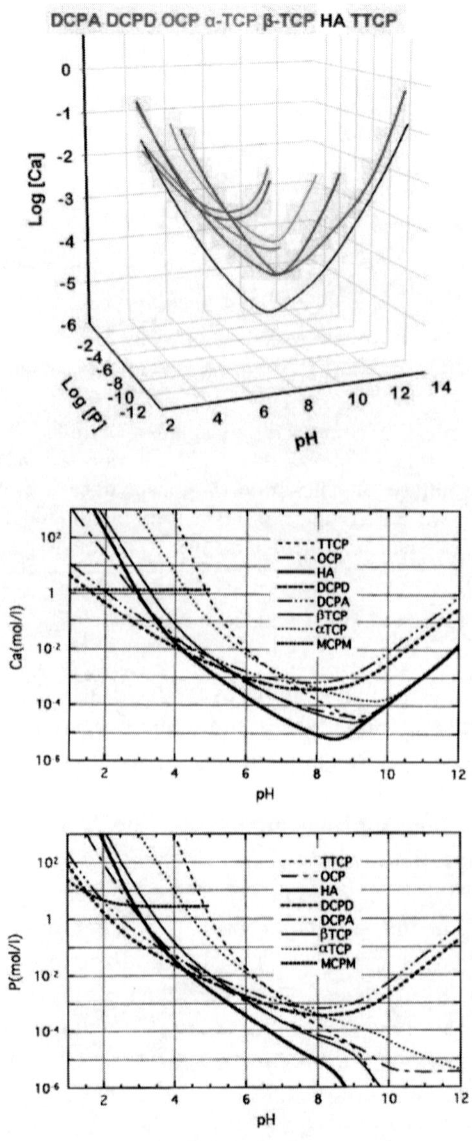

Figure 1.6 Top: a 3D version of the classical solubility phase diagrams for the ternary system $Ca(OH)_2$–H_3PO_4–H_2O. Reprinted from Ref. [122] with permission. Middle and bottom: solubility phase diagrams in 2D graphs, showing two logarithms of the concentrations of (a) calcium and (b) orthophosphate ions as a function of the pH in solutions saturated with various salts. Reprinted from Ref. [123] with permission.

pastes. Besides, MCPM might be added to ceramics and glasses, while agriculture is the main consumer of a technical grade MCPM, where it is used as a fertilizer [36, 134].

1.3.2 MCPA (or MCP)

Monocalcium phosphate anhydrous ($Ca(H_2PO_4)_2$; the chemically correct name is calcium dihydrogen orthophosphate anhydrous) is the anhydrous form of MCPM. It crystallizes under the same conditions as MCPM but at temperatures above ~100°C (e.g., from highly concentrated mother liquors during fertilizer production). Like MCPM, MCPA never appears in calcified tissues and is not biocompatible due to its acidity. There is no current application of MCPA in medicine. Due to the similarity with MCPM, in many cases, MCPA might be used instead of MCPM [36, 134]; however, highly hydroscopic properties of MCPA reduce its commercial application.

1.3.3 DCPD

Dicalcium phosphate dihydrate ($CaHPO_4 \cdot 2H_2O$; the chemically correct name is calcium hydrogen orthophosphate dihydrate; the mineral brushite [136]) can be easily crystallized from aqueous solutions at $\sim 2.0 < pH < \sim 6.5$. Interestingly, but precipitation of DCPD by mixing a $Ca(OH)_2$ suspension and a H_3PO_4 solution in the equimolar quantities was found to occur in five stages, being HA the first precipitated phase [137, 138]. Besides, DCPD might be prepared in gels [139, 140]. DCPD transforms into DCPA at temperatures above ~80°C and this transformation is accompanied by ~11% increase in volume [141] and structural changes [142]. The value for $\Delta_r G^0$ for DCPD → DCPA transformation is −1.032 kJ/mol [142]. Briefly, DCPD crystals consist of $CaPO_4$ chains arranged parallel to each other, while lattice water molecules are interlayered between them. Using surface X-ray diffraction, Arsic et al. determined the atomic structure of the {010} interface of DCPD with water [143, 144]. Since DCPD contains water layers as part of its crystal structure, special ordering properties at the interface are expected. This interface consists of two water bilayers with different ordering properties. The first is highly ordered and can be considered as part of the DCPD crystal structure. Surprisingly, the second water bilayer exhibits no in-plane order but shows only layering in the perpendicular direction. It has been proposed that the low level of water ordering at the interface is correlated with the low solubility of DCPD in water [144]. Recently, data on DCPD solubility have been updated [145]. The optical properties of DCPD are well described in Ref. [146], while many additional data on DCPD, as well as a good picture of DCPD atomic structure are available in literature [147].

DCPD is of biological importance because it is often found in pathological calcifications (dental calculi, crystalluria, chondrocalcinosis, and urinary stones) and some carious lesions [27, 83–85]. It has been proposed as an intermediate in both bone mineralization and dissolution of enamel in acids (dental erosion) [27, 83, 84]. In medicine, DCPD is used in calcium orthophosphate cements [131, 148–151] and as an intermediate for tooth remineralization. DCPD is added to toothpaste both for caries protection (in this case, it is coupled with F-containing compounds such as NaF and/or Na_2PO_3F) and as a gentle polishing agent [152–156]. Other applications include a flame retardant [157], a slow release fertilizer, using in glass production, as well as calcium supplement in food, feed and cereals [134]. The importance of DCPD as a constituent of infant's food was discovered as early as in 1917 [158]. In food industry, it serves as a texturizer, bakery improver and water retention additive. In diary industry, DCPD is used as a mineral supplement. If added to food products, DCPD should be marked as E341 according to the European classification of food additives. In addition, plate-like crystals of DCPD might be used as a non-toxic, anticorrosive, and passivating pigment for some ground coat paints.

1.3.4 DCPA or DCP

Dicalcium phosphate anhydrous ($CaHPO_4$; the chemically correct name is calcium hydrogen orthophosphate anhydrate; the mineral monetite [159]) is the anhydrous form of DCPD. It is less soluble than DCPD due to the absence of water inclusions. Like DCPD, DCPA can be crystallized from aqueous solutions but at temperatures ~100°C. Furthermore, it might be prepared at room temperature in gels [139], ethanol [160], as well as in oil-in-water and water-in-oil systems [161]. DCPA is physically stable and resisted hydration even when dispersed in water for over 7 months in the temperature range of 4–50°C [162]. A calcium-deficient DCPA was prepared recently. It might be sintered at ~300°C [163]. Unlike DCPD, DCPA occurs in neither normal nor pathological calcifications. It is used in calcium orthophosphate cements [150, 164–171]. Besides, DCPA might be implanted [172]. Other applications include using as a polishing agent, a source of calcium and phosphate in nutritional supplements (e.g., in prepared breakfast cereals, enriched flour and noodle products), a tabletting aid [173] and a toothpaste component [134]. In addition, it is used as a dough conditioner in food industry.

1.3.5 OCP

Octacalcium phosphate ($Ca_8(HPO_4)_2(PO_4)_4 \cdot 5H_2O$; the chemically correct name is octacalcium bis(hydrogenphosphate) tetrakis(phosphate) pentahydrate) is often found as an unstable transient intermediate during the precipitation of the thermodynamically more stable calcium orthophosphates (e.g., CDHA) in aqueous solutions. Its preparation technique might be found elsewhere [174–179]. A partially hydrolyzed form of OCP with Ca/P molar ratio of 1.37 might be prepared as well [179, 180]. The full hydrolysis of OCP into CDHA occurs within ~6 hours [178]. Furthermore, OCP might be non-stoichiometric and be either Ca-deficient (Ca/P = 1.26) or include excessive calcium (up to Ca/P = 1.48) in the structure [179]. Ion-substituted OCP might be prepared as well [181]. Crystals of OCP are typically small, extremely platy and almost invariably twinned.

The triclinic structure of OCP displays apatitic layers (with atomic arrangements of calcium and orthophosphate ions similar to those of HA) separated by hydrated layers (with atomic arrangements of calcium and orthophosphate ions similar to those in DCPD) [27–29, 182, 183]. A similarity in crystal structure between OCP and HA [184, 185] is one reason that the epitaxial growth of these phases is observed. Morphologically, OCP crystallizes as {100} blades of triclinic pinacoidal symmetry, elongated along the *a*-axis and bordered by the forms {010}, {001} and {011}. It is generally assumed that, in solutions, the hydrated layer of the (100) face is the layer most likely exposed to solution. The water content of OCP crystals is about 20% that of DCPD and this is partly responsible for its lower solubility. New data on OCP solubility have been published recently [186].

OCP is of a great biological importance because it is one of the stable components of human dental and urinary calculi [187–190].

OCP was first proposed by W. E. Brown to participate as the initial phase in enamel mineral formation and bone formation through subsequent precipitation and stepwise hydrolysis of OCP [184, 185, 191]. It plays an important role in *in vivo* formation of apatitic biominerals. A "central OCP inclusion" (also known as "central dark line") is seen by transmission electron microscopy in many biological apatites and in some synthetically precipitated HA [192–196]. Although OCP has not been observed in vascular calcifications, it has been strongly suggested as a precursor phase to biological apatite found in natural and prosthetic heart valves [197, 198]. In surgery, OCP is used for implantation into bone defects [199–205]. For the comprehensive information on OCP, the readers are referred to other reviews [179, 189].

1.3.6 β-TCP

β-tricalcium phosphate (β-$Ca_3(PO_4)_2$; the chemically correct names are calcium orthophosphate tribasic beta or tricalcium bis(orthophosphate) beta) cannot be precipitated from aqueous solutions. It is a high temperature phase, which only can be prepared at temperatures above 800°C by thermal decomposition of CDHA or by solid-state interaction of acidic calcium orthophosphates, e.g., DCPA, with a base, e.g., CaO. Apart from the chemical preparation routes, ion-substituted β-TCP can be prepared by calcining of bones [206]: such type of β-TCP is occasionally called "bone ash". In β-TCP, there are three types of crystallographically nonequivalent PO_4^{3-} groups located at general points of the crystal, each type with different intratetrahedral bond lengths and angles. At temperatures above ~1125°C, β-TCP is transformed into a high-temperature phase α-TCP. Being the stable phase at room temperature, β-TCP is less soluble in water than α-TCP (Table 1.1). Furthermore, the ideal β-TCP structure contains calcium ion vacancies that are too small to accommodate calcium ions but allow for the inclusion of magnesium ions, which thereby stabilize the structures [207, 208]. Both ion-substituted [209–212] and organically modified [213–215] forms of β-TCP can be synthesized as well. The maximum substitution of Mg^{2+} in β-TCP was found to correspond to the $Ca_{2.61}(Mg(1)_{0.28},Mg(2)_{0.11})(PO_4)_2$ stoichiometric formula [212]. The modern structural data on β-TCP are available in Refs. [216–218], those on Vicker's and Knoop microhardness studies might be found if Ref. [219], while solubility data in Ref. [220]. Furthermore, an ability of β-TCP to store an electrical charge by electrical polarization was studied and this material was found to have a suitable composition and structure for both ion conduction and charge storage [221].

Pure β-TCP never occurs in biological calcifications. Only the Mg-substituted form called whitlockite [222] (β-TCMP—β-tricalcium magnesium phosphate, β-$(Ca,Mg)_3(PO_4)_2$) is found in dental calculi and urinary stones, dentineal caries, salivary stones, arthritic cartilage, as well as in some soft-tissue deposits [27,83–85,229–232]. However, it has not been observed in enamel, dentine, or bone. In biomedicine, β-TCP is used in calcium orthophosphate bone cements [233–238] and other types of bone substitution bioceramics [206, 239–246]. Dental applications of β-TCP are also known [247]. Pure β-TCP is added to some brands of toothpaste as a gentle polishing agent. Multivitamin complexes with calcium orthophosphate are widely available in the market and β-TCP is used as the calcium phosphate there. In addition, β-TCP serves as a texturizer, bakery improver and anti-clumping agent for dry powdered food (flour, milk powder, dried cream, cocoa powder). Besides, β-TCP is added as a dietary or mineral supplement to food and feed, where it is marked as E341 according to the European classification of food additives. A prenatal development of rats during gestation was found to be sensitive to E341 (TCP) exposure [248]. There is a good review on the toxicological properties of inorganic phosphates, where the interested readers are referred to [249]. Occasionally, β-TCP might be used as inert filler in pelleted drugs. Other applications comprise porcelains, pottery, enamel, using as a component for mordants and ackey, as well as a polymer stabilizer [134]. β-TCP of a technical grade (as either calcined natural phosphorites or bone dust) is used as a slow release fertilizer for acidic soils [36].

1.3.7 α-TCP

α-tricalcium phosphate (α-$Ca_3(PO_4)_2$; the chemically correct names are calcium orthophosphate tribasic alpha or tricalcium bis(orthophosphate) alpha) is usually prepared from β-TCP at heating above ~1125°C [250] and it might be considered as a high temperature phase of β-TCP. However, at the turn of the millennium, the previously forgotten data that the presence of silicates stabilized α-TCP at lower temperatures of 800–1000°C [251] have been rediscovered again. Such type of α-TCP is called "silicon stabilized α-TCP" [252–257].

Although α-TCP and β-TCP have exactly the same chemical composition, they differ by the crystal structure (Table 1.3) and solubility (Table 1.1). In addition, β-TCP is more stable than the α-phase [217]. Therefore, of them, α-TCP is more reactive in aqueous systems, has a higher specific energy and it can be hydrolyzed to CDHA [258–260]. Milling was found to increase the α-TCP reactivity even more [261]. Although, α-TCP never occurs in biological calcifications, in medicine, it is used as a component of calcium orthophosphate cements [131, 148–151, 166–168, 262–265]. On the other hand, the chemically pure α-TCP has received not much interest in the biomedical field [244]. The disadvantage for using α-TCP is its quick resorption rate, which limits its application in this area. However, the silicon stabilized α-TCP (more precisely as a biphasic composite with HA) has been commercialized as a starting material to produce bioresorbable porous ceramic scaffolds to be used as artificial bone grafts [239, 252–256]. Theoretical insights into bone grafting properties of the silicon-stabilized α-TCP might be found in Ref. [266]. The structure of α-TCP is well described in literature [217, 218, 267], while the surface and adsorption properties are available in Ref. [268]. Similar to β-TCP, α-TCP of a technical grade might be used slow release fertilizer for acidic soils [134].

1.3.8 ACP

Amorphous calcium phosphates (ACPs) represent a special class of calcium orthophosphate salts, having variable chemical but rather identical glass-like physical properties, in which there are neither translational nor orientational long-range orders (LRO) of the atomic positions. Until recently [269], ACP has been considered as an individual chemical compound; however, this is just an amorphous state of other calcium orthophosphates. Therefore, in principle, all compounds mentioned in Table 1.1 might be somehow fabricated in an amorphous state but, currently, only few of them (e.g., an amorphous TCP) are known [269]. Thus, strictly speaking, ACP should be excluded from Table 1.1.

Depending on the production temperatures, ACPs are divided into two major groups: low-temperature ACPs (prepared in aqueous solutions) and high-temperature ACPs. Low-temperature ACPs (described by the chemical formula $Ca_xH_y(PO_4)_z{\cdot}nH_2O$, n = 3–4.5; 15–20% H_2O) are often encountered as a transient precursor phase during precipitation of other calcium orthophosphates in aqueous systems. Usually, an ACP is the first phase precipitated from a supersaturated solution prepared by rapid mixing of solutions containing ions of calcium and orthophosphate [28, 270–275]; however, other production techniques are known. ACPs are thought to be formed at the beginning of the precipitation due to a lower surface energy than that of OCP and apatites [271]. The amorphization degree of ACPs increases with the concentration increasing of Ca- and PO_4-containing solutions, as well as at a high solution pH and a low crystallization temperature. A continuous gentle agitation of as precipitated ACPs in the mother solution, especially at elevated temperatures, results in a slow recrystallization and formation of better crystalline calcium orthophosphates, such as CDHA [27, 28]. The lifetime of ACPs in aqueous solution was reported to be a function of the presence of additive molecules and ions, pH, ionic strength and temperature. Thus, ACPs may persist for appreciable periods and retain the amporphous state under some specific experimental conditions [276]. The chemical composition of ACPs strongly depends on the solution pH and the concentrations of mixing solutions. For example, ACPs with Ca/P ratios in the range of 1.18 (precipitated at solution pH = 6.6) to 1.53 (precipitated at solution pH = 11.7) [28, 277] and even to 2.5 [27, 83, 84] have been described. The presence of poly(ethylene glycol) [278], ions of pyrophosphate, carbonate and/or magnesium in solution during the crystallization promotes formation of ACPs and slows down their further transformation into more crystalline calcium orthophosphates, while the presence of fluoride has the opposite effect [27–29, 107, 279]. The solution-mediated transformation of an ACP to CDHA, which can be described by a "first-order" rate law, is a function only of the solution pH and depends upon the experimental conditions which regulate both the dissolution of ACP and the formation of early HA nuclei [280].

High-temperature ACPs might be prepared using high energy processing at elevated temperatures. This method is based on a rapid quenching of melted calcium orthophosphates occurring, e.g., during plasma spraying of HA [281–283]. A plasma jet, possessing very high temperatures (5000–20000°C), partly decomposes HA. That results in formation of a complicated mixture of products, some of which would be ACPs. Obviously, all types of high-temperature ACPs are definitively anhydrous contrary to the precipitated ACPs. Unfortunately, no adequate chemical formula is available to describe the high-temperature ACPs.

In general, as all amorphous compounds are characterized by a lack of LRO, it is problematic to discuss the structure of ACPs (they are X-ray amorphous). Concerning a short-range order (SRO) in ACPs, it exists, just due to the nature of chemical bonds. Unfortunately, in many cases, the SRO in ACPs is uncertain either, because it depends on many variables, such as Ca/P ratio, preparation conditions, storage, admixtures, etc. It is well known that freshly precipitated ACPs contain 10–20% by weight of tightly bound water, which is removed by vacuum drying at elevated temperature [284]. Infrared spectra of ACPs show broad featureless phosphate absorption bands. Electron microscopy of freshly precipitated ACPs usually shows featureless nearly spherical particles with diameters in the range of 20 to 200 nm. However, there is a questionable opinion that ACPs might have an apatitic structure but with a crystal size so small, that they are X-ray amorphous. This is supported by X-ray absorption spectroscopic data (EXAFS) on biogenic and synthetic samples [285–288]. On the other hand, it was proposed that the basic structural unit of the precipitated ACPs is a 9.5 Å diameter, roughly spherical cluster of ions with the composition of $Ca_9(PO_4)_6$ (Fig. 1.7) [28, 277, 289, 290]. These clusters were found experimentally as first nuclei during the crystallization of CDHA and a model was developed to describe the crystallization of HA as a step-wise assembly of these units [291] (see section 1.3.10. *HA (or HAp, or OHAp)* below). Biologically, ion-substituted ACPs (always containing ions of Na, Mg, carbonate and pyrophosphate) are found in soft-tissue pathological calcifications (e.g., heart valve calcifications of uremic patients) [27, 83–85].

In medicine, pure ACPs are used in calcium orthophosphate cements [148–150] and as a filling material in dentistry [269]. Bioactive composites of ACPs with polymers have properties suitable for use in dentistry [292–295] and surgery [296–299]. Due to a reasonable solubility and physiological pH of aqueous solutions, ACP appeared to be consumable by some microorganisms and, due to this reason, it might be added as a mineral supplement to culture media. Non-biomedical applications of ACPs comprise their using as a component for mordants and ackey. In food industry, ACPs are used for syrup clearing. Occasionally, they might be used as inert filler in pelleted drugs. In addition, ACPs are used in glass and pottery production and as a raw material for production of some organic phosphates. To get further details on ACPs, the readers are referred to both special reviews [269, 290, 300, 301] and Chapter 2 of this book.

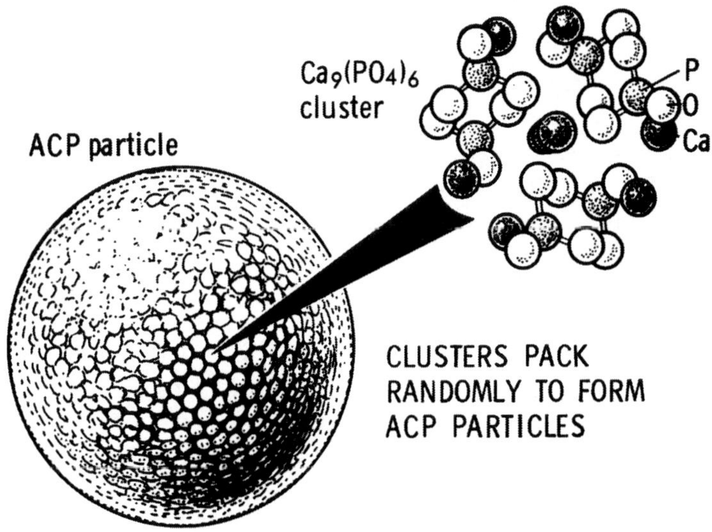

Figure 1.7 A model of ACP structure. Reprinted from Ref. [289] with permission.

1.3.9 CDHA (or Ca-def HA)

Calcium-deficient hydroxyapatite ($Ca_{10-x}(HPO_4)_x(PO_4)_{6-x}(OH)_{2-x}$ ($0 < x < 1$)) can be easily prepared by simultaneous addition of calcium- and orthophosphate-containing solutions into boiling water followed by boiling the suspension for several hours (an ageing stage). That is why in literature it might be called as "precipitated HA (PHA)" [302, 303]. Besides, it might be prepared by hydrolysis of α-TCP [258–260]. Other preparation techniques of CDHA are known as well [304–306]. During ageing, initially precipitated ACPs are restructured and transformed into CDHA [307]. Therefore, there are many similarities in the structure, properties and application between the precipitated in alkaline solutions (pH > 8) ACPs and CDHA. Recent data indicated on presence of intermediate phases during further hydrolysis of CDHA to a more stable HA-like phase [311]. CDHA crystals are poorly crystalline and of submicron dimensions. They have a very large specific surface area, typically 25–100 m^2/g. On heating above ~700°C, dry CDHA with Ca/P = 1.5 will convert to β-TCP and that with 1.5 < Ca/P < 1.67 will convert into a biphasic composite of HA and β-TCP (see section 1.3.14 Biphasic (BCP) and triphasic calcium (ortho)phosphates below) [312–323]. A reasonable solid-state mechanism of a high-temperature transformation of CDHA into BCP has been proposed [324, 325].

The variability in Ca/P molar ratio of CDHA has been explained through different models: surface adsorption, lattice substitution and intercrystalline mixtures of HA and OCP [326]. Due to a lack of stoichiometry, CDHA usually contains other ions [82]. The extent depends on the counter-ions of the chemicals used for preparation (e.g., Na^+, Cl^-). Direct determinations of the CDHA structures are still missing and the unit cell parameters remain uncertain. However, unlike that in ACPs (see section *1.3.8. ACP* above), a LRO exists in CDHA. The following lattice parameters were reported for formate (HCO_2^-) containing CDHA with Ca/P = 1.596 (ionic): $a = 9.4729(20)$ and $c = 6.8855(9)$ Å. A loss of Ca^{2+} ions happened exclusively from Ca(2) sites, while the PO_4 tetrahedron volume and P–O bonds were ~4.4% and ~1.4% smaller, respectively, than those in HA [327].

A systematic study of defect constellations in CDHA is available in literature [328]. As a first approximation, CDHA may be considered as HA with some ions missing [329]. The more calcium is deficient, the more disorder and imperfections are in CDHA structure [330]. Furthermore, a direct correlation between Ca deficiency and the mechanical properties of the crystals was found: calcium deficiency leads to an 80% reduction in the hardness and elastic modulus and at least a 75% reduction in toughness in plate-shaped HA crystals [331]. According to the chemical formula of CDHA (Table 1.1), there are vacancies of Ca^{2+} (mainly on Ca(2) sites) and OH^- ions in crystal structure of this compound [327, 329–334]. However, due to Ca^{2+} vacancies in CDHA, the resulting negative charge might be compensated by protonation of both an OH^- ion within the deficient calcium-triangle and a PO_4^{3-} ion in the nearest neighborhood of the vacant calcium site. This results in the presence of some water in CDHA structure: $Ca_{10-x}(HPO_4)_x(PO_4)_{6-x}(OH)_{2-x}(H_2O)_x$ $(0 < x < 1)$ [328]. According to this approach, there are no hydroxide vacancies in CDHA, just a portion of OH^- ions are substituted by water molecules. Concerning possible vacancies of orthophosphate ions, nothing is known about their presence in CDHA. It is considered that a portion of PO_4^{3-} ions is either protonated (as HPO_4^{2-}) or substituted by other ions (e.g., CO_3^{2-}) [335]. Theoretical investigations of the defect formation mechanism relevant to non-stoichiometry in CDHA are available elsewhere [336].

Unsubstituted CDHA (i.e., that containing ions of Ca^{2+}, PO_4^{3-}, HPO_4^{2-} and OH^- only) does not exist in biological systems. However, the ion substituted CDHA: Na^+, K^+, Mg^{2+}, Sr^{2+} for Ca^{2+}; CO_3^{2-} for PO_4^{3-} or HPO_4^{2-}; F^-, Cl^-, CO_3^{2-} for OH^-, plus some water forms biological apatite — the main inorganic part of animal and human normal and pathological calcifications [27, 82, 83]. Therefore, CDHA is a very promising compound for industrial manufacturing of artificial bone substitutes [337], including drug delivery applications [338]. Non-biomedical applications of CDHA are similar to those of ACP and HA. Interestingly, but CDHA was found to possess a catalytic activity to produce biogasoline [339].

1.3.10 HA (or HAp, or OHAp)

Hydroxyapatite [340] ($Ca_5(PO_4)_3(OH)$, but is usually written as $Ca_{10}(PO_4)_6(OH)_2$ to denote that the crystal unit cell comprises two molecules; the chemically correct name is pentacalcium hydroxide tris(phosphate)) is the second most stable and least soluble calcium orthophosphate after FA. Chemically pure HA crystallizes in the monoclinic space group $P2_1/b$ [341]. However, at temperatures above ~250°C, there is a monoclinic to hexagonal phase transition in HA (space group $P6_3/m$) [28, 115, 277, 342, 343]. The detailed description of the HA structure was first reported in 1964 [344] and its interpretation in terms of aggregation of $Ca_9(PO_4)_6$ clusters, the so-called Posner's clusters, has been widely used since publication of the article by Posner and Betts [284]. The $Ca_9(PO_4)_6$ clusters appeared to be energetically favored in comparison to alternative candidates including $Ca_3(PO_4)_2$ and $Ca_6(PO_4)_4$ clusters [345]. In hexagonal HA, the hydroxide ions are more disordered within each row, when compared with the monoclinic form, pointing either upward or downward in the structure. This induces strains that are compensated for by substitutions or ion vacancies. Some impurities, like partial substitution of hydroxide by fluoride or chloride, stabilize the hexagonal structure of HA at ambient temperature. Due to this reason, hexagonal HA is seldom the stoichiometric phase and very rare single crystals of natural HA always exhibit the hexagonal space group. The crystal structure of HA is well described elsewhere [28, 114–116], the detailed analysis of the electronic structure, bonding, charge transfer, optical and elastic properties are also available [346–350], while the readers interested in Posner's clusters are referred to other papers [345, 351–353]. A shell model was developed to study the lattice dynamics of HA [354], while a cluster growth model was created to illustrate its growth [291]. Polarization characteristics [355, 356] and pyroelectrical properties [357] of HA bioceramics have been investigated. First-principles calculations for the elastic properties of doped HA [358] and vacancy formation in HA [359] were performed. Computer simulations of the structures and properties of HA are well described in recent feature articles [360, 361].

Several techniques might be utilized for HA preparation; they can be divided into solid-state reactions and wet methods [362], which include precipitation, hydrothermal synthesis and hydrolysis of other calcium orthophosphates. Even under the ideal stoichiometric conditions, the precipitates are generally non-stoichiometric, suggesting intermediate formation of precursor phases, such as ACP and CDHA. HA can be prepared in aqueous solutions by mixing exactly stoichiometric quantities of Ca- and PO_4-containing solutions at pH > 9, followed by boiling for several days in CO_2-free atmosphere (the ageing or maturation stage), filtration, drying and, usually, sintering at about 1000°C [363]. As the first precipitates are rich in non-apatitic environments (see ACP and CDHA), the ageing stage appears to be very important: the Ca/P molar ratio of 1.67 was found to attain in as little as 5 hours after the completion of the reaction at 90°C [364]. The surface of freshly precipitated HA is composed of a structured hydrated layer containing easily exchangeable mobile ionic species [365]. Usually unsintered HA is poorly crystalline and often non-stoichiometric, resembling the aforementioned CDHA. However, well crystalline HA can be prepared from an aqueous solution [366]. Microcrystalline samples of HA can also be prepared by solid-state reaction of other calcium phosphates (e.g., MCPM, DCPA, DCPD, OCP) with CaO, $Ca(OH)_2$, or $CaCO_3$ at temperatures above ~1200°C in an atmosphere of equal volumes of water and nitrogen. HA can be prepared by hydrothermal synthesis [28, 277, 367, 368]. A water-free synthesis can be performed in ethanol from $Ca(OEt)_2$ (Et = ethyl) and H_3PO_4 [369, 370]. In addition, HA might be prepared by mechanochemical synthesis of a dry mixture of CaO and DCPD [362, 371] or from coral skeletal carbonate by hydrothermal exchange [372–374]. Relatively large single crystals of HA might be prepared from those of chlorapatite [375] or by recently developed controlled homogeneous precipitation method [376]. Lower sized particles of HA might be prepared by a pyrosol technique, where an aerosol, containing calcium and orthophosphate ions in the adequate ratio, is transported to a furnace where the pyrolisis takes place [377]. Synthesis of nano-sized HA has also been described [378, 379], while the chronological development of nano-sized HA synthesis might be found in another paper [380]. Two-dimensional nanocrystalline HA might be also synthesized [381]. Space-grown and terrestrial HA crystals were found to differ in size: the former appeared to be at least 1–1.5 orders of magnitude bigger in length [382, 383]. Transparent HA ceramics is also known [384–387]. The detailed information on HA synthesis is available elsewhere [388–396]. In addition, there are good reviews on HA solubility, crystal growth and intermediate phases of HA crystallization [397], as well as on HA dissolution (Ref. [398] and Chapter 7 of this book).

Pure HA never occurs in biological systems. However, due to the chemical similarities to bone and teeth mineral (Table 1.2), HA is widely used as a coating on orthopedic (e.g., hip joint prosthesis) and dental implants [399–406]. HA particles might be implanted as well [407]. Due to a great similarity to biological apatite, over a long time HA has been used in liquid chromatography of nucleic acids, proteins and other biological compounds [408–417] and for drug delivery purposes [418–421]. Also, HA is added to some brands of toothpaste as a gentle polishing agent instead of calcium carbonate [422, 423]. Non-biomedical applications of HA include its using as an environmentally friendly filler for elastomers [424], a sorbent of poisonous chemical elements [425, 426] and a carrier for various catalysts [427–429]. Furthermore, HA by itself might act as a catalyst for formaldehyde combustion at room temperature [430]. To conclude this topic, one should mention on other reviews devoted to HA and its biomedical applications [431–435].

1.3.11 FA (or FAp)

Fluorapatite ($Ca_5(PO_4)_3F$, but is usually written as $Ca_{10}(PO_4)_6F_2$ to denote that the crystal unit cell comprises two molecules; the chemically correct name is pentacalcium fluoride tris(phosphate)) is the only ion-substituted calcium orthophosphate, considered in this review. It is the hardest (5 according to the Mohs' scale of mineral hardness), most stable, and least soluble compound among all calcium orthophosphates (Table 1.1). Perhaps, such "extreme" properties of FA are related to the specific position of F^- ions in the center of Ca(2) triangles of the crystal structure [115]. Due to its properties, FA is the only calcium orthophosphate that naturally forms large deposits suitable for the commercial use [35–38] (see also Fig. 1.2). Preparation techniques of the chemically pure FA are similar to the aforementioned ones for HA but the synthesis must be performed in presence of the necessary amount of F^- ions (usually, NaF or NH_4F is added). Unlike that for HA (see CDHA), no data are available on existence of calcium-deficient FA. Under some special crystallization conditions (e.g., in presence of gelatin or citric acid), FA might form unusual dumbbell-like fractal morphology that finally are closed to spheres (Fig. 1.8) [436–442]. A hierarchical structure for FA was proposed [443]. The crystal structure of FA for the first time was studied in 1930 [444, 445] and is well described elsewhere [28, 114–116, 446]. The detailed analysis of the electronic structure, bonding, charge transfer and optical properties is available as well [348]. In addition, there are reviews on FA solubility [397] and the dissolution mechanism (Ref. [398] and Chapter 7 of this book).

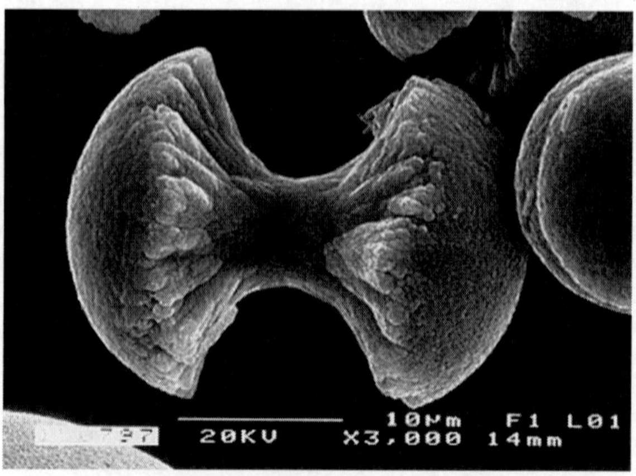

Figure 1.8 A biomimetically grown aggregate of FA that was crystallized in a gelatin matrix. Its shape can be explained and simulated by a fractal growth mechanism. Scale bar: 10 μm. Reprinted from Ref. [437] with permission.

FA easily forms solid solutions with HA with any desired F/OH molar ratio. Such compounds are called fluorhydroxyapatites (FHA) or hydroxyfluorapatites (HFA) and described with a chemical formula $Ca_{10}(PO_4)_6(OH)_{2-x}F_x$, where $0 < x < 2$. If the F/OH ratio is either uncertain or not important, the chemical formula of FHA and HFA is often written as $Ca_{10}(PO_4)_6(F,OH)_2$. The lattice parameters, crystal structure, solubility and other properties of FHA and HFA lay in between of those for the chemically pure FA and HA [447–451].

Similar to pure HA, pure FA never occurs in biological systems. Obviously, a lack of the necessary amount of toxic fluorides (the acute toxic dose of fluoride is ~5 mg/kg of body weight) in living organisms is the main reason of this fact (pure FA contains 3.7% mass. F). Enameloid of shark teeth [107, 452–457] and some exoskeletons of mollusks [458] seem to be the only exclusions because they contain substantial amounts of FA. Among all normal calcified tissues of humans, the highest concentration of fluorides is found in bones and the lowest — in dental enamel [459]. However, even in bones, the total amount of fluorides is not enough to form FA; it is generally considered that the inorganic part of bones consists of ion-substituted CDHA. Due to the lowest solubility, good chemical stability and toxicity of high amounts of fluorides, chemically pure FA is rarely used as a bone substituting material [466]. However, various FA-containing composites [467–469], FHA [470, 471] and porous FA bioceramics [472] seem to be better candidates for biomedical applications. Furthermore, due to the ability to form FHA and/or HFA, minor amounts of fluorides might be intentionally added to calcium orthophosphate biomaterials [473–479]. The effect of fluoride contents in FHA on both osteoblast behavior [480, 481] and leukemia cells proliferation [482] has been described. Non-biomedical applications of FA include its application as a catalyst [483].

1.3.12 OA (or OAp, or OXA)

Oxyapatite ($Ca_{10}(PO_4)_6O$; the chemically correct name is decacalcium oxide hexakis(phosphate)) is the least studied calcium orthophosphate. To the best of my knowledge, pure OA has never been prepared; therefore, its properties are not well established. Furthermore, still there are doubts that OA exists. However, a mixture of OA and HA (oxy-HA) might be prepared by dehydration of HA at temperatures exceeding ~900°C (e.g., during plasma-spray of HA) only in the absence of water vapor [28, 29, 484, 485]. It also might be crystallized in glass-ceramics [486]. Computer modeling techniques have been employed to qualitatively and quantitatively investigate the dehydration of HA to OA [487]. OA has the hexagonal space group symmetry $P\overline{6}$ (174) of cesanite type [114], while the space group symmetry for partially dehydrated HA was found to change from hexagonal P63/m to triclinic $P\overline{1}$ when more than *ca.* 35% of the structurally bound water had been removed [485]. OA has no stability field in aqueous conditions [488]; it is very reactive and transforms to HA in contact with water vapor [484]. Due to the aforementioned problems with OA preparation, no information on biomedical applications of pure OA is available. Plasma-sprayed coatings of HA, in which OA might be present as an admixture phase, seems to be the only application.

1.3.13 TTCP (or TetCP)

Tetracalcium phosphate or tetracalcium orthophosphate monoxide ($Ca_4(PO_4)_2O$; the chemically correct name is tetracalcium oxide bis(phosphate); the mineral hilgenstockite [489]) is the most basic calcium orthophosphate. However, its solubility in water is higher than that of HA (Table 1.1). TTCP cannot be precipitated from aqueous solutions. It can be prepared only by a solid-state reaction at temperatures above 1300°C, e.g., by heating homogenized equimolar quantities of DCPA and $CaCO_3$ in dry air, or in a flow of dry nitrogen [28, 277, 490, 491]. These reactions should be carried out in a dry atmosphere, in vacuum or with rapid cooling (to prevent uptake of water and formation of HA). Easily DCPA might be replaced by ammonium orthophosphates [492, 493], while calcium carbonate might be replaced by calcium acetate [493]. Furthermore, TTCP often appears as an unwanted by-product in plasma-sprayed HA coatings, where it is formed as a result of the thermal decomposition of HA to a mixture of high-temperature phases of α-TCP, TTCP and CaO [494]. TTCP is metastable: in both wet environment and aqueous solutions it slowly hydrolyses to HA and calcium hydroxide [28, 277, 495]. Consequently, TTCP is never found in biological calcifications. In medicine, TTCP is widely used for preparation of various self-setting calcium orthophosphate cements [120, 132, 148, 164, 170, 171, 263, 494, 496]; however, to the best of my knowledge, there is no commercial bone-substituting product consisting solely of TTCP. For the comprehensive information on TTCP, the readers are referred to a recent review [494], while the structure [497], spectra [498] and solubility [220] of TTCP are well described elsewhere.

There is an opinion [115, 189], that all calcium orthophosphates listed in Table 1.1 might be classified into three major structural types: (i) the apatite type, $Ca_{10}(PO_4)_6X_2$, which includes HA, FA, OA, CDHA, OCP and TTCP; (ii) the glaserite type, named after the mineral glaserite, $K_3Na(SO_4)_2$, which includes all polymorphs of TCP and, perhaps, ACP; (iii) the Ca–PO_4 sheet-containing compounds, which include DCPD, DCPA, MCPM and MCPA. According to the authors, a closer examination of the structures revealed that all available calcium orthophosphates could be included into distorted glaserite type structures, but with varying degrees of distortion [115, 189].

1.3.14 Multiphasic and Polyphasic Calcium Orthophosphate Compositions

Calcium orthophosphates might form multiphasic and/or polyphasic compositions, in which the individual components cannot be separated from each other. Presumably, the individual phases of such compositions are homogeneously "mixed" at a far submicron level (< 0.1 μm) and strongly integrated with each other. Nevertheless, the presence of all individual phases is easily seen by X-ray diffraction technique.

The main idea of the multiphasic concept is determined by a balance of more stable calcium orthophosphate phases (e.g., HA) and more soluble calcium orthophosphate phases (e.g., TCP). The usual way to prepare multiphasic calcium orthophosphates consist of sintering of non-stoichiometric calcium orthophosphates, such as ACP and CDHA, at temperatures above ~700°C. Furthermore, a thermal decomposition of the stoichiometric calcium orthophosphates at temperatures above ~1300°C might be used as well [499, 500]; however, this approach often results in formation of complicated mixtures of various products including admixtures of CaO, calcium pyrophosphates, etc. Namely, transformation of HA into polyphasic calcium orthophosphates by annealing in a vacuum occurs as this: the outer part of HA is transformed into α-TCP and TTCP, while the α-TCP phase of the surface further transforms into CaO. Besides, in the boundary phase, HA is transformed into TTCP [499].

Historically, Nery and Lynch with co-workers first used the term biphasic calcium phosphate (BCP) in 1986 to describe a bioceramic, that consisted of a mixture of HA and β-TCP [237]. Based on the results of X-ray diffraction analysis, these authors found that the "tricalcium phosphate" preparation material used in their early publication [238] was in fact a mixture of ~20% HA and ~80% β-TCP. Currently, only biphasic and triphasic calcium orthophosphate formulations are known; perhaps, more complicated formulations would be manufactured in future. Furthermore, nowadays, multiphasic and/or polyphasic compositions consisting of high-temperature phases of calcium orthophosphates, such as α-TCP, β-TCP, HA and, perhaps, high-temperature ACP, OA and TTCP, are known only. No information on multiphasic compositions, containing MCPM, MCPA, DCPD, DCPA, low-temperature ACP, OCP and CDHA has been found in literature. Perhaps, such formulations will be produced in future.

All BCP formulations might be subdivided into two major groups: those consisting of calcium orthophosphates having either the same (e.g., α-TCP and β-TCP) or different (e.g., β-TCP and HA) molar Ca/P ratios. Among all known BCP formulations, BCP consisting of HA and β-TCP is both the most known and the best investigated [312–323]. In 1986, LeGeros in USA and Daculsi in France initiated the basic studies on preparation of this type of BCP and its *in vitro* properties. This material is soluble and gradually dissolves in the body, seeding new bone formation as it releases calcium and orthophosphate ions into the biological medium. Presently, commercial BCP products of different or similar HA/β-TCP ratios are manufactured in many parts of the world (see partly overlapping Tables 4.2, 5.2 and 6.4) as bone-graft or bone substitute materials for orthopaedic and dental applications under various trade marks and several manufacturers [322]. A similar combination of α-TCP with HA forms BCP as well [252, 253, 255, 501–504].

Recently the concept of BCP has been extended by preparation and characterization of biphasic TCP (BTCP), consisting of α-TCP and β-TCP phases [505–509]. The biphasic TCP is usually prepared by heating ACP precursors [507–509], in which the α-TCP/β-TCP ratio can be controlled by aging time and pH value during synthesis of the amorphous precursor [508]. Furthermore, triphasic formulations, consisting of HA, α-TCP and β-TCP [510] or HA, α-TCP and TTCP [499, 500] have been prepared.

It is important to recognize, that the major biomedical properties (such as bioactivity, bioresorbability, osteoconductivity and osteoinductivity) of the multiphasic and/or polyphasic compositions might be adjusted by changing the ratio among the calcium orthophosphate phases. When compared to both α- and β-TCP, HA is a more stable phase under the physiological conditions, as it has a lower solubility (Table 1.1) and, thus, slower resorption kinetics. Therefore, due to a higher biodegradability of the α- or β-TCP component, the reactivity of BCP increases with the TCP/HA ratio increasing. Thus, *in vivo* bioresorbability of BCP can be adjusted through the phase composition. Similar conclusions are also valid for both the biphasic TCP (in which α-TCP is a more soluble phase) and the triphasic (HA, α-TCP and β-TCP) formulation.

A phase transition from α-TCP into β-TCP in three types of BCPs (HA + TCP) was investigated and the experimental results indicated that a sintering temperature for the complete phase transition from α-TCP into β-TCP increased with HA content increasing in BCP [511].

1.3.15 Ion-substituted Calcium Orthophosphates

Finally, one should very briefly mention on existence of carbonateapatite [512–518], chlorapatite [519–521], as well as on a great number of various ion-substituted calcium orthophosphates [82, 522, 523]. Usually, they are of a non-stoichiometric nature and there are too many of them to be mentioned in one review. Currently this is a hot investigation topic; therefore, the readers are referred to other books and reviews [27–29, 35, 37, 47, 107, 277, 432]. In addition, there is a very good review, in which the structures of more than 75 chemically different apatites have been discussed [114].

To conclude this topic, it is interesting to note, that chemical elements not found in natural bones can be intentionally incorporated into calcium orthophosphate biomaterials to get special properties. For example, addition of Ag^{+} [524–526], Zn^{2+} [526, 527] and Cu^{2+} [526–528] has been used for imparting antimicrobial effect, while radioactive isotopes of ^{90}Y [529], ^{153}Sm [530–532] and ^{186}Re [530] have been incorporated into HA bioceramics and injected into knee joints to treat rheumatoid joint synovitis [529, 530, 532]. More to the point, apatites were found to incorporate individual molecules, such as water, oxygen and carbon dioxide [82].

1.4 Biological Hard Tissues of Calcium Orthophosphates

Biological mineralization (or biomineralization) is the process of *in vivo* formation of inorganic minerals (so-called, biominerals). One should stress, that the term "biomineral" refers not only to a mineral produced by organisms but also to the fact that almost all of these mineralized products are composite materials comprised of both inorganic and bioorganic components. Furthermore, having formed *in vivo* under well-controlled conditions, the biomineral phases often have properties, such as shape, size, crystallinity, isotopic and trace element compositions, quite unlike its inorganically formed counterpart (please, compare Figs. 1.2, 1.8, 1.10 and 1.14). Thus, the term "biomineral" reflects all this complexity [454, 455].

As shown in Table 1.2 and discussed above, in the body of mammals the vast majority of both normal and pathological calcifications consist of non-stoichiometric and ion-substituted calcium orthophosphates, mainly of apatitic structure [87, 533]. At the element scale, nano-sized crystals bone apatite exhibit a variety of substitutions and vacancies that make the Ca/P molar ratio distinct from the stoichiometric HA ratio of 1.67. Their chemical composition is complicated and varies in relatively wide ranges. This depends on what the animal has ingested [534]. Occasionally, attempts are performed to compose chemical formulas of biological apatites. For example, the following formula $Ca_{8.856}Mg_{0.088}Na_{0.292}K_{0.010}(PO_4)_{5.312}(HPO_4)_{0.280}(CO_3)_{0.407}(OH)_{0.702}Cl_{0.078}(CO_3)_{0.050}$ was proposed to describe the chemical composition of the inorganic part of dental enamel [535].

The impurities in biological apatite of bones and teeth introduce significant stresses into the crystal structure, which make it less stable and more reactive. Among all substituting ions, the presence of 4–8% of carbonates instead of orthophosphate anions (so called, B-type substitution [27–29, 516]) and of 0.5–1.5% of Mg is of the special importance because it leads to large lattice strain and significantly increases the solubility [533, 535, 536]. Higher concentrations of magnesium and carbonates in bone or dentine compared to those in enamel (Table 1.2) may explain a higher solubility and a lower crystallinity (smaller crystal size) of bone or dentine compared to enamel.

In addition, the crystals of biological apatite are always very small which also increases its solubility when compared with that for the chemically pure HA and even CDHA [82]. However, biologic apatites of enamel have considerably larger crystal size (about 2000 nm) compared to that of either bone or dentine apatite, as indicated by the well-defined diffraction peaks in the X-ray diffraction profile of enamel apatite and much broader diffraction peaks of either bone or dentine apatites (Fig. 1.9, center). Small dimensions and a low crystallinity are two distinct features of biological apatites, which, combined with their non-stoichiometric composition, inner crystalline disorder and presence of other ions in the crystal lattice, allow explaining their special behavior. For example, the small crystal size means that a large percentage of the atoms are on the surface of the crystals, providing a large specific surface area for sorption of ions, proteins and drugs [536, 537]. The major physical properties of biological apatite are summarized in Fig. 1.9. It is interesting to note, that the solubility and equilibrium phenomena of calcium orthophosphates related to the calcification process have been studied, at least, since 1925 [538, 539].

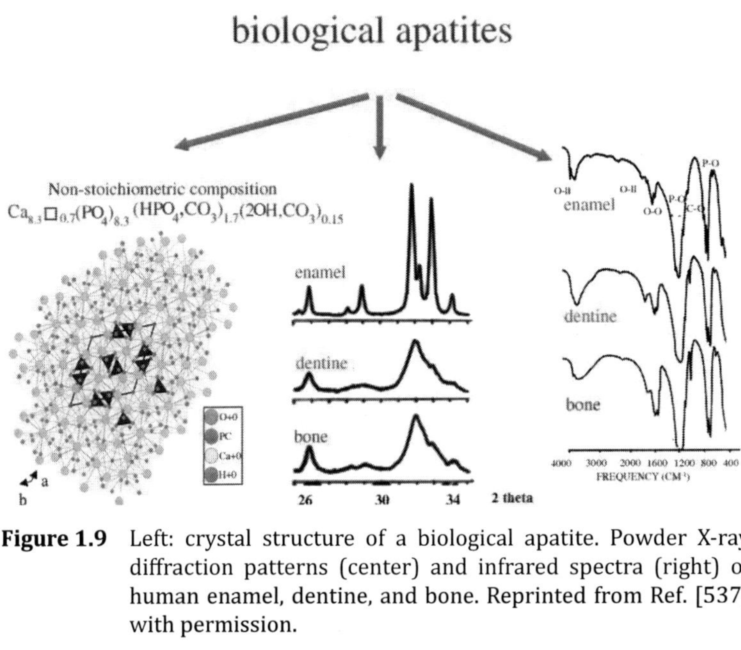

Figure 1.9 Left: crystal structure of a biological apatite. Powder X-ray diffraction patterns (center) and infrared spectra (right) of human enamel, dentine, and bone. Reprinted from Ref. [537] with permission.

Attempts to mimic the calcium orthophosphate nature of bones were first performed in 1913 [540]. This discovery was clarified afterwards, suggesting that the bone mineral could be carbonated apatite [541, 542]. Further optical and X-ray analysis of bones and other mineralized tissues matched analyses of two apatites: FA and dahllite [543]. Additional historical data on this point are available in literature [41]. Nowadays, according to Weiner and Wagner: "the term bone refers to a family of materials, all of which are built up of mineralized collagen fibrils" [544, 545]. For mammals, this family of materials includes dentine — the material that constitutes the inner layers of teeth, cementum — the thin layer that binds the roots of teeth to the jaw, deer antlers and some other materials [544, 546]. It is worth noting, that bones and teeth contain almost 99% of the total body calcium and about 85% of the total body phosphorus that amounts to a combined mass of approximately 2 kg in an average person [547, 548]. In addition, it is important to recognize that calcium orthophosphates of bones are by no means inert; they play an important role in the metabolic functions of the body. The recent data on the physico-chemical and crystallographic study of biological apatite have been reviewed elsewhere [535]. Besides, there is a comprehensive review on the application of surface science methods to study the properties of dental materials and related biomaterials [549].

1.4.1 Bone

Bone, also called osseous tissue (Latin: *os*), is a type of hard endoskeletal connective tissue found in many vertebrate animals. All bones of a single animal are, collectively, known as the skeleton. True bones are present in bony fish (osteichthyes) and all tetrapods. Bones support body structures, protect internal organs and, in conjunction with muscles, facilitate movement [550]. In addition, bones are also involved with blood cell formation, calcium metabolism and act for mineral storage. From the material point of view, bone is a dynamic, highly vascularized tissue that is formed from a complicated biocomposite containing both inorganic (Table 1.2) and bioorganic (chiefly, collagen) compounds [533, 551–557]. More than 20 human collagens have been reported in the human body, among which type I collagen is the most abundant protein and provides much of the structural integrity for connective tissue, particularly in bones, tendons and ligaments. Furthermore, there is a cellular phase, which consists of three different types of cells: osteoblasts, osteoclasts, and osteocytes; however, that is another story. The inorganic to bioorganic ratio is approximately 75% to 25% by dry weight and about 65% to 35% by volume. This ratio not only differs among animals, among bones in the same animal and over time in the same animal but also it exerts a major control on the material properties of bone, such as its toughness, ultimate strength and stiffness. In general, load-bearing ability of bones depends on not only architectural properties, such as cortical thickness and bone diameter, but also intrinsic, size-independent, material properties such as porosity, level of mineralization, crystal size and properties derived from the organic phase of bone [558]. A higher mineral to collagen ratio typically yields stronger, but more brittle, bones [559–561]. For example, bone from the leg of a cow has a relatively high concentration of calcium orthophosphates (for support), whereas bone from the antler of a deer has a relatively high concentration of collagen (for flexibility) [124]. It is interesting to note, that bone exhibits several physical properties such as piezoelectricity [562] and pyroelectricity [563].

Stability of the mineral composition of bones has a very long history: calcium orthophosphates were found in dinosaur fossils [52, 99, 564–567]. Therefore, organisms have had a great deal of time to exploit the feedback between composition and structure in apatite, on the one hand, and benefit from its biological functionality, on the other. Bones of modern animals is a relatively hard and lightweight porous composite material, formed mostly of biological apatite (i.e., poorly crystalline CDHA with ionic substitutions). It has relatively high compressive strength but poor tensile strength [568]. While bone is essentially brittle, it has a degree of significant plasticity contributed by its organic components.

The distribution of the inorganic and bioorganic phases depends on a highly complex process that takes place during bone formation. Each of these components may be assembled in different proportions creating two different architectural structures depending on the bone type and function. They are characterized by different structural features that strongly correlate with the mechanical performance of the tissue. These two types of bones are the following: the cortical bone (or compact bone), which is a dense structure and the cancellous bone (also known as trabecular or spongious bone), which is less dense and less stiff than compact bone. Usually bone is composed of a relatively dense outer layer of cortical bone covering an internal mesh-like structure (average porosity of 75–95%) of cancellous bone, the density of which is about 0.2 g/cm^3 but it may vary at different points (Fig. 1.10). Cortical bone makes up a large portion of the skeletal mass; but due to its high density (~1.80 g/cm^3) it has a low surface area. Cancellous bone has an open meshwork or honeycomb-like structure. It has a relatively high surface area but forms a smaller portion of the skeleton. Bone is a porous material with the pore sizes range from 1 to 100 μm in normal cortical bones and 200–400 μm in trabecular bones. About 55–70% of the pores in trabecular bones are interconnected. The porosity reduces the strength of bones but also reduces their weight [27, 83, 84, 105–107, 454, 544, 552–556, 569–573].

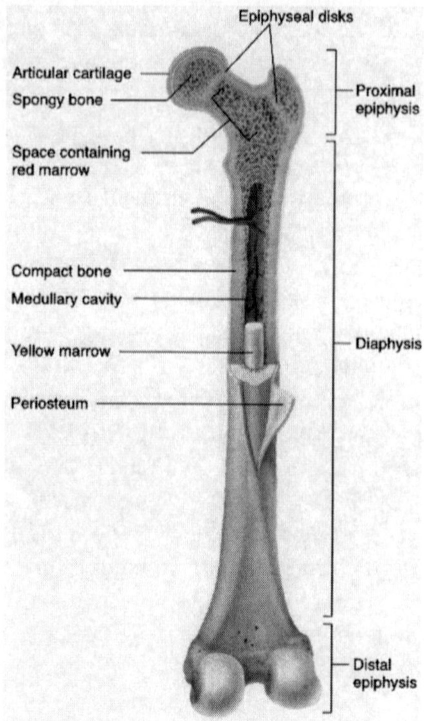

Figure 1.10 General structure of a mammalian bone. Other very good graphical sketches of the mammalian bone structure are available in Refs. [87, 537].

Bones can be either woven or lamellar. The fibers of woven bones are randomly aligned and as the result have a low strength. In contrast, lamellar bones have parallel fibers and are much stronger. Woven bones are put down rapidly during growth or repair [574] but as growth continues, they are often replaced by lamellar bones. The replacement process is called "secondary bone formation" and described in details elsewhere [575 and references therein]. In addition, bones might be long, short, flat, and irregular. The sizes and shapes of bones reflect their function. Namely, broad and flat bones, such as scapulae, anchor large muscle masses, flat skull bones protect the brain, ribs protect the lungs, pelvis protects other internal organs, short tubular bones in the digits of hands and feet provide specific grasping functions, hollow and thick-walled tubular bones, such as femur or radius, support weight and long bones enable locomotion [576, 577]. Long bones are tubular in structure (e.g., the tibia). The central shaft of a long bone is called the diaphysis and has a medullar cavity filled with bone marrow (Fig. 1.10). Surrounding the medullar cavity is a thin layer of cancellous bone that also contains marrow. The extremities of the bone are called the epiphyses and are mostly cancellous bone covered by a relatively thin layer of compact bone. Short bones (e.g., finger bones) have a similar structure to long bones, except that they have no medullar cavity. Flat bones (e.g., the skull and ribs) consist of two layers of compact bone with a zone of cancellous bone sandwiched between them. Irregular bones (e.g., vertebrae) do not conform to any of the previous forms. Thus, bones are shaped in such a manner that strength is provided only where it is needed. All bones contain living cells embedded in a mineralized organic matrix that makes up the main bone material [576–578]. The structure of bones is most easily understood by differentiating between seven levels of organization because bones exhibit a strongly hierarchical structure (Fig. 1.11) [432, 454, 533, 544, 551–556, 562–567, 569–572, 579–584].

The mechanical properties of bones reconcile high stiffness and high elasticity in a manner that is not yet possible with synthetic materials [584]. Cortical bone specimens have been found to have tensile strength in the range of 79–151 MPa in longitudinal direction and 51–56 MPa in transversal direction. Bone's elasticity is also important for its function giving the ability to the skeleton to withstand impact. Estimates of modulus of elasticity of bone samples are of the order of 17–20 GPa in longitudinal direction and of 6–13

GPa in the transversal direction [585]. The elastic properties of bone were successfully modeled at the level of mineralized collagen fibrils via step-by-step homogenization from the staggered arrangement of collagen molecules up to an array of parallel mineralized fibrils [586]. Recent investigations revealed that bone deformation was not homogeneous but distributed between a tensile deformation of the fibrils and a shearing in the interfibrillar matrix between them [587, 588]. Readers, who are interested in further details, are addressed to a good review on the effects of the microscopic and nano-scale structure on bone fragility [589].

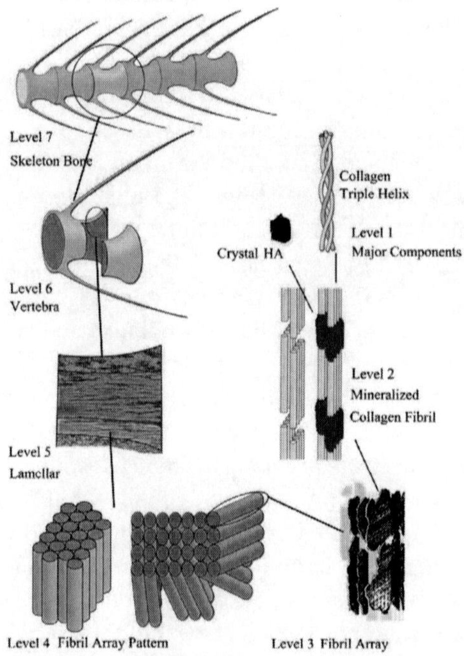

Figure 1.11 The seven hierarchical levels of organization of the zebrafish skeleton bone. Level 1: Isolated crystals and part of a collagen fibril with the triple helix structure. Level 2: Mineralized collagen fibrils. Level 3: The array of mineralized collagen fibrils with a cross-striation periodicity of nearly 60–70 nm. Level 4: Two fibril array patterns of organization as found in the zebrafish skeleton bone. Level 5: The lamellar structure in one vertebra. Level 6: A vertebra. Level 7: Skeleton bone. Reprinted from Ref. [602] with permission. Other good graphical sketches of the hierarchical structure of bones are available in Refs. [544, 583, 584].

The smallest level of the bone hierarchy consists of the molecular components: water, biological apatite, collagen and other proteins [533]. The second smallest hierarchical level is formed by mineralization of collagen fibrils, which are of 80 to 100 nm thickness and a length of a few to tens of microns (Fig. 1.11). Thus, biocomposites of biological apatite and molecules of type I collagen are formed [87, 544, 551, 557, 590]. Some evidences for direct physical bonding between the collagen fibers and apatite crystals in bone were found [591]. Eppell et al., used atomic force microscopy to measure the crystallites of mature cow bone [592]. They are always platelet-like (elongated along the crystallographic *c*-axis) and very thin [86, 593–595], with remarkably uniform thicknesses (determined in transmission electron microscopy) of 2–4 nm [596] (just a few unit cells thick — see Table 1.2). The nano-sized crystals of biological apatite exist in bones not as discrete aggregates but rather as a continuous phase, which is indirectly evidenced by a very good strength of bones. This results in a very large surface area facing extracellular fluids, which is critically important for the rapid exchange of ions with these fluids. The nano-sized crystals of biological apatite are inserted in a nearly parallel way into the collagen fibrils, while the latter are formed by self-assembly [597] of collagen triple helices [544, 551, 599–602] using the self-organization mechanism [603, 604]. Recent data from electron diffraction studies revealed that that the mineral plates of biological apatite are not quite as ordered as previously assumed [575]. This imperfect arrangement of nearly parallel crystals has been supported by recent SAXS and transmission electron microscopy studies [605].

The lowest level of hierarchical organization of bone has successfully been simulated by CDHA precipitation on peptide-amphiphile nanodimensional fibers [604]. However, apatite platelets nucleating on the surface of peptide tubules are not similar to the nanostructure of bone and they are only an example of surface induced nucleation (and not accurately characterized either), while the nanostructure of bone consists of intra-fibrillar platelets intercalated within the collagen fibrils. Olszta and Gower were the first to truly duplicate the bone nanostructure [575]. Unfortunately, the interface between collagen and crystals of biological apatite is still poorly understood; for the available details, the readers are referred to a review devoted to the structure and mechanical quality of the collagen/mineral nanodimensional biocomposite of bones

[590]. There is still no clear idea why the crystals of biological apatite are platelet-shaped even though dahllite has hexagonal crystal symmetry [454, 544, 552–556, 562–567, 569–573]. One possible reason is that they grow via an OCP transition phase, which crystals are plate-shaped [544].

The processes of bone formation (ossification) and growth are very complicated ones and it is difficult to describe them without making a deep invasion into biology. It has been studied for decades [574] but still there are missing points. Very briefly, it is considered that bones appear and grow as the result of calcification (or biomineralization) of connective tissues, mainly cartilage [533, 575]. The ossified tissue is invaginated with blood vessels, which bring ions of calcium and orthophosphate to be deposited in the ossifying tissue. The biomineralization process is controlled to some extent by cells and the organic matrices made by those cells facilitate the deposition of crystals [578]. There is an opinion, that, initially, the mineral crystals are formed in an environment rich in the so-called SIBLING (Small Integrin-Binding Ligand N-linked Glycoprotein) proteins. As bone crystals grow, there is greater association with proteins, such as osteocalcin, that regulate remodeling [606]. Thus, *in vivo* formation of hard tissues always occurs by mineral reinforcement of the previously formed network of soft tissues [533, 575–577, 602].

Cartilage is composed of cells (chondrocytes and their precursor forms known as chondroblasts), fibers (collagen and elastic fibers) and extracellular matrix (proteoglycans, which are a special class of heavily glycosylated glycoproteins) [607–609]. The initial stage involves the synthesis and extracellular assembly of the collagen matrix framework of fibrils. At the second stage, the chondrocytes calcify the matrix before undergoing the programmed cell death (apoptosis). At this point, blood vessels penetrate this calcified matrix, bringing in osteoblasts (they are mononuclear cells primarily responsible for bone formation), which use the calcified cartilage matrix as a template to build bone, thus completing ossification [607–609].

During ossification, the crystals of biological apatite grow with a specific crystalline orientation — the *c*-axes of the crystals are roughly parallel to the long axes of the collagen fibrils within which they are deposited [533, 534, 544, 546–549, 552–554, 557, 575]. Earlier, it was believed that this process occurred via epitaxial growth mechanism [610]. The same was suggested for dentine and enamel [611, 612] (see section *1.4.2. Teeth* below), as well as for more primitive living organisms. For example, in the shell of the fossil marine animal *Lingula brachiopod unguis* that consists of a biological apatite, the crystal *c*-axes are oriented parallel to the β-chitin fibrils [458, 613–616]. Therefore, the orientation of biological apatite crystals parallel to the long axes of the organic framework could be a general feature of calcium orthophosphate biomineralization. However, the degree of biological apatite orientation appears to be a useful parameter to evaluate *in vivo* stress distribution, nano-scale microstructure and the related mechanical function, the regenerative process of the regenerated bone and to diagnose bone diseases such as osteoarthritis [617, 618]. It is interesting to note, that contrary to what might be expected in accordance with possible processes of dissolution, formation and remineralization of hard tissues, no changes in phase composition of mineral part, crystal sizes (length, width and thickness) and arrangement of crystals on collagen fibers were detected in abnormal (osteoporotic) human bones compared to the normal ones [619].

Some animals, such as newts, are able to regenerate amputated limbs. This is, of course, of a high interest in regenerative medicine. Bone regeneration in the forelimbs of mature newts was studied by noninvasive X-ray microtomography to image regenerating limbs from 37 to 85 days. The missing limb skeletal elements were restored in a proximal-to-distal direction, which reiterated the developmental patterning program. However, in contrast to this proximal-distal sequence, the portion of the humerus distal to the amputation site was found to fail to ossify in synchrony with the regenerating radius and ulna. This finding suggests that the replacement of cartilage with mineralized bone close to the amputation site is delayed with respect to other regenerating skeletal elements [620].

Unlike other mineralized tissues, bone continuously undergoes a remodeling process, as it is resorbed by specialized cells called osteoclasts and formed by another type of cells called osteoblasts (so called "bone lining cells") in a delicate equilibrium [533, 575, 578, 621, 622]. The purpose of remodeling is the release of calcium and the repair of micro-damaged bones from everyday stress. Osteoblasts are mononuclear cells primarily responsible for bone formation. They contain alkaline phosphatase, which enzymatically produces orthophosphate anions needed for the mineralization. In addition, there is one more type of the cells called osteocytes that originate from osteoblasts, which have migrated into, become trapped and surrounded by bone matrix, which they themselves produce [533, 552–555, 575–578].

If osteoblasts are bone-forming cells, osteoclasts are multinuclear, macrophage-like cells, which can be described as bone destroying cells because they mature and migrate to discrete bone surfaces [578, 621, 622]. Upon arrival, active enzymes, such as acid phosphatase, are secreted against the mineral substrate that causes dissolution. This process, called bone resorption, allows stored calcium to be released into systemic circulation and is an important process in regulating calcium balance [621, 622]. The iteration of remodeling events at the cellular level is influential on shaping and sculpting the skeleton both during growth and afterwards. That is why, mature bones always consist of a very complex mesh of bone patches, each of which has both a slightly different structure and a different age [454, 533–535, 544, 546–549, 552–554, 575]. The interested readers are suggested to read a review on the interaction between biomaterials and osteoclasts [623].

Still there is no general agreement on the chemical mechanism of bone formation. It is clear that the inorganic part of bone consists of biological apatite, i.e. CDHA with ionic substitutions but without the detectable amounts of hydroxide [624–628]. However, the recent results of solid-state nuclear magnetic resonance on fresh-frozen and ground whole bones of several mammalian species revealed that the bone crystal OH^- was readily detectable; a rough estimate yielded an OH^- content of human cortical bone of about 20% of the amount expected in stoichiometric HA [629]. Various *in vitro* experiments on precipitation of CDHA and HA revealed that none of these compounds is directly precipitated from supersaturated aqueous solutions containing calcium and orthophosphate ions: some intermediate phases (precursors) are always involved [27, 83, 84, 192–198, 270–274]. Depending on both the solution pH and crystallization conditions, three calcium orthophosphates (DCPD, ACP and OCP) have been discussed as possible precursors of CDHA precipitation *in vitro*. Due to this reason, the same calcium orthophosphates are suggested as possible precursors of biological apatite formation *in vivo*.

The transient nature of the precursor phase of bone, if it exists at all, makes it very difficult to detect, especially *in vivo* [93]. However, in 1966 W. E. Brown proposed that OCP was the initial precipitate that then acted as a template upon which biological apatite nucleates [191]. This idea was extended in his further investigations [630–633]. The principal support for this concept is derived from the following: (i) the close structural similarity of OCP and HA [183, 184]; (ii) formation of interlayered single crystals of OCP and HA (pseudomorphs of OCP); (iii) the easier precipitation of OCP compared with HA; (iv) the apparent plate- or lath-like habit of biological apatites that does not conform to hexagonal symmetry, but looks like a pseudomorph of triclinic OCP; and (v) the presence of HPO_4^{2-} in bone mineral, particularly in newly formed bones [535]. Some evidences supporting this idea were found using high-resolution transmission electron microscopy: computer-simulated lattice images of the "central dark line" in mineralized tissues revealed that it consisted of OCP [192–196]. Recently, Raman spectroscopic indication for an OCP precursor phase was found during intra-membranous bone formation [634]. Other evidences of OCP to HA transformation, including a mechanistic model for the central dark line formation, might be found in literature [635].

Simultaneously with Brown, the research group led by A. S. Posner proposed that ACP was the initially precipitated phase of bone and dentine mineral formation *in vivo*, thus explaining the non-stoichiometric Ca/P ratio in bones and teeth [636–638]. This conclusion was drawn from the following facts: (i) when calcium orthophosphates are prepared by rapid precipitation from aqueous solutions containing ions of calcium and orthophosphate at pH > 8.5, the initial solid phase is amorphous; (ii) mature bone mineral is composed of a mixture of ion-substituted ACP and poorly crystallized ion-substituted CDHA; (iii) early bone mineral has a lower crystallinity than mature bone and the observed improvement in crystallinity with the age of the bone mineral is a result of a progressive reduction in the ACP content [535, 636–644]. However, there are thermodynamic data proving that the transition of freshly precipitated ACP into CDHA involves intermediate formation of OCP [645, 646]. Recently the discovery of a stable amorphous calcium carbonate in sea urchin spines [647] reawakened the suggestion that a transient amorphous phase might also exist in bones [575, 648–652]. Even more recently, evidences of an abundant ACP phase in the continuously forming fin bones of zebrafish were found [653, 654]. The new bone mineral was found to be delivered and deposited as packages of nanodimensional spheres of ACP, which further transformed into platelets of crystalline apatite within the collagen matrix [654].

Furthermore, to investigate how apatite crystals form inside collagen fibrils, researchers carried out a time-resolved study starting from the earliest stages of mineral formation [655]. After 24 h of mineralization calcium orthophosphate particles were found outside the fibril, associated with the overlap region, in close proximity to the gap zone. Cryogenic energy-dispersive X-ray spectroscopy confirmed that these precipitates were composed of calcium orthophosphate, while a low-dose selected-area electron diffraction technique showed a diffuse band characteristic of ACP. After 48 h, apatite crystals started to develop within a bed of ACP and after 72 h, elongated electron-dense crystals were abundant within the fibril, in many cases still embedded within a less dense matrix. A low-dose selected-area electron diffraction technique demonstrated that the mineral phase consisted of both ACP and oriented apatite, the latter identical to bone apatite [655]. This process is schematically shown in Fig. 1.12 [656]. The modern points of view on the bone formation mechanisms have been summarized in a recent excellent review [575], where the interested readers are referred.

The maturation mechanism of bone minerals is not well established, mainly because of the difficulty involved in the nanostructural analyses of bone minerals [575, 657]. Only indirect evidences for the *in vivo* bone mineral maturation are available. For example, X-ray diffraction patterns of bones from animals of different age show that the reflections become sharper with age increasing [98, 658]. This effect is more pronounced in the crystallographic *a*-axis [(310) reflections] as compared to the *c*-axis [(002) reflections] [659, 660]. The most comprehensive report describing how normal human bone mineral changes in composition and crystal size as a function of age was based on X-ray diffraction analyses by Hanschin and Stern [661], who examined 117 homogenized iliac crest biopsies from patients aged 0–95 years. They found that the bone mineral crystal size and perfection increased during the first 25–30 years and then decreased thereafter, slightly increasing in the oldest individuals. The same 117 homogenized biopsy samples were analyzed by wavelength-dispersive X-ray fluorescence to quantify the carbonate substitution in biological apatite as a function of age. Although the changes observed in carbonate substitution were relatively slight (at most 10%), there was a general increase from 0 to 90 years that is distinct from the absence of a change in crystallinity after age 30 in these samples [583]. In addition, other changes, like an increase of Ca^{2+} content and a decrease of HPO_4^{2-}, occur in bone mineral with age [662–665]. Both the crystal sizes and carbonate content were found to increase during aging in rats and cows [663, 664]. The increase in carbonate content with age has also been reported in still other studies [666–668]. From a chemical point of view, these changes indicate to a slow transformation of poorly crystallized non-apatitic calcium orthophosphates into a better-crystallized ion-substituted carbonate-containing CDHA. While there are still many gaps in our knowledge, the researchers seem to be comfortable in stating that in all but the youngest bone and dentine, the only phase present is a highly disordered, highly substituted biological apatite.

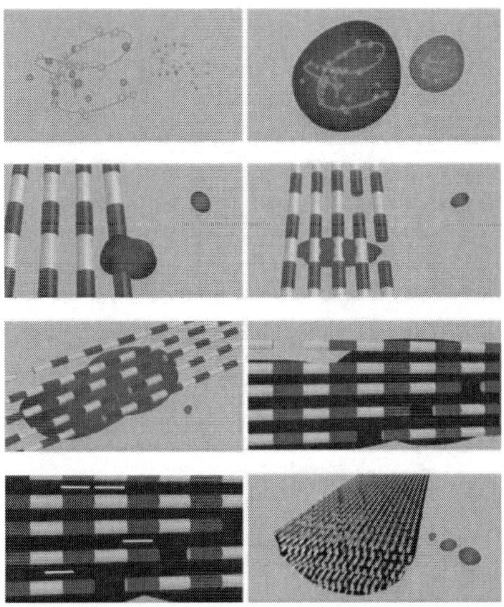

Figure 1.12 A schematic illustration of *in vivo* mineralization of a collagen fibril: top layer — calcium orthophosphate clusters (green) form complexes with biopolymers (orange line), forming stable mineral droplets; second top layer — mineral droplets bind to a distinct region on the collagen fibers and enter the fibril; second bottom layer — once inside the collagen, the mineral in a liquid state diffuses through the interior of the fibril and solidifies into a disordered phase of ACP (black); bottom layer — finally, directed by the collagen, ACP is transformed into oriented crystals of biological apatite (yellow). Reprinted from Ref. [656] with permission.

In general, the biomineralization process (therefore, bone formation) can happen in two basic ways: either the mineral phase develops from the ambient environment as it would from a supersaturated solution of the requisite ions, but requires the living system to nucleate and localize mineral deposition, or the mineral phase is developed under the direct regulatory control of the organism, so that the mineral deposits are not only localized but may be directed to form unique crystal habits not normally developed by a saturated solution of the requisite ions. In a very famous paper [669] and two extended elaborations [454, 670] the first type of biomineralization was called "biologically induced" mineralization and the second "(organic) matrix-mediated" biomineralization. In some papers, the former process is called "passive" and the latter one — "active" biomineralization [34]. Briefly, an "active process" means the assembly of nano-sized crystals of calcium orthophosphate into bones due to an activity of the suitable cells (e.g., osteoblasts), i.e. within a matrix vesicle. Such structures have been discovered by transmission electron microscopy for bone and teeth formation [671, 672]. A "passive process" does not require involvement of cells and means mineralization from supersaturated solutions with respect to the precipitation of biological apatite. In the latter case, thermodynamically, biomineralization might occur at any suitable nucleus. The collagen fibrils have a specific structure with a 67 nm periodicity and 35–40 nm gaps or holes between the ends of the collagen molecules where bone mineral is incorporated in the mineralized fibril [454, 544, 545, 557, 576, 577]. Such a nucleation within these holes would lead to discrete crystals with a size related to the nucleating cavity in the collagen fibril (Fig. 1.11). It was proposed that a temporary absence of the specific inhibitors might regulate the process of bone formation [673–675].

To conclude the bone subject, let me briefly mention on the practical application of bones. In the Stone Age, bones were used to manufacture art, weapons, needles, catchers, amulets, pendants, headdresses, etc. Nowadays, cut and polished bones from a variety of animals are sometimes used as a starting material for jewelry and other crafts. Ground cattle bone is occasionally used as a fertilizer. Furthermore, in medicine, bones are used for bone graft substitutes, e.g., allografts from cadavers.

1.4.2 Teeth

Teeth (singular: tooth) are dense structures found in the jaws of many vertebrates. They have various structures to allow them to fulfill their different purposes. The primary function of teeth is to tear, smell and chew food, while for carnivores it is also a weapon. Therefore, teeth have to withstand a range of physical and chemical processes, including compressive forces (up to ~700 N), abrasion and chemical attack due to acidic foods or products of bacterial metabolism [549]. The roots of teeth are covered by gums. From the surface teeth are covered by enamel of up to ~2 mm thick at the cutting edges of the teeth, which helps to prevent cavities on the teeth. The biggest teeth of some gigantic animals (elephants, hippopotamuses, walruses, mammoths, narwhals, etc.) are known as tusks or ivory.

Similar to the various types of bones, there are various types of teeth. The shape of the teeth is related to the animal's food, as well as its evolutionary descent. For example, plants are hard to digest, so herbivores have many molars for chewing. Carnivores need canines to kill and tear and since meat is easy to digest, they can swallow without the need for molars to chew the food well. Thus, the following types of teeth are known: molars (used for grinding up food), carnassials (used for slicing food), premolars (small molars), canines (used for tearing apart food) and incisors (used for cutting food). While humans only have two sets of teeth, some animals have many more: for example, sharks grow a new set of teeth every two weeks. Some other animals grow just one set during the life, while teeth of rodents grow and wear away continually through the animal gnawing, maintaining constant length [676, 677].

Similar to bones, the inorganic part of teeth also consist of biological apatite [678]. The stability of the mineral composition of teeth also has a very long history: namely, calcium orthophosphates were found in fossil fish teeth [679]. Recent investigations of biological apatite from fossil human and animal teeth revealed its similarity to the modem biological apatite [680].

The structure of teeth appears to be even more complicated than that of bones (see Fig. 1.13). Unlike bones, teeth consist of at least two different materials: enamel, which is a hard outer layer consisting of calcium orthophosphates, and dentine, which is a bone-like magnesium-rich tissue that forms the bulk of vertebrate teeth. In addition, there is a thin layer around the tooth roots called cementum. It is a thin layer of a bone-like calcified tissue that covers dentine at the roots of teeth and anchors them to the jaw [681–684]. Finally, there is the core called pulp (commonly called "the nerve") — it is a remnant of the embryologic organ for tooth development and contains nerves and blood vessels necessary for tooth function (Fig. 1.13) [576, 577, 676, 677]. Both dentine and cementum are mineralized connective tissues with an organic matrix of collagenous proteins, while the inorganic component of them consists of biological apatite. As shown in Table 1.2, dentine, cementum and bone are quite similar and for general purposes of material scientists they can be regarded as being essentially the same material [454, 535, 544, 546–549, 552–556, 562–567, 569–571, 590, 594, 595, 599, 662, 663]. Thus, most statements made in the previous section for bones are also valid for dentine and cementum; however, unlike bones, both dentine and cementum lack vascularization [685].

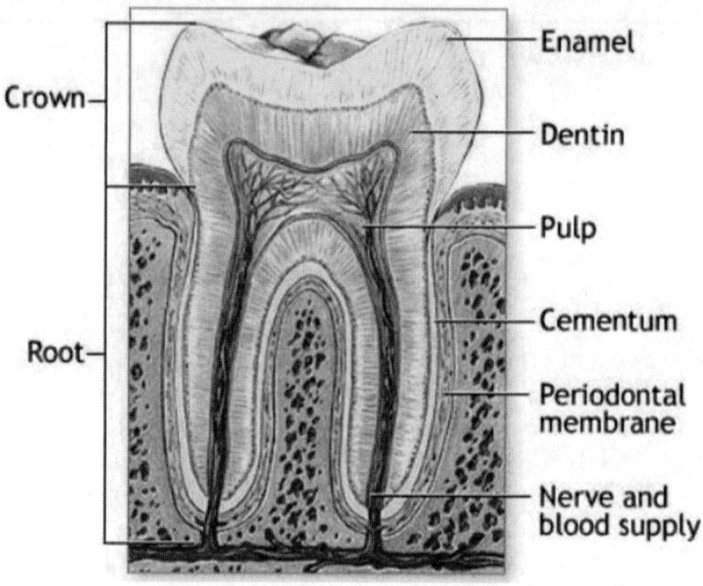

Figure 1.13 A schematic drawing of a tooth. Other very good graphical sketches of the mammalian tooth structure, including the hierarchical levels, are available in Refs. [533, 584].

Dental enamel is the outermost layer of teeth. It is white and translucent and its true color might be observed at the cutting edges of the teeth only. Enamel is highly mineralized and acellular, so it is not a living tissue. Nevertheless, it is sufficiently porous for diffusion and chemical reactions to occur within its structure, particularly acidic dissolution (dental caries) and remineralization from saliva (possible healing of caries lesions). Enamel is the hardest substance in the body [568] and forms a solid, tough and wear-resistant surface for malaxation. In the mature state, it contains up to 98% of inorganic phase (Table 1.2). The crystals of biological apatite of enamel are much larger as evidenced by higher crystallinity (reflecting greater crystal size and perfection) demonstrated in their X-ray diffraction patterns, than those of bone and dentine. Besides, enamel apatite has fewer ionic substitutions than bone or dentine mineral and more closely approximates the stoichiometric HA [576]. The organic phase of enamel does not contain collagen. Instead, enamel has two unique classes of proteins called amelogenins and enamelins. While the role of these proteins is not fully understood yet, it is believed that both classes of proteins aid in the enamel development by serving as a framework support [676, 677, 686]. The large amount of minerals in enamel accounts not only for its strength but also for its brittleness. Dentine, which is less mineralized and less brittle, compensates for enamel and is necessary as a support [676, 677]. Shark enameloid is an intermediate form bridging enamel and dentine. It has enamel-like crystals of fluoridated biological apatite associated with collagen fibrils [82, 452–457]. Due to the presence of fluorides, biological apatite of shark enameloid shows both higher crystal sizes and a more regular hexagonal symmetry if compared to non-fluoridated biological apatite of bones and teeth [107]. Similar correlation between the presence of fluorides and crystal dimensions was found for enamel [687].

Like that for bones, seven levels of structural hierarchy have been also discovered in human enamel; moreover, the analysis of the enamel and bone hierarchical structure suggests similarities of the scale distribution at each level [533, 585, 688]. On the mesoscale level, there are three main structural components: a rod, an interrod and aprismatic enamel. Among them, the enamel rod (formerly called an enamel prism) is the basic unit of enamel. It is a tightly packed mass of biological apatite in an organized pattern. Each rod traverses uninterrupted through the thickness of enamel. They number 5 to 12 million rods per crown. The rods increase in diameter (4 up to 8 microns) as they flare outward from the dentine-enamel junction

(DEJ). Needle-like enamel rods might be tens of microns long (up to 100 μm) but sometimes only 50 nm wide and 30 nm thick (Fig. 1.14) [676, 677, 689–696]. They are quite different from the much smaller crystals of dentine and bone (Table 1.2), but all of them consist of biological apatite [438, 697, 698]. In cross section, an enamel rod is best compared to a keyhole, with the top, or head, oriented toward the crown of the tooth and the bottom, or tail, oriented toward the root of the tooth.

Figure 1.14 Scanning electron micrograph of the forming enamel of a continuously growing rat incisor showing ordered rods of calcium orthophosphates. Scale bar: 10 μm. Reprinted from Ref. [454] with permission.

The arrangement of the crystals of biological apatite within each enamel rod is highly complex. Enamel crystals in the head of the enamel rod are oriented parallel to the long axis of the rod. When found in the tail of the enamel rod, the crystals' orientation diverges slightly from the long axis [676, 677]. The arrangement of the enamel rods is understood more clearly than their internal structure. Enamel rods are found in rows along the tooth (Fig. 1.14) and, within each row, the long axis of the enamel rod is generally perpendicular to the underlying dentine [676, 677, 689–693]. Recent AFM study indicated that CDHA crystals in enamel exhibited regular sub-domains or subunits with distinct chemical properties related to topographical features and gave rise to patterned behavior in terms of the crystal surface itself and the manner in which it responded to low pH [699].

The second structural component of the enamel matrix is the interrod (or interprismatic) enamel, which surrounds and packs between the rods. The difference between the rod and the interrod is the orientation of apatite crystals; the rod contains aligned crystallites, whereas the mineral in the interrod is less ordered. These structures coalesce to form the tough tissue of enamel, which can withstand high forces and resist damage by crack deflection. The third structure, aprismatic enamel, refers to the structures containing apatite crystals that show no mesoscale or macroscale alignment [533]. Enamel is a selectively permeable membrane, allowing water and certain ions to pass via osmosis [676, 677].

The *in vivo* formation and development of teeth appears to be even more complicated when compared with the afore-described process of bone formation. It is a very complex biological process, by which teeth are formed from embryonic cells, grow and erupt into the mouth [578]. For human teeth enamel, dentine and cementum must all be developed during the appropriate stages of fetal development. Primary (baby) teeth start to form between the sixth and eighth weeks *in utero*, while the permanent teeth begin to form in the twentieth week *in utero* [676, 677]. Recent data confirmed the necessity of calcium orthophosphates in the diet of pregnant and nursing mother to prevent early childhood dental caries [700].

As teeth consist of at least two materials with different properties (enamel and dentine), the tooth bud (sometimes called "the tooth germ" — that is an aggregation of cells that eventually forms a tooth) is organized into three parts: the enamel organ, the dental papilla and the dental follicle. The enamel organ is composed of at least four other groups of cells (for the biological details see Refs. [676, 677]). Altogether, these groups of cells give rise to ameloblasts, which secret enamel matrix proteins. The protein gel adjacent to ameloblasts is supersaturated with calcium orthophosphates, which leads to the precipitation of biological apatite. Similarly, the dental papilla contains cells that develop into odontoblasts, which are dentine-forming cells. The dental follicle gives rise to three important entities: cementoblasts, osteoblasts, and fibroblasts. Cementoblasts form the cementum of a tooth [682]. Osteoblasts give rise to the alveolar bone around the roots of teeth (see bone formation above). Fibroblasts develop the periodontal ligaments that connect teeth to the alveolar bone through cementum [576–578, 676, 677].

The first detectable crystals in enamel formation are flat thin ribbons [691–693], that were reported to be OCP [569, 701–703], β-$(Ca,Mg)_3(PO_4)_2$ [702], DCPD [624, 627], or ACP [704]. The formation process of enamel is different from that for bone or dentine: amelogenin being hydrophobic self-assembles into nano-sized spheres that guide the growth of the ribbon-like dental enamel crystals. During maturation of enamel, the mineral content increases from initially ~45 wt.% up to ~98–99 wt.% [624, 676, 677]. The enamel crystal rods widen and thicken by additional growth [624, 627, 705] with a simultaneous increase of the Ca/P molar ratio [705] and a decrease in carbonate content [706–708], finally resulting in the most highly mineralized and hardest substance produced by vertebrates. It is interesting to note that in the radular teeth of chitons, ACP was found to be the first-formed calcium orthophosphate mineral, which over a period of weeks was transformed to dahllite [709].

The crystal faces expressed in enamel are always (100) face and at the ends presumably (001) [710, 711], which are the ones usually found in HA. The centers of enamel crystals contain a linear structure known as the "central dark line" (this line was also observed in bone and dentine), which consists of OCP [192–196, 635]. As described above for bones, X-ray diffraction shows that the crystals of younger dentine are less crystalline than those of more mature dentine [662]. Therefore, maturation of dentine also means a slow transformation (re-crystallization?) of biological calcium orthophosphates from ion-substituted ACP to a better-crystallized ion-substituted CDHA.

The development of individual enamel and dentine crystals was studied by high-resolution transmission electron microscopy [712–714]. Both processes appear to be roughly comparable and were described in a four-step process. The first two steps include the initial nucleation and formation of nano-sized particles of biological apatite. They are followed by ribbon-like crystal formation, which until recently was considered as the first step of biological crystal formation [712–714]. These complicated processes, starting with the heterogeneous nucleation of inorganic calcium orthophosphates on an organic extracellular matrix, are controlled in both tissues by the organic matrix and are under cellular control [715]. To complicate the process even further, regular and discrete domains of various charges or charge densities on the surface of apatite crystals derived from the maturation stage of enamel development were recently discovered by a combination of atomic and chemical force microscopy [716]. Binding of organic molecules (e.g., amelogenin [716]) at physiological solution pH appears to occur on the charged surface domains of apatite. The modern visions on dental tissue research have been reviewed recently [717].

As teeth consist of several materials, there are mutual junctions among them. For example, a dentine-enamel junction (DEJ) is the interface between dentine and enamel. It is a remnant of the onset of enamel formation because enamel grows outwards from this junction [677, 718, 719]. DEJ plays an important role in preventing crack propagation from enamel into dentine [720]. The major steps of enamel crystal growth at the junction have been described above but the mechanism of the junction formation is still debatable. Some authors claim that enamel crystals grow epitaxially on the pre-existing dentine crystals because of a high continuity between enamel and dentine crystals [721–723]. Others have shown that enamel crystals are formed at a given distance from the dentine surface [701–703, 724] and could either reach dentine crystals by a subsequent growth [725] or remain distant [724, 726]. In addition, there are a cementum–enamel junction (CEJ) [727], which is quite similar to DEJ, and a cementum–dentine junction (CDJ) [681–683, 728].

Enamel formation, or amelogenesis, is a highly regulated process involving precise genetic control as well as protein–protein interactions, protein–mineral interactions and interactions involving the cell membrane. Much is still unknown about the interactions among proteins present in enamel matrix and the final crystalline phase of biological apatite [533, 729]. At some point before a tooth erupts into the mouth, the ameloblasts are broken down. Consequently, enamel, unlike bones, has no way to regenerate itself using the process of "active mineralization" (see bone formation) because there is no biological process that repairs degraded or damaged enamel [676, 677]. In addition, certain bacteria in the mouth feed on the remains of foods, especially sugars. They produce lactic acid, which dissolves the biological apatite of enamel in a process known as enamel demineralization that takes place below the critical pH of about 5.5. Similar process called enamel erosion occurs when a person consumes acid-containing (citric, lactic, phosphoric, etc.) soft drinks [689, 730–733]. Evidences exist that there is a preferential loss of carbonates and Mg during acidic dissolution of mineral in dental caries. Luckily, saliva gradually neutralizes the acids that cause pH on teeth surface to rise above the critical pH. This might cause partial enamel remineralization, i.e. a return of the dissolved calcium orthophosphates to the enamel surface. Until recently, it was generally agreed, that if there was sufficient time between the intake of foods (generally, two to three hours) and damage was very limited, teeth could repair themselves by the "passive mineralization" process [734]. Data on increased remineralization of tooth enamel by milk containing added casein phosphopeptide — ACP nanodimensional complexes [735] are in support of this hypothesis.

However, studies performed by using atomic force microscopy nano-indentation technique revealed that previously demineralized samples of dental enamel further exposed to remineralizing solutions did show a crystalline layer of calcium orthophosphates formed on their surface. Unfortunately, the re-precipitated deposits of calcium orthophosphates always consisted of loosely packed crystals and did not protect the underlying enamel from a subsequent acid attack. Furthermore, these surface deposits were completely removed by either a toothbrush or a short exposure to an erosive acidic solution [689, 736–738]. In this context, it should be emphasized that the term "remineralization", which is often misused in the literature, should imply the process of mineral growth that goes hand in hand with a strengthening effect of the weakened enamel surface. Since no strengthening of an exposure to remineralizing solutions was observed, it might be considered that no "passive mineralization" was found (in spite of the real evidence of the re-precipitated surface deposits of calcium orthophosphates) [689, 737, 738].

An interesting hypothesis that nano-sized apatite crystallites occur in the oral cavity during extensive physiological wear of the hierarchical structured enamel surface due to dental abrasion and attrition has been published recently [739]. These nano-scaled apatite enamel crystallites might promote remineralization at the tooth surface. However, this idea should be verified experimentally. Thus, according to the current knowledge, the enamel self-repairing ability by a passive remineralization appears to be doubtful, while an active remineralization is impossible. Nevertheless, investigations in this field keep going [213–215, 740–751]. For example, ACP-containing orthodontic biocomposite resins might reduce the enamel decalcification found in patients with poor oral hygiene [751].

A content of fluoride added to either toothpaste or mouthwash lowers the solubility of calcium orthophosphates (by formation of FHA on the surface) and therefore improves the acid-resistance of dental enamel [438, 460–465, 752]. Furthermore, fluorides also reduce production of acids by bacteria in the mouth by reducing their ability to metabolize sugars. However, dental treatment by fluorides must be used with care because an improper treatment results in formation of CaF_2 globules deposited on the enamel surface [753].

To conclude the teeth subject, let me briefly mention on the practical application of teeth. Due to relatively small dimensions of normal teeth, only tusks and ivory of giant animals are used. For example, both the Greek and Roman civilizations used large quantities of ivory to make high value works of art, precious religious objects and decorative boxes for costly objects. Ivory was often used to form the whites of the eyes of statues. Prior to introduction of plastics, it was used for billiard balls, piano keys, buttons and ornamental items. The examples of modern carved ivory objects are small statuary, netsukes, jewelry, flatware handles and furniture inlays.

1.4.3 Antlers

Deer antlers (Fig. 1.15) are unique biological structures since their growth rate is without parallel in vertebrates and because they are the only bony appendages in mammals capable of complete regeneration. This allows for basic research in bone biology without the interference of surgical procedures and their adverse effects in animals where samples are obtained. In addition, antlers also allow for the gathering of a large amount of samples from different populations to assess nutritional and ecological effects on bone composition and structure [754–757]. They are costly sexual secondary characters of male deer and constitute 1 to 5% of the body weight [758]. Recent studies suggest that antler regeneration is a stem cell-based process and that these stem cells are located in the pedicle periosteum [759, 760].

Antlers are not true horns; they are a simple extension of bone, so they have a matrix of biological apatite similar to that of mammalian bones [761]. Antlers are large and complex horn-like appendages of deer consisting of bony outgrowths from the head with no covering of keratin as is found in true horns. Usually, they begin growing in March and reach maturity in August. In winter, antlers fall off; this is known as shedding. Similar to bones, antlers contain pores and can withstand applied stresses of over 300 MPa [762–766], which is even higher than that of bones (Table 1.2). Therefore, antlers are occasionally considered an almost unbreakable bone [561]. Each antler grows from an attachment point on the skull called a pedicle. While an antler is growing, it is covered with highly vascular skin called velvet, which supplies oxygen and nutrients to the growing bone. Once the antler has achieved its proper size, the velvet starts to dry out, cracks and breaks off, while the antler's bone dies. Fully developed antlers consist of dead bone only [767–776]. It was found that food processing cannot supply the mineral needs required for antler growth and thus, male deer must temporary resorb calcium orthophosphate minerals from their own skeleton for antler growth [777–779]. Detailed studies revealed that daily food intake provided between 25 and 40% of calcium needed for antler mineralization, which resulted in a temporary skeleton demineralization [778, 779]. Interestingly, but antlers may act as large hearing aids, namely, moose with antlers have far more sensitive hearing than moose without [780].

Figure 1.15 Red deer stag at velvet shedding. The bare bone of the hard antlers is exposed. Reprinted from Ref. [760] with permission. A good cross-sectional image of a deer antler is available in Ref. [584].

Antlers are a good model to study bone biology because they are accessible, shed after mating season and cast every year [781]. However, people seldom come across the antlers in the woods.

Rabbits and rodents such as mice and chipmunks eat antlers (and bones of wild animals after they die) for calcium. Rodents and rabbits also gnaw bones and antlers to sharpen their incisors. Due to an extremely high growth rate, which can achieve 2–4 cm per day [767], combined with a very fast biomineralization, these unique appendages might be a well-suited animal model for studying the disturbances of bone formation induced by additives (e.g., by excess of fluoride) [769]. Antler size and external characteristics were found to be influenced by nutrition, climatic variability and other factors. Thus, since antlers are periodically replaced, the analysis of naturally cast antlers offers the opportunity for a continuous and a noninvasive monitoring of the environmental pollution by these additives [769]. Recently, the first attempt to evaluate a potential use of deer antlers as a bone regeneration biomaterial has been performed [782].

To conclude this part, let me briefly mention on the practical application of antlers. Associated with aristocracy, antlers have adorned European castles and hunting lodges for centuries. Today, furnishings and accessories made from antlers are featured in fine homes throughout the world and are a reflection of grace and elegance (Fig. 1.16).

Figure 1.16 Fallen antlers used to make a chandelier.

1.5 Pathological Calcification of Calcium Orthophosphates

In the body of mammals, osteoblasts and odontoblasts fix ions of calcium and orthophosphate and then precipitate biological apatite onto an organic matrix. This is the process of physiological biomineralization that is restricted to the specific sites in skeletal tissues, including growth plate cartilage, bones, teeth and antlers [107, 454]. Normally, mammals are supposed to die with calcium orthophosphates located in bones and teeth (and antlers for male deer) only and nowhere else, because under the normal conditions soft tissues are not mineralized. Unfortunately, owing to ageing, various diseases and under certain pathological conditions blood vessels, muscles, extracellular matrix of articular cartilaginous tissues of the joints and some internal organs are calcified as well. This process is called pathological calcification or ectopic (bio) mineralization and leads to a morbidity and a mortality [107, 454, 783]. In general, any type of abnormal accumulation of calcium orthophosphates in wrong places is accounted for by a disruption of systemic defense mechanism against calcification [784].

To the best of my knowledge, the first paper on a negative influence of unwanted depositions of calcium orthophosphates in the body was published as early as in 1911 [785]. This finding was confirmed in later studies [786, 787]. Unwanted depositions always lead to various diseases, for instance: soft tissue calcification (in damaged joints, blood vessels, dysfunctional areas in the brain, diseased organs, scleroderma, prostate stones) [230, 231, 232, 788–793], kidney and urinary stones [27, 794–797], dental pulp stones and dental calculus [187, 188, 190, 223, 229, 798–800], salivary stones [801], gall stones, pineal gland calcification, atherosclerotic arteries and veins [85, 802–805], coronary calcification [806], cardiac skeleton, damaged cardiac valves [807], calcification on artificial heart valves [808–812], carpal tunnel [813], cataracts [814], malacoplakia, calcified menisci [815, 816], dermatomyositis [817, 818], and still other diseases [107]. In addition, there is a metastatic calcification of nonosseous viable tissue occurring throughout the body [819, 820], but it primarily affects the interstitial tissue of the blood vessels, kidney, lungs and gastric mucosa. A metastatic calcification is defined as a deposition of calcium orthophosphates in previously normal tissue due to an abnormal biochemistry with disturbances in the calcium or phosphorus metabolism [821]. Common causes of the metastatic calcification include hyperparathyroidism, chronic renal disease, massive bone destruction in widespread bone metastases and increased intestinal calcium absorption. One author has mentioned on "apatite diseases" which are characterized by the appearance of needle-like crystals comparable to those of bone apatite in the fibrous connective tissue [822]. All these cases are examples of a calcinosis [823–825], which might be described as a formation of calcium orthophosphate deposits in any soft tissue. In dentistry, a calculus or a tartar refers to a hardened plaque on the teeth, formed by the presence of saliva, debris and minerals [826]. Its rough surface provides an ideal medium for bacterial growth, threatening the health of the gums and absorbing unaesthetic stains far more easily than natural teeth [27].

Calcifying nanodimensional particles are the first calcium orthophosphate mineral containing particles isolated from human blood and were detected in numerous pathologic calcification related diseases [827]. Interestingly, but contrary to the mineral phases of normal calcifications (bone, dentine, enamel, cementum, antlers), which consist of only one type of calcium orthophosphate (namely, biological apatite), the mineral phases of abnormal and/or pathological calcifications are found to occur as single or mixed phases of other types of calcium orthophosphates (ACP, DCPD, OCP, β-$(Ca,Mg)_3(PO_4)_2$) and/or other phosphatic and non-phosphatic compounds (e.g., magnesium orthophosphates, calcium pyrophosphates, calcium oxalates, etc.) in addition to or in place of biological apatite (Table 1.4) [27, 29, 84, 107, 147, 223, 229–232, 313, 828–832]. However, precipitation of biological apatite in wrong places is also possible; this is so-called "HA deposition disease" [833–836].

Occurrence of non-apatite phases in the pathological calcifications may indicate that they were crystallized under the conditions different from homeostasis or crystallization of the apatite structures was inhibited and less stable phases crystallized instead, without further change to the more stable one. Furthermore, in the places of pathological calcifications the solution pH is often relatively low. Given that nucleation and crystal growth is not a highly regulated process in any pathological deposits, there is not likely just one fundamental formation mechanism for all possible calcification types.

Furthermore, various bioorganic impurities in the local environment undoubtedly influence the crystallization process, resulting in a great variety of pathological deposits. Thus, it is a highly complex problem. In some cases, the chemical composition of an unwanted inorganic phase might depend on the age of the pathological calcification and its location. For example, DCPD is more frequently found in young (3 months or younger) calculus, biological apatite is present in all ages of calculus, while β-$(Ca,Mg)_3(PO_4)_2$ occurs more frequently in sub-gingival calculus. In mature calculus, the relative abundance of OCP, β-$(Ca,Mg)_3(PO_4)_2$ and biological apatite also differ between the inner and outer layers [84]. It is interesting to note that the mineral phases of animal calculus (e.g., from dog) was found to consist of calcium carbonate and biological apatite, while human calculi do not contain calcium carbonate [84, 837].

Table 1.4 Occurrence of calcium phosphates in biological systems (human) [84].

Calcium phosphate	Occurrence
Biological apatite	Enamel, dentine, bone, dental calculi, stones, urinary stones, soft-tissue deposits
OCP	Dental calculi and urinary stones
DCPD	Dental calculi, crystalluria, chrondrocalcinosis, in some carious lesions
β-$(Ca,Mg)_3(PO_4)_2$	Dental calculi, salivary stones, arthritic cartilage, soft-tissue deposits
$Ca_2P_2O_7\cdot 2H_2O$	Pseudo-gout deposits in synovium fluids
ACP	Heart calcifications in uremic patients, kidney stones

The nucleation process is the main step in both normal and pathological calcifications. *In vitro* experiments conducted by Grases and Llobera [838] to simulate the formation of sedimentary urinary stones, demonstrated that in the absence of organic matter no calcium orthophosphates crystallized in cavities with scarce liquid renovation, but regular CDHA layers appeared on the wall around the cavity. Visible deposits of calcified organic materials (mixtures of organic matter and spherulites of CDHA) were formed when a glycoprotein (mucin) was present. In this case, the walls of the cavity as well as the glycoproteins had the capacity to act as heterogeneous nucleators of calcium orthophosphates. CDHA microcrystal nucleation on the surface of epithelial cells can be a critical step in the formation of kidney stones [839] and identical mechanisms can be thought for unwanted calcifications in other soft tissues of the body, such as cardiac valves or vascular ducts. Monolayers of CDHA crystals can bind to epithelial cells. A large amount of kidney stones contains CDHA as the crystallization nuclei.

In general, formation of crystals in pathological mineralizations follows the same principles as normal calcifications [840–842]. Namely, local conditions for nucleation require a certain degree of local supersaturation induced by biochemical processes, which can be promoted by deficiency of inhibitors (like diphosphate, Mg^{2+} or even citrate ions) and/or the presence of matrix of a bioorganic material (such as cholesterol) or other crystals of different solids, those might act as heterogeneous nuclei. In addition, other regulators (activators and inhibitors) of physiological biomineralization have been identified and characterized [840–847]. What's more, the biological fluids (e.g., serum, saliva, synovial fluids) are normally supersaturated with respect to biological apatite precipitation [27, 84, 454]; therefore, in principle, calcification is thermodynamically feasible in any part of the body. However, normally it is not the case. Therefore, in the healthy body, the appropriate inhibitory mechanisms must be at work to prevent a superfluous calcification of soft tissues. These inhibition mechanisms are a hot research topic in molecular medicine but this subject is beyond the scope of current review. The interested readers are forwarded, for example, to a very interesting review on molecular recognition at the protein/HA interface [848]. More to the point, molecular, endocrine and genetic mechanisms of arterial calcification have been reviewed in another paper [849].

Very recently, an arachidic acid Langmuir monolayer system has been reported as a model for pathological mineralization of ion-substituted carbonateapatites from simulated body fluid [850]. The authors have demonstrated that the surface-induced formation of carbonateapatite starts with aggregation of prenucleation clusters of yet unknown calcium orthophosphates leading to nucleation of ACP before further development of oriented apatite crystals. This process is schematically shown in Fig. 1.17 [656, 850].

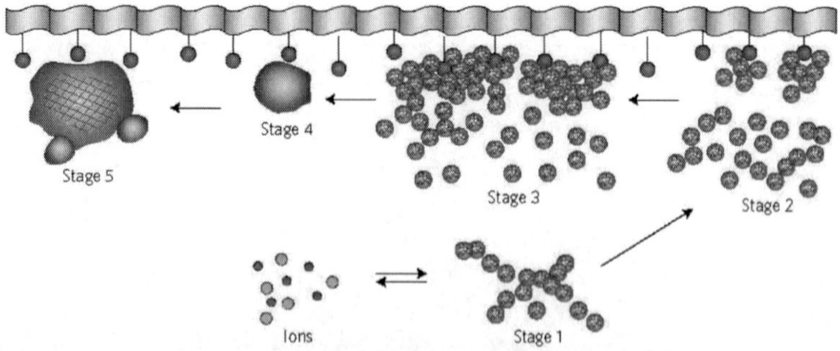

Figure 1.17 A schematic representation of the different stages of a surface-directed mineralization of calcium orthophosphates. In stage 1, aggregates of pre-nucleation clusters are in equilibrium with ions in solution. The clusters approach a surface with chemical functionality. In stage 2, pre-nucleation clusters aggregate near the surface, with loose aggregates still in solution. In stage 3, further aggregation causes densification near the surface. In stage 4, nucleation of spherical particles of ACP occurs at the surface only. In stage 5, crystallization occurs in the region of the ACP particles directed by the surface. Reprinted from Refs. [656, 850] with permission.

To conclude this part, it is worth reminding that calcium orthophosphates of biological origin are sparingly soluble in aqueous solutions. Removing them from the places of unwanted deposition would be an equivalent of bone demineralization; that is a challenge. Therefore, the majority of therapeutic approaches are directed at preventing the progression of pathological calcifications. Among them, a chelation therapy might be of some interest to chemists and materials researchers because it deals with chemical processes [851, 852]. The general principles of demineralization and decalcification (i.e., removing the mineral Ca-containing compounds (phosphates and carbonates) from the bioorganic matrix) have been extensively reviewed [853, 854], where the interested readers are referred to.

1.6 Biomimetic Crystallization of Calcium Orthophosphates

The term "biomimetics" ("the mimicry of life") was coined by an American inventor, engineer and biophysicist Otto Herbert Schmitt

(1913–1998) in the 1950s. Biomimetics (also known as bionics, biognosis, and/or biomimicry) might be defined as application of the methods and systems found in nature to the study, design and construction of new engineering systems, materials, chemical compounds and modern technology. Another definition describes biomimetics as a micro-structural process that mimics or inspires the biological mechanism, in part or as a whole [855]. This biological process generates highly ordered materials with a hybrid composition, a complex texture and ultra-fine crystallites through a hierarchical self-assembly and begins by designing and synthesizing molecules that have an ability to self-assemble or self-organize spontaneously to higher order structures.

Historically, the biomimetic concept is very old. (e.g., the Chinese wanted to make artificial silk ~3000 years ago; Daedalus' wings was one of the early design failure) but the implementation is gathering momentum only recently. The first papers with the term "biomimetics" in the title were published in 1972 [856, 857]. In spite of the tremendous achievements of modern science and technology, the nature's ability to assemble inorganic compounds into hard tissues (shells, spicules, teeth, bones, antlers, skeletons, etc.) is still not achievable by the synthetic procedures. This is not surprising — designs found in nature are the result of millions of years of evolution and competition for survival. The models that failed are fossils; those that survived are the success [858]. In the frames of this review, biomimetics is considered as mimicking natural manufacturing methods to generate artificial calcified tissues (grafts, implants, prostheses) those might be used as temporary or permanent replacements of the missing, lost, injured or damaged bones and teeth. It is important to notice, that precipitation of calcium orthophosphates and calcium carbonates have been considered to correlate with bone formation, at least, since 1923 [859].

A key step in the biomimetic bone graft production is attributed to the crystal growth of apatite phase onto a collagen matrix. Therefore, the matter of choosing the correct experimental conditions and well-mimicking solutions is of the primary importance. The easiest way to perform the crystallization would be mixing of aqueous solutions containing the ions of calcium and orthophosphate [27–29]. Unfortunately, such type of crystallization provides precipitates with the properties (chemical composition, Ca/P ratio, crystallinity level, particle size distribution, etc.) far different from those of biological apatite. This can be explained by the following paramount differences between the *in vivo* biological and *in vitro* chemical crystallization conditions [860]:

(i) *In vitro* crystallization normally occurs at permanently depleting concentrations of calcium and orthophosphate ions, while the concentrations of all ions and molecules are kept strictly constant during biological mineralization (the same is valid for the solution pH);

(ii) Chemical crystallization is a fast process (time scale of minutes to days), while the biological process is a slow one (time scale of weeks to years);

(iii) Many inorganic, bioorganic, biological and polymeric compounds are present in biological liquids (blood plasma, serum, saliva). Each of these compounds might act as an inhibitor, promoter, nucleator or even as a template for the growth of biological apatite [537]. In addition, each of them somehow influences the crystallization kinetics and might be either incorporated into the solid structure or co-precipitated with calcium orthophosphates.

(iv) Chemical crystallization is, by all means, a "passive" process, while the biological mineralization is strongly influenced by cells and occurs by the self-organization mechanisms [578, 603, 604]. Still there are no good ways to overcome this difference.

The first and the second differences might be overcome by using the appropriate crystallization techniques. The details are available elsewhere [860] but, briefly, the first problem might be overcome by either a continuous flow of a supersaturated solution [861, 862] or using a constant-composition (CC) technique [198, 863, 864]. The second difference might be surpassed by a restrained diffusion of calcium and orthophosphate ions from the opposite directions in, for example, a double-diffusion (DD) crystallization device or in viscous gels [436–438, 440, 441, 865–868]. The CC and DD techniques have been combined into a single constant-composition double-diffusion (CCDD) device, which currently seems to be the most advanced experimental tool to perform biomimetic crystallization [860, 869–873]. However, in no case the CCDD device should be considered as the final construction; it still has much room for further improvement, e.g., by upgrading the design of the crystallization chamber [874].

Other constructions, e.g., to study calcification of biological heart valve prostheses [875], are also possible. In addition, one should keep in mind that the potential of the standard CC technique has not reached its limit yet: for example, recently a good mimicking of the self-organized microstructure of tooth enamel has been achieved [876].

The third major difference between the *in vivo* and *in vitro* crystallization conditions might be overcome by using the appropriate crystallization solutions [860]. The presence of calcium and orthophosphate ions in some biological fluids has been known, at least, since 1921 [877, 878]. Therefore, the best way would be to perform experiments using natural liquids (blood serum, saliva, lymph, etc.), but this is not easy due to great variability of the chemical and biochemical compositions of natural liquids and problems with their collection and storage. As stated before, using supersaturated aqueous solutions containing only the ions of calcium and orthophosphate appears to be unable to mimic the crystallization of biological apatite; therefore, more advanced solutions have been elaborated. To the best of my knowledge, Hanks' balanced salt solution (HBSS) [879] was the first successful simulating medium, containing the ions of calcium and orthophosphate together with other inorganic ions and glucose. HBSS is commercially available and still used in biomimetic experiments [880–882]; its chemical composition might be taken, e.g., from Refs. [883, 884]. Other popular physiological solutions include α-modified Eagle's [885] medium (α-MEM) and its variation Dulbecco's [886] modified Eagle's medium (DMEM), which contain numerous bioorganic (alanine, aspartic acid, glycine, biotin, vitamin C, folic acid, riboflavin) and inorganic ($CaCl_2$, KCl, NaCl, NaH_2PO_4) components [887–891], phosphate buffered saline (PBS) that contains only inorganic ($CaCl_2$, $MgCl_2$, KCl, KH_2PO_4, NaCl, NaH_2PO_4) components [892, 893]. Furthermore, artificial saliva [894–896], synthetic urine [838, 897] and simulated milk ultrafiltrate (SMUF) [898–901] solutions are available. They contain both bioorganic (e.g., xantan gum or sodium carboxymethylcellulose, sorbitol, etc.) and inorganic (e.g., $CaCl_2$, $MgCl_2$, KCl, KH_2PO_4, NaCl, KH_2PO_4) compounds. Additional media used for mineralization studies are listed in Table 3 of Ref. [578]. All these simulating solutions are commercially available.

However, the most popular biomimetic solution is a protein-free acellular simulated body fluid (SBF). It was introduced by Kokubo et al., [902] and occasionally named as Kokubo's SBF. It is a metastable aqueous solution with pH ~7.40, supersaturated with respect to the precipitation of OCP, β-TCP, CDHA and HA [903], containing only inorganic ions in concentrations nearly equal to those in human blood plasma. However, the standard SBF formulation, firstly, contains the tris/HCl buffer, and, secondly, the concentration of hydrogencarbonate (4.2 mM) is only a fraction of that in blood plasma (27 mM) [902]. The problem of a low concentration of hydrogencarbonate ions has been overcome by first introducing a "synthetic body fluid" [904–906] and later a revised SBF (rSBF) [907, 908]. Due to the chemical similarity with human blood plasma, rSBF currently seems to be the best simulating solution. However, it contains Hepes buffer, loses CO_2 in open vessels and does not contain any organic and/or biological molecules [907, 908]. Other types of SBF are also available [909–912] and the interested readers are referred to a leading opinion co-authored by the SBF inventor [913], where the entire history and the preparation techniques of various SBF formulations are well described. Recently, another leading opinion on the suitability of SBF for the *in vitro* bioactivity tests was published [914]. The authors demonstrated that (i) there is presently no enough scientific data to support the SBF suitability and (ii) even though bioactivity tests with SBFs are valid, the way the tests are generally conducted leaves room for further improvements. Furthermore, the preparation protocol of SBF solutions was reconsidered and a new procedure was suggested to improve the reproducibility of bioactivity tests [914]. The application of SBF for the surface mineralization of various materials *in vitro* has been reviewed in Ref. [915], while the theoretical analysis of calcium orthophosphate precipitation (the driving force and the nucleation rate based on the classical crystallization theory) in SBF is also available [903]. It is important to note that nanometer-sized prenucleation clusters in SBF solutions have been discovered [850]; those clusters are believed to be the initial building blocks of crystallized calcium orthophosphates (e.g., CDHA [291]), while the crystallization process itself occurs via intermediate formation of ACP (Fig. 1.17).

Further attempts to improve the biomimetic properties of SBF and rSBF have been performed [913, 914]. Efforts were made to replace artificial buffers (tris/HCl, Hepes) with simultaneous increasing the concentration of hydrogencarbonates for SBF [916–918] or avoiding losses of CO_2 from open vessels for rSBF [860, 869–873] by means of permanent bubbling of gaseous CO_2 through the solutions. Addition of the most important organic and biological compounds like glucose [871], albumin [869, 918], lactates [919] and collagen [920] is another direction to improve biomimetic properties of various types of SBF. Once a cow milk-based rSBF has been prepared [921]. Further improvements of all biomimetic solutions are to be made in future. Occasionally, condensed solutions of SBF (e.g., 1.5-fold, 2-fold [920, 922, 923], 5-fold [924, 925], and even 10-fold [926]) are used to accelerate precipitation and increase the amount of precipitates. However, whenever possible this should be avoided because the application of condensed solutions of SBF leads to changes in the chemical composition of the precipitates; namely, the concentration of carbonates increases, while the concentration of orthophosphates decreases [927].

To conclude this part, one should note on difficulties in mimicking the calcification process that occurs in bones and teeth. A reasonable mechanism of the induction of CDHA nucleation and crystallization by carboxylate groups on the bioorganic matrices looks as this. At first, calcium and orthophosphate ions are combined with carboxylate groups. By using this as seeds, CDHA crystals then grow to generate interfaces that contain the most stable structure of the {100} faces. Such a crystallization mechanism explains why the *c*-axes of biological apatite are parallel to the organic matrices. Collagen fibers can be regarded as axis-like organic matrices: when CDHA is formed on the surface of collagen fibers parallel to the *c*-axes, the *c*-axes are oriented parallel to the fiber orientation [928]. A step further would be to perform the precipitation from the simulating solutions on templates of biomineralization proteins for the control of crystal organization and properties. For example, there are successful attempts to crystallize calcium orthophosphates on collagen in order to obtain bone-like composites [572, 929–938]. Such collagen/calcium orthophosphate biocomposites are currently under investigation for clinical use (see section *6.4.4. Biocomposites with collagen* for details). Other popular biomimetic matrixes to perform calcium orthophosphate crystallization comprise gelatin [436–441, 939–941], chitosan [939, 942, 943], organic polyelectrolytes [944–947], metals and alloys [948–954], polymers [955], cellulose [956], self-assembled monolayers [957] and many other materials (Chapter 6). Such biomimetically prepared calcium orthophosphate precipitates are occasionally called "organoapatites" [533, 958].

1.7 Conclusions and Outlook

By the end of the twentieth century, it became clear that calcium orthophosphate biomaterials and bioceramics by themselves could not give a complete response to the clinical needs for artificial implants. Biomaterials with more demanding properties were required. Namely, in 1998, Prof. Larry L. Hench published a forecast for the future of biomaterials development [959], where he noted that available that time bioactive materials (calcium orthophosphates, bioactive glasses and glass ceramics) had already improved prostheses lifetime but, unfortunately, any type of prosthesis had mechanical limitations. As the solution, he proposed that biomaterial researchers would need to focus on tissue regeneration instead of tissue replacement. A working hypothesis was announced: "Long-term survivability of prosthesis will be increased by the use of biomaterials that enhance the regeneration of natural tissues" [959]. One path to follow is the regeneration of bone using calcium orthophosphate scaffolds that mimic the structure of biological apatite, bond to bone and in some cases activate the genes within bone cells to stimulate new bone growth [960–962]. Thus, more than 10 years ago Prof. Hench predicted a rapid development of tissue engineering field, where calcium orthophosphates play an auxiliary role. The history has shown that tissue engineering, indeed, is a very rapidly developed field of science and research [963].

However, what can be said about calcium orthophosphates themselves? The major questions on chemistry, crystallization, ion-substitution, crystallography, thermodynamics and phase relationships for the chemically pure calcium orthophosphates have been answered in the twentieth century. Some important topics for DCPD and CDHA have been additionally investigated in the field of self-setting calcium orthophosphate formulations (Chapter 5). Conversely, calcium orthophosphates of biological origin, including the control of their morphology and interaction of calcium orthophosphate bioceramics with various bioorganic compounds are not well investigated yet. The same is valid for the nanocrystalline (Chapter 3) and amorphous (Chapter 2) samples of calcium orthophosphates. Small amounts of bone-like apatite might be easily prepared by crystallization from SBF and rSBF but what can be said about larger quantities? A standard way of the concentration increasing causes chemical changes in the precipitates [927]. After a necessary technology is developed, one will have to think on scaffold preparation from this material, keeping in mind that any thermal treatment would destroy this material. A spark plasma sintering approach based on the use of pulsed current and enabling very fast heating and cooling rates seemed to be a first hint to achieve this goal [964]. However, a rapid development of the self-setting calcium orthophosphate formulations (Chapter 5), which can be easily doped by the necessary chemical elements, seems to be a better solution of this problem. Furthermore, the existence of OA remains to be questionable, as well as the bioactivity mechanism of calcium orthophosphates requires better understanding.

To date, although calcium orthophosphate biomaterials and bioceramics have been extensively studied for over 50 years, their ability to trigger bone formation is still incomparable with other biomaterials. Naturally, the biomaterials' field is shifting towards biologically active systems in order to improve their performance and to expand their use [965]. Due to this case, tissue engineering is the strongest direction of current research, which, in the case of calcium orthophosphates, means fabrication of proper substrates and/or scaffolds to carry cells, hormones and biochemical factors to be further used in surgery and medicine. Presumably, a synthesis of various types of calcium orthophosphate-based biocomposites and hybrid biomaterials (Chapter 6) occupies the second important place. For example, even composites with carbon nanotubes already exist [966–968]! The third important place is occupied by investigations devoted to the synthesis and characterization of various nano-sized particles and nanodimensional crystals of calcium orthophosphates (Chapter 3), as well as by synthesis of calcium orthophosphates with controlled particle geometry [537]. In general, the geometry of crystal phases can be varied by controlling the precipitation conditions, such as temperature, solution pH, concentration of the reagents, hydrodynamics, presence of various admixtures, inhibitors or promoters, ultrasonication, etc. All these approaches might be useful in preparation of calcium orthophosphate fibers, whiskers, hollow microspheres, etc. In addition, a great attention is paid to manufacturing of the self-setting calcium orthophosphate formulations (Chapter 5) and multiphase [969] mixtures mimicking as closely as possible the mineral component of biological apatite. A work along the ecological ways of synthesis of calcium orthophosphates might be of a great importance as well [971]. A deeper study of the fascinating growth rate of deer antlers and the ability of some animals, such as newts, to regenerate amputated limbs might provide new and unexpected approaches to the bone-healing concept, as well as this will be important for further development of both biomimetics and biomineralization fields. Unfortunately, no currently available grafting biomaterials can substitute the bones' mechanical function, illustrating yet unmet medical need that would entirely substitute and regenerate a damaged tissue or organ. In a close future, the foreseeable application of calcium orthophosphates will be as a component of the third generation biomaterials [959, 962], where they will support cells and/or other biologically active substances (peptides, growth factors, hormones, drugs, etc.) to guide regeneration of hard tissues [972–982].

To finalize this chapter, one should note that, in spite of a long history of the calcium orthophosphate research (Chapter 8) and many important discoveries, still many gaps remain in our knowledge to be investigated in future.

References and Notes

1. As a mineral species, apatite was first recognized by the father of German geology Abraham Gottlob Werner (1750–1817) in 1786 and named by him from the ancient Greek απατάω (apatao) – "to mislead" or "to deceive", because it had previously been mistaken for other minerals, such as beryl, tourmaline, chrysolite, amethyst, fluorite, etc. Currently, apatite is the name for a group of minerals with the same crystallographic structure and does not indicate one chemical composition. That is why, the term "calcium apatite" is used in this book.
2. Shepperd, J. (2004) The early biological history of calcium phosphates, in *Fifteen Years of Clinical Experience with Hydroxyapatite Coatings in Joint Arthroplasty* (ed. Epinette, J. A., and Manley, M. T.), Springer, France, pp. 3–8.
3. Berzelius, J. (1845) Ueber basische phosphorsaure kalkerde, *Ann. Chem. Pharmac.*, **53**, 286–288.

4. Warington, Jr., R. (1866) Researches on the phosphates of calcium, and upon the solubility of tricalcic phosphate, *J. Chem. Soc.*, **19**, 296–318.

5. Fresenius, R. (1867) Ueber die Bestimmung der Phosphorsäure im Phosphorit nebst Mittheilung der Analysen des Phosphorits und Staffelits aus dem Lahnthal, *Zeitschrift für Analytische Chemie*, **6**, 403–409.

6. Church, A. H. (1873) New analyses of certain mineral arseniates and phosphates. 1. Apatite; 2. Arseniosiderite; 3. Childrenite; 4. Ehlite; 5. Tyrolite; 6. Wavellite, *J. Chem. Soc.*, **26**, 101–111.

7. L., W. J. (1883) On a fine specimen of apatite from Tyrol, lately in the possession of Mr. Samuel Henson, *Nature*, **27**, 608–609.

8. Tereg, A., and Das V. der (1883) Calciumphosphate im Organismus der Fleischfresser. Pflüger, Archiv für die Gesammte Physiologie des Menschen und der Thiere, *Eur. J. Physiology*, **32**, 122–170.

9. Mohr, C. (1884) Ueber die quantitative Bestimmung der zurückgegangenen Phosphorsäure und der Phosphorsäure im Dicalciumphosphat, *Zeitschrift für Analytische Chemie*, **23**, 487–491.

10. Glaser, C. (1885) Bemerkungen zu der Abhandlung des Herrn Carl Mohr über die quantitative Bestimmung der zurückgegangenen Phosphorsäure und der Phosphorsäure im Dicalciumphosphat, *Zeitschrift für Analytische Chemie*, **24**, 180.

11. Hutchings, W. M. (1887) Occurrence of apatite in slag, *Nature*, **36**, 460.

12. Hilgenstock, G. (1883) Eine neue Verbindung von P_2O_5 und CaO, *Stahl und Eisen*, **3**, 498.

13. Hilgenstock, G. (1887) Das vierbasische Kalkphosphat und die Basicitätsstufe des Silicats in der Thomas-Schlacxke, *Stahl und Eisen*, **7**, 557–560.

14. Scheibler, C. (1886) Ueber die Herstellung reicher Kalkphosphate in Verbindung mit einer Verbesserung des Thomasprocesses, *Ber. Dtsch. Chem. Ges.*, **19**, 1883–1893.

15. Georgievics, G. V. (1891) Über das Verhalten des Tricalciumphosphats gegen Kohlensäure und Eisenhydroxyd, *Monatshefte für Chemie*, **12**, 566–581.

16. Dreesmann, H. (1892) Ueber Knochenplombierung, *Beitr. Klin. Chir.*, **9**, 804–810.

17. Cameron, F. K. (1904) The action of water upon the phosphates of calcium, *J. Am. Chem. Soc.*, **26**, 1454–1463.

18. Cameron, F. K., and Seidell, A. (1905) The phosphates of calcium, I. *J. Am. Chem. Soc.*, **27**, 1503–1512.

19. Cameron, F. K., and Bell, J. M. (1905) The phosphates of calcium, II. *J. Am. Chem. Soc.*, **27**, 1512–1514.

20. Cameron, F. K., and Bell, J. M. (1906) The phosphates of calcium, III; Superphosphate, *J. Am. Chem. Soc.*, **28**, 1222–1229.

21. Cameron, F. K., and Bell, J. M. (1910) The phosphates of calcium, IV. *J. Am. Chem. Soc.*, **32**, 869–873.

22. Bassett, H., Jr. (1907) Beiträge zum Studium der Calciumphosphate. I. Die Hydrate der Calcium-Hydroorthophosphate, *Z. Anorg. Chem.*, **53**, 34–48.

23. Bassett, H., Jr. (1907) Beiträge zum Studium der Calciumphosphate. II. Die Einwirkung von Ammoniakgas auf Calcium-Hydroorthophosphate, *Z. Anorg. Chem.*, **53**, 49–62.

24. Bassett, H., Jr. (1908) Beiträge zum Studium der Calciumphosphate. III. Das System CaO – P_2O_5 – H_2O, *Z. Anorg. Chem.*, **59**, 1–55.

25. Bassett, H., Jr. (1917) The phosphates of calcium. Part IV. The basic phosphates, *J. Chem. Soc.*, **111**, 620–642.

26. Lide, D. R. (2005) *The CRC Handbook of Chemistry and Physics*, 86th edn., CRC Press, Boca Raton, p. 2544.

27. LeGeros, R. Z. (1991) Calcium phosphates in oral biology and medicine, *Monographs in Oral Science*, vol. 15, Karger, Basel, p. 201.

28. Elliott, J. C. (1994) Structure and chemistry of the apatites and other calcium orthophosphates, *Studies in Inorganic Chemistry*, vol. 18, Elsevier, Amsterdam, p. 389.

29. Amjad, Z. (ed.) (1997) *Calcium Phosphates in Biological and Industrial Systems*, Kluwer Academic Publishers, Boston, p. 529.

30. Dorozhkin, S. V. (2007) Calcium orthophosphates, *J. Mater. Sci.*, **42**, 1061–1095.

31. Dorozhkin, S. V. (2009) Calcium orthophosphates in nature, biology and medicine, *Materials*, **2**, pp. 399–498.

32. Cantelar, E., Lifante, G., Calderón, T., Meléndrez, R., Millán, A., Alvarez, M. A., and Barboza-Flores, M. (2001) Optical characterisation of rare earths in natural fluorapatite, *J. Alloy Compd.*, pp. 323–324, pp. 851–854.

33. Ribeiro, H. B., Guedes, K. J., Pinheiro, M. V. B., Greulich-Weber, S., and Krambrock, K. (2005) About the blue and green colours in natural fluorapatite, *Physica Status Solidi C*, **2**, pp. 720–723.

34. Dorozhkin, S. V., and Epple, M. (2002) Biological and medical significance of calcium phosphates, *Angew. Chem. Int. Ed. Engl.*, **41**, 3130–3146.

35. McConnell, D. (1973) Apatite: Its crystal chemistry, mineralogy, utilization, and geologic and biologic occurrences, *Applied Mineralogy*, vol. 5., Springer-Verlag, Vienna and New York, p. 111.

36. Becker, P. (1989) Phosphates and phosphoric acid: raw materials technology and economics of the wet process, *Fertilizer Science and Technology Series*, 2nd edn., Marcel Dekker, New York, p. 760.

37. Smith, D. K. (1994) Calcium phosphate apatites in nature, in *Hydroxyapatite and Related Materials* (ed. Brown, P. W., and Constantz, B.), CRC Press, Boca Raton, pp. 29–44.

38. Angelov, A. I., Levin, B. V., and Chernenko, Y. D. (2000) *Phosphate Ore* (in Russian), Nedra Busyness Centre, Moscow, p. 120.

39. Cook, P. J., Shergold, J. H., and Davidson, D. F. (eds.) (2005) *Phosphate Deposits of the World: Phosphate Rock Resources*, Cambridge University Press, Cambridge, vol. 2, p. 600.

40. Zhang, J. Z., Guo, L., and Fischer, C. J. (2010) Abundance and chemical speciation of phosphorus in sediments of the Mackenzie river delta, the Chukchi sea and the Bering sea: importance of detrital apatite, *Aquatic Geochem*, **16**, 353–371.

41. Omelon, S. J., and Grynpas, M. D. (2008) Relationships between polyphosphate chemistry, biochemistry and apatite biomineralization, *Chem. Rev.*, **108**, 4694–4715.

42. Jarvis, I. (1994) Phosphorite geochemistry: state-of-the-art and environmental concerns, *Eclogae Geologicae Helvetiae*, **87**, 643–700.

43. Glenn, C. R. (1994) Phosphorus and phosphorites: sedimentology and environments of formation, *Eclogae Geologicae Helvetiae*, **87**, 747–788.

44. McClellan, G. H. (1980) Mineralogy of carbonate fluorapatites (Francolites), *J. Geological Soc.*, **137**, 675–681.

45. Mcarthur, J. M. (1985) Francolite geochemistry-compositional controls during formation, diagenesis, metamorphism and weathering, *Geochim. Cosmochim. Acta*, **49**, 23–35.

46. Schuffert, J. D., Kastner, M., Emanuele, G., and Jahnke, R. A. (1990) Carbonate-ion substitution in francolite: a new equation, *Geochim. Cosmochim. Acta*, **54**, 2323–2328.

47. Pan, Y., and Fleet, M. E. (2002) Compositions of the apatite-group minerals: substitution mechanisms and controlling factors, in

Phosphates: Geochemical, Geobiological and Materials Importance. Series: Reviews in Mineralogy and Geochemistry, (ed. Hughes, J. M., Kohn, M., and Rakovan, J.), vol. 48, *Mineralogical Society of America*, Washington, D.C., pp. 13–49.

48. Zanin, Y. N. (2004) The classification of calcium phosphates of phosphorites, *Lithol. Miner. Resour.*, **39**, 281–282.

49. Rogers, A. F. (1922) Collophane, a much neglected mineral, *Am. J. Sci.*, **3**, 269–276.

50. Kozlovsky, Y. G., Yarygin, V. I., and Shokarev, M. M. (1977) Colloid calcium phosphate (collophan) in the composition of urinary calculi, *Urologiya i Nefrologiya*, **42**, 27–32.

51. *http://www.mindat.org/min-10072.html* (accessed in September 2010).

52. Elorza, J., Astibia, H., Murelaga, X., and Pereda-Suberbiola, X. (1999) Francolite as a diagenetic mineral in dinosaur and other upper cretaceous reptile bones (Lano, Iberian peninsula): microstructural, petrological and geochemical features, *Cretaceous Res.*, **20**, 169–187.

53. Hubert, B., Álvaro, J. J., and Chen, J. Y. (2005) Microbially mediated phosphatization in the Neoproterozoic Doushantuo Lagerstätte, *South China. Bull. Soc. Géol. Fr.*, **176**, 355–361.

54. Xiao, S., Zhang, Y., and Knoll, A. H. (1998) Three-dimensional preservation of algae and animal embryos in a neoproterozoic phosphorite, *Nature*, **391**, 553–558.

55. Xiao, S., Yuan, X., and Knoll, A. H. (2000) Eumetazoan fossils in terminal Proterozoic phosphorites? *Proc. Natl. Acad. Sci. USA*, **97**, 13684–13689.

56. Chakhmouradian, A. R., and Medici, L. (2006) Clinohydroxylapatite: a new apatite-group mineral from northwestern Ontario (Canada), and new data on the extent of Na-S substitution in natural apatites, *Eur. J. Mineral.*, **18**, 105–112.

57. *http://www.mindat.org/gallery.php?min=9293* (accessed in October 2010).

58. Klein, C. (1901) Brushite from the island of Mona (between Haiti and Puerto Rico), *Sitzber. K. Preuss. Aka.*, 720–725.

59. Merrill, G. P. (1917) On the calcium phosphate in meteoric stones, *Am. J. Sci.*, **43**, 322–324.

60. Tyrrell, G. W. (1938) Apatite, nepheline, and rare-earth mining in the Kola Peninsula, *Nature*, **141**, 354–355.

61. Kogarko, L. N. (1999) Problems of the genesis of giant apatite and rare metal deposits of the Kola Peninsula, Russia, *Geology of Ore Deposits*, **41**, 351–366.

62. Ford, A. K. (1917) A remarkable crystal of apatite from Mt. Apatite, Auburn, Maine, *Am. J. Sci.*, **44**, 245–246.

63. Hogarth, D. D. (1974) The discovery of apatite on the Lievre River, *Quebec. Mineral. Rec.*, **5**, 178–182.

64. van Velthuizen, J. (1992) Giant fluorapatite crystals: a question of locality, *Mineral. Rec.*, **23**, 459–463.

65. Trueman, N. A. (1966) Substitutions for phosphate ions in apatite, *Nature*, **210**, 937–938.

66. Gilinskaya, L. G. (2010) Organic radicals in natural apatites according to EPR data: potential genetic and paleoclimatic indicators, *J. Struct. Chem.*, **51**, 471–481.

67. Gilinskaya, L. G. (2010) A stable perinaphthenyl radical in natural apatites, *J. Struct. Chem.*, **51**, 761–764.

68. Sánchez-Salcedo, S., Vila, M., Izquierdo-Barba, I., Cicuéndez, M., and Vallet-Regí, M. (2010) Biopolymer-coated hydroxyapatite foams: a new antidote for heavy metal intoxication, *J. Mater. Chem.*, **20**, 6956–6961.

69. White, T., Ferraris, C., Kim, J., and Madhavi, S. (2005) Apatite – an adaptive framework structure, in *Micro- and Mesoporous Mineral Phases. Series: Reviews in Mineralogy and Geochemistry* (ed. Ferraris, G., and Merlino, S.), vol. 57, Mineralogical Society of America, Washington, D.C., pp. 307–401.

70. Jacob, K. D., and Reynolds, D. S. (1928) Reduction of tricalcium phosphate by carbon, *Ind. Eng. Chem.*, **20**, 1204–1210.

71. Emsley, J. (2001) The Shocking History of Phosphorus: A Biography of the Devil's Element, Pan Books, Macmillan, p. 336.

72. van der Sluis, S., Meszaros, Y., Marchee, W. G. J., Wesselingh, H. A., and van Rosmalen, G. M. (1987) The digestion of phosphate ore in phosphoric acid, *Ind. Eng. Chem. Res.*, **26**, 2501–2505.

73. Dorozhkin, S. V. (1996) Fundamentals of the wet-process phosphoric acid production. 1. Kinetics and mechanism of the phosphate rock dissolution, *Ind. Eng. Chem. Res.*, **35**, 4328–4335.

74. Dorozhkin, S. V. (1997) Fundamentals of the wet-process phosphoric acid production. 2. kinetics and mechanism of $CaSO_4 \cdot 0.5H_2O$ surface crystallization and coating formation, *Ind. Eng. Chem. Res.*, **36**, 467–473.

75. Dorozhkin, S. V. (1998) Ecological principles of wet-process phosphoric acid technology, *J. Chem. Technol. Biotechnol.*, **71**, 227–233.

76. Copson, R. L., Newton, R. H., and Lindsay, J. D. (1936) Superphosphate manufacture – mixing phosphate rook with concentrated phosphoric acid, *Ind. Eng. Chem.*, **28**, 923–927.

77. Newton, R. H., and Copson, R. L. (1936) Superphosphate manufacture – composition of superphosphate made from phosphate rock and concentrated phosphoric acid, *Ind. Eng. Chem.*, **28**, 1182–1186.

78. Rossete, A. L. R. M., Carneiro, J. M. T., Bendassolli, J. A., Tavares, C. R. O., and Sant'Ana Filho, C. R. (2008) Production of single superphosphate labeled with 34S, *Scientia Agricola*, **65**, 91–94.

79. Magda, A., Pode, V., Niculescu, M., Muntean, C., Bandur, G., and Iovi, A. (2008) Studies on process of obtaining the fertilizers based on ammonium phosphates with addition of boric acid, *Rev. Chim. Bucharest*, **59**, 1340–1344.

80. Abouzeid, A. Z. M. (2007) Upgrading of phosphate ores – a review, *Powder Handling and Processing*, **19**, 92–109.

81. Abouzeid, A. Z. M. (2008) Physical and thermal treatment of phosphate ores – an overview, *Int. J. Mineral Process.*, **85**, 59–84.

82. Rey, C., Combes, C., Drouet, C., and Sfihi, H. (2006) Chemical diversity of apatites, *Adv. Sci. Technol.*, **49**, 27–36.

83. O'Neill, W. C. (2007) The fallacy of the calcium – phosphorus product, *Kidney Int.*, **72**, 792–796.

84. LeGeros, R. Z. (2001) Formation and transformation of calcium phosphates: relevance to vascular calcification, *Z. Kardiol.*, **90**(3), III116–III125.

85. Becker, A., Epple, M., Müller, K. M., and Schmitz, I. (2004) A comparative study of clinically well-characterized human atherosclerotic plaques with histological, chemical, and ultrastructural methods, *J. Inorg. Biochem.*, **98**, 2032–2038.

86. Wopenka, B., and Pasteris, J. D. (2005) A mineralogical perspective on the apatite in bone, *Mater. Sci. Eng. C*, **25**, 131–143.

87. Pasteris, J. D., Wopenka, B., and Valsami-Jones, E. (2008) Bone and tooth mineralization: why apatite? *Elements*, **4**, 97–104.

88. Sun, Y., and Hanley, E. N., Jr. (2007) Calcium-containing crystals and osteoarthritis, *Curr. Opin. Orthoped.*, **18**, 472–478.

89. Daculsi, G., LeGeros, R. Z., and Mitre, D. (1989) Crystal dissolution of biological and ceramic apatites, *Calcif. Tiss. Int.*, **45**, 95–103.

90. Lee-Thorp, J. A., and van der Merwe, N. J. (1991) Aspects of the chemistry of modern and fossil biological apatites, *J. Archaeological Sci.*, **18**, 343–354.

91. Bigi, A., Foresti, E., Gregorini, R., Ripamonti, A., Roveri, N., and Shah, J. S. (1992) The role of magnesium on the structure of biological apatites, *Calcif. Tiss. Int.*, **50**, 439–444.

92. Nakano, T., Kaibara, K., Tabata, Y., Nagata, N., Enomoto, S., Marukawa, E., and Umakoshi, Y. (2002) Unique alignment and texture of biological apatite crystallites in typical calcified tissues analyzed by microbeam x-ray diffractometer system, *Bone*, **31**, 479–487.

93. Grynpas, M. D., and Omelon, S. (2007) Transient precursor strategy or very small biological apatite crystals? *Bone*, **41**, 162–164.

94. Lee, J. W., Sasaki, K., Ferrara, J. D., Akiyama, K., Sasaki, T., and Nakano, T. (2009) Evaluation of preferential alignment of biological apatite (BAp) crystallites in bone using transmission X-ray diffraction optics, *J. Jpn. Ins. Metal.*, **73**, 786–793.

95. Ishimoto, T., Sakamoto, T., and Nakano, T. (2010) Orientation of biological apatite in rat calvaria analyzed by microbeam X-ray diffractometer, *Mater. Sci. Forum*, 638–642, 576–581.

96. Danil'chenko, S. N., Kulik, A. N., Bugai, A. N., Pavlenko, P. A., Kalinichenko, T. G., Ul'yanchich, N. V., and Sukhodub, L. F. (2005) Thermally activated diffusion of magnesium from bioapatite crystals, *J. Appl. Spectrosc.*, **72**, 899–905.

97. Passey, B. H., Robinson, T. F., Ayliffe, L. K., Cerling, T. E., Sphonheimer, M., Dearing, M. D., Roeder, B. L., and Ehleringer, J. R. (2005) Carbon isotopic fractionation between diet, breath and bioapatite in different mammals, *J. Arch. Sci.*, **32**, 1459–1470.

98. Meneghini, C., Dalconi, M. C., Nuzzo, S., Mobilio, S., and Wenk, R. H. (2003) Rietveld refinement on X-ray diffraction patterns of bioapatite in human fetal bones, *Biophys. J.*, **84**, 2021–2029.

99. Eagle, R. A., Schauble, E. A., Tripati, A. K., Tütken, T., Hulbert, R. C., and Eiler, J. M. (2010) Body temperatures of modern and extinct vertebrates from ^{13}C-^{18}O bond abundances in bioapatite, *Proc. Natl. Acad. Sci. USA*, **107**, 10377–10382.

100. There are reports that dahllite belongs to the francolite group. Natural dahllite might be a rock-forming mineral [101]. For example, it was found in some phosphorite concretions of Podolia [102, 103]. In addition, it was found in both massive and accretionary crustal phosphorites [48].

101. Rogers, A. F. (1912) Dahllite from Tonopah, Nevada; Voelckerite, a new basic calcium phosphate; comments on the chemical composition of apatite and phosphorite, *Z. Krystallogr. Mineral.*, **52**, 209–217.

102. Schaller, W. T. (1910) The probable identity of podolite with dahllite, *Am. J. Sci.*, **30**, 309–310.

103. Schaller, W. T. (1910) On the likely identity of podolite and dahllite, *Z. Krystallogr. Mineral.*, **48**, 559–561.

104. Collagens are fibrous, insoluble proteins found in the connective tissues, including skin, bone, ligaments and cartilage.

105. Skinner, H. C. W. (2005) *Biominerals. Mineral. Mag.*, **69**, 621–641.

106. Daculsi, G. (1992) Physicochemical and ultrastructural analysis of bone bioactive interface, *Biomater. Tissue Int.*, **10**, 296–304.

107. Daculsi, G., Bouler, J. M., and LeGeros, R. Z. (1997) Adaptive crystal formation in normal and pathological calcifications in synthetic calcium phosphate and related biomaterials, *Int. Rev. Cytology*, **172**, 129–191.

108. Driessens, F. C. M., and Verbeeck, R. M. H. (1990) *Biominerals*, CRC Press, Boca Raton, p. 440.

109. Clark, N. A. (1931) The system $P_2O_5 - CaO - H_2O$ and the recrystallization of monocalcium phosphate, *J. Phys. Chem.*, **35**, 1232–1238.

110. Brown, P. W. (1992) Phase relationships in the ternary system $CaO - P_2O_5 - H_2O$ at 25°C, *J. Am. Ceram. Soc.*, **75**, 17–22.

111. Martin, R. I. and Brown, P. W. (1997) Phase equilibria among acid calcium phosphates, *J. Am. Ceram. Soc.*, **80**, 1263–1266.

112. Kreidler, E. R., and Hummel, F. A. (1967) Phase relationships in the system $SrO-P_2O_5$ and the influence of water vapor on the formation of $Sr_4P_2O_9$, *Inorg. Chem.*,**6**, 884–891.

113. Carayon, M. T., and Lacout, J. L. (2003) Study of the Ca/P atomic ratio of the amorphous phase in plasma-sprayed hydroxyapatite coatings, *J. Solid State Chem.*,**172**, 339–350.

114. White, T. J., and Dong, Z. L. (2003) Structural derivation and crystal chemistry of apatites, *Acta Crystallogr. B*, **B59**, 1–16.

115. Mathew, M., and Takagi, S. (2001) Structures of biological minerals in dental research, *J. Res. Natl. Inst. Stand. Technol.*, **106**, 1035–1044.

116. Hughes, J. M., and Rakovan, J. (2002) The crystal structure of apatite, $Ca_5(PO_4)_3(F,OH,Cl)$, in *Phosphates: Geochemical, Geobiological and Materials Importance. Series: Reviews in Mineralogy and Geochemistry* (ed. Hughes, J. M., Kohn, M., and Rakovan, J.), vol. 48, Mineralogical Society of America, Washington, D.C., pp. 1–12.

117. Wang, L., and Nancollas, G. H. (2008) Calcium orthophosphates: crystallization and dissolution, *Chem. Rev.*, **108**, 4628–4669.

118. Lynn, A. K., and Bonfield, W. (2005) A novel method for the simultaneous, titrant-free control of pH and calcium phosphate mass yield, *Acc. Chem. Res.*, **38**, 202–207.

119. León, B., and Jansen, J. J. (eds.) (2009) *Thin Calcium Phosphate Coatings for Medical Implants*, Springer, New York, p. 326.

120. Fernández, E., Gil, F. J., Ginebra, M. P., Driessens, F. C. M., Planell, J. A., and Best, S. M. (1999) Calcium phosphate bone cements for clinical applications. Part I: Solution chemistry, *J. Mater. Sci. Mater. Med.*, **10**, 169–176.

121. McDowell, H., Gregory, T. M., and Brown, W. E. (1977) Solubility of $Ca_5(PO_4)_3OH$ in the system $Ca(OH)_2$ - H_3PO_4 - H_2O at 5, 15, 25, and 37 degree C. *J. Res. Natl. Bur. Stand. Sect. A Phys. Chem.*, **81A**, 273–281.

122. Chow, L. C. (2009) Next generation calcium phosphate-based biomaterials, *Dent. Mater. J.*, **28**, 1–10.

123. Ishikawa, K. (2010) Bone substitute fabrication based on dissolution-precipitation reactions, *Materials*, **3**, 1138–1155.

124. Pan, H. B., and Darvell, B. W. (2009) Calcium phosphate solubility: the need for re-evaluation, *Cryst. Growth Des.*, **9**, 639–645.

125. Boonchom, B. (2009) Parallelogram-like microparticles of calcium dihydrogen phosphate monohydrate ($Ca(H_2PO_4)_2{\cdot}H_2O$) obtained by a rapid precipitation route in aqueous and acetone media, *J. Alloys and Compounds*, **482**, 199–202.

126. Biocompatibility is the ability of a material to perform with an appropriate host response in a specific application [127]. For further details on this topic, the interested readers are referred to Ref. [128].

127. Williams, D. F. (1999) *The Williams Dictionary of Biomaterials*, Liverpool University Press, Liverpool, p. 368.

128. Kohane, D. S., and Langer, R. (2010) Biocompatibility and drug delivery systems, *Chem. Sci.*, **1**, 441–446.

129. Köster, K., Heide, H., and König, R. (1977) Resorbierbare Calciumphosphatkeramik im Tierexperiment unter Belastung, *Langenbecks Arch. Chir.*, **343**, 173–181.

130. Bermúdez, O., Boltong, M. G., Driessens, F. C. M., and Planell, J. A. (1994) Optimization of a calcium orthophosphate cement formulation occurring in the combination of monocalcium phosphate monohydrate with calcium oxide, *J. Mater. Sci. Mater. Med.*, **5**, 67–71.

131. Bermúdez, O., Boltong, M. G., Driessens, F. C. M., Planell, J. A. (1994) Development of some calcium phosphate cements from combinations of α-TCP, MCPM and CaO, *J. Mater. Sci. Mater. Med.*, **5**, 160–163.

132. Driessens, F. C. M., Boltong, M. G., Bermúdez, O., Planell, J. A., Ginebra, M. P., and Fernández, E. (1994) Effective formulations for the preparation of calcium phosphate bone cements, *J. Mater. Sci. Mater. Med.*, **5**, 164–170.

133. Huan, Z., and Chang, J. (2009) Novel bioactive composite bone cements based on the β-tricalcium phosphate – monocalcium phosphate monohydrate composite cement system, *Acta Biomater.*, **5**, 1253–1264.

134. Budavari, S., O'Neil, M. J., Smith, A., Heckelman, P. E., and Kinneary, J. F. (eds.) (1996) The Merck Index: An Encyclopedia of Chemicals, Drugs, and Biologicals, 12th edn., Chapman & Hall, USA, p. 1741.

135. Stein, H. H., Kadzere, C. T., Kim, S. W., and Miller, P. S. (2008) Influence of dietary phosphorus concentration on the digestibility of phosphorus in monocalcium phosphate by growing pigs, *J. Animal Sci.*, **86**, 1861–1867.

136. To honor Prof. George Jarvis Brush (1831–1912), an American mineralogist, Yale University, New Haven, Connecticut, USA.

137. Ferreira, A., Oliveira, C., and Rocha, F. (2003) The different phases in the precipitation of dicalcium phosphate dihydrate, *J. Cryst. Growth*, **252**, 599–611.

138. Oliveira, C., Ferreira, A., and Rocha, F. (2007) Dicalcium phosphate dihydrate precipitation: Characterization and crystal growth, *Chem. Eng. Res. Design*, **85**, 1655–1661.

139. Sivkumar, G. R., Girija, E. K., Kalkura, S. N., and Subramanian, C. (1997) Crystallization and characterization of calcium phosphates: brushite and monetite, *Crystal Res. Technol.*, **33**, 197–205.

140. Madhurambal, G., Subha, R., and Mojumdar, S. C. (2009) Crystallization and thermal characterization of calcium hydrogen phosphate dihydrate crystals, *J. Thermal Analysis and Calorimetry*, **96**, 73–76.

141. MacDowell, H., Brown W. E., and Sutter, J. R. (1971) Solubility study of calcium hydrogenphosphate. Ion pair formation, *Inorg. Chem.*, **10**, 1638–1643.

142. Landin, M., Rowe, R. C., and York, P. (1994) Structural changes during the dehydration of dicalcium phosphate dihydrate, *Eur. J. Pharmac. Sci.*, **2**, 245–252.

143. Curry, N. A., and Jones, D. W. (1971) Crystal structure of brushite, calcium hydrogen orthophosphate dihydrate: a neutron-diffraction investigation, *J. Chem. Soc. A Inorg. Phys. Theoret. Chem.*, 3725–3729.

144. Arsic, J., Kaminski, D., Poodt, P., and Vlieg, E. (2004) Liquid ordering at the brushite – {010} – water interface, *Phys. Rev. B*, **69**, 245–406.

145. Pan, H. B., and Darvell, B. W. (2009) Solubility of dicalcium phosphate dihydrate by solid titration, *Caries Res.*, **43**, 254–260.

146. Lundager-Madsen, H. E. (2008) Optical properties of synthetic crystals of brushite ($CaHPO_4 \cdot 2H_2O$), *J. Cryst. Growth*, **310**, 617–623.

147. Qiu, S. R., and Orme, C. A. (2008) Dynamics of biomineral formation at the near-molecular level, *Chem. Rev.*, **108**, 4784–4822.

148. Kurashina, K., Kurita, H., Hirano, M., Kotani, A., Klein, C. P., and de Groot, K. (1997) *In vivo* study of calcium phosphate cements: implantation of an α-tricalcium phosphate / dicalcium phosphate dibasic / tetracalcium phosphate monoxide cement paste, *Biomaterials*, **18**, 539–543.

149. Driessens, F. C. M., Planell, J. A., Boltong, M. G., Khairoun, I., Ginebra, M. P. (1998) Osteotransductive bone cements, *Proc. Inst. Mech. Eng. H: J. Eng. Med.*, **212**, 427–435.

150. Takagi, S., Chow, L. C., and Ishikawa, K. (1998) Formation of hydroxyapatite in new calcium phosphate cements, *Biomaterials*, **19**, 1593–1599.

151. Yamamoto, H., Niwa, S., Hori, M., Hattori, T., Sawai, K., Aoki, S., Hirano, M., and Takeuchi, H. (1998) Mechanical strength of calcium phosphate cement *in vivo* and *in vitro*, *Biomaterials*, **19**, 1587–1591.

152. Crall, J. J., and Bjerga, J. M. (1987) Effects of DCPD/APF application and prolonged exposure to fluoride on caries-like lesion formation *in vitro*, *J. Oral Pathol.*, **16**, 488–491.

153. Wefel, J. S. and Harless, J. D. (1987) The use of saturated DCPD in remineralization of artificial caries lesions *in vitro*, *J. Dent. Res.*, **66**, 1640–1643.

154. Hoppenbrouwers, P. M., Groenendijk, E., Tewarie, N. R., Driessens, F. C. M. (1988) Improvement of the Caries Resistance of human dental roots by a two-step conversion of the root mineral into fluoridated hydroxylapatite, *J. Dent. Res.*, **67**, 1254–1256.

155. Gaffar, A., Blake-Haskins, J., and Mellberg, J. (1993) *In vivo* studies with a dicalcium phosphate / MFP system for caries prevention, *Int. Dent. J.*, **43**(1), 81–88.

156. Sullivan, R. J., Charig, A., Blake-Haskins, J., Zhang, Y. P., Miller, S. M., Strannick, M., Gaffar, A., and Margolis, H. C. (1997) *In vivo* detection

of calcium from dicalcium phosphate dihydrate dentifrices in demineralized human enamel and plaque, *Adv. Dent. Res.*, **11**, 380–387.

157. Mostashari, S. M., Haddadi, H., and Hashempoor, Z. (2006) Effect of deposited calcium hydrogen phosphate dihydrate on the flame retardancy imparted to cotton fabric, *Asian J. Chem.*, **18**, 2388–2390.

158. Powditch, H. I., and Bosworth, A. W. (1917) Studies on infant feeding. IX. The availability of dicalcium phosphate when present as a constituent of infant's food, *Boston Med. Surg. J.*, **177**, 864–867.

159. For Moneta (now Monito) Island (archipelago of Puerto Rico), which contains a notable occurrence.

160. Tas, A. C. (2009) Monetite ($CaHPO_4$) synthesis in ethanol at room temperature, *J. Am. Ceram. Soc.*, **92**, 2907–2912.

161. Chen, G. G., Luo, G. S., Yang, L. M., Xu, J. H., Sun, Y., and Wang, J. D. (2005) Synthesis and size control of $CaHPO_4$ particles in a two-liquid phase micro-mixing process, *J. Cryst. Growth*, **279**, 501–507.

162. Miyazaki, T., Sivaprakasam, K., Tantry, J., Suryanarayanan, R. (2009) Physical characterization of dibasic calcium phosphate dihydrate and anhydrate, *J. Pharmac. Sci.*, **98**, 905–916.

163. Eshtiagh-Hosseini, H., Houssaindokht, M. R., Chahkandhi, M., and Youssefi, A. (2008) Preparation of anhydrous dicalcium phosphate, DCPA, through sol-gel process, identification and phase transformation evaluation, *J. Non-Cryst. Solids*, **354**, 3854–3857.

164. Fukase, Y., Eanes, E. D., Takagi, S., Chow, L. C., and Brown, W. E. (1990) Setting reactions and compressive strengths of calcium phosphate cements, *J. Dent. Res.*, **69**, 1852–1856.

165. TenHuisen, K. S., and Brown, P. W. (1994) The formation of hydroxyapatite-ionomer cements at 38 degrees, *C. J. Dent. Res.*, **73**, 598–606.

166. Fernández, E., Ginebra, M. P., Boltong, M. G., Driessens, F. C. M., Ginebra, J., de Maeyer, E. A., Verbeeck, R. M. H., and Planell, J. A. (1996) Kinetic study of the setting reaction of a calcium phosphate bone cement, *J. Biomed. Mater. Res.*, **32**, 367–374.

167. Fernández, E., Gil, F. J., Best, S. M., Ginebra, M. P., Driessens, F. C. M., and Planell, J. A. (1998) The cement setting reaction in the $CaHPO_4$ – α-$Ca_3(PO_4)_2$ system: an X-ray diffraction study, *J. Biomed. Mater. Res.*, **42**, 403–406.

168. Fernández, E., Gil, F. J., Ginebra, M. P., Driessens, F. C. M., Planell, J. A., and Best, S. M. (1999) Production and characterization of new calcium

phosphate bone cements in the $CaHPO_4$ – α-$Ca_3(PO_4)_2$ system: pH, workability and setting times, *J. Mater. Sci. Mater. Med.*, **10**, 223–230.

169. Desai, T. R., Bhaduri, S. B., and Tas, A. C. (2006) A self-setting, monetite ($CaHPO_4$) cement for skeletal repair, *Ceram. Eng. Sci. Proc.*, **27**, 61–69.

170. Sturgeon, J. L., and Brown, P. W. (2009) Effects of carbonate on hydroxyapatite formed from $CaHPO_4$ and $Ca_4(PO_4)_2O$, *J. Mater. Sci. Mater. Med.*, **20**, 1787–1794.

171. Chen, G., Li, W., Yu, X., and Sun, K. (2009) Study of the cohesion of TTCP/DCPA phosphate cement through evolution of cohesion time and remaining percentage, *J. Mater. Sci.*, **44**, 828–834.

172. Tamimi, F., Torres, J., Bassett, D., Barralet, J., and Cabarcos, E. L. (2010) Resorption of monetite granules in alveolar bone defects in human patients, *Biomaterials*, **31**, 2762–2769.

173. Takami, K., Machimura, H., Takado, K., Inagaki, M., and Kawashima, Y. (1996) Novel preparation of free flowing spherically granulated dibasic calcium phosphate anhydrous for direct tabletting, *Chem. Pharm. Bull.*, **44**, 868–870.

174. LeGeros, R. Z. (1985) Preparation of octacalcium phosphate (OCP): a direct fast method, *Calcif. Tissue Int.*, **37**, 194–197.

175. Bigi, A., Boanini, E., Borghi, M., Cojazzi, G., Panzavolta, S., and Roveri, N. (1999) Synthesis and hydrolysis of octacalcium phosphate: effect of sodium polyacrylate, *J. Inorg. Biochem.*, **75**, 145–151.

176. Nakahira, A., Aoki, S., Sakamoto, K., and Yamaguchi, S. (2001) Synthesis and evaluation of various layered octacalcium phosphates by wet-chemical processing, *J. Mater. Sci. Mater. Med.*, **12**, 793–800.

177. Shelton, R. M., Liu, Y., Cooper, P. R., Gbureck, U., German, M. J., and Barralet, J. E. (2006) Bone marrow cell gene expression and tissue construct assembly using octacalcium phosphate microscaffolds, *Biomaterials*, **27**, 2874–2881.

178. Arellano-Jiménez, M. J., García-García, R., and Reyes-Gasga, J. (2009) Synthesis and hydrolysis of octacalcium phosphate and its characterization by electron microscopy and X-ray diffraction, *J. Phys. Chem. Solids*, **70**, 390–395.

179. Suzuki, O. (2010) Octacalcium phosphate: osteoconductivity and crystal chemistry, *Acta Biomater.*, **6**, 3379–3387.

180. Miyatake, N., Kishimoto, K. N., Anada, T., Imaizumi, H., Itoi, E., and Suzuki, O. (2009) Effect of partial hydrolysis of octacalcium phosphate on its osteoconductive characteristics, *Biomaterials*, **30**, 1005–1014.

181. Boanini, E., Gazzano, M., Rubini, K., and Bigi, A. (2010) Collapsed octacalcium phosphate stabilized by ionic substitutions, *Cryst. Growth Des.*, **10**, 3612–3617.

182. Brown, W. E., Mathew, M., and Tung, M. S. (1981) Crystal chemistry of octacalcium phosphate, *Prog. Cryst. Growth Charact.*, **4**, 59–87.

183. Mathew, M., Brown, W. E., Schroeder, L., and Dickens, B. (1988) Crystal structure of octacalcium bis(hydrogenphosphate) tetrakis(phosphate) pentahydrate, $Ca_8(HPO_4)_2(PO_4)_4{\cdot}5H_2O$, *J. Crystallogr. Spectrosc. Res.*, **18**, 235–250.

184. Brown, W. E. (1962) Octacalcium phosphate and hydroxyapatite: crystal structure of octacalcium phosphate, *Nature*, **196**, 1048–1050.

185. Brown, W. E., Smith, J. P., Lehr, J. R., and Frazier, A. W. (1962) Octacalcium phosphate and hydroxyapatite: crystallographic and chemical relations between octacalcium phosphate and hydroxyapatite, *Nature*, **196**, 1050–1055.

186. Pan, H. B., and Darvell, B. W. (2009) Solid titration of octacalcium phosphate, *Caries Res.*, **43**, 322–330.

187. LeGeros, R. Z. (1974) Variations in the crystalline components of human dental calculus: I. Crystallographic and spectroscopic methods of analysis, *J. Dent. Res.*, **53**, 45–50.

188. Schroeder, H. (1969) Formation and inhibition of dental calculus, *J. Periodontol.*, **40**, 643–646.

189. Chow, L. C., and Eanes, E. D. (eds.) (2001) Octacalcium phosphate, *Monographs in Oral Science*, vol. 18, Karger, Basel, 2001, p. 167.

190. Kakei, M., Sakae, T., and Yoshikawa, M. (2009) Electron microscopy of octacalcium phosphate in the dental calculus, *J. Electron Microscopy*, **58**, 393–398.

191. Brown, W. E. (1966) Crystal growth of bone mineral, *Clin. Orthop. Relat. Res.*, **44**, 205–220.

192. Nelson, D. G. A., Wood, G. J., Barry, J. C., and Featherstone, J. D. B. (1986) The structure of (100) defects in carbonated apatite crystallites: a high-resolution electron microscope study, *Ultramicroscopy*, **19**, 253–265.

193. Iijima, M., Nelson, D. G. A., Pan, Y., Kreinbrink, A. T., Adachi, M., Goto, T., and Moriwaki, Y. (1996) Fluoride analysis of apatite crystals with a central planar OCP inclusion: concerning the role of F^- ions on apatite / OCP / apatite structure formation, *Calcif. Tissue Int.*, **59**, 377–384.

194. Bodier-Houllé, P., Steuer, P., Voegel, J. C., and Cuisinier, F. J. G. (1998) First experimental evidence for human dentine crystal formation

involving conversion of octacalcium phosphate to hydroxyapatite, *Acta Crystallogr. D: Biol. Crystallogr.*,**54**, 1377–1381.

195. Aoba, T., Komatsu, H., Shimazu, Y., Yagishita, H., and Taya, Y. (1998) Enamel mineralization and an initial crystalline phase, *Connect. Tissue Res.*, **38**, 129–137.

196. Rodríguez-Hernández, A. G., Fernández, M. E., Carbajal-de-la-Torre, G., García-García, R., and Reyes-Gasga, J. (2005) Electron microscopy analysis of the central dark line defect of the human tooth enamel, *Mater. Res. Soc. Symp. Proc.*, **839**, 157–162.

197. Tomazic, B. B., Brown, W. E., and Shoen, F. J. (1994) Physicochemical properties of calcific deposits isolated from porcine bioprosthetic heart valves removed from patients following 2–13 years function, *J. Biomed. Mater. Res.*, **28**, 35–47.

198. Nancollas, G. H., and Wu, W. (2000) Biomineralization mechanisms: a kinetics and interfacial energy approach, *J. Cryst. Growth*, **211**, 137–142.

199. Kamakura, S., Sasano, Y., Homma, H., Suzuki, O., Kagayama, M., and Motegi, K. (1999) Implantation of octacalcium phosphate (OCP) in rat skull defects enhances bone repair, *J. Dent. Res.*, **78**, 1682–1687.

200. Kamakura, S., Sasano, Y., Homma, H., Suzuki, O., Kagayama, M., and Motegi, K. (2001) Implantation of octacalcium phosphate nucleates isolated bone formation in rat skull defects, *Oral Dis.*, **7**, 259–265.

201. Sargolzaei-Aval, F., Sobhani, A., Arab, M. R., Sarani, S. A., and Heydari, M. H. (2004) The efficacy of implant of octacalcium phosphate in combination with bone matrix gelatin (BMG) on bone regeneration in skull defects in rat, *Iran J. Med. Sci.*, **29**, 124–129.

202. Suzuki, O., Kamakura, S., Katagiri, T., Nakamura, M., Zhao, B., Honda, Y., and Kamijo, R. (2006) Bone formation enhanced by implanted octacalcium phosphate involving conversion into Ca-deficient hydroxyapatite, *Biomaterials*, **27**, 2671–2681.

203. Suzuki, O., Imaizumi, H., Kamakura, S., and Katagiri, T. (2008) Bone regeneration by synthetic octacalcium phosphate and its role in biological mineralization, *Curr. Med. Chem.*, **15**, 305–313.

204. Kikawa, T., Kashimoto, O., Imaizumi, H., Kokubun, S., and Suzuki, O. (2009) Intramembranous bone tissue response to biodegradable octacalcium phosphate implant, *Acta Biomater.*, **5**, 1756–1766.

205. Murakami, Y., Honda, Y., Anada, T., Shimauchi, H., and Suzuki, O. (2010) Comparative study on bone regeneration by synthetic octacalcium phosphate with various granule sizes, *Acta Biomater.*, **6**, 1542–1548.

206. Hou, X. J., Mao, K. Y., and Chen, D. F. (2008) Bone formation performance of beta-tricalcium phosphate sintered bone, *J. Clin. Rehabil. Tiss. Eng. Res.*, **12**, 9627–9630.

207. Dickens, B., Schroeder, L. W., and Brown, W. E. (1974) Crystallographic studies of the role of Mg as a stabilizing impurity in β-$Ca_3(PO_4)_2$. I. The crystal structure of pure β- $Ca_3(PO_4)_2$, *J. Solid State Chem.*, **10**, 232–248.

208. Schroeder, L. W., Dickens, B., and Brown, W. E. (1977) Crystallographic studies of the role of Mg as a stabilizing impurity in β-$Ca_3(PO_4)_2$. II. Refinement of Mg-containing β- $Ca_3(PO_4)_2$, *J. Solid State Chem.*, **22**, 253–262.

209. Ito, A., and LeGeros, R. Z. (2008) Magnesium- and zinc-substituted beta-tricalcium phosphates as potential bone substitute biomaterials, *Key Eng. Mater.*, **377**, 85–98.

210. Kannan, S., Goetz-Neunhoeffer, F., Neubauer, J., and Ferreira, J. M. F. (2009) Synthesis and structure refinement of zinc-doped β-tricalcium phosphate powders, *J. Am. Ceram. Soc.*, **92**, 1592–1595.

211. Kannan, S., Goetz-Neunhoeffer, F., Neubauer, J., Pina, S., Torres, P. M. C., and Ferreira, J. M. F. (2010) Synthesis and structural characterization of strontium- and magnesium-co-substituted β-tricalcium phosphate, *Acta Biomater.*, **6**, 571–576.

212. Araújo, J. C., Sader, M. S., Moreira, E. L., Moraes, V. C. A., LeGeros, R. Z., and Soares, G. A. (2009) Maximum substitution of magnesium for calcium sites in Mg-β-TCP structure determined by X-ray powder diffraction with the Rietveld refinement, *Mater. Chem. Phys.*, **118**, 337–340.

213. Karlinsey, R. L., and Mackey, A. C. (2009) Solid-state preparation and dental application of an organically modified calcium phosphate, *J. Mater. Sci.*, **44**, 346–349.

214. Karlinsey, R. L., Mackey, A. C., Walker, E. R., and Frederick, K. E. (2010) Preparation, characterization and *in vitro* efficacy of an acid-modified β-TCP material for dental hard-tissue remineralization, *Acta Biomater.*, **6**, 969–978.

215. Karlinsey, R. L., Mackey, A. C., Walker, E. R., and Frederick, K. E. (2010) Surfactant-modified β-TCP: structure, properties, and *in vitro* remineralization of subsurface enamel lesions, *J. Mater. Sci. Mater. Med.*, **21**, 2009–2020.

216. Yashima, M., Sakai, A., Kamiyama, T., and Hoshikawa, A. (2003) Crystal structure analysis of β-tricalcium phosphate $Ca_3(PO_4)_2$ by neutron powder diffraction, *J. Solid State Chem.*, **175**, 272–277.

217. Yin, X., Stott, M. J., and Rubio, A. (2003) α- and β-tricalcium phosphate: a density functional study, *Phys. Rev. B*, **68**, 205–211 (7 pages).

218. Liang, L., Rulis, P., and Ching, W. Y. (2010) Mechanical properties, electronic structure and bonding of α- and β-tricalcium phosphates with surface characterization, *Acta Biomater.*, **6**, 3763–3771.

219. Kumar, A. R., and Kalainathan, S. (2010) Microhardness studies on calcium hydrogen phosphate (brushite) crystals, *Mater. Res. Bull.*, **45**, 1664–1667.

220. Pan, H. B., and Darvell, B. W. (2009) Solubility of TTCP and β-TCP by solid titration, *Arch. Oral Biol.*, **54**, 671–677.

221. Wang, W., Itoh, S., Yamamoto, N., Okawa, A., Nagai, A., and Yamashita, K. (2010) Electrical polarization of β-tricalcium phosphate ceramics, *J. Am. Ceram. Soc.*, **93**, 2175–2177.

222. In 1941, to honor Mr. Herbert Percy Whitlock (1868 – 1948), an American mineralogist, the curator of the American Museum of Natural History, New York City, New York, USA, the term whitlockite was coined as a synonym for β-TCP identified by its X-ray diffraction pattern in phosphate rocks [223-225]. Therefore, strictly speaking, β-TCMP should be called as a "magnesium whitlokite". Its solubility is less than that of β-TCP [226]. An iron-containing whitlockite with chemical formula $Ca_9(Mg,Fe^{2+})(PO_4)_6(PO_3,OH)$ exists in nature: is a relatively rare natural mineral but is found in granitic pegmatite and has also been found in meteorites. It can form small, but distinct and well-formed crystals [227, 228].

223. Jensen, A. T., and Rowles, S. L. (1957) Magnesium whitlockite, a major constituent of dental calculus, *Acta Odont. Scand.*, **16**, 121–139.

224. Frondel, C. (1941) Whitlockite: a new calcium phosphate, $Ca_3(PO_4)_2$, *Am. Mineral.*, **26**, 145–152.

225. Frondel, C. (1943) Mineralogy of the calcium phosphates in insular phosphate rock, *Am. Mineral.*, **28**, 215–232.

226. Li, X., Ito, A., Sogo, Y., Wang, X., and LeGeros, R. Z. (2009) Solubility of Mg-containing β-tricalcium phosphate at 25°C, *Acta Biomater.*, **5**, 508–517.

227. Calvo, C., and Gopal, R. (1975) The crystal structure of whitlockite from the Palermo Quarry, *Am. Mineral.*, **60**, 120–133.

228. *http://rruff.geo.arizona.edu/doclib/hom/whitlockite.pdf* (accessed in October 2010).

229. Kodaka, T., Debari, K., and Higashi, S. (1988) Magnesium-containing crystals in human dental calculus, *J. Electron Microsc. (Tokyo)*, **37**, 73–80.

230. Reid, J. D., and Andersen, M. E. (1993) Medial calcification (whitlockite) in the aorta, *Atherosclerosis*, **101**, 213–224.

231. Scotchford, C. A., and Ali, S. Y. (1995) Magnesium whitlockite deposition in articular cartilage: a study of 80 specimens from 70 patients, *Annals of the Rheumatic Diseases*, **54**, 339–344.

232. P'ng, C. H., Boadle, R., Horton, M., Bilous, M., and Bonar, F. (2008) Magnesium whitlockite of the aorta, *Pathology*, **40**, 539–540.

233. Mirtchi, A. A., Lemaître, J., and Munting, E. (1990) Calcium phosphate cements: study of the β-tricalcium phosphate – dicalcium phosphate – calcite cements, *Biomaterials*, **11**, 83–88.

234. Mirtchi, A. A., Lemaître, J., and Munting, E. (1991) Calcium phosphate cements: effect of fluorides on the setting and hardening of β-tricalcium phosphate – dicalcium phosphate – calcite cements, *Biomaterials*, **12**, 505–510.

235. Lemaître, J., Munting, E., and Mirtchi, A. A. (1992) Setting, hardening and resorption of calcium phosphate ionic cements, *Rev. Stomatol. Chir. Maxillofac.*, **93**, 163–165.

236. Ohura, K., Bohner, M., Hardouin, P., Lemaître, J., Pasquier, G., and Flautre, B. (1996) Resorption of, and bone formation from, new β-tricalcium phosphate-monocalcium phosphate cements: an *in vivo* study, *J. Biomed. Mater. Res.*, **30**, 193–200.

237. Ellinger, R. F., Nery, E. B., and Lynch, K. L. (1986) Histological assessment of periodontal osseous defects following implantation of hydroxyapatite and biphasic calcium phosphate ceramics: a case report, *Int. J. Periodont. Restor. Dent.*, **3**, 22–33.

238. Nery, E. B., Lynch, K. L., Hirthe, W. M., and Mueller, K. H. (1975) Bioceramic implants in surgically produced infrabony defects, *J. Periodontol.*, **46**, 328–347.

239. Metsger, D. S., Driskell, T. D., and Paulsrud, J. R. (1982) Tricalcium phosphate ceramic – a resorbable bone implant: review and current status, *J. Am. Dent. Assoc.*, **105**, 1035–1038.

240. Galois, L., Mainard, D., and Delagoutte, J. P. (2002) Beta-tricalcium phosphate ceramic as a bone substitute in orthopaedic surgery, *Int. Orthopaedics*, **26**, 109–115.

241. Ogose, A., Hotta, T., Kawashima, H., Kondo, N., Gu, W., Kamura, T., and Endo, N. (2005) Comparison of hydroxyapatite and beta tricalcium phosphate as bone substitutes after excision of bone tumors, *J. Biomed. Mater. Res. B* (*Appl. Biomater.*), **72B**, 94–101.

242. Horch, H. H., Sader, R., Pautke, C., Neff, A., Deppe, H., and Kolk, A. (2006) Synthetic, pure-phase beta-tricalcium phosphate ceramic granules (Cerasorb®) for bone regeneration in the reconstructive surgery of the jaws, *Int. J. Oral Maxillofac. Surg.*, **35**, 708–713.

243. Ogose, A., Kondo, N., Umezu, H., Hotta, T., Kawashima, H., Tokunaga, K., Ito, T., Kudo, N., Hoshino, M., Gu, W., and Endo, N. (2006) Histological assessment in grafts of highly purified beta-tricalcium phosphate (OSferion®) in human bones, *Biomaterials*, **27**, 1542–1549.

244. Kamitakahara, M., Ohtsuki, C., and Miyazaki, T. (2008) Review paper: Behavior of ceramic biomaterials derived from tricalcium phosphate in physiological condition, *J. Biomaterials Applications*, **23**, 197–212.

245. Liu, Y., Pei, G. X., Shan, J., and Ren, G. H. (2008) New porous beta-tricalcium phosphate as a scaffold for bone tissue engineering, *J. Clin. Rehabil. Tiss. Eng. Res.*, **12**, 4563–4567.

246. Epstein, N. E. (2009) Beta tricalcium phosphate: observation of use in 100 posterolateral lumbar instrumented fusions, *Spine J.*, **9**, 630–638.

247. Shayegan, A., Petein, M., and Abbeele, A. V. (2009) The use of beta-tricalcium phosphate, white MTA, white Portland cement and calcium hydroxide for direct pulp capping of primary pig teeth, *Dental Traumatology*, **25**, 413–419.

248. Güngörmüş, C., Kiliç, A., Akay, M. T., and Kolankaya, D. (2010) The effects of maternal exposure to food additive E341 (tricalcium phosphate) on foetal development of rats, *Environmental Toxicology and Pharmacology*, **29**, 111–116.

249. Weiner, M. L., Salminen, W. F., Larson, P. R., Barter, R. A., Kranetz, J. L., and Simon, G. S. (2001) Toxicological review of inorganic phosphates, *Food and Chemical Toxicology*, **39**, 759–786.

250. Jokic, B., Jankovic-Castvan, I., Veljovic, D. J., Bucevac, D., Obradovic-Djuricic, K., Petrovic, R., and Janackovic, D. J. (2007) Synthesis and settings behavior of α-TCP from calcium deficient hyroxyapatite obtained by hydrothermal method, *J. Optoelectronics Adv. Mater.*, **9**, 1904–1910.

251. Nurse, R. W., Welch, J. B., and Gun, W. (1959) High-temperature phase equilibria in the system dicalcium silicate – tricalcium phosphate, *J. Chem. Soc.*, 1077–1083.

252. Langstaff, S. D., Sayer, M., Smith, T. J. N., Pugh, S. M., Hesp, S. A. M., and Thompson, W. T. (1999) Resorbable bioceramics based on stabilized calcium phosphates. Part I: Rational design, sample preparation and material characterization, *Biomaterials*, **20**, 1727–1741.

253. Langstaff, S. D., Sayer, M., Smith, T. J. N., and Pugh, S. M. (2001) Resorbable bioceramics based on stabilized calcium phosphates. Part II: Evaluation of biological response, *Biomaterials*, **22**, 135–150.

254. Sayer, M., Stratilatov, A. D., Reid, J. W., Calderin, L., Stott, M. J., Yin, X., MacKenzie, M., Smith, T. J .N., Hendry, J. A., and Langstaff, S. D. (2003) Structure and composition of silicon-stabilized tricalcium phosphate, *Biomaterials*, **24**, 369–382.

255. Reid, J. W., Pietak, A. M., Sayer, M., Dunfield, D., and Smith, T. J. N. (2005) Phase formation and evolution in the silicon substituted tricalcium phosphate / apatite system, *Biomaterials*, **26**, 2887–2897.

256. Reid, J. W., Tuck, L., Sayer, M., Fargo, K., and Hendry, J. A. (2006) Synthesis and characterization of single-phase silicon substituted α-tricalcium phosphate, *Biomaterials*, **27**, 2916–2925.

257. Astala, R., Calderin, L., Yin, X., and Stott, M. J. (2006) Ab initio simulation of Si-doped hydroxyapatite, *Chem. Mater.*, **18**, 413–422.

258. TenHuisen, K. S., and Brown, P. W. (1998) Formation of calcium-deficient hydroxyapatite from α-tricalcium phosphate, *Biomaterials*, **19**, 2209–2217.

259. Durucan, C., and Brown, P. W. (2000) α-tricalcium phosphate hydrolysis to hydroxyapatite at and near physiological temperature, *J. Mater. Sci. Mater. Med.*, **11**, 365–371.

260. Durucan, C., and Brown, P. W. (2002) Kinetic model for α-tricalcium phosphate hydrolysis, *J. Am. Ceram. Soc.*, **85**, 2013–2018.

261. Camiré, C. L., Gbureck, U., Hirsiger, W., and Bohner, M. (2005) Correlating crystallinity and reactivity in an α-tricalcium phosphate, *Biomaterials*, **26**, 2787–2794.

262. Constantz, B. R., Ison, I. C., Fulmer, M. T., Poser, R. D., Smith, S. T., Vanwagoner, M., Ross, J., Goldstein, S. A., Jupiter, J. B., and Rosenthal, D. I. (1995) Skeletal repair by *in situ* formation of the mineral phase of bone, *Science*, **267**, 1796–1799.

263. Tagaya, M., Goto, H., Iinuma, M., Wakamatsu, N., Tamura, Y., and Doi, Y. (2005) Development of self-setting Te-CP/α-TCP cement for pulpotomy, *Dent. Mater. J.*, **24**, 555–561.

264. Oda, M., Takeuchi, A., Lin, X., Matsuya, S., and Ishikawa, K. (2008) Effects of liquid phase on basic properties of α-tricalcium phosphate-based apatite cement, *Dent. Mater. J.*, **27**, 672–677.

265. Camiré, C.L., Nevsten, P., Lidgren, L., and McCarthy, I. (2006) The effect of crystallinity on strength development of α-TCP bone substitutes, *J. Biomed. Mater. Res. B* (*Appl. Biomater.*), **79B**, 159–165.

266. Yin, X., and Stott, M. J. (2005) Theoretical insights into bone grafting Si-stabilized α-tricalcium phosphate, *J. Chem. Phys.*, **122**, 024709 (9 pages).

267. Mathew, M., Schroeder, L. W., Dickens, B., and Brown, W. E. (1977) The crystal structure of α-$Ca_3(PO_4)_2$, *Acta Crystallogr. B*, **B33**, 1325–1333.

268. Yin, X., and Stott, M. J. (2006) Surface and adsorption properties of α-tricalcium phosphate, *J. Chem. Phys.*, **124**, 124701 (9 pages).

269. Dorozhkin S. V. (2010) Amorphous calcium (ortho)phosphates, *Acta Biomater.*, **6**, 4457–4475.

270. Termine, J. D., and Eanes, E. D. (1972) Comparative chemistry of amorphous and apatitic calcium phosphate preparations, *Calcif. Tissue Res.*, **10**, 171–197.

271. Eanes, E. D., Termine, J. D., and Nylen, M. U. (1973) An electron microscopic study of the formation of amorphous calcium phosphate and its transformation to crystalline apatite, *Calcif. Tissue Res.*, **12**, 143–158.

272. Meyer, J. L., and Eanes, E. D. (1978) A thermodynamic analysis of the amorphous to crystalline calcium phosphate transformation, *Calcif. Tissue Res.*, **28**, 59–68.

273. Meyer, J. L., and Eanes, E. D. (1978) A thermodynamic analysis of the secondary transition in the spontaneous precipitation of calcium phosphate, *Calcif. Tissue Res.*, **28**, 209–216.

274. Wuthier, R. E., Rice, G. S., Wallace, J. E., Weaver, R. L., LeGeros, R. Z., and Eanes, E. D. (1985) *In vitro* precipitation of calcium phosphate under intracellular conditions: formation of brushite from an amorphous precursor in the absence of ATP, *Calcif. Tissue Int.*, **37**, 401–410.

275. Sinyaev, V. A., Shustikova, E. S., Levchenko, L. V., Sedunov, A. A. (2001) Synthesis and dehydration of amorphous calcium phosphate, *Inorg. Mater.*, **37**, 619–622.

276. Termine, J. D., Peckauskas, R. A., and Posner, A. S. (1970) Calcium phosphate formation *in vitro*. II. Effects of environment on amorphous-crystalline transformation, *Arch. Biochem. Biophys.*, **140**, 318–325.

277. Elliott, J. C. (1998) Recent studies of apatites and other calcium orthophosphates, in *Les matériaux en phosphate de calcium. Aspects fondamentaux. / Calcium phosphate materials. Fundamentals* (ed. Brès, E., and Hardouin, P.), Sauramps Medical, Montpellier, pp. 25–66.

278. Li, Y., and Weng, W. (2007) *In vitro* synthesis and characterization of amorphous calcium phosphates with various Ca/P atomic ratios, *J. Mater. Sci. Mater. Med.*, **18**, 2303–2308.

279. Tadic, D., Peters, F., and Epple, M. (2002) Continuous synthesis of amorphous carbonated apatites, *Biomaterials*, **23**, 2553–2559.

280. Boskey, A. L., and Posner, A. S. (1973) Conversion of amorphous calcium phosphate to microcrystalline hydroxyapatite. A pH-dependent, solution-mediated, solid-solid conversion, *J. Phys. Chem.*, **77**, 2313–2317.

281. Keller, L., and Dollase, W. A. (2000) X-ray determination of crystalline hydroxyapatite to amorphous calcium-phosphate ratio in plasma sprayed coatings, *J. Biomed. Mater. Res.*, **49**, 244–249.

282. Carayon, M. T., and Lacout, J. L. (2003) Study of the Ca/P atomic ratio of the amorphous phase in plasma-sprayed hydroxyapatite coatings, *J. Solid State Chem.*, **172**, 339–350.

283. Kumar, R., Cheang, P., and Khor, K. A. (2004) Phase composition and heat of crystallization of amorphous calcium phosphate in ultra-fine radio frequency suspension plasma sprayed hydroxyapatite powders, *Acta Mater.*, **52**, 1171–1181.

284. Posner, A. S., and Betts, F. (1975) Synthetic amorphous calcium phosphate and its relation to bone mineral structure, *Acc. Chem. Res.*, **8**, 273–281.

285. Harries, J. E., Hukins, D. W. L., and Hasnain, S. S. (1986) Analysis of the EXAFS spectrum of hydroxyapatite, *J. Phys. C: Solid State Phys.*, **19**, 6859–6872.

286. Harries, J. E., Hukins, D. W. L., Holt, C., and Hasnain, S. S. (1987) Conversion of amorphous calcium phosphate into hydroxyapatite investigated by EXAFS spectroscopy, *J. Cryst. Growth*, **84**, 563–570.

287. Taylor, M. G., Simkiss, K., Simmons, J., Wu, L. N. Y., and Wuthier, R. E. (1998) Structural studies of a phosphatidyl serine-amorphous calcium phosphate complex, *Cell. Mol. Life Sci.*, **54**, 192–202.

288. Peters, F., Schwarz, K., and Epple, M. (2000) The structure of bone studied with synchrotron X-ray diffraction, X-ray absorption spectroscopy and thermal analysis, *Thermochim. Acta*, **361**, 131–138.

289. Posner, A. S., Betts, F., and Blumenthal, N. C. (1980) Formation and structure of synthetic and bone hydroxyapatite, *Progr. Cryst. Growth Char.*, **3**, 49–64.

290. Boskey, A. L. (1997) Amorphous calcium phosphate: the contention of bone, *J. Dent. Res.*, **76**, 1433–1436.

291. Onuma, K., and Ito, A. (1998) Cluster growth model for hydroxyapatite, *Chem. Mater.*, **10**, 3346–3351.

292. Skrtic, D., Hailer, A. W., Takagi, S., Antonucci, J. M., and Eanes, E. D. (1996) Quantitative assessment of the efficacy of amorphous calcium phosphate / methacrylate composites in remineralizing caries-like lesions artificially produced in bovine enamel, *J. Dent. Res.*, **75**, 1679–1686.

293. Skrtic, D., Antonucci, J. M., and Eanes, E. D. (1996) Improved properties of amorphous calcium phosphate fillers in remineralizing resin composites, *Dent. Mater.*, **12**, 295–301.

294. Skrtic, D., Antonucci, J. M., and Eanes, E. D. (2003) Amorphous calcium phosphate-based bioactive polymeric composites for mineralized tissue regeneration, *J. Res. Natl. Inst. Stand. Technol.*, **108**, 167–182.

295. Skrtic, D., Antonucci, J. M., Eanes, E. D., Eichmiller, F. C., and Schumacher, G. E. (2000) Physicochemical evaluation of bioactive polymeric composites based on hybrid amorphous calcium phosphates, *J. Biomed. Mater. Res. (Appl. Biomater.)*, **53**, 381–391.

296. Schiller, C., Siedler, M., Peters, F., and Epple, M. (2001) Functionally graded materials of biodegradable polyesters and bone-like calcium phosphates for bone replacement, *Ceram. Transact.*, **114**, 97–108.

297. Linhart, W., Peters, F., Lehmann, W., Schilling, A. F., Schwarz, K., Amling, M., Rueger, J. M., and Epple, M. (2001) Biologically and chemically optimized composites of carbonated apatite and polyglycolide as bone substitution materials, *J. Biomed. Mater. Res.*, **54**, 162–171.

298. Tadic, D., Beckmann, F., Schwarz, K., and Epple, M. (2004) A novel method to produce hydroxylapatite objects with interconnecting porosity that avoids sintering, *Biomaterials*, **25**, 3335–3340.

299. Tadic, D., and Epple, M. (2002) Amorphous calcium phosphates as bone substitution materials, *Eur. J. Trauma*, **28**, 136–137.

300. Eanes, E. D. (2001) Amorphous calcium phosphate, in *Octacalcium Phosphate. Monographs in Oral Science* (ed. Chow, L.C., and Eanes, E.D.), vol. 18, Karger, Basel, pp. 130–147.

301. Combes, C., and Rey, C. (2010) Amorphous calcium phosphates: synthesis, properties and uses in biomaterials, *Acta Biomater*, **6**, 3362–3378.

302. Sinha, A., Nayar, S., Agrawal, A., Bhattacharyya, D., and Ramachandrarao, P. (2003) Synthesis of nanosized and microporous precipitated hydroxyapatite in synthetic polymers and biopolymers, *J. Am. Ceram. Soc.*, **86**, 357–359.

303. Mayer, I., Jacobsohn, O., Niazov, T., Werckmann, J., Iliescu, M., Richard-Plouet, M., Burghaus, O., and Reinen, D. (2003) Manganese in precipitated hydroxyapatites, *Eur. J. Inorg. Chem.*, 1445–1451.

304. Vallet-Regí, M., Rodríguez-Lorenzo, L. M., and Salinas, A. J. (1997) Synthesis and characterisation of calcium deficient apatite, *Solid State Ionics*, 101–103, 1279–1285.

305. Siddharthan, A., Seshadri, S. K., and Kumar, T. S. S. (2004) Microwave accelerated synthesis of nanosized calcium deficient hydroxyapatite, *J. Mater. Sci. Mater. Med.*, **15**, 1279–1284.

306. Hutchens, S. A., Benson, R. S., Evans, B. R., O'Neill, H. M., and Rawn, C. J. (2006) Biomimetic synthesis of calcium-deficient hydroxyapatite in a natural hydrogel, *Biomaterials*, **27**, 4661–4670.

307. In some research papers, CDHA is defined as "precipitated HA" [308–310].

308. Knowles, J. C., Callcut, S., and Georgiou, G. (2000) Characterisation of the rheological properties and zeta potential of a range of hydroxyapatite powders, *Biomaterials*, **21**, 1387–1392.

309. Kumar, R., Prakash, K. H., Cheang, P., and Khor, K. A. (2004) Temperature driven morphological changes of chemically precipitated hydroxyapatite nanoparticles, *Langmuir*, **20**, 5196–5200.

310. Sung, Y. M., Lee, J. C., and Yang, J. W. (2004) Crystallization and sintering characteristics of chemically precipitated hydroxyapatite nanopowder, *J. Cryst. Growth*, **262**, 467–472.

311. Brès, E. F., Duhoo, T., Leroy, N., and Lemaitre, J. (2005) Evidence of a transient phase during the hydrolysis of calcium-deficient hydroxyapatite, *Zeitschrift fuer Metallkunde / Mater. Res. Adv. Tech.*, **96**, 503–506

312. Lecomte, A., Gautier, H., Bouler, J. M., Gouyette, A., Pegon, Y., Daculsi, G., and Merle, C. (2008) Biphasic calcium phosphate: a comparative study of interconnected porosity in two ceramics, *J. Biomed. Mater. Res. B* (*Appl. Biomater.*), **84B**, 1–6.

313. Tancret, F., Bouler, J. M., Chamousset, J., Minois, L. M. (2006) Modelling the mechanical properties of microporous and macroporous biphasic calcium phosphate bioceramics, *J. Eur. Ceram. Soc.*, **26**, 3647–3656.

314. Bouler, J. M., Trecant, M., Delecrin, J., Royer, J., Passuti, N., and Daculsi, G. (1996) Macroporous biphasic calcium phosphate ceramics: Influence of five synthesis parameters on compressive strength, *J. Biomed. Mater. Res.*, **32**, 603–609.

315. Wang, J., Chen, W., Li, Y., Fan, S., Weng, J., and Zhang, X. (1998) Biological evaluation of biphasic calcium phosphate ceramic vertebral laminae, *Biomaterials*, **19**, 1387–1392.

316. Daculsi, G. (1998) Biphasic calcium phosphate concept applied to artificial bone, implant coating and injectable bone substitute, *Biomaterials*, **19**, 1473–1478.

317. Daculsi, G., Weiss, P., Bouler, J. M., Gauthier, O., Millot, F., and Aguado, E. (1999) Biphasic calcium phosphate / hydrosoluble polymer composites: a new concept for bone and dental substitution biomaterials, *Bone*, **25**(2), 59S–61S.

318. LeGeros, R. Z. Lin, S., Rohanizadeh, R., Mijares, D., and LeGeros, J. P. (2003) Biphasic calcium phosphate bioceramics: preparation, properties and applications, *J. Mater. Sci. Mater. Med.*, **14**, 201–209.

319. Daculsi, G., Laboux, O., Malard, O., and Weiss, P. (2003) Current state of the art of biphasic calcium phosphate bioceramics, *J. Mater. Sci. Mater. Med.*, **14**, 195–200.

320. Alam, I., Asahina, I., Ohmamiuda, K., and Enomoto, S. (2001) Comparative study of biphasic calcium phosphate ceramics impregnated with rhBMP-2 as bone substitutes, *J. Biomed. Mater. Res.*, **54**, 129–138.

321. Daculsi, G. (2006) Biphasic calcium phosphate granules concept for injectable and mouldable bone substitute, *Adv. Sci. Technol.*, **49**, 9–13.

322. Daculsi, G., Baroth, S., and LeGeros, R. Z. (2010) 20 years of biphasic calcium phosphate bioceramics development and applications, *Ceram. Eng. Sci. Proc.*, **30**, 45–58.

323. Lobo, S. E., and Arinzeh, T. L. (2010) Biphasic calcium phosphate ceramics for bone regeneration and tissue engineering applications, *Materials*, **3**, 815–826.

324. Dorozhkina, E. I., and Dorozhkin, S. V. (2002) Mechanism of the solid-state transformation of a calcium-deficient hydroxyapatite (CDHA) into biphasic calcium phosphate (BCP) at elevated temperatures, *Chem. Mater.*, **14**, 4267–4272.

325. Dorozhkin, S. V. (2003) Mechanism of solid-state conversion of non-stoichiometric hydroxyapatite to diphase calcium phosphate, *Russ. Chem. Bull.* (*Int. Ed.*), **52**, 2369–2375.

326. Rodríguez-Lorenzo, L. (2005) Studies on calcium deficient apatites structure by means of MAS-NMR spectroscopy, *J. Mater. Sci. Mater. Med.*, **16**, 393–398.

327. Wilson, R. M., Elliott, J. C., and Dowker, S. E. P. (2003) Formate incorporation in the structure of Ca-deficient apatite: Rietveld structure refinement, *J. Solid State Chem.*, **174**, 132–140.

328. Zahn, D., and Hochrein, O. (2008) On the composition and atomic arrangement of calcium-deficient hydroxyapatite: an *ab-initio* analysis, *J. Solid State Chem.*, **181**, 1712–1716.

329. Brown, P. W., and Martin, R. I. (1999) An analysis of hydroxyapatite surface layer formation, *J. Phys. Chem. B*, **103**, 1671–1675.

330. Honghui, Z., Hui, L., and Linghong, G. (2007) Molecular and crystal structure characterization of calcium-deficient apatite, *Key Eng. Mater.*, 330–332, 119–122.

331. Viswanath, B., Shastry, V. V., Ramamurty, U., and Ravishankar, N. (2010) Effect of calcium deficiency on the mechanical properties of hydroxyapatite crystals, *Acta Mater.*, **58**, 4841–4848.

332. Mortier, A., Lemaître, J., Rodrique, L., and Rouxhet, P. G. (1989) Synthesis and thermal behavior of well-crystallized calcium-deficient phosphate apatite, *J. Solid State Chem.*, **78**, 215–219.

333. Jeanjean, J., McGrellis, S., Rouchaud, J. C., Fedoroff, M., Rondeau, A., Perocheau, S., and Dubis, A. (1996) A crystallographic study of the sorption of cadmium on calcium hydroxyapatites: incidence of cationic vacancies, *J. Solid State Chem.*, **126**, 195–201.

334. Wilson, R. M., Elliott, J. C., Dowker, S. E. P., and Rodriguez-Lorenzo, L. M. (2005) Rietveld refinements and spectroscopic studies of the structure of Ca-deficient apatite, *Biomaterials*, **26**, 1317–1327.

335. Ivanova, T. I., Frank-Kamenetskaya, O. V., Kol'tsov, A. B., and Ugolkov, V. L. (2001) Crystal structure of calcium-deficient carbonated hydroxyapatite thermal decomposition, *J. Solid State Chem.*, **160**, 340–349.

336. Matsunaga, K. (2008) Theoretical investigation of the defect formation mechanism relevant to nonstoichiometry in hydroxyapatite, *Phys. Rev. B*, **77**, 104106 (14 pages).

337. Bourgeois, B., Laboux, O., Obadia, L., Gauthier, O., Betti, E., Aguado, E., Daculsi, G., and Bouler, J. M. (2003) Calcium-deficient apatite: a first *in vivo* study concerning bone ingrowth, *J. Biomed. Mater. Res. A*, **65A**, 402–408.

338. Liu, T. Y., Chen, S. Y., Liu, D. M., and Liou, S. C. (2005) On the study of BSA-loaded calcium-deficient hydroxyapatite nano-carriers for controlled drug delivery, *J. Control. Release*, **107**, 112–121.

339. Tsuchida, T., Yoshioka, T., Sakuma, S., Takeguchi, T., and Ueda, W. (2008) Synthesis of biogasoline from ethanol over hydroxyapatite catalyst, *Ind. Eng. Chem. Res.*, **47**, 1443–1452.

340. It is worth noting that *hydroxylapatite* would be a more accurate description (perhaps, *hydroxideapatite* would be even better because it relates to calcium hydroxide) while by both the medical and material communities it is usually called as *hydroxyapatite*.

341. Elliott, J. C., Mackie, P. E., and Young, R. A. (1973) Monoclinic hydroxyapatite, *Science*, **180**, 1055–1057.

342. Rangavittal, N., Landa-Cánovas, A. R., González-Calbet, J. M., and Vallet-Regí, M. (2000) Structural study and stability of hydroxyapatite and β-tricalcium phosphate: two important bioceramics, *J. Biomed. Mater. Res.*, **51**, 660–668.

343. Kim, J. Y., Fenton, R. R., Hunter, B. A., and Kennedy, B. J. (2000) Powder diffraction studies of synthetic calcium and lead apatites, *Austr. J. Chem.*, **53**, 679–686.

344. Kay, M. I., Young, R. A., and Posner, A. S. (1964) Crystal structure of hydroxyapatite, *Nature*, **204**, 1050–1052.

345. Treboux, G., Layrolle, P., Kanzaki, N., Onuma, K., and Ito, A. (2000) Existence of Posner's cluster in vacuum, *J. Phys. Chem. A*, **104**, 5111–5114.

346. Gilmore, R. S., and Katz, J. L. (1982) Elastic properties of apatites, *J. Mater. Sci.*, **17**, 1131–1141.

347. Calderin, L., Stott, M. J., and Rubio, A. (2003) Electronic and crystallographic structure of apatites, *Phys. Rev. B*, **67**, 134106 (6 pages).

348. Rulis, P., Ouyang, L., and Ching, W. Y. (2004) Electronic structure and bonding in calcium apatite crystals: hydroxyapatite, fluorapatite, chlorapatite and bromapatite, *Phys. Rev. B*, **70**, 155104 (8 pages).

349. Snyders, R., Music, D., Sigumonrong, D., Schelnberger, B., Jensen, J., and Schneider, J. M. (2007) Experimental and ab initio study of the mechanical properties of hydroxyapatite, *Appl. Phys. Lett.*, 90, 193902 (3 pages).

350. Ching, W. Y., Rulis, P., and Misra, A. (2009) *Ab initio* elastic properties and tensile strength of crystalline hydroxyapatite, *Acta Biomater.*, **5**, 3067–3075.

351. Treboux, G., Layrolle, P., Kanzaki, N., Onuma, K., and Ito, A. (2000) Symmetry of Posner's cluster, *J. Am. Chem. Soc.*, **122**, 8323–8324.

352. Yin, X., and Stott, M. J. (2003) Biological calcium phosphates and Posner's cluster, *J. Chem. Phys.*, **118**, 3717–3723.

353. Kanzaki, N., Treboux, G., Onuma, K., Tsutsumi, S., and Ito, A. (2001) Calcium phosphate clusters, *Biomaterials*, **22**, 2921–2929.

354. Calderin, L., Dunfield, D., and Stott, M. J. (2005) Shell-model study of the lattice dynamics of hydroxyapatite, *Phys. Rev. B*, **72**, 224304 (12 pages).

355. Tanaka, Y., Iwasaki, T., Nakamura, M., Nagai, A., Katayama, K., and Yamashita, K. (2010) Polarization and microstructural effects of ceramic hydroxyapatite electrets, *J. Appl. Phys.*, **107**, 014107 (10 pages).

356. Tanaka, Y., Iwasaki, T., Katayama, K., Hojo, J., and Yamashita, K. (2010) Effect of ionic polarization on crystal structure of hydroxyapatite ceramic with hydroxide nonstoichiometry, *J. Jpn. Soc. Powder and Powder Metall.*, **57**, 520–528.

357. Tofail, S. A. M., Baldisserri, C., Haverty, D., McMonagle, J. B., and Erhart, J. (2009) Pyroelectric surface charge in hydroxyapatite ceramics, *J. Appl. Phys.*, **106**, 106104 (3 pages).

358. Kawabata, K., and Yamamoto, T. (2010) First-principles calculations of the elastic properties of hydroxyapatite doped with divalent ions, *J. Ceram. Soc. Jpn.*, **118**, 548–549.

359. Matsunaga, K., and Kuwabara, A. (2007) First-principles study of vacancy formation in hydroxyapatite, *Phys. Rev. B*, **75**, 014102 (9 pages).

360. de Leeuw, N. H. (2010) Computer simulations of structures and properties of the biomaterial hydroxyapatite, *J. Mater. Chem.*, **20**, 5376–5389.

361. Corno, M., Rimola, A., Bolis, V., and Ugliengo, P. (2010) Hydroxyapatite as a key biomaterial: quantum-mechanical simulation of its surfaces in interaction with biomolecules, *Phys. Chem. Chem. Phys.*, **12**, 6309–6329.

362. Briak-Ben, El, Abdeslam, H., Ginebra, M. P., Vert, M., and Boudeville, P. (2008) Wet or dry mechanochemical synthesis of calcium phosphates? Influence of the water content on DCPD – CaO reaction kinetics, *Acta Biomater.*, **4**, 378–386.

363. LeGeros, R. Z., and LeGeros, J. P. (1993) Dense hydroxyapatite, in *An Introduction to Bioceramics* (ed. Hench, L. L., and Wilson, J.), World Scientific, London, pp. 139–180.

364. Rodriguez-Lorenzo, L. M., and Vallet-Regí, M. (2000) Controlled crystallization of calcium phosphate apatites, *Chem. Mater.*, **12**, 2460–2465.

365. Cazalbou, S., Combes, C., Eichert, D., and Rey, C. (2004) Adaptive physico-chemistry of bio-related calcium phosphates, *J. Mater. Chem.*, **14**, 2148–2153.

366. Markovic, M., Fowler, B. O., and Tung, M. S. (2004) Preparation and comprehensive characterization of a calcium hydroxyapatite reference material, *J. Res. Natl. Inst. Stand. Technol.*, **109**, 553–568.

367. Ioku, K., Kawachi, G., Sasaki, S., Fujimori, H., and Goto, S. (2006) Hydrothermal preparation of tailored hydroxyapatite, *J. Mater. Sci.*, **41**, 1341–1344.

368. Ito, N., Kamitakahara, M., Murakami, S., Watanabe, N., and Ioku, K. (2010) Hydrothermal synthesis and characterization of hydroxyapatite from octacalcium phosphate, *J. Ceram. Soc. Jpn.*, **118**, 762–766.

369. Layrolle, P., and Lebugle, A. (1994) Characterization and reactivity of nanosized calcium phosphates prepared in anhydrous ethanol, *Chem. Mater.*, **6**, 1996–2004.

370. Layrolle, P., and Lebugle, A. (1996) Synthesis in pure ethanol and characterization of nanosized calcium phosphate fluoroapatite, *Chem. Mater.*, **8**, 134–144.

371. Yeong, B., Junmin, X., and Wang, J. (2001) Mechanochemical synthesis of hydroxyapatite from calcium oxide and brushite, *J. Am. Ceram. Soc.*, **84**, 465–467.

372. Roy, D. M., and Linnehan, S. K. (1974) Hydroxyapatite formed from coral skeletal carbonate by hydrothermal exchange, *Nature*, **247**, 220–222.

373. Ben-Nissan, B. (2003) Natural bioceramics: from coral to bone and beyond, *Curr. Opin. Solid State Mater. Sci.*, **7**, 283–288.

374. Ben-Nissan, B., Milev, A., and Vago, R. (2004) Morphology of sol-gel derived nano-coated coralline hydroxyapatite, *Biomaterials*, **25**, 4971–4975.

375. Elliott, J. C., and Young, R. A. (1967) Conversion of single crystals of chlorapatite into single crystals of hydroxyapatite, *Nature*, **214**, 904–906.

376. Tao, J., Jiang, W., Pan, H., Xu, X., and Tang, R. (2007) Preparation of large-sized hydroxyapatite single crystals using homogeneous releasing controls, *J. Cryst. Growth*, **308**, 151–158.

377. Vallet-Regí, M., Gutièrrez Rios, M. T., Alonso, M. P., de Frutos, M. I., and Nicolopoulos, S. (1994) Hydroxyapatite particles synthesized by pyrolysis of an aerosol, *J. Solid State Chem.*, **112**, 58–64.

378. Montero, M. L., Sáenz, A., Rodríguez, J. G., Arenas, J., and Castaño, V. M. (2006) Electrochemical synthesis of nanosized hydroxyapatite, *J. Mater. Sci.*, **41**, 2141–2144.

379. Kumar, A. R., and Kalainathan, S. (2008) Growth and characterization of nano-crystalline hydroxyapatite at physiological conditions, *Crystal Res. Technol.*, **43**, 640–644.

380. Kalita, S. J., Bhardwaj, A., and Bhatt, H. A. (2007) Nanocrystalline calcium phosphate ceramics in biomedical engineering, *Mater. Sci. Eng. C*, **27**, 441–449.

381. Melikhov, I. V., Komarov, V. F., Severin, A. V., Bozhevolnov, V. E., and Rudin, V. N. (2000) Two-dimensional crystalline hydroxyapatite, *Doklady Phys. Chem.*, **373**, 125–128.

382. Suvorova, E. I., and Buffat, P. A. (2001) Electron diffraction and HRTEM characterization of calcium phosphate precipitation from aqueous solutions under biomineralization conditions, *Eur. Cell Mater.*, **1**, 27–42.

383. Suvorova, E. I., and Buffat, P. A. (2002) Model of the mechanism of Ca loss by bones under microgravity and earth conditions, *J. Biomed. Mater. Res.*, **63**, 424–432.

384. Uematsu, K., Takagi, M., Honda, T., Uchida, N., and Saito, K. (1989) Transparent hydroxyapatite prepared by hot isostatic pressing of filter cake, *J. Am. Ceram. Soc.*, **72**, 1476–1478.

385. Takikawa, K., and Akao, M. (1996) Fabrication of transparent hydroxyapatite and application to bone marrow derived cell / hydroxyapatite interaction observation *in vivo*, *J. Mater. Sci. Mater. Med.*, **7**, 439–445.

386. Watanabe, Y., Ikoma, T., Monkawa, A., Suetsugu, Y., Yamada, H., Tanaka, J., and Moriyoshi, Y. (2005) Fabrication of transparent hydroxyapatite sintered body with high crystal orientation by pulse electric current sintering, *J. Am. Ceram. Soc.*, **88**, 243–245.

387. Kotobuki, N., Ioku, K., Kawagoe, D., Fujimori, H., Goto, S., and Ohgushi, H. (2005) Observation of osteogenic differentiation cascade of living mesenchymal stem cells on transparent hydroxyapatite ceramics, *Biomaterials*, **26**, 779–785.

388. Perloff, A., and Posner, A. S. (1956) Preparation of pure hydroxyapatite crystals, *Science*, **124**, 583–584.

389. LeGeros, R. Z., LeGeros, J. P., Daculsi, G., and Kijkowska, R. (1995) Calcium phosphate biomaterials: preparation, properties and biodegradation, in *Encyclopedic Handbook of Biomaterials and Bioengineering. Part A: Materials*, (ed. Wise, D. L., Trantolo, D. J., Altobelli, D. E., Yaszemski, M. J., Gresser, J. D., and Schwartz, E. R.), Marcel Dekker, New York, pp. 1429–1463.

390. Narasaraju, T. S. B., and Phebe, D. E. (1996) Some physico-chemical aspects of hydroxylapatite, *J. Mater. Sci.*, **31**, 1–21.

391. Orlovskii, V. P., and Barinov, S. M. (2001) Hydroxyapatite and hydroxyapatite-matrix materials: a survey, *Russ. J. Inorg. Chem.*, **46**(2), S129–S149.

392. Riman, R. E., Suchanek, W. L., Byrappa, K., Chen, C. W., Shuk, P., and Oakes, C. S. (2002) Solution synthesis of hydroxyapatite designer particulates, *Solid State Ionics*, **151**, 393–402.

393. Orlovskii, V. P., Komlev, V. S., and Barinov, S. M. (2002) Hydroxyapatite and hydroxyapatite-based ceramics, *Inorg. Mater.*, **38**, 973–984.

394. Peňa, J., and Vallet-Regí, M. (2003) Hydroxyapatite, tricalcium phosphate and biphasic materials prepared by a liquid mix technique, *J. Eur. Ceram. Soc.*, **23**, 1687–1696.

395. Koutsopoulos, S. (2002) Synthesis and characterization of hydroxyapatite crystals: a review study on the analytical methods, *J. Biomed. Mater. Res.*, **62**, 600–612.

396. Norton, J., Malik, K. R., Darr, J. A., and Rehman, I. U. (2006) Recent developments in processing and surface modification of hydroxyapatite, *Adv. Appl. Ceram.*, **105**, 113–139.

397. Rakovan, J. (2002) Growth and surface properties of apatite, in *Phosphates: Geochemical, Geobiological and Materials Importance. Series: Reviews in Mineralogy and Geochemistry* (ed. Hughes, J. M., Kohn, M., and Rakovan, J.), vol. 48, Mineralogical Society of America, Washington, D.C., pp. 51–86.

398. Dorozhkin, S. V. (2002) A review on the dissolution models of calcium apatites, *Prog. Cryst. Growth Charact.*, **44**, 45–61.

399. Suchanek, W., and Yoshimura, M. (1998) Processing and properties of hydroxyapatite-based biomaterials for use as hard tissue replacement implants, *J. Mater. Res.*, **13**, 94–117.

400. Willmann, G. (1999) Coating of implants with hydroxyapatite – material connections between bone and metal, *Adv. Eng. Mater.*, **1**, 95–105.

401. Sun, L., Berndt, C. C., Gross, K. A., and Kucuk, A. (2001) Review: Material fundamentals and clinical performance of plasma sprayed hydroxyapatite coatings, *J. Biomed. Mater. Res.* (*Appl. Biomater.*), **58**, 570–592.

402. Ong, J. L., and Chan, D. C. N. (1999) Hydroxyapatites and their use as coatings in dental implants: a review, *Crit. Rev. Biomed. Eng.*, **28**, 667–707.

403. Geesink, R. G. (2002) Osteoconductive coatings for total joint arthroplasty, *Clin. Orthop. Rel. Res.*, **395**, 53–65.

404. Hench, L. L. (1991) Bioceramics: from concept to clinic, *J. Am. Ceram. Soc.*, **74**, 1487–1510.

405. Hench, L. L. (1998) Bioceramics, *J. Am. Ceram. Soc.*, **81**, 1705–1728.

406. Dey, A., Mukhopadhyay, A. K., Gangadharan, S., Sinha, M. K., and Basu, D. (2009) Characterization of microplasma sprayed hydroxyapatite coating, *J. Thermal Spray Technol.*, **18**, 578–592.

407. Mangano, C., Piattelli, A., Perrotti, V., and Iezzi, G. (2008) Dense hydroxyapatite inserted into postextraction sockets: a histologic and histomorphometric 20-year case report, *J. Periodontol.*, **79**, 929–933.

408. Dickinson, W. (1956) Chromatography of insulin on calcium phosphate columns, *Nature* **178**, 994–995.

409. Bernardi, G. (1965) Chromatography of nucleic acids on hydroxyapatite, *Nature*, **206**, 779–783.

410. Xia, Z., Duan, X., Locklin, R. M., Quijano, M., Dobson, R. L., Triffitt, J. T., Ebetino, F. H., and Russell, G. R. (2008) Evaluation of the relative mineral-binding affinities of clinically-relevant bisphosphonates by using hydroxyapatite-column chromatography and adsorption isotherms combined with mass spectrometric analysis, *Bone*, **42**(1), S90–S91.

411. Brand, M., Rampalli, S., Chaturvedi, C. P., and Dilworth, F. J. (2008) Analysis of epigenetic modifications of chromatin at specific gene loci by native chromatin immunoprecipitation of nucleosomes isolated using hydroxyapatite chromatography, *Nat. Protoc.*, **3**, 398–409.

412. Yoshitake, T., Kobayashi, S., Ogawa, T., and Okuyama, T. (2006) Hydroxyapatite chromatography of guanidine denatured proteins: 1. Guanidine containing phosphate buffer system, *Chromatography*, **27**, 19–26.

413. Smith, G. P., and Gingrich, T. R. (2005) Hydroxyapatite chromatography of phage-display virions, *Biotechniques*, **39**, 879–884.

414. Jungbauer, A., Hahn, R., Deinhofer, K., and Luo, P. (2004) Performance and characterization of a nanophased porous hydroxyapatite for protein chromatography, *Biotechnol. Bioeng.*, **87**, 364–375.

415. Doonan, S. (2004) Chromatography on hydroxyapatite, *Methods Mol. Biol.*, **244**, 191–194.

416. Cummings, L. J., Snyder, M. A., and Brisack, K. (2009) Chapter 24. Protein chromatography on hydroxyapatite columns, *Methods in Enzymology*, **463**, 387–404.

417. Gagnon, P. (2009) Monoclonal antibody purification with hydroxyapatite, *New Biotechnology*, **25**, 287–293.

418. Palazzo, B., Sidoti, M. C., Roveri, N., Tampieri, A., Sandri, M., Bertolazzi, L., Galbusera, F., Dubini, G., Vena, P., and Contro, R. (2005) Controlled drug delivery from porous hydroxyapatite grafts: an experimental and theoretical approach, *Mater. Sci. Eng. C*, **25**, 207–213.

419. Palazzo, B., Iafisco, M., Laforgia, M., Margiotta, N., Natile, G., Bianchi, C. L., Walsh, D., Mann, S., and Roveri, N. (2007) Biomimetic hydroxyapatite-drug nanocrystals as potential bone substitutes with antitumor drug delivery properties, *Adv. Func. Mater.*, **17**, 2180–2188.

420. Tang, S. H., Jin, A. M., Lü, R. F., Wang, X. D., Wang, Y. F., Yu, B., and Zhang, B. P. (2007) Preparation and *in vivo* release experiment of nano-hydroxyapatite/gentamicin drug delivery system, *J. Clin. Rehabil. Tiss. Eng. Res.*, **11**, 3573–3576.

421. Ye, F., Guo, H., Zhang, H., and He, X. (2010) Polymeric micelle-templated synthesis of hydroxyapatite hollow nanoparticles for a drug delivery system, *Acta Biomater.*, **6**, 2212–2218.

422. Niwa, M., Sato, T., Li, W., Aoki, H., Aoki, H., and Daisaku, T. (2001) Polishing and whitening properties of toothpaste containing hydroxyapatite, *J. Mater. Sci. Mater. Med.*, **12**, 277–281.

423. Kim, B. I., Jeong, S. H., Jang, S. O., Kim, K. N., Kwon, H. K., and Park, Y. D. (2006) Tooth whitening effect of toothpastes containing nano-hydroxyapatite, *Key Eng. Mater.*, 309–311, 541–544.

424. Pietrasik, J., Szustakiewicz, K., Zaborski, M., and Haberko, K. (2008) Hydroxyapatite: an environmentally friendly filler for elastomers, *Mol. Cryst. Liq. Cryst.*, **483**, 172–178.

425. Bailliez, S., Nzihou, A., Bèche, E., and Flamant, G. (2004) Removal of lead (Pb) by hydroxyapatite sorbent, *Process Safety and Environmental Protection*, **82**, 175–180.

426. Corami, A., Mignardi, S., and Ferrini, V. (2008) Cadmium removal from single- and multi-metal (Cd plus Pb plus Zn plus Cu) solutions by sorption on hydroxyapatite, *J. Colloid Interf. Sci.*, **317**, 402–408.

427. Phonthammachai, N., Ziyi, Z., Jun, G., Fan, H. Y., and White, T. J. (2008) Synthesis of high performance hydroxyapatite-gold catalysts for CO oxidation, *Gold Bull.*, **41**, 42–50.

428. Chen, W., Huang, Z. L., Liu, Y., and He, Q. J. (2008) Preparation and characterization of a novel solid base catalyst hydroxyapatite loaded with strontium, *Catalysis Comm.*, **9**, 516–521.

429. Domínguez, M. I., Romero-Sarria, F., Centeno, M. A., and Odriozola, J. A. (2009) Gold/hydroxyapatite catalysts. Synthesis, characterization and catalytic activity to CO oxidation, *Appl. Catalysis B*, **87**, 245–251.

430. Xu, J., White, T., Li, P., He, C., and Han, Y. F. (2010) Hydroxyapatite foam as a catalyst for formaldehyde combustion at room temperature, *J. Am. Chem. Soc.*, **132**, 13172–13173.

431. de Groot, K., Wolke, J. G. C., and Jansen, J. A. (1998) Calcium phosphate coatings for medical implants, *Proc. Inst. Mech. Eng. H: J. Eng. Med.*, **212**, 137–147.

432. Gross, K. A., and Berndt C. C. (2002) Biomedical application of apatites, in *Phosphates: Geochemical, Geobiological and Materials Importance. Series: Reviews in Mineralogy and Geochemistry*, (ed. Hughes, J. M., Kohn, M., and Rakovan, J.), vol. 48., Mineralogical Society of America, Washington, D.C., pp. 631–672.

433. Ferraz, M. P., Monteiro, F. J., and Manuel, C. M. (2004) Hydroxyapatite nanoparticles: a review of preparation methodologies, *J. Appl. Biomater. Biomech.*, **2**, 74–80.

434. Damien, E., and Revell, P. A. (2004) Coralline hydroxyapatite bone graft substitute: a review of experimental studies and biomedical applications, *J. Appl. Biomater. Biomech.*, **2**, 65–73.

435. Aoki, H. (1991) *Science and Medical Applications of Hydroxyapatite*, JAAS, Tokyo, p. 245.

436. Kniep, R., and Busch, S. (1996) Biomimetic growth and self-assembly of fluorapatite aggregates by diffusion into denatured collagen matrices, *Angew. Chem. Int. Ed. Engl.*, **35**, 2624–2626.

437. Busch, S., Dolhaine, H., Duchesne, A., Heinz, S., Hochrein, O., Laeri, F., Podebrad, O., Vietze, U., Weiland, T., and Kniep, R. (1999) Biomimetic morphogenesis of fluorapatite-gelatin composites: fractal growth, the question of intrinsic electric fields, core / shell assemblies, hollow spheres and reorganization of denatured collagen, *Eur. J. Inorg. Chem.*, 1643–1653.

438. Göbel, C., Simon, P., Buder, J., Tlatlik, H., and Kniep, R. (2004) Phase formation and morphology of calcium phosphate-gelatine-composites grown by double diffusion technique: the influence of fluoride, *J. Mater. Chem.*, **14**, 2225–2230.

439. Prymak, O., Sokolova, V., Peitsch, T., and Epple, M. (2006) The crystallization of fluoroapatite dumbbells from supersaturated aqueous solution, *Cryst. Growth Des.*, **6**, 498–506.

440. Tlatlik, H., Simon, P., Kawska, A., Zahn, D., and Kniep, R. (2006) Biomimetic fluorapatite-gelatine nanocomposites: pre-structuring of gelatine matrices by ion impregnation and its effect on form development, *Angew. Chem. Int. Ed. Engl.*, **45**, 1905–1910.

441. Simon, P., Zahn, D., Lichte, H., and Kniep, R. (2006) Intrinsic electric dipole fields and the induction of hierarchical form developments in fluorapatite-gelatine nanocomposites: a general principle for morphogenesis of biominerals? *Angew. Chem. Int. Ed. Engl.*, **45**, 1911–1915.

442. Wu, Y. J., Tseng, Y. H., and Chan, J. C. C. (2010) Morphology control of fluorapatite crystallites by citrate ions, *Cryst. Growth Des.*, **10**, 4240–4242.

443. Dorozhkin, S. V. (2007) A hierarchical structure for apatite crystals, *J. Mater. Sci. Mater. Med.*, **18**, 363–366.

444. Mehmel, M. (1930) On the structure of apatite, *I. Z. Kristallogr.*, **75**, 323–331.

445. Naray-Szabo, S. (1930) The structure of apatite $(CaF)Ca_4(PO_4)_3$, *Z. Kristallogr.*, **75**, 387–398.

446. Sudarsanan, K., Mackie, P. E., and Young, R. A. (1972) Comparison of synthetic and mineral fluorapatite, $Ca_5(PO_4)_3F$, in crystallographic detail, *Mater. Res. Bull.*, **7**, 1331-337.

447. Barinov, S. M., Shvorneva, L. I., Ferro, D., Fadeeva, I. V., and Tumanov, S. V. (2004) Solid solution formation at the sintering of hydroxyapatite-fluorapatite ceramics, *Sci. Technol. Advanc. Mater.*, **5**, 537–541.

448. Nikcevic, I., Jokanovic, V., Mitric, M., Nedic, Z., Makovec, D., and Uskokovic, D. (2004) Mechanochemical synthesis of nanostructured fluorapatite / fluorhydroxyapatite and carbonated fluorapatite/ fluorhydroxyapatite, *J. Solid State Chem.*, **177**, 2565–2574.

449. Cheng, K., Weng, W., Qu, H., Du, P., Shen, G., Han, G., Yang, J., and Ferreira, J. M. F. (2004) Sol-gel preparation and *in vitro* test of fluorapatite / hydroxyapatite films, *J. Biomed. Mater. Res. B* (*Appl. Biomater.*), **69B**, 33–37.

450. Rodriguez-Lorenzo, L. M., Hart, J. N., and Gross, K. A. (2003) Influence of fluorine in the synthesis of apatites. Synthesis of solid solutions of hydroxy-fluorapatite, *Biomaterials*, **24**, 3777–3785.

451. Wei, M., Evans, J. H., Bostrom, T., and Grondahl, L. (2003) Synthesis and characterization of hydroxyapatite, fluoride-substituted hydroxyapatite and fluorapatite, *J. Mater. Sci. Mater. Med.*, **14**, 311–320.

452. Daculsi, G., and Kerebel, L. M. (1980) Ultrastructural study and comparative analysis of fluoride content of enameloid in sea-water and fresh-water sharks, *Arch. Oral Biol.*, **25**, 145–151.

453. Daculsi, G., Kerebel, L. M., and Kerebel, B. (1981) Effects of fluoride on human enamel and selachian enameloid *in vitro*: a high-resolution TEM and electron diffraction study, *Calcif. Tissue Int.*, **33**, 9–13.

454. Lowenstam, H. A., and Weiner, S. (1989) *On Biomineralization*, *Oxford University Press*, p. 324.

455. Weiner, S., and Dove, P. M. (2003) An overview of biomineralization processes and the problem of the vital effect, in *Biomineralization. Series: Reviews in Mineralogy and Geochemistry*, (ed. Dove, P. M., de Yoreo, J. J., and Weiner, S.), vol. 54., Mineralogical Society of America, Washington, D.C., pp. 1–29.

456. Dahm, S., and Risnes, S. (1999) A comparative infrared spectroscopic study of hydroxide and carbonate absorption bands in spectra of shark enameloid, shark dentine, and a geological apatite, *Calcif. Tissue Int.*, **65**, 459–465.

457. Carr, A., Kemp, A., Tibbetts, I., Truss, R., and Drennan, J. (2006) Microstructure of pharyngeal tooth enameloid in the parrotfish Scarus rivulatus (Pisces: Scaridae), *J. Microscopy*, **221**, 8–16.

458. Leveque, I., Cusack, M., Davis, S. A., and Mann, S. (2004) Promotion of fluorapatite crystallization by soluble-matrix proteins from Lingula Anatina shells, *Angew. Chem. Int. Ed. Engl.*, **43**, 885–888.

459. The amount of fluorides on the very surface of dental enamel might be increased by using fluoride-containing toothpastes and mouthwashes [460-463]. Fluoride-containing toothpastes and mouthwashes are widely used in practice due to the well-known anti-cariogenic effect of fluorides that is related to the solubility decreasing [464, 465].

460. Fonteles, C. S. R., Zero, D. T., Moss, M. E., and Fu, J. (2005) Fluoride concentrations in enamel and dentine of primary teeth after pre- and postnatal fluoride exposure, *Caries Res.*, **39**, 505–508.

461. Wiegand, A., Krieger, C., Attin, R., Hellwig, E., and Attin, T. (2005) Fluoride uptake and resistance to further demineralisation of demineralised enamel after application of differently concentrated acidulated sodium fluoride gels, *Clin. Oral. Investig.*, **9**, 52–57.

462. Waszkiel, D., Opalko, K., Lagocka, R., and Chlubek, D. (2004) Fluoride and magnesium content in superficial enamel layers of teeth with erosions, *Fluoride*, **37**, 271–277.

463. Vieira, A. P. G. F., Hancock, R., Dumitriu, M., Schwartz, M., Limeback, H., and Grynpas, M. D. (2005) How does fluoride affect dentine microhardness and mineralization? *J. Dental Res.*, **84**, 951–957.

464. Driessens, F. C. M. (1973) Relation between apatite solubility and anti-cariogenic effect of fluoride, *Nature*, **243**, 420–421.

465. Moreno, E. C., Kresak, M., and Zahradnik, R. T. (1974) Fluoridated hydroxyapatite solubility and caries formation, *Nature*, **247**, 64–65.

466. Heling, L., Heindel, R., and Merin, B. (1981) Calcium-fluorapatite. A new material for bone implants, *J. Oral Implantol.*, **9**, 548–555.

467. Busch, S., Schwarz, U., and Kniep, R. (2001) Morphogenesis and structure of human teeth in relation to biomimetically grown fluorapatite-gelatine composites, *Chem. Mater.*, **13**, 3260–3271.

468. Kniep, R., and Simon, P. (2006) Fluorapatite-gelatine-nanocomposites: self-organized morphogenesis, real structure and relations to natural hard materials, *Topics Curr. Chem.*, **270**, 73–125.

469. Bogdanov, B. I., Pashev, P. S., Hristov, J. H., and Markovska, I. G. (2009) Bioactive fluorapatite-containing glass ceramics, *Ceram. Int.*, **35**, 1651–1655.

470. Savarino, L., Fini, M., Ciapetti, G., Cenni, E., Granchi, D., Baldini, N., Greco, M., Rizzi, G., Giardino, R., and Giunti, A. (2003) Biologic effects of surface roughness and fluorhydroxyapatite coating on osteointegration in external fixation systems: an *in vivo* experimental study, *J. Biomed. Mater. Res. A*, **66 A**, 652–661.

471. Vitkovič, M., Noaman, M. S. M., Palou, M. T., and Jantová, S. (2009) Potential applications of fluorhydroxyapatite as biomaterials in medicine, *Central Eur. J. Chem.*, **7**, 246–251.

472. Chaari, K., Ayed, F. B., Bouaziz, J., and Bouzouita, K. (2009) Elaboration and characterization of fluorapatite ceramic with controlled porosity, *Mater. Chem. Phys.*, **113**, 219–226.

473. Bibby, J. K., Bubb, N. L., Wood, D. J., and Mummery, P. M. (2005) Fluorapatite-mullite glass sputter coated Ti6Al4V for biomedical applications, *J. Mater. Sci. Mater. Med.*, **16**, 379–385.

474. Yoon, B. H., Kim, H. W., Lee, S. H., Bae, C. J., Koh, Y. H., Kong, Y. M., and Kim, H. E. (2005) Stability and cellular responses to fluorapatite-collagen composites, *Biomaterials*, **26**, 2957–2963.

475. Gineste, L., Gineste, M., Ranz, X., Ellefterion, A., Guilhem, A., Rouquet, N., and Frayssinet, P. (1999) Degradation of hydroxylapatite, fluorapatite and fluorhydroxyapatite coatings of dental implants in dogs, *J. Biomed. Mater. Res.*, **48**, 224–234.

476. Bhadang, K. A., and Gross, K. A. (2004) Influence of fluorapatite on the properties of thermally sprayed hydroxyapatite coatings, *Biomaterials*, **25**, 4935–4945.

477. Gross, K. A., and Bhadang, K. A. (2004) Sintered hydroxyfluorapatites. Part III: Sintering and resultant mechanical properties of sintered blends of hydroxyapatite and fluorapatite, *Biomaterials*, **25**, 1395–1405.

478. Barinov, S. M., Rustichelli, F., Orlovskii, V. P., Lodini, A., Oscarsson, S., Firstov, S. A., Tumanov, S. V., Millet, P., and Rosengren, A. (2004) Influence of fluorapatite minor additions on behavior of hydroxyapatite ceramics, *J. Mater. Sci. Mater. Med.*, **15**, 291–296.

479. Agathopoulos, S., Tulyaganov, D. U., Marques, P. A. A. P., Ferro, M. C., Fernandes, M. H., and Correia, R. N. (2003) The fluorapatite – anorthite system in biomedicine, *Biomaterials*, **24**, 1317–1331.

480. Qu, H., and Wei, M. (2006) The effect of fluoride contents in fluoridated hydroxyapatite on osteoblast behavior, *Acta Biomater.*, **2**, 113–119.

481. Bhadang, K. A., Holding, C. A., Thissen, H., McLean, K. M., Forsythe, J. S., and Haynes, D. R. (2010) Biological responses of human osteoblasts and osteoclasts to flame-sprayed coatings of hydroxyapatite and fluorapatite blends, *Acta Biomater.*, **6**, 1575–1583.

482. Theiszova, M., Jantova, S., Letasiova, S., Palou, M., and Cipak, L. (2008) Cytotoxicity of hydroxyapatite, fluorapatite and fluor-hydroxyapatite: a comparative *in vitro* study, *Neoplasma*, **55**, 312–316.

483. An, L., Zhang, L., and Zou, J. (2009) Fluorapatite: an effective solid base catalyst for Michael addition of indole/pyrrole to nitroalkenes under solventless condition, *Chinese J. Chem.*, **27**, 2223–2228.

484. Gross, K. A., Berndt, C. C., Dinnebier, R., and Stephens, P. (1998) Oxyapatite in hydroxyapatite coatings, *J. Mater. Sci. Mater. Med.*, **33**, 3985–3991.

485. Alberius-Henning, P., Adolfsson, E., Grins, J., and Fitch, A. (2001) Triclinic oxy-hydroxyapatite, *J. Mater. Sci.*, **36**, 663–668.

486. van't Hoen, C., Rheinberger, V., Höland, W., and Apel, E. (2007) Crystallization of oxyapatite in glass-ceramics, *J. Eur. Ceram. Soc.*, **27**, 1579–1584.

487. de Leeuw, N. H., Bowe, J. R., and Rabone, J. A. L. (2007) A computational investigation of stoichiometric and calcium-deficient oxy- and hydroxy-apatites, *Faraday Disc.*, **134**, 195–214.

488. Duff, E. J. (1972) Orthophosphates – VII. Thermodynamical considerations concerning the stability of oxyapatite, $Ca_{10}O(PO_4)_6$, in aqueous media, *J. Inorg. Nuclear Chem.*, **34**, 853–857.

489. To honor Gustav Hilgenstock (1844 – 1913), a German metallurgist, who first discovered it in Thomas slags nearly 120 years ago [12, 13].

490. Kai, D., Fan, H., Li, D., Zhu, X., and Zhang, X. (2009) Preparation of tetracalcium phosphate and the effect on the properties of calcium phosphate cement, *Mater. Sci. Forum*, 610–613, 1356–1359.

491. Brown, W. E., and Epstein, E. F. (1965) Crystallography of tetracalcium phosphate, *J. Res. Nat. Bur. Stand. A*, **69A**, 547–551.

492. Romeo, H. E., and Fanovich, M. A. (2008) Synthesis of tetracalcium phosphate from mechanochemically activated reactants and assessment as a component of bone cements, *J. Mater. Sci. Mater. Med.*, **19**, 2751–2760.

493. Jalota, S., Tas, A. C., and Bhaduri, S. B. (2005) Synthesis of HA-seeded TTCP ($Ca_4(PO_4)_2O$) powders at 1230°C from $Ca(CH_3COO)_2{\cdot}H_2O$ and $NH_4H_2PO_4$, *J. Am. Ceram. Soc.*, **88**, 3353–3360.

494. Moseke, C., and Gbureck, U. (2010) Tetracalcium phosphate: synthesis, properties and biomedical applications, *Acta Biomater.*, **6**, 3815–3823.

495. Martin, I., and Brown, P. W. (1993) Hydration of tetracalcium phosphate, *Adv. Chem. Res.*, **5**, 115–125.

496. Fernández, E., Gil, F. J., Ginebra, M. P., Driessens, F. C. M., Planell, J. A., and Best, S. M. (1999) Calcium phosphate bone cements for clinical applications. Part II: Precipitate formation during setting reactions, *J. Mater. Sci. Mater. Med.*, **10**, 177–183.

497. Dickens, B., Brown, W. E., Kruger, G. J., and Stewart, J. M. (1973) $Ca_4(PO_4)_2O$, tetracalcium diphosphate monoxide, crystal structures and relationships to $Ca_5(PO_4)_3OH$ and $K_3Na(SO_4)_2$, *Acta Crystallogr. B*, **29**, 2046–2056.

498. Jillavenatesa, A., and Condrate, Sr. R. A. (1997) The infrared and Raman spectra of tetracalcium phosphate ($Ca_4(PO_4)_2O$), *Spectrosc. Lett.*, **30**, 1561–1570.

499. Nakano, T., Kaibara, K., Umakoshi, Y., Imazato, S., Ogata, K., Ehara, A., Ebisu, S., and Okazaki, M. (2002) Change in microstructure and solubility improvement of HAp ceramics by heat-treatment in a vacuum, *Mater. Transact.*, **43**, 3105–3111.

500. Kiba, W., Imazato, S., Takahashi, Y., Yoshioka, S., Ebisu, S., and Nakano, T. (2010) Efficacy of polyphasic calcium phosphates as a direct pulp capping material, *J. Dentistry*, **38**, 828–837.

501. Pan, L., Li, Y., Weng, W., Cheng, K., Song, C., Du, P., Zhao, G., Shen, G., Wang, J., and Han, G. (2006) Preparation of submicron biphasic α-TCP/HA powders, *Key Eng. Mater.*, 309–311, 219–222.

502. Kui, C. (2007) Slip casting derived α-TCP/HA biphasic ceramics, *Key Eng. Mater.*, 330–332, 51–54.

503. Sanchez-Sálcedo, S., Arcos, D., and Vallet-Regí, M. (2008) Upgrading calcium phosphate scaffolds for tissue engineering applications, *Key Eng. Mater.*, **377**, 19–42.

504. Li, Y., Kong, F., and Weng, W. (2009) Preparation and characterization of novel biphasic calcium phosphate powders (α-TCP/HA) derived from carbonated amorphous calcium phosphates, *J. Biomed. Mater. Res. B* (*Appl. Biomater.*), **89B**, 508–517.

505. Oishi, M., Ohtsuki, C., Kitamura, M., Kamitakahara, M., Ogata, S., Miyazaki, T., and Tanihara, M. (2004) Fabrication and chemical durability of porous bodies consisting of biphasic tricalcium phosphates, *Phosphorus Res. Bull.*, **17**, 95–100.

506. Kamitakahara, M., Ohtsuki, C., Oishi, M., Ogata, S., Miyazaki, T., and Tanihara, M. (2005) Preparation of porous biphasic tricalcium phosphate and its *in vivo* behavior, *Key Eng. Mater.*, 284–286, 281–284.

507. Wang, R., Weng, W., Deng, X., Cheng, K., Liu, X., Du, P., Shen, G., and Han, G. Dissolution behavior of submicron biphasic tricalcium phosphate powders, *Key Eng. Mater.*, 309–311, 223–226.

508. Li, Y., Weng, W., and Tam, K. C. (2007) Novel highly biodegradable biphasic tricalcium phosphates composed of α-tricalcium phosphate and β-tricalcium phosphate, *Acta Biomater.*, **3**, 251–254.

509. Li, Y., Li, D., and Weng, W. (2008) *In vitro* dissolution behavior of biphasic tricalcium phosphate composite powders composed of α-tricalcium phosphate and β-tricalcium phosphate, *Key Eng. Mater.*, 368–372, 1206–1208.

510. Vani, R., Girija, E. K., Elayaraja, K., Prakash Parthiban, S., Kesavamoorthy, R., and Narayana Kalkura, S. (2009) Hydrothermal synthesis of porous triphasic hydroxyapatite/(α and β) tricalcium phosphate, *J. Mater. Sci. Mater. Med.*, **20**, S43–S48.

511. Huang, Y., Huang, W., Sun, L., Wang, Q., He, A., and Han, C. C. (2010) Phase transition from α-TCP into β-TCP in TCP/HA composites, *Int. J. Appl. Ceram. Technol.*, **7**, 184–188.

512. McConnell, D., and Posner, A. S. (1961) Carbonate in apatites, *Science*, **134**, 213–215.

513. LeGeros, R. Z. (1965) Effect of carbonate on the lattice parameters of apatite, *Nature*, **206**, 403–404.

514. LeGeros, R. Z., Trautz, O. R., LeGeros, J. P., Klein, E., and Shirra, W. P. (1967) Apatite crystallites: effects of carbonate on morphology, *Science*, **155**, 1409–1411.

515. Astala, R., and Stott, M. J. (2005) First principles investigation of mineral component of bone: CO_3 substitutions in hydroxyapatite, *Chem. Mater.*, **17**, 4125–4133.

516. Lafon, J. P., Champion, E., and Bernache-Assollant, D. (2008) Processing of AB-type carbonated hydroxyapatite $Ca_{10-x}(PO_4)_{(6-x)}(CO_3)_{(x)}(OH)_{(2-x-2y)}(CO_3)_{(y)}$ ceramics with controlled composition, *J. Eur. Ceram. Soc.*, **28**, 139–147.

517. Tonegawa, T., Ikoma, T., Yoshioka, T., Hanagata, N., and Tanaka, J. (2010) Crystal structure refinement of A-type carbonate apatite by X-ray powder diffraction, *J. Mater. Sci.*, **45**, 2419–2426.

518. Yahia, F. B. H., and Jemal, M. (2010) Synthesis, structural analysis and thermochemistry of B-type carbonate apatites, *Thermochim. Acta*, **50**, 22–32.

519. Kannan, S., Rebelo, A., Lemos, A. F., Barba, A., and Ferreira, J. M. F. (2007) Synthesis and mechanical behaviour of chlorapatite and chlorapatite/β-TCP composites, *J. Eur. Ceram. Soc.*, **27**, 2287–2294.

520. Teshima, K., Yubuta, K., Ooi, S., Suzuki, T., Shishido, T., and Oishi, S. (2006) Environmentally friendly growth of calcium chlorapatite whiskers from a sodium chloride flux, *Cryst. Growth Des.*, **6**, 2538–2542.

521. García-Tuñón, E., Franco, J., Dacuña, B., Zaragoza, G., and Guitián, F. (2010) Chlorapatite conversion to hydroxyapatite under high temperature hydrothermal conditions, *Mater. Sci. Forum*, 636–637, 9–14.

522. Kannan, S., Goetz-Neunhoeffer, F., Neubauer, J., and Ferreira, J. M. F. (2008) Ionic substitutions in biphasic hydroxyapatite and β-tricalcium phosphate mixtures: structural analysis by Rietveld refinement, *J. Am. Ceram. Soc.*, **91**, 1–12.

523. Boanini, E., Gazzano, M., and Bigi, A. (2010) Ionic substitutions in calcium phosphates synthesized at low temperature, *Acta Biomater.*, **6**, 1882–1894.

524. Pushpakanth, S., Srinivasan, B., Sastry, T. P., and Mandal, A. B. (2008) Biocompatible and antibacterial properties of silver-doped hydroxyapatite, *J. Biomed. Nanotechnol.*, **4**, 62–66.

525. Zhang, Y., Yin, Q. S., Zhang, Y., Xia, H., Ai, F. Z., Jiao, Y. P., and Chen, X. Q. (2010) Determination of antibacterial properties and cytocompatibility

of silver-loaded coral hydroxyapatite, *J. Mater. Sci. Mater. Med.*, **21**, 2453–2462.

526. Kim, T. N., Feng, Q. L., Kim, J. O., Wu, J., Wang, H., Chen, G. C., and Cui, F. Z. Antimicrobial effects of metal ions (Ag^+, Cu^{2+}, Zn^{2+}) in hydroxyapatite, *J. Mater. Sci. Mater. Med.*, **9**, 129–134.

527. Stanić, V., Dimitrijević, S., Antić-Stanković, J., Mitrić, M., Jokić, B., Plećaš, I. B., and Raičević, S. (2010) Synthesis, characterization and antimicrobial activity of copper and zinc-doped hydroxyapatite nanopowders, *Appl. Surf. Sci.*, **256**, 6083–6089.

528. Li, Y., Ho, J., and Ooi, C. P. (2010) Antibacterial efficacy and cytotoxicity studies of copper (II) and titanium (IV) substituted hydroxyapatite nanoparticles, *Mater. Sci. Eng. C*, **30**, 1137–1144.

529. Thomas, S., Assi, P., Marycel, B., Correa, M., Liberato, W., and Brito, V. (2008) Yttrium 90-hydroxyapatite, a new radioisotope for chronic synovitis in hemophilia, *Haemophilia*, **14**, 77.

530. Chinol, M., Vallabhajosula, S., Goldsmith, S. J., Klein, M. J., Deutsch, K. F., Chinen, L. K., Brodack, J. W., Deutsch, E. A., Watson, B. A., and Tofe, A. J. (1993) Chemistry and biological behavior of samarium-153 and rhenium-186-labeled hydroxyapatite particles: potential radiopharmaceuticals for radiation synovectomy, *J. Nucl. Med.*, **34**, 1536–1542.

531. Argüelles, M. G., Berlanga, I. S. L., and Torres, E. A. (1999) Preparation of ^{153}Sm-particles for radiosynovectomy, *J. Radioanal. Nucl. Chem.*, **240**, 509–511.

532. O'Duffy, E., Clunie, G., Lui, D., Edwards, J., and Ell, P. (1999) Double blind glucocorticoid controlled trial of samarium-153 particulate hydroxyapatite radiation synovectomy for chronic knee synovitis, *Ann. Rheum. Dis.*, **58**, 554–558.

533. Palmer, L. C., Newcomb, C. J., Kaltz, S. R., Spoerke, E. D., and Stupp, S. I. (2008) Biomimetic systems for hydroxyapatite mineralization inspired by bone and enamel, *Chem. Rev.*, **108**, 4754–4783.

534. Grynpas, M. D., Hancock, R. G., Greenwood, C., Turnquist, J., and Kessler, M. J. (1993) The effects of diet, age, and sex on the mineral content of primate bones, *Calcif Tissue Int*, **52**, 399–405.

535. Elliott, J. C. (2002) Calcium phosphate biominerals, in *Phosphates: Geochemical, Geobiological and Materials Importance. Series: Reviews in Mineralogy and Geochemistry*, (ed. Hughes, J. M., Kohn, M., and Rakovan, J.), vol. 48,Mineralogical Society of America, Washington, D.C., pp. 13–49.

536. Boskey, A. L. (2006) Assessment of bone mineral and matrix using backscatter electron imaging and FTIR imaging, *Curr. Osteoporos. Rep.*, **4**, 71–75.

537. Vallet-Regí, M., and González-Calbet, J. M. (2004) Calcium phosphates as substitution of bone tissues, *Progr. Solid State Chem.*, **32**, 1–31.

538. Holt, L. E., la Mer, V. K., and Chown, H. B. (1925) Studies in calcification. I. The solubility product of secondary and tertiary calcium phosphate under various conditions, *J. Biol. Chem.*, **64**, 509–565.

539. Holt, L. E., la Mer, V. K., and Chown, H. B. (1925) Studies in calcification. II. Delayed equilibrium between the calcium phosphates and its biological significance, *J. Biol. Chem.*, **64**, 567–578.

540. Gassmann, T. (1913) The preparation of a complex salt corresponding to apatite-typhus and its relations to the constitution of bones, *H.-S. Z. Physiol. Chem.*, **83**, 403–408.

541. de Jong, W. F. (1926) La substance minérale dans les os, *Recl. Trav. Chim. Pays-Bas*, **45**, 445–449.

542. Bredig, M. A. (1933) The apatite structure of inorganic bone and tooth substance, *H.-S. Z. Physiol. Chem.*, **216**, 239–243.

543. Taylor, N. W., and Sheard, C. (1929) Microscopic and X-ray investigations on the calcification of tissue, *J. Biol. Chem.*, **81**, 479–493.

544. Weiner, S., and Wagner, H. D. (1998) Material bone: structure-mechanical function relations, *Ann. Rev. Mater. Sci.*, **28**, 271–298.

545. Weiner, S., Traub, W., and Wagner, H. D. (1999) Lamellar bone: structure-function relations, *J. Struct. Biol.*, **126**, 241–255.

546. Limeback, H. (1991) Molecular mechanisms in dental hard tissue mineralization, *Curr. Opin. Dent.*, **1**, 826–835.

547. Matkovic, V. (1991) Calcium metabolism and calcium requirements during skeletal modeling and consolidation of bone mass, *Am. J. Clin. Nutr.*, **54**, 245S–260S.

548. Power, M. L., Heaney, R. P., Kalkwarf, H. J., Pitkin, R. M., Repke, J. T., Tsang, R. C., and Schulkin, J. (1999) The role of calcium in health and disease, *Am. J. Obstetrics Gynecol.*, **181**, 1560–1569.

549. Jones, F. H. (2001) Teeth and bones: application of surface science to dental materials and related biomaterials, *Surf. Sci. Rep.*, **42**, 75–205.

550. Loveridge, N. (1999) Bone: more than a stick, *J. Anim. Sci.*, **77**(2), 190–196.

551. Nightingale, J. P., and Lewis, D. (1971) Pole figures of the orientation of apatite in bones, *Nature*, **232**, 334–335.

552. Currey, J. D. (2002) Bones: structure and mechanics, *Princeton Univercity Press*, Princeton, p. 456.

553. Rho, J. Y., Kuhn-Spearing, L., and Zioupos, P. (1998) Mechanical properties and the hierarchical structure of bone, *Med. Eng. Phys.*, **20**, 92–102.

554. Mann, S. (1996) *Biomimetic Materials Chemistry*, VCH, Weinheim, p. 400.

555. Hancox, N. M. (1972) *Biology of Bone*, Cambridge University Press, Cambridge, 1972, p. 199.

556. Martin, R. B. (1999) Bone as a ceramic composite material, *Mater. Sci. Forum*, **7**, 5–16.

557. Tzaphlidou, M. (2008) Bone architecture: collagen structure and calcium/phosphorus maps, *J. Biol. Phys.*, **34**, Spec. Iss. 1–2, 39–49.

558. Davison, K. S., Siminoski, K., Adachi, J. D., Hanley, D. A., Goltzman, D., Hodsman, A. B., Josse, R., Kaiser, S., Olszynski, W. P., Papaioannou, A., Ste-Marie, L. G., Kendler, D. L., Tenenhouse, A., and Brown, J. P. (2006) Bone strength: the whole is greater than the sum of its parts, *Semin. Arthritis Rheum.*, **36**, 22–31.

559. Turner, C. H., and Burr, D. B. (1993) Basic biomechanical measurements of bone: a tutorial, *Bone*, **14**, 595–608.

560. Currey, J. D. (2004) Tensile yield in compact bone is determined by strain, post-yield behaviour by mineral content, *J. Biomech.*, **37**, 549–556.

561. Currey, J. D., Brear, K., and Zioupos, P. (2004) Notch sensitivity of mammalian mineralized tissues in impact, *Proc. R. Soc. London B*, **217**, 517–522.

562. Anderson, J. C., and Eriksson, C. (1970) Piezoelectric properties of dry and wet bone, *Nature*, **227**, 491–492.

563. Lang, S. B. (1966) Pyroelectric effect in bone and tendon, *Nature*, **212**, 704–705.

564. Haynes, V. (1968) Radiocarbon: analysis of inorganic carbon of fossil bone and enamel, *Science*, **161**, 687–688.

565. Rensberger, J. M., and Watabe, M. (2000) Fine structure of bone in dinosaurs, birds and mammals, *Nature*, **406**, 619–622.

566. Kolodny, Y., Luz, B., Sander, M., and Clemens, W. A. (1996) Dinosaur bones: fossils or pseudomorphs? The pitfalls of physiology reconstruction from apatitic fossils, *Palaeo*, **126**, 161–171.

567. Trueman, C. N., and Tuross, N. (2002) Trace elements in recent and fossil bone apatite, in *Phosphates: Geochemical, Geobiological and*

Materials Importance. Series: Reviews in Mineralogy and Geochemistry, (ed. Hughes, J. M., Kohn, M., and Rakovan, J.), vol. 48, Mineralogical Society of America, Washington, D.C., 2002, pp. 489–522.

568. Currey, J. D., and Brear, K. (1990) Hardness, Young's modulus and yield stress in mammalian mineralized tissues, *J. Mater. Sci. Mater. Med.*, **1**, 14–20.

569. Brown, W. E., and Chow, L. C. (1976) Chemical properties of bone mineral, *Ann. Rev. Mater. Sci.*, **6**, 213–236.

570. Lakes, R. (1993) Materials with structural hierarchy, *Nature*, **361**, 511–515.

571. Meyers, M. A., Lin, A. Y. M., Seki, Y., Chen, P. Y., Kad, B. K., and Bodde, S. (2006) Structural biological composites: an overview, *JOM*, **58**, 36–43.

572. Wahl, D. A., and Czernuszka, J. T. (2006) Collagen – hydroxyapatite composites for hard tissue repair, *Eur. Cell Mater.*, **11**, 43–56.

573. Rey, C., Combes, C., Drouet, C., and Glimcher, M. J. (2009) Bone mineral: update on chemical composition and structure, *Osteoporosis Int.*, **20**, 1013–1021, Erratum: *ibid.*, p. 2155.

574. Watt, J. C. (1925) The deposition of calcium phosphate and calcium carbonate in bone and in areas of calcification, *Arch. Surg. Chicago*, **10**, 983–990.

575. Olszta, M. J., Cheng, X., Jee, S. S., Kumar, R., Kim, Y. Y., Kaufman, M. J., Douglas, E. P., and Gower, L. B. (2007) Bone structure and formation: a new perspective, *Mater. Sci. Eng. R*, **58**, 77–116.

576. Boskey, A. L. (2007) Mineralization of bones and teeth, *Elements*, **3**, 385–391.

577. Glimcher, M. J. (2006) Bone: nature of the calcium phosphate crystals and cellular, structural and physical chemical mechanisms in their formation, in *Medical Mineralogy and Geochemistry, Series: Reviews in Mineralogy and Geochemistry*, (ed. Sahai, N., and Schoonen M. A. A.), vol. 64, Mineralogical Society of America, Washington, D.C., pp. 223–282.

578. Boskey, A. L., and Roy, R. (2008) Cell culture systems for studies of bone and tooth mineralization, *Chem. Rev.*, **108**, 4716–4733.

579. McKee, M. D., Addison, W. N., and Kaartinen, M. T. (2005) Hierarchies of extracellular matrix and mineral organization in bone of the craniofacial complex and skeleton, *Cell Tiss. Organ*, **181**, 176–188.

580. Currey, J. D. (2005) Hierarchies in biomineral structures, *Science*, **309**, 253–254.

581. Fantner, G. E., Rabinovych, O., Schitter, G., Thurner, P., Kindt, J. H., Finch, M. M., Weaver, J. C., Golde, L. S., Morse, D. E., Lipman, E. A., Rangelow, I. W., and Hansma, P. K. (2006) Hierarchical interconnections in the nano-composite material bone: fibrillar cross-links resist fracture on several length scales, *Composite Sci. Technol.*, **66**, 1202–1208.

582. Fratzl, P., and Weinkamer, R. (2007) Nature's hierarchical materials, *Progress Mater. Sci.*, **52**, 1263–1334.

583. Boskey, A. L., and Coleman, R. (2010) Aging and bone, *J. Dent. Res.*, **89**, 1333–1348.

584. Meyers, M. A., Chen, P. Y., Lin, A. Y. M., and Seki, Y. (2008) Biological materials: structure and mechanical properties, *Prog. Mater. Sci.*, **53**, 1–206.

585. Athanasiou, K. A., Zhu, C., Lanctot, D. R., Agrawal, C. M., and Wang, X. (2000) Fundamentals of biomechanics in tissue engineering of bone, *Tissue Eng.*, **6**, 361–381.

586. Nikolov, S., and Raabe, D. (2008) Hierarchical modeling of the elastic properties of bone at submicron scales: the role of extrafibrillar mineralization, *Biophys. J.*, **94**, 4220–4232.

587. Gupta, H. S., Wagermaier, W., Zickler, G. A., Aroush, D. R. B., Funari, S. S., Roschger, P., Wagner, H. D., and Fratzl, P. (2005) Nanoscale deformation mechanisms in bone, *Nano Lett.*, **5**, 2108–2111.

588. Peterlik, H., Roschger, P., Klaushofer, K., and Fratzl, P. (2006) From brittle to ductile fracture of bone, *Nature Mater.*, **5**, 52–55.

589. Ruppel, M. E., Miller, L. M., and Burr, D. B. (2008) The effect of the microscopic and nanoscale structure on bone fragility, *Osteop. Int.*, **19**, 1251–1265.

590. Fratzl, P., Gupta, H. S., Paschalis, E. P., and Roschger, P. (2004) Structure and mechanical quality of the collagen-mineral nano-composite in bone, *J. Mater. Chem.*, **14**, 2115–2123.

591. Marino, A. A., and Becker, R. O. (1967) Evidence for direct physical bonding between the collagen fibres and apatite crystals in bone, *Nature*, **213**, 697–698.

592. Eppell, S. J., Tong, W., Katz, J. L., Kuhn, L., and Glimcher, M. J. (2001) Shape and size of isolated bone mineralites measured using atomic force microscopy, *J. Orthop. Res.*, **19**, 1027–1034.

593. Clark, S. M., and Iball, J. (1954) Orientation of apatite crystals in bone, *Nature*, **174**, 399–400.

594. Rubin, M. A., Jasiuk, I., Taylor, J., Rubin, J., Ganey, T., and Apkarian, R. P. (2003) TEM analysis of the nanostructure of normal and osteoporotic human trabecular bone, *Bone*, **33**, 270–282.

595. Su, X., Sun, K., Cui, F. Z., and Landis, W. J. (2003) Organization of apatite crystals in human woven bone, *Bone*, **32**, 150–162.

596. Due to the nanoscopic dimensions, biological apatite is occasionally called "nano-apatite" [544].

597. Self-assembling is the autonomous organization of components into patterns or structures without human intervention. It is considered that self-assembling processes are common throughout nature and technology [598].

598. Whitesides, G. M., and Grzybowski, B. (2002) Self-assembly at all scales, *Science*, **295**, 2418–2421.

599. Fratzl, P., Fratzl-Zelman, N., Klaushofer, K., Vogl, G., and Koller, K. (1991) Nucleation and growth of mineral crystals in bone studied by small-angle X-ray scattering, *Calcif. Tissue Int.*, **48**, 407–413.

600. Landis, W. J., Hodgens, K. J., Arena, J., Song, M. J., and McEwen, B. F. (1996) Structural relations between collagen and mineral in bone as determined by high voltage electron microscopic tomography, *Microsc. Res. Techn.*, **33**, 192–202.

601. Rosen, V. B., Hobbs, L. W., and Spector, M. (2002) The ultrastructure of anorganic bovine bone and selected synthetic hyroxyapatites used as bone graft substitute materials, *Biomaterials*, **23**, 921–928.

602. Cui, F. Z., Li, Y., and Ge, J. (2007) Self-assembly of mineralized collagen composites, *Mater. Sci. Eng. R*, **57**, 1–27.

603. Sato, K. (2006) Inorganic-organic interfacial interactions in hydroxyapatite mineralization processes, *Topics Curr. Chem.*, **270**, 127–153.

604. Hartgerink, J. D., Beniash, E., and Stupp, S. I. (2001) Self-assembly and mineralization of peptide-amphiphile nanofibers, *Science*, **294**, 1684–1688.

605. Burger, C., Zhou, H., Wang, H., Sics, I., Hsiao, B. S., Chu, B., Graham, L., and Glimcher, M. (2008) Lateral packing of mineral crystals in bone collagen fibrils, *Biophys. J.*, **95**, 1985–1992.

606. George, A., and Veis, A. (2008) Phosphorylated proteins and control over apatite nucleation, crystal growth, and inhibition, *Chem. Rev.*, **108**, 4670–4693.

607. Hall, B. K. (ed.) (1983) *Cartilage: Structure, Function and Biochemistry*, Academic Press, New York, p. 385.

608. Robins, S. P., Bilezikian, J. P., and Seibel, M. J. (1999) Dynamics of bone and cartilage metabolism, *Academic Press*, New York, p. 672.

609. Hall, B. K. (ed.), *Bones and Cartilage: Developmental Skeletal Biology*, Academic Press, p. 792.

610. Marino, A. A., and Becker, R. O. (1970) Evidence for epitaxy in the formation of collagen and apatite, *Nature*, **226**, 652–653.

611. Jodaikin, A., Weiner, S., Talmon, Y., Grossman, E., and Traub, W. (1988) Mineral-organic matrix relations in tooth enamel, *Int. J. Biol. Macromol.*, **10**, 349–352.

612. Fincham, A. G., Moradian-Oldak, J., Diekwisch, T. G. H., Lyaruu, D. M., Wright, J. T., Bringas, P., and Slavkin, H. C. (1995) Evidence for amelogenin "nanospheres" as functional components of secretory-stage enamel matrix, *J. Struct. Biol.*, **115**, 50–59.

613. LeGeros, R. Z. Pan, C. M., Suga, S., and Watabe, N. (1985) Crystallo-chemical properties of apatite in atremate brachiopod shells, *Calcif. Tissue Int.*, **37**, 98–100.

614. Iijima, M., and Moriwaki, Y. (1990) Orientation of apatite and organic matrix in Lingula unguis shell, *Calcif. Tissue Int.*, **47**, 237–242.

615. Williams, A., Cusack, M., Buckman, J. O., and Stachel, T. (1998) Siliceous tablets in the larval shells of apatitic discinid *Brachiopods*, *Science*, **297**, 2094–2096.

616. Rohanizadeh, R., and LeGeros, R. Z. (2007) Mineral phase in linguloid brachiopod shell: *Lingula adamsi*, *Lethaia*, **40**, 61–68.

617. Nakano, T., Ishimoto, T., Lee, J. W., and Umakoshi, Y. (2008) Preferential orientation of biological apatite crystallite in original, regenerated and diseased cortical bones, *J. Ceram. Soc. Jpn.*, **116**, 313–315.

618. Nakano, T., Ishimoto, T., Umakoshi, Y., and Tabata, Y. (2007) Variation in bone quality during regenerative process, *Mater. Sci. Forum*, 539–543, 675–680.

619. Suvorova, E. I., Petrenko, P. P., and Buffat, P. A. (2007) Scanning and transmission electron microscopy for evaluation of order/disorder in bone structure, *Scanning*, **29**, 162–170.

620. Stock, S. R., Blackburn, D., Gradassi, M., and Simon, H. G. (2003) Bone formation during forelimb regeneration: a microtomography (microCT) analysis, *Dev. Dyn.*, **226**, 410–417.

621. Teitelbaum, S. L. (2000) Bone resorption by osteoclasts, *Science*, **289**, 1504–1508.

622. Rodan, G. A., and Martin, T. J. (2000) Therapeutic approaches to bone diseases, *Science*, **289**, 1508–1514.

623. Schilling, A. F., Filke, S., Brink, S., Korbmacher, H., Amling, M., and Rueger, J. M. (2006) Osteoclasts and biomaterials, *Eur. J. Trauma*, **32**, 107–113.

624. Bonar, L. C., Shimizu, M., Roberts, J. E., Griffin, R. G., and Glimcher, M. J. (1991) Structural and composition studies on the mineral of newly formed dental enamel: a chemical, X-ray diffraction, and ^{31}P and proton nuclear magnetic resonance study, *J. Bone Miner. Res.*, **6**, 1167–1176.

625. Biltz, R. M., and Pellegrino, E. D. (1971) The hydroxyl content of calcified tissue mineral, *Calcif. Tissue Res.*, **7**, 259–263.

626. Rey, C., Renugopalakrishnan, V., Collins, B., and Glimcher, M. J. (1991) Fourier transform infrared spectroscopic study of the carbonate ions in bone mineral during aging, *Calcif. Tissue Int.*, **49**, 251–258.

627. Rey, C., Miquel, J. L., Facchini, L., Legrand, A. P., and Glimcher, M. J. (1995) Hydroxyl groups in bone mineral, *Bone*, **16**, 583–586.

628. Loong, C. K., Rey, C., Kuhn, L. T., Combes, C., Wu, Y., Chen, S. H., and Glimcher, M.J. (2000) Evidence of hydroxyl-ion deficiency in bone apatites: an inelastic neutron scattering study, *Bone*, **26**, 599–602.

629. Cho, G., Wu, Y., and Ackerman, J. L. (2003) Detection of hydroxyl ions in bone mineral by solid-state NMR spectroscopy, *Science*, **300**, 1123–1127.

630. Tung, M. S., and Brown, W. E. (1983) An intermediate state in hydrolysis of amorphous calcium phosphate, *Calcif. Tissue Int.*, **35**, 783–790.

631. Tung, M. S., and Brown, W. E. (1985) The role of octacalcium phosphate in subcutaneous heterotopic calcification, *Calcif. Tissue Int.*, **37**, 329–331.

632. Brown, W. E., Eidelman, N., and Tomazic, B. B. (1987) Octacalcium phosphate as a precursor in biomineral formation, *Adv. Dent. Res.*, **1**, 306–313.

633. Siew, C., Gruninger, S. E., Chow, L. C., and Brown, W. E. (1992) Procedure for the study of acidic calcium phosphate precursor phases in enamel mineral formation, *Calcif. Tissue Int.*, **50**, 144–148.

634. Crane, N. J., Popescu, V., Morris, M. D., Steenhuis, P., Ignelzi, J. M. A. (2006) Raman spectroscopic evidence for octacalcium phosphate and other transient mineral species deposited during intramembranous mineralization, *Bone*, **39**, 434–442.

635. Tseng, Y. H., Mou, C. Y., and Chan, J. C. C. (2006) Solid-state NMR study of the transformation of octacalcium phosphate to hydroxyapatite: a mechanistic model for central dark line formation, *J. Am. Chem. Soc.*, **128**, 6909–6918.

636. Eanes, E. D., Gillessen, I. H., and Posner, A. S. (1965) Intermediate states in the precipitation of hydroxyapatite, *Nature*, **208**, 365–367.

637. Termine, J. D., and Posner, A. S. (1966) Infrared analysis of rat bone: age dependency of amorphous and crystalline mineral fractions, *Science*, **153**, 1523–1525.

638. Termine, J. D., and Posner, A. S. (1966) Infra-red determination of the percentage of crystallinity in apatitic calcium phosphates, *Nature*, **211**, 268–270.

639. Harper, R. A., and Posner, A. S. (1966) Measurement of non-crystalline calcium phosphate in bone mineral, *Proc. Soc. Exptl. Biol. Med.*, **122**, 137–142.

640. Posner, A. S. (1969) Crystal chemistry of bone mineral, *Physiol. Rev.*, **49**, 760–792.

641. Posner, A. S. (1973) Bone mineral on the molecular level, *Fed. Proc.*, **32**, 1933–1937.

642. Boskey, A. L., and Posner, A. S. (1976) Formation of hydroxyapatite at low supersaturation, *J. Phys. Chem.*, **80**, 40–44.

643. Posner, A. S. (1978) The chemistry of bone mineral, *Bull. Hosp. Joint Dis.*, **39**, 126–144.

644. Glimcher, M. J., Bonar, L. C., Grynpas, M. D., Landis, W. J., and Roufosse, A. H. (1981) Recent studies of bone mineral: Is the amorphous calcium phosphate theory valid?, *J. Cryst. Growth*, **53**, 100–119.

645. Meyer, J. L., and Eanes, E. D. (1978) A thermodynamic analysis of the amorphous to crystalline calcium phosphate transformation, *Calcif. Tissue Res.*, **25**, 59–68.

646. Meyer, J. L., and Eanes, E. D. (1978) A thermodynamic analysis of the secondary transition in the spontaneous precipitation of calcium phosphate, *Calcif. Tissue Res.*, **25**, 209–216.

647. Politi, Y., Arad, T., Klein, E., Weiner, S., and Addadi, L. (2004) Sea urchin spine calcite forms via a transient amorphous calcium carbonate phase, *Science*, **306**, 1161–1164.

648. Weiner, S., Sagi, I., and Addadi, L. (2005) Choosing the crystallization path less traveled, *Science*, **309**, 1027–1028.

649. Weiner, S. (2006) Transient precursor strategy in mineral formation of bone, *Bone*, **39**, 431–433.

650. Pekounov, Y., and Petrov, O. E. (2008) Bone resembling apatite by amorphous-to-crystalline transition driven self-organisation, *J. Mater. Sci. Mater. Med.*, **19**, 753–759.

651. Gower, L. B. (2008) Biomimetic model systems for investigating the amorphous precursor pathway and its role in biomineralization, *Chem. Rev.*, **108**, 4551–4627.

652. Weiner, S., Mahamid, J., Politi, Y., Ma, Y., and Addadi, L. (2009) Overview of the amorphous precursor phase strategy in biomineralization, *Frontiers of Materials Science in China*, **3**, 104–108.

653. Mahamid, J., Sharir, A., Addadi, L., and Weiner, S. (2008) Amorphous calcium phosphate is a major component of the forming fin bones of zebrafish: indications for an amorphous precursor phase, *Proc. Natl. Acad. Sci. USA*, **105**, 12748–12753.

654. Mahamid, J., Aichmayer, B., Shimoni, E., Ziblat, R., Li, C., Siegel, S., Paris, O., Fratzl, P., Weiner, S., and Addadi, L. (2010) Mapping amorphous calcium phosphate transformation into crystalline mineral from the cell to the bone in zebrafish fin rays, *Proc. Natl. Acad. Sci. USA*, **107**, 6316–6321.

655. Nudelman, F., Pieterse, K., George, A., Bomans, P. H. H., Friedrich, H., Brylka, L. J., Hilbers, P. A. J., de With, G., Sommerdijk, N. A. J. M. (2010) The role of collagen in bone apatite formation in the presence of hydroxyapatite nucleation inhibitors, *Nature Mater.*, **9**, 1004–1009.

656. Cölfen, H. (2010) A crystal-clear view, *Nature Mater.*, **9**, 960–961.

657. Sahar, N. D., Hong, S. I., and Kohn, D. H. (2005) Micro- and nano-structural analyses of damage in bone, *Micron*, **36**, 617–629.

658. Bilezikian, J. P., Raisz, L. G., and Martin, T. J. (2008) *Principles of Bone Biology*, 3rd edn., Academic Press, New York, p. 1900.

659. Burnell, J. M., Teubner, E. J., and Miller, A. G. (1980) Normal maturational changes in bone matrix, mineral, and crystal size in the rat, *Calcif. Tissue Int.*, **31**, 13–19.

660. Weiner, S., and Traub, W. (1986) Organization of hydroxyapatite crystals within collagen fibrils, *FEBS Lett.*, **206**, 262–266.

661. Hanschin, R. G., and Stern, W. B. (1995) X-ray diffraction studies on the lattice perfection of human bone apatite (*Crista iliaca*), *Bone*, **16**(4), 355S–363S.

662. Pellegrino, E. D., and Blitz, R. M. (1972) Mineralization in the chick embryo I. Monohydrogen phosphate and carbonate relationships during maturation of the bone crystal complex, *Calcif. Tissue Res.*, **10**, 128–135.

663. LeGeros, R. Z. Balmain, N., and Bonel, G. (1987) Age-related changes in mineral of rat and bovine cortical bone, *Calcif. Tissue Int.*, **41**, 137–144.

664. Rey, C., Hina, A., Tofighi, A., and Glimcher, M. J. (1995) Maturation of poorly crystalline apatites: chemical and structural aspect *in vivo* and *in vitro* behavior, *Cell Mater.*, **5**, 345–356.

665. Verdelis, K., Lukashova, L., Wright, J. T., Mendelsohn, R., Peterson, M. G. E., Doty, S., and Boskey, A. L. (2007) Maturational changes in dentine mineral properties, *Bone*, **40**, 1399–1407.

666. Yerramshetty, J. S., Lind, C., and Akkus, O. (2006) The compositional and physicochemical homogeneity of male femoral cortex increases after the sixth decade, *Bone*, **39**, 1236–1243.

667. Kuhn, L. T., Grynpas, M. D., Rey, C., Wu, Y., Ackerman, J. L., and Glimcher, M. J. (2008) A comparison of the physical and chemical differences between cancellous and cortical bovine bone mineral at two ages, *Calcif. Tissue Int.*, **83**, 146–154.

668. Donnelly, E., Boskey, A. L., Baker, S. P., and van der Meulen, M. C. (2009) Effects of tissue age on bone tissue material composition and nanomechanical properties in the rat cortex, *J. Biomed. Mater. Res. A*, **92A**, 1048–1056.

669. Lowenstam, H. A. (1981) Minerals formed by organisms, *Science*, **211**, 1126–1131.

670. Lowenstam, H. A., and Weiner, S. (1983) Mineralization by organisms and the evolution of biomineralization, in *Biomineralization and Biological Metal Accumulation*, (ed. Westbroek, P., and de Jong, E. W.), D. Reidel Pub Co., Dordrecht, pp. 191–203.

671. Plate, U., Tkotz, T., Wiesmann, H. P., Stratmann, U., Joos, U., and Höhling, H. J. (1996) Early mineralization of matrix vesicles in the epiphyseal growth plate, *J. Microsc.*, **183**, 102–107.

672. Stratmann, U., Schaarschmidt, K., Wiesmann, H. P., Plate, U., and Höhling, H. J. (1996) Mineralization during matrix-vesicle-mediated mantle dentine formation in molars of albino rats: a microanalytical and ultrastructural study, *Cell Tiss. Res.*, **284**, 223–230.

673. Jahnen-Dechent, W., Schinke, T., Trindl, A., Muller-Esterl, W., Sablitzky, F., Kaiser, S., and Blessing, M. (1997) Cloning and targeted deletion of the mouse fetuin gene, *J. Biol. Chem.*, **272**, 31496–31503.

674. Schinke, T., McKnee, M. D., and Karsenty, G. (1999) Extracellular matrix calcification: where is the action?, *Nature Genetics*, **21**, 150–151.

675. Jahnen-Dechent, W., Schäfer, G., Heiss, A., and Grötzinger, J. (2001) Systemic inhibition of spontaneous calcification by the serum protein a_2-HS glycoprotein/fetuin, *Z. Kardiol.*, **90**(3), III/47–III/56.

676. Avery, J. K. (2001) *Oral Development and Histology*, 3rd edn., Thieme Medical Publishers, New York, p. 435.

677. Cate, A. R. (1998) Oral Histology: Development, Structure, and Function, 5th edn., Mosby-Year Book, Saint Louis, p. 497.

678. Xue, J., Zhang, L., Zou, L., Liao, Y., Li, J., Xiao, L., and Li, W. (2008) High-resolution X-ray microdiffraction analysis of natural teeth, *J. Synchrotron Rad.*, **15**, 235–238.

679. Gaft, M., Shoval, S., Panczer, G., Nathan, Y., Champagnon, B., and Garapon, C. (1996) Luminescence of uranium and rare-earth elements in apatite of fossil fish teeth, *Palaeogeogr. Palaeocl.*, **126**, 187–193.

680. Huang, C. M., Zhang, Q., Bai, S., and Wang, C. S. (2007) FTIR and XRD analysis of hydroxyapatite from fossil human and animal teeth in Jinsha Relict, Chengdu, *Spectrosc. Spect. Anal.*, **27**, 2448–2452.

681. Ho, S. P., Yu, B., Yun, W., Marshall, G. W., Ryder, M. I., and Marshall, S. J. (2009) Structure, chemical composition and mechanical properties of human and rat cementum and its interface with root dentine, *Acta Biomater.*, **5**, 707–718.

682. Bosshardt, D. D., and Selvig, K. A. (1997) Dental cementum: the dynamic tissue covering of the root, *Periodontology*, **14**, 41–75.

683. Ho, S. P., Senkyrikova, P., Marshall, G. W., Yun, W., Wang, Y., Karan, K., Li, C., and Marshall, S. J. (2009) Structure, chemical composition and mechanical properties of coronal cementum in human deciduous molars, *Dental Mater.*, **25**, 1195–1204.

684. Yamamoto, T., Li, M., Liu, Z., Guo, Y., Hasegawa, T., Masuki, H., Suzuki, R., and Amizuka, N. (2010) Histological review of the human cellular cementum with special reference to an alternating lamellar pattern, *Odontology*, **98**, 102–109.

685. Strictly speaking, there are some differences among these biological materials. For example, the hardness of live dentine is less than that of enamel but is greater than that of bone or cementum [681]. When pulp of the tooth dies or is removed by a dentist, the properties of dentine change: it becomes brittle, liable to fracture and looses a reparative capability.

686. Margolis, H. C., Beniash, E., and Fowler, C. E. (2006) Role of macromolecular assembly of enamel matrix proteins in enamel formation, *J. Dent. Res.*, **86**, 775–793.

687. Vieira, A., Hancock, R., Limeback, H., Schwartz, M., and Grynpas, M. D. (2003) How does fluoride concentration in the tooth affect apatite crystal size? *J. Dent. Res.*, **82**, 909–913.

688. Cui, F. Z., and Ge, J. (2007) New observations of the hierarchical structure of human enamel, from nanoscale to microscale, *J. Tissue Eng. Regener. Med.*, **1**, 185–191.

689. Jandt, K. D. (2006) Probing the future in functional soft drinks on the nanometre scale – towards tooth friendly soft drinks, *Trends Food Sci. Technol.*, **17**, 263–271.

690. Chen, H., Chen, Y., Orr, B. G., Banaszak-Holl, M., Majoros, I., and Clarkson, B. H. (2004) Nanoscale probing of the enamel nanorod surface using polyamidoamine dendrimers, *Langmuir*, **20**, 4168–4171.

691. Rönnholm, E. (1962) The amelogenesis of human teeth as revealed by electron microscopy. II. The development of the enamel crystallites, *J. Ultrastruct. Res.*, **6**, 249–303.

692. Nylen, M. U., Evans, E. D., and Omnel, K. A. (1963) Crystal growth in rat enamel, *J. Cell. Biol.*, **18**, 109–123.

693. Miake, Y., Shimoda, S., Fukae, M., and Aoba, T. (1993) Epitaxial overgrowth of apatite crystals on the thin-ribbon precursor at early stages of porcine enamel mineralization, *Calcif. Tissue Int.*, **53**, 249–256.

694. Daculsi, G., Menanteau, J., Kerebel, L. M., and Mitre, D. (1984) Length and shape of enamel crystals, *Calcif. Tissue Int.*, **36**, 550–555.

695. Jodaikin, A., Traub, W., and Weiner, S. (1984) Enamel rod relations in the developing rat incisor, *J. Ultrastruct. Res.*, **89**, 324–332.

696. Bres, E. F., and Hutchison, J. L. (2002) Surface structure study of biological calcium phosphate apatite crystals from human tooth enamel, *J. Biomed. Mater. Res.*, **63**, 433–440.

697. Schroeder, I., and Frank, R. M. (1985) High-resolution transmission electron microscopy of adult human peritubular dentine, *Cell Tissue Res.*, **242**, 449–451.

698. Brès, E. F., Voegel, J. C., and Frank, R. M. (1990) High-resolution electron microscopy of human enamel crystals, *J. Microsc.*, **160**, 183–201.

699. Robinson, C., Connell, S., Kirkham, J., Shorea, R., and Smith, A. (2004) Dental enamel – a biological ceramic: regular substructures in enamel hydroxyapatite crystals revealed by atomic force microscopy, *J. Mater. Chem.*, **14**, 2242–2248.

700. Warf, R. D., and Watson, R. R. (2008) Calcium phosphate – nutrition in prevention of early childhood dental caries, in *Wild-Type Food in Health Promotion and Disease Prevention: The Columbus Concept*, (ed. de Meester, F., and Watson, R. R.), Humana Press, Totowa, pp. 343–353.

701. Simmer, J. P., and Fincham, A. G. (1995) Molecular mechanisms of dental enamel formation, *Crit. Rev. Oral Biol. Med.*, **6**, 84–108.

702. Diekwisch, T. G., Berman, B. J., Genters, S., and Slavkin, H. C. (1995) Initial enamel crystals are not spatially associated with mineralized dentine, *Cell Tissue Res.*, **279**, 149–167.

703. Aoba, T. (1996) Recent observations on enamel crystal formation during mammalian amelogenesis, *Anat. Rec.*, **245**, 208–218.

704. Beniash, E., Metzler, R. A., Lam, R. S. K., Gilbert, P. U. P. A. (2009) Transient amorphous calcium phosphate in forming enamel, *J. Struct. Biol.*, **166**, 133–143. Corrigendum *ibid*. **167**, p. 95.

705. Smith, C. E. (1998) Cellular and chemical events during enamel maturation, *Crit. Rev. Oral. Biol. Med.*, **9**, 128–161.

706. Sydney-Zax, M., Mayer, I., and Deutsch, D. (1991) Carbonate content in developing human and bovine enamel, *J. Dent. Res.*, **70**, 913–916.

707. Rey, C., Renugopalakrishnan, V., Shimizu, M., Collins, B., and Glimcher, M. J. (1991) A resolution-enhanced Fourier transform infrared spectroscopic study of the environment of the CO_3^{2-} ion in the mineral phase of enamel during its formation and maturation, *Calcif. Tissue Int.*, **49**, 259–268.

708. Takagi, T., Ogasawara, T., Tagami, J., Akao, M., Kuboki, Y., Nagai, N., and LeGeros, R. Z. (1998) pH and carbonate levels in developing enamel, *Connect. Tissue Res.*, **38**, 181–187.

709. Lowenstam, H. A., and Weiner, S. (1985) Transformation of amorphous calcium phosphate to crystalline dahillite in the radular teeth of chitons, *Science*, **227**, 51–53.

710. Selvig, K. A. (1970) Periodic lattice images of hydroxyapatite crystals in human bone and dental hard tissues, *Calcif. Tissue Res.*, **6**, 227–238.

711. Selvig, K. A. (1973) Electron microscopy of dental enamel: analysis of crystal lattice images, *Cell Tissue Res.*, **137**, 271–280.

712. Cuisinier, F. J. G., Steuer, P., Senger, B., Voegel, J. C., and Frank, R. M. (1993) Human amelogenesis: High resolution electron microscopy of nanometer-sized particles, *Cell Tissue Res.*, **273**, 175–182.

713. Cuisinier, F. J. G., Steuer, P., Brisson, A., and Voegel, J. C. (1995) High resolution electron microscopy study of crystal growth mechanisms in chicken bone composites, *J. Cryst. Growth*, **156**, 443–453.

714. Houllé, P., Voegel, J. C., Schultz, P., Steuer, P., and Cuisinier, F. J. G. (1997) High resolution electron microscopy: structure and growth mechanisms of human dentine crystals, *J. Dent. Res.*, **76**, 895–904.

715. Mann, S. (1993) Molecular tectonics in biomineralization and biomimetic materials chemistry, *Nature*, **365**, 499–505.

716. Kirkham, J., Zhang, J., Brookes, S. J., Shore, R. C., Wood, S. R., Smith, D. A., Wallwork, M. L., Ryu, O. H., and Robinson, C. (2000) Evidence for charge domains on developing enamel crystal surfaces, *J. Dent. Res.*, **79**, 1943–1947.

717. Smith, T. M., and Tafforeau, P. (2008) New visions of dental tissue research: tooth development, chemistry, and structure, *Evol. Anthropol.*, **17**, 213–226.

718. Warshawsky, H., and Nanci, A. (1982) Stereo electron microscopy of enamel crystallites, *J. Dental Res.*, **61**, 1504–1514.

719. Warshawsky, H. (1989) Organization of crystals in enamel, *Anat. Rec.*, **224**, 242–262.

720. Xu, C., Yao, X., Walker, M. P., and Wang, Y. (2009) Chemical / molecular structure of the dentine-enamel junction is dependent on the intratooth location, *Calcif. Tiss. Int.*, **84**, 221–228.

721. Arsenault, A. L., and Robinson, B. W. (1989) The dentineo-enamel junction: a structural and microanalytical study of early mineralization, *Calcif. Tissue Int.*, **45**, 111–121.

722. Hayashi, Y. (1992) High resolution electron microscopy in the dentineo-enamel junction, *J. Electron Microsc.* (Tokyo), **41**, 387–391.

723. Hayashi, Y. (1993) High resolution electron microscopic study on the human dentine crystal, *J. Electron Microsc.* (Tokyo), **42**, 141–146.

724. Bodier-Houllé, P., Steuer, P., Meyer, J. M., Bigeard, L., and Cuisinier, F. J. G. (2000) High-resolution electron-microscopic study of the relationship between human enamel and dentine crystals at the dentineo-enamel junction, *Cell Tissue Res.*, **301**, 389–395.

725. Takano, Y., Hanaizumi, Y., and Oshima, H. (1996) Occurrence of amorphous and crystalline mineral deposits at the epithelial-mesenchymal interface of incisors in the calcium-loaded rat: Implication of novel calcium binding domains, *Anat. Rec.*, **245**, 174–185.

726. Dong, W., and Warshawsky, H. (1996) Lattice fringe continuity in the absence of crystal continuity in enamel, *Adv. Dent. Res.*, **10**, 232–237.

727. Wang, R., Hu, Y., and Ng, C. (2006) Microstructure and interfacial fracture at the cementum-enamel junctions in equine and bovine teeth, *J. Mater. Res.*, **21**, 2146–2155.

728. Ho, S. P., Goodis, H., Balooch, M., Nonomura, G., Marshall, S. J., and Marshall, G. W. (2004) The effect of sample preparation technique on

determination of structure and nanomechanical properties of human cementum hard tissue, *Biomaterials*, **25**, 4847–4857.

729. Paine, M. L., White, S. N., Luo, W., Fong, H., Sarikaya, M., and Snead, M. L. (2001) Regulated gene expression dictates enamel structure and tooth function, *Matrix Biol.*, **20**, 273–292.

730. Finke, M., Parker, D. M., and Jandt, K. D. (2002) Influence of soft drinks on the thickness and morphology of *in situ* acquired pellicle layer on enamel, *J. Colloid Interf. Sci.*, **251**, 263–270.

731. Barbour, M. E., Parker, D. M., and Jandt, K. D. (2003) Enamel dissolution as a function of solution degree of saturation with respect to hydroxyapatite: a nanoindentation study, *J. Colloid Interf. Sci.*, **265**, 9–14.

732. Lippert, F., Parker, D. M., and Jandt, K. D. (2004) Susceptibility of deciduous and permanent enamel to dietary acid-induced erosion studied with atomic force microscopy nanoindentation, *Eur. J. Oral Sci.*, **112**, 61–66.

733. Barbour, M. E., Parker, D. M., Allen, G. C., and Jandt, K. D. (2005) Human enamel erosion in constant composition citric acid solutions as a function of degree of saturation with respect to hydroxyapatite, *J. Oral Rehabil.*, **32**, 16–21.

734. LeGeros, R. Z. (1999) Calcium phosphates in demineralization and remineralization processes, *J. Clin. Dent.*, **10**, 65–73.

735. Walker, G., Cai, F., Shen, P., Reynolds, C., Ward, B., Fone, C., Honda, S., Koganei, M., Oda, M., and Reynolds, E. (2006) Increased remineralization of tooth enamel by milk containing added casein phosphopeptide-amorphous calcium phosphate, *J. Dairy Res.*, **73**, 74–78.

736. Lippert, F., Parker, D. M., and Jandt, K. D. (2004) *In situ* remineralisation of surface softened human enamel studied with AFM nanoindentation, *Surface Sci.*, **553**, 105–114.

737. Lippert, F., Parker, D. M., and Jandt, K. D. (2004) *In vitro* demineralization / remineralization cycles at human tooth enamel surfaces investigated by AFM and nanoindentation, *J. Colloid Interf. Sci.*, **280**, 442–448.

738. Lippert, F., Parker, D. M., and Jandt, K. D. (2004) Toothbrush abrasion of surface softened enamel studied with tapping mode AFM and AFM nanoindentation, *Caries Res.*, **38**, 464–472.

739. Hannig, C., and Hannig, M. (2010) Natural enamel wear – a physiological source of hydroxylapatite nanoparticles for biofilm management and tooth repair? *Medical Hypotheses*, **74**, 670–672.

740. Busch, S. (2004) Regeneration of human tooth enamel, *Angew. Chew. Int. Ed.*, **43**, 1428–1431.

741. Onuma, K., Yamagishi, K., and Oyane, A. (2005) Nucleation and growth of hydroxyapatite nanocrystals for nondestructive repair of early caries lesions, *J. Cryst. Growth*, **282**, 199–207.

742. He, L., and Feng, Z. (2007) Preparation and characterization of dicalcium phosphate dihydrate coating on enamel, *Mater. Lett.*, **61**, 3923–3926.

743. Li, L., Pan, H., Tao, J., Xu, X., Mao, C., Gu, X., and Tang, R. (2008) Repair of enamel by using hydroxyapatite nanoparticles as the building blocks, *J. Mater. Chem.*, **18**, 4079–4084.

744. Cochrane, N. J., Saranathan, S., Cai, F., Cross, K. J., and Reynolds, E. C. (2008) Enamel subsurface lesion remineralisation with casein phosphopeptide stabilised solutions of calcium, phosphate and fluoride, *Caries Res.*, **42**, 88–97.

745. Wang, X., Xia, C., Zhang, Z., Deng, X., Wei, S., Zheng, G., and Chen, H. (2009) Direct growth of human enamel-like calcium phosphate microstructures on human tooth, *J. Nanosci. Nanotechnol.*, **9**, 1361–1364.

746. Roveri, N., Battistella, E., Bianchi, C. L., Foltran, I., Foresti, E., Iafisco, M., Lelli, M., Naldoni, A., Palazzo, B., and Rimondini, L. (2009) Surface enamel remineralization: biomimetic apatite nanocrystals and fluoride ions different effects, *J. Nanomater.*, **746383** (9 pages).

747. Roveri, N., Foresti, E., Lelli, M., and Lesci, I. G. (2009) Recent advancements in preventing teeth health hazard: the daily use of hydroxyapatite instead of fluoride, *Recent Patents on Biomed. Eng.*, **2**, 197–215.

748. Yin, Y., Yun, S., Fang, J., and Chen, H. (2009) Chemical regeneration of human tooth enamel under near-physiological conditions, *Chem. Commun.*, **21**, 5892–5894.

749. Peters, M. C., Bresciani, E., Barata, T. J. E., Fagundes, T. C., Navarro, R. L., Navarro, M. F. L., and Dickens, S. H. (2010) *In vivo* dentine remineralization by calcium-phosphate cement, *J. Dent. Res.*, **89**, 286–291.

750. Orsini, G., Procaccini, M., Manzoli, L., Giuliodori, F., Lorenzini, A., and Putignano, A. (2010) A double-blind randomized-controlled trial comparing the desensitizing efficacy of a new dentifrice containing carbonate/hydroxyapatite nanocrystals and a sodium fluoride/potassium nitrate dentifrice, *J. Clin. Periodontology*, **37**, 510–517.

751. Uysal, T., Amasyali, M., Koyuturk, A. E., Ozcan, S., and Sagdic, D. (2010) Amorphous calcium phosphate-containing orthodontic composites. Do they prevent demineralisation around orthodontic brackets? *Austral. Orthodontic J.*, **26**, 10–15.

752. McClendon, J. F. (1966) Fluorapatite and teeth, *Science*, **151**, p. 151.

753. Wang, X., Klocke, A., Mihailova, B., Tosheva, L., and Bismayer, U. (2008) New insights into structural alteration of enamel apatite induced by citric acid and sodium fluoride solutions, *J. Phys. Chem. B*, **112**, 8840–8848.

754. Price, J., Faucheux, C., and Allen, S. (2005) Deer antlers as a model of mammalian regeneration, *Curr. Top. Dev. Biol.*, **67**, 2–49.

755. Yue, Z., Deng, X., and Feng, H. (2005) The mechanisms of deer antlers development and regeneration, *J. Econ. Animal*, **9**, 46–49.

756. Zhao, L., Yue, Z., Zhang, X., and Deng, X. (2006) The deer antlers endochondral ossification and its regulation mechanisms, *J. Econ. Animal*, **10**, 238–241.

757. Landete-Castilleijos, T., Garcia, A., and Gallego, L. (2007) Body weight, early growth and antler size influence antler bone mineral composition of Iberian Red Deer (Cervus elaphus hispanicus), *Bone*, **40**, 230–235.

758. Huxley, J. (1931) The relative size of antlers of deer, *Proc. Zool. Soc. London*, **72**, 819–864.

759. Kierdorf, U., Li, C., and Price, J. S. (2009) Improbable appendages: deer antler renewal as a unique case of mammalian regeneration, *Seminars in Cell and Developmental Biology*, **20**, 535–542.

760. Kierdorf, U., and Kierdorf, H. (2011) Deer antlers – a model of mammalian appendage regeneration: an extensive review, *Gerontology*, **57**, 53–65.

761. Chen, P. Y., Stokes, A. G., and McKittrick, J. (2009) Comparison of the structure and mechanical properties of bovine femur bone and antler of the North American elk (Cervus elaphus canadensis), *Acta Biomater.*, **5**, 693–706.

762. Zioupos, P., Wang, X. T., and Currey, J. D. (1996) Experimental and theoretical quantification of the development of damage in fatigue tests of bone and antler, *J. Biomech.*, **29**, 989–1002.

763. Landete-Castilleijos, T., Currey, J. D., Estevez, J. A., Gaspar-Lopez, E., Garcia, A., and Gallego, L. (2007) Influence of physiological effort of growth and chemical composition on antler bone mechanical properties, *Bone*, **41**, 794–803.

764. Evans, L. A., McCutcheon, A. L., Dennis, G. R., Mulley, R. C., and Wilson, M. A. (2005) Pore size analysis of fallow deer (Dama dama) antler bone, *J. Mater. Sci.*, **40**, 5733–5739.

765. Akhtar, R., Daymond, M. R., Almer, J. D., and Mummery, P. M. (2008) Elastic strains in antler trabecular bone determined by synchrotron X-ray diffraction, *Acta Biomater.*, **4**, 1677–1687.

766. Currey, J. D., Landete-Castillejos, T., Estevez, J., Ceacero, F., Olguin, A., Garcia, A., and Gallego, L. (2009) The mechanical properties of red deer antler bone when used in fighting, *J. Exp. Biol.*, **212**, 3985–3993.

767. Goss, R. J. (1983) Deer antlers regeneration, function and evolution, Academic Press, New York, 1983, p. 316.

768. Bubenik, G. A., and Bubenik, A. B. (eds.) (1990) Horns, pronghorns, and antlers: evolution, morphology, physiology, and social significance, Springer, New York, p. 562.

769. Kierdorf, U., and Kierdorf, H. (2005) Antlers as biomonitors of environmental pollution by lead and fluoride: a review, Eur. J. Wildlife Res., **51**, 137–150.

770. Barling, P. M., Gupta, D. K., and Lim, C. E. L. (1999) Involvement of phosphodiesterase I in mineralization: histochemical studies using antler from red deer (*Cervus elaphus*) as a model, *Calcif. Tissue Int.*, **65**, 384–389.

771. Barling, P. M., and Chong, K. W. (1999) The involvement of phosphohydrolases in mineralization: studies on enzymatic activities extracted from red deer antler, *Calcif. Tissue Int.*, **65**, 232–236.

772. Kierdorf, U., and Kierdorf, H. (2000) The fluoride content of antlers as an indicator of fluoride exposure in red deer (*Cervus elaphus*): a historical biomonitoring study, *Fluoride*, **33**, 92–94.

773. Pathak, N. N., Pattanaik, A. K., Patra, R. C., and Arora, B. M. (2001) Mineral composition of antlers of three deer species reared in captivity, *Small Rumin. Res.*, **42**, 61–65.

774. Yuxia, Y., Rui, D., Wang, Y., and Wang, S. (2002) Calcium and phosphors contents of three-branched and two-branched antler and ossificational antler from sika deer, *J. Econ. Animal*, **6**, 6–8.

775. Kim, H. Y., and Rhyu, M. R. (2000) Sectional composition of minerals in domestic deer antler, *Korean J. Food Sci. Technol.*, **32**, 31–36.

776. Li, C., Suttie, J. M., and Clarck, D. E. (2005) Histological examination of antler regeneration in red deer (*Cervus elaphus*), *Anat. Rec.*, **282A**, 163–174.

777. Meister, W. (1956) Changes in biological structure of the long bones of white-tailed deer during the growth of antlers, *Anat. Rec.*, **124**, 709–721.

778. Muir, P. D., Sykes, A. R., and Barrell, G. K. (1987) Calcium metabolism in red deer (*Cervus elaphus*) offered herbages during antlerogenesis: kinetic and stable balance studies, *J. Agric. Sci. Camb.*, **109**, 357–364.

779. Baxter, B. J., Andrews, R. N., and Barrell, G. K. (1999) Bone turnover associated with antler growth in red deer (*Cervus elaphus*), *Anat. Rec.*, **256**, 14–19.

780. Bubenik, G. A., and Bubenik, P. G. (2008) Palmated antlers of moose may serve as a parabolic reflector of sounds, *Eur. J. Wildlife Res.*, **54**, 533–535.

781. Landete-Castilleijos, T., Estevez, J. A., Martinez, A., Ceacero, F., Garcia, A., and Gallego, L. (2007) Does chemical composition of antler bone reflect the physiological effort made to grow it? *Bone*, **40**, 1095–1102.

782. Baciut, M., Baciut, G., Simon, V., Albon, C., Coman, V., Prodan, P., Florian, S., and Bran, S. (2007) Investigation of deer antler as a potential bone regenerating biomaterial, *J. Optoelectronics Adv. Mater.*, **9**, 2547–2550.

783. Block, G. A., Hulbert-Shearon, T. E., Levin, N. W., and Port, F. K. (1998) Association of serum phosphorus and calcium x phosphate product with mortality risk in chronic hemodialysis patients: a national study. *Am. J. Kidney Dis.*, **31**, 607–617.

784. Kazama, J. J., Amizuka, N., and Fukagawa, M. (2006) Ectopic calcification as abnormal biomineralization, *Ther. Apher. Dial.*, **10**(1), S34–S38.

785. Mackarell, W. W., Moore, B., and Thomas, W. T. (1911) On the presence of insoluble salts of calcium (oxalate and phosphate) in renal calculi in large amount in a preponderating number of cases, and the bearing of this finding upon calcium metabolism in gout and allied conditions, *Biochem. J.*, **5**, 161–180.

786. Frondel, C., and Prien, E. L. (1942) Carbonate-apatite and hydroxylapatite in urinary calculi, *Science*, **95**, 431.

787. Frondel, C., and Prien, E. L. (1946) Deposition of calcium phosphates accompanying senile degeneration and disease, *Science*, **103**, 326.

788. Brancaccio, D., and Cozzolino, M. (2005) The mechanism of calcium deposition in soft tissues, *Contrib. Nephrol.*, **149**, 279–286.

789. Goff, A. K., and Reichard, R. (2006) A soft-tissue calcification: differential diagnosis and pathogenesis, *J. Forensic Sci.*, **51**, 493–497.

790. Bittmann, S., Gunther, M. W., and Ulus, H. (2003) Tumoral calcinosis of the gluteal region in a child: case report with overview of different soft-tissue calcifications, *J. Pediatric Surg.*, **38**, E4–E7.

791. Molloy, E. S., and McCarthy, G. M. (2006) Basic calcium phosphate crystals: pathways to joint degeneration, *Curr. Opin. Rheumatol.*, **18**, 187–192.

792. Giachelli, C. M. (2004) Vascular calcification mechanisms, *J. Am. Soc. Nephrol.*, **15**, 2959–2964.

793. Kazama, J. J., Amizuka, N., and Fukagawa, M. (2007) The making of a bone in blood vessels: from the soft shell to the hard bone, *Kidney Int.*, **72**, 533–534.

794. Achilles, W. (1987) Crystallization in gel matrices: a new experimental model of calcium stone formation, *Contrib. Nephrol.*, **58**, 59–64.

795. Achilles, W., Jockel, U., Schaper, A., Burk, M., and Riedmiller, H. (1995) *In vitro* formation of "urinary stones": generation of spherulites of calcium phosphate in gel and overgrowth with calcium oxalate using a new flow model of crystallization, *Scan. Microscc.*, **9**, 577–585; discussion 585–586.

796. Ross, A. E., Handa, S., Lingeman, J. E., and Matlaga, B. R. (2008) Kidney stones during pregnancy: an investigation into stone composition, *Urol. Res.*, **36**, 99–102.

797. Ringdén, I., and Tiselius, H. G. (2007) Composition and clinically determined hardness of urinary tract stones, *Scand. J. Urol. Nephrol.*, **41**, 316–323.

798. Le May, O., Kaqueler, J. C., and Kodaka, T. (1993) Electron probe micro-analysis of human dental pulp stones, *Scan. Microscc.*, **7**, 267–272.

799. Kodaka, T., Hirayama, A., Mori, R., and Sano, T. (1998) Spherulitic brushite stones in the dental pulp of a cow, *J. Electron Microsc.*, **47**, 57–65.

800. Hayashizaki, J., Ban, S., Nakagaki, H., Okumura, A., Yoshii, S., and Robinson, C. (2008) Site specific mineral composition and microstructure of human supra-gingival dental calculus, *Arch. Oral Biology*, **53**, 168–174.

801. Zelentsov, E. L., Moroz, T. N., Kolmogorov, Y. P., Tolmachev, V. E., Dragun, G. N., Palchik, N. A., and Grigorieva, T. N. (2001) The elemental SRXRF analysis and mineral composition of human salivary stones, *NIM A*, **470**, 417–421.

802. Ortlepp, J. R., Schmitz, F., Mevissen, V., Wei, S., Huster, J., Dronskowski, R., Langebartels, G., Autschbach, R., Zerres, K., Weber, C., Hanrath, P.,

and Hoffmann, R. (2004) The amount of calcium-deficient hexagonal hydroxyapatite in aortic valves is influenced by gender and associated with genetic polymorphisms in patients with severe calcific aortic stenosis, *Eur. Heart J.*, **25**, 514–522.

803. Kim, K. M. (2001) Cellular mechanism of calcification and its prevention in glutaraldehyde treated vascular tissue, *Z. Kardiol.*, **90**(3), III99–III105.

804. Tomazic, B. B. (2001) Physicochemical principles of cardiovascular calcification, *Z. Kardiol.*, **90**(3), III68–III80.

805. Marra, S. P., Daghlian, C. P., Fillinger, M. F., and Kennedy, F. E. (2006) Elemental composition, morphology and mechanical properties of calcified deposits obtained from abdominal aortic aneurysms, *Acta Biomater.*, **2**, 515–520.

806. Fitzpatrick, L. A., Turner, R. T., and Ritman, E. R. (2003) Endochondral bone formation in the heart: a possible mechanism of coronary calcification, *Endocrinology*, **144**, 2214–2219.

807. Suvorova, E. I., and Buffat, P. A. (2005) Pathological mineralization of cardiac valves: causes and mechanism, *J. Long Term Eff. Med. Implants*, **15**, 355–367.

808. Giachelli, C. (2001) Ectopic calcification: new concepts in cellular regulation, *Z. Kardiol.*, **90**(3), III31–III37.

809. Glasmacher, B., Nellen, E., Reul, H., and Rau, G. (1999) *In vitro* hemocompatibility testing of new materials for mechanical heart valves, *Mat.-Wiss. u. Werkstofftech.*, **30**, 806–808.

810. Deiwick, M., Glasmacher, B., Pettenazzo, E., Hammel, D., Castellon, W., Thiene, G., Reul, H., Berendes, E., and Scheld, H. H. (2001) Primary tissue failure of bioprostheses: new evidence from *in vitro* tests, *Thorac. Cardiovasc. Surg.*, **49**, 78–83.

811. Pettenazzo, E., Deiwick, M., Thiene, G., Molin, G., Glasmacher, B., Martignago, F., Bottio, T., Reul, H., and Valente, M. (2001) Dynamic *in vitro* calcification of bioprosthetic porcine valves: evidence of apatite crystallization, *J. Thorac. Cardiovasc. Surg.*, **121**, 500–509.

812. Schoen, F. J., and Levy, R. J. (2005) Calcification of tissue heart valve substitutes: progress toward understanding and prevention, *Ann. Thorac. Surg.*, **79**, 1072–1080.

813. Sensui, K., Saitoh, S., Kametani, K., Makino, K., Ohira, M., Kimura, T., Cheng, G. A., and Hata, Y. (2003) Property analysis of ectopic calcification in the carpal tunnel identification of apatite crystals: a case report, *Arch. Orthop. Trauma Surg.*, **123**, 442–445.

814. Kim, C. J., and Choi, S. K. (2007) Analysis of aqueous humor calcium and phosphate from cataract eyes with and without diabetes mellitus, *KJO*, **21**, 90–94.

815. McCarty, D. J., Hogan, H. M., Gatter, R. A., and Grossmann, M. (1966) Studies on pathological calcifications in human cartilage, *J. Bone Joint Surg.*, **48**, 309–325.

816. Laird, D. F., Mucalo, M. R., and Yokogawa, Y. (2006) Growth of calcium hydroxyapatite (Ca-HAp) on cholesterol and cholestanol crystals from a simulated body fluid: a possible insight into the pathological calcifications associated with atherosclerosis, *J. Colloid Interf. Sci.*, **295**, 348–363.

817. Stock, S. R., Ignatiev, K., Lee, P. L., Abbott, K., and Pachman, L. M. (2004) Pathological calcification in juvenile dermatomyositis (JDM): microCT and synchrotron X-ray diffraction reveal hydroxyapatite with varied microstructures, *Connect. Tissue Res.*, **45**, 248–256.

818. Pachman, L. M., and Boskey, A. L. (2006) Clinical manifestations and pathogenesis of hydroxyapatite crystal deposition in juvenile dermatomyositis, *Curr. Rheumatol. Rep.*, **8**, 236–243.

819. Hale, E. K. (2003) Metastatic calcification, *Dermatology Online J.*, **9**, 2 (http://dermatology.cdlib.org/94/NYU/Nov2001/3.html) – accessed in October 2010.

820. Alkan, O., Tokmak, N., Demir, S., and Yildirim, T. (2009) Metastatic pulmonary calcification in a patient with chronic renal failure, *J. Radiology Case Reports*, **3**, 14–17.

821. Grech, P., Ell, P. J., Martin, T. J., Barrington, N. A., and Martin, T. J. (1998) Diagnosis in metabolic bone disease, *Hodder Arnold H&S: USA*, 300.

822. Mohr, W. (2003) Apatite diseases. The pathological substrate in dependence of tissue processing and in consideration of pathogenesis, *Aktuel Rheumatol.*, **28**, 53–58.

823. Gorospe, M., and Fadare, O. (2007) Gastric mucosal calcinosis: clinicopathologic considerations, *Adv. Anat. Pathology*, **14**, 224–228.

824. Dubols, L. A., Gray, D. K., and Tweedie, E. J. (2007) Surgical images: soft tissue calcinosis cutis, *Can. J. Surg.*, **50**, 217–218.

825. Sprecher, E. (2010) Familial tumoral calcinosis: from characterization of a rare phenotype to the pathogenesis of ectopic calcification, *J. Investigative Dermatology*, **130**, 652–660.

826. White, D. J. (1997) Dental calculus: recent insights into occurrence, formation, prevention, removal and oral health effects of supragingival and subgingival deposits, *Eur. J. Oral Sci.*, **105**, 508–522.

827. ÇiftÇioğlu, N., and McKay, D. S. (2010) Pathological calcification and replicating calcifying-nanoparticles: general approach and correlation, *Pediatric Res.*, **67**, 490–499.

828. Tomazic, B. B. (1994) Characterization of mineral phases in cardiovascular calcification, in *Hydroxyapatite and Related Materials*, (ed. Brown, P. W., and Constantz, B.), CRC Press, Boca Raton, pp. 93–116.

829. Lee, R. S., Kayser, M. V., and Ali, S. Y. (2006) Calcium phosphate microcrystal deposition in the human intervertebral disc, *J. Anat.*, **208**, 13–19.

830. Rosenthal, A. K. (2007) Update in calcium deposition diseases, *Curr. Opin. Rheumatol.*, **19**, 158–162.

831. Lagier, R., and Baud, C. A. (2003) Magnesium whitlockite, a calcium phosphate crystal of special interest in pathology, *Pathol. Res. Pract.*, **199**, 329–335.

832. Wesson, J. A., and Ward, M. D. (2007) Pathological biomineralization of kidney stones, *Elements*, **3**, 415–421.

833. Hayes, C. W., and Conway, W. F. (1990) Calcium hydroxyapatite deposition disease, *Radiographics*, **10**, 1031–1048.

834. Best, J. A., Shapiro, R. D., Kalmar, J., and Westsson, P. L. (1997) Hydroxyapatite deposition disease of the temporomandibular joint in a patient with renal failure, *J. Oral Maxillofac. Surg.*, **55**, 1316–1322.

835. Garcia, G. M., McCord, G. C., and Kumar, R. (2003) Hydroxyapatite crystal deposition disease, *Seminars in Musculoskeletal Radiology*, **7**, 187–193.

836. Melrose, J., Burkhardt, D., Taylor, T. K. F., Dillon, C. T., Read, R., Cake, M., and Little, C. B. (2009) Calcification in the ovine intervertebral disc: a model of hydroxyapatite deposition disease, *Eur. Spine J.*, **18**, 479–489.

837. LeGeros, R. Z. Orly, I., LeGeros, J. P., Gomez, C., Kazimiroff, J., Tarpley, T., and Kerebel, B. (1988) Scanning electron microscopy and electron probe microanalyses of the crystalline components of human and animal dental calculi, *Scan. Microsc.*, **2**, 345–356.

838. Grases, F., and Llobera, A. (1998) Experimental model to study sedimentary kidney stones, *Micron*, **29**, 105–111.

839. Lieske, J. C., Norris, R., and Toback, F. G. (1997) Adhesion of hydroxyapatite crystals to anionic sites on the surface of renal epithelial cells, *Am. J. Physiol. - Renal Physiol.*, **273**, F224–F233.

840. Kirsch, T. (2006) Determinants of pathological mineralization, *Curr. Opin. Rheumatol.*, **18**, 174–180.

841. Giachelli, C. M. (1999) Ectopic calcification: gathering hard facts about soft tissue mineralization, *Am. J. Pathol.*, **154**, 671–675.

842. Kirsch, T. (2007) Physiological and pathological mineralization: a complex multifactorial process, *Curr. Opin. Orthop.*, **18**, 425–427.

843. Speer, M. Y., and Giachelli, C. M. (2004) Regulation of cardiovascular calcification, *Cardiovasc. Pathol.*, **13**, 63–70.

844. Giachelli, C. M. (2005) Inducers and inhibitors of biomineralization: lessons from pathological calcification, *Orthod. Craniofac. Res.*, **8**, 229–231.

845. Giachelli, C. M., Speer, M. Y., Li, X., Rajachar, R. M., and Yang, H. (2005) Regulation of vascular calcification: roles of phosphate and osteopontin, *Circ. Res.*, **96**, 717–722.

846. Schmitt, C. P., Odenwald, T., and Ritz, E. (2006) Calcium, calcium regulatory hormones, and calcimimetics: impact on cardiovascular mortality, *J. Am. Soc. Nephrol.*, **17**, 78–80.

847. Azari, F., Vali, H., Guerquin-Kern, J. L., Wu, T. D., Croisy, A., Sears, S. K., Tabrizian, M., and McKee, M. D. (2008) Intracellular precipitation of hydroxyapatite mineral and implications for pathologic calcification, *J. Struct. Biology*, **162**, 468–479.

848. Stayton, P. S., Drobny, G. P., Shaw, W. J., Long, J. R., and Gilbert, M. (2003) Molecular recognition at the protein-hydroxyapatite interface, *Crit. Rev. Oral. Biol. Med.*, **14**, 370–376.

849. Doherty, T. M., Fitzpatrick, L. A., Inoue, D., Qiao, J. H., Fishbein, M. C., Detrano, R. C., Shah, P. K., and Rajavashisth, T. B. (2004) Molecular, endocrine and genetic mechanisms of arterial calcification, *Endocr. Rev.*, **25**, 629–672.

850. Dey, A., Bomans, P. H. H., Müller, F. A., Will, J., Frederik, P. M., de With G., and Sommerdijk, N. A. J. M. (2010) The role of prenucleation clusters in surface-induced calcium phosphate crystallization, *Nature Mater.*, **9**, 1010–1014.

851. Lamas, G. A., and Ackermann, A. (2000) Clinical evaluation of chelation therapy: is there any wheat amidst the chaff? *Am. Heart J.*, **140**, 4–5.

852. Knudtson, M. L., Wyse, D. G., Galbraith, P. D., Brant, R., Hildebrand, K., Paterson, D., Richardson, D., Burkart, C., and Burgess, E. (2002) Chelation therapy for ischemic heart disease: a randomized controlled trial, *JAMA*, **287**, 481–486.

853. Ehrlich, H., Koutsoukos, P. G., Demadis, K. D., and Pokrovsky, O. S. (2008) Principles of demineralization: modern strategies for the isolation of organic frameworks. Part I. Common definitions and history, *Micron*, **39**, 1062–1091.

854. Ehrlich, H., Koutsoukos, P. G., Demadis, K. D., and Pokrovsky, O. S. (2009) Principles of demineralization: modern strategies for the isolation of organic frameworks. Part II. Decalcification, *Micron*, **40**, 169–193.

855. Green, D., Walsh, D., Mann, S., and Oreffo, R. O .C. (2002) The potential of biomimesis in bone tissue engineering: Lessons from the design and synthesis of invertebrate skeletons, *Bone*, **30**, 810–815.

856. Burke, D. E., de Markey, C. A., le Quesne, P. W., and Cook, J. M. (1972) Biomimetic synthesis of the bis-indole alkaloid macralstonine, *J. Chem. Soc. Chem. Communic.*, 1346–1347.

857. Breslow, R. (1972) Centenary lecture: biomimetic chemistry, *Chem. Soc. Rev.*, **1**, 553–580.

858. Benyus, J. M. (1997) *Biomimicry: Innovation Inspired by Nature*, William Morrow, New York, p. 308.

859. Watt, J. C. (1923) The behavior of calcium phosphate and calcium carbonate (bone salts) precipitated in various media, with applications to bone formation, *Biol. Bull.*, **44**, 280–288.

860. Dorozhkin, S. V., Dorozhkina, E. I., and Epple, M. (2004) A model system to provide a good *in vitro* simulation of biological mineralization, *Cryst. Growth Des.*, **4**, 389–395.

861. Izquierdo-Barba, I., Salinas, A. J., and Vallet-Regí, M. (2000) Effect of the continuous solution exchange on the *in vitro* reactivity of a $CaO–SiO_2$ sol-gel glass, *J. Biomed. Mater. Res.*, **51**, 191–199.

862. Vallet-Regí, M., Pėrez-Pariente, J., Izquierdo-Barba, I., and Salinas, A. J. (2000) Compositional variations in the calcium phosphate layer growth on gel glasses soaked in a simulated body fluid, *Chem. Mater.*, **12**, 3770–3775.

863. Koutsoukos, P., Amjad, Z., Tomson, M. B., and Nancollas, G. H. (1980) Crystallization of calcium phosphates: a constant composition study, *J. Am. Chem. Soc.*, **102**, 1553–1557.

864. Tomson, M. B., and Nancollas, G. H. (1978) Mineralization kinetics: a constant composition approach, *Science*, **200**, 1059–1060.

865. Manjubala, I., Scheler, S., Bössert, J., and Jandt, K. D. (2006) Mineralisation of chitosan scaffolds with nano-apatite formation by double diffusion technique, *Acta Biomater.*, **2**, 75–84.

866. Cai, H. Q., Li, Q. L., Zhou, J., Tang, J., and Chen, H. (2010) Biomimetic synthesis and cytocompatibility of agar-hydroxyapatite composites, *J. Clin. Rehabil. Tiss. Eng. Res.*, **14**, 410–414.

867. Yokoi, T., Kawashita, M., Kikuta, K., and Ohtsuki, C. (2010) Biomimetic mineralization of calcium phosphate crystals in polyacrylamide hydrogel: effect of concentrations of calcium and phosphate ions on crystalline phases and morphology, *Mater. Sci. Eng. C*, **30**, 154–159.

868. Sadjadi, M. S., Meskinfam, M., Sadeghi, B., Jazdarreh, H., and Zare, K. (2010) *In situ* biomimetic synthesis, characterization and *in vitro* investigation of bone-like nanohydroxyapatite in starch matrix, *Mater. Chem. Phys.*, **124**, 217–222.

869. Dorozhkin, S. V., and Dorozhkina, E. I. (2003) The influence of bovine serum albumin on the crystallization of calcium phosphates from a revised simulated body fluid, *Colloids Surf. A*, **215**, 191–199.

870. Dorozhkina, E. I., and Dorozhkin, S. V. (2003) *In vitro* crystallization of carbonateapatite on cholesterol from a modified simulated body fluid, *Colloids Surf. A*, **223**, 231–237.

871. Dorozhkin, S. V., Dorozhkina, E. I., and Epple, M. (2003) Precipitation of carbonateapatite from a revised simulated body fluid in the presence of glucose, *J. Appl. Biomater. Biomech.*, **1**, 200–208.

872. Dorozhkin, S. V., and Dorozhkina, E. I. (2005) *In vitro* simulation of vascular calcification by the controlled crystallization of amorphous calcium phosphates onto porous cholesterol, *J. Mater. Sci.*, **40**, 6417–6422.

873. Dorozhkin, S. V. (2007) *In vitro* mineralization of silicon containing calcium phosphate bioceramics, *J. Am. Ceram. Soc.*, **90**, 244–249.

874. Becker, A., and Epple, M. (2005) A high-throughput crystallisation device to study biomineralisation *in vitro*, *Mater. Res. Soc. Symp. Proc.*, **873E**, K12.1.1–K12.1.10.

875. Krings, M., Kanellopoulou, D., Mavrilas, D., and Glasmacher, B. (2006) *In vitro* pH-controlled calcification of biological heart valve prostheses, *Mat.-Wiss. u. Werkstofftech.*, **37**, 432–435.

876. Wang, L. J., Guan, X. Y., Yin, H. Y., Moradian-Oldak, J., and Nancollas, G. H. (2008) Mimicking the self-organized microstructure of tooth enamel, *J. Phys. Chem. C*, **112**, 5892–5899.

877. Howland, J., and Kramer, B. (1921) Calcium and phosphorus in the serum in relation to rickets, *Am. J. Dis. Child.*, **22**, 105–119.

878. Tisdall, F. F. (1922) The effects of ultra violet rays on the calcium and inorganic phosphate content of the blood serum of rachitic infants, *Can. Med. Assoc. J.*, **7**, 536–538.

879. Hanks, J. H., and Wallace, R. E. (1949) Relation of oxygen and temperature in the preservation of tissues by refrigeration, *Proc. Soc. Exp. Biol. Med.*, **71**, 196–200.

880. Shibata, Y., Takashima, H., Yamamoto, H., and Miyazaki, T. (2004) Functionally gradient bonelike hydroxyapatite coating on a titanium metal substrate created by a discharging method in HBSS without organic molecules, *Int. J. Oral Max. Impl.*, **19**, 177–183.

881. Marques, P. A. A. P., Serro, A. P., Saramago, B. J., Fernandes, A. C., Magalhães, M. C. F., and Correia, R. N. (2003) Mineralisation of two phosphate ceramics in HBSS: role of albumin, *Biomaterials*, **24**, 451–460.

882. Gopikrishna, V., Baweja, P. S., Venkateshbabu, N., Thomas, T., and Kandaswamy, D. (2008) Comparison of coconut water, propolis, HBSS, and milk on PDL cell survival, *J. Endod.*, **34**, 587–589.

883. Hanawa, T., Asami, K., and Asaoka, K. (1998) Repassivation of titanium and surface oxide film regenerated in simulated bioliquid, *J. Biomed. Mater. Res.*, **40**, 530–538.

884. Rocha, L. A., Anza, E., Costa, A. M., Oliveira, F. J., and Silva, R. F. (2003) Electrochemical behavior of Ti / Al_2O_3 interfaces produced by diffusion bonding, *Mater. Res.*, **6**, 439–444.

885. Named after Harry Eagle (1905 – 1992), an American physician and pathologist.

886. Named after Renato Dulbecco (born February 22, 1914), an Italian-born US virologist, who shared the 1975 Nobel Prize in Physiology or Medicine for his work on reverse transcriptase.

887. Meuleman, N., Tondreau, T., Delforge, A., Dejeneffe, M., Massy, M., Libertalis, M., Bron, D., and Lagneaux, L. (2006) Human marrow mesenchymal stem cell culture: serum-free medium allows better expansion than classical α-MEM medium, *Eur. J. Haematol.*, **76**, 309–316.

888. Okazaki, Y., and Gotoh, E. (2002) Corrosion fatigue properties of metallic biomaterials in Eagle's medium, *Mater. Trans.*, **43**, 2949–2955.

889. Coelho, M. J., Cabral, A. T., and Fernandes, M. H. (2000) Human bone cell cultures in biocompatibility testing. Part I: Osteoblastic differentiation of serially passaged human bone marrow cells cultured in α-MEM and in DMEM, *Biomaterials*, **21**, 1087–1094.

890. Mandel, S., and Tas, A. C. (2010) Brushite ($CaHPO_4 \cdot 2H_2O$) to octacalcium phosphate ($Ca_8(HPO_4)_2(PO_4)_4 \cdot 5H_2O$) transformation in DMEM solutions at 36.5°C, *Mater. Sci. Eng. C*, **30**, 245–254.

891. Chen, C., Lee, I. S., Zhang, S. M., and Yang, H. C. (2010) Biomimetic apatite formation on calcium phosphate-coated titanium in Dulbecco's phosphate-buffered saline solution containing $CaCl_2$ with and without fibronectin, *Acta Biomater.*, **6**, 2274–2281.

892. Gao, Y. B., Weng, W. J., Deng, X. L., Cheng, K., Liu, X. G., Du, P. Y., Shen, G., and Han, G. R. (2006) Surface morphology variations of porous nano-calcium phosphate / poly(L-lactic acid) composites in PBS, *Key Eng. Mater.*, **309–311**, 569–572.

893. Lewis, A. C., Kilburn, M. R., Papageorgiou, I., Allen, G. C., and Case, C. P. (2005) Effect of synovial fluid, phosphate-buffered saline solution, and water on the dissolution and corrosion properties of CoCrMo alloys as used in orthopedic implants, *J. Biomed. Mater. Res. A*, **73A**, 456–467.

894. Humphrey, S. P., and Williamson, R. T. (2001) A review of saliva: normal composition, flow, and function, *J. Prosthet. Dent.*, **85**, 162–169.

895. Preetha, A., and Banerjee, R. (2005) Comparison of artificial saliva substitutes, *Trends Biomater. Artif. Organs*, **18**, 178–186.

896. Sato, Y., Sato, T., Niwa, M., and Aoki, H. (2006) Precipitation of octacalcium phosphates on artificial enamel in artificial saliva, *J. Mater. Sci. Mater. Med.*, **17**, 1173–1177.

897. Grases, F., Ramis M., and Costa-Bauzá, A. (2000) Effects of phytate and pyrophosphate on brushite and hydroxyapatite crystallization (comparison with the action of other polyphosphates), *Urol. Res.*, **28**, 136–140.

898. Jenness, R., and Koops, J. (1962) Preparation and properties of a salt solution which simulates milk ultrafiltrate, *Neth. Milk Dairy J.*, **16**, 153–164.

899. Spanos, N., Patis, A., Kanellopoulou, D., Andritsos, N., and Koutsoukos, P. G. Precipitation of calcium phosphate from simulated milk ultrafiltrate solutions, *Cryst. Growth Des.*, **7**, 25–29.

900. Gao, R., van Halsema, F. E. D., Temminghoff, E. J. M., van Leeuwen, H. P., van Valenberg, H. J. F., Eisner, M. D., Giesbers, M., and van Boekel, M. A. J. S. (2010) Modelling ion composition in simulated milk ultrafiltrate (SMUF). I: Influence of calcium phosphate precipitation, *Food Chem.*, **122**, 700–709.

901. Gao, R., van Halsema, F. E. D., Temminghoff, E. J. M., van Leeuwen, H. P., van Valenberg, H. J. F., Eisner, M. D., and van Boekel, M. A. J. S. (2010) Modelling ion composition in simulated milk ultrafiltrate (SMUF). II. Influence of pH, ionic strength and polyphosphates, *Food Chem.*, **122**, 710–715.

902. Kokubo, T., Kushitani, H., Sakka, S., Kitsugi, T., and Yamamuro, T. (1990) Solutions able to reproduce *in vivo* surface-structure changes in bioactive glass-ceramic A-W3, *J. Biomed. Mater. Res.*, **24**, 721–734.

903. Lu, X., and Leng, Y. (2005) Theoretical analysis of calcium phosphate precipitation in simulated body fluid, *Biomaterials*, **26**, 1097–1108.

904. Tas, A. C. (2000) Synthesis of biomimetic Ca-hydroxyapatite powders at 37°C in synthetic body fluids, *Biomaterials*, **21**, 1429–1438.

905. Landi, E., Tampieri, A., Celotti, G., Langenati, R., Sandri, M., and Sprio, S. (2005) Nucleation of biomimetic apatite in synthetic body fluids: dense and porous scaffold development, *Biomaterials*, **26**, 2835–2845.

906. Jalota, S., Bhaduri, S. B., and Tas, A. C. (2008) Using a synthetic body fluid (SBF) solution of 27 mM HCO_3^- to make bone substitutes more osteointegrative, *Mater. Sci. Eng. C*, **28**, 129–140.

907. Kim, H. M., Miyazaki, T., Kokubo, T., and Nakamura, T. (2001) Revised simulated body fluid, in *Bioceramics 13* (ed. Giannini, S., and Moroni, A.), *Trans Tech Publishing*, Switzerland, **192–195**, pp. 47–50.

908. Oyane, A., Kim, H. M., Furuya, T., Kokubo, T., Miyazaki, T., and Nakamura, T. (2003) Preparation and assessment of revised simulated body fluids, *J. Biomed. Mater. Res. A*, **65A**, 188–195.

909. Müller, L., and Müller, F. A. (2006) Preparation of SBF with different HCO_3^- content and its influence on the composition of biomimetic apatites, *Acta Biomater.*, **2**, 181–189.

910. Hu, K., Yang, X. J., Cai, Y. L., Cui, Z. D., and Wei, Q. (2006) Preparation of bone-like composite coating using a modified simulated body fluid with high Ca and P concentrations, *Surf. Coat. Technol.*, **201**, 1902–1906.

911. Wen, Z., Wu, C., Dai, C., and Yang, F. (2009) Corrosion behaviors of Mg and its alloys with different Al contents in a modified simulated body fluid, *J. Alloys and Compounds*, **488**, 392–399.

912. Gemelli, E., Resende, C. X., and de Soares, G. D. A. (2010) Nucleation and growth of octacalcium phosphate on treated titanium by immersion in a simplified simulated body fluid, *J. Mater. Sci. Mater. Med.*, **21**, 2035–2047.

913. Kokubo, T., and Takadama, H. (2006) How useful is SBF in predicting *in vivo* bone bioactivity? *Biomaterials*, **27**, 2907–2915.

914. Bohner, M., and Lemaitre, J. (2009) Can bioactivity be tested *in vitro* with SBF solution? *Biomaterials*, **30**, 2175–2179.

915. Kim, H. M. (2003) Ceramic bioactivity and related biomimetic strategy, *Curr. Opin. Solid State Mater. Sci.*, **7**, 289–299.

916. Marques, P. A. A. P., Magalhães, M. C. F., and Correia, R. N. (2003) Inorganic plasma with physiological CO_2 / HCO_3^- buffer, *Biomaterials*, **24**, 1541–1548.

917. Dorozhkina, E. I., and Dorozhkin, S. V. (2002) Surface mineralisation of hydroxyapatite in modified simulated body fluid (mSBF) with higher amounts of hydrogencarbonate ions, *Colloids Surf. A*, **210**, 41–48.

918. Marques, P. A. A. P., Cachinho, S. C. P., Magalhães, M. C. F., Correia, R. N., and Fernandes, M. H. V. (2004) Mineralization of bioceramics in simulated plasma with physiological CO_2/HCO_3^- buffer and albumin, *J. Mater. Chem.*, **14**, 1861–1866.

919. Pasinli, A., Yuksel, M., Celik, E., Sener, S., and Tas, A. C. (2010) A new approach in biomimetic synthesis of calcium phosphate coatings using lactic acid – Na lactate buffered body fluid solution, *Acta Biomater.*, **6**, 2282–2288.

920. Sun, T., and Wang, M. (2010) Electrochemical deposition of apatite/collagen composite coating on NiTi shape memory alloy and coating properties, *Mater. Res. Soc. Symp. Proc.*, **1239**, 141–146.

921. Dorozhkin, S. V., and Dorozhkina, E. I. (2007) Crystallization from a milk-based revised simulated body fluid, *Biomed. Mater.*, **2**, 87–92.

922. Miyaji, F., Kim, H. M., Handa, S., Kokubo, T., and Nakamura, T. (1999) Bonelike apatite coating on organic polymers: novel nucleation process using sodium silicate solution, *Biomaterials*, **20**, 913–919.

923. Kim, H. M., Kishimoto, K., Miyaji, F., Kokubo, T., Yao, T., Suetsugu, Y., Tanaka, J., and Nakamura, T. (2000) Composition and structure of apatite formed on organic polymer in simulated body fluid with a high content of carbonate ion, *J. Mater. Sci. Mater. Med.*, **11**, 421–426.

924. Barrere, F., van Blitterswijk, C. A., de Groot, K., and Layrolle, P. (2002) Influence of ionic strength and carbonate on the Ca-P coating formation from SBF×5 solution, *Biomaterials*, **23**, 1921–1930.

925. Barrere, F., van Blitterswijk, C. A., de Groot, K., and Layrolle, P. (2002) Nucleation of biomimetic Ca-P coatings on Ti6Al4V from a SBF×5 solution: influence of magnesium, *Biomaterials*, **23**, 2211–2220.

926. Tas, A. C., and Bhaduri, S. B. (2004) Rapid coating of Ti6Al4V at room temperature with a calcium phosphate solution similar to 10x simulated body fluid, *J. Mater. Res.*, **19**, 2742–2749.

927. Dorozhkina, E. I., and Dorozhkin, S. V. (2003) Structure and properties of the precipitates formed from condensed solutions of the revised simulated body fluid, *J. Biomed. Mater. Res. A*, **67A**, 578–581.

928. Sato, K. (2007) Mechanism of hydroxyapatite mineralization in biological systems, *J. Ceram. Soc. Jpn.*, **115**, 124–130.

929. Pompe, W., Lampenscherf, S., Rößler, S., Scharnweber, D., Weis, K., Worch, H., and Hofinger, J. (1999) Functionally graded bioceramics, *Mater. Sci. Forum*, 308–311, 325–330.

930. Fan, Y., Duan, K., and Wang, R. (2005) A composite coating by electrolysis-induced collagen self-assembly and calcium phosphate mineralization, *Biomaterials*, **26**, 1623–1632.

931. Kikuchi, M., Itoh, S., Ichinose, S., Shinomiya, K., and Tanaka, J. (2001) Self-organization mechanism in a bone-like hydroxyapatite/collagen nanocomposite synthesized *in vitro* and its biological reaction *in vivo*, *Biomaterials*, **22**, 1705–1711.

932. Yamauchi, K., Goda, T., Takeuchi, N., Einaga, H., and Tanabe, T. (2004) Preparation of collagen / calcium phosphate multilayer sheet using enzymatic mineralization, *Biomaterials*, **25**, 5481–5489.

933. Zhang, W., Huang, Z. L., Liao, S. S., and Cui, F. Z. (2003) Nucleation sites of calcium phosphate crystals during collagen mineralization, *J. Am. Ceram. Soc.*, **86**, 1052–1054.

934. Tampieri, A., Celotti, G., Landi, E., Sandri, M., Roveri, N., and Falini, G. (2003) Biologically inspired synthesis of bone-like composite: self-assembled collagen fibers / hydroxyapatite nanocrystals, *J. Biomed. Mater. Res. A*, **67A**, 618–625.

935. Wang, Y., Yang, C., Chen, X., and Zhao, N. (2006) Biomimetic formation of hydroxyapatite / collagen matrix composite, *Adv. Eng. Mater.*, **8**, 97–100.

936. Lickorish, D., Ramshaw, J. A. M., Werkmeister, J. A., Glattauer, V., and Howlett, C. R. (2004) Collagen – hydroxyapatite composite prepared by biomimetic process, *J. Biomed. Mater. Res. A*, **68A**, 19–27.

937. Yunoki, S., Ikoma, T., Monkawal, A., Ohtal, K., and Tanaka, J. (2007) Preparation and characterization of hydroxyapatite/collagen nanocomposite gel, *J. Nanosci. Nanotechnol.*, **7**, 818–821.

938. Nassif, N., Gobeaux, F., Seto, J., Belamie, E., Davidson, P., Panine, P., Mosser, G., Fratzl, P., and Giraud Guille, M. M. (2010) Self-assembled collagen-apatite matrix with bone-like hierarchy, *Chem. Mater.*, **22**, 3307–3309.

939. Zhao, F., Yin, Y., Lu, W. W., Leong, J. C., Zhang, W., Zhang, J., Zhang, M., and Yao, K. (2002) Preparation and histological evaluation of biomimetic three-dimensional hydroxyapatite/chitosan-gelatin network composite scaffolds, *Biomaterials*, **23**, 3227–3234.

940. Kim, H. W., Kim, H. E., and Salih, V. (2005) Stimulation of osteoblast responses to biomimetic nanocomposites of gelatin – hydroxyapatite for tissue engineering scaffolds, *Biomaterials*, **26**, 5221–5230.

941. Kim, H. W., Knowles, J. C., and Kim, H. E. (2005) Porous scaffolds of gelatin-hydroxyapatite nanocomposites obtained by biomimetic approach: characterization and antibiotic drug release, *J. Biomed. Mater. Res. B* (*Appl. Biomater.*), **74B**, 686–698.

942. Chen, Z., Li, Q. L., Zen, Q., Li, G., Jiang, H., Liu, L., and Darvell, B. W. (2005) Biomimetic mineralization and bioactivity of phosphorylated chitosan, *Key Eng. Mater.*, **288–289**, 429–432.

943. Li, Q. L., Chen, Z., Ou, G., Liu, L., Jiang, H., Zeng, Q., Li, G., He, G., Mo, A., and Darvell, B. W. (2005) Biomimetic synthesis of apatite – polyelectrolyte complex (chitosan – phosphorylated chitosan) hydrogel as an osteoblast carrier, *Key Eng. Mater.*, 288–289, 75–78.

944. Stupp, S. I., and Ciegler, G. W. (1992) Organoapatites: materials for artificial bone. I. Synthesis and microstructure, *J. Biomed. Mater. Res.*, **26**, 169–183.

945. Stupp, S. I., and Braun, P. V. (1997) Molecular manipulation of microstructures: biomaterials, ceramics, and semiconductors, *Science*, **277**, 1242–1248.

946. Stupp, S. I., Mejicano, G. C., and Hanson, J. A. (1993) Organoapatites: materials for artificial bone. II. Hardening reactions and properties, *J. Biomed. Mater. Res.*, **27**, 289–299.

947. Stupp, S. I., Hanson, J. A., Eurell, J. A., Ciegler, G. W., and Johnson, A. (1993) Organoapatites: materials for artificial bone. III. Biological testing, *J. Biomed. Mater. Res.*, **27**, 301–311.

948. Liu, Y., Layrolle, P., de Bruijn, J., van Blitterswijk, C. A., and de Groot, K. (2001) Biomimetic coprecipitation of calcium phosphate and bovine serum albumin on titanium alloy, *J. Biomed. Mater. Res.*, **57**, 327–335.

949. Wang, J., Layrolle, P., Stigter, M., and de Groot, K. (2004) Biomimetic and electrolytic calcium phosphate coatings on titanium alloy: physicochemical characteristics and cell attachment, *Biomaterials*, **25**, 583–592.

950. Zhang, Q., and Leng, Y. (2005) Electrochemical activation of titanium for biomimetic coating of calcium phosphate, *Biomaterials*, **26**, 3853–3859.

951. Bigi, A., Boanini, E., Bracci, B., Facchini, A., Panzavolta, S., Segatti, F., Sturba, L. (2005) Nanocrystalline hydroxyapatite coatings on titanium: a new fast biomimetic method, *Biomaterials*, **26**, 4085–4089.

952. Kim, H. M., Himeno, T., Kawashita, M., Lee, J. H., Kokubo, T., and Nakamura, T. (2003) Surface potential change in bioactive titanium metal during the process of apatite formation in simulated body fluid, *J. Biomed. Mater. Res. A*, **67A**, 1305–1309.

953. Allegrini, S., Rumpel, E., Kauschke, E., Fanghänel, J., and König, B. (2006) Hydroxyapatite grafting promotes new bone formation and osseointegration of smooth titanium implants, *Ann. Anat.*, **188**, 143–151.

954. Arnould, C., Delhalle, J., and Mekhalif, Z. (2008) Multifunctional hybrid coating on titanium towards hydroxyapatite growth: electrodeposition of tantalum and its molecular functionalization with organophosphonic acids films, *Electrochim. Acta*, **53**, 5632–5638.

955. Iwatsubo, T., Kusumocahyo, S. P., Kanamori, T., and Shinbo, T. (2006) Mineralization of hydroxyapatite on a polymer substrate in a solution supersaturated by polyelectrolyte, *J. Appl. Polymer Sci.*, **100**, 1465–1470.

956. Bodin, A., Gustafsson, L., and Gatenholm, P. (2006) Surface-engineered bacterial cellulose as template for crystallization of calcium phosphate, *J. Biomater. Sci. Polym. Ed.*, **17**, 435–447.

957. Toworfe, G. K., Composto, R. J., Shapiro, I. M., and Ducheyne, P. (2006) Nucleation and growth of calcium phosphate on amine-, carboxyl- and hydroxyl-silane self-assembled monolayers, *Biomaterials*, **27**, 631–642.

958. Storrie, H., and Stupp, S. I. (2005) Cellular response to zinc-containing organoapatite: an *in vitro* study of proliferation, alkaline phosphatase activity and biomineralization, *Biomaterials*, **26**, 5492–5499.

959. Hench, L. L. (1998) Biomaterials: a forecast for the future, *Biomaterials*, **19**, 1419–1423.

960. Jones, J. R., and Hench, L. L. (2003) Regeneration of trabecular bone using porous ceramics, *Curr. Opin. Solid State Mater. Sci.*, **7**, 301–307.

961. Griffith, L. G., and Naughton, G. (2002) Tissue engineering – current challenges and expanding opportunities, *Science*, **295**, 1009–1014.

962. Hench, L. L., and Polak, J. M. (2002) Third-generation biomedical materials, *Science*, **295**, 1014–1017.

963. Ratner, B. D., and Bryant, S. J. (2004) Biomaterials: where we have been and where we are going, *Ann. Rev. Biomed. Eng.*, **6**, 41–75.

964. Drouet, C., Largeot, C., Raimbeaux, G., Estournès, C., Dechambre, G., Combes, C., and Rey, C. (2006) Bioceramics: spark plasma sintering (SPS) of calcium phosphates, *Adv. Sci. Technol.*, **49**, 45–50.

965. Anderson, J. M. (2006) The future of biomedical materials, *J. Mater. Sci. Mater. Med.*, **17**, 1025–1028.

966. White, A. A., Best, S. M., and Kinloch, I. A. (2007) Hydroxyapatite – carbon nanotube composites for biomedical applications: a review, *Int. J. Appl. Ceram. Technol.*, **4**, 1–13.

967. Kealley, C., Elcombe, M., van Riessen, A., and Ben-Nissan, B. (2006) Development of carbon nanotube-reinforced hydroxyapatite bioceramics, *Physica B*, **385–386**, 496–498.

968. Balani, K., Anderson, R., Laha, T., Andara, M., Tercero, J., Crumpler, E., and Agarwal, A. (2007) Plasma-sprayed carbon nanotube reinforced hydroxyapatite coatings and their interaction with human osteoblasts *in vitro*, *Biomaterials*, **28**, 618–624.

969. For multiphase compositions of various calcium orthophosphates, the problem of accurate phase quantification often arises. The problem is usually solved by the Rietveld refinement and the readers are referred to a recent paper on this subject [970].

970. Reid, J. W., and Hendry, J. A. (2006) Rapid, accurate phase quantification of multiphase calcium phosphate materials using Rietveld refinement, *J. Appl. Cryst.*, **39**, 536–543.

971. Dorozhkin, S. V. (2008) Green chemical synthesis of calcium phosphate bioceramics, *J. Appl. Biomater. Biomech.*, **6**, 104–109.

972. Salinas, A. J., and Vallet-Regí, M. (2007) Evolution of ceramics with medical applications, *Z. Anorg. Allg. Chem.*, **633**, 1762–1773.

973. Matsumoto, T., Okazaki, M., Nakahira, A., Sasaki, J., Egusa, H., and Sohmura, T. (2007) Modification of apatite materials for bone tissue engineering and drug delivery carriers, *Curr. Med. Chem.*, **14**, 2726–2733.

974. Mizushima, Y., Ikoma, T., Tanaka, J., Hoshi, K., Ishihara, T., Ogawa, Y., and Ueno, A. (2006) Injectable porous hydroxyapatite microparticles as a new carrier for protein and lipophilic drugs, *J. Control. Release*, **110**, 260–265.

975. Ginebra, M. P., Traykova, T., and Planell, J. A. (2006) Calcium phosphate cements as bone drug delivery systems: a review, *J. Control. Release*, **113**, 102–110.

976. Ginebra, M. P., Traykova, T., and Planell, J. A. (2006) Calcium phosphate cements: competitive drug carriers for the musculoskeletal system? *Biomaterials*, **27**, 2171–2177.

977. Fan, J., Lei, J., Yu, C., Tu, B., and Zhao, D. (2007) Hard-templating synthesis of a novel rod-like nanoporous calcium phosphate bioceramics and their capacity as antibiotic carriers, *Mater. Chem. Phys.*, **103**, 489–493.

978. Barrère, F., Mahmood, T. A., de Groot, K., van Blitterswijk, C. A. (2008) Advanced biomaterials for skeletal tissue regeneration: instructive and smart functions, *Mater. Sci. Eng. R*, **59**, 38–71.

979. Sudo, A., Hasegawa, M., Fukuda, A., and Uchida, A. (2008) Treatment of infected hip arthroplasty with antibiotic-impregnated calcium hydroxyapatite, *J. Arthroplasty*, **23**, 145–150.

980. Verron, E., Khairoun, I., Guicheux, J., and Bouler, J. M. (2010) Calcium phosphate biomaterials as bone drug delivery systems: a review, *Drug Discovery Today*, **15**, 547–552.

981. Wernike, E., Montjovent, M. O., Liu, Y., Wismeijer, D., Hunziker, E. B., Siebenrock, K. A., Hofstetter, W., and Klenke, F. M. (2010) VEGF incorporated into calcium phosphate ceramics promotes vascularisation and bone formation *in vivo*, *Eur. Cell Mater.*, **19**, 30–40.

982. Jiang, P. J., Wynn-Jones, G., and Grover, L. M. (2010) A calcium phosphate cryogel for alkaline phosphatase encapsulation, *J. Mater. Sci.*, **45**, 5257–5263.

Chapter 2

Amorphous Calcium (Ortho) Phosphates

2.1 Introduction

In nature, amorphous phases exist extensively with readily moldable isotropic properties and of structure materials. For example, amorphous structures represent ~20% of approximately 60 different inorganic compounds and minerals formed by living organisms. These biologically formed minerals are often called biominerals, while the process of their formation is called biomineralization [1]. A recent review on the subject indicates that many biominerals are formed from amorphous precursors and, furthermore, the amorphous phases may possess fluidic properties that impart new processing capabilities to the system [2]. Among the existing biogenic amorphous minerals, those composing from calcium orthophosphates are most abundant in teeth and exoskeletal structures of marine invertebrates [2–6]. On the other hand, the existence of similar amorphous calcium phosphate (ACP) minerals in vertebrate organisms has not been well established experimentally except in highly specialized locations such as the inner ear structures of embryonic sharks [1], mammalian milk [7, 8], dental enamel [9], as well as in mitochondria [1] and sarcoplasmic reticulum [10] of some cells. Despite the intensive efforts, accumulated evidences for ACP as an integral mineral component in major hard tissues, such as bones and teeth, are equivocal and for many years have been the subject of considerable debates [11–24]. However, recent studies on bone and teeth formation have suggested a presence of transient amorphous

Calcium Orthophosphates: Applications in Nature, Biology, and Medicine
Sergey Dorozhkin

ISBN 978-981-4316-62-0 (Hardcover), 978-981-4364-17-1 (eBook)
www.panstanford.com

mineral precursors and a universal strategy for calcium carbonate-based and calcium orthophosphate-based biomineralization in both vertebrates and invertebrates [21–29]. See Section 1.4.1 for additional details on this topic.

An interesting study on a potential role of ACP in facilitating assembly of nano-sized particles of HA into highly ordered structures has been published recently [30]. Higher order HA architectures were detected only when the starting particles were aggregates of nanodimensional spheres with HA cores and ACP shells. Surface ACP initially linked HA nano-sized particles in a way that allowed a parallel orientation of the nano-sized particles of HA and then was incorporated into HA by phase transformation to produce more ordered architectures with the characteristic features of apatites in biologic structures. Further, it was demonstrated that enamel- and bone-like apatites could be prepared by using nanodimensional HA and ACP under the controls of different biological additives [30]. This study points to an important role of ACP might play in bone formation. However, due to a lack of undeniable proofs, the question on the occurrence of ACP phases in newly mineralized tissues of vertebrates still remains unanswered [31].

The complete list of the available calcium orthophosphates combined with their standard abbreviations, chemical formulae, and major properties is given in Table 1.1 [32, 33]. As the majority of calcium orthophosphates are crystalline, this chapter is devoted to the detailed description of ACP, which is of a great biomedical importance due to its chemical and structural similarities to the calcified tissues of mammals. Furthermore, with a few important exceptions, neither ion-substituted forms of ACP [34–47], nor ACP-containing biocomposites [48–75] are considered and discussed. The readers interested in either these topics and/or other types of amorphous calcium phosphates (e.g., amorphous calcium polyphosphate [76–80] and amorphous calcium metaphosphate [81]) are referred to the original publications.

2.2 Basic Definitions and Knowledge on the Amorphous State of Solids

According to the thermodynamic laws, the perfect infinite crystals cannot exist in the real world. Various disorders in the forms of vacancies, interstitial atoms, impurities, dislocations, grain boundaries, surfaces, and other interfaces disrupt the periodicity of otherwise "perfect" crystals and in many cases determine their physical properties. By contrast, highly disordered solids are those solids that are so irregular that the concept of a reference crystal lattice must be abandoned. Such highly disordered materials are called amorphous materials [82]. As Wikipedia, the free encyclopedia, has it: an *amorphous* (from the Greek term αμορφος, which means "shapeless" or "without form") solid is a solid, in which there is no translational and orientational long-range order (LRO) of the atomic positions [83]. Early researchers categorized solids as amorphous or crystalline materials based on the macroscopic properties such as their external shapes, fracture mechanisms, and optical properties long before X-ray diffraction techniques and other methods became available to reveal their atomic structures. Only in the past century, an understanding of the microscopic nature of amorphous materials has become possible [82]. However, there is still much debate concerning the exact nature of these materials. For example, in a recent article, Sheng et al., [84] have mentioned: "the atomic arrangements in amorphous alloys remain mysterious at present."

An amorphous structure is distinctly different from a densely packed assembly of microcrystals and is closely related to the structure of a liquid phase. Ideally, an amorphous solid should be described by the model of a perfectly random structure [85]; however, this is the boundary condition. As such, the structure of amorphous solids is normally described in terms of statistical distributions. Nevertheless, prior a further description, one must specify the existing atomic length scales. The shortest length scale usually used to describe the structure of a material consists of an atom and its nearest neighbors, out to perhaps two or three atoms distant. All solids and liquids have some structure on this scale, which is called a short-range order (SRO). For crystalline solids, structural order persists over much longer distances (at least, tens or hundreds of atomic distances), such that the atoms occupy sites in a periodic three-dimensional array. Such materials are said to have a LRO and include most metals and many covalently bonded solids. Non-crystalline solids, including glasses, lack a LRO, and are said to be amorphous even though they can have a SRO that is quite well defined [86].

According to the available literature, for each particular atom of any solid there are a SRO of 2–5 Å, a medium-range order (MRO) of 5–20 Å, and a LRO at distances exceeding ~20 Å [87, 88]. In the case of covalent materials, in which a directed chemical bonding is dominant, a SRO can be characterized in terms of the well-defined coordination polyhedra, which, in many cases, appear to concur with the unit cells. The definition of a MRO is more contentious and is helpful to subdivide MRO into 3 subcategories. At the shortest length scale (~5 Å), a near-MRO describes the connections among the coordination polyhedra. At the next length scale (~5–8 Å), an intermediate MRO can be associated with correlations between pairs of preferred dihedral angles for neighboring bonds. Finally, on a yet larger length scale (~8–20 Å), a far-MRO can be associated with the total dimensionality of the covalently bonded amorphous network. Thus, characterizing the nature of MRO in disordered solids is very important for understanding their structure. Figure 2.1 represents an excellent visual demonstration of the differences between a SRO and a MRO [89]. Farther details on this topic are available in literature [87, 88].

Figure 2.1 A formation of skydivers illustrates disorders on an intermediate length scale. Each skydiver has a simple set of rules for bonding to the next skydiver (SRO) but there is a sufficient flexibility for different patterns of ordering to be created on the scale of a few body lengths (MRO).

In covalent solids, bond angles and bond lengths, as well as a number of the nearest neighbors, are all part of the appropriate bonding scheme. Thus, due the nature of chemical bonding, even the truly amorphous materials have some structural SRO and, perhaps, some MRO. For example, MRO regions of ~15 Å in dimensions and comprising about 100 atoms have been directly observed in amorphous carbon [90]. Some order in two-dimensional projections of thin amorphous three-dimensional structures was found [91]. Besides, covalent amorphous solids were found to exhibit a MRO at length scales up to 20 Å or so [88]. Such MRO clusters are called paracrystals [86]. These paracrystals have a crystalline topology but the atomic positions are highly distorted from those of a perfect crystal. However, in solids there is a serious problem of very small particles. Specifically, if the crystal sizes are extremely small, it is difficult to make a distinction between the truly amorphous and crystalline solids. Namely, if a powder consists of tiny perfect crystals with dimensions of 2 nm × 2 nm × 2 nm (8 nm^3) or less, both this powder and any bulk materials prepared from this powder (e.g., by compaction) will be amorphous, just due to the case their sizes are below the minimal value of LRO. Additionally, in very small crystals a large fraction of the atoms are located at/or near surface. Relaxation of the surface and various interfacial effects distort the atomic positions, decreasing the structural order. Thus, even the most advanced structural characterization techniques, such as X-ray, neutron and electron diffraction, as well as transmission electron microscopy (TEM), have difficulties in distinguishing between the amorphous and crystalline structures on these length scales [83, 92].

Many studies revealed that the majority of solids could be found or prepared in an amorphous state. For example, cooling strongly reduces atomic and/or molecular mobility. Thus, in principle, given a sufficiently high cooling rate, any liquid can be transformed into an amorphous solid. As cooling is performed, the material changes from a super-cooled liquid, with properties one would expect from a liquid state material, to a solid. The temperature at which this transition occurs is called glass transition temperature. If a cooling rate is faster than the rate at which atoms and/or molecules can be organized into a more thermodynamically favorable crystalline state, then an amorphous solid will be formed. In contrast, if atoms and/or molecules have a sufficient time to be organized into structures with two- or three-dimensional order, then a crystalline (at least, a semi-crystalline) solid will be formed. Furthermore, in many cases amorphous materials can be produced by additives, which interfere with the ability of the primary constituent to crystallize. For example, addition of soda to melted silicon dioxide results in amorphous window glass and addition of glycols to water results in a vitrified solid [83]. More to the point, amorphization of many solids might be achieved by applying mechanical forces, e.g., by intensive milling [93, 94], as well as by irradiation [95, 96].

From a thermodynamic point of view, amorphous materials are at best metastable. Given a sufficient time, they tend to transform to crystalline phases that are thermodynamically more stable. Interestingly, but when an unstable crystalline solid is transformed to an amorphous phase, this transformation frequently exhibits features that are associated with ordinary melting. Namely, amorphization frequently begins at grain boundaries, surfaces or other defect sites, as does ordinary melting. Further, as the transformation proceeds, a sharp interface that separates the amorphous materials from the untransformed crystalline material is always observed [82].

2.3 Amorphous Calcium Phosphates (ACPs)

2.3.1 History

According to Eanes [20], the history of ACPs looks as this: "In 1955, Robinson and Watson [97] were the first to suggest that a substantial portion of newly formed mineral in young bone was not crystalline. Instead, they described this early mineral as being more similar in character to an amorphous-like precipitate they had prepared in a study on synthetic HA [98]. This precipitate, which appeared initially in their synthesis when sufficiently concentrated solutions of $CaCl_2$ and Na_2HPO_4 were mixed at room temperature and neutral pH, had as its most distinctive features an extremely fine, non-crystalline texture when examined by TEM and no discernable electron diffraction pattern. This latter feature led them to infer that the considerably more diffuse electron diffraction pattern of newly formed bone mineral as compared to more mature bone mineral, although still apatitic in character, indicated the presence also of an amorphous component" [20].

However, Boskey [19] has reported another story: "In 1964, Dr. Paul Tannenbaum, a graduate student in periodontics at Columbia, and a research assistant in Dr. Posner's laboratory at hospital for special surgery, was studying the effect of fluoride on apatite crystal size. He prepared a synthetic apatite by mixing high concentrations (~30 mM) of calcium chloride and (~20 mM) sodium acid phosphate in buffer, and, being anxious to confirm that the precipitate which formed was apatite, pelleted it by centrifugation, dried it with acetone and placed it on a holder for analysis by wide-angle X-ray diffraction. The pattern obtained (Fig. 2.2, bottom) was broad and diffuse, with a maximum at ~30° 2 theta, had no features, and was clearly not apatite. Dr. Posner suggested that Dr. Tannenbaum did not have the settings correct on the X-ray diffractometer, but since it was late on Friday, decided to correct the settings on Monday. On Monday, the sample, which had been left on the diffractometer over the weekend, was again subjected to X-ray diffraction analysis, but now the pattern had the appearance of a poorly crystalline apatite (Fig. 2.2, middle). Dr. Tannenbaum was certain that the settings on the diffractometer were not different from those he had used previously. Instructed by Dr. Posner to repeat the experiment, he observed the same phenomenon. Immediately after being mixed, the precipitate formed was amorphous, while after several hours, it converted to poorly crystalline apatite. It seemed plausible to Dr. Posner that were such an "amorphous" material (i.e., one that did not give a crystalline diffraction pattern) present in bone, along with the apatite, it might account for the broad diffraction pattern of bone mineral" [19]. One should stress, that both Chow et al., [99] and Eanes [100] published corrigenda to this story by Boskey.

In 1960s, both X-ray diffraction and infrared spectroscopic techniques were used to obtain a quantitative estimate of the amorphous content of bone mineral and then, based on the methods used in polymer chemistry, an algorithm to estimate the ACP amount in bones was developed [101–103]. Early X-ray diffraction estimates indicated the presence of ~30% or more of a non-crystalline mineral in bones of several animal species. Later estimates by X-ray radial distribution analysis placed the upper limit of ~10% ACP in bones and brought into question whether all X-ray amorphous mineral of bones was truly non-crystalline [17, 104–107]. However, further studies by higher-resolution techniques have shown that ~99% of the mineral in bone is a poorly crystalline ion-substituted CDHA of a biological origin [19].

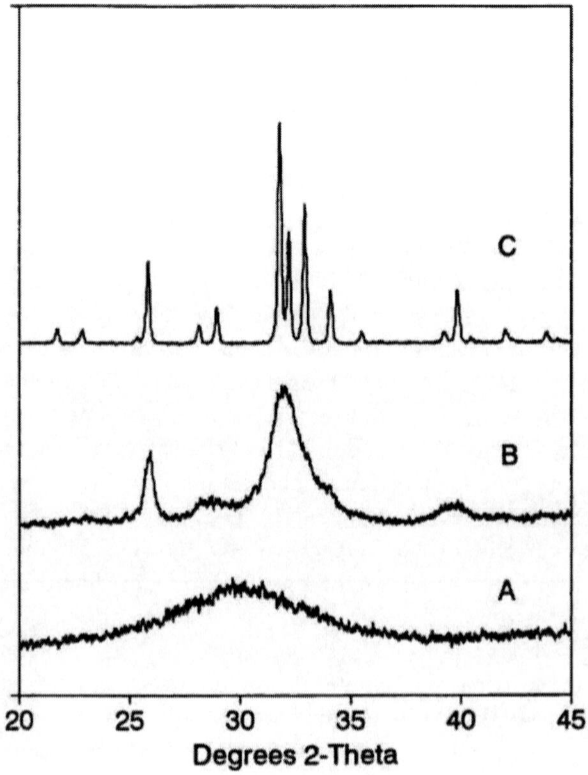

Figure 2.2 Comparison of X-ray diffraction patterns (Cu K_{α} radiation, = 0.154 nm) of synthetic ACP (bottom), poorly crystalline CDHA (middle), and well-crystalline HA (top). The intensity values of the top pattern have been multiplied by a factor of 10, accounting for a high noise level.

Morphological evidence establishing the extent of ACP in skeletal tissues of mammals is equally ambiguous. Although some studies [97, 108–112] report a presence of small spheroidal particles atypical of crystalline material, primarily in actively metabolizing regions, most TEM studies of bones do not even mention finding such possibly amorphous structures. Furthermore, during ageing, the amount of ACP in bones and teeth of mammals decreases while the crystalline forms of biologically formed ion-substituted CDHA increases during early stages of formation [3, 11] (see Section 1.4.1 for additional details on this topic). Since both physical and morphological evidences for ACP in skeletal tissues of mammals have been difficult to establish directly, much of our progress in clarifying the possible roles of ACP in biogenic calcification has come from both synthetic and *in vitro* studies [20].

2.3.2 Preparation

2.3.2.1 Wet-chemistry

Already in early 1970s, researchers established that the final and stable product of a reaction between calcium and orthophosphate salts in neutral or basic aqueous solutions was crystalline stoichiometric HA. However, the stoichiometric well crystalline HA might be prepared at elevated temperatures only; thus, in the vast majority cases, in aqueous solutions CDHA is formed instead (Chapter 1). Furthermore, during CDHA precipitation, over a broad range of the solution conditions, an ACP precursor phase is often formed [113–123], in some cases, via a short intermediate stage of OCP formation [124, 125]. Data are available that CDHA crystallization from ACP simply involves a LRO increase in the structure [126]. One should stress, that already in the mid of 1970s ACP was found to be not the mandatory precursor to CDHA. Namely, in dilute aqueous solutions CDHA was found to precipitate without going through an ACP precursor [126]. Afterward, a model was developed to illustrate factors influencing the nature of non-stoichiometric amorphous precursor phases precipitating in highly supersaturated solutions [127].

The basic approach to synthesize ACP still consists of a spontaneous precipitation by mixing concentrated aqueous solutions of calcium and orthophosphate ions, first developed in 1953 by Watson and Robinson [98]. Another commonly used method is to prepare an acidic (pH within 1–5) sub-saturated aqueous solution of a calcium orthophosphate salt (e.g., DCPD, MCPM) and afterward to induce precipitation by a rapid addition of a strong base (e.g., NaOH, KOH, NH_4OH) to reach the desired solution pH [128]. Vigorous mixing is highly desirable. By means of both approaches, various types of ACPs have been prepared from solutions encompassing a wide range of pH (from ~6.5 to ~13), Ca/P ionic ratios (from ~0.1 to ~10), calcium and orthophosphate concentrations (from ~0.002 to ~1 mol/l), as well as at temperatures within 0°C to 50°C. However, the Ca/P ratio of the mixing reagents (classically, $Ca(NO_3)_2 \cdot 4H_2O$ and $(NH_4)_2HPO_4$) is typically kept within 1.50–1.67, while a basicity of the mixing solutions is frequently created by NH_4OH addition [48, 116, 128–132]. At acidic pH, crystalline calcium orthophosphates normally are precipitated. However, in presence of stabilizers (magnesium and/or citrates), ACP could be precipitated at solution pH within 6.0–6.5 [133]. No information on ACP precipitation from even more acidic aqueous solutions has been found in literature. The obtained precipitates should be collected shortly after the preparation (the sooner, the better), because in aqueous media ACP is spontaneously converted to the crystalline calcium orthophosphates, mainly to CDHA [116, 134]. Furthermore, it was shown that the final calcium orthophosphate (a dry powder) would be amorphous if, beside the appropriate key factors of the synthesis (a high concentration of reagents, a basic solution pH, a rapid mixing, and a low temperature), both a high addition rate and a mandatory freeze-drying of the precipitates were employed [40, 116, 132, 135].

In all wet-precipitation techniques, the amorphous precursors, although related to the final CDHA phase, are differed from the final phase in atomic structure, particle morphology, and stoichiometry. For example, the X-ray diffraction pattern of ACP (Fig. 2.2, bottom), if compared to those of CDHA (Fig. 2.2, middle) and HA (Fig. 2.2, top), shows a single and a very broad diffraction peak, typical for amorphous materials, which lack the atomic LRO characteristics of all crystalline materials, including HA [104]. The precipitated ACP phases appear to be spherical (Fig. 2.3a,b) in an electron microscope (diameter ca. 30–100 nm), unlike the needle-like crystals of CDHA (Fig. 2.3c,d). The solution pH, concentration of the mixing reagents, and a preparation temperature all affect the ACP particle sizes; namely, a higher supersaturation produces smaller ACP particles [129]. Although ACP can be prepared with a Ca/P molar ratio as low as ~1.2 (at low pH — see Fig. 2.4) or as high as 1.7 (at high supersaturation), the departure from a Ca/P of ~1.5 has been shown to be due to surface-adsorbed soluble phases those can be washed away or to occluded Ca, respectively [104].

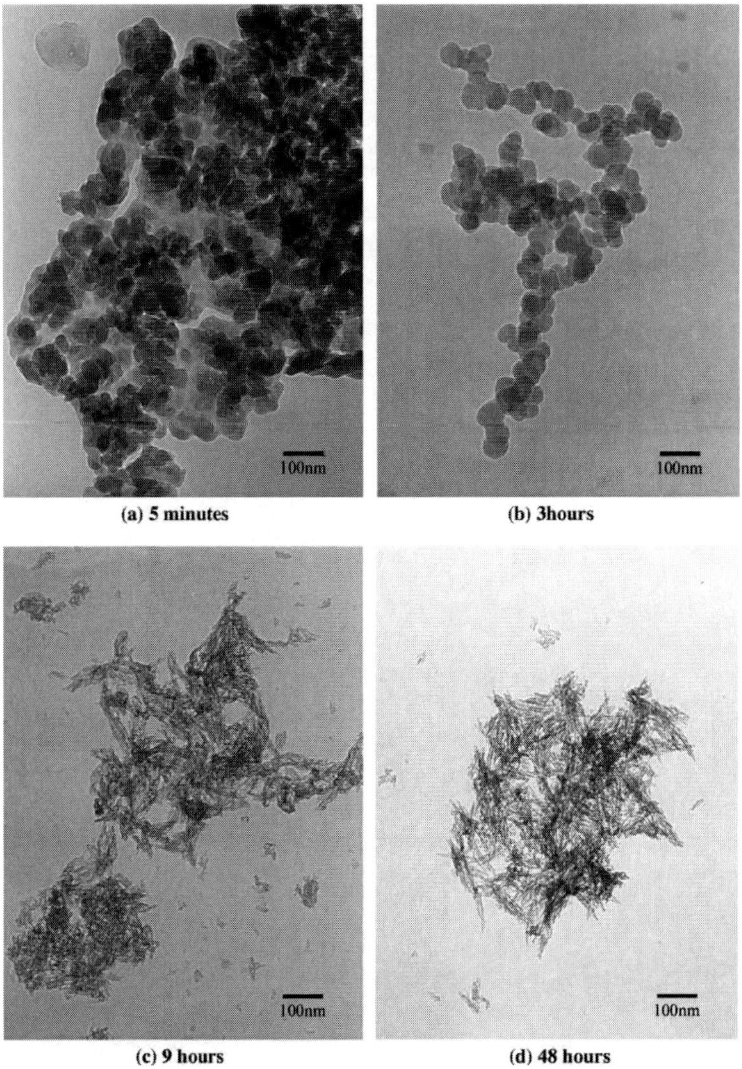

Figure 2.3 Bright field transmission electron micrographs of ACP → CDHA transformation at reaction times of (a) 5 min, (b) 3 h, (c) 9 h, (d) 48 h. Reprinted from Ref. [136] with permission. Another set of electron micrographs of ACP → CDHA transformation might be found in Ref. [253].

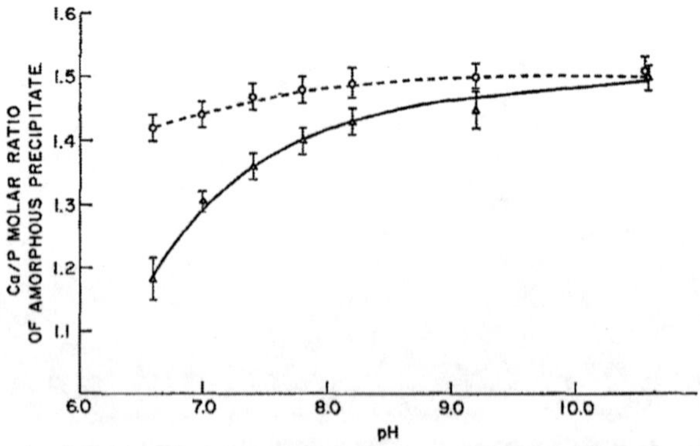

Figure 2.4 Ca/P molar ratios of washed (dashed line) and unwashed (solid line) ACP precipitates as a function of their formation pH. Reprinted from Ref. [137] with permission.

2.3.2.2 Non-aqueous solutions and solvents (sol-gel)

Besides, ACPs might be easily prepared in either non-aqueous or solvent + water media [72, 131, 138–152]. The presence of organic compounds and/or solvents results in decreasing of a dielectric constant. Therefore, all ions in solutions appear to be less solvated than those in water. The consequence of this is a strong decrease of solubility and an increase in precipitation kinetics, which simplifies amorphization [153]. Furthermore, in such systems, complexes of calcium with organic agents can be formed. This favors ACP formation, which is attributed to the coordinated complexing agents remaining in the structure of ACP [144]. The influence of the presence of organic solvents to the amorphization degree of precipitated calcium orthophosphates is well illustrated in Fig. 2.5a. In some cases, incorporation of organic compounds into ACPs has been detected [131, 150]. Interestingly, but a replacement of a freeze-drying stage of a wet ACP precipitate by an oven drying at 80°C resulted in its transformation to CDHA [139]. This process was ascribed to an internal hydrolysis of a part of orthophosphate ions of ACP to those of CDHA according to the schemes [139, 154]:

$$PO_4^{3-} + H_2O \rightarrow HPO_4^{2-} + OH^- \quad (2.1)$$

$$Ca_9(PO_4)_6 + H_2O \rightarrow Ca_9(HPO_4)(PO_4)_5OH \quad (2.2)$$

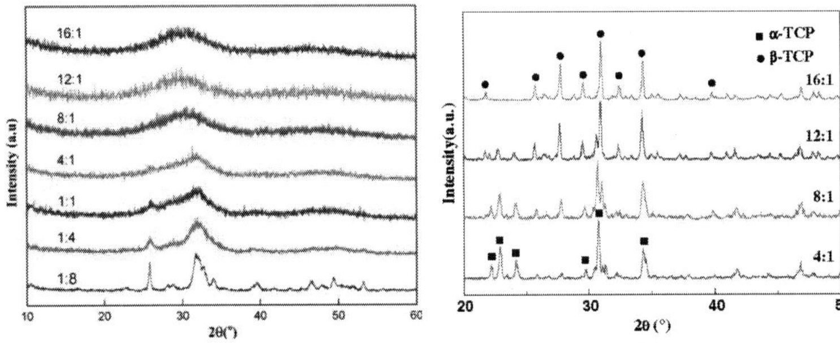

(a) (b)

Figure 2.5 X-ray diffraction patterns of: (a) freeze-dried precipitates prepared from aqueous solutions containing different amounts of polyethylene glycol. Please, note a shift of the center of the broadened peak from ~32° toward ~31° with increasing polyethylene glycol/Ca molar ratio, which implies some structural differences in the resulting ACPs; (b) heat-treated (800°C) ACPs prepared from aqueous solutions containing different amounts of polyethylene glycol. Reprinted from Ref. [150] with permission.

2.3.2.3 Mechanical and pressure-induced techniques

In addition to the aforementioned solution-based methods, various types of ACPs might be prepared by dry chemical techniques. For example, an ACP was prepared using a mechano-chemical method involving a dry mixture of DCPD and $Ca(OH)_2$ reactants with a Ca/P ratio of 1.67 [155]. Other authors have shown that a prolonged high-energy ball milling of either α-TCP, β-TCP powder in ethanol or a dry mixture of ACP and DCPD powders lead to ACP formation after 24 h [156–158]. Furthermore, prolonged high-energy ball milling of TTCP was found to result in a mechanical activation with the formation of undisclosed nanocrystalline and/or amorphous domains within the compound [159]. However, there is a non-negligible risk of powder contamination (ball wear) when using this processing route over extended periods to obtain an ACP [31]. In addition, a crystalline to amorphous transition has been detected for various calcium orthophosphates at very high (up to 10 GPa) pressures (Fig. 2.6) [160, 161].

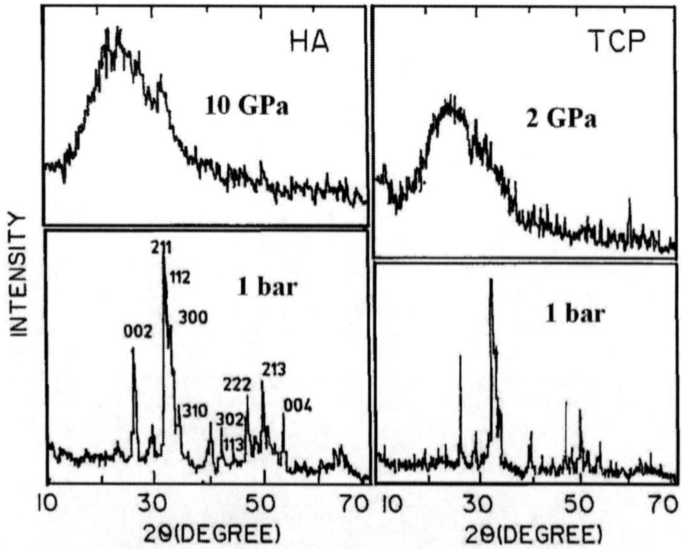

Figure 2.6 Effect of pressure on X-ray diffraction patterns of HA and hydrated TCP at 1 bar and high pressures. Reprinted from Ref. [161] with permission.

2.3.2.4 Thermal

Furthermore, ACPs might be prepared using high energy processing at elevated temperatures. This method is based on a rapid quenching of melted calcium orthophosphates occurring, e.g., during plasma spraying of HA [162–174]. A plasma jet, which possesses very high temperatures (5000–20000°C), partly decomposes HA. Furthermore, there is an opinion, that thermal spraying produces the amorphous phase, not only due to the high cooling rate but also to the removal of hydroxyl ions which make it more difficult for the crystalline phase to form [167]. This generally leads to a mixture of calcium orthophosphate phases with variable compositions, often containing impurities, which is not convenient for preparation of pure ACPs. Interestingly, but the amorphization degree of the plasma sprayed HA coatings appeared to correlate with the presence of vacancies of hydroxyl ions in the structure of HA: the more vacancies were present in the apatite structure due to missing hydroxyl sites, the more amount of ACP was present in the resultant coatings [166]. This might be due to the fact that particles resident in plasma for a longer period of time lose more structural water. Other studies have shown more amorphous phase located adjacent to the substrate and a gradient tending toward a lower ACP content at the top of the coating [31].

To summarize, the amorphous phases in plasma sprayed HA coatings are in intimated mixtures with both crystalline calcium orthophosphates and other compounds, such as CaO [9], and up to now nobody has ever succeed to extract ACPs from the blend. However, the amorphous regions in plasma sprayed HA coatings might be mapped using a scanning cathodoluminescence microscopy technique [170]. Furthermore, due to a number of uncertainties, a reproducibility of such experiments is poor; thus, the plasma spaying technique is not considered as a valuable method to produce ACPs. A flame spray synthesis, when a liquid precursor solution is fed through a capillary into a burning methane/oxygen supporting flame, seems to be more preferable to produce ACPs at high temperatures [175, 176]. Further details on and additional examples of ACP preparation might be found in literature [31].

2.3.2.5 Irradiation

Just a few studies are available on the amorphization of calcium orthophosphates (up to now, apatites only) by irradiation [177, 178]. However, due to the obvious risks caused by the induced radioactivity, this amorphization approach is highly unlikely to be ever used to prepare calcium orthophosphates for biomedical applications. Unfortunately, it remains unclear in what extent the structures, compositions, and properties of ACPs prepared by various production approaches might be mutually compared.

To conclude the preparation part, one should briefly mention on an interesting attempt to precipitate separately hydroxyapatites of Mg, Ca, Sr, and Ba from basic supersaturated orthophosphate solutions containing a 10 : 6 divalent cation / PO_4 molar ratio [179]. In the cases of Mg, Ca, and Sr, the first precipitated phase had a 3 : 2 ratio (i.e., that of a TCP), while only Ba went directly to the stable 10 : 6 HA phase. Furthermore, the precipitated magnesium orthophosphate was amorphous and remaining in the mother solution did not convert to a Mg-deficient HA. The precipitated calcium orthophosphate formed ACP, which converted to CDHA by a solution-mediated autocatalytic mechanism [130]. The precipitated strontium orthophosphate was not amorphous but poorly crystallized and readily converted in solution to a strontium HA. Thus, the smaller alkaline earth cation systems tended to form the more stable amorphous 3 : 2 compounds [179].

2.3.3 Morphology of Precipitated ACPs

When viewed by TEM, ACP solids precipitated from aqueous solutions usually have a curvilinear appearance rather than the faceted and angular shape of crystalline calcium orthophosphates [180, 181]. However, this curvilinear aspect has only been clearly established for dried ACP. The morphological form of highly hydrated flocculent solids that appear initially in freshly precipitated ACP suspensions is not known. What is observed when drops of these suspensions are placed on carbon-coated grids, excess solution removed, and air-dried are irregularly shaped, anastomosing aggregates of low-contrast, disk-shaped particles varying widely in lateral dimensions (from ~0.01 μm to 5+ μm) [181, 182]. These highly flattened particles represent collapsed, de-solvated residues of the initial wet ACP flocculates.

As ACP suspensions age, high contrast particles with a more spherical aspect begin to appear, initially evolving as bud-like extensions from the disks [181, 182]. With time, these spherical forms become the dominant shape for ACP. Although generally smaller (20–300 nm in diameter) than the disks they supplant, the spherical forms, like the disks, frequently aggregate into irregularly shaped and branching clusters. The progression from disk-shaped to ball-like particles most probably represents a spontaneous desolvation *in situ* or the initial gel-like flocculates into smaller, denser, less hydrated structures [183]. That the spherules are formed in suspension and are not a drying artifact is supported by the crystallization behavior of ACP preparations. Although the evolution of a spherical morphology would be favored during consolidation as this shape minimizes interfacial tension with the surrounding solution, it also requires that the contracting surface be isotropic. This is possible for uniformly curved surfaces only when the enclosed structure remains non-crystalline while desolvating [20].

2.3.4 Chemical Composition

Nowadays, ACPs should be recognized as a special class of biomedically relevant calcium orthophosphate salts having variable chemical but rather identical glass-like physical properties. Presumably, all calcium orthophosphates mentioned in Table 1.1 might somehow be fabricated in an amorphous state. Therefore, perhaps, sometime in the future people will deal with an amorphous phase corresponding to the chemical composition of MCPM ("amorphous MCPM"), an amorphous phase corresponding to the chemical composition of DCPA ("amorphous DCPA"), an amorphous phase corresponding to the chemical composition of TTCP ("amorphous TTCP"), etc. (in most cases, stabilization procedures will become necessary), as well as with various mixtures thereof. Currently the majority of such compounds remain unknown and, in the available literature, a variety of ACPs is distinguished by the Ca/P ratio only. Since the greater part of ACPs has the Ca/P ratio close to 1.5 (see below), a term "amorphous TCP" (ATCP) becomes usual in literature [31, 140, 154, 158, 175, 184–191]. Other terms such as "amorphous carbonated apatite" [40, 192, 193], "amorphous $CaHPO_4$" [140], which is equal to "amorphous DCPA" (Table 1.1), "amorphous $Ca_8(HPO_4)_2(PO_4)_4$" [140], which is equal to "amorphous OCP" (Table 1.1), "amorphous OCP" [142], "amorphous $Ca_{10}(PO_4)_6(OH)_2$" [194], which is equal to "amorphous HA" (Table 1.1) and "amorphous dicalcium phosphate" [195] are rare but have been already mentioned. One should note, that ACPs with the Ca/P ratio < ~1.0 currently remain unknown.

2.3.4.1 Precipitated ACPs

Although first described in 1953 [98], the quantitative chemical studies on the precipitated ACPs were not reported until 1965, when methods were devised to isolate large amounts of unstable solids for analysis. To minimize changing during sample drying, those methods utilized filtration and/or centrifugation to wash excess ions from ACP slurries, then freezing wet ACPs under high vacuum to remove any remaining entrapped solvent by sublimation [128]. Early chemical studies [134, 196] on ACPs prepared at pH ~10.5, filtered, washed, and lyophilized, showed that the Ca/PO_4 molar ratio was very close to 1.5, suggesting a TCP composition (as $Ca_3(PO_4)_2{\cdot}nH_2O$ [103, 197]). No OH^- ions were found in it. Furthermore, the electron spin resonance spectra of vanadyl (VO^{2+})

ions adsorbed on ACP formed under varying conditions also revealed that ACP is either a non-crystalline form of hydrated TCP or a solid solution with the composition of $Ca_3(PO_4)_{1.87}(HPO_4)_{0.2}$ [198]. Other researchers reported a stoichiometry of the precipitated ACP more akin to OCP [138, 140, 142, 199–204], DCPA [140], and DCPD [205]; however, with a few exceptions [140, 142] the subsequent terms "an amorphous OCP" and "an amorphous DCPD" were not introduced. Namely, although the Ca/P ratio of early formed ACP phases was varied between 1.35 and 1.38, which was close to that of OCP (Table 1.1), the authors of Refs. [202–204] used a term ACP2 to explain initial variations in solution pH during the transformation of ACP into more crystalline phases of OCP and/or CDHA. Further, based on the results of TEM analysis, ACP2 was identified as a separate amorphous phase with a floccular morphology and no electron diffraction pattern, if compared with a spherular morphology of the initially precipitated amorphous phase ACP1 [202–204]. The same terms ACP1 and ACP2 were used in another study [145], in which the authors studied ageing of calcium orthophosphate precipitates in methanol at room temperature, finally leading to formation of a nano-sized β-TCP. Furthermore, at precipitation experiments from aqueous solutions containing polyethylene glycol, two types of ACPs were detected [150]: one with a broadened peak centered at ~31° and another with a broadened peak centered at ~32° (Fig. 2.5a, spectra 16:1 and 4:1, respectively). According to the authors, the first one was similar to the basic structure of β-TCP, while the second one was similar to the basic structure of HA [150]. Perhaps, one can mention on amorphous β-TCP and amorphous HA, respectively. A similar shift of the position of the amorphous maximum but obtained at different aging time of ACP precipitates (Table 2.1) has been detected in another study [132].

Table 2.1 Position of the X-ray diffraction amorphous maximum at different aging time of ACP precipitates [132].

Aging time	2 min	1 h	2 h	3 h	4 h	5 h	6 h
2θ, degree	29.5	30.1	30.4	30.3	30.5	30.8	29.0

Note: The 2θ values were derived from the profile analysis of the scattering curves. Negligible changes were found for longer aged ACPs. The experimental error ± 0.5 degrees 2θ [132].

Additional analyzes of ACPs prepared from aqueous solutions at pH ~7.4 at wide variations (from 5 : 1 to 1 : 5) in starting Ca/PO_4 molar ratios showed that the compositional Ca/P ratio decreased only slightly from ~1.5 due to the presence of small amounts (<15%) of HPO_4^{2-} ions [137, 206–208]. This is an indirect confirmation of the existence of an amorphous TCP. The presence of HPO_4^{2-} in ACP is easy to detect by infrared spectroscopy most notably by an absorption peak at about 890 cm^{-1}, due to a P–O(H) stretching mode of protonated orthophosphate species (Fig. 2.7) [133, 140]. Furthermore, in infrared spectra of HPO_4^{2-}-contained ACPs there is a large band of a weak intensity near 2300 cm^{-1}, which corresponds to H–O(P) stretching in HPO_4^{2-} ions [140]. More to the point, the amount of HPO_4^{2-} in ACP strongly depends on the solution pH: the lower the Ca/P ratio for any given ACP precipitate, the greater is its HPO_4^{2-} content (Fig. 2.8).

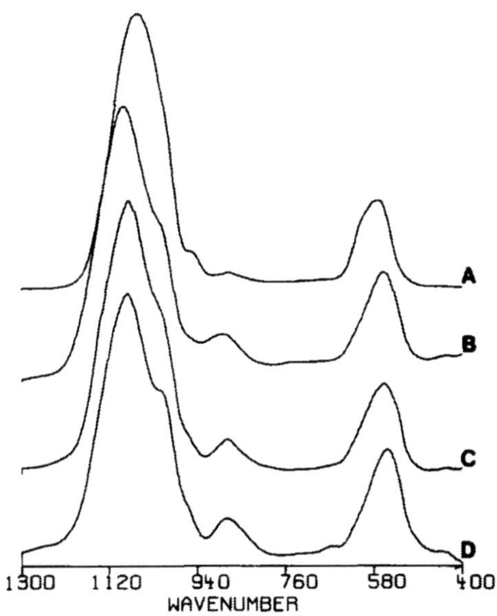

Figure 2.7 Infrared spectra after baseline subtraction for ACPs prepared at: A – pH 10.0, B – pH 7.0, C – pH 6.5, and D – pH 6.0. Reprinted from Ref. [133] with permission. The complete (from 400 cm^{-1} to 4000 cm^{-1}) infrared spectrum of ACP is available in Ref. [31].

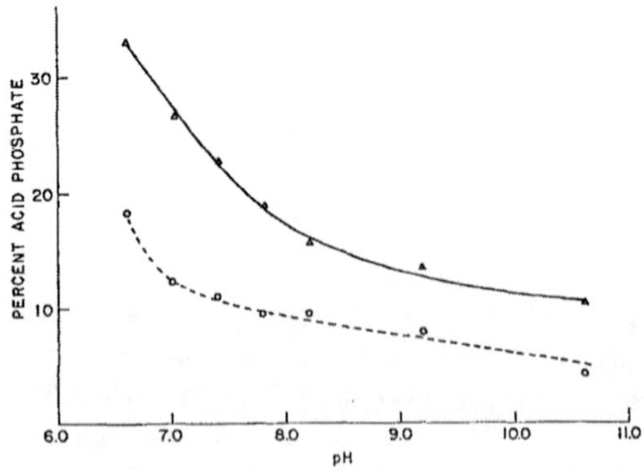

Figure 2.8 Acid phosphate content (percent of total P as HPO_4^{2-}) as a function of pH for washed (dashed line) and unwashed (solid line) ACP precipitates. Reprinted from Ref. [137] with permission.

The presence of HPO_4^{2-} ions in ACP has been confirmed by solid-state nuclear magnetic resonance (NMR) technique: the results revealed the presence of ~20% of HPO_4^{2-} in two ACP samples [209]. Furthermore, orthophosphate units close to water and a third (remained unknown) type of orthophosphate groups were found in the NMR spectra of the ACPs. What's more, substantial differences were noticed between the NMR spectra of two ACP samples, donated by two different research groups [209]. In other studies, two ACP samples of different chemical composition (since they had been prepared at solution pH = 6.5 and 10.0, respectively) were found to give very similar EXAFS spectra [133, 210]. The latter means that a SRO around calcium ions of both ACPs appeared to be very similar; however, the reasons why the pH differences were not reflected in the calcium environment remained unclear. These experiments confirm the fact that ACPs are not a single chemical compound but represent a special class of amorphous calcium orthophosphate salts. Thus, precipitated ACPs cannot be described by a single chemical formula; a sketch $Ca_xH_y(PO_4)_z{\cdot}nH_2O$ (Table 1.1) seems to be the most reasonable element demonstration of this class of calcium orthophosphate salts. Presumably, ACPs with the Ca/P ratio exceeding ~1.6 should also contain some amount hydroxide anions; however, no information on this point has been found in literature. However, a presence of CaO is frequently detected in ACP-containing calcium orthophosphate coatings prepared by plasma spraying.

Nevertheless, in slightly alkaline aqueous solutions, ACP might have a well-defined chemical composition. For example, there is a finding that ACP slurries over the pH range 7.4–9.25 appeared to have a nearly constant solution ion activity product of ~1.6×10^{-25} when the solid-phase composition is postulated to be $Ca_3(PO_4)_{1.87}(HPO_4)_{0.2}$, i.e. when the local chemical unit in ACP is postulated to have a Ca/P molar ratio of 1.45, with ~10% HPO_4^{2-}, but without OH^- ions [207]. Thus, in spite of the absence of a LRO, this relative constancy in the composition over such a wide pH range indirectly suggests that ACP should have some well-defined local chemical units. However, at solution pH exceeding 9.25, ACP does not appear to have a nearly constant solution ion activity product. This breakdown in the solution constancy suggests that the solubility-controlling ions of ACP are subtly dependent compositionally on the preparation conditions. Perhaps, at solution pH > 9.25 the content of HPO_4^{2-} ions no more remains constant and gradually decreases with pH increasing [128]. Furthermore, at more acidic pH = 6.9, ACP precipitates with Ca/P molar ratios as low as 1.15 have been reported [211]. These latter precipitates are extremely unstable and rapidly change over into crystalline DCPD. Again, a term "amorphous DCPD" has not been introduced in that study.

Even after a lyophilization, solution-matured, spheroidal ACP solids still retain ~15% water by weight [197, 212]. A temperature programmed description analysis by Sedlak and Beebe indicated that the most part (~75%) of this retained water was tightly bound inside the solid, while the rest was a more loosely held surface water with different activation energies of 20.0 and 10.5 kcal/mole, respectively [213]. These results suggest that ACPs do not completely desolvate in solutions but remain partially hydrated with about 3 water molecules per formula unit. Other researchers found that water occurred in regions those were only loosely associated with calcium cations in ACP [5]. Furthermore, when prepared from carbonate-containing solutions, ACPs can readily incorporate carbonate anions [35–40, 214, 215]. The amount of carbonate incorporated at any given pH increases with solution carbonate concentration. At a given concentration, carbonate uptake also increases with pH. Incorporating carbonate into ACP does not affect the HPO_4^{2-} content but raises the Ca/P molar ratio. At physiological pH, the carbonate content of ACP precipitated from solutions containing 30 mmol/l carbonates is ~3% by weight [182]. These data suggest that ACP, if present in skeletal tissues, would contain appreciable amounts of carbonates, although less than those present in the apatitic phases of bones [215]. Two other ions that readily incorporate into the ACP structure are Mg^{2+} [37, 44, 133, 179, 216–218] and $P_2O_7^{4-}$ [215, 219, 220]. Ions such as $P_2O_7^{4-}$, carbonate, and Mg^{2+} increase the solution stability of ACP and, in the case of the latter two ions, could possibly play an important role in maintaining the presence of ACP in skeletal tissue. In addition, other ionic substitutions are possible; however, such inorganic additives alter the ACP composition, which would enhance the negative effects in the biomedical application of ACP. Besides, with a few important exceptions, ion-substituted forms of ACP [34–47] are not discussed here.

2.3.4.1 Other types of ACPs

Little is known on the chemical composition of ACPs, prepared by other amorphization techniques. For example, various ACP samples prepared by compressing of several calcium orthophosphates at very high pressures revealed collapses of their initial crystal structures but possible changes in their chemical compositions were not investigated [160, 161]. Interestingly, but the authors found that in the region below 550 cm^{-1} the infrared spectra of DCPD in amorphous phase resembled that of HA in the crystalline phase and conversely the spectra of DCPD in the crystalline phase resembles that of HA in the amorphous phase [161]. In the case of milling, calcium orthophosphates were found to become amorphous; however, no additional phases were detected [157, 158]. Presumably, this means that during amorphization their chemical composition remained unchanged.

Concerning the ACPs formed in plasma-sprayed HA coatings, the authors of one study reported that "the amorphous phase mostly consists of a dehydroxylated calcium phosphate" [174], which, presumably, meant dehydroxylated HA. If so, the chemical composition of that particular ACP should be close to amorphous OA. The authors of another study considered "that the amorphous phase substance consists of HA molecules" (Ref. [163], p. 227). However, in the next study, the same authors mentioned that "the plasma-sprayed amorphous phase is an oxyapatite" [164]. No further clarification has been provided; however, all these authors have come to the conclusion on the apatitic chemical composition of the plasma-sprayed amorphous phases. Besides, these ACPs are definitively anhydrous contrary to the precipitated ACPs.

To conclude the chemical part, one should mention on solubility of ACPs. Due to the chemical variations, this value cannot be measured precisely (Table 1.1). Several different solubility products have been proposed for various ACPs and the interested readers are referred to Table 1 of Ref. [31] for the details.

2.3.5 Structure

In general, determination of the atomic structure of amorphous solids is a non-trivial task. As the structure can be defined essentially only in terms of unit-cells containing an infinitely large amount of atoms (as there is no LRO periodic symmetry), a statistical description appears to be unavoidable. Thus, the structure of a particular amorphous solid can never be determined unambiguously and this uncertainty is compounded by the fact that the structure of a non-crystalline material often depends on the specific details of preparation techniques [88]. Furthermore, the chemical composition, namely the Ca/P ratio, of ACPs varies a lot ($1.2 < Ca/P < 2.2$ — see Table 1.1), which makes the task even more complicated. The latter results in the fact that different samples of ACP possess diverse properties and the data found for one particular sample appear to be inapplicable to other ACP samples. A possible solution of this problem seems to be in a wide recognition of the fact that ACP is not a single chemical compound but represents a group of calcium orthophosphates having diverse physical and chemical properties. In other words, depending on the Ca/P ratio and/or other properties, first of all, any particular sample of ACP must be ascribed to the existing crystalline phase from Table 1.1 and only afterward any structural investigations should be performed. This is the only way to succeed in clarification of the ACP structures from experimental measurements in future. Current state-of-the-art on the structure of ACPs is given below.

The first quantitative studies on a synthetic ACP were done in the mid of 1960s on a material precipitated at pH ~10 [134, 196]. As Watson and Robinson found at neutral pH, the initial phase that spontaneously formed immediately upon mixing concentrated alkaline Ca- and PO_4-containing solutions was structurally non-crystalline [98]. The X-ray diffraction pattern of this rapidly precipitated phase showed only two very broad and diffuse peaks, typical for substances that lack the periodic LRO [196]. The extreme diffuseness of the synthetic ACP pattern also provided a basis for interpreting the reduced intensity of the apatitic X-ray diffraction patterns of bones as being due to the mineral having an amorphous component [101]. The diffracted X-ray energy from this component was so uniformly dispersed that it could not be separated from the subtracted background intensity [20].

Initially, there were suggestions that a synthetic ACP was, in fact, HA of such small crystal dimensions that its X-ray diffraction pattern was widely broadened to appear amorphous in character. However, calculated X-ray diffraction patterns assuming that ACP consisted of small groups of HA unit cells, or even a single HA unit cell, did not match the observed ACP diffraction data [221]. The NMR spectra of ACP are also sufficiently different from those of HA to suggest that they do not have the same structural motif [222]. Then, a probability that ACP was structurally distinct from HA led to a study of this material by the X-ray radial distribution method, which showed that its characteristic diffuseness resulted from a breakdown in atomic LRO within the interior of the synthetic ACP [223–225]. The major peak positions were found to occur at distances of 2.55, 3.75, and 6.40 Å with an evidence of a peak at 2.9 Å [223]. Furthermore, any orderly atomic arrangements did not extend beyond ~0.95 nm in diameter that corresponded to the smallest values of a far-MRO. A contiguous periodic regularity in the distribution of these domains typical of crystalline materials was absent. Extended X-ray absorption fine structure (EXAFS) spectroscopy [116, 226] indicated that possible regularities in the local environment around individual Ca^{2+} ions were even more circumscribed, not extending beyond distances of ~0.3 nm (Fig. 2.9). This is a SRO scale. In addition, an infrared analysis showed a similar lack of crystalline order for orthophosphate anions in the ACP structure [31, 129, 140, 228]. The method is based on the observation that a splitting of the P–O anti-symmetric bending mode at 550–600 cm^{-1} (Fig. 2.10) increases as crystallinity increases. This apparent lack of crystalline regularity is one of the striking features of ACP that distinguishes it from other calcium orthophosphates and provides the structural basis for its name [20].

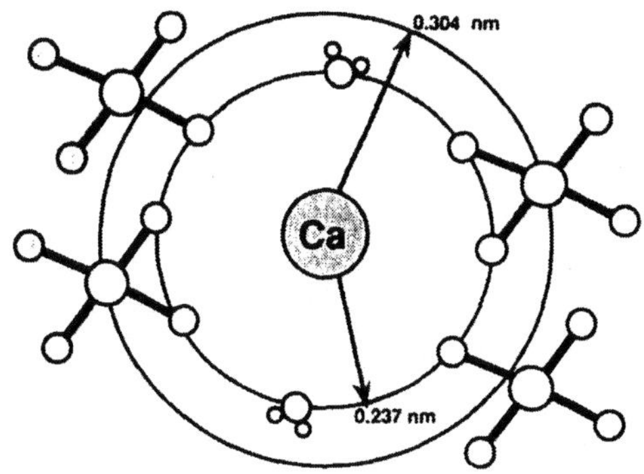

Figure 2.9 Diagrammatic representation of the SRO structure of four phosphate tetrahedra and two water molecules about calcium ions in an acidic ACP, calculated using the DCPD shell model. The positions of hydrogen atoms are not determined. Reprinted from Ref. [227] with permission.

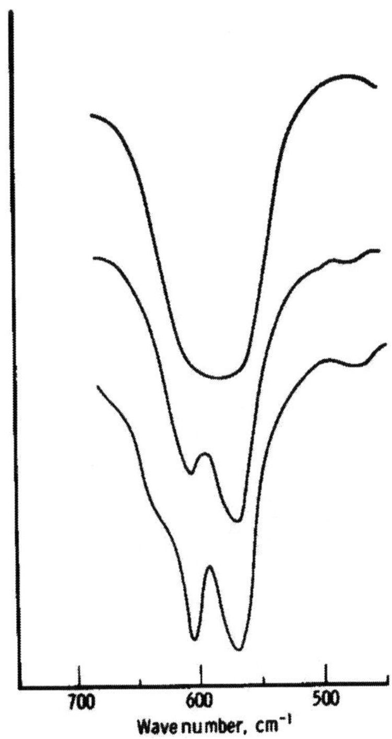

Figure 2.10 Fragments of the infrared spectra of ACP (upper), CDHA (middle), and HA (lower). Reprinted from Ref. [129] with permission.

The apparent absence of observable crystalline features in ACPs does not rule out a possibility that ACPs might have a well-defined local structural unit. The compositional constancy of ACPs over a relatively wide range of the preparative conditions suggests an existence of such a core structure [104]. Several lines of evidence, however, indicate that the local unit of ACPs is not a crypto-structural variant of one of the crystalline calcium orthophosphates. For example, although in the majority of cases precipitated ACPs appear to be closest in composition to TCP, in aqueous solutions ACPs transform into either OCP or CDHA but not into TCP. If ACP were a cryptocrystalline TCP, direct growth into observably crystalline TCP by crystal ripening would be expected to occur. However, a possibility that ACP might be a cryptocrystalline forms of CDHA or HA cannot be as easily precluded. Only below pH ~ 9 CDHA can be ruled out for the same reason as TCP. However, even above this pH, the finding that CDHA forms primarily as an outgrowth from the surface of ACP suggests that an *in situ* ripening process does not occur. Furthermore, dissimilarity in composition rules out ACP as being made up of highly disordered arrays of OCP unit cells. It is equally unlikely that ACP is an orthophosphate-deficient OCP as the structural integrity of the latter depends on a full complement of orthophosphate groups in the unit cell. The finding that the Ca/P molar ratio remains relatively constant at ~1.5 over a relatively wide range of pH (7.4–9.25) appears to preclude ACP as a cryptocrystalline mixture of OCP and CDHA. Instead of remaining constant, a rise in pH would expect to increase the overall Ca/P molar ratio of such a mixture as the relative proportion of the OCP and CDHA components would shift in favor of the latter [20].

Based on the relatively constant Ca/P ratio of ACPs formed under varying precipitation conditions, such as different concentrations of calcium and orthophosphate ions, solution pH and different temperatures, as well as spectroscopic and structural analyzes, Posner and Betts hypothesized that the initial solid phase of ACPs (more precisely, of an amorphous TCP) precipitated *in vitro* consisted of spherical particles ranging within 300–1000 Å in diameter with water molecules in the interstices [104, 224, 225]. Namely, the defining structural unit of ACP is a spatial subset of the HA unit cell consisting of a central Ca^{2+} ion coordinated by the oxygens of six surrounding orthophosphate groups, which, in turn, are stabilized by another 8 Ca^{2+} ions spherically distributed around the outer boundary of the subset (Fig. 1.7). This appears to be a neutral ion cluster of ~ 9.5 Å in diameter, whose composition is expressed by the formula $Ca_9(PO_4)_6$, i.e., 3 times of TCP. Since then, in literature it is defined as a Posner's cluster. It was later suggested that these clusters in fact possessed a S_6 symmetry [229, 230]. This subset is linearly expanded by ~ 3% and its radial distribution function is similar to that calculated from the diffuse X-ray diffraction profile of ACP. The authors postulated that at the far-MRO level these slightly enlarged spherical subsets are randomly clustered in ACP particles (Fig. 1.7) with water filling the intervening spaces [104, 224, 225]. No data have been found in the references that plasma-spayed ACPs cannot contain Posner's clusters as the structural units; however, undoubtedly, they cannot contain water molecules in the intervening spaces. Interestingly, but the atomic arrangement of Posner's clusters appears to be analogous to that existing in several types of crystalline calcium orthophosphates, such as HA, OCP, and β-TCP. For example, the relationship between the original Posner's cluster and the atomic structure of HA is represented in Fig. 2.11. However, one must stress, that this is a model only because the real arrangements of the clusters into larger structures are not known yet.

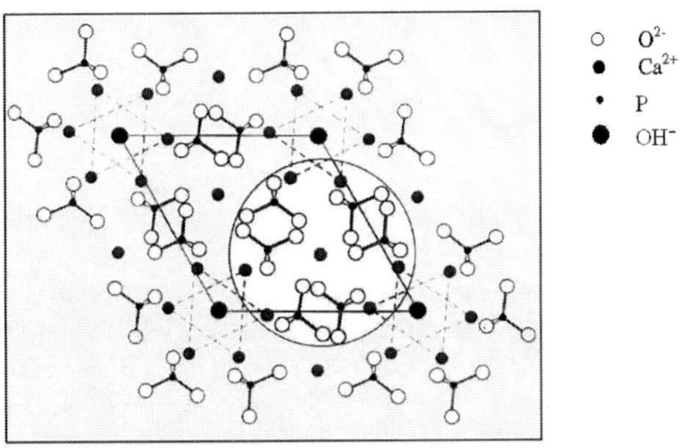

Figure 2.11 A model of Posner's cluster (in a circle) showing its relationship with the HA crystal structure. Black lines forming a rhomb delimit the borders of one unit-cell of HA. Reprinted from Ref. [31] with permission. A similar relationship is also available in Ref. [254]. This correlation between Posner's clusters and HA structure was first published in Ref. [104]. One should note, that in Ref. [230] another set of atoms of the HA crystal structure has been chosen to represent a Posner's cluster.

In general, an electrostatic interaction, a hydrophobic interaction and a "cementing" effect of water molecules are among the main factors holding ions of a cluster altogether [231]. Again, a "cementing" effect of water molecules is applied for ACPs precipitated from aqueous solutions only. Theoretical investigations on the stability of different calcium orthophosphate clusters with an increasing number of ions have confirmed that Posner's clusters are the most stable arrangement [232]. The potential energy surfaces associated with $[Ca_3(PO_4)_2]_n$ clusters were analyzed in details using *ab initio* calculations for *n* ranging from 1 to 4 [233]. The energy criteria were found to favor the Posner's cluster, which is the core of the actual structural model of ACP (more precisely, of amorphous TCP). Moreover, the calculations showed that an aggregation of the clusters corresponded to a large energy stabilization whatever the cluster considered [233], which is in agreement with the Posner's hypothesis in which clusters are proposed to be closely packed in the ACP structure. Furthermore, the vibration spectra of ACP have been modeled as well [234]. In spite of the aforementioned, it is not quite clear from what experimental, spectroscopic, and/or structural data the cluster model (Fig. 1.7) has been derived, except the TCP composition. Further, it is not entirely clear whether there are volumetric regions within the large 300–1000 Å particles that are free from the 9.5 Å clusters. In other words, are there some volumetric regions within the large 300–1000 Å particles where the atoms are completely random with no SRO? Moreover, the results of EXAFS analysis indicated that the range of orderly Ca–Ca and Ca–P interactions in ACP were much shorter than would be predicted from the Betts and Posner model [116, 226]. Another weakness in the model is that water only serves to fill the interstices between the HA subsets and is not an essential part of the ACP structure, contrary to the assumptions by Sedlak and Beebe [213]. Both of these weaknesses could possibly be remedied if a smaller spatial domain is carved from a portion of the OCP unit cell that contains a part of the hydration layer. However, this suggestion has not been examined in details. It is also possible, however, that the defining structural unit for ACP is a truly unique entity with no satisfactory crystallographic model. In fact, there is no *a priori* need for such a model, as the defining unit for ACP would not be constrained structurally by the symmetry requirements for crystalline arrangements. Furthermore, it will also depend on the Ca/P ratio. Therefore, ACP could have a structural motif not found or even permissible in any of the crystalline calcium orthophosphates [20].

More to the point, mechanisms of calcium and orthophosphate ion association in aqueous solutions were elucidated by means of quantum and classical molecular mechanics simulations [235]. A special focus was dedicated to the role of the protonation state of orthophosphate ions and depronation of the hydrogen orthophosphate ions appeared to be necessary during crystal growth. According to the simulation results, a triple ion $[Ca^{2+}\cdot\cdot(HPO_4)^{2-}\cdot\cdot Ca^{2+}]^{2+}$ could form in aqueous solutions and subsequently yield another triple ion $[Ca^{2+}\cdot\cdot(PO_4)^{3-}\cdot\cdot Ca^{2+}]^{+}$ by releasing a proton. The latter ion suggested to be the smallest stable aggregate, which contains an entirely deprotonated orthophosphate ion [235]. Some indirect experimental evidences in support of this hypothesis have been obtained recently [236]. Nevertheless, a verification of the aforementioned structural models of ACPs will require details of the chemistry and the processing conditions together with viewing techniques at the atomic scale. Unfortunately, no additional information on the structure of high-temperature ACP phases has been found in the references. Thus, for want of anything better, one is forced to assume, that except of water molecules and possible presence of HPO_4^{2-} ions, all the aforementioned is valid for ACPs presented in plasma-sprayed coatings. Undoubtedly, this assumption should be verified and corrected in future.

To conclude the structural part, determination of the specific surface area of ACPs generally leads to surprisingly low numbers. This might be related to the larger spherical associations of Posner's clusters and the apparent hindrance to nitrogen adsorption of these inner surfaces [189, 212].

2.3.6 Thermal Properties

All types of ACP (presumably, except of ones presented in plasma-sprayed coatings) are thermally unstable and sustain neither calcining nor sintering. For example, if precipitated ACPs are heated, first of all, they lose water. Two types of water loss occur, corresponding to loosely bound water molecules adsorbed on the surface of ACP agglomerates and more strongly bound internal water molecules, respectively. The first loss is essentially reversible, whereas the second is mostly irreversible [213, 237]. Furthermore, in experiments, where water interactions were minimized by allowing for the escape of volatile components, crystallization of ACP was found to begin at about 530°C [197]. Below this temperature, the non-crystalline features of ACPs seemed thermally stable. The first crystalline phase to appear was invariably β-TCP. However, between 600°C and 800°C, depending upon the preparation, α-TCP was found to become generally the favored ignition product, even though β-TCP is normally the stable phase up to ~ 1200°C. Neither washing + drying procedures employed to isolate the amorphous material, nor the choice of soluble orthophosphate salt used in its preparation, were found to have any significant effect on the thermo-crystallization properties of ACP [197]. However, in other studies, both α-TCP [147, 175, 176, 189] and carbonated CDHA [141] appeared to be the first detectable crystalline phase at heating of various ACPs to 550–600°C. Interestingly, but the presence of organic solvents (in that case, polyethylene glycol) at the ACP preparation stage was found to influence the products formed at elevated temperatures [150]. Namely, when the amount of polyethylene glycol was small, α-TCP was formed at heating; when the amount of polyethylene glycol was big, β-TCP was formed at heating; and biphasic (α-TCP + β-TCP) formulations were formed when the amount of polyethylene glycol was average (Fig. 2.5b). Similar effects of both aging time and the solution pH were also detected [151].

ACPs with Ca/P ratios of 1.00 ("amorphous DCPA") and 1.34 ("amorphous OCP") were found to remain amorphous at heating up to 600°C, while crystalline compounds (β-$Ca_2P_2O_7$ in the case of Ca/P = 1.00 and α-TCP + β-$Ca_2P_2O_7$ in the case of Ca/P = 1.34) started to appear at 620°C [140]. In the same study, ACP with Ca/P ratio of 1.51 ("amorphous TCP") were found to remain amorphous at heating up to ~ 550°C, while crystalline compounds (β-TCP) started to appear at ~ 600°C. Interestingly, but heating of a crystalline DCPA (monetite) leads to γ-$Ca_2P_2O_7$ and then this phase is transformed at ~ 750°C to β-$Ca_2P_2O_7$ [238]. Thus, "amorphous DCPA" showed a thermal behavior different from that of crystalline DCPA [140].

Furthermore, crystallization of ACPs is an exothermic process. The heat produced was found to be ~21 kJ/mol, while the activation energy was ~450 kJ/mol [168]. Other researchers reported the activation energy values of 440 kJ/mol for crystallization of hydroxyl-depleted areas of the amorphous phase to OA and 230 kJ/mol for crystallization of hydroxyl-rich areas of the amorphous phase to HA [174]. Therefore, hydroxylated ACP regions were found to crystallize more readily compared to the dehydroxylated ones. A heat of re-crystallization of ACP to HA was calculated to be ~43 kJ/mol in still another study [172]. Presumably, such variations are due to the differences in the chemical composition of various ACP samples.

To conclude the thermal part, the results mentioned in this section appear to be additional indirect confirmations of the fact that ACP is not a single chemical compound but represents a special class of calcium orthophosphate salts. Furthermore, as the first approximation and for want of anything better, the first crystalline product formed at heating of a specific ACP sample might become an indicator of its type. Namely, if the first crystalline product was β-TCP [197], the authors of that study dealt with an "amorphous β-TCP." Similarly, if the first crystalline product was α-TCP [147, 175, 176, 189], the authors of those studies dealt with an "amorphous α-TCP," etc. Certainly, this is a hypothesis only, which needs to be investigated and/or verified both theoretically and experimentally in future.

2.3.7 Amorphous-to-Crystalline Transformations in Aqueous Solutions

As previously stated, in the vast majority of cases, ACPs are the first solid phases to appear upon mixing of calcium- and orthophosphate-containing aqueous solutions at pH > 7 and concentrations sufficiently high to produce an immediate precipitation. The spontaneous formation of ACPs is a kinetically driven process. A rapid mixing of highly concentrated solutions creates sufficiently strong stochastic interactions among the ions and they quickly coalesce into irregularly coordinated highly hydrated clusters large enough to separate from solution in a gel-like state before they had a chance to be rearranged into orderly nuclei capable of growth as crystals. This structural arrangement, however, is inherently unstable. In addition to desolvating, ACPs kept in solution eventually disappear, being supplanted by more stable crystalline phases such as OCP or CDHA [20].

Watson and Robinson in their pioneering study on ACP [98] were the first, who observed a transient nature of ACP when kept in contact with its preparative medium. They found that electron diffraction patterns of ACP taken later in the precipitation reaction no longer were diffuse but resembled patterns of a poorly crystalline CDHA. Further investigations revealed that this amorphous-to-crystalline transition was not gradual but occurred rather precipitously. Initially, there is a period of a relative stability, where surfaces of the high-contrast spherules generally remain smooth and regular [181]. However, as seen from the data of Table 2.1 [132], some yet unknown changes occur with the precipitated ACP during this time. Afterward, the transition follows a sigmoid evolution by the solid phase rapidly progressing from being barely crystalline to where the amorphous features disappear. Once the first crystals appear on the surface of the spherules, the transition proceeds rapidly to completion. Simultaneously, dramatic declines in ionic concentrations of calcium and orthophosphate ions occur in the mother solution. The time it takes to reach this amorphous-to-crystalline boundary various considerably with the preparation conditions, being particularly sensitive to temperature and solution pH [68, 130]. For example, at pH ~ 7.4, ACP converts five times faster at 37°C than at 20°C [239]. The pH dependency is somewhat more complex than that for temperature. Namely, at 25°C the aqueous lifetime of freshly precipitated ACP is less than 0.3 h at pH ~ 7.4. It increases to a maximum lifetime of over 9 h between pH ~ 10.0 and pH ~ 10.5, then rapidly decreases until at pH ~ 12.8 the lifetime is nearly as short as that at pH ~ 7.4 [207, 240]. The solution lifetime of ACP can be greatly extended by inclusion of simple inorganic ions such as Mg^{2+}, Zr^{2+}, silicates, carbonates, and pyrophosphates [40, 41, 43, 133, 179, 204, 216–219, 241, 242]. As an extreme example, ACP prepared from Mg^{2+}-containing solutions at pH = 10.0 and 32.5°C remained in a gel-like amorphous state for up to 20 weeks when the reactant Mg/Ca molar ratio was set at 0.2 [179]. Other substances those can increase the stability of ACP in aqueous solutions include F^- [239, 240], various polyelectrolytes [67, 68], polyalcohols and polyglycols [131, 144, 146], phospholipids [243], dentin phosphoprotein [244], phosvitin [220, 244], glycochenodeoxycholic acid [245], biomacromolecules such as casein phosphopeptide [69], as well as adenosine di- and triphosphates (but not the monophosphate) [246, 247]. On the other hand, excess of Ca^{2+} ions in the solution accelerates the transformation of ACP into a crystalline CDHA [136].

Interestingly, but collagen, which is the principal matrix protein in skeletal tissues, has no effect on ACP stability [239, 244], while presence of gelatin promotes transformation of ACP into crystalline phases [49]. One should note, that in wet atmosphere solid ACP with Ca/P ratio of 1.33 ("amorphous OCP") was found to re-crystallize into a mixture of CDHA + DCPA [142].

Conversion of ACPs to solid phases with X-ray diffraction characteristics of CDHA has been studied largely in aqueous suspensions [67, 68, 72, 116, 119, 121, 130, 132, 136, 154, 185, 190, 196, 218, 248–253]. Transmission electron micrographs of this transformation are shown in Fig. 2.3. Kinetics of this process can be described by an empirical equation: $dC/dt = k_1 + k_2C$, where C is the fraction of ACP converted into the crystalline phase by time t, k_1 is a rate constant associated with the nucleation of the first crystals and k_2 is an autocatalytic rate constant indicative of the observation that the transition rate is proportional to the mass fraction already crystallized and not to the fraction of remaining ACP [134, 196]. Numerical values for k_2 appeared to be much larger than those for k_1, reflective of the exponential rapidity of the transition once started [196].

The amorphous-to-crystalline transformation mechanisms of ACPs have not been well elucidated. Chemically, this process is described by Eqs. (2.1) and (2.2); while in the reality it might proceed along several pathways. Namely, it might occur as dissolution of ACP and re-precipitation of crystalline phases (e.g., CDHA — see Fig. 2.12 [72]) [20, 72, 118, 121, 128, 130, 154, 190, 248], internal structure rearrangements [136, 190, 218, 229, 254, 255], development of a LRO without changing the immediate environment of Ca [116], formation of the crystalline phases either directly within the ACP phases [149, 236] or on the surface of ACP particles [253], as well as by self-aggregation and surface-mediated transformations [31]. For example: "... When the density reached a critical value, the random arrangement of growth units became disadvantageous in terms of total free energy, resulting in a sudden regularization of the structure, which was deposited as HA." (Ref. [229], page 241). Furthermore, Yin and Stott suggested that, in the transformation from ACP to CDHA, ACP needed only to dissociate into clusters rather than undergo complete ionic salvation [256].

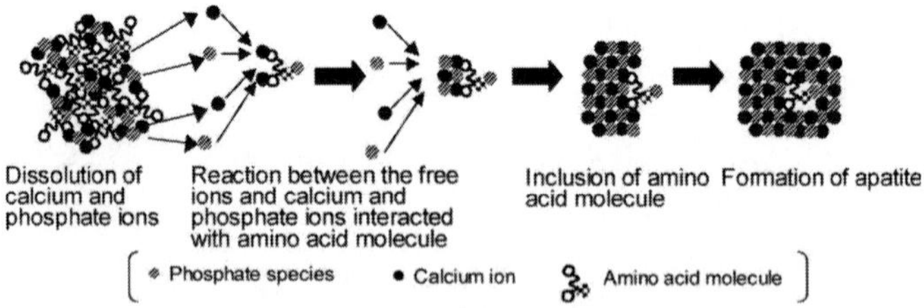

Figure 2.12 A proposed mechanism of a CDHA/amino acid biocomposite from an ACP/amino acid biocomposite. Reprinted from Ref. [72] with permission.

Having summarized findings of their own and previously published results by other researchers, Wang et al., [236] have depicted a following picture of the main events that take place during the induction period and finally trigger the rapid precipitation of calcium orthophosphates from supersaturated aqueous solutions: "Calcium and orthophosphate ions form pairs and clusters successively in the first few seconds. These ions, pairs, and clusters then compose the initial solid phase that is heavily hydrated and contains hydrogenorthophosphates. Growing and aggregating, the solid increases in size and quantity without affecting the solution pH. During the induction period, the solid particles exhibit a steady size distribution, which is around 300 nm at the early stage and shifts toward 1000 nm with time. These particles are agglomerates of primary particles of 60–100 nm in diameter and are originally amorphous in structure. At multiple sites inside a particle, crystalline domains develop from ion pairs and/or clusters by taking up calcium and releasing hydrated proton, possibly through a stage at which the more compact cluster $Ca_9(PO_4)_6$ presents. Since the expansion of crystalline domains consumes surrounding calcium and orthophosphate ions (or their pairs and clusters) and releases hydrated protons, the mechanic strength decreases in the inter-domain regions. Finally, under the action of the shearing strength of the fluid, these primary particles collapse and the liberated crystallites induce the rapid precipitation of calcium orthophosphates, together with the previously trapped hydrated protons in primary particles, resulting in the abrupt pH drop. Indeed, it is the crystallization at multiple sites inside amorphous particles that finally triggered the rapid precipitation of calcium phosphate from the supersaturated solution." (Ref. [236], pp. 2625–2626).

To conclude this part, as ACPs represent a special class of calcium orthophosphates, it is reasonably to presume that various amorphous-to-crystalline transformation mechanisms might exist due to the initial differences of the investigated ACPs. Furthermore, various competitive processes might occur simultaneously and their importance might depend on the conversion conditions. Further details on this topic are available in literature [31, 123].

2.4 ACP *in vivo*

As stated in the introduction, both physical and morphological evidences for the presence of ACPs in skeletal tissue of mammals have been difficult to establish directly and the validity of inferential evidences for their presence and amounts has been the subject of considerable debates. Indirect assessments, such as X-ray diffraction methods, have produced widely varying estimates of the amorphous content of bone mineral, placing it at less than ~1% [19] to more than ~30% of the total mineral mass, the rest presumably being poorly crystalline ion-substituted CDHA (biological apatite). Even if ACPs only occur at a lower percentage, one would expect TEM to reveal some evidences [9], but as stated earlier, most TEM studies of bones and teeth do not even mention the existence of amorphous-like structures that could be ascribed to ACPs. However, the absence of such structures could have been a negative artifact caused by aqueous dissolution of more labile ACPs during the sample preparation for examination. A few early TEM studies [108, 109] that avoided aqueous processing of bone specimens by directly embedding and sectioning freeze-dried material revealed a zone of electron dense, 6– 20 μm diameter spheroidal bodies adjacent to crystal-rich areas of bone tissue. Electron diffraction of these sites revealed a hazy, diffuse pattern similar to that observed by Watson and Robinson [98] in their synthetic, amorphous-like precipitates. Whether or not these spheroidal particles represented ACPs has been debatable but a labile nature of ACPs suggests the need to maintain carefully the anhydrous conditions in preserving this phase in calcified material for examinations [20].

Another possibility that could account for the inability to establish with certainty the existence of ACPs in mammalian bones is that the ACP does not exist as separate particles but, instead, as an amorphous layer on crystals of the poorly crystalline phase of biological apatite. Although there has been no direct evidence for this possibility, similar coatings appear to form in physiological-like solutions seeded with HA [257, 258]. Under these *in vitro* conditions, the initial accretions appeared to form an amorphous calcium carbonate-ACP coating on the seed crystals. The initial phase also incorporated a small amount of Mg^{2+} ions from solution that was subsequently released upon the ancillary formation of CDHA crystals. This Mg^{2+} behavior, consistent with the amorphous-to-crystalline transition, is the most compelling compositional evidence from these studies for the initial coat being an amorphous layer [257].

A proper assessment of the possibility that some of the mineral in skeletal tissues is in a free-standing amorphous state is further complicated by the fact that the minimum ion activity product needed to form ACP *de novo* in physiological-like synthetic solutions at pH 7.4 is considerably greater than that calculated for serum [126, 259]. If extracellular skeletal fluids were in electrolyte equilibrium with serum, it would appear unlikely that ACPs could form *in vivo* except possibly as a coating on crystals of biological apatite of bones. Without stronger evidence than that described above, even this possibility, though, is highly speculative. However, there is some evidence that suggests that calcifying bone matrix may be compartmentalized with the establishment of an interior milieu different from that of serum [260]. Unfortunately, it is not known whether such compartmentalization results in an extracellular fluid space capable of initiating *de novo* ACP formation [20].

Although the general compartmentalization of bone-producing conditions favorable for ACP needs to be further established, a considerable body of evidence suggests that local micro-compartments exist in bones that could allow for ACP development. The most thoroughly studied of these micro-spaces are the membrane-enclosed aqueous cores of matrix vesicles [261, 262]. Most commonly found near osteoblasts in the extracellular regions of rapidly mineralizing embryonic bone, these spherical bodies of cellular origin are the sites of initial mineral formation [263]. Preceding appearance of the first crystals at these sites is an accumulation of calcium and orthophosphate ions within the aqueous cores of the vesicles to levels that far exceed the threshold level for *de novo* ACP formation [264, 265]. Studies with synthetic liposomes confirm that ACP should readily form under such compartmentalized conditions

[266]. However, Raman spectroscopic data indicate that ACP in matrix vesicles is not in a pure chemical state but instead calcium and orthophosphate ions are combined in a single-phase complex with lipid and protein moieties found within the vesicles [267]. Also consistent with formation of an amorphous precursor phase are infrared and Raman spectroscopic findings that the first crystals in matrix vesicles are OCP and not CDHA [267, 268]. These crystals, in turn, penetrate the enclosing membrane and initiate a chain of crystallization events that appears to trigger the mineralization of the collagenous matrix. Thus ACP, even when present in small amounts, may be an important initiating factor in the calcification of skeletal tissues [20].

Besides the questionable cases of calcified tissues of mammals, ACPs are found in mammalian milk [7, 8, 73]. Clearly, the presence of the most easily biodegradable calcium orthophosphates in the form of ACPs in milk is necessary to construct skeletons of young organisms.

2.5 Biomedical Application of ACPs

Currently, biomaterials and bioceramics of calcium orthophosphates are available in various physical forms: powders, particles, granules, dense blocks, porous scaffolds, injectable formulations, self-setting cements and concretes, implant coatings, as well as composite components of different origin (natural, biological or synthetic) often with the specific shapes, such as implants, prostheses or prosthetic devices (see Refs. [32, 33, 269] and Chapter 4). In principle, all these physical forms should apply to ACPs; however, not all of them have been realized yet. Namely, it is easy to prepare ACPs in a powder form by a wet-precipitation technique (see section 2.3.2); however, manufacturing of other physical forms of ACPs is not so simple. Furthermore, to be used in surgery, all implantable 3D constructions must possess the necessary mechanical properties, which is difficult to achieve in the case of ACPs. For example, both dense blocks and 3D porous scaffolds made of calcium orthophosphates reach their mechanical properties only after sintering at temperatures, exceeding ~ 1000°C, which is impossible in the case of ACPs (see section 2.3.6). Furthermore, an ACP powder might be easily added as a component to self-setting calcium orthophosphate formulations

(Chapter 5); however, no cement formulation is known, which results in ACP formation, as the major end product [270, 271]. More to the point, ACPs are present as components of various calcium orthophosphate coatings; however, just a few studies are known, in which ACP coatings have been fabricated [272–275].

Not many examples of the biomedical applications of ACPs are currently available. First of all, the majority of cases of plasma sprayed calcium orthophosphate coatings should be excluded because the coatings consist of a complicated mixture of various phases, where ACP is just one of them [9, 162–174]. Besides, both the amount and the composition of these amorphous phases cannot be well controlled. Even in the case, when ACP was plasma sprayed, the coating was found to be a mixture of a crystalline carbonate-containing apatite phase with considerable amounts of TTCP and trace amounts of α-TCP [276]. Thus, these cases may hardly be mentioned as the biomedical applications of ACPs.

There is another type of calcium orthophosphate coatings, obtained by a biomimetic route. Some of them contain ACP as the major component [277–280]. For example, an ACP coating on polyethersulphone plates was prepared by precipitation from a simulated body fluid (SBF) at 35°C. It took authors 12 days to obtain a ~20 μm thick ACP coating and 28 days for a ~50 μm thick one [277]. Similar results were obtained in other studies [278, 279]. One should stress, that in the vast majority cases, the biomimetic approach presumes application of SBF, which contains ions of sodium, magnesium, potassium, sulfate, and hydrogencarbonate; thus, ion-substituted ACPs are always formed as a result.

Furthermore, ACP coatings can be deposited on titanium using an electrochemical technique at 36°C and solution pH 6.4 [280]. The obtained ACP coatings appeared to be unstable and transformed into those of CDHA. By choice of different electrochemical parameters, a homogeneous coating of ACP, CDHA or some intermediate phases could be achieved, thus allowing formation of the coatings with different morphology and solubility [280].

Among the self-setting calcium orthophosphate formulations (Chapter 5), there is only one ACP-based composition currently on the market, namely Biobon® (α-BSM®) [281, 282]. This cement comprises a mixture of ACP (50 wt. %) and DCPD (50 wt. %) which is mixed with an appropriate amount of aqueous medium (de-ionized water or saline) with a liquid to solid ratio of 0.8 ml/g at room temperature. An injectable paste is obtained which sets in less than 20 min at 37°C. After hardening, the cement is constituted of nanocrystalline CDHA with crystal dimensions close to those in human bones [31]. A similar formulation but additionally containing mechano-activated β-cyclodextrins has been studied as well [155]. Furthermore, a self-setting ACP + DCPA formulation is known [283]. Besides, ACP might be added as an admixture phase to the standard calcium orthophosphate cement formulations [44, 284]. In such cases, addition of ACP resulted in cements exhibited shorter setting times, a compressive strength suitable for non-load-bearing applications and the full conversion to nanocrystalline CDHA. Moreover, ACP-containing formulations demonstrated good cell viability, making them suitable candidates for biomedical applications [284]. More to the point, nano-sized particles of 13, 19, and 40 nm diameter of amorphous TCP were found to be highly reactive and set to CDHA within minutes, which is much faster if compared with microcrystalline powders of both α-TCP and β-TCP [187, 188]. A similar approach was reported in other studies [156–158].

Since ACPs do not sustain heating above ~ 600°C (see section 2.3.6), a possibility of preparing dense ACP bioceramics was studied using spark plasma sintering technique at temperatures ranging from 150 to 200°C with the processing time < 15 min. Unfortunately, the observed mechanical strength of the prepared consolidated disks (sintered for 6 min at 150°C) was poor [285]. However, an increase in the sintering processing time to 13 min indicated to a possibility of improving the mechanical properties. The low temperature conditions appeared to be more adaptive to the processing of ACP when compared with the experimental conditions of conventional sintering. However, the physicochemical characterization of the prepared consolidated bioceramics indicated a crystallization of the initial ACP to an apatitic phase with no other detectable crystalline phase [31].

However, the majority of cases of the biomedical applications of ACPs comprise various biocomposites and hybrid biomaterials containing ACP as one of the phases [48–75, 155, 286–300]. Several ACP-containing formulations (e.g., Recaldent™ and Enamelon™) are now commercially available [299]. As can be seen from the references, most of the applications are in dentistry field. In all cases, addition of calcium orthophosphates, including ACPs, imparts both biocompatibility and bioactivity of the biocomposites [290].

For example, to improve cell adhesion, hydrophilic array of ACP was fabricated on a surface of hydrophobic polystyrene [301]. A similar effect was found when coatings composed of ACP and hyaluronic acid were used [302]. Furthermore, in the acidic oral environment the ACP-containing biocomposites take advantage of the ability of ACPs to release calcium and orthophosphate ions, which potentially can take part in enamel remineralization [51–66, 188, 288, 289, 303–324]. Such ACP-containing biocomposites and hybrid biomaterials might be coatings [292] and cements [155, 286, 287]. In dentistry, the ACP-containing formulations are used mainly as anti-cariogenic and/or remineralizing agents [303–324], e.g., in chewing gums [306–308], sugar confections [74], various tooth mousses [309–311], bleaching gels [314, 315], various drinks [316, 317] or even in milk [321, 322]. Furthermore, ACP-containing formulations are used in orthodontics [291–298].

2.6 Conclusions

ACPs can be both found in living organisms (especially invertebrates) and synthesized in a laboratory. Due to their chemical variability, they should be recognized as a special class of calcium orthophosphates offering a wide variety of compositions. Presumably, all known calcium orthophosphates (see Table 1.1) might be prepared in an amorphous state; however, not all of them (especially, those with Ca/P ratio < 1.1) have been prepared. Furthermore, in the available literature nothing has been found on existence of amorphous TTCP. The vast majority of the available publications on ACPs in fact describe wet-precipitated compounds with Ca/P ratio close to ~1.5, i.e. amorphous TCP. As Poster's cluster has a chemical composition of TCP, it might be located in the atomic structures of HA (Fig. 2.11), TCPs, and OCP. Since TTCP and all acidic calcium orthophosphates have the Ca/P ionic ratios far away from 1.5, presumably, at SRO and MRO scales, all of them in an amorphous state should contain structural units, which will be quite different from Posner's clusters. Nothing on this point has been found in the available literature. Obviously, this topic needs to be investigates in future.

In aqueous solutions, all currently known ACPs are easily converted into crystalline phases of calcium orthophosphates, especially into poorly crystalline CDHA. Therefore, advantages can be taken of ACPs' high reactivity to prepare various bioactive biomaterials. Currently, ACPs are involved as transient or constitutive phases in several commercial substitute bone materials, such as plasma sprayed coatings on metal prostheses and injectable cements for orthopedic applications. ACPs are also used for dental applications as fillers in ionomer cements to fill cavities or as colloidal suspensions in toothpastes, chewing gums or mouthwashes to promote remineralization of carious lesions and/or to prevent tooth demineralization [31]. Obviously, these examples are just the initial steps of the biomedical applications of ACPs.

As written above, the atomic structure of amorphous compounds is difficult to investigate due to the lack of LRO. Thus, in some of the aforementioned studies on ACPs, an "amorphous" character of the phases might be due to formation of nanodimensional and/or nanocrystalline calcium orthophosphates [325, 326] with crystal dimensions of 2 nm × 2 nm × 2 nm (8 nm^3) or so (see section 2.2). Namely, employment of novel high resolution techniques resulted in doubt of the existence of several calcium orthophosphate samples in an amorphous state [327, 328]. Therefore, there is still an unanswered question concerning the structure of the substantial amounts of nanodimensional and/or nanocrystalline calcium orthophosphates: whether they appear to be almost amorphous (according to numerous results of X-ray diffraction studies) due to their exceedingly small dimensions of well-crystallized structures or due to a really amorphous (i.e., retaining only a SRO) matter? A good attempt to discuss this topic is available in literature [329], where the interested readers are referred to.

To conclude the chapter, the biomedical applications of ACPs nowadays are limited due to great difficulties in their consolidations to various 3D structures, which have to possess the sufficient mechanical properties. As ACPs do not sustain heating at temperatures exceeding ~ 600°C, they must be consolidated by low temperature techniques only. Spark plasma sintering approach is one of them [31, 285]. Another one comprises mixing of thermally unstable calcium orthophosphates with water soluble porogens, followed by cold isostatic pressing of the prepared mixture and dissolving the porogens [330]. Obviously, still other "mild" consolidation approaches are possible, those need to be developed. For example, a very interesting approach to fabricate hollow ACP-based nano-sized tubes has been proposed recently. These nano-sized tubes appear to be non-toxic to fibroblasts and can be taken up easily by mammalian cells [331]. Presumably, they may serve as vehicles for drugs and growth factors, as well as for tissue repair including bone regeneration.

References and Notes

1. Lowenstam, H. A., and Weiner, S. (1989) *On Biomineralization*, Oxford University Press, New York, p. 324.
2. Gower, L. B. (2008) Biomimetic model systems for investigating the amorphous precursor pathway and its role in biomineralization, *Chem. Rev.*, **108**, 4551–4627.
3. Lowenstam, H. A., and Weiner, S. (1985) Transformation of amorphous calcium phosphate to crystalline dahillite in the radular teeth of chitons, *Science*, **227**, 51–53.
4. Stricker, S. A., and Weiner, S. (1985) Amorphous calcium phosphate in the stylets produced by a marine worm (*Nemertea*), *Experientia*, **41**, 1557–1559.
5. Mitchell, P. C. H., Parker, S. F., Simkiss, K., Simmons, J., and Taylor, M. G. (1996) Hydrated sites in biogenic amorphous calcium phosphates: an infrared, Raman, and inelastic neutron scattering study, *J. Inorg. Biochem.*, **62**, 183–197.
6. Becker, A., Ziegler, A., and Epple, M. (2005) The mineral phase in the cuticles of two species of *Crustacea* consists of magnesium calcite, amorphous calcium carbonate, and amorphous calcium phosphate, *Dalton Transactions*, **10**, 1814–1820.
7. McGann, T. C. A., Buchheim, W., Kearney, R. D., and Richardson, T. (1983) Composition and ultrastructure of calcium phosphate-citrate complexes in bovine milk systems, *Biochim. Biophys. Acta*, **760**, 415–420.
8. McGann, T. C. A., Kearney, R. D., and Buckheim, W. (1983) Amorphous calcium phosphate in casein micelles of bovine milk, *Calcif. Tiss. Int.*, **35**, 821–823.
9. Brès, E. F., Moebus, G., Kleebe, H. J., Pourroy, G., Werkmann, J., and Ehret, G. (1993) High resolution electron microscopy study of amorphous calcium phosphate, *J. Cryst. Growth*, **129**, 149–162.
10. Raeymaekers, L., Agostini, B., and Hasselbach, W. (1981) The formation of intravesicular calcium phosphate deposits in microsomes of smooth muscle: a comparison with sarcoplasmic reticulum of skeletal muscle, *Histochemistry*, **70**, 139–150.

11. Termine, J. D., and Posner, A. S. (1996) Infrared analysis of rat bone: age dependency of amorphous and crystalline mineral fractions, *Science*, **153**, 1523–1525.

12. Termine, J. D., Wuthier, R. E., and Posner, A. S. (1967) Amorphous-crystalline mineral changes during endochondral and periosteal bone formation, *Proceedings of the Society for Experimental Biology and Medicine*, **125**, 4–9.

13. Eanes, E. D., Termine, J. D., and Posner, A. S. (1967) Amorphous calcium phosphate in skeletal tissues, *Clin. Orthop. Relat. Res.*, **53**, 223–235.

14. Tannenbaum, P. J., Schraer, H., and Posner, A. S. (1974) Crystalline changes in avian bone related to the reproductive cycle. II. Percent crystallinity changes, *Calcif. Tiss. Int.*, **14**, 83–86.

15. Glimcher, M. J., Bonar, L. C., Grynpas, M. D., Landis, W. J., and Roufosse, A. H. (1981) Recent studies of bone mineral: is the amorphous calcium phosphate theory valid?, *J. Cryst. Growth*, **53**, 100–119.

16. Grynpas, M. D., Bonar, L. C., and Glimcher, M. J. (1981) On the question of amorphous tricalcium phosphate in bone mineral, *Dev. Biochem.*, **22**, 279–283.

17. Grynpas, M. D., Bonar, L. C., and Glimcher, M. J. (1984) Failure to detect an amorphous calcium-phosphate solid phase in bone mineral: a radial distribution function study, *Calcif. Tiss. Int.*, **36**, 291–301.

18. Aoba, T., and Moreno, E. (1990) Changes in the nature and composition of enamel mineral during porcine amelogenesis, *Calcif. Tiss. Int.*, **47**, 356–364.

19. Boskey, A. L. (1997) Amorphous calcium phosphate: the contention of bone, *J. Dent. Res.*, **76**, 1433–1436.

20. Eanes, E. D. (2001) Amorphous calcium phosphate, in *Octacalcium Phosphate* (ed. Chow, L. C., and Eanes, E. D.), *Monographs Oral Sci.*, vol. 18, Karger, Basel, pp. 130–147.

21. Weiner, S., Sagi, I., and Addadi, L. (2005) Choosing the crystallization path less travelled, *Science*, **309**, 1027–1028.

22. Weiner, S. (2006) Transient precursor strategy in mineral formation of bone, *Bone*, **39**, 431–433.

23. Suvorova, E. I., Petrenko, P. P., and Buffat, P. A. (2007) Scanning and transmission electron microscopy for evaluation of order/disorder in bone structure, *Scanning*, **29**, 162–170.

24. Rey, C., Combes, C., Drouet, C., and Glimcher, M. J. (2009) Bone mineral: an update on chemical composition and structure, *Osteoporos. Int.*, **20**, 1013–1321.

25. Olszta, M. J., Odom, D. J., Douglas, E. P., and Gower, L. B. (2003) A new paradigm for biomineral formation: mineralization via an amorphous liquid phase precursor, *Connect. Tiss. Res.*, **44**, 326–334.

26. Mahamid, J., Sharir, A., Addadi, L., and Weiner, S. (2008) Amorphous calcium phosphate is a major component of the forming fin bones of zebrafish: indications for an amorphous precursor phase, *Proc. Natl. Acad. Sci. USA*, **105**, 12748–12753.

27. Mahamid, J., Aichmayer, B., Shimoni, E., Ziblat, R., Li, C., Siegel, S., Paris, O., Fratzl, P., Weiner, S., and Addadi, L. (2010) Mapping amorphous calcium phosphate transformation into crystalline mineral from the cell to the bone in zebrafish fin rays, *Proc. Natl. Acad. Sci. USA*, **107**, 6316–6321.

28. Tsuji, T., Onuma, K., Yamamoto, A., Iijima, M., and Shiba, K. (2008) Direct transformation from amorphous to crystalline calcium phosphate facilitated by motif-programmed artificial proteins, *Proc. Natl. Acad. Sci. USA*, **105**, 16866–16870.

29. Beniash, A., Metzler, R. A., Lam, R. S. K., and Gilbert, P. U. P. A. (2009) Transient amorphous calcium phosphate in forming enamel, *J. Struct. Biol.*, **166**, 133–143.

30. Tao, J., Pan, H., Zeng, Y., Xu, R., and Tang, R. (2007) Roles of amorphous calcium phosphate and biological additives in the assembly of hydroxyapatite nanoparticles, *J. Phys. Chem. B*, **111**, 13410–13418.

31. Combes, C., and Rey, C. (2010) Amorphous calcium phosphates: synthesis, properties and uses in biomaterials, *Acta Biomater.*, **6**, 3362–3378.

32. Dorozhkin, S. V. (2007) Calcium orthophosphates, *J. Mater. Sci.*, **42**, 1061–1095.

33. Dorozhkin, S. V. (2009) Calcium orthophosphates in nature, biology and medicine, *Materials*, **2**, 399–498.

34. Holt, C., van Kemenade, M. J. J. M., Harries, J. E., Nelson, L. S., Jr., Bailey, R. T., Hukins, D. W. L., Hasnain, S. S., and de Bruyn, P. L. (1988) Preparation of amorphous calcium-magnesium phosphates at pH 7 and characterization by X-ray absorption and Fourier transform infrared spectroscopy, *J. Cryst. Growth*, **92**, 239–252.

35. Bachra, B. N. (1963) Precipitation of calcium carbonates and phosphates from metastable solutions, *Ann. NY Acad. Sci.*, **109**, 251–255.

36. Bachra, B. N., Trautz, O. R., and Simon, S. L. (1963) Precipitation of calcium carbonates and phosphates under physiological conditions, *Arch. Biochem. Biophys.*, **103**, 124–138.

37. Bachra, B. N., Trautz, O. R., and Simon, S. L. (1965) Precipitation of calcium carbonates and phosphates. III. The effect of magnesium and fluoride ions on the spontaneous precipitation of calcium carbonates and phosphates, *Arch. Oral Biol.*, **10**, 731–738.

38. Greenfield, D. J., and Eanes, E. D. (1972) Formation chemistry of amorphous calcium phosphates prepared from carbonate containing solutions, *Calcif. Tiss. Res.*, **9**, 152–162.

39. Olesen, P. T., Steenberg, T., Christensen, E., and Bjerrum, N. J. (1998) Electrolytic deposition of amorphous and crystalline zinc-calcium phosphates, *J. Mater. Sci.*, **33**, 3059–3063.

40. Tadic, D., Peters, F., and Epple, M. (2002) Continuous synthesis of amorphous carbonated apatite, *Biomaterials*, **23**, 2553–2559.

41. LeGeros, R. Z., Mijares, D., Park, J., Chang, X. F., Khairoun, I., Kijkowska, R., Dias, R., and LeGeros, J. P. (2005) Amorphous calcium phosphates (ACP): formation and stability, *Key Eng. Mater.*, 284–286, 7–10.

42. Aimanova, O. J., LeGeros, R. Z., and Sinyayev, V. A. (2005) Antimicrobiologic property hydrated amorphous calcium phosphates containing silver, *Key Eng. Mater.*, 284–286, 439–442.

43. Skrtic, D., Antonucci, J. M., Eanes, E. D., and Brunworth, R. T. (2002) Silica- and zirconia-hybridized amorphous calcium phosphate: effect on transformation to hydroxyapatite, *J. Biomed. Mater. Res.*, **59**, 597–604.

44. Julien, M., Khairoun, I., LeGeros, R. Z., Delplace, S., Pilet, P., Weiss, P., Daculsi, G., Bouler, J. M., and Guicheux, J. (2007) Physico-chemical–mechanical and *in vitro* biological properties of calcium phosphate cements with doped amorphous calcium phosphates, *Biomaterials*, **28**, 956–965.

45. Sinyaev, V. A., LeGeros, R. Z., Levchenko, L. V., Shustikova, E. S., and Karzhaubaeva, R. A. (2008) State of water in amorphous calcium and calcium-magnesium phosphates, *Russ. J. General Chem.*, **78**, 864–867.

46. Sinyaev, V. A., Shustikova, E. S., Levchenko, L. V., Karzhaubaeva, R. A., and Tokseitova, G. A. (2008) Amorphous calcium barium monophosphate and its dehydration in air at room temperature, *Russ. J. Appl. Chem.*, **81**, 1899–1903.

47. Lee, D., and Kumta, P. N. (2010) Chemical synthesis and characterization of magnesium substituted amorphous calcium phosphate (MG-ACP), *Mater. Sci. Eng. C*, **30**, 1313–1317.

48. Taylor, M. G., Simkiss, K., Simmons, J., Wu, L. N. Y., and Wuthier, R. E. (1998) Structural studies of a phosphatidyl serine-amorphous calcium phosphate complex, *Cell. Mol. Life Sci.*, **54**, 196–202.

49. Brečević, L., Hlady, V., and Füredi-Milhofer, H. (1987) Influence of gelatin on the precipitation of amorphous calcium phosphate, *Colloids Surf.*, **28**, 301–313.

50. Ambrosio, A. M. A., Sahota, J. S., Khan, Y., and Laurencin, C. T. (2001) A novel amorphous calcium phosphate polymer ceramic for bone repair: I. Synthesis and characterization, *J. Biomed. Mater. Res.*, **58**, 295–301.

51. Skrtic, D., Antonucci, J. M., Eanes, E. D., Eichmiller, F. C., and Schumacher, G. E. (2000) Physicochemical evaluation of bioactive polymeric composites based on hybrid amorphous calcium phosphates, *J. Biomed. Mater. Res.* (*Appl. Biomater.*), **53**, 381–391.

52. Skrtic, D., Antonucci, J. M., and Eanes, E. D. (2001) Effect of the monomer and filler system on the remineralizing potential of bioactive dental composites based on amorphous calcium phosphate, *Polym. Adv. Technol.*, **12**, 369–379.

53. Skrtic, D., Antonucci, J. M., and Eanes, E. D. (2003) Amorphous calcium phosphate-based bioactive polymeric composites for mineralized tissue regeneration, *J. Res. Natl. Inst. Stands. Technol.*, **108**, 167–182.

54. Skrtic, D., Antonucci, J. M., Eanes, E. D., and Eidelman, N. (2004) Dental composites based on hybrid and surface-modified amorphous calcium phosphates, *Biomaterials*, **25**, 1141–1150.

55. Skrtic, D., and Antonucci, J. M. (2005) Matrix resin effects on selected physicochemical properties of amorphous calcium phosphate composites, *J. Bioactive Compatible Polym.*, **20**, 29–49.

56. Skrtic, D., Antonucci, J. M., and Eanes, E. D. (1996) Improved properties of amorphous calcium phosphate fillers in remineralizing resin composites, *Dent. Mater.*, **12**, 295–301.

57. Skrtic, D., and Antonucci, J. M. (2007) Dental composites based on amorphous calcium phosphate – resin composition/physicochemical properties study, *J. Biomater. Applic.*, **21**, 375–393.

58. Skrtic, D., Antonucci, J. M., and Liu, D. W. (2006) Ethoxylated bisphenol dimethacrylate-based amorphous calcium phosphate composites, *Acta Biomater.*, **2**, 85–94.

59. Skrtic, D., Hailer, A. W., Takagi, S., Antonucci, J. M., and Eanes, E. D. (1996) Quantitative assessment of the efficacy of amorphous calcium phosphate/methacrylate composites in remineralizing caries-like lesions artificially produced in bovine enamel, *J. Dental Res.*, **75**, 1679–1686.

60. Park, M. S., Eanes, E. D., Antonucci, J. M., and Skrtic, D. (1998) Mechanical properties of bioactive amorphous calcium phosphate/methacrylate composites, *Dent. Mater.*, **14**, 137–141.

61. Lee, S. Y., Regnault, W. F., Antonucci, J. M., and Skrtic, D. (2007) Effect of particle size of an amorphous calcium phosphate filler on the mechanical strength and ion release of polymeric composites, *J. Biomed. Mater. Res. B* (*Appl. Biomater.*), **80B**, 11–17.

62. Antonucci, J. M., Liu, D. W., and Skrtic, D. (2007) Amorphous calcium phosphate based composites: effect of surfactants and poly(ethylene oxide) on filler and composite properties, *J. Dispersion Sci. Technol.*, **28**, 819–824.

63. Skrtic, D., Lee, S. Y., Antonucci, J. M., and Liu, D. W. (2005) Amorphous calcium phosphate based polymeric composites: effects of polymer composition and filler's particle size on composite properties, *Key Eng. Mater.*, 284–286, 737–740.

64. O'Donnell, J. N. R., Schumacher, G. E., Antonucci, J. M., and Skrtic, D. (2009) Adhesion of amorphous calcium phosphate composites bonded to dentin: a study in failure modality, *J. Biomed. Mater. Res. B* (*Appl. Biomater.*), **90B**, 238–249.

65. Antonucci, J. M., O'Donnell, J. N. R., Schumacher, G. E., and Skrtic, D. (2009) Amorphous calcium phosphate composites and their effect on composite–adhesive–dentin bonding, *J. Adhes. Sci. Technol.*, **23**, 1133–1147.

66. Reynolds, E. C., Cai, F., Cochrane, N. J., Shen, P., Walker, G. D., Morgan, M. V., and Reynolds, C. (2008) Fluoride and casein phosphopeptide-amorphous calcium phosphate, *J. Dent. Res.*, **87**, 344–348.

67. Amjad, Z. (1990) Influence of polyelectrolytes on the precipitation of amorphous calcium phosphate, *Colloids Surf.*, **48**, 95–106.

68. Bar-Yosef, O. P., Govrin-Lippman, R., Garti, N., and Füredi-Milhofer, H. (2004) The influence of polyelectrolytes on the formation and phase transformation of amorphous calcium phosphate, *Cryst. Growth Des.*, **4**, 177–183.

69. Cross, K. J., Huq, N. L., Palamara, J. E., Perich, J. W., and Reynolds, E. C. (2005) Physicochemical characterisation of casein phosphopeptide-amorphous calcium phosphate nanocomplexes, *J. Biological Chem.*, **280**, 15362–15369.

70. Gutiérrez, M. C., Jobbágy, M., Ferrer, M. L., and del Monte, F. (2008) Enzymatic synthesis of amorphous calcium phosphate-chitosan nanocomposites and their processing into hierarchical structures, *Chem. Mater.*, **20**, 11–13.

71. Cushnie, E. K., Khan, Y. M., and Laurencin, C. T. (2008) Amorphous hydroxyapatite-sintered polymeric scaffolds for bone tissue

regeneration: physical characterization studies, *J. Biomed. Mater. Res. A*, **84A**, 54–62.

72. Ikawa, N., Kimura, T., Oumi, Y., and Sano, T. (2009) Amino acid containing amorphous calcium phosphates and the rapid transformation into apatite, *J. Mater. Chem.*, **19**, 4906–4913.

73. Lin, Q., Li, Y., Lan, X., Lu, C., and Xu, Z. (2009) Preparation of amorphous calcium phosphate/tricalcium silicate composite powders, *Adv. Mater. Res.*, 79–82, 1643–1646.

74. Walker, G. D., Cai, F., Shen, P., Adams, G. G., Reynolds, C., and Reynolds, E. C. (2010) Casein phosphopeptide-amorphous calcium phosphate incorporated into sugar confections inhibits the progression of enamel subsurface lesions *in situ*, *Caries Res.*, **44**, 33–40.

75. Antonucci, J. M., Regnault, W. F., and Skrtic, D. (2010) Polymerization shrinkage and stress development in amorphous calcium phosphate / urethane dimethacrylate polymeric composites, *J. Composite Mater.*, **44**, 355–367.

76. Sinyaev, V. A., Shustikova, E. S., Levchenko, L. V., and Sedunov, A. A. (2001) Synthesis and dehydration of amorphous calcium phosphate, *Inorg. Mater.*, **37**, 619–622.

77. Dion, A., Berno, B., Hall, G., and Filiaggi, M. J. (2005) The effect of processing on the structural characteristics of vancomycin-loaded amorphous calcium phosphate matrices, *Biomaterials*, **26**, 4486–4494.

78. Dion, A., Langman, M., Hall, G., and Filiaggi, M. J. (2005) Vancomycin release behaviour from amorphous calcium polyphosphate matrices intended for osteomyelitis treatment, *Biomaterials*, **26**, 7276–7285.

79. Lee, B., Kim, M., Choi, S., and Lee, Y. K. (2009) Amorphous calcium polyphosphate bone regenerative materials based on calcium phosphate glass, *Key Eng. Mater.*, 396–398, 209–212.

80. Chen, G., Li, W., Zhao, B., and Sun, K. (2009) A novel biphasic bone scaffold: β-calcium phosphate and amorphous calcium polyphosphate, *J. Am. Ceram. Soc.*, **92**, 945–948.

81. Chun, S., Jeong, J. H., Kim, K. M., and Kim, S. (2001) Biodegradation study of amorphous and crystalline calcium metaphosphate in the SBF and tris-buffer solution, *Key Eng. Mater.*, 192–195, 131–134.

82. Cheng, Y. T., and Johnson, W. L. (1987) Disordered materials: a survey of amorphous solids, *Science*, **235**, 997–1002.

83. http://en.wikipedia.org/wiki/Amorphous (accessed in October 2010).

84. Sheng, H. W., Luo, W. K., Alamgir, F. M., Bai, J. M., and Ma, E. (2006) Atomic packing and short-to-medium-range order in metallic glasses, *Nature*, **439**, 419–425.

85. Lee, C. Y., Stachurski, Z. H., and Welberry, T. R. (2010) The geometry, topology and structure of amorphous solids, *Acta Mater.*, **58**, 615–625.

86. Hufnagel, T. C. (2004) Amorphous materials: finding order in disorder, *Nature Mater.*, **3**, 666–667.

87. Elliott, S. R. (1990) *Physics of Amorphous Materials*, 2nd edn., Longman, London, p. 481.

88. Elliott, S. R. (1991) Medium-range structural order in covalent amorphous solids, *Nature*, **354**, 445–452.

89. Salmon, P. S. (2002) Amorphous materials: order in disorder, *Nature Mater.*, **1**, 87–88.

90. Krivanek, O. L., Gaskell, P. H., and Howie, A. (1976) Seeking order in 'amorphous' materials, *Nature*, **362**, 454–457.

91. Mountjoy, G. (1999) Order in two-dimensional projections of thin amorphous three-dimensional structures, *J. Phys. Cond. Matter*, **11**, 2319–2336.

92. Simon, V., Lazăr, D., Turcu, R. V. F., Mocuta, H., Magyari, K., Prinz, M., Neumann, M., and Simon, S. (2009) Atomic environment in sol-gel derived nanocrystalline hydroxyapatite, *Mater. Sci. Eng. B*, **165**, 247–251.

93. Weeber, A. W., and Bakker, H. (1988) Amorphization by ball milling. A review, *Physica B*, **153**, 93–135.

94. Maier, G., Zipper, P., Stubičar, M., and Schurz, J. (2005) Amorphization of different cellulose samples by ball milling, *Cellulose Chem. Technol.*, **39**, 167–177.

95. Motta, A. T. (1997) Amorphization of intermetallic compounds under irradiation – a review, *J. Nucl. Mater.*, **244**, 227–250.

96. Edmondson, P. D., Riley, D. J., Birtcher, R. C., and Donnelly, S. E. (2009) Amorphization of crystalline Si due to heavy and light ion irradiation, *J. Appl. Phys.*, **106**, 043505 (8 pages).

97. Robinson, R. A., and Watson, M. L. (1955) Crystal-collagen relationships in bone as observed in the electron microscope. III. Crystal and collagen morphology as a function of age, *Ann. NY Acad. Sci.*, **60**, 596–660.

98. Watson, M. L., and Robinson, R. A. (1953) Collagen-crystal relationships in bone. II. Electron microscope study of basic calcium phosphate crystals, *Am. J. Anat.*, **93**, 25–59.

99. Chow, L. C., Takagi, S., and Vogel, G. L. (1998) Letter to the Editor, *J. Dent. Res.*, **77**, 6.

100. Eanes, E. D. (1998) Letter to the Editor, *J. Dent. Res.*, **77**, 6.

101. Harper, R. A., and Posner, A. S. (1966) Measurement of non-crystalline calcium phosphate in bone mineral, *Proc. Soc. Exp. Biol. Med.*, **122**, 137–142.

102. Termine, J. D., and Posner, A. S. (1967) Amorphous/crystalline interrelationships in bone mineral, *Calcif. Tiss. Res.*, **1**, 8–23.

103. Posner, A. S. (1969) Crystal chemistry of bone mineral, *Physiol. Rev.*, **40**, 760–792.

104. Posner, A. S., and Betts, F. (1975) Synthetic amorphous calcium phosphate and its relation to bone mineral structure, *Acc. Chem. Res.*, **8**, 273–281.

105. Betts, F., Blumenthal, N. C., Posner, A. S., Becker, G. L., and Lehninger, A. L. (1975) Atomic structure of intracellular amorphous calcium phosphate deposits, *Proc. Nat. Acad. Sci. USA*, **72**, 2088–2090.

106. Posner, A. S., Betts, F., and Blumenthal, N. C. (1980) Formation and structure of synthetic and bone hydroxyapatite, *Progr. Cryst. Growth Char.*, **3**, 49–64.

107. Bonar, L. C., Roufousse, A. H., Sabine, W. K., Grynpass, M. D., and Glimcher, M. J. (1983) X-ray diffraction studies of the crystallinity of bone mineral in newly synthesized and density fractionated bone, *Calcif. Tiss. Int.*, **35**, 202–209.

108. Molnar, Z. (1959) Development of the parietal bone of young mice I. Crystals of bone mineral in frozen-dried preparations, *J. Ultrastruct. Res.*, **3**, 39–45.

109. Molnar, Z. (1960) Additional observations on bone crystal dimensions, *Clin. Orthop.*, **17**, 38–42.

110. Thyberg, J. (1974) Electron microscopic studies on the initial phases of calcification in guinea pig epiphyseal cartilage, *J. Ultrastruct. Res.*, **46**, 206–218.

111. Gay, C. V. (1977) The ultrastructure of the extracellular phase of bone as observed in frozen thin sections, *Calcif. Tiss. Res.*, **23**, 215–223.

112. Schraer, H., and Gay, C. V. (1977) Matrix vesicles in newly synthesizing bone observed after ultracryotomy and ultramicroincineration, *Calcif. Tiss. Res.*, **23**, 185–188.

113. Nancollas, G. H., and Mohan, M. S. (1970) The growth of hydroxyapatite crystals, *Arch. Oral Biol.*, **15**, 731–745.

114. Blumenthal, N. C., Betts, F., and Posner, A. S. (1981) Formation and structure of Ca-deficient hydroxyapatite, *Calcif. Tiss. Int.*, **33**, 111–117.

115. Meyer, J. L. (1983) Phase transformations in the spontaneous precipitation of calcium phosphate, *Croatica Chim. Acta*, **56**, 753–767.

116. Harries, J. E., Hukins, D. W. L., Holt, C., and Hasnain, S. S. (1987) Conversion of amorphous calcium phosphate into hydroxyapatite investigated by EXAFS spectroscopy, *J. Cryst. Growth*, **84**, 563–570.

117. Lopez-Valero, I., Gomez-Lorente, C., and Boistelle, R. (1992) Effects of sodium and ammonium ions on occurrence., evolution and crystallinity of calcium phosphates, *J. Cryst. Growth*, **121**, 297–304.

118. Lazić, S. (1995) Microcrystalline hydroxyapatite formation from alkaline solutions, *J. Cryst. Growth*, **147**, 147–154.

119. Gadaleta, S. J., Paschalis, E. P., Betts, F., Mendelson, R., and Boskey, A. L. (1996) Fourier transform infrared spectroscopy of the solution-mediated conversion of amorphous calcium phosphate to hydroxyapatite: new correlations between X-ray diffraction and infrared data, *Calcif. Tiss. Int.*, **58**, 9–16.

120. Tarasevich, B. J., Chusuei, C. C., and Allara, D. L. (2003) Nucleation and growth of calcium phosphate from physiological solutions onto self-assembled templates by a solution-formed nucleus mechanism, *J. Phys. Chem. B*, **107**, 10367–10377.

121. Kim, S., Ryu, H. S., Jung, H. S., and Hong, K. S. (2004) Influence of Ca/P ratios of starting solutions on the crystallization of amorphous calcium phosphate to hydroxyapatite, *Metals Mater. Int.*, **10**, 171–175.

122. Kim, S., Ryu, H. S., Shin, H., Jung, H. S., and Hong, K. S. (2004) Direct observation of hydroxyapatite nucleation from amorphous phase in a stoichiometric calcium/phosphate aqueous solution, *Chem. Lett.*, **33**, 1292–1293.

123. Wang, L., and Nancollas, G. H. (2008) Calcium orthophosphates: crystallization and dissolution, *Chem. Rev.*, **108**, 4628–4669.

124. Brečević, L. J., and Füredi-Milhofer, H. (1972) Precipitation of calcium phosphates from electrolyte solutions – II. The formation and transformation of the precipitates, *Calcif. Tiss. Res.*, **10**, 82–90.

125. Liu, C., Huang, Y., Shen, W., and Cui, J. (2001) Kinetics of hydroxyapatite precipitation at pH 10 to 11, *Biomaterials*, **22**, 301–306.

126. Boskey, A. L., and Posner, A. S. (1976) Formation of hydroxyapatite at low supersaturation, *J. Phys. Chem.*, **80**, 40–45.

127. Feenstra, T. P., and de Bruyn, P. L. (1981) The Ostwald rule of stages in precipitation from highly supersaturated solutions: a model and

its application to the formation of the nonstoichiometric amorphous calcium phosphate precursor phase, *J. Coll. Interf. Sci.*, **84**, 66–72.

128. Eanes, E. D. (1998) Amorphous calcium phosphate: thermodynamic and kinetic considerations, in Calcium Phosphates in Biological and Industrial Systems (ed. Amjad, Z.), Kluwer, Massachusetts, pp. 21–39.

129. Blumenthal, N. C., Posner, A. S., and Holmes, J. M. (1972) Effect of preparation conditions on the properties and transformation of amorphous calcium phosphate, *Mater. Res. Bull.*, **7**, 1181–1189.

130. Boskey, A. L., and Posner, A. S. (1973) Conversion of amorphous calcium phosphate to microcrystalline hydroxyapatite. A pH-dependent., solution-mediated., solid-solid conversion, *J. Phys. Chem.*, **77**, 2313–2317.

131. Li, Y., and Weng, W. (2007) *In vitro* synthesis and characterization of amorphous calcium phosphates with various Ca/P atomic ratios, *J. Mater. Sci. Mater. Med.*, **18**, 2303–2308.

132. Zyman, Z. Z., Rokhmistrov, D. V., and Glushko, V. I. (2010) Structural and compositional features of amorphous calcium phosphate at the early stage of precipitation, *J. Mater. Sci. Mater. Med.*, **21**, 123–130.

133. Holt, C., van Kemenade, M. J. J. M., Nelson, L. S., Jr., Hukins, D. W. L., Bailey, R. T., Harries, J. E., Hasnain, S. S., and de Bruyn, P. L. (1989) Amorphous calcium phosphates prepared at pH 6.5 and 6.0, *Mater. Res. Bull.*, **23**, 55–62.

134. Eanes, E. D., Gillessen, I. H., and Posner, A. S. (1965) Intermediate states in the precipitation of hydroxyapatite, *Nature*, **208**, 365–367.

135. Urch, H., Vallet-Regi, M., Ruiz, L., Gonzalez-Calbet, J. M., and Epple, M. (2009) Calcium phosphate nanoparticles with adjustable dispersability and crystallinity, *J. Mater. Chem.*, **19**, 2166–2171.

136. Kim, S., Ryu, H. S., Shin, H., Jung, H. S., and Hong, K. S. (2005) *In situ* observation of hydroxyapatite nanocrystal formation from amorphous calcium phosphate in calcium-rich solutions, *Mater. Chem. Phys.*, **91**, 500–506.

137. Termine, J. D., and Eanes, E. D. (1972) Comparative chemistry of amorphous and apatitic calcium phosphate preparations, *Calcif. Tiss. Res*, **10**, 171–197.

138. Zahidi, E., Lebugle, A., and Bonel, G. (1985) Sur une nouvelle classe de materiaux pour protheses osseuses ou dentaires. [A new class of materials for bone or dental prostheses], *Bull. Sot. Chim. Fr.*, **4**, 523–527.

139. Lebugle, A., Zahidi, E., and Bonel, G. (1986) Effect of structure and composition on the thermal decomposition of calcium phosphates (Ca/P = 1.33), *React Solids*, **2**, 151–161.

140. Layrolle, P., and Lebugle, A. (1994) Characterization and reactivity of nanosized calcium phosphates prepared in anhydrous ethanol, *Chem. Mater.*, **6**, 1996–2004.

141. Layrolle, P., Ito, A., and Tateishi, T. (1998) Sol-gel synthesis of amorphous calcium phosphate and sintering into microporous hydroxyapatite bioceramics, *J. Am. Ceram. Soc.*, **81**, 1421–1428.

142. Rodrigues, A., and Lebugle, A. (1999) Behavior in wet atmosphere of an amorphous calcium phosphate with an atomic Ca/P ratio of 1.33, *J. Solid State Chem.*, **148**, 308–315.

143. Ohta, M., Honma, T., Umesaki, M., and Nakahira, A. (2006) Synthesis and evaluation of amorphous calcium phosphate (ACP) by various synthesis methods, *Key Eng. Mater.*, 309–311, 175–178.

144. Li, Y. B., Weng, W. J., Cheng, K., Du, P. Y., Shen, G., and Han, G. R. (2004) Complexes of Ca(II) with polymers as precursors for preparation of amorphous calcium phosphate, *Mater. Sci. Technol.*, **20**, 1075–1078.

145. Bow, J. S., Liou, S. C., and Chen, S. Y. (2004) Structural characterization of room-temperature synthesized nano-sized β-tricalcium phosphate, *Biomaterials*, **25**, 3155–3161.

146. Li, Y. B., Weng, W. J., Cheng, K., Du, P. Y., Shen, G., Wang, J., and Han, G. R. (2003) Preparation of amorphous calcium phosphate in the presence of poly(ethylene glycol), *J. Mater. Sci. Lett.*, **22**, 1015–1016.

147. Wang, R., Weng, W., Deng, X., Cheng, K., Liu, X., Du, P., Shen, G., and Han, G. (2006) Dissolution behavior of submicron biphasic tricalcium phosphate powders, *Key Eng. Mater.*, 309–311, 223–226.

148. Li, Y., Wiliana, T., and Tam, K. C. (2007) Synthesis of amorphous calcium phosphate using various types of cyclodextrins, *Mater. Res. Bull.*, **42**, 820–827.

149. Tao, J., Pan, H., Zhai, H., Wang, J., Li, L., Wu, J., Jiang, W., Xu, X., and Tang, R. (2009) Controls of tricalcium phosphate single-crystal formation from its amorphous precursor by interfacial energy, *Cryst. Growth Des.*, **9**, 3154–3160.

150. Liu, S., Weng, W., Li, Z., Pan, L., Cheng, K., Song, C., Du, P., Shen, G., and Han, G. (2009) Effect of PEG amount in amorphous calcium phosphate on its crystallized products, *J. Mater. Sci. Mater. Med.*, **20**, 359–363.

151. Li, Y., Weng, W., and Tam, K. C. (2007) Novel highly biodegradable biphasic tricalcium phosphates composed of α-tricalcium phosphate and β-tricalcium phosphate, *Acta Biomater.*, **3**, 251–254.

152. Li, Y., Li, D., and Weng, W. (2008) *In vitro* dissolution behavior of biphasic tricalcium phosphate composite powders composed of α-tricalcium phosphate and β-tricalcium phosphate, *Key Eng. Mater.*, 368–372, 1206–1208.

153. Rodrigues, A., and Lebugle, A. (1998) Influence of ethanol in the precipitation medium on the composition., structure and reactivity of tricalcium phosphate, *Colloids Surf. A*, **145**, 191–204.

154. Heughebaert, J. C., and Montel, G. (1982) Conversion of amorphous tricalcium phosphate into apatitic tricalcium phosphate, *Calcif. Tiss. Int.*, **34**, S103–S108.

155. Yu, T., Ye, J., and Wang, Y. (2009) Synthesis and property of a novel calcium phosphate cement, *J. Biomed. Mater. Res. B* (*Appl. Biomater.*), **90B**, 745–751.

156. Tofighi, A., and Palazzolo, R. (2005) Calcium phosphate bone cement preparation using mechano-chemical process, *Key Eng. Mater.*, 284–286, 101–104.

157. Gbureck, U., Grolms, O., Barralet, J. E., Grover, L. M., and Thull, R. (2003) Mechanical activation and cement formation of β-tricalcium phosphate, *Biomaterials*, **24**, 4123–4131.

158. Gbureck, U., Barralet, J. E., Radu, L., Klinger, H., and Thull, R. (2004) Amorphous α-tricalcium phosphate: preparation and aqueous setting reaction, *J. Am. Ceram. Soc.*, **87**, 1126–1132.

159. Gbureck, U., Hofmann, M. P., and Barralet, J. E. (2005) Thermal performance of mechanically activated tetracalcium phosphate, *J. Am. Ceram. Soc.*, **88**, 1327–1330.

160. Vaidya, S. N., Karunakaran, C., Pande, B. M., Gupta, N. M., Iyer, R. K., and Karweer, S. B. (1997) Pressure-induced crystalline to amorphous transition in hydroxylapatite, *J. Mater. Sci.*, **32**, 3213–3217.

161. Vaidya, S. N., and Sugandhi, V. (1999) Pressure induced amorphization in calcium phosphates, *J. Mater. Sci.*, **34**, 3769–3778.

162. Lemons, J. E. (1988) Hydroxyapatite coatings, *Clin. Orthop.*, **235**, 220–223.

163. Zyman, Z. Z., Weng, J., Liu, X., Zhang, X., and Ma, Z. (1993) Amorphous phase and morphological structure of hydroxyapatite plasma coatings, *Biomaterials*, **14**, 225–228.

164. Weng, J., Liu, X., Zhang, X., Ma, Z., Ji, X., and Zyman, Z. Z. (1993) Further studies on the plasma-sprayed amorphous phase in hydroxyapatite coatings and its deamorphization, *Biomaterials*, **14**, 578–582.

165. Tong, W., Chen, J., and Zhang, X. (1995) Amorphorization and recrystallization during plasma spraying of hydroxyapatite, *Biomaterials*, **16**, 829–832.

166. Weng, J., Liu, X. G., Li, X. D., and Zhang, X. D. (1995) Intrinsic factors of apatite influencing its amorphization during plasma-spray coating, *Biomaterials*, **16**, 39–44.

167. Gross, K. A., Berndt, C. C., and Herman, H. (1998) Amorphous phase formation in plasma-sprayed hydroxyapatite coatings, *J. Biomed. Mater. Res.*, **39**, 407–414.

168. Feng, C. F., Khor, K. A., Kweh, S. W. K., and Cheang, P. (2000) Thermally induced crystallization of amorphous calcium phosphate in plasma-spheroidized hydroxyapatite powders, *Mater. Lett.*, **46**, 229–233.

169. Tong, W., and Li, P. (2007) *In vitro* dissolution of amorphous calcium phosphate (Acp) increased the wear particle generation of plasma-sprayed HA coatings, *Key Eng. Mater.*, 330–332, 561–564.

170. Gross, K. A., and Phillips, M. R. (1998) Identification and mapping of the amorphous phase in plasma-sprayed hydroxyapatite coatings using scanning cathodoluminescence microscopy, *J. Mater. Sci. Mater. Med.*, **9**, 797–802.

171. Carayon, M. T., and Lacout, J. L. (2003) Study of the Ca/P atomic ratio of the amorphous phase in plasma-sprayed hydroxyapatite coatings, *J. Solid State Chem.*, **172**, 339–350.

172. Kumar, R., Cheang, P., and Khor, K. A. (2004) Phase composition and heat of crystallization of amorphous calcium phosphate in ultra-fine radio frequency suspension plasma sprayed hydroxyapatite powders, *Acta Mater.*, **52**, 1171–1181.

173. Keller, L., and Dollase, W. A. (2000) X-ray determination of crystalline hydroxyapatite to amorphous calcium-phosphate ratio in plasma sprayed coatings, *J. Biomed. Mater. Res.*, **49**, 244–249.

174. Gross, K. A., Gross, V., and Berndt, C. C. (1998) Thermal analysis of amorphous phases in hydroxyapatite coatings, *J. Am. Ceram. Soc.*, **81**, 106–112.

175. Döbelin, N., Brunner, T. J., Stark, W. J., Eggimann, M., Fisch, M., and Bohner, M. (2009) Phase evolution of thermally treated amorphous tricalcium phosphate nanoparticles, *Key Eng. Mater.*, 396–398, 595–598.

176. Maciejewski, M., Brunner, T. J., Loher, S. F., Stark, W. J., and Baiker, A. (2008) Phase transitions in amorphous calcium phosphates with different Ca/P ratios, *Thermochim. Acta*, **468**, 75–80.

177. Tisserand, R., Rebetez, M., Grivet, M., Bouffard, S., Benyagoub, A., Levesque, F., and Carpena, J. (2004) Comparative amorphization quantification of two apatitic materials irradiated with heavy ions using XRD and RBS results, *Nucl. Instrum. Methods Phys. Res. B*, **215**, 129–136.

178. Weikusat, C., Glasmacher, U. A., Schuster, B., Trautmann, C., Miletich, R., and Neumann, R. (2011) Raman study of apatite amorphised with swift heavy ions under various irradiation conditions, *Phys. Chem. Minerals*, 38, 293–303.

179. Eanes, E. D., and Posner, A. S. (1968) Intermediate phases in the basic solution preparation of alkaline earth phosphates, *Calcif. Tiss. Res.*, **2**, 38–48.

180. Nylen, M. U., Eanes, E. D., and Termine, J. D. (1972) Molecular and ultrastructural studies of noncrystalline calcium phosphates, *Calcif. Tiss. Res.*, **9**, 95–108.

181. Eanes, E. D., Termine, J. D., and Nylen, M. U. (1973) An electron microscopic study of the formation of amorphous calcium phosphate and its transformation to crystalline apatite, *Calcif. Tiss. Res.*, **12**, 143–158.

182. Eanes, E. D. (1975) Amorphous intermediates in the formation of biological apatites. Physico-chimie et cristallographie des apatites d'intérêt biologique, *Coll. Int. CNRS*, **230**, 295–301.

183. Barton, S. S., and Harrison, B. H. (1976) Surface and bulk properties of amorphous calcium phosphate, in *Colloid and Interface Science* (ed. Kerker, M.), vol. 3, 50th Proceeding Int'l Conf., Academic Press, New York, p. 71.

184. Lundager-Madsen, H. E., Lopez-Valero, I., and Lopez-Acevedo, V. (1986) The formation product of amorphous tricalcium phosphate at 37°C, *J. Cryst. Growth*, **75**, 429–434.

185. Roberts, J. E., Heughebaert, M., Heughebaert, J. C., Bonar, L. C., Glimcher, M. J., and Griffin, R. G. (1991) Solid state ^{31}NMR studies of the conversion of amorphous tricalcium phosphate to apatitic tricalcium phosphate, *Calcif. Tiss. Int.*, **49**, 378–382.

186. Gbureck, U., Barralet, J. E., and Thull, R. (2004) Thermodynamic study of formation of amorphous β-tricalcium phosphate for calcium phosphate cements, *Key Eng. Mater.*, 254–256, 249–252.

187. Brunner, T. J., Bohner, M., Dora, C., Gerber, C., and Stark, W. J. (2007) Comparison of amorphous TCP nanoparticles to micron-sized α-TCP as starting materials for calcium phosphate cements, *J. Biomed. Mater. Res. B* (*Appl. Biomater.*), **83B**, 400–407.

188. Brunner, T. J., Grass, R. N., Bohner, M., and Stark, W. J. (2007) Effect of particle size, crystal phase and crystallinity on the reactivity of tricalcium phosphate cements for bone reconstruction, *J. Mater. Chem.*, **17**, 4072–4078.

189. Somrani, S., Rey, C., and Jemal, M. (2003) Thermal evolution of amorphous tricalcium phosphate, *J. Mater. Chem.*, **13**, 888–892.

190. Somrani, S., Banu, M., Jemal, M., and Rey, C. (2005) Physico-chemical and thermochemical studies of the hydrolytic conversion of amorphous tricalcium phosphate into apatite, *J. Solid State Chem.*, **178**, 1337–1348.

191. Mohn, D., Ege, D., Feldman, K., Schneider, O. D., Imfeld, T., Boccaccini, A. R., and Stark, W. J. (2010) Spherical calcium phosphate nanoparticle fillers allow polymer processing of bone fixation devices with high bioactivity, *Polym. Eng. Sci.*, **50**, 952–960.

192. Dekker, R. J., de Bruijn, J. D., Stigter, M., Barrere, F., Layrolle, P., and van Blitterswijk, C. A. (2005) Bone tissue engineering on amorphous carbonated apatite and crystalline octacalcium phosphate-coated titanium discs, *Biomaterials*, **26**, 5231–5239.

193. Amin, M. S., Randeniya, L. K., Bendavid, A., Martin, P. J., and Preston, E. W. (2009) Amorphous carbonated apatite formation on diamond-like carbon containing titanium oxide, *Diamond and Related Materials*, **18**, 1139–1144.

194. Imai, H., Kusunoki, M., Hashimoto, Y., Nishikawa, H., and Hontsu, S. (2007) Evaluation of biological molecular adsorption on hydroxyapatite and amorphous $Ca_{10}(PO_4)_6(OH)_2$ thin films using QCM method, *IEEJ Trans. EIS*, **127**, 1839–1842.

195. Holt, C., Wahlgren, N. M., and Drakenberg, T. (1996) Ability of a β-casein phosphopeptide to modulate the precipitation of calcium phosphate by forming amorphous dicalcium phosphate nanoclusters, *Biochem. J.*, **314**, 1035–1039.

196. Eanes, E. D., and Posner, A. S. (1965) Kinetics and mechanism of conversion of noncrystalline calcium phosphate to crystalline hydroxyapatite, *Trans. NY Acad. Sci.*, **28**, 233–241.

197. Eanes, E. D. (1970) Thermochemical studies on amorphous calcium phosphate, *Calcif. Tiss. Res.*, **5**, 133–145.

198. Oniki, T., and Oyamada, M. (1983) Surface structure of amorphous calcium phosphate by ESR of VO^{2+} adsorbed on it, *Calcif. Tiss. Int.*, **35**, 477–480.

199. Boulet, M., and Marier, J. R. (1961) Precipitation of calcium phosphates from solutions at near physiological concentrations, *Arch. Biochem.*, **98**, 157–165.

200. Newesely, H. (1961) Changes in crystal types of low solubility calcium phosphates in the presence of accompanying ions, *Arch. Oral Biol.*, spec. suppl. 6, 174–180.

201. Vuilleumier, C., and Lerch, P. (1996) Étude de la structure, par la diffraction des rayons X et la microscopic é1ectronique, de l'hydroxylapatite calcique et des orthophosphates dits tri- et octacalcique, *Helv. Chim. Acta*, **49**, 663–670.

202. Christoffersen, J., Christoffersen, M. R., Kibalczyc, W., and Andersen, F. A. (1989) A contribution to the understanding of the formation of calcium phosphates, *J. Cryst. Growth*, **94**, 767–777.

203. Christoffersen, J., Christoffersen, M. R., and Kibalczyc, W. (1990) Apparent solubilities of two amorphous calcium phosphates and of octacalcium phosphate in the temperature range 30-42°C, *J. Cryst. Growth*, **106**, 349–354.

204. Kibalczyc, W., Christoffersen, J., Christoffersen, M. R., Zielenkiewicz, A., and Zielenkiewicz, W. (1990) The effect of magnesium ions on the precipitation of calcium phosphates, *J. Cryst. Growth*, **106**, 355–366.

205. Francis, M. D., and Webb, N. C. (1971) Hydroxyapatite formation from a hydrated calcium monohydrogen phosphate precursor, *Calcif. Tiss. Res.*, **6**, 335–342.

206. Walton, A. G., Bodin, W. J., Furedi, H., and Sehwartz, A. (1967) Nucleation of calcium phosphate from solution, *Can. J. Chem.*, **45**, 2695–2701.

207. Meyer, J. L., and Eanes, E. D. (1978) A thermodynamic analysis of the amorphous to crystalline calcium phosphate transformation, *Calcif. Tiss. Res.*, **25**, 59–68.

208. Meyer, J. L. (1979) Hydroxyl content of solution-precipitated calcium phosphates, *Calcif. Tiss. Int.*, **27**, 153–160.

209. Jaeger, C., Maltsev, S., and Karrasch, A. (2006) Progress of structural elucidation of amorphous calcium phosphate (ACP) and hydroxyapatite (HAp): disorder and surfaces as seen by solid state NMR, *Key Eng. Mater.*, 309–311, 69–72.

210. Nelson, L. S. Jr., Holt, C., Harries, J. E., and Hukins, D. W. L. (1989) Amorphous calcium phosphates of different composition give very similar EXAFS spectra, *Physica B*, **158**, 105–106.

211. Wuthier, R. E., Rice, G. S., Wallace, J. E. B. Jr., Weaver, R. L., LeGeros, R. Z., and Eanes, E. D. (1985) *In vitro* precipitation of calcium phosphate

under intracellular conditions: formation of brushite from an amorphous precursor in the absence of ATP, *Calcif. Tiss. Int.*, **37**, 401–410.

212. Holmes, J. M., and Beebe, R. A. (1971) Surface areas by gas adsorption on amorphous calcium phosphate and crystalline hydroxyapatite, *Calcif. Tiss. Res.*, **7**, 163–174.
213. Sedlak, J. M., and Beebe, R. A. (1974) Temperature programmed dehydration of amorphous calcium phosphate, *J. Coll. Interf. Sci.*, **47**, 483–489.
214. LeGeros, R. Z., Shirra, W. P., Miravite, M. A., and LeGeros, J. P. (1975) Amorphous calcium phosphates: synthetic and biological. Physico-chimie et cristallographie des apatites d'intérêt biologique, *Coll. Int. CNRS*, **230**, 105–115.
215. LeGeros, R. Z. (1991) *Calcium Phosphates in Oral Biology and Medicine* (*Monographs in Oral Science*), vol. 15, Karger, Basel, p. 201.
216. Boskey, A. L., and Posner, A. S. (1974) Magnesium stabilization ob amorphous calcium phosphate: a kinetic study, *Mater. Res. Bull.*, **9**, 907–916.
217. Blumenthal, N. C., Betts, F., and Posner, A. S. (1977) Stabilization of amorphous calcium phosphate by Mg and ATP, *Calcif. Tiss. Res.*, **23**, 245–250.
218. Abbona, F., and Baronnet, A. (1996) A XRD and TEM study on the transformation of amorphous calcium phosphate in the presence of magnesium, *J. Cryst. Growth*, **165**, 98–105.
219. Fleisch, H., Russell, R. G. G., Bisaz, S., Termine, J. D., and Posner, A. S. (1968) Influence of pyrophosphate on the transformation of amorphous to crystalline calcium phosphate, *Calcif. Tiss. Res.*, **2**, 49–59.
220. Termine, J. D., Peckauskas, R. A., and Posner, A. S. (1970) Calcium phosphate formation *in vitro*. II. Effects of environment on amorphous-crystalline transformation, *Arch. Biochem. Biophys.*, **140**, 318–325.
221. Bienenstock, A., and Posner, A. S. (1968) Calculation of the X-ray intensities from arrays of small crystallites of hydroxyapatite, *Arch. Biochem. Biophys.*, **124**, 604–607.
222. Tropp, J., Blumenthal, N. C., and Waugh, J. S. (1983) Phosphorus NMR study of solid amorphous calcium phosphate, *J. Am. Chem. Soc.*, **105**, 22–26.
223. Fawcett, R. W. (1973) A radial distribution function analysis of an amorphous calcium phosphate with calcium to phosphorus molar ratio of 1.42, *Calcif. Tiss. Res.*, **13**, 319–325.

224. Betts, F., and Posner, A. S. (1974) An X-ray radial distribution study of amorphous calcium phosphate, *Mater. Res. Bull.*, **9**, 353–360.

225. Betts, F., and Posner, A. S. (1974) A structural model for amorphous calcium phosphate, *Trans. Am. Crystal Assoc.*, **10**, 73–84.

226. Eanes, E. D., Powers, L., and Costa, J. L. (1981) Extended X-ray absorption fine structure (EXAFS) studies on calcium in crystalline and amorphous solids of biological interest, *Cell Calcium*, **2**, 251–262.

227. Holt, C., and Hukins, D. W. L. (1991) Structural analysis of the environment of calcium ions in crystalline and amorphous calcium phosphates by X-ray absorption spectroscopy and a hypothesis concerning the biological function of the casein micelle, *Int. Dairy J.*, **1**, 151–165.

228. Termine, J. D., and Lundy, D. R. (1974) Vibrational spectra of some phosphate salts amorphous to X-ray diffraction, *Calcif. Tiss. Res.*, **15**, 55–70.

229. Onuma, K. (2006) Recent research on pseudobiological hydroxyapatite crystal growth and phase transition mechanism, *Prog. Cryst. Growth Charact. Mater.*, **52**, 223–245.

230. Treboux, G., Layrolle, P., Kanzaki, N., Onuma, K., and Ito, A. (2000) Symmetry of Posner's cluster, *J. Am. Chem. Soc.*, **122**, 8323–8324.

231. Georgalis, Y., Kierzek, A. M., and Saenger, W. (2000) Cluster formation in aqueous electrolyte solutions observed by dynamic light scattering, *J. Phys. Chem. B*, **104**, 3405–3406.

232. Treboux, G., Layrolle, P., Kanzaki, N., Onuma, K., and Ito, A. (2000) Existence of Posner's cluster in vacuum, *J. Phys. Chem. A*, **104**, 5111–5114.

233. Kanzaki, N., Treboux, G., Onuma, K., Tsutsumi, S., and Ito, A. (2001) Calcium phosphate clusters, *Biomaterials*, **22**, 2921–2929.

234. Boldeskul, I. E., Sukhodub, L. F., Kalinkevich, A. N., and Khavryutchenko, V. D. (2006) *Ab initio* modelling of calcium phosphate clusters and their vibrational spectra, *Cond. Matter Phys.*, **9**, 669–679.

235. Zahn, D. (2004) Mechanisms of calcium and phosphate ion association in aqueous solution, *Zeitschrift fur Anorganische und Allgemeine Chemie*, **630**, 1507–1511.

236. Wang, C. G., Liao, J. W., Gou, B. D., Huang, J., Tang, R. K., Tao, J. H., Zhang, T. L., and Wang, K. (2009) Crystallization at multiple sites inside particles of amorphous calcium phosphate, *Cryst. Growth Des.*, **9**, 2620–2626.

237. Kojima, Y., Sakama, K., Toyama, T., Yasue, T., and Arai, Y. (1994) Dehydration of water molecules in amorphous calcium phosphate, *Phosphorus Res. Bull.*, **4**, 47–52.

238. Wikholm, N. W., Beebe, R. A., and Kittelberger, J. S. (1975) Kinetics of the conversion of monetite to calcium pyrophosphate, *J. Phys. Chem.*, **79**, 853–856.

239. Eanes, E. D., and Meyer, J. L. (1977) The maturation of crystalline calcium phosphates in aqueous suspensions at physiologic pH, *Calcif. Tiss. Res.*, **23**, 259–269.

240. Meyer, J. L., and Weatherall, C. C. (1982) Amorphous to crystalline calcium phosphate phase transformation at elevated pH, *J. Coll. Interf. Sci.*, **89**, 257–267.

241. Ajibola, V. O., and Thomas, S. A. (1997) Transformation of amorphous calcium phosphate hydroxyapatite in the presence of some ions, *Bull. Chem. Soc. Ethiopia*, **11**, 19–24.

242. Root, M. J. (1990) Inhibition of the amorphous calcium phosphate phase transition reaction by polyphosphates and metal ions, *Calcif. Tiss. Int.*, **47**, 112–116.

243. Wuthier, R. E., and Eanes, E. D. (1975) Effect of phospholipids on the transformation of amorphous calcium phosphate to hydroxyapatite *in vitro*, *Calcif. Tiss. Int.*, **19**, 197–210.

244. Termine, J. D., Eanes, E. D., and Conn, K. M. (1980) Phosphoprotein modulation of apatite crystallization, *Calcif. Tiss. Int.*, **31**, 247–251.

245. Qiu, S. M., Wen, G., Hirakawa, N., Soloway, R. D., Hong, N. K., and Crowther, R. S. (1991) Glycochenodeoxycholic acid inhibits calcium phosphate precipitation *in vitro* by preventing the transformation of amorphous calcium phosphate to calcium hydroxyapatite, *J. Clin. Investigation*, **88**, 1265–1271.

246. Biumenthal, N. C., Betts, F., and Posner, A. S. (1975) Nucleotide stabilization of amorphous calcium phosphate, *Mater. Res. Bull.*, **10**, 1055–1060.

247. Termine, J. D., and Conn, K. M. (1976) Inhibition of apatite formation by phosphorylated metabolites and macromolecules, *Calcif. Tiss. Res.*, **22**, 149–157.

248. Tung, M. S., and Brown, W. E. (1983) An intermediate state in hydrolysis of amorphous calcium phosphate, *Calcif. Tiss. Int.*, **35**, 783–790.

249. Kazanci, M., Fratzl, P., Klaushofer, K., and Paschalis, E. P. (2006) Complementary information on *in vitro* conversion of amorphous (precursor) calcium phosphate to hydroxyapatite from Raman microspectroscopy and wide-angle X-ray scattering, *Calcif. Tiss. Int.*, **79**, 354–359.

250. Pekounov, Y., and Petrov, O. E. (2008) Bone resembling apatite by amorphous-to-crystalline transition driven self-organisation, *J. Mater. Sci. Mater. Med.*, **19**, 753–759.

251. Tao, J., Pan, H., Wang, J., Wu, J., Wang, B., Xu, X., and Tang, R. (2008) Evolution of amorphous calcium phosphate to hydroxyapatite probed by gold nanoparticles, *J. Phys. Chem. C*, **112**, 14929–14933.

252. Rabadjieva, D., Gergulova, R., Titorenkova, R., Tepavitcharova, S., Dyulgerova, E., Balarew, C., and Petrov, O. (2010) Biomimetic transformations of amorphous calcium phosphate: kinetic and thermodynamic studies, *J. Mater. Sci. Mater. Med.*, **21**, 2501–2509.

253. Pan, H., Liu, X. Y., Tang, R., and Xu, H. Y. (2010) Mystery of the transformation from amorphous calcium phosphate to hydroxyapatite, *Chem. Comm.*, **46**, 7415–7417.

254. Onuma, K., and Ito, A. (1998) Cluster growth model for hydroxyapatite, *Chem. Mater.*, **10**, 3346–3351.

255. Oyane, A., Onuma, K., Kokubo, T., and Ito, A. (1999) Clustering of calcium phosphate in the system $CaCl_2$–H_3PO_4–KCl–H_2O, *J. Phys. Chem. B*, **103**, 8230–8235.

256. Yin, X., and Stott, M. J. (2003) Biological calcium phosphates and Posner's cluster, *J. Chem. Phys.*, **118**, 3717–3723.

257. Eanes, E. D. (1976) The interaction of supersaturated calcium phosphate solutions with apatitic substrates, *Calcif. Tiss. Res.*, **20**, 75–89.

258. Eanes, E. D. (1980) Crystal growth of mineral phases in skeletal tissues, *Progr. Cryst. Growth Character.*, **3**, 3–15.

259. Termine, J. D., and Eanes, E. D. (1974) Calcium phosphate deposition from balanced salt solutions, *Calcif. Tiss. Res.*, **15**, 81–84.

260. Soares, A. M. V., Arana-Chavez, V. E., Reid, A. R., and Katchburian, E. (1992) Lanthanum tracer and freeze-fracture studies suggest that compartmentalization of early bone matrix may be related to initial mineralization, *J. Anat.*, **181**, 345–356.

261. Bonucci, E. (1967) Fine structure of early cartilage calcification, *J. Ultrastruct. Res.*, **20**, 33–50.

262. Anderson, H. C. (1969) Vesicles associated with calcification in the matrix of epiphyscal cartilage, *J. Cell Biol.*, **41**, 59–72.

263. Bernard, G. W., and Pease, D. C. (1969) An electron microscopic study of initial intramembranous osteogenesis, *Am. J. Anat.*, **125**, 271–290.

264. Wuthier, R. E. (1977) Electrolytes of isolated epiphyseal chondrocytes, matrix vesicles, and extracellular fluid, *Calcif. Tiss. Res.*, **23**, 125–133.

265. Wuthier, R. E., and Gore, S. T. (1977) Partition of inorganic ions und phospholipids in isolated cell, membrane and matrix vesicle fractions: evidence for Ca-P_i-acidic phospholipid complexes, *Calcif. Tiss. Res.*, **24**, 163–171.

266. Eanes, E. D., Hailer, A. W., and Costa, J. L. (1984) Calcium phosphate formation in aqueous suspensions of multilamellar liposomes, *Calcif. Tiss. Int.*, **36**, 421–430.

267. Sauer, G. R., Zunie, W. B., Durig, J. R., and Wuthier, R. E. (1994) Fourier-transform Raman-spectroscopy of synthetic and biological calcium phosphates, *Calcif. Tiss. Int.*, **54**, 414–420.

268. Sauer, G. R., and Wuthier, R. E. (1988) Fourier-transform infrared characterization of mineral phases formed during induction of mineralization by collagenase-released matrix vesicles *in vitro*, *J. Biol. Chem.*, **263**, 13718–13724.

269. Dorozhkin, S. V. (2010) Bioceramics of calcium orthophosphates, *Biomaterials*, **31**, 1465–1485.

270. Dorozhkin, S. V. (2008) Calcium orthophosphate cements for biomedical application, *J. Mater. Sci.*, **43**, 3028–3057.

271. Dorozhkin, S. V. (2009) Calcium orthophosphate cements and concretes, *Materials*, **2**, 221–291.

272. Maxian, S. H., Zawadsky, J. P., and Dunn, M. G. (1993) *In vitro* evaluation of amorphous calcium phosphate and poorly crysiallized hydroxyapatite coatings on titanium implants, *J. Biomed. Mater. Res.*, **27**, 111–117.

273. Maxian, S. H., Zawadsky, J. P., and Dunn, M. G. (1993) Mechanical and histological evaluation of amorphous calcium phosphate and poorly crystallized hydroxyapatite coatings on titanium implants, *J. Biomed. Mater. Res.*, **27**, 717–728.

274. Garcia, F., Arias, J. L., Mayor, B., Pou, J., Rehman, I., Knowles, J., Best, S. M., León, B., Pérez-Amor, M., and Bonfield, W. (1998) Effect of heat treatment on pulsed laser deposited amorphous calcium phosphate coatings, *J. Biomed. Mater. Res.* (*Appl. Biomater.*), **43**, 69–76.

275. Heimann, R. B., and Wirth, R. (2006) Formation and transformation of amorphous calcium phosphates on titanium alloy surfaces during atmospheric plasma spraying and their subsequent *in vitro* performance, *Biomaterials*, **27**, 823–831.

276. Liu, D. M., Chou, H. M., Wu, J. D., and Tung, M. S. (1994) Hydroxyl apatite coating via amorphous calcium phosphate, *Mater. Chem. Phys.*, **37**, 39–44.

277. Nagano, M., Nakamura, T., Kokubo, T., Tanahashi, M., and Ogawa, M. (1996) Differences of bone bonding ability and degradation behaviour *in vivo* between amorphous calcium phosphate and highly crystalline hydroxyapatite coating, *Biomaterials*, **17**, 1771–1777.

278. Leeuwenburgh, S. C. G., Layrolle, P., Barrère, F., de Bruijn, J., Schoonman, J., van Blitterswijk, C. A., and de Groot, K. (2001) Osteoclastic resorption of biomimetic calcium phosphate coatings *in vitro*, *J. Biomed. Mater. Res.*, **56**, 208–215.

279. Habibovic, P., Barrère, F., van Blitterswijk, C. A., de Groot, K., and Layrolle, P. (2002) Biomimetic hydroxyapatite coating on metal implants, *J. Am. Ceram. Soc.*, **85**, 517–522.

280. Rössler, S., Sewing, A., Stolzel, M., Born, R., Scharnweber, D., Dard, M., and Worch, H. (2003) Electrochemically assisted deposition of thin calcium phosphate coatings at near-physiological pH and temperature, *J. Biomed. Mater. Res. A*, **64A**, 655–663.

281. Lee, D. D., Tofighi, A., Aiolova, M., Chakravarthy, P., Catalano, A., Majahad, A., and Knaack, D. (1999) α-BSM®: a biomimetic bone substitute and drug delivery vehicle, *Clin. Orthop. Relat. Res.*, **367**, S396–S405.

282. Tofighi, A., Mounic, S., Chakravarthy, P., Rey, C., and Lee, D. (2000) Setting reactions involved in injectable cements based on amorphous calcium phosphate, *Key Eng. Mater.*, **192**, 769–772.

283. Wang, X., Ye, J., Wang, Y., Wu, X., Bai, B. (2007) Control of crystallinity of hydrated products in a calcium phosphate bone cement, *J. Biomed. Mater. Res. A*, **81A**, 781–790.

284. van den Vreken, N. M. F., Pieters, I. Y., Declercq, H. A., Cornelissen, M. J., and Verbeeck, R. M. H. (2010) Characterization of calcium phosphate cements modified by addition of amorphous calcium phosphate, *Acta Biomater.*, **6**, 617–625.

285. Drouet, C., Largeot, C., Raimbeaux, G., Estournès, C., Dechambre, G., Combes, C., and Rey, C. (2006) Bioceramics: spark plasma sintering (SPS) of calcium phosphates, *Adv. Sci. Technol.*, **49**, 45–50.

286. Mazzaoui, S. A., Burrow, M. F., Tyas, M. J., Dashper, S. G., Eakins, D., and Reynolds, E. C. (2003) Incorporation of casein phosphopeptide – amorphous calcium phosphate into a glass-ionomer cement, *J. Dent. Res.*, **82**, 914–918.

287. Uysal, T., Amasyali, M., Koyuturk, A. E., and Sagdic, D. (2009) Efficiency of amorphous calcium phosphate-containing orthodontic composite and resin modified glass ionomer on demineralization evaluated by a new laser fluorescence device, *Eur. J. Dent.*, **3**, 127–134.

288. Reynolds, E. C. (1998) Anticariogenic complexes of amorphous calcium phosphate stabilized by casein phosphopeptides: a review, *Special Care in Dentistry*, **18**, 8–16.

289. Tung, M. S., and Eichmiller, F. C. (1999) Dental applications of amorphous calcium phosphates, *J. Clinical Dentistry*, **10**, 1–6.

290. Dorozhkin, S. V. (2009) Calcium orthophosphate-based biocomposites and hybrid biomaterials, *J. Mater. Sci.*, **44**, 2343–2387.

291. Uysal, T., Ustdal, A., Nur, M., and Catalbas, B. (2010) Bond strength of ceramic brackets bonded to enamel with amorphous calcium phosphate-containing orthodontic composite, *Eur. J. Orthodontics*, **32**, 281–284.

292. Wei, D., and Zhou, Y. (2009) Characteristic and biocompatibility of the TiO_2-based coatings containing amorphous calcium phosphate before and after heat treatment, *Appl. Surf. Sci.*, **255**, 6232–6239.

293. Dunn, W. J. (2007) Shear bond strength of an amorphous calcium-phosphate-containing orthodontic resin cement, *Am. J. Orthod. Dentofac. Orthoped.*, **131**, 243–247.

294. Keçik, D., Çehreli, S. B., Şar, Ç., and Ünver, B. (2008) Effect of acidulated phosphate fluoride and casein phosphopeptide-amorphous calcium phosphate application on shear bond strength of orthodontic brackets, *Angle Orthod.*, **78**, 129–133.

295. Foster, J. A., Berzins, D. W., and Bradley, T. G. (2008) Bond strength of an amorphous calcium phosphate-containing orthodontic adhesive, *Angle Orthod.*, **78**, 339–344.

296. Uysal, T., Ulker, M., Akdogan, G., Ramoglu, S. I., and Yilmaz, E. (2009) Bond strength of amorphous calcium phosphate-containing orthodontic composite used as a lingual retainer adhesive, *Angle Orthod.*, **79**, 117–121.

297. Uysal, T., Amasyali, M., Koyuturk, A. E., Ozcan, S., and Sagdic, D. (2010) Amorphous calcium phosphate-containing orthodontic composites. Do they prevent demineralisation around orthodontic brackets? *Austral. Orthodontic J.*, **26**, 10–15.

298. Bröchner, A., Christensen, C., Kristensen, B., Tranæus, S., Karlsson, L., Sonnesen, L., Twetman, S. (2011) Treatment of post-orthodontic white spot lesions with casein phosphopeptide-stabilised amorphous calcium phosphate. Clin. Oral Invest. 15, 369–373.

299. Reynolds, E. C. (2008) Calcium phosphate-based remineralization systems: scientific evidence? *Australian Dent. J.*, **53**, 268–273.

300. Al Zraikat, H., Palamara, J. E., Messer, H. H., Burrow, M. F., and Reynolds, E. C. (2011) The incorporation of casein phosphopeptide-amorphous calcium phosphate into a glass ionomer cement, *Dent Mater.* 27, 235–243.

301. Kim, I., Kim, H. J., and Kim, H. M. (2010) Array of amorphous calcium phosphate particles improves cellular activity on a hydrophobic surface, *J. Biomed. Mater. Res. B* (*Appl. Biomater.*), **93B**, 113–121.

302. Sun, W., Zhang, F., Guo, J., Wu, J., and Wu, W. (2008) Effects of amorphous calcium phosphate on periodontal ligament cell adhesion and proliferation *in vitro*, *J. Medical and Biological Eng.*, **28**, 107–112.

303. Llena, C., Forner, L., and Baca, P. (2009) Anticariogenicity of casein phosphopeptide-amorphous calcium phosphate: a review of the literature, *J. Contemp. Dent. Pract.*, **10**, 1–9.

304. Cai, F., Shen, P., Morgan, M. V., and Reynolds, E. C. (2003) Remineralization of enamel subsurface lesions *in situ* by sugar-free lozenges containing casein phosphopeptide-amorphous calcium phosphate, *Australian Dental J.*, **48**, 240–243.

305. Langhorst, S. E., O'Donnell, J. N. R., and Skrtic, D. (2009) *In vitro* remineralization of enamel by polymeric amorphous calcium phosphate composite: quantitative microradiographic study, *Dent. Mater.*, **25**, 884–891.

306. Shen, P., Cai, F., Nowicki, A., Vincent, J., and Reynolds, E.C. (2001) Remineralization of enamel subsurface lesions by sugar-free chewing gum containing casein phosphopeptide-amorphous calcium phosphate, *J. Dent. Res.*, **80**, 2066–2070.

307. Iijima, Y., Cai, F., Shen, P., Walker, G., Reynolds, C., and Reynolds, E. C. (2004) Acid resistance of enamel subsurface lesions remineralized by a sugar-free chewing gum containing casein phosphopeptide-amorphous calcium phosphate, *Caries Res.*, **38**, 551–556.

308. Cai, F., Manton, D. J., Shen, P., Walker, G. D., Cross, K. J., Yuan, Y., Reynolds, C., and Reynolds, E. C. (2007) Effect of addition of citric acid and casein phosphopeptide-amorphous calcium phosphate to a sugar-free chewing gum on enamel remineralization *in situ*, *Caries Res.*, **41**, 377–383.

309. Kumar, V. L. N., Itthagarun, A., and King, N. M. (2008) The effect of casein phosphopeptide-amorphous calcium phosphate on remineralization of artificial caries-like lesions: an *in vitro* study, *Australian Dental J.*, **53**, 34–40.

310. Ranjitkar, S., Rodriguez, J. M., Kaidonis, J. A., Richards, L. C., Townsend, G. C., and Bartlett, D. W. (2009) The effect of casein phosphopeptide-

amorphous calcium phosphate on erosive enamel and dentine wear by toothbrush abrasion, *J. Dentistry*, **37**, 250–254.

311. Ranjitkar, S., Narayana, T., Kaidonis, J. A., Hughes, T. E., Richards, L. C., and Townsend, G. C. (2009) The effect of casein phosphopeptide-amorphous calcium phosphate on erosive dentine wear, *Australian Dental J.*, **54**, 101–107.

312. Wegehaupt, F. J., and Attin, T. (2010) The role of fluoride and casein phosphopeptide/amorphous calcium phosphate in the prevention of erosive/abrasive wear in an *in vitro* model using hydrochloric acid, *Caries Res.*, **44**, 358–363.

313. Al-Mullahi, A. M., and Toumba, K. J. (2010) Effect of slow-release fluoride devices and casein phosphopeptide/amorphous calcium phosphate nanocomplexes on enamel remineralization *in vitro*, *Caries Res.*, **44**, 364–371.

314. Giniger, M., MacDonald, J., Spaid, M., and Felix, H. (2005) A 180-day clinical investigation of the tooth whitening efficacy of a bleaching gel with added amorphous calcium phosphate, *J. Clinical Dentistry*, **16**, 11–16.

315. Giniger, M., MacDonald, J., Ziemba, S., and Felix, H. (2005) The clinical performance of professionally dispensed bleaching gel with added amorphous calcium phosphate, *J. Am. Dental Association*, **136**, 383–392.

316. Ramalingam, L., Messer, L. B., and Reynolds, E. C. (2005) Adding casein phosphopeptide-amorphous calcium phosphate to sports drinks to eliminate *in vitro* erosion, *Pediatric Dentistry*, **27**, 61–67.

317. Panich, M., and Poolthong, S. (2009) The effect of casein phosphopeptide-amorphous calcium phosphate and a cola soft drink on *in vitro* enamel hardness, *J. Am. Dental Assoc.*, **140**, 455–460.

318. Silva, K. G., Pedrini, D., Delbem, A. C. B., Ferreira, L., and Cannon, M. (2010) *In situ* evaluation of the remineralizing capacity of pit and fissure sealants containing amorphous calcium phosphate and/or fluoride, *Acta Odontologica Scandinavica*, **68**, 11–18.

319. Bayrak, S., Tunc, E. S., Sonmez, I. S., Egilmez, T., and Ozmen, B. (2009) Effects of casein phosphopeptide-amorphous calcium phosphate (CPP-ACP) application on enamel microhardness after bleaching, *Am. J. Dentistry*, **22**, 393–396.

320. Yengopal, V., and Mickenautsch, S. (2009) Caries preventive effect of casein phosphopeptide-amorphous calcium phosphate (CPP-ACP): a meta-analysis, *Acta Odontologica Scandinavica*, **67**, 321–332.

321. Walker, G. D., Cai, F., Shen, P., Reynolds, C., Ward, B., Fone, C., Honda, S., Koganei, M., Oda, M., and Reynolds, E. C. (2006) Increased remineralization of tooth enamel by milk containing added casein phosphopeptide-amorphous calcium phosphate, *J. Dairy Res.*, **73**, 74–78.

322. Walker, G. D., Cai, F., Shen, P., Bailey, D. L., Yuan, Y., Cochrane, N. J., Reynolds, C., and Reynolds, E. C. (2009) Consumption of milk with added casein phosphopeptide-amorphous calcium phosphate remineralizes enamel subsurface lesions *in situ*, *Australian Dental J.*, **54**, 245–249.

323. Willershausen, B., Schulz-Dobrick, B., and Gleissner, C. (2009) *In vitro* evaluation of enamel remineralisation by a casein phosphopeptide-amorphous calcium phosphate paste, *Oral Health & Preventive Dentistry*, **7**, 13–21.

324. Mei, H. L., Chen, L. Y., Zhang, D., Zhang, P. L., Liu, B., Zhao, W., and Qi, H. Y. (2009) Effects of casein phosphopeptide-stabilized amorphous calcium phosphate solution on enamel remineralization, *J. Clin. Rehabil. Tiss. Eng. Res.*, **13**, 4825–4828.

325. Dorozhkin, S. V. (2009) Nanodimensional and nanocrystalline apatites and other calcium orthophosphates in biomedical engineering, biology and medicine, *Materials*, **2**, 1975–2045.

326. Dorozhkin, S. V. (2010) Nanosized and nanocrystalline calcium orthophosphates, *Acta Biomater.*, **6**, 715–734.

327. Suvorova, E. I., and Buffat, P. A. (2001) Size effect in X-ray and electron diffraction patterns from hydroxyapatite particles, *Crystallogr. Rep.*, **46**, 722–729.

328. Suvorova, E. I., and Buffat, P. A. (2001) Electron diffraction and high resolution transmission electron microscopy in the characterization of calcium phosphate precipitation from aqueous solutions under biomineralization conditions, *Eur. Cell Mater.*, **1**, 27–42.

329. Celotti, G., Tampieri, A., Sprio, S., Landi, E., Bertinetti, L., Martra, G., and Ducati, C. (2006) Crystallinity in apatites: how can a truly disordered fraction be distinguished from nanosize crystalline domains? *J. Mater. Sci. Mater. Med.*, **17**, 1079–1087.

330. Tadic, D., Beckmann, F., Schwarz, K., and Epple, M. (2004) A novel method to produce hydroxyapatite objects with interconnecting porosity that avoids sintering, *Biomaterials*, **25**, 3335–3340.

331. Chandanshive, B., Dyondi, D., Ajgaonkar, V. R., Banerjee, R., and Khushalani, D. (2010) Biocompatible calcium phosphate based tubes, *J. Mater. Chem.*, **20**, 6923–6928.

Chapter 3

Nanodimensional and Nanocrystalline Calcium Orthophosphates

3.1 Introduction

Living organisms can create the amazing ways to produce various high-performance materials and over 60 different inorganic minerals of biological origin have already been revealed [1]. Among them, calcium orthophosphates are of a special importance since they are the most important inorganic constituents of hard tissues in vertebrates [2, 3]. In the form of a poor crystalline, non-stoichiometric, ion-substituted CDHA (commonly referred to as "biological apatite"), calcium orthophosphates are present in bones, teeth, deer antlers, and tendons of mammals to give these organs stability, hardness, and function [2, 4, 5]. Through we still do not exactly know why the highly intelligent animals use conformable calcium orthophosphates as their crucial biomineral for survival [6], current biomedical questions of persistent pathological and physiological mineralization in the body force people to focus on the processes, including the occurrence, formation, and degradation of calcium orthophosphates in living organisms [7–9].

Biological mineralization (or biomineralization) is a process of *in vivo* formation of inorganic minerals [1, 2]. In the biomineralization processes, organized assemblies of organic macromolecules regulate nucleation, growth, morphology, and assembly of inorganic crystals. Biologically formed calcium orthophosphates (biological apatite)

Calcium Orthophosphates: Applications in Nature, Biology, and Medicine
Sergey Dorozhkin

ISBN 978-981-4316-62-0 (Hardcover), 978-981-4364-17-1 (eBook)
www.panstanford.com

are always nanodimensional and nanocrystalline, which have been formed *in vivo* under mild conditions. According to many reports, dimensions of biological apatite in the calcified tissues always possess a range of a few to hundreds of nanometers with the smallest building blocks on the nanometer size scale [2, 4, 5, 10, 11]. For example, tens to hundreds of nanometer-sized apatite crystals in a collagen matrix are combined into self-assembled structures during bone and teeth formation [2, 4, 5]. Recent advances suggest that this is a natural selection, since the nanostructured materials provide a better capability for the specific interactions with proteins [12].

Due to the aforementioned, nanodimensional and nanocrystalline forms of calcium orthophosphates are able to mimic both the composition and dimensions of constituent components of the calcified tissues. Thus, they can be utilized in biomineralization and as biomaterials due to the excellent biocompatibility [13, 14]. Further development of calcium orthophosphate-based biomaterials obviously will stand to benefit mostly from nanotechnology [15], which offers unique approaches to overcome shortcomings of many conventional materials. For example, nano-sized ceramics can exhibit significant ductility before failure contributed by the grain-boundary phase. Namely, already in 1987, Karch et al. reported that, with nanodimensional grains, a brittle ceramic could permit a large plastic strain up to 100% [16]. In addition, nanostructured ceramics can be sintered at lower temperatures; thereby major problems associated with a high temperature sintering are also decreased. Thus, nanodimensional and nanocrystalline forms of bioceramics clearly represent a promising class of orthopedic and dental implant formulations with improved biological and biomechanical properties [17].

Many other advances have been made in biomaterial field due to a rapid growth of nanotechnology [18]. For example, a recent theory of "aggregation-based crystal growth" [19] and a new concept of "mesocrystals" [20, 21] highlighted the roles of nano-sized particles in biological crystal engineering. In this aspect, the study of calcium orthophosphates is a specific area in nanotechnology, because they might be applied readily to repair hard skeletal tissues of mammals [22–24].

Herein, an overview of nanodimensional and nanocrystalline apatites and other calcium orthophosphates in studies on biomineralization and biomaterials is given. The available calcium orthophosphates are listed in Table 1.1. To narrow the subject of this chapter, with a few important exceptions, undoped and un-substituted calcium orthophosphates are considered and discussed only. The readers interested in various nanodimensional and nanocrystalline ion-substituted calcium orthophosphates [25–52] are referred to the original publications. Furthermore, details on calcium orthophosphate-based nanodimensional biocomposites [53–71] or nanodimensional calcium orthophosphate-based biocomposites [72–87] are available in both Ref. [88] and section 6.4.3 of this book.

This chapter is organized into several sections. After a brief introduction (current section), general information on "nano" is provided in the second section. The third section briefly compares the micron-sized and nanodimensional calcium orthophosphates. The forth section briefly discusses the presence of nano-sized and nanocrystalline calcium orthophosphates in normal calcified tissues of mammals. The structure of nano-sized and nanocrystalline apatites is described in the fifth section. Synthesis of nanodimensional and nanocrystalline calcium orthophosphates of various dimensions and shapes is reviewed in the sixth section, while the biomedical applications are examined in the seventh section. Finally, the summary and reasonable future perspectives in this active research area are given in the last section.

3.2 General Information on "Nano"

The prefix "nano" specifically means a measure of 10^{-9} units. Although it is widely accepted that the prefix "nano" specifically refers to 10^{-9} units, in the context of nano-sized and nanocrystalline materials, the units should only be those of dimensions, rather than of any other unit of the scientific measurements. Besides, for practical purposes, it appears to be unrealistic to consider the prefix "nano" to solely and precisely refer to 10^{-9} m, just as it is not considered that "micro" specifically and solely concerns something with a dimension of precisely 10^{-6} m [89]. Currently, there is a general agreement that the subject of nanoscience and nanotechnology started after the famous talk: "There's plenty of room at the bottom" given by the Nobel Prize winner in physics Prof. Richard P. Feynman on December 26, 1959, at the annual meeting of the American Physical Society held at California Institute of Technology. This well-known talk has been widely published in various media (e.g., [90]).

In a recent extensive discussion about a framework for definitions presented to the European Commission, the nano-scale has been defined as being of the order of 100 nm or less. Similarly, a nanomaterial has been defined as "any form of a material that is composed of discrete functional parts, many of which have one or more dimensions of the order of 100 nm or less" [91]. Other definitions logically follow this approach such as: a nanocrystalline material is "a material that is comprised of many crystals, the majority of which have one or more dimensions of the order of 100 nm or less" (normally, with presence of neither the micron-sized crystals nor an intergranular amorphous phase) and a nanocomposite is a "multi-phase material in which the majority of the dispersed phase components have one or more dimensions of the order of 100 nm or less" [89]. Similarly, nanostructured materials are defined as the materials containing structural elements (e.g., clusters, crystallites or molecules) with dimensions in the 1–100 nm range [92], nanocoatings represent individual layers or multilayer surface coatings of 1–100 nm thick, nanopowders are extremely fine powders with an average particle size in the range of 1–100 nm and nanofibers are the fibers with a diameter within 1–100 nm [93, 94]. It also has been proposed to extend the lower size limit to 0.1 nm [95], which would include all existing organic molecules, allowing chemists to rightly claim they have been working on nanotechnology for very many years [96].

Strictly speaking, there are serious doubts that the term "nanomaterial" has a reasonable meaning. For example, let me cite Prof. David F. Williams, the Editor-in-Chief of *Biomaterials*: "... some words which have no rational basis whatsoever become part of everyday language so rapidly, even if so illogically, that it is impossible to reverse the process and their common use has to be accepted, or perhaps, accommodated. Nanomaterial is one such word, where I have argued that it should not exist, but accept that it does through common usage and have to recognize its existence [89]. The discussion about nanomaterial provides a hint of the analysis of a biomaterial that follows, since a prefix, which is an indicator of scale, cannot specify the integer that follows (in this case a material) unless that integer can be qualified by that scale. In other words, it is very clear what a nanometer is because nano means 10^{-9} and a meter is a measure of length. In the case of nanomaterial, what is it about the material that is 10^{-9}. Is it the dimension of a crystal within the material, or of a grain boundary, a domain, or a molecule, or is it a parameter of a surface feature of the sample, or perhaps of the resistivity or thermal conductivity of the material? Clearly this is nonsense, but one has to accept that nanomaterials are here to stay, with even some journal titles containing the word" [97, p. 5898, left column]. Following this logic, such terms as "nanocomposite," "nanocoatings," "nanopowders," "nanofibers," and "nanocrystals" are senseless either and should be replaced, for example, by "composites with nano-sized (or nanodimensional) dispersed phase(s)," "coatings of nano-sized (or nanodimensional) thickness," "nano-sized (or nanodimensional) powders," "fibers of nano-sized (or nanodimensional) thickness," and "nano-sized (or nanodimensional) crystals," respectively. At least, this has been done in this book.

According to their geometry, all nanodimensional materials can be divided into three major categories: equiaxed, one-dimensional (or fibrous), and two-dimensional (or lamellar) forms. Selected examples and typical applications of each category of nanodimensional materials and their use in biomedical applications are available in literature [98]. It is important to note, that in literature on calcium orthophosphates there are cases, when the prefix "nano" has been applied for the structures, with the minimum dimensions exceeding 100 nm [42, 71, 99–108].

As a rule, nanodimensional materials can be manufactured from nearly any substance. Of crucial importance, there are two major characteristics conferring the special properties of any nanodimensional material. These are the quantum effects associated with the very small dimensions (currently, this is not applicable to the biomaterials field) and a large surface-to-volume ratio that is encountered at these dimensions. For instance, specific surface areas for submicron-sized particles are typically 60–80 m^2/g, while decreasing particle diameter to tens of nanometers increases the specific surface area up to 5 times more — an amazing amount of surface area per mass! Furthermore, all nanophase materials have the unique surface properties, such as an increased number of grain boundaries and defects on the surface, huge surface area and altered electronic structure, if compared to the micron-sized materials [89, 109]. While less than ~1% of a micron-sized particle's atoms occupy the surface positions, over a tenth of the atoms in a 10 nm diameter particle reside on its surface and ~60% in a 2 nm particle [110]. This very high surface-to-volume ratio of nanodimensional materials provides a tremendous driving force for diffusion, especially at elevated temperatures, as well as causes a self-aggregation into larger particles. Besides, solubility of many substances increases with particle size decreasing [111, 112]. What's more, nanophase materials could have surface features (e.g., a higher amount of nano-scale pores) to influence the type and amount of adsorption of selective proteins that could enhance specific osteoblast adhesion [113]. Finally and yet importantly, the nanodimensional and nanocrystalline materials have different mechanical, electrical, magnetic, and optical properties if compared to the larger grained materials of the same chemical composition [114–117].

The nanostructured materials can take the form of powders, dispersions, coatings or bulk materials. In general, nanostructured materials contain a large volume fraction (greater than 50%) of defects such as grain boundaries, interphase boundaries, and dislocations, which strongly influence their chemical and physical properties. The great advantages of nanostructuring were first understood in electronic industry with the advent of thin film deposition processes. Other application areas have followed. For example, nanostructured bioceramics was found to improve friction and wear problems associated with joint replacement components because it was tougher and stronger than coarser-grained bioceramics [118]. Furthermore, nanostructuring has allowed chemical homogeneity and structural uniformity to an extent, which was once thought to impossible to achieve [92]. In calcium orthophosphate bioceramics, the major target of nanostructuring is to mimic the architecture of bones and teeth [119, 120].

3.3 Micron- and Submicron-Sized Calcium Orthophosphates Versus the Nanodimensional Ones

The micron-sized calcium orthophosphate-based bioceramic powders suffer from poor sinterability, mainly due to a low surface area (typically 2–5 m^2/g), while the specific surface area of nanodimensional calcium orthophosphates exceeds 100 m^2/g [121]. In addition, the resorption process of synthetic micron-sized calcium orthophosphates was found to be quite different from that of bone mineral [122].

Although the nanodimensional and nanocrystalline features of natural calcium orthophosphates of bones and teeth had been known earlier [2, 123–128], the history of the systematic investigations of this field has started only in 1994. Namely, a careful search in scientific databases using various combinations of keywords "nano" + "calcium phosphate," "nano" + "apatite," "nano" + "hydroxyapatite," etc., in the article title revealed 5 papers published in 1994 [129–133]. No papers published before 1994 with the aforementioned keywords in the title have been found.

Nanodimensional (size ~67 nm) HA was found to have a higher surface roughness of 17 nm if compared to 10 nm for the submicron-sized (~180 nm) HA, while the contact angles (a quantitative measure of the wetting of a solid by a liquid) were significantly lower for nano-sized HA (6.1) if compared to the submicron-sized HA (11.51). Additionally, the diameter of individual pores in nanodimensional HA compacts is several times smaller (pore diameter ~6.6 Å) than that in the submicron grain-sized HA compacts (pore diameter within 19.8–31.0 Å) [134]. A surface roughness is known to enhance the osteoblast functions while a porous structure improves the osteoinduction compared with smooth surfaces and nonporpous structure, respectively [113]. Furthermore, nanophase HA appeared to have ~11% more proteins of fetal bovine serum adsorbed per 1 cm^2 than submicron-sized HA [135]. Interfacial interactions between calcined HA nano-sized crystals and various substrates were studied and a bonding strength appeared to be influenced not only by the nature of functional groups on the substrate but also by matching of surface roughness between the nano-sized crystals and the substrate [136]. More to the point, incorporating of nanodimensional particles of HA into polyacrylonitrile fibers were found to result in their crystallinity degree rising by about 5% [137].

In general, nanostructured biomaterials [138] offer much improved performances than their larger particle sized counterparts due to their huge surface-to-volume ratio and unusual chemical synergistic effects. Such nanostructured systems constitute a bridge between single molecules and bulk material systems [139]. For instance, powders of nanocrystalline apatites [140–146] and β-TCP [147] were found to exhibit an improved sinterability and enhanced densification due to a greater surface area. This is explained by the fact that the distances of material transport during the sintering becomes shorter for ultrafine powders with a high specific surface area, resulting in a densification at a low temperature. Therefore, due to low grain growth rates, a low-temperature sintering appears to be effective to produce fine-grained apatite bioceramics [148]. Furthermore, the mechanical properties (namely, hardness and toughness) of HA bioceramics appeared to increase as the grain size decreased from sub-micrometers to nanometers [149].

More to the point, nano-sized HA is also expected to have a better bioactivity than coarser crystals [150–152]. Namely, Kim et al. found that osteoblasts (bone-forming cells) attached to the nano-sized HA/gelatin biocomposites to a significantly higher degree than to micrometer size analog did [153]. An increased osteoblast and decreased fibroblast (fibrous tissue-forming cells) adhesion on nanophase ceramics [134, 154–158], as well as on nanocrystalline HA coatings on titanium, if compared to traditionally used plasma-sprayed HA coatings, was also discovered by other researchers [159–161]. Scientists also observed enhanced osteoclast (bone-resorbing cells) functions to show healthy remodeling of bone at the simulated implant surface [151]. Besides, the proliferation and osteogenic differentiation of periodontal ligament cells were found to be promoted when a nanophase HA was used, if compared to dense HA bioceramics [162]. Thus, the underlying material property, responsible for this enhanced osteoblast function, is the surface roughness of the nanostructured surface [18]. Interestingly, but an increased osteoblast adhesion was discovered on nano-sized calcium orthophosphate powders with higher Ca/P ratios [163], which points out to some advantages of apatites over other calcium orthophosphates. Furthermore, a histological analysis revealed a superior biocompatibility and osteointegration of bone graft substitutes when nano-sized HA was employed in biocomposites [164, 165]. However, data are available that nano-sized HA could inhibit growth of osteoblasts in a dose-dependent manner [166].

Obviously, the volume fraction of grain boundaries in nanodimensional calcium orthophosphates is increased significantly leading to improved osteoblast adhesion, proliferation, and mineralization. Therefore, a composition of these biomaterials at the nano-scale emulates the bone's hierarchic organization, to initiate the growth of an apatite layer and to allow for the cellular and tissue response of bone remodeling. These examples emphasize that nanophase materials deserve more attention in improving orthopedic implant failure rates. However, to reduce surface energy, all nano-sized materials tend to agglomerate, and to avoid self-aggregation of calcium orthophosphate nano-sized particles [167–169], special precautions might be necessary [102, 170–172].

Finally yet importantly, nano-sized crystals of CDHA obtained by precipitation methods in aqueous solutions were shown to exhibit physico-chemical characteristics rather similar to those of bone apatite [173]. In particular, their chemical composition departs from stoichiometry by calcium and hydroxide ions deficiency, leading to an increased solubility, and in turn bioresorption rate *in vivo* [123, 174–176]. The nano-sized crystals of CDHA have also a property to evolve in solution (maturation) like bone crystals. Namely, freshly precipitated CDHA has been shown to be analogous to embryonic bone mineral crystals whereas aged precipitates resemble bone crystals of old vertebrates [173].

3.4 Nanodimensional and Nanocrystalline Calcium Orthophosphates in Calcified Tissues of Mammals

3.4.1 Bones

Bone is the most typical calcified tissue of mammals and comes in all sorts of shapes and sizes in order to achieve various functions of protection and mechanical support for the body. The major inorganic component of bone mineral is a biological apatite, which might be defined as a poorly crystalline, non-stoichiometric, and ion-substituted CDHA [2–5, 174–177]. From the material point of view, bone can be considered as an assembly of distinct levels of seven hierarchical structural units from macro- to micro- and to nano-scale (Fig. 1.11) to meet numerous functions [2, 5, 109, 178–180]. Furthermore, all these levels of bones permanently interact with cells and biological macromolecules. At the nanostructural level, tiny plate-like crystals of biological apatite in bone occur within the discrete spaces within the collagen fibrils and grow with specific crystalline orientation along the *c*-axes, which are roughly parallel to the long axes of the collagen fibrils [181]. Type I collagen molecules are self-assembled into fibrils with a periodicity of ~67 nm and ~40 nm gaps between the ends of their molecules, into which the apatite nano-sized crystals are placed. A biocomposite of these two constituents forms mineralized fibers. The fibers also may be cross-linked, which provides a highly dynamic system capable of modification through the selection of different amino acids to allow for different mechanical properties for different biomaterial applications [182]. This is why bone is usually termed a fiber-reinforced composite of a biological origin, in which nanometer-sized hard inclusions are embedded into a soft protein matrix [183]. Though dimensions of biological apatite crystals reported in the literature vary due to different treatment methods and analytical techniques, it is generally around the nanometric level with values in the ranges of 30–50 nm (length), 15–30 nm (width), and 2–10 nm (thickness) [184]. Why does the nanometer scale appear to be so important to bones? It was recently demonstrated that natural biocomposites exhibit a generic mechanical structure in which the nanometer sizes of mineral particles are used to ensure the optimum strength and maximum tolerance of flaws [185, 186]. Furthermore, nanodimensional apatite has another crucial function for organisms. It is a huge reservoir of calcium and orthophosphate ions necessary for a wide variety of metabolic functions, which offer or consume calcium and orthophosphate ions through a so-called remodeling process because of a continuous resorption and formation of nanodimensional apatite by osteoclasts and osteoblasts, respectively, in a delicate equilibrium [2, 5]. Additional details on the structure, properties, and composition of bones are available in section 1.4.1 of this book, while even more comprehensive information might be found in special literature [5, 177, 187].

3.4.2 Teeth

Teeth are another normal calcium orthophosphate-based calcified tissue of vertebrates. Unlike bone, teeth consist of at least two different biominerals: enamel (a crown, the part above the gum line) and dentin (root, the part below the gum line) [188]. Dental enamel contains up to 98% of biological apatite, ~1% of bioorganic compounds, and up to 2% of water. Typical rods in enamel are composed of rod-like apatite crystals measuring 25–100 nm and an undetermined length of 100 nm to 100 μm or longer along the c-axis [189–191]. However, the apatite crystals in enamel were found to exhibit regular sub-domains or subunits with distinct chemical properties [192]. This subunit structure reflects an assembly mechanism for such biological crystals [193, 194]. Like that for bones (Fig. 1.11), seven levels of structural hierarchy have been also discovered in human enamel; moreover, the analysis of the enamel and bone hierarchical structures suggests similarities of the scale distribution at each level [195]. In enamel, nano-sized crystals of biological apatite at first form mineral nanodimensional fibrils; the latter always align lengthways, aggregating into fibrils and afterwards into thicker fibers; further, prism/interprism continua are formed from the fibers. At the micro-scale, prisms are assembled into prism bands, which present different arrangements across the thickness of the enamel layer. These compositional and structural characteristics endow enamel special properties such as anisotropic elastic modulus, effective viscoelastic properties, much higher fracture toughness and stress–strain relationships more similar to metals than ceramics [196].

Dentin contains ~50% of biological apatite, ~30% of bioorganic compounds, and ~20% of water. In dentin, the nanodimensional building blocks (~25 nm width, ~4 nm thickness, and ~35 nm length) of biological apatite are smaller than those of enamel. Dentin is analogous to bone in many aspects, for example, it has a similar composition and a hierarchical structure up to the level of the bone lamellae [174, 175]. Additional details on the structure, properties, and composition of teeth are available in section 1.4.2 of this book, while even more comprehensive information might be found in special literature [197].

3.5 The Structure of the Nanodimensional and Nanocrystalline Apatites

Due to the apatitic structure on natural calcified tissues, apatites appear to be the best investigated compounds among the available calcium orthophosphates (Table 1.1). Thus, nanodimensional and nanocrystalline apatites have been extensively studied by various physico-chemical techniques and chemical analysis methods [169, 198–210] with a special attention to the "nano" effect (i.e., an enhanced contribution of the surface against the volume). Unfortunately, no publications on the structure of other nanodimensional and/or nanocrystalline calcium orthophosphates were found in the available literature.

Due to a nanocrystalline nature, various diffraction techniques have not yet given much information on the fine structural details related to apatite nano-sized crystals (assemblies of nano-sized particles give only broad diffraction patterns, similar to ones from an amorphous material — see Chapter 2) [198, 199]. Nevertheless, the diffraction studies with electron microprobes of 35 ± 10 nm in diameter clearly indicated a crystalline character of the nano-sized particles in these assemblies. Furthermore, high-resolution transmission electron microscopy results revealed that nano-sized particles of HA behaved a fine monocrystalline grain structure [169, 198].

`Therefore, a recent progress on the structure of nanodimensional and nanocrystalline apatites has relied mainly on diverse spectroscopic methods, which are sensitive to disturbances of the closest environments of various ions. Namely, the structure analysis revealed an existence of structural disorder at the particle surface, which was explained by chemical interactions between the orthophosphate groups and either adsorbed water molecules or hydroxyl groups located at the surface of nano-sized apatites [200]. More to the point, infrared (FTIR) spectra of nanocrystalline apatites, in the ν_4 PO_4 domain, revealed the existence of additional bands of orthophosphate ions which could not be assigned to an apatitic environment and which were not present in well-crystallized apatites (Fig. 3.1). These bands were assigned to non-apatitic environments of PO_4^{3-} and HPO_4^{2-} ions of the nano-sized crystals. Thus, FTIR spectra can be used to provide a sufficiently accurate evaluation of the amounts of such environments. Furthermore, the non-apatitic environments were found to correspond to hydrated domains of the nano-sized crystals, which were distinct from the apatite domains [202]. Hence, precipitated crystals of nano-sized apatite appeared to have a hydrated surface layer containing labile ionic species, which easily and rapidly could be exchanged by ions and/or macromolecules from the surrounding fluids [201, 202, 209]. For the as-precipitated apatites, such a layer appears to constitute mainly by water molecules coordinated to surface Ca^{2+} ions, approximately in the 1 : 1 ratio, while the OH groups account only for ~20% of the surface hydration species. The FTIR data indicated that water molecules, located on the surface of nanodimensional apatites, are coordinated to surface cations and experience hydrogen bonding significantly stronger than that in liquid water [208]. The surface hydrated layer is very delicate and becomes progressively transformed into a more stable apatitic lattice upon ageing in aqueous media. Furthermore, it irreversibly altered upon drying [202]. Outgassing at increasing temperatures up to ~300°C resulted in a complete surface dehydration, accompanied by a decrease of the capability to re-adsorb water. Combination of these data with rehydration tests suggested that a significant part of the surface Ca^{2+} ions, once dehydrated, could undergo a relaxation inward the surface, more irreversibly as the outgassing temperature increased [207].

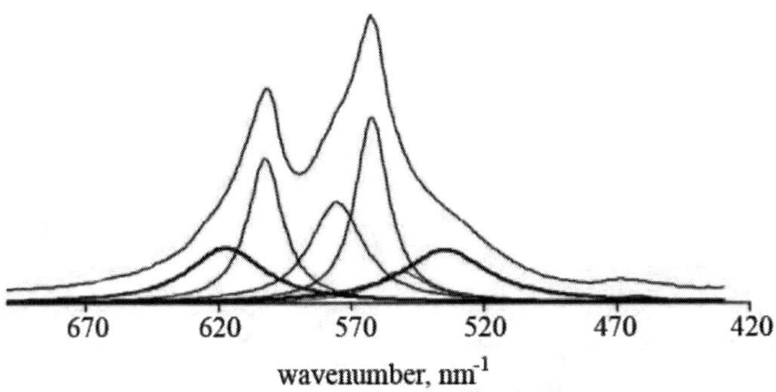

Figure 3.1 FTIR spectra of poorly crystalline apatites showing the non-apatitic environments of the orthophosphate ions (bold lines with peaks at 617 and 534 cm^{-1}) and the apatitic PO_4^{3-} (thin lines with peaks at 600, 575 and 560 cm^{-1}) and HPO_4^{2-} (thin line with peak at 550 cm^{-1}) in the ν_4 PO_4 domain. Reprinted from Ref. [202] with permission.

In another study, elongated nano-sized crystals of CDHA of ~10 nm thick and of ~30–50 nm length were synthesized followed by investigations with X-ray diffraction and nuclear magnetic resonance techniques. The nano-sized crystals of CDHA were shown to consist of a crystalline core with the composition close to the stoichiometric HA and a disordered (amorphous) surface layer of 1–2 nm thick [206, 207] with the composition close to DCPD [205]. Based on the total

Ca/P ratio, on the one hand, and the crystal shape, on another hand, a thickness of the DCPD surface layer along the main crystal axis was estimated to be ~1 nm [205], which is close to dimensions of the unit-cells (Table 1.3). A similar structure of a crystalline core with the composition of the stoichiometric HA and a disordered (amorphous) surface layer was found by other researchers [211]; however, in yet another study devoted to nanodimensional carbonateapatites [212], the model of a crystalline core and an outer amorphous layer was not confirmed. Perhaps, this discrepancy could be explained by the presence of carbonates. A lack of hydroxide in nanodimensional apatites was detected; an extreme nanocrystallinity was found to place an upper bound on OH^- possible in apatites [213].

However, it is possible to address the structure of surface terminations of HA nano-sized particles to be amorphous or crystalline by properly selecting the preparation parameters and, in particular, the temperature; thus, nanodimensional HA without the amorphous layer on the surface has been prepared [214]. The two types of surfaces (amorphous or crystalline) of nanodimensional HA appeared to be quite similar in terms of their first hydration layer, as well as Lewis acid strength of exposed Ca^{2+} ions. Both features have a strong dependence on the local structure of surface sites (well probed by small molecules, such as H_2O and CO) that appeared essentially unaffected by the organization at a longer range. Interestingly, but once treated at 573 K, the crystalline surfaces of nanodimensional HA was found to adsorb multilayers of water in a larger extent than the amorphous ones [214].

Nevertheless, after summarizing the available data, the following statements on the structure of nano-sized crystals of apatites have been made:

(1) they involve non-apatitic anionic and cationic chemical environments (in another study, the researchers mentioned on "ordered and disordered HA" [206]),
(2) at least part of these environments are located on the surface of the nano-sized crystals and are in strong interaction with hydrated domains,
(3) immature samples show FTIR band fine substructure that is altered upon drying without leading to long-range order (LRO) modifications,
(4) this fine substructure shows striking similarities with the FTIR spectrum of OCP [203].

All these elements favor a model in which nano-sized crystals of apatites are covered with a rather fragile but structured surface hydrated layer containing relatively mobile ions (mainly, bivalent anions and cations: Ca^{2+}, HPO_4^{2-}, CO_3^{2-}) in "non-apatitic" sites (Fig. 3.2), which is supposed to be of either OCP or DCPD structure. Unfortunately, both the exact structure and the chemical composition of this hydrated layer are still uncertain (regrettably, as the hydrated layer cannot be isolated, it is not possible to standardize the methods for detailed studies) [203, 205–207]. Nevertheless, it is known that the surface layer might adsorb considerable amounts of foreign compounds (molecules and ions) in the percent mass range [215]. Strictly speaking, all the aforementioned apply to both biological apatite of calcified tissues [216] and micron-sized apatites as well [217]; nonetheless, in nano-sized crystals, the composition of the hydrated surface layer contributes to the global composition for a non-negligible proportion. The results of electron states spectroscopy of nanostructural HA bioceramics are available elsewhere [218, 219].

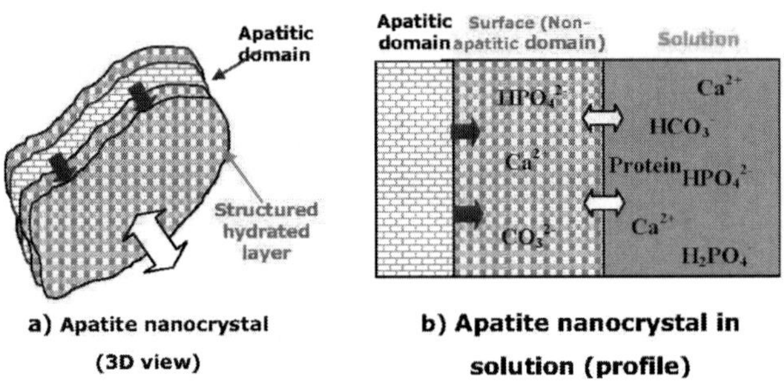

Figure 3.2 A schematic representation of the "surface hydrated layer model" for poorly crystalline apatite nanocrystals. Reprinted from Ref. [203] with permission.

The hydrated surface layer confers unexpected properties to nano-sized apatite, is responsible for most of the properties of apatites, and, for example, can help to explain the regulation by biological apatites of the concentration in mineral ions in body fluids (homeostasis). These properties are important for living organisms; therefore, they need to be used in both material science and biotechnology [202]. The consideration of this type of surface state can help understanding and explaining the behavior of biological apatites in participating in homeostasis due to a very high specific surface area of bone crystals and in constituting an important ion reservoir with an availability that depends on the maturation state. The important consequences are that the surface of nanodimensional apatites has nothing in common with the bulk composition and that the chemistry of such materials (e.g., binding of protein molecules) must be reconsidered [203, 205]. Interestingly, but, in response to an electrical potential, the surface of nano-sized HA bioceramics was found to exhibit dynamic changes in interfacial properties, such as wettability. The wettability modification enabled both a sharp switching from hydrophilic to hydrophobic states and a microscopic wettability patterning of the HA surface, which may be used for fabrication of spatially arrayed HA for biological cells immobilization or gene transfer [220].

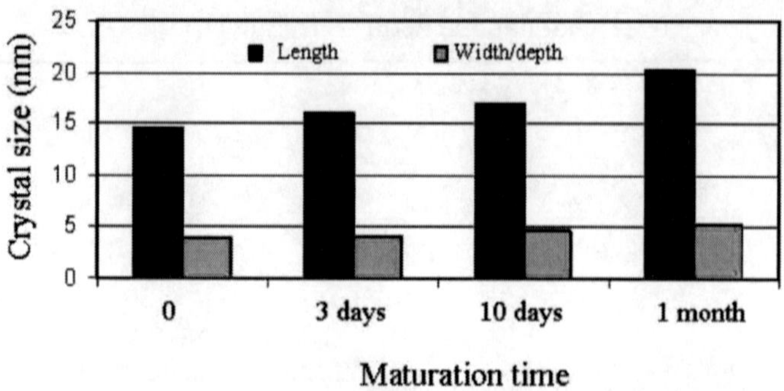

Figure 3.3 Variation of nanocrystalline apatite dimensions with maturation time. Reprinted from Ref. [203] with permission.

Furthermore, dry powders of nanodimensional HA were found to contain an X-ray amorphous portion with an unspecified location [221]. After mixing of an initial nano-sized HA powder with a physiological solution (aqueous isotonic 0.9% NaCl solution for injections), this amorphous portion was fully converted into the crystalline phase of HA. The initial crystallite average size (~35 nm) was enlarged by a factor of about four within the first 100 min after mixing the powder with the physiological solution and no more structural changes were detected during the following period [221]. In the light of the aforementioned studies, presumably, the discovered X-ray amorphous component of the initial powder was located on the surface of nanodimensional HA.

3.6 Synthesis of the Nanodimensional and Nanocrystalline Calcium Orthophosphates

3.6.1 General Nanotechnological Approaches

The synthesis of nano-scale materials has received considerable attention and their novel properties can find numerous applications, for example, in the biomedical field. This has encouraged the invention of chemical, physical, and biomimetic methods by which such nano-sized materials can be obtained [109]. Generally, all approaches for preparation of nanodimensional and nanocrystalline materials can be categorized as "bottom-up" and "top-down" ones [117, 222]. The bottom-up approach refers to the build up of a material from the bottom, i.e., atom by atom, molecule by molecule or cluster by cluster and then assembles them into the final nanostructured material. An example is production of a nano-sized powder and its compaction into the final product (e.g., hot-pressed or sintered nanostructured ceramics). The top–down approach starts from a bulk material and then, via different dimension decreasing techniques, such as milling, slicing, or successive cutting, leads to the formation of nanodimensional materials [109]. Using this approach, a novel two-dimensional carbon material graphene of just 1 atom thick has been prepared from bulk graphite.

Concerning calcium orthophosphates, presumably, all of them (see Table 1.1) might be manufactured in a nanodimensional and/or a nanocrystalline state; however, not all of them (especially those with low Ca/P ionic ratios) have been prepared yet. The details on the available preparation techniques are given below:

3.6.2 Nanodimensional and Nanocrystalline Apatites

First of all, one should stress that the stoichiometric HA with well resolved X-ray diffraction patterns might be prepared mostly at temperatures exceeding ~700°C either by calcining of CDHA with the Ca/P molar ratio very close to 1.67 or by solid-state reactions of other calcium orthophosphates with various chemicals (e.g., DCPA + CaO). Thus, with the exception of a hydrothermal synthesis [223–225], in aqueous solutions only CDHA might be prepared [123, 174–176, 226–230]. As apatites (CDHA, HA, and FA) belong to the sparingly soluble compounds (Table 1.1), simple mixing of calcium- and orthophosphate-containing aqueous solutions at pH >9 results in formation of extremely supersaturated solutions and, therefore, a very fast precipitation of the tremendous amounts of very fine crystals [231], initially of ACP (see Chapter 2), that afterwards is re-crystallized into apatites [174–176, 232–235]. The dimensions of the precipitated nano-sized crystals might be slightly increased by the Ostwald ripening approach (maturation), that is, by boiling and/or ambient aging in the mother liquid (Fig. 3.3) [131, 143, 173, 203, 224, 234–239]. Heat treatment of ACP might be applied as well [240]. Therefore, preparation of nanodimensional and/or nanocrystalline apatites is not a problem at all and has been known for many years [131, 132, 241–243]; however, prefix "nano" had not been used before 1994. On the contrary, with the exception of a thermally stable FA (thus, big crystals of FA might be produced by a melt-growth process [244, 245]), manufacturing of big crystals of both CDHA and HA still is a challenge.

Many different methodologies have been proposed to prepare nanodimensional and/or nanocrystalline structures [246–253]. Prior to describing them, it is important to stress that in the vast majority of the available literature on apatites, the authors do not tell the difference between CDHA and HA. Therefore, getting through scientific papers, an attentive reader often finds statements, as: "Because natural bone is composed of both organic components (mainly type I collagen) and inorganic components (HA), ..." [98, p. 357], "The HA nanorods are synthesized via a wet precipitation process ..." [141, p. 2364], "... (TTCP) has been shown previously to be an essential component of self-setting calcium phosphate cements that form hydroxyapatite (HA) as the only end-product. ..." [254, abstract], etc. The matter with distinguishing between CDHA and HA becomes even much more complicated, when researchers deal with nanodimensional and/or nanocrystalline apatites because the assemblies of nano-sized particles give only broad diffraction patterns, similar to ones from an amorphous material [198, 199]. While composing this review, I always tried to specify whether each cited study dealt with CDHA or HA; unfortunately, the necessary data were found in just a few papers. Therefore, in many cases, I was forced to mention just "apatites" without a further clarification. Thus, the readers are requested to be understandable on this uncertainty.

The greater part of the published reports on synthesizing of nanodimensional and/or nanocrystalline apatites is focused on the bottom-up approach. Among the available preparation techniques, a wet chemical precipitation is the most popular one [63, 72, 74, 87, 101, 103, 141–143, 146, 152, 172, 173, 232, 241, 255–292]. Various authors discussed the effects of synthesis parameters, such as temperature [267–270, 286], time [268], calcium ion concentration [270], presents of surfactants [273], calcination [268], and the use of different reagents on the morphological properties of nanodimensional apatites. In general, the shape, stoichiometry, dimensions, and specific surface area of nano-sized apatites appeared to be very sensitive to both the reaction temperature (Fig. 3.4) and the reactant addition rate [267, 280, 286]. Namely, particle sizes of nanodimensional apatites were observed to increase in a linear correlation with temperature [269, 286], which is a good indication that sizes of nanodimensional apatites can possibly be tailored. Furthermore, the initial pH values and reaction temperatures both play important roles in the morphology of the precipitated apatites, as well as on the phase formation and degree of crystallinity [290]. For example, significant differences in the chemical composition, morphology, and amorphous character of nano-sized CDHA produced through the reaction between aqueous solutions of $Ca(NO_3)_2$ and $(NH_4)_2HPO_4$ can be induced, simply by changing the pH of the reactant hydrogen phosphate solution [291]. Thus, the solvent systems, dispersant species, and drying methods appear to have effects on the particle size and dispersibility. However, some conflicting results have been obtained on how certain synthesis parameters can affect the morphological properties of these nano-sized particles. Nevertheless, it was commonly observed that nano-sized crystals of apatites synthesized through the chemical precipitation were often highly agglomerated; however, these agglomerates could be clusters of ultra-fine primary particles [271].

A hydrothermal synthesis [58, 61, 131, 132, 224, 225, 255, 286, 287, 293–313] seems to be the second most popular preparation technique of the nanodimensional and/or nanocrystalline apatites. The term "hydrothermal" refers to a chemical reaction of substances in a sealed heated solution above ambient temperature and pressure [314] and this process allows synthesis of highly pure fine-grained single crystals, with controlled morphology and narrow size distribution [293]. Extraneous additives, such as EDTA [311], surfactants [312], anionic starburst dendrimer [313], etc., might be utilized to modify the morphology of nanodimensional and/or nanocrystalline apatites during the hydrothermal synthesis. Most of these techniques produced rod-like crystals or whiskers, while plate-like shapes were obtained in just a few studies [295, 305, 307].

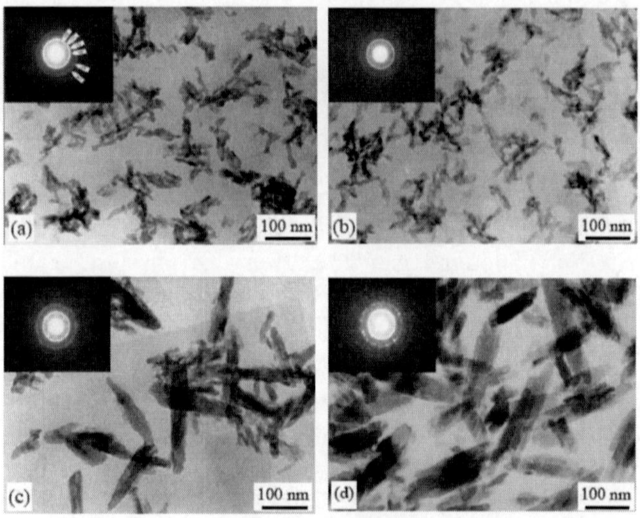

Figure 3.4 The influence of the reaction temperature on the crystal dimensions of precipitated CDHA: (a) 25°C, (b) 37°C, (c) 55°C, (d) 75°C.

Other preparation methods include sol-gel [30, 162, 199, 200, 237, 292, 315–326], co-precipitation [238, 293, 294, 327–329], mechanochemical approach [54, 218, 303, 308, 330–335], mechanical alloying [336, 337], ball milling [308, 332, 338, 339], radio frequency induction plasma [340, 341], vibro-milling of bones [342], flame spray pyrolysis [343], liquid-solid-solution synthesis [344], electrocrystallization [133, 345, 346], microwave processing

[32, 58, 255, 293, 294, 302, 347–355], hydrolysis of other calcium orthophosphates [356–358], double step stirring [359], emulsion-based [275, 309, 360–372], steam-assistant [373], and solvothermal [374] syntheses. Besides, still other preparation techniques are known [31, 43, 122, 129, 242, 375–387]. Continuous preparation procedures are also available [171, 388]. Application of both ultrasound [389–391] and viscous systems [392] might be helpful. Furthermore, nanodimensional HA might be manufactured by a laser-induced fragmentation of HA microparticles in water [393–395] and in solvent-containing aqueous solutions [304, 323, 396], while dense nanocrystalline HA films might be produced by radio frequency magnetron sputtering [397, 398]. An interesting approach using sitting drop vapor diffusion technique should be mentioned as well [399]. A comparison between the sol-gel synthesis and wet chemical precipitation technique was performed and both methods appeared to be suitable for synthesis of nanodimensional apatite [292]. By means of these methods, a variety of nanodimensional calcium orthophosphate building blocks with various structures and morphologies have been synthesized, including needle-like, spherical, fibrous, and mesoporous nano-sized crystals, as well as nano-sized rods, hollow spheres, layered structures, and flowers as shown in Fig. 3.5 [387].

Table 3.1 presents some data on the chronological development of synthesis of nanodimensional apatites for the period of 1995–2004 [122]. Among the methods described, the thinnest crystals of apatite (60 × 15 × 0.69 or 0.84 nm) have been prepared by Melikhov et al.; they have been called "two dimensional crystalline HA" [263], while the smallest ones (size between 2.1 and 2.3 nm, i.e. around two times the HA unit cell parameters) have been found by Biggemann et al. [169], Liu et al. [405, 406], and Han et al. [404] synthesized nano-sized HA via a template-mediated and a non-template-mediated sol-gel techniques, respectively. Both triethylphosphate [405, 406] and other alkylphosphates [104] might be used to produce nanocrystalline apatites. Besides, nanodimensional ion-substituted CDHA might be precipitated from both a synthetic [328] and a simulated [407] body fluids. A relatively simple sol-gel process using ethanol and/or water as a solvent has also been reported to obtain the stoichiometric, nanocrystalline single phase HA [323].

Table 3.1 Synthesis of nanodimensional apatites: a chronological development [122].

Year	Process	Reference
1995	Synthesis of nanocrystalline HA (particle size ~20 nm) for the first time using calcium nitrate and diammonium hydrogen orthophosphate as precursors by solution spray dry method.	[400]
2000	Synthesis of biomimetic nanosized CDHA powders (~50 nm) at 37°C and pH of 7.4 from calcium nitrate tetrahydrate and diammonium hydrogen orthophosphate salts in synthetic body fluid using a novel chemical precipitation technique.	[328]
2002	Preparation of nanosized HA particles and HA/chitosan nanocomposite.	[401]
2002	Direct precipitation from dilute calcium chloride and sodium orthophosphate solutions.	[402]
2003	Radio frequency plasma spray process employing fine spray dried HA powders (average size ~15 μm) as a feedstock.	[340]
2003	Sol-gel process using equimolar solutions of calcium nitrate and diammonium hydrogen orthophosphate dissolved in ethanol.	[323]
2003	Chemical precipitation through aqueous solutions of calcium chloride and ammonium hydrogen orthophosphate.	[268]
2003	Mechanochemical synthesis of nanosized HA and β-TCP powders using DCPD and CaO as starting materials.	[333]
2003	Synthesis of nano-powders via sucrose-templated sol-gel method using calcium nitrate and diammonium hydrogen orthophosphate as precursor chemicals.	[403]
2004	Hydrolysis method of DCPD and $CaCO_3$ by 2.5 M NaOH (aq).	[356]
2004	Citric acid sol-gel combustion process using calcium nitrate tetrahydrate, diammonium hydrogen orthophosphate and citric acid.	[404]

Nanocrystalline HA powder was synthesized at a low calcination temperature of 750°C by the citric acid sol-gel combustion method [404]. The attractive features of this method were to synthesize materials with a high purity, a better homogeneity, and a high surface area in a single step [404, 408]. An array of highly ordered HA nano-sized tubes of uniform length and diameter was synthesized by sol-gel auto-combustion method with porous anodic aluminum oxide template [322]. Varma et al. synthesized nano-sized HA by polymeric combustion method and self-propagating combustion synthesis by using novel body fluid solutions [409]. Furthermore, nano-sized particles of both FA and β-TCP might be synthesized by a simultaneous combustion of calcium carboxylate and tributylphosphate based precursors in a flame spray reactor [410]. Both a flame-based technique [411] and a spray drying approach [106, 412] might be applied as well. Furthermore, crystalline and phase pure nano-sized HA and CDHA were synthesized in a continuous hydrothermal flow system using supercritical water at $t < 400°C$ and 24 MPa pressure [298].

Nanopowders of the stoichiometric HA of ~20 nm particle size were synthesized by hydrolysis of a mixture of DCPD and $CaCO_3$ performed with 2.5 M aqueous solution of NaOH at 75°C for 1 h. The only product synthesized was nanocrystalline HA and its crystallinity was improved with increasing annealing temperature [356]. Similar results were obtained in other studies [357, 358]. Furthermore, Xu et al. used radio frequency plasma spray process to synthesize nanodimensional HA powders with particle size in the range of 10–100 nm [340]. Kuriakose et al. synthesized nanocrystalline HA of size ~1.3 nm that was thermally stable until 1200°C [323]. Nanocrystalline plate-shaped particles of HA were directly precipitated at ambient temperature and pH ~7.4 from dilute aqueous solutions of calcium chloride and sodium orthophosphate. The direct precipitation of nano-sized HA was achieved by submitting the aqueous suspension to microwave irradiation immediately after mixing [402]. A simple and easy approach for synthesizing thermally stable nanostructured stoichiometric HA powder under invariant pH conditions of 7.5, known as the NanoCaP process, was developed. Under these conditions, the synthesized HA not only remained in the nanostructured state but also did not exhibit any compositional fluctuations that were observed in conventional approaches for synthesizing HA [12]. Other preparation techniques of nano-sized apatite might be found elsewhere [242]. Bulk bioceramics made of nanocrystalline HA with a grain size of no more than 50 nm and a near-theoretical density might be prepared by application of a high (~3.5 GPa) pressure in uniaxial compaction of nanodimensional powders with subsequent sintering at 640°C [142]. A similar approach was reported by another research group [351].

Figure 3.5 A variety of nano-scale calcium orthophosphates with different structures and morphologies synthesized by: (A and B) sol-gel processing, (C) co-precipitation, (D) emulsion technique, (E) hydrothermal process, (F) ultrasonic technique, (G) mechano-chemical method, (H–L) template method, (M) microwave processing, (N) emulsion-hydrothermal combination, (O) microwave-hydrothermal combination. Reprinted from Ref. [387] with permission.

Mechanochemical processing is another compelling method to produce nanostructured apatites in solid state [54, 206, 303, 330–334]. For example, Yeong et al. used the appropriate amounts of DCPA and calcium oxide. The initial stage of mechanical activation resulted in a significant refinement in crystallite and particle sizes, together with a degree of amorphization in the starting powder mixture. This was followed by steady formation and subsequent growth of HA crystallites with increasing degree of mechanical activation. Finally, a single-phase HA of an average particle size of ~25 nm, a specific surface area of ~76 m^2/g and a high crystallinity was attained after 20 h of mechanical activation [331].

The use of macromolecules as templating agents to manipulate the growth of inorganic crystals has been realized in many biological systems. Namely, in the presence of biological macromolecules (such as collagen), nucleation, and growth of nanocrystalline apatite to form highly organized bone minerals is one of the most fascinating processes in nature. These processes might be simulated. For example, layers of nanocrystalline apatite were formed *in situ* on the surface of various films at soaking them in aqueous solutions containing ions of calcium and orthophosphate. The *in situ* synthesized particles were found to be less agglomerated which was believed to be the result of nucleation of apatite crystallites on the regularly arranged side groups located on polymer chains [413, 414]. Another approach comprises precipitation of nanodimensional apatites from aqueous solutions in the presence of dissolved high molecular weight polyacrylic acid [415, 416] that acts as an inhibitor for the crystallization of apatite crystals [417, 418]. A similar inhibiting effect was found for dimethyl acetamide [419], polyvinyl alcohol [258], and several other (bio) polymers [420, 421]. This type of synthesis is expected to lead to formation of nanodimensional composites, which might be structurally more comparable to bones with closely related mechanical and biological properties. Furthermore, a control of particle size of aqueous colloids of apatite nano-sized particles was described involving a presence of amino acids [422, 423]. The amino acids ensured effective growth inhibition by a predominant adsorption onto the Ca-rich surfaces during the initial stages of crystallization. Thus, the nano-sized particles were formed by an oriented aggregation of primary crystallite domains along the *c*-axis direction. The size of the domains was shown to be governed by the interactions with the amino acid additives, which restricted a growth of the primary crystallites [422, 423]. Furthermore, nanodimensional apatites might be precipitated from aqueous solutions of gelatin [59, 424]. The development of nano-sized apatite in aqueous gelatin solutions was highly influenced by the concentration of gelatin: namely, a higher concentration of gelatin induced formation of tiny (4 × 9 nm) nano-sized crystals, while a lower concentration of gelatin contributed to the development of bigger (30 × 70 nm) nano-sized crystals. In this experiment, a higher concentration of gelatin supplied abundant reaction sites containing groups such as carboxyl, which could bind with calcium ions. This leads to formation of a very large number of nuclei and creation of a large number of tiny nano-sized crystals [59].

Although each of the reported approaches to produce nanodimensional apatites has both a scientific and a practical relevance, a little attention has been dedicated to the physicochemical details involved in the careful control of the particle size distribution and particle shape. Indeed, in the case of particle size distribution, most of the reported ways to synthesize nanodimensional apatites really produced a particle mixture with a wide size distribution from tens to hundreds of nanometers. Moreover, the control of particle shape is another problem for these methods, which commonly result in pin-like or irregular particles. It is well known that bone consists of homogeneous plate-like crystals of biological apatite of 15–30 nm wide and 30–50 nm long, while enamel consists of rod-like crystals of biological apatite of 25–100 nm thick and lengths of 100 nm to microns (Fig. 1.14) [2, 5, 177, 178, 180, 187, 195, 197]. The study of higher-level biomineralization and biomimetic assembly involves a search for advanced methods so that the synthesis of nano-sized apatite can be accurately controlled [425]. Namely, the size-controlled synthesis of materials can be achieved by using limited reaction spaces. For example, microemulsions have been shown to be one of the few techniques, which is able to produce particle sizes in the range of nanometers and with minimum agglomeration [426]. Thus, microemulsions [309, 363–371], micelles [427], and reverse micelles [299, 428, 429] have been successfully applied to synthesize nanodimensional apatites with minimal agglomeration. It was found that experimental conditions, such as aqueous/organic phase volume ratio, pH, aging time, aging temperature, and ion concentration in the aqueous phase can affect the crystalline phase, surface area, particle size, and morphology of nanodimensional apatites.

In some cases, special polymers can be used as spatial reaction vessels for fabrication of CDHA. For example, Shchukin et al. employed a poly (allylamine hydrochloride) / PO_4^{3-} complex as a source of orthophosphate anions to capture calcium cations and make them react in the capsule volume [430]. Bose and Saha synthesized spherical-like nanocrystalline CDHA powder with particle diameters of ~30 and ~50 nm using the emulsion route [365]. Furthermore, nano-sized crystals of apatite might be aggregated into microspheres [376, 431]. Hexadecyl (cetyl) trimethylammonium bromide (CTAB) was selected as an efficient agent to modulate the formation of CDHA nano-sized particles [428, 432]. The particle size can be regulated feasibly by changing the concentration of CTAB in the supersaturated by calcium orthophosphates solutions. For example, three different types of spherical particles of nano-sized CDHA with average diameters of 20 ± 5, 40 ± 10, and 80 ± 12 nm were fabricated using a series of CTAB concentrations to control the particle size. The experimental results revealed that the dimensions of the prepared nano-sized CDHA were relatively uniform. In contrast, nano-sized CDHA grown in the absence of organic additives are typical, rod-like particles with lengths of hundreds of nanometers and width of tens of nanometers [432].

To conclude this part, the surface of nano-sized particles of apatite might be functionalized by various compounds (even by quantum dots [433, 434]) to provide new important properties [422, 435–440], e.g., fluorescence [27–29, 439] and luminescence [434, 440]. Both fluorescence and luminescence can be used as a tracking property for the nano-sized particles to give an observable indication of agent delivery, while the particles are served to protect the agent *in vivo* until it has reached the destination.

3.6.3 Nanodimensional and Nanocrystalline TCP

Many researchers have formulated synthesis of nanodimensional β-TCP. For example, Bow et al. synthesized β-TCP powders of ~50 nm particle diameter at room temperature in anhydrous methanol as a solvent [441]. With increase in aging time, the phase transformation was found to take place from initial DCPA, to intermediate ACP phases, then to final β-TCP. The authors observed that incorporation of carbonates helped in suppressing formation of ACP phases with apatitic structure and its transformation into poorly crystalline (almost amorphous) CDHA and favored the formation of β-TCP phase [441]. Nano-sized particles of both FA and β-TCP were synthesized by a simultaneous combustion of calcium carboxylate and tributylphosphate based precursors in a flame spray reactor [410]. The same technique was used to synthesize nano-sized particles of amorphous TCP of 25–60 nm size [442–445], those after calcinations transformed into α-TCP or β-TCP. Nanodimensional β-TCP powders with an average grain size of ~100 nm [147, 446] and less [447] were prepared by wet precipitation methods, followed by calcining at elevated temperatures. Furthermore, a sol-gel technique [448], reverse micelle-mediated synthesis [449], and a polystyrene template method [450] are also applicable. In wet precipitation techniques, dialysis might be applied as a separation method [446]. When wet-precipitation methods were used, initially nanodimensional CDHA with Ca/P ratio of ~1.50 was precipitated, that was transformed into nano-sized β-TCP at calcination.

To synthesize nano-sized TCP, both milling [451, 452] and a high temperature flame spray pyrolysis [453] techniques might be employed as well. Afterwards, the nanodimensional β-TCP powders can be compacted into 3D specimens, followed by sintering to achieve the appropriate mechanical strength [147]. The maximal values of the bending strength, elastic modulus, Vickers hardness, and compressive strength of the samples fabricated from nano-sized β-TCP powders were more than two times higher as compared to those of bioceramics obtained from micron-sized β-TCP powders. However, the degradability of bioceramics sintered from nanodimensional powders was just about one fourth of that sintered from micron-sized powders. Thus, the degradability of β-TCP bioceramics could be additionally regulated by the particle dimensions [147].

Nano-sized whiskers of several calcium orthophosphates (HA, β-TCP, and BCP (HA + β-TCP)) were produced by using a novel microwave-assisted "combustion synthesis (auto ignition)/molten salt synthesis" hybrid route. Aqueous solutions containing $NaNO_3$, $Ca(NO_3)_2$, and KH_2PO_4 (with or without urea) were irradiated in a household microwave oven for 5 min at 600 watts of power. The as-synthesized precursors were then simply stirred in water at room temperature for 1 h to obtain the nano-sized whiskers of the desired calcium orthophosphate [454]. Furthermore, nanostructured biphasic (HA + β-TCP) bioceramics was successfully prepared by microwave synthesis [455, 456] and a polymer matrix mediated process [457] in other studies. Good cellular activities of the biphasic bioceramics have been reported.

Layrolle and Lebugle developed a synthesis route of nano-sized FA and other calcium orthophosphates, using calcium diethoxide ($Ca(OEt)_2$) and H_3PO_4 [129] (+ NH_4F to prepare FA [458]) as the initial reagents and anhydrous ethanol as a solvent. By a simple variance of the ratio of reagents, calcium orthophosphates of various chemical compositions were precipitated in ethanol. The precipitates were characterized and the results indicated that those calcium orthophosphates were amorphous and nanodimensional. Furthermore, they had large specific surface areas and possessed a high reactivity [129, 458].

3.6.4 Other Nanodimensional and Nanocrystalline Calcium Orthophosphates

Nano-sized particles of DCPD (with some amount of CDHA and ACP) of a relatively high monodispersity could be synthesized from aqueous solutions of calcium nitrate and H_3PO_4 in the presence of 2-carboxyethylphosphonic acid. They are produced in a discoid shape with a diameter of 30–80 nm and a height of less than ~5 nm. They form stable colloidal solutions displaying minimal agglomeration [459]. Nano-sized rods and nanodimensional fibers of DCPD with average diameters of 25 ± 5 nm (aspect ratio ~6) and 76 ± 20 nm (aspect ratio ~40), respectively, were synthesized by sucrose ester based reverse microemulsion technique [460]. A similar approach was used in another study [371]. Nanodimensional crystals of both DCPD and DCPA were prepared by EDTA-assisted hydrothermal method [311]. An interesting approach comprises precipitation of calcium orthophosphates inside nano-sized pores of another material. For example, nanodimensional clusters DCPD were immobilized into pores of an oxide network by immersion of this network into an acidic (pH = 2.7) calcium orthophosphate solution at 50°C [461]. The acid-base reaction between the calcium orthophosphate solution and the hydroxyl groups of the oxide network resulted in formation of nanodimensional clusters of DCPD immobilized inside the oxide pores. Interestingly, but the immobilized nanodimensional clusters of DCPD were further converted into those of ACP and CDHA by supplementary treatment of the oxide network in alkaline solutions [461]. Hollow nano-sized shells of undisclosed calcium orthophosphates (presumably, of ACP) with a size distribution of (120–185) ± 50 nm and predictable mean shell thickness from 10 to 40 nm were prepared by crystallization onto the surface of nanodimensional liposomes [462, 463]. Both the suspension stability and shell thickness control were achieved through the introduction of carboxyethylphosphoric acid. Variation of shell thickness and stoichiometry may be a way of manipulating the dissolution kinetics of ACP coating to control the release of encapsulated materials, necessary for drug delivery purposes [462, 463]. Roughly spherical DCPA particles of approx. 50–100 nm in sizes were synthesized via a spray-drying technique [106, 464–466], while ribbon-like fibers of nano-sized DCPA might be prepared upon hydrolysis in urea [357]. Furthermore, nanodimensional calcium orthophosphate powders with DCPD as the major phase have been synthesized by an inverse microemulsion system using kerosene as the oil phase, a cationic surfactant and a non-ionic surfactant [467]. Microskeletal constructions might be synthesized as well [468].

When it comes to ACP, it is nanodimensional in the vast majority cases (Chapter 2). Approximately spherical nano-sized particles of ACP with a diameter of about 50 nm can be prepared by rapid precipitation from water and subsequent colloidal stabilization by coating with polymers [469]. Nano-sized clusters of ACP [470] or those comprising a spherical core of 355 ± 20 DCPD units with density of 2.31 g/cm^3 and radius of 2.30 ± 0.05 nm surrounded by 49 ± 4 peptide chains with a partial specific volume of 0.7 cm^3/g, forming a tightly packed shell with an outer radius of 4.04 ± 0.15 nm were prepared by precipitation using 10 mg/ml of the 25-amino-acid N-terminal tryptic phosphopeptide of bovine β-casein as a stabilizing agent [471]. Nano-sized particles of ACP were prepared by mixing of solutions of $Ca(NO_3)_2 \cdot 4H_2O$ (450 mmol/L) in acetone and $(NH_4)_2HPO_4$ (30 mmol/L) in deionized water at pH within 10.0–11.0 [472]. Furthermore, nanodimensional powders of ACP might be prepared by both an electrostatic spray pyrolysis [473, 474] and spray drying [106] techniques.

Self-assembled shell cross-linked poly(acrylic acid-b-isoprene) micelles and/or cross-linked poly(acrylic) acid nano-sized cages in aqueous solutions might be used as templates for preparation of polymer/calcium orthophosphate nanodimensional capsules of 50–

70 nm in diameter, which consisted of spherical polymer nano-sized particles enclosed within a continuous 10–20 nm thick surface layer of ACP [475]. Synthesis of hollow spherical calcium orthophosphate nano-sized particles using polymeric templates has been also reported by other researchers [476]. Furthermore, bundles of surfactant-coated ACP nanodimensional filaments of ~2 nm in width and >300 μm in length were synthesized in reverse micelles [477]. Bundles of the nanodimensional filaments were stable in the reverse micelle phase up to around 5 days, after which they transformed into 5 nm-wide surfactant-coated CDHA rods. Discrete filaments of 100–500 × 10–15 nm in size and a linear superstructure based on the side-on stacking of surfactant-coated ACP nano-sized rods were also prepared [417]. A double reverse-micelle strategy was realized to synthesize amine, carboxylate-, and polyethylene glycol surface functionalized calcium orthophosphate nano-sized particles of an undisclosed nature [478]. Furthermore, the reverse micelle technique might be applied to prepare nanodimensional DCPA [428, 479].

Concerning OCP, an oriented growth of nanodimensional belts of OCP with a clean surface has been achieved by wet-chemical approach using cetyltrimethylammonium bromide [480]. Pulsed laser deposition technique was employed to obtain thin films of nanocrystalline OCP on pure Ti substrates [481]. The deposition was performed by a pulsed UV laser source in a flux of hot water vapors. High-resolution electron microscopy and X-ray diffraction at grazing incidence investigations indicated that the coatings were made of nanocrystalline OCP (unfortunately, the dimensions were not indicated). *In vitro* tests proved that both fibroblasts and osteoblasts adhered, reached a normal morphology, proliferated and remained viable when cultured on the nanocrystalline OCP coatings, supporting a good biocompatibility and absence of any toxicity [481].

Nanodimensional powders of BCP (both HA + β-TCP [482, 483] and HA + α-TCP [484]) have been fabricated as well. To get the details, the interested readers are referred to the original publications.

Similar to that for apatites (see above), the surface of nano-sized particles of both TCP and other calcium orthophosphate might be functionalized by various compounds to provide new important properties [171, 478, 484–489], such as fluorescence [487, 488] or a good disperseability in organic solvents [484]. Furthermore, nano-sized calcium orthophosphates might be used as templates to manufacture nanodimensional capsules [490].

3.6.5 Biomimetic Construction Using Nanodimensional Particles

Morphological control of bioinorganic materials is another interested issue in biomineralization, by which inorganic materials with complex morphologies can be produced. Complex forms or patterns with a hierarchical structure over several length scales are important features of biomineralization. Pattern formation in biomineralization is a process in which self-assembled organic templates are transformed by a material's replication into organized inorganic structures. Needless to mention, that researchers try to reproduce these processes in laboratories. For example, Chen et al. reported a way to create enamel-like structures by modifying synthetic nano-sized rods of apatite with a surfactant, bis(2-ethylhexyl)sulfosuccinate salt, that allowed the nano-sized rods to self-assemble into prism-like structures at the water/air interface [193]. A nanometer-scale rod array of apatite having preferred orientation to the *c*-axis was successfully prepared simply by soaking calcium-containing silicate glass substrates in Na_2HPO_4 aqueous solution at 80°C for various periods [491]. A biomimetic bottom-up route to obtain the first hierarchical level of bone was reported [182]. A pH-induced self-assembly of peptide-amphiphile to make a nanostructured fibrous scaffold reminiscent of extracellular bone matrix was obtained. After the cross-linking of the scaffold, the fibers were able to direct mineralization of CDHA to form a biocomposite, in which the crystallographic *c*-axes of the nano-sized crystals of CDHA were aligned with the long axes of the fibers. This alignment was similar to that observed between collagen fibrils and crystals of biological apatite in bones [182]. Other attempts to fabricate artificial materials having bone-like both nanostructure and chemical composition were performed and several significant achievements were obtained [492, 493].

The classical model of biomineralization considers mineral formation as an amplification process in which individual atoms or molecules are added to existing nuclei or templates [1, 2, 494]. This process occurs in the presence of various bioorganic molecules, which deterministically modify nucleation, growth, and facet stability. A model involving aggregation-based growth [495] recently challenged this conventional concept for the crystal growth. Inorganic nano-sized crystals were found to aggregate into ordered solid phases via oriented attachment to control the reactivity of nanophase materials in nature [19, 496]. A model of "bricks and mortar" was suggested to explain the biological aggregation of nano-sized apatite [497]. In this model, ACP acts as "mortar" to cement the crystallized "bricks" of nano-sized HA. Meanwhile, biological molecules control the construction process. By using nanodimensional spheres of HA as the building blocks, highly ordered enamel-like and bone-like apatites were hierarchically constructed in the presence of glycine and glutamate, respectively. It is interesting that, during the evolution of biological apatite, the amorphous "mortar" can be eventually turned into the "brick" by phase-to-phase transformation to ensure the integrity of biominerals [497].

3.7 Biomedical Applications of the Nanodimensional and Nanocrystalline Calcium Orthophosphates

3.7.1 Bone Repair

Due to advances in surgical practice and a fast aging of the population, there is a permanently increasing demand for bone grafts [498]. Modern grafts should not only replace the missing bones, but also should be intrinsically osteoinductive by acting as scaffolds for guided bone growth. Furthermore, an ability to form a biologically active apatite layer to bond to living bone it is an essential requirement to modern biomaterials [499]. In addition, a good graft should provide a framework to support new blood vessels and soft tissues in forming a bridge to existing bones [498].

Calcium orthophosphate bioceramics of micron dimensions have been used in dentistry, orthopedics, and surgery for over 30 years (Chapters 1, 4, 5, 6) because of their chemical similarity to calcified tissues of mammals and, therefore, excellent biocompatibility [123, 174–176, 226, 227]. Due to a rapid development of nanotechnology, the potential of nanodimensional and nanocrystalline forms of calcium orthophosphates has received a considerable attention [18] because they produce favorable results in repair of bone defects [500]. For example, due to an improved sinterability, an enhanced densification, and a better bioactivity than coarser crystals, they might be chosen as the major components of self-setting bone cements [14, 25, 442, 443, 501–505]. However, there is a study in which an increase of particle and crystallite sizes of TCP did not prolong but shortened the induction time until the cement setting reaction started [445], which was against the common physical rules (generally, smaller particles or crystallites should enhance reactivity). Nevertheless, two general directions of the biomedical application of nanodimensional and nanocrystalline calcium orthophosphates can be outlined: (i) using them in powder form as filling materials to impart bioactivity to various biocomposites and hybrid biomaterials [53–87, 153, 506] and Chapter6; (ii) manufacturing of either dense compacts or porous scaffolds, possessing the sufficient mechanical properties [63, 80, 261, 262, 492, 463, 507, 508]. As the nanodimensional and nanocrystalline calcium orthophosphates tend to agglomerate at heating (Fig. 3.6) [278, 509–511], normally a low-temperature [143, 323] and/or a rapid consolidation [143, 225, 287, 512–517] techniques must be employed. The low-temperature approach comprises gel hardening (at 4°C) [323] and uni-axial pressing at 150–200°C [143]. The rapid consolidation techniques comprise spark plasma sintering [143, 225, 287, 512–514], pressure sintering [513], and microwave sintering over the temperature range 1000–1300°C, using a rapid sintering schedule [515–517]. Furthermore, nanodimensional crystals of calcined HA might be fabricated by calcination at 800°C for 1 h with an anti-sintering agent surrounding the original nano-sized CDHA particles and the agent is subsequently removed by washing after the calcination [518–520]. These consolidation approaches provided a limited alteration of the initial nano-sized crystals, while the final bioceramics possessed the mechanical properties similar to those reached with sintered stoichiometric HA.

Already in 1990s, implants prepared from nanodimensional apatites, as well as biocomposites of nanodimensional apatite with organic compounds were tested *in vivo* [521–523]. Cylinders made of both pure nanodimensional apatite and organoapatite containing a synthetic peptide were analyzed 28 days after implantation into spongy bones of Chinchilla rabbits. Both implant types were well incorporated and interface events were found to be similar to those observed on human bone surfaces with regard to resorption by osteoclast-like cells and bone formation by osteoblasts. That study revealed a suitability of such materials for both bone replacement and drug release purposes [521]. Similar results were obtained in other studies [522, 523].

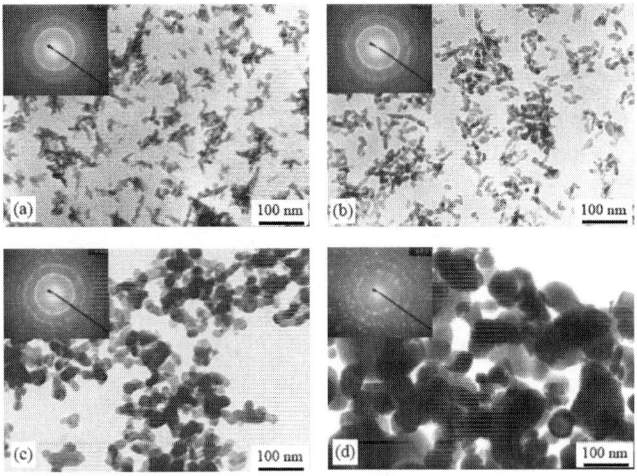

Figure 3.6 Particle sizes and crystallinity of HA powders after a heat treatment at various temperatures: (a) 300°C, (b) 500°C, (c) 700°C, (d) 900°C.

Among the available commercial formulations, NanOss™ bone void filler from Angstrom Medica, Inc. [524], is considered as the first nanotechnological medical device received the clearance by the US Food and Drug Administration (FDA) in 2005. It is prepared by precipitation of nano-sized calcium orthophosphates from aqueous solutions and the resulting white powder is then compressed and heated to form a dense, transparent, and nanocrystalline material. NanOss™ mimics the microstructure, composition, and performance of human bone, as well as it is mechanically strong and osteoconductive. It is remodeled over time into human bone with applications in the sports medicine, trauma, spine, and general orthopedics markets [524].

Ostim® (Osartis GmbH & Co. KG, Obernburg, Germany) is another popular commercial formulation. It is a ready-to-use injectable paste that received CE (Conformite Europeenne) approval in 2002. Ostim® is a suspension of synthetic nanocrystalline HA (average crystal dimensions: 100 × 20 × 3 nm^3 (a needle-like appearance); specific surface area ~100 m^2/g) in water, prepared by a wet chemical reaction [525]. After completion, the HA content in the paste is ~35%. Ostim® does not harden when mixed with blood or spongiosa, so it is highly suitable for increasing the volume of autologous or homologous material. Simultaneously, its viscosity enables its applications to form-fit in close contact with the bone. Ostim® can be used in metaphyseal fractures and cysts, alveolar ridge augmentation, acetabulum reconstruction, and periprosthetic fractures during hip prosthesis exchange operations, osteotomies, filling cages in spinal column surgery, etc. [524, 526–536]. It might be incorporated into bones and a new bone formation is visible after only three months [537]. For a number of clinical applications, Ostim® might be combined with other types of calcium orthophosphate bioceramics, e.g., with a HA bioceramic core (Cerabone®) [525, 538] or with biphasic (β-TCP + HA) granules (BoneSaves®) [539]. Application of such combinations of a nanocrystalline Ostim® with the microcrystalline calcium orthophosphate bioceramics appeared to be an effective method for treatment of both tibia head compression fractures [525] and metaphyseal osseous volume defects in the metaphyseal spongiosa [538]. Besides, such combinations might be used for acetabular bone impaction grafting procedures [539].

Cui et al. developed nano-sized HA/collagen biocomposites, which mimicked the nanostructure of bones [178, 540]. After implantation, such biocomposites can be incorporated into bone metabolism. Due to processing difficulties and poor mechanical properties of bulk calcium orthophosphates, their applications are currently confined to non-load-bearing implants and porous bodies/scaffolds. Porous 3D biocomposites of nanodimensional HA and collagen/polymer mimic bones in composition and microstructure and can be employed as a matrix for the tissue engineering of bone [75].

Owing to their low mechanical properties, the use of calcium orthophosphates in load-bearing applications is rather limited: calcium orthophosphates are too stiff and brittle for such use. Today's solutions for weight-bearing applications rely mostly on biologically friendly metals, like cobalt–chromium alloys, titanium and its alloys, as well as stainless steel 316L, but problems with stress-shielding and long-term service can cause failures. All these metals, although nontoxic, are always bioinert and cannot bond to bone directly. In order to improve the biological properties of the metallic implants, nanostructured calcium orthophosphates (mainly, apatites) are generally used as a coating material to accelerate bone growth and enhance bone fixation [160, 161, 281, 377, 481, 541–562]. The coating techniques include thermal spraying, sputter coating, pulsed laser deposition, dynamic mixing method, dip coating, sol-gel method, electrophoretic deposition, biomimetic process, hot isostatic pressing, and some other methods [563]. In the majority cases, the coatings are composed of uniform nanocrystalline apatites (Fig. 3.7). They are capable in performing bone formation and promoting direct osseointegration with juxtaposed bone [566–569]. For example, an enhanced new bone formation can be clearly seen on nanophase HA-coated tantalum compared to micro-scale HA-coated tantalum and non-coated tantalum (see Fig. 2 in Ref. [98]). Furthermore, nanostructured calcium orthophosphates might be used as a coating material to impart surface bioactivity to other materials, e.g., glasses [570] and polymers [571, 572]. Finally but yet importantly, such coatings might be patterned, e.g., by laser direct writing [491] or electrohydrodynamic atomization spraying technique [573].

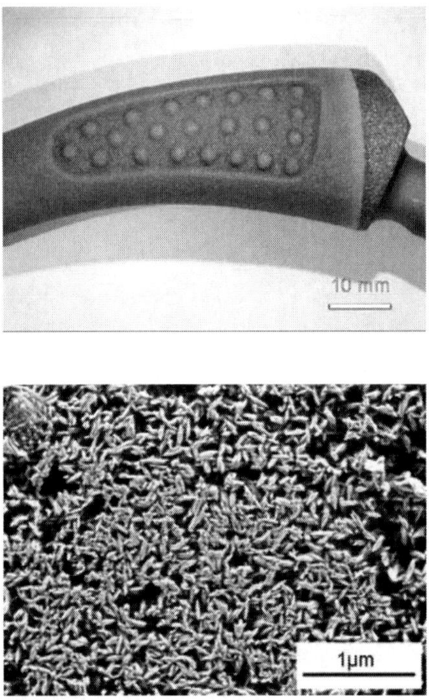

Figure 3.7 (a) A photo of a titanium implant coated with electrochemically deposited HA at 37°C (Cenos® BoneMaster); (b) A micrograph of a titanium implant surface coated with electrochemically deposited HA at 37°C. Reprinted from Ref. [564] with permission. Other micrographs of nano-CDHA coatings biomimetically deposited on NaOH-treated Ti6Al4V surfaces might be found in Ref. [565].

3.7.2 Nanodimensional and Nanocrystalline Calcium Orthophosphates and Bone-related Cells

It is well accepted that bone-related cells (especially, osteoblasts and osteoclasts) play the key roles in the physiological formation of calcified tissues. Bone-related cells not only are speculated to take part in the formation of biominerals and macrostructure constructions of bones, but they also continuously modulate the density, regeneration, and degradation of bones. Therefore, understanding the relationship between the bone-related cells and nano-sized calcium orthophosphates has been paid much attention in order to elucidate the formation mechanism of bones, to prevent and cure bone-related diseases, and to design novel biomaterials. Better structural biomimicity and osteoconductivity can be achieved using nanodimensional and nanocrystalline calcium orthophosphates [152, 153, 159, 160, 574–576]. Biocompatibility of such biomaterials is the key question for their application possibility for clinical use. For example, adhesion, proliferation, and differentiation of mesenchymal stem cells were studied on nano-sized HA/polyamide biocomposite scaffolds. The results indicated that such biocomposites exhibited a good biocompatibility and an extensive osteoconductivity with host bone *in vitro* and *in vivo* and proved that nano-sized HA/polyamide scaffolds had a potential to be used in orthopedic, reconstructive, and maxillofacial surgery [577–579].

Most results demonstrate that nanostructured HA can improve cell attachment and mineralization *in vivo*, which suggests that nano-sized HA may be a better candidate for clinical use in terms of bioactivity [159, 160, 164, 580–582]. The size effects of nanodimensional HA on bone-related cells, as well as the influence of crystallinity of nano-sized HA were studied [432, 583]. Different nano-sized particles of HA, typically of 20 ± 5, 40 ± 10, and 80 ± 12 nm in diameter, were prepared and their effects on the proliferation of two types of bone-related cells, bone marrow mesenchymal stem cells (MSCs) and osteosarcoma cells (U2OS and MG63) were studied. The cell culture experiments showed an improved cytophilicity of the nanophase HA if compared to the submicron-sized HA. A greater cell viability and proliferation of MSCs were measured for nano-sized HA, remarkably for 20 nm-sized particles. However, the opposite phenomenon occurred for bone tumour cells when nano-sized HA were co-cultured with cells. Nano-sized HA can inhibit proliferation of U2OS and MG63 cells and the inhibited strengths were inversely proportion to the particle size, i.e. smaller particles possessed a greater ability to prevent cell proliferation. This suggests that nano-sized HA can exhibit favorable cell proliferation to optimize biological functionality, in which the particle dimensions are believed to play a key role. These *in vitro* findings are of a great significance for the understanding of cytophilicity and biological activity of nano-sized particles during biomineralization [432].

Studies confirmed that nano-sized ACP had an improved bioactivity if compared to nano-sized HA since a better adhesion and proliferation of osteogenic cells had been observed on the ACP substrates [584]. However, in order to understand the influence of crystallinity of the nano-sized calcium orthophosphates on the osteogenic cells correctly, it was critical to use nano-sized ACP and HA of the same size distribution [583]. Thus, ACP and HA particles of ~20 nm size were synthesized and the effects of crystallinity were studied. The adhesion, proliferation, and differentiation of MSC cells were measured on both ACP and HA films and compared at the same size scale. Surprisingly, more cells were adsorbed and proliferated on the films of the well crystallized nano-sized HA than those on the films of nano-sized ACP. Alkaline phosphatase activity assay and RT-PCR assay were also used to evaluate the differentiation of MSC cells. The results showed that the differentiation of MSC cells from osteoblasts was promoted significantly by nano-sized HA. These experimental phenomena clearly demonstrate that the crystallized phase of HA provides a better substrate for MSC cells than ACP, when the factor of size effect is removed. This new view on the relationship between the crystallinity of calcium orthophosphates and the responses of MSCs emphasized the importance of both size and phase control in the application of biomedical materials [583, 584].

On the other hand, the chemical composition of the samples appears to be important. Interestingly, but in spite on the fact that the biological apatite of bones contains the substantial amount of carbonates, among investigated samples of nanocrystalline apatites, osteoclastic differentiation was found to be constrained on carbonate-rich samples, leading to smaller numbers of osteoclast-like cells and fewer resorption pits. Furthermore, the highest resorption rate was found for nanodimensional HA with a low carbonate content, which strongly stimulated the differentiation of osteoclast-like cells on its surface [585].

Cells are sufficiently sensitive and nano-scale alterations in topography might elicit diverse cell behavior [586–588]. How cells can recognize the particle size and other very small differences in the properties of nano-sized HA in these experiments remains unclear. Actually, determining the mechanisms whereby nano-sized particles of calcium orthophosphates and their sizes exert effects on bone-related cells will require further systematic studies.

3.7.3 Dental Applications

Dental caries is a ubiquitous and worldwide oral disease. At the initial stage of caries lesions, bacteria cause damage of dental enamel, which is the exterior coating of teeth and possesses remarkable hardness and resistance. As the most highly mineralized structure in vertebrate bodies, enamel is composed of numerous needle-like apatite crystals of nanodimensional sizes, which are bundled in parallel ordered prisms to ensure unique mechanical strength and biological protection (see section 1.4.2 of this book). As a non-living tissue, the main constituent (~97 wt. %) of mature enamel is inorganic nanodimensional apatite so that enamel is scarcely self-repaired by living organisms after substantial mineral loss. Filling with artificial materials is a conventional treatment to repair damaged enamel. However, secondary caries frequently arise at the interfaces between the tooth and foreign materials [589].

Nanodimensional HA and CDHA are often considered as model compounds of dental enamel due to the chemical and phase similarities [123, 174, 175]. Therefore, enamel remineralization by using nanodimensional apatite or other calcium orthophosphates is suggested in dental research. For example, toothpastes containing nanodimentional apatite could promote a partial remineralization of demineralized enamel [590–592], as well as possess some whitening effect [593]. Furthermore, nano-sized HA might be added to a fluoride-containing mouthrinse [594]. A remineralization potential of sports drink, containing nano-sized HA, was also investigated [595]. A positive influence of addition of nanodimensional β-TCP against acid demineralization and promoted remineralization of enamel surface was detected as well [596]. Unfortunately, these chemically analogous compounds of enamel are not widely applied in clinical practices. The native structure of dental enamel is too complex to be remodeled and the synthesized apatite crystallites often have different dimensions, morphologies, and orientations from the natural ones, which result in a poor adhesion and mechanical strength during dental restoration. Recent advances in biomineralization also indicate that features of smaller particles of nano-sized HA might approximate features of biological apatite more closely than features of the larger HA particles that are conventionally used [13]. For example, it has been demonstrated that nano-sized HA can be self-assembled to form enamel-like structures in the laboratory [193]. Therefore, a biomimetic technique is suggested as follows: the localized repair of the enamel surface can be improved by nano-sized HA (dimension of ~20 nm), analogues to the basic building blocks of enamel rods. Furthermore, it is found that nano-sized HA can adsorb onto the enamel surface strongly and can even be integrated into the natural enamel structure [597].

It is surprising but nano-sized HA of ~20 nm can inhibit significantly a mineral loss from the enamel surface [196]. Without any treatment, the demineralization of the natural enamel surface was remarkable in acidic solution (pH ~4.5 ± 0.1, experimental period of 2 days) and damaged sites were observed. The mass loss rate was about 0.12 ± 0.04 mg/mm^2 per day. In contrast, a layer of nano-sized HA on the treated enamel surface was almost unchanged in acidic solution. The rate of mass loss of enamel coated by nano-sized HA approached zero (< 0.02 mg/mm^2 per day), which was beyond the sensitivity of the detection methods. Since the coating by nano-sized HA appeared to be insensitive to dissolution, the underlying enamel surface was well protected under slightly acidic conditions. Furthermore, the enamel surface coated by ~20 nm-sized HA had a hardness of 4.6 ± 0.4 GPa and an elastic modulus of 95.6 ± 8.4 GPa. These data appeared to be very similar to those of natural enamel samples, which are 4.2 ± 0.2 and 94.1 ± 5.4 GPa, respectively [196].

The similarity between ~20 nm-sized HA and building blocks of dental enamel results in a good fixation of artificial biomaterials to natural tissues. Moreover, the enamel structure appears to be reinforced by nano-sized HA since secondary caries formation is suppressed and hardness is retained [589, 598, 599]. This strategy may have prospective applications in dentistry as it offers an easy but effective method to reconstruct tooth enamel that is suffering from mineral losses. Generally, these studies also suggest that analogues of nanodimensional building blocks of biominerals should be highlighted in the entire subject of biomineralization.

In the case of nanodimensional DCPA, decreasing of DCPA particle dimensions were found to increase the Ca- and PO_4-ions releases from DCPA-based biocomposites. Therefore, biocomposites based on nano-sized DCPA, possessing both a high strength and good release of Ca- and PO_4-ions, may provide the needed and unique combination of stress-bearing and caries-inhibiting capabilities suitable for dental applications [466].

3.7.4 Other Applications

Several other applications of nanodimensional and nanostructured calcium orthophosphates are in progress, some of which are described here.

Surface modification of nanodimensional calcium orthophosphates was performed in order to modulate their colloid stability, prevent dissolution in the case of low pH, avoid inflammation, serve as an intermediate layer to allow strong bond formation between HA/polymer matrices and potentially enhance its bioactivity or improves its conjugation ability with special functional groups [12, 600–603]. Such surface modified nano-sized particles might be applied for oral insulin delivery [604]

In another aspect, many strategies have been employed to load various agents, i.e. therapeutic, bio-imaging, etc., to nanodimensional calcium orthophosphates (mainly, apatites). In summary, these strategies can be broadly categorized into two main approaches. One approach is to load these agents during the synthesis — so-called *in situ* loading. This is done by adding the desired agent(s) to the reaction mixture before the formation of a nanodimensional calcium orthophosphate is completed. The second approach is to load the agent(s) only after a nanodimensional calcium orthophosphate has been fully synthesized or, in other words, after the synthesis process — so-called *ex situ* loading. This is mainly done through surface adsorption where the agents are adsorbed onto the surfaces of pre-synthesized nanodimensional particles [605]. Therefore, due to established biocompatibility, ease of handling, and notorious adsorption affinity, nano-sized calcium orthophosphates have been applied as non-viral carriers for drug delivery and gene therapy [173, 256, 282, 289, 435, 463, 478, 606–615]. After loading with genes or drugs, nanodimensional calcium orthophosphates provide a protective environment that shields them from degradation while providing a convenient pathway for cell membrane penetration and controlled release of the genes or drugs [436]. The experimental results proved that nanodimensional calcium orthophosphates possessed a higher penetration rate into cell membranes and their transfection efficiency could be 25-fold higher than that of the micron-sized particles. Furthermore, due to the larger specific surface areas, nanodimensional calcium orthophosphates can hold larger load amounts of drugs than coarser particles. These results indicate the potential of nano-sized calcium orthophosphates in gene delivery and as drug carriers [436, 616–619]. Furthermore, nanodimensional calcium orthophosphates can be stably loaded with radioisotopes [282, 620].

A transfer of functional foreign nucleic acids (DNA or RNA) into nuclei of living cells (transfection) with the aim of repairing missing cell function and to provide means to enhance or silence gene expression is currently used extensively in the laboratory and is fast becoming a therapeutic reality. As nucleic acids alone are unable to penetrate the cell wall, efficient carriers are required [621, 622]. Nanodimensional calcium orthophosphates can be represented as a unique class of the non-viral vectors, which can serve as efficient and alternative DNA carriers for targeted delivery of genes [256, 610, 623–635] and cells [485, 636–642]. The standard transfection method using calcium orthophosphates, first introduced by Graham and van der Eb in 1973 [641], is still used in biochemistry. It involves a straightforward *in situ* precipitation of calcium orthophosphate/DNA aggregates. Interestingly, but the transfection efficiency of nanodimensional calcium orthophosphates were found to depend on Ca/P ionic ratio: namely, calcium orthophosphates with Ca/P = 1.30 ratio exhibited a fourfold increase in the transfection efficiency over the ones with Ca/P = 1.65 ratio composition [256]. This data emphasize the importance of understanding the interaction between calcium orthophosphates and DNA to optimize the DNA uptake and its channeling to the nucleus of the cell. Besides, it has been demonstrated that surface modified particles of nano-sized calcium orthophosphates can be used *in vivo* to target genes specifically to a liver [643]. Attachment of galactose moiety onto the particle surface has increased the targetability of the nano-sized particles. Furthermore, this surface modification makes it possible for site-specific gene delivery [643, 644]. Assemblies of block-copolymer/nano-sized calcium orthophosphate were prepared and used for cell transfection; a high biocompatibility of this system was emphasized [645]. Structures that are even more complex are known as well [646–648]. A schematic drawing of a functionalized nano-sized particle suitable for both imaging and drug delivery purposes is shown in Fig. 3.8 [649]. Furthermore, vaccination to protect against human infectious diseases may be enhanced by using adjuvants that can selectively stimulate immuno-regulatory responses and nano-sized particles of calcium orthophosphates were found to be suitable for such purposes [650, 651].

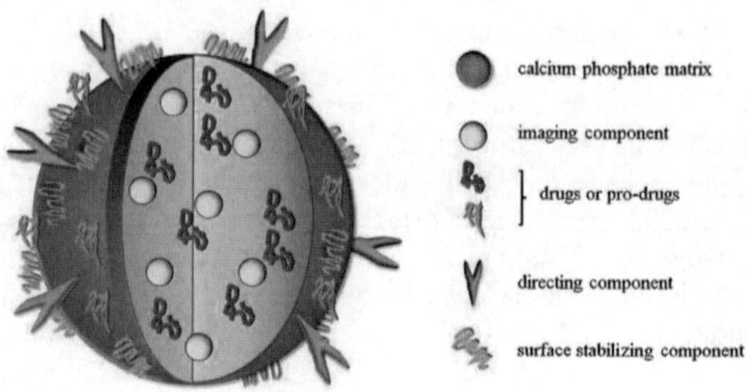

Figure 3.8 A generalized schematic setup of a nanodimensional particle of a calcium orthophosphate suitable for both imaging and drug delivery purposes. Reprinted from Ref. [649] with permission.

In all these new applications of nano-sized calcium orthophosphates, knowledge of the exact internalization pathway into the cells represents the first necessary step towards the detailed investigation and optimization of the functional mechanism. The main groups of pathways into the cell are diffusion, passive and active transport, as well as a number of endocytic mechanisms [612]. Bigger particles of far above 10 nm are internalized by eukaryotic cells through the endocytic pathways including phagocytosis, macropinocytosis, clathrin-mediated endocytosis, and non-clathrin-mediated endocytosis such as internalization via caveolae. To date, the exact internalization pathway of nano-sized calcium orthophosphates into cells has not been determined and there are many questions that remain to be answered, particularly, concerning possible interactions of calcium orthophosphates with nucleic acids. Furthermore, the mechanisms of cellular uptake and transport to the cell nucleus of calcium orthophosphate/DNA nanodimensional complexes remain unclear either. Therefore, there is a need to conduct a focused study on the synthesis of various forms of nano-sized calcium orthophosphates that could elucidate the mechanisms of binding, transport, and release of attached plasmid DNA for understanding the gene delivery method. Research is also warranted to understand the tracking of DNA intracellularly [638] to understand the release and transport of DNA into cellular nuclei. Already, some data are available that clathrin-mediated endocytosis might be responsible for the uptake of nano-sized HA [612].

Concerning the healing abilities of nano-sized calcium orthophosphates, an *in vitro* inhibiting effect and even apoptotic action of un-functionalized nano-sized HA of about 50 nm diameter on a hepatoma cell line in the concentration range of 50–200 mg/1 was reported [652]. A similar inhibiting effect was discovered for discrete nano-sized particles of HA, which appeared to cause apoptosis of leukemia P388 cells [101] and rat macrophages [653]. This effect might be due to a harmful increase in the intracellular calcium concentration. However, the correlation between the particle dimensions and the apoptotic action of nano-sized calcium orthophosphates appears not to be straightforward. Namely, the apoptosis efficacy of nanodimensional particles of HA of various sizes was found to decrease in the order of 45 > 26 > 78 > 175 nm [654].

Hollow nano-sized spheres are extremely attractive constructions because they can greatly enhance the load quantity. Though these novel biomaterials can improve the total intake of drugs, they also bring new problems, e.g., uncontrolled release kinetics and unreasonable metabolism pathway of the carriers [655]. In order to solve these problems, calcium orthophosphates were selected as suitable biomaterials to construct nanodimensional spheres hollow inside [462, 463, 614, 656, 657]. Such spheres with dimensions ranged from 110 to 180 nm were synthesized by an ultrasonic-assisted wet chemical reaction in the presence of a modifier [658]. In addition, they might be prepared through emulsions [659]. Transmission electron microscopy investigations revealed that the uniform nanodimensional spheres were formed and they were well dispersed in the solutions. Thickness of the shells was about 45 nm; thus, they always had ~60 nm sized internal cavities, which could be used to load drugs. The hollow spheres appeared to be stable in both air and aqueous solutions without ultrasonic application. However, when an ultrasonic treatment (40 kHz, 150 W) was applied, the hollow structures deconstructed to form pin-like nano-sized crystals of calcium orthophosphates [658]. During this transformation, the encapsulated drugs and chemicals are released [478, 658]. Different from a free and slow diffusion of encapsulated drugs from the cavity through the shells [170], the release kinetics in this system was triggered and controlled by ultrasound. Furthermore, the power density of ultrasound can conveniently regulate the release dynamics. Besides, the formed pin-like nano-sized crystals of calcium orthophosphates had similar behavior to the biological apatite of bones. Thus, a combination of the hollow calcium orthophosphate nanospheres and ultrasonic treatment might provide a good system for drug delivery and release [658].

Interestingly (although this is beyond the subject of biomaterials), but nano-sized particles of calcium orthophosphates with a mean size of 150 ± 20 nm filled with a solution containing luminol, haematin, and fluorescein were found to improve the ease and accuracy of H_2O_2 sensing [660].

3.8 Summary and Perspectives

As the basic building blocks of calcified tissues of mammals, nano-sized calcium orthophosphates of the apatitic structure play an important role in the construction of these biominerals. Therefore, they appear to be almost the ideal biomaterials due to their good biocompatibility and bioresorbability. Even more enhanced applications are expected in drug delivery systems [661]. However, there is still an unanswered question concerning their structure: whether the majority of nanodimensional calcium orthophosphates appear to be almost amorphous (according to numerous results of X-ray diffraction studies) due to their nanoscopic dimensions of well-crystallized structures or due to a really amorphous (i.e., retaining only a short-range order at the scale of few atomic neighbors — see Chapter 2) matter? A good attempt to discuss this topic is available in literature [662], where the interested readers are referred to.

In future, an ability to functionalize surfaces with different molecules of varying nature and dimensions by means of their attachment to cells will enable them to act selectively on biological species such as proteins and peptides. The capability of synthesizing and processing of nanodimensional and nanocrystalline calcium orthophosphates with the controlled structures and topographies, in attempts to simulate the basic units of bones and teeth, will provide a possibility of designing novel proactive bioceramics necessary for enhanced repair efficacy. The various primary positive results on the biocompatibility and biomimicity of novel nanostructured bioceramics merit further confirmations. Namely, much work remains to be undertaken to address the following key challenges and critical issues of nanodimensional and nanocrystalline calcium orthophosphates [663]:

- Consistency of the processing technologies;
- Optimization the structure and properties mimicking bones;
- Matching the strength of nanodimensional and nanocrystalline constructs with those of bones in order to provide a uniform distribution of stresses (load sharing);
- Optimizing bioresorption without comprising the mechanical properties;
- Assessing the inflammatory response to validate their biosafety.

Furthermore, substantial research efforts are required in the analysis of cells and their different behaviors with regard to their interactions with nanodimensional and nanocrystalline calcium orthophosphates [663]. An important but still unsolved question is how the cells can recognize the particle dimensions and crystallinity of nano-sized calcium orthophosphates. What is the signal for nanodimensional biomaterials to promote cell proliferation and differentiation and how can the pathways be found out? According to the experiments results on transfection, nano-sized particles can enter into cells readily but many details of this process remain unclear. Namely, the pathways for the nano-sized particles to enter the cells through the membranes should be revealed [664]. A greater influence of the hydrated surface layer with labile ionic species of smaller particles and crystals (see section 3.5) might be another possible option, to be confirmed experimentally. Then, it is important to examine the metabolism process of nano-sized calcium orthophosphates inside cells, so the existing forms of these particles during the biological processes could be understood. Further, a critical step will be the investigation of possible changes of gene or protein expression in the absence and presence of various nano-sized calcium orthophosphates, which may directly be related to cell proliferation and differentiation [13].

An understanding of the interactions between nano-sized particles and living cells is still a great challenge [663]. Future studies will focus on (1) the detailed interfacial structure of nanodimensional calcium orthophosphates and the specific adsorption of proteins or other matrices; (2) an uptake processes of the nano-sized particles by cells; (3) metabolism of nano-sized calcium orthophosphates inside the cells and its possible interference with physiological reactions. Another important topic is a biological security of nano-sized particles in general [119, 120, 665, 666] and those of calcium orthophosphates particularly [667]. For example, toxicity of nano-sized HA was found to vary considerably, which was related to their physico-chemical properties. Furthermore, cell death correlate strongly with the load of nano-sized particles. Namely, the biological effects of rod-shaped apatite, 50–80 nm in length, were investigated on human monocyte-derived macrophages [152]. High concentrations of apatite (200 nano-sized particles per cell) were incubated for 24 h with the macrophages in both serum and serum-free conditions. This induced high levels of lactate dehydrogenase release, which is an indicator of cellular damage. However, lower concentrations (20 and 2 nano-sized particles per cell) of the rod-shaped apatite did not affect the cell viability similarly to the control group that did not contain nano-sized apatite [152]. Similarly, intracellular dissolution of nano-sized HA as a function of time suggests that increased cytoplasmic calcium load is likely to be the cause of cell death [667]. Furthermore, nano-sized calcium orthophosphates were found to interfere with cell cycle of cultured human ovarian granulosa cells thus increasing cell apoptosis [668]. That pilot study suggested that effects of nano-sized particles on ovarian function should be extensively investigated. Additional examples of cytotoxicity experiments of nanodimensional calcium orthophosphates are well described in a recent review [605].

To finalize this topic, one should stress that, *in vivo* evaluation of nano-sized particles includes the particle's activity, biodistribution, and pharmacokinetic properties. Ultimately, all these properties are determined by dimensions, surface charge, morphology, and surface chemistry. Furthermore, nano-sized particles penetrate and leave biological organisms more readily using a number of pathways. Namely, very small (<10 nm) particles are generally eliminated from the body via renal clearance, i.e. being filtered through the kidneys and eliminated through urine, while nano-sized particles of larger dimensions are phagocytized by tissue macrophages of the reticuloendothelial system in the liver and spleen [609]. For example, intravenously administered nanodimensional (~40 nm and ~200 nm) rod-shaped crystals of apatite showed clearance from the bloodstream within 2 h, with ~90% of them being cleared in the first 10 min post injection; those nanodimensional crystals of apatite were observed primarily in the liver with a minority seen in the spleen [282]. These results indicate that bloodstream clearance occurs rapidly for a wide range of nanodimensional sizes. The accumulation of nanodimensional (50–100 nm in size) apatite in the liver was also noted in another study [434].

Thus, understanding the biological influence of nano-sized and nanocrystalline calcium orthophosphates is essential for a future development of bionanotechnology [670]. This interdisciplinary approach is very complicated and the effective collaboration of scientists from different disciplines is the key [13].

3.9 Conclusions

With a high surface area, un-agglomerated nanodimensional and nanocrystalline bioceramic particles are of interest for many applications including injectable or controlled setting bone cements, high strength porous or non-porous synthetic bone grafts, and the reinforcing phase in biocomposites that attempt to mimic both the complex structure and superior mechanical properties of bone. Therefore, nano-sized and nanocrystalline calcium orthophosphates have already gained much regard in the biomedical field due to their superior biocompatibility and biomechanical properties. This is easily seen from a permanent increasing of the amount of publications. At present, apatites (HA and CDHA) and β-TCP are the major calcium orthophosphates used in clinics. Currently, nanodimensional apatites are used primarily as bioactive coatings on bioinert materials like titanium and its alloys, in bone tissue repairs and implants, as well as for drug delivery purposes. The nano-sized β-TCP exhibits a significant biological affinity and activity and responds very well to the physiological environment. A lot of research is expected for much enhanced applications of the nanodimensional and nanocrystalline calcium orthophosphates for both drug delivery systems and as resorbable scaffolds that can be replaced by the endogenous hard tissues with the passage of time [122, 672].

Although the nanostructured biomaterials may have many potential advantages in the context of promoting bone cell responses [455–457, 588], it is important to remember that studies on nanophase materials have only just begun; there are still many other issues regarding human health that must be answered. Since particles of very low size have higher reactivity and effectiveness, a rapid technical development of nanometer-scaled particles in the biomedical field leads to concerns regarding the unknown risks of such materials [665, 666]. These nano-sized particles might induce inflammatory reactions, cytotoxicity, oxidative stresses or thrombogenesis when injected for drug delivery purposes. Namely, nano-sized particles may enter the human body through pores and may accumulate in the cells of the respiratory or other organ systems (when becoming dislodged through wear debris) and the health effects are yet to be largely known. This could happen during commercial-scale processing of the nano-sized particles as well as using these materials as implants [673]. Besides, nano-sized particles might be the objects whose existence has not been assumed by living body defense system [18, 119, 120]. Up to now, only a small number of short-term and small-scale health effects of single nanodimensional materials have been examined in toxicological studies, usually of the lungs [666]. Therefore, prior to clinical applications, any toxicity concerns of the nanophase materials [674–679] need to be overcome.

In summary, despite the challenges that lie ahead, significant evidences now exist elucidating that nanophase biomaterials represent an important growing area of research that may improve bonding between the implants and the surrounding tissues. It has proven to be a versatile approach that can increase bone cell functions on a wide range of orthopedic implant chemistries. Even if the nanodimensional and nanocrystalline calcium orthophosphates do not provide the ultimate answer for increasing bone cell responses (due to some potential problems as mentioned above), researchers have learned a tremendous amount of information concerning bone cell recognition with nanostructured surfaces that will most certainly aid in improving orthopedic implant efficacy [119, 120].

3.10 Post-Conclusion Remarks

According to Prof. D. F. Williams [97], the term "nanomaterial" should not exist because it is senseless (see section 3.2). Following this logic, the term "nanoapatite" is senseless as well. However, it is presented in the titles of several publications, namely Refs. [521, 543, 567, 580]. In a slightly modified form, the term "nano-apatite" is presented in the titles of several other publications, namely Refs. [75, 84, 194, 435, 545, 630, 672]. Furthermore, similar terms "nano-HA" [87, 492, 523, 573, 591, 598], "nano-hydroxyapatite" [40, 73, 76, 77, 80, 81, 86, 101, 164, 165, 261, 262, 279, 298, 389, 413, 419, 495, 500, 521, 541, 542, 555, 577, 579, 587, 592-594, 599, 601, 603] and "nanohydroxyapatite" [72, 78, 83, 137, 172, 207, 208, 210, 214, 221, 241, 284, 392, 493, 570, 577, 657] are presented in the titles of still other publications. Presumably, it is wiser not to use these terms anymore.

References and Notes

1. Mann, S. (2001) *Biomineralization Principles and Concepts in Bioinorganic Materials Chemistry*, Oxford University Press, New York, p. 216.
2. Lowenstam, H. A., and Weiner, S. (1989) *On Biomineralization*, Oxford University Press,New York, p. 324.
3. Vallet-Regí, M., and González-Calbet, J. M. (2004) Calcium phosphates as substitution of bone tissues, *Prog. Solid State Chem*, **32**, 1–31.
4. Weiner, S., and Addadi, L. (1997) Design strategies in mineralized biological materials, *J. Mater. Chem.*, **7**, 689–702.
5. Weiner, S., and Wagner, H. D. (1998) The material bone: structure-mechanical function relations, *Ann. Rev. Mater. Sci.*, **28**, 271–298.
6. Pasteris, J. D., Wopenka, B., and Valsami-Jones, E. (2008) Bone and tooth mineralization: why apatite? *Elements*, **4**, 97–104.
7. Giachelli, C. M. (1999) Ectopic calcification: gathering hard facts about soft tissue mineralization, *Am. J. Pathol.*, **154**, 671–675.

8. Kirsch, T. (2006) Determinants of pathological mineralization: crystal deposition diseases, *Curr. Opin. Rheumatol.*, **18**, 174–180.
9. Christian, R. C., and Fitzpatrick, L. A. (1999) Vascular calcification, *Curr. Opin. Nephrol. Hypertens.*, **8**, 443–448.
10. Boskey, A. (2003) Bone mineral crystal size, *Osteoporosis Int.*, **14**(5), S16-S20; discussion S20–S21.
11. Alivisatos, A. P. (2000) Enhanced naturally aligned nanocrystals, *Science*, **289**, 736–737.
12. Narayan, R. J., Kumta, P. N., Sfeir, C., Lee, D. H., Choi, D., and Olton, D. (2004) Nanostructured ceramics in medical devices: applications and prospects, *JOM*, **56**, 38–43.
13. Cai, Y., and Tang, R. (2008) Calcium phosphate nanoparticles in biomineralization and biomaterials, *J. Mater. Chem.*, **18**, 3775–3787.
14. Ginebra, M. P., Driessens, F. C. M., and Planell, J. A. (2004) Effect of the particle size on the micro and nanostructural features of a calcium phosphate cement: a kinetic analysis, *Biomaterials*, **25**, 3453–3462.
15. Nanotechnology is an application of science and engineering at the nanoscale.
16. Karch, J., Birringer, R., and Gleiter, H. (1987) Ceramics ductile at low temperature, *Nature*, **330**, 556–558.
17. Webster, T. J. (2001) Nanophase ceramics: the future of orthopedic and dental implant material, in *Nanostructured Materials*, (ed. Ying, J. Y.), Academic Press, New York, pp. 125–166.
18. Tasker, L. H., Sparey-Taylor, G. J., and Nokes, L. D. (2007) Applications of nanotechnology in orthopaedics, *Clin. Orthop. Relat. Res.*, **456**, 243–249.
19. Banfield, J. F., Welch, S. A., Zhang, H., Ebert, T. T., and Penn, R. L. (2000) Aggregation-based crystal growth and microstructure development in natural iron oxyhydroxide biomineralization products, *Science*, **289**, 751–754.
20. Cölfen, H. (2007) Bio-inspired mineralization using hydrophilic polymers, *Top. Curr. Chem.*, **271**, 1–77.
21. Oaki, Y., and Imai, H. (2005) Nanoengineering in echinoderms: the emergence of morphology from nanobricks, *Small*, **2**, 66–70.
22. Lee, S. H., and Shin, H. (2007) Matrices and scaffolds for delivery of bioactive molecules in bone and cartilage tissue engineering, *Adv. Drug Delivery Rev.*, **59**, 339–359.
23. Ben-Nissan, B. (2004) Nanoceramics in biomedical applications, *MRS Bulletin*, **29**, 28–32.

24. Rehman, I. (2004) Nano bioceramics for biomedical and other applications, *Mater. Technol.*, **19**, 224–233.

25. Driessens, F. C. M., Boltong, M. G., de Maeyer, E. A. P., Wenz, R., Nies, B., and Planell, J. A. (2002) The Ca/P range of nanoapatitic calcium phosphate cements, *Biomaterials*, **23**, 4011–4017.

26. Doat, A., Fanjul, M., Pellé, F. Hollande, E., and Lebugle, A. (2003) Europium-doped bioapatite: a new photostable biological probe, internalizable by human cells, *Biomaterials*, **24**, 3365–3371.

27. Doat, A., Pellé, F., Gardant, N., and Lebugle, A. (2004) Synthesis of luminescent bioapatite nanoparticles for utilization as a biological probe, *J. Solid State Chem.*, **177**, 1179–1187.

28. Lebugle, A., Pellé, F., Charvillat, C., Rousselot, I., and Chane-Ching, J. Y. (2006) Colloidal and monocrystalline Ln^{3+} doped apatite calcium phosphate as biocompatible fluorescent probes, *Chem. Commun.*, pp. 606–608.

29. Mondejar, S. P., Kovtun, A., and Epple, M. (2007) Lanthanide-doped calcium phosphate nanoparticles with high internal crystallinity and with a shell of DNA as fluorescent probes in cell experiments, *J. Mater. Chem.*, **17**, 4153–4159.

30. Kalita, S. J., and Bhatt, H. A. (2007) Nanocrystalline hydroxyapatite doped with magnesium and zinc: synthesis and characterization, *Mater. Sci. Eng. C*, **27**, 837–848.

31. Huang, J., Jayasinghe, S. N., Best, S. M., Edirisinghe, M. J., Brooks, R. A., Rushton, N., and Bonfield, W. (2005) Novel deposition of nanosized silicon substituted hydroxyapatite by electrostatic spraying, *J. Mater. Sci. Mater. Med.*, **16**, 1137–1142.

32. Pon-On, W., Meejoo, S., and Tang, I. M. (2007) Incorporation of iron into nano hydroxyapatite particles synthesized by the microwave process, *Int. J. Nanosci.*, **6**, 9–16.

33. Predoi, D., Barsan, M., Andronescu, E., Vatasescu-Balcan, R. A., and Costache, M. (2007) Hydroxyapatite–iron oxide bioceramic prepared using nano-size powders, *J. Optoelectronics Adv. Mater.*, **9**, 3609–3613.

34. Bakunova, N. V., Fomin, A. S., Fadeeva, I. V., Barinov, S. M., and Shvorneva, L. I. (2007) Silicon-containing hydroxylapatite nanopowders, *Russ. J. Inorg. Chem.*, **52**, 1492–1497.

35. Miao, S., Weng, W., Cheng, K., Du, P., Shen, G., and Han, G. (2007) Preparation of nano-sized strontium containing tricalcium phosphate particles, *Key Eng. Mater.*, pp. 330–332, 263–266.

36. Liu, Y., Zhou, R., Mo, A., Chen, Z., and Wu, H. (2007) Synthesis and characterization of yttrium/hydroxyapatite nanoparticles, *Key Eng. Mater.*, pp. 330–332, 295–298.

37. Wu, H. C., Wang, T. W., Sun, J. S., Wang, W. H., and Lin, F. H. (2007) A novel biomagnetic nanoparticle based on hydroxyapatite, *Nanotechnology*, **18**, 165601 (9 pages).

38. Rameshbabu, N., Kumar, T. S. S., Prabhakar, T. G., Sastry, V. S., Murty, K. V. G. K., and Rao, K. P. (2007) Antibacterial nanosized silver substituted hydroxyapatite: synthesis and characterization, *J. Biomed. Mater. Res. A*, **80A**, 581–591.

39. Fujii, E., Ohkubo, M., Tsuru, K., Hayakawa, S., Osaka, A., Kawabata, K., Bonhomme, C., and Babonneau, F. (2006) Selective protein adsorption property and characterization of nano-crystalline zinc-containing hydroxyapatite, *Acta Biomater*, **2**, 69–74.

40. Zhang, S. M., Hu, W., Zhou, W., Li, J., Liu, Y. H., and Qiu, Z. Y. (2007) Dialysis preparation of zinc-substituted nano-hydroxyapatite and its characterization, *Key Eng. Mater.*, pp. 330–332, 219–222.

41. Pon-On, W., Meejoo, S., and Tang, I. M. (2008) Substitution of manganese and iron into hydroxyapatite: core/shell nanoparticles, *Mater. Res. Bull.*, **43**, 2137–2144.

42. Zou, C., Weng, W., Cheng, K., Du, P., Shen, G., and Han, G. (2008) Preparation of nanosized β-tricalcium phosphate particles with Zn substitution, *J. Mater. Sci. Mater. Med.*, **19**, 1133–1136.

43. Hwang, K. S., Hwangbo, S., and Kim, J. T. (2008) Silver-doped calcium phosphate nanopowders prepared by electrostatic spraying, *J. Nanoparticle Res.*, **10**, 1337–1341.

44. Lee, D., Sfeir, C., and Kumta, P. N. (2009) Novel in-situ synthesis and characterization of nanostructured magnesium substituted β-tricalcium phosphate (β-TCMP), *Mater. Sci. Eng. C*, **29**, 69–77.

45. Petchsang, N., Pon-On, W., Hodak, J. H., and Tang, I. M. (2009) Magnetic properties of Co-ferrite-doped hydroxyapatite nanoparticles having a core/shell structure, *J. Magnetism Magnetic Mater.*, **321**, 1990–1995.

46. Hou, C. H., Hou, S. M., Hsueh, Y. S., Lin, J., Wu, H. C., and Lin, F. H. (2009) The *in vivo* performance of biomagnetic hydroxyapatite nanoparticles in cancer hyperthermia therapy, *Biomaterials*, **30**, 3956–3960.

47. Petchsang, N., Pon-On, W., Hodak, J. H., and Tang, I. M. (2009) Magnetic properties of Co-ferrite-doped hydroxyapatite nanoparticles having a core/shell structure, *J. Magnetism Magnetic Mater.*, **321**, 1990–1995.

48. Cacciotti, I., Bianco, A., Lombardi, M., and Montanaro, L. (2009) Mg-substituted hydroxyapatite nanopowders: synthesis, thermal stability and sintering behaviour, *J. Eur. Ceram. Soc.*, **29**, 2969–2978.

49. Bianco, A., Cacciotti, I., Lombardi, M., and Montanaro, L. (2009) Si-substituted hydroxyapatite nanopowders: synthesis, thermal stability and sinterability, *Mater. Res. Bull.*, **44**, 345–354.

50. Capuccini, C., Torricelli, P., Boanini, E., Gazzano, M., Giardino, R., and Bigi, A. (2009) Interaction of Sr-doped hydroxyapatite nanocrystals with osteoclast and osteoblast-like cells, *J. Biomed. Mater. Res. A*, **89A**, 594–600.

51. Hou, C. H., Chen, C. W., Hou, S. M., Li, Y. T., and Lin, F. H. (2009) The fabrication and characterization of dicalcium phosphate dihydrate-modified magnetic nanoparticles and their performance in hyperthermia processes *in vitro*, *Biomaterials*, **30**, 4700–4707.

52. Hanifi, A., Fathi, M. H., Sadeghi, H. M. M., and Varshosaz, J. (2010) Mg^{2+} substituted calcium phosphate nano particles synthesis for non viral gene delivery application, *J. Mater. Sci. Mater. Med.*, **21**, 2393–2401.

53. Li, W., and Gao, L. (2003) Fabrication of Hap-ZrO_2 (3Y) nano-composite by SPS, *Biomaterials*, **24**, 937–940.

54. Wang, L., Nemoto, R., and Senna, M. (2002) Microstructure and chemical states of hydroxyapatite/silk fibroin nanocomposites synthesized via a wet-mechanochemical route, *J. Nanopart. Res.*, **4**, 535–540.

55. Nemoto, R., Wang, L., Ikoma, T., Tanaka, J., and Senna, M. (2004) Preferential alignment of hydroxyapatite crystallites in nanocomposites with chemically disintegrated silk fibroin, *J. Nanopart. Res.*, **6**, 259–265.

56. Fang, L. M., Leng, Y., and Gao, P. (2006) Processing and mechanical properties of HA/UHMWPE nanocomposites, *Biomaterials*, **27**, 3701–3707.

57. Sugawara, A., Yamane, S., and Akiyoshi, K. (2006) Nanogel-templated mineralization: polymer-calcium phosphate hybrid nanomaterials. *Macromol. Rapid Commun.*, **27**, 441–446.

58. Pushpakanth, S., Srinivasan, B., Sreedhar, B., and Sastry, T.P. (2008) An *in situ* approach to prepare nanorods of titania–hydroxyapatite (TiO_2–HAp) nanocomposite by microwave hydrothermal technique, *Mater. Chem. Phys.*, **107**, 492–498.

59. Chang, M. C., Ko, C. C., and Douglas, W. H. (2003) Preparation of hydroxyapatite-gelatin nanocomposite, *Biomaterials*, **24**, 2853–2862.

60. Hao, J., Liu, Y., Zhou, S., Li, Z., and Deng, X. (2003) Investigation of nanocomposites based on semi-interpenetrating network of [L-poly(ε-caprolactone)]/[net-poly(ε-caprolactone)] and hydroxyapatite nanocrystals, *Biomaterials*, **24**, 1531–1539.

61. Deng, X. M., Hao, J. Y., and Wang, C. S. (2001) Preparation and mechanical properties of nanocomposites of poly(D,L-lactide) with Ca-deficient hydroxyapatite nanocrystals, *Biomaterials*, **22**, 2867–2873.

62. Hong, Z., Zhang, P., He, C., Qiu, X., Liu, A., Chen, L., Chena, X., and Jing, X. (2005) Nanocomposite of poly(L-lactide) and surface grafted hydroxyapatite: mechanical properties and biocompatibility, *Biomaterials*, **26**, 6296–6304.

63. Ramay, H. R. R., and Zhang, M. (2004) Biphasic calcium phosphate nanocomposite porous scaffolds for load-bearing bone tissue engineering, *Biomaterials*, **25**, 5171–5180.

64. Cross, K. J., Huq, N. L., Palamara, J. E., Perich, J. W., and Reynolds, E. C. (2005) Physicochemical characterization of casein phosphopeptide — amorphous calcium phosphate nanocomplexes, *J. Biol. Chem.*, **280**, 15362–15369.

65. Murugan, R., and Ramakrishna, S. (2005) Development of nanocomposites for bone grafting, *Comp. Sci. Tech.*, **65**, 2385–2406.

66. Liou, S. C., Chen, S. Y., and Liu, D. M. (2004) Phase development and structural characterization of calcium phosphate ceramics — polyacrylic acid nanocomposites at room temperature in water-methanol mixtures, *J. Mater. Sci. Mater. Med.*, **15**, 1261–1266.

67. Sung, Y. M., Shin, Y. K., and Ryu, J. J. (2007) Preparation of hydroxyapatite/zirconia bioceramic nanocomposites for orthopaedic and dental prosthesis applications, *Nanotechnology*, **18**, 065602 (6 pages).

68. Sreedhar, B., Aparna, Y., Sairam, M., and Hebalkar, N. (2007) Preparation and characterization of HAP / carboxymethyl chitosan nanocomposites, *J. Appl. Polym. Sci.*, **105**, 928–934.

69. Pramanik, N., Biswas, S. K., and Pramanik, P. (2008) Synthesis and characterization of hydroxyapatite/poly(vinyl alcohol phosphate) nanocomposite biomaterials, *Int. J. Appl. Ceram. Technol.*, **5**, 20–28.

70. Li, X., and Chang, J. (2008) Preparation of bone-like apatite — collagen nanocomposites by a biomimetic process with phosphorylated collagen, *J. Biomed. Mater. Res. A*, **85A**, 293–300.

71. Ohsawa, H., Ito, A., Sogo, Y., Yamazaki, A., and Ohno, T. (2007) Synthesis of albumin/DCP nano-composite particles, *Key Eng. Mater.*, pp. 330–332, 239–242.

72. Degirmenbasi, N., Kalyon, D. M., and Birinci, E. (2006) Biocomposites of nanohydroxyapatite with collagen and poly(vinyl alcohol), *Colloids Surf. B*, **48**, 42–49.

73. Zhang, X., Li, Y. B., Zuo, Y., Lv, G. Y., Mu, Y. H., and Li, H. (2007) Morphology, hydrogen-bonding and crystallinity of nano-hydroxyapatite/polyamide 66 biocomposites, *Composites A*, **38**, 843–848.

74. Wei, J., Li, Y. B., and Lau, K. T. (2007) Preparation and characterization of a nano apatite/polyamide$_6$ bioactive composite, *Composites B*, **38**, 301–305.

75. Wei, J., and Li, Y. B. (2004) Tissue engineering scaffold material of nano-apatite crystals and polyamide composite, *Eur. Polym. J.*, **40**, 509–515.

76. Szaraniec, B., Rosół, P., and Chłopek, J. (2005) Carbon composite material and polysulfone modified by nano-hydroxyapatite, *e-Polymers*, **030**.

77. Pramanik, N., Mohapatra, S., and Pramanik, P. (2007) Processing and properties of nano-hydroxyapatite (n-HAp) / poly(ethylene-co-acrylic acid) (EAA) composite using a phosphonic acid coupling agent for orthopedic applications, *J. Am. Ceram. Soc.,* **90**, 369–375.

78. Ren, Y. J., Sun, X. D., Cui, F. Z., Wei, Y. T., Cheng, Z. J., and Kong, X. D. (2007) Preparation and characterization of Antheraea pernyi silk fibroin based nanohydroxyapatite composites, *J. Bioact. Compat. Polym*, **22**, 465–474.

79. Xu, H. H. K., Sun, L., Weir, M. D., Takagi, S., Chow, L. C., and Hockey, B. (2007) Effects of incorporating nanosized calcium phosphate particles on properties of whisker-reinforced dental composites, *J. Biomed. Mater. Res. B* (*Appl. Biomater.*), **81B**, 116–125.

80. Liu, L., Liu, J., Wang, M., Min, S., Cai, Y., Zhu, L., and Yao, J. (2008) Preparation and characterization of nano-hydroxyapatite/silk fibroin porous scaffolds, *J. Biomater. Sci. Polymer Edn.*, **19**, 325–338.

81. Xu, F., Li, Y. B., Deng, Y., and Xiong, J. (2008) Porous nano-hydroxyapatite/ poly(vinyl alcohol) composite hydrogel as artificial cornea fringe: characterization and evaluation *in vitro*, *J. Biomater. Sci. Polymer Edn.*, **19**, 431–439.

82. Huang, J., Lin, Y. W., Fu, X. W., Best, S. M., Brooks, R. A., Rushton, N., and Bonfield, W. (2007) Development of nanosized hydroxyapatite reinforced composites for tissue engineering scaffolds, *J. Mater. Sci. Mater. Med.*, **18**, 2151–2157.

83. Yusong, P., Dangsheng, X., and Xiaolin, C. (2007) Mechanical properties of nanohydroxyapatite reinforced poly(vinyl alcohol) gel composites as biomaterial, *J. Mater. Sci.*, **42**, 5129–5134.

84. Deng, C., Weng, J., Lu, X., Zhou, S. B., Wan, J. X., Qu, S. X., Feng, B., Li, X. H., and Cheng, Q. Y. (2008) Mechanism of ultrahigh elongation rate of poly(D, L-lactide)-matrix composite biomaterial containing nano-apatite fillers, *Mater. Lett.*, **62**, 607–610.

85. Sundaraseelan, J., and Sastry, T. P. (2007) Fabrication of a biomimetic compound containing nano hydroxyapatite — demineralised bone matrix, *J. Biomed. Nanotechnol.*, **3**, 401–405.

86. Teng, S., Chen, L., Guo, Y., and Shi, J. (2007) Formation of nano-hydroxyapatite in gelatin droplets and the resulting porous composite microspheres, *J. Inorg. Biochem.*, **101**, 686–691.

87. Meng, Y. H., Tang, C. Y., Tsui, C. P., and Chen, D. Z. (2008) Fabrication and characterization of needle-like nano-HA and HA/MWNT composites, *J. Mater. Sci. Mater. Med.*, **19**, 75–81.

88. Dorozhkin, S. V. (2009) Calcium orthophosphate-based biocomposites and hybrid biomaterials, *J. Mater. Sci.*, **44**, 2343–2387.

89. Williams, D. F. (2008) The relationship between biomaterials and nanotechnology, *Biomaterials*, **29**, 1737–1738.

90. Feynman, R. P. (1992) There's plenty of room at the bottom, *J. Microelectromechanical Systems*, **1**, 60–66.

91. European Commission, Scientific Committee on Emerging and Newly Identified Health Risks (SCENIHR) Opinion on "the scientific aspects of the existing and proposed definitions relating to products of nanoscience and nanotechnologies". *Adopted Brussels: European Commission.*, 29 November 2007.

92. Moriarty, P. (2001) Nanostructured materials, *Rep. Prog. Phys.*, **64**, 297–381.

93. Webster, T. J., and Ahn, E. S. (2006) Nanostructured biomaterials for tissue engineering bone, *Adv. Biochem. Eng. Biotechnol.*, **103**, 275–308.

94. Streicher, R. M., Schmidt, M., and Fiorito, S. (2007) Nanosurfaces and nanostructures for artificial orthopedic implants, *Nanomedicine*, **2**, 861–874.

95. Havancsak, K. (2003) Nanotechnology at present and its promises in the future. *Mater. Sci. Forum*, pp. 414–415, 85–94.

96. Duncan, R. (2004) Nanomedicines in action, *Pharm. J.*, **273**, 485–488.

97. Williams, D. F. (2007) On the nature of biomaterials, *Biomaterials*, **30**, 5897–5909.

98. Liu, H., and Webster, T. J. (2007) Nanomedicine for implants: a review of studies and necessary experimental tools, *Biomaterials*, **28**, 354–369.

99. Murugan, R., and Ramakrishna, S. (2004) Bioresorbable composite bone paste using polysaccharide based nano hydroxyapatite, *Biomaterials*, **25**, 3829–3835.

100. Murugan, R., and Ramakrishna, S. (2005) Aqueous mediated synthesis of bioresorbable nanocrystalline hydroxyapatite, *J. Cryst. Growth*, **274**, 209–213.

101. Li, G., Huang, J., Li, Y., Zhang, R., Deng, B., Zhang, J., and Aoki, H. (2007) *In vitro* study on influence of a discrete nano-hydroxyapatite on leukemia P388 cell behavior, *Biomed. Mater. Eng.*, **17**, 321–327.

102. Ganesan, K., Kovtun, A., Neumann, S., Heumann, R., and Epple, M. (2008) Calcium phosphate nanoparticles: colloidally stabilized and made fluorescent by a phosphate-functionalized porphyrin, *J. Mater. Chem.*, **18**, 3655–3661.

103. Kim, H. W., and Kim, H. E. (2005) Nanofiber generation of hydroxyapatite and fluor-hydroxyapatite bioceramics, *J. Biomed. Mater. Res. B* (*Appl. Biomater.*), **77B**, 323–328.

104. Cihlar, J., and Castkova, K. (2002) Direct synthesis of nanocrystalline hydroxyapatite by hydrothermal hydrolysis of alkylphosphates, *Monatshefte für Chemie*, **133**, 761–771.

105. Lak, A., Mazloumi, M., Mohajerani, M., Kajbafvala, A., Zanganeh, S., Arami, H., and Sadrnezhaad, S. K. (2008) Self-assembly of dandelion-like hydroxyapatite nanostructures via hydrothermal method, *J. Am. Ceram. Soc.*, **91**, 3292–3297.

106. Sun, L., Chow, L. C., Frukhtbeyn, S. A., and Bonevich, J. E. (2010) Preparation and properties of nanoparticles of calcium phosphates with various Ca/P ratios, *J. Res. Natl. Inst. Stand. Technol.*, **115**, 243–255.

107. Sylvie, J., Sylvie, T. D., Pascal, P. M., Fabienne, P., Hassane, O., and Guy, C. (2010) Effect of hydroxyapatite and β-tricalcium phosphate nanoparticles on promonocytic U937 cells, *J. Biomed. Nanotechnol*, **6**, 158–165.

108. Wang, J., Chen, X., Yang, X., Xu, S., Zhang, X., and Gou, Z. (2011) A facile pollutant-free approach toward a series of nutritionally effective calcium phosphate nanomaterials for food and drink additives, *J. Nanopart. Res.* **13**, 1039–1048.

109. Traykova, T., Aparicio, C., Ginebra, M. P., and Planell, J. A. (2006) Bioceramics as nanomaterials, *Nanomedicine*, **1**, 91–106.

110. Grainger, D. W., and Castner, D. G. (2008) Nanobiomaterials and nanoanalysis: opportunities for improving the science to benefit biomedical technologies, *Adv. Mater.*, **20**, 867–877.

111. Nelson, K. G. (1972) The Kelvin equation and solubility of small particles, *J. Pharmac. Sci.*, **61**, 479–480.

112. Fan, C., Chen, J., Chen, Y., Ji, J., and Teng, H. H. (2006) Relationship between solubility and solubility product: the roles of crystal sizes and crystallographic directions, *Geochim. Cosmochim. Acta*, **70**, 3820–3829.

113. Sato, M., and Webster, T. J. (2004) Nanobiotechnology: implications for the future of nanotechnology in orthopedic applications, *Expert Rev. Med. Dev.*, **1**, 105–114.

114. Hahn, H. (2003) Unique features and properties of nanostructured materials, *Adv. Eng. Mater.*, **5**, 277–284.

115. Aronov, D., Karlov, A., and Rosenman, G. (2007) Hydroxyapatite nanoceramics: basic physical properties and biointerface modification, *J. Eur. Ceram. Soc.*, **27**, 4181–4186.

116. Ramsden, J. J., and Freeman, J. (2009) The nanoscale, *Nanotechnol. Percept.*, **5**, 3–25.

117. Rempel, A. A. (2007) Nanotechnologies. Properties and applications of nanostructured materials, *Russ. Chem. Rev.*, **76**, 435–461.

118. Catledge, S. A., Fries, M. D., Vohra, Y. K., Lacefield, W. R., Lemons, J. E., Woodard, S., and Venugopalan, R. (2002) Nanostructured ceramics for biomedical implants, *J. Nanosci. Nanotechnol.*, **2**, 1–20.

119. Balasundarama, G., and Webster, T. J. (2006) A perspective on nanophase materials for orthopedic implant applications, *J. Mater. Chem.*, **16**, 3737–3745.

120. Balasundarama, G., and Webster, T. J. (2006) Nanotechnology and biomaterials for orthopedic medical applications, *Nanomedicine*, **1**, 169–176.

121. Padilla, S., Izquierdo-Barba, I., and Vallet-Regí, M. (2008) High specific surface area in nanometric carbonated hydroxyapatite, *Chem. Mater.*, **20**, 5942–5944.

122. Kalita, S. J., Bhardwaj, A., and Bhatt, H. A. (2007) Nanocrystalline calcium phosphate ceramics in biomedical engineering, *Mater. Sci. Eng. C*, **27**, 441–449.

123. LeGeros, R. Z. (1991) *Calcium Phosphates in Oral Biology and Medicine*, Karger, Basel, p. 210.

124. Mann, S. (1986) The study of biominerals by high resolution transmission electron microscopy, *Scan. Electron. Microsc.*, Pt. 2, 393–413.

125. Katsura, N. (1990) Nanospace theory for biomineralization, *Dent. Jpn.*, **27**, 57–63.

126. Cuisinier, F. J. G., Voegel, J. C., Yacaman, J., and Frank, R. M. (1992) Structure of initial crystals formed during human amelogenesis, *J. Cryst. Growth*, **116**, 314–318.

127. Cuisinier, F. J. G., Steuer, P., Senger, B., Voegel, J. C., and Frank, R. M. (1993) Human amelogenesis: high resolution electron microscopy of nanometer-sized particles, *Cell Tissue Res.*, **273**, 175–182.

128. Brès, E. F., Moebus, G., Kleebe, H. J., Pourroy, G., Werkmann, J., and Ehret, G. (1993) High resolution electron microscopy study of amorphous calcium phosphate, *J. Cryst. Growth*, **129**, 149–162.

129. Layrolle, P., and Lebugle, A. (1994) Characterization and reactivity of nanosized calcium phosphate prepared in anhydrous ethanol, *Chem. Mater.*, **6**, 1996–2004.

130. Cui, F. Z., Wen, H. B., Zhang, H. B., Ma, C. L., and Li, H. D. (1994) Nanophase hydroxyapatite-like crystallites in natural ivory, *J. Mater. Sci. Lett.*, **13**, 1042–1044.

131. Li, Y. B., de Wijn, J., Klein, C. P. A. T., de Meer, S. V., and de Groot, K. (1994) Preparation and characterization of nanograde osteoapatite-like rod crystals, *J. Mater. Sci. Mater. Med.*, **5**, 252–255.

132. Li, Y. B., de Groot, K., de Wijn, J., Klein, C. P. A. T., and de Meer, S. V. (1994) Morphology and composition of nanograde calcium phosphate needle-like crystals formed by simple hydrothermal treatment, *J. Mater. Sci. Mater. Med.*, **5**, 326–331.

133. Shirkhanzadeh, M. (1994) X-ray diffraction and Fourier transform infrared analysis of nanophase apatite coatings prepared by electrocrystallization, *Nanostruct. Mater.*, **4**, 677–684.

134. Webster, T. J., Ergun, C., Doremus, R. H., Siegel, R. W., and Bizios, R. (2000) Specific proteins mediate enhanced osteoblast adhesion on nanophase ceramics, *J. Biomed. Mater. Res.*, **51**, 475–483.

135. Chan, C. K., Kumar, T. S. S., Liao, S., Murugan, R., Ngiam, M., and Ramakrishnan, S. (2006) Biomimetic nanocomposites for bone graft applications, *Nanomedicine*, **1**, 177–188.

136. Okada, M., Furukawa, K., Serizawa, T., Yanagisawa, Y., Tanaka, H., Kawai, T., and Furuzono, T. (2009) Interfacial interactions between calcined hydroxyapatite nanocrystals and substrates, *Langmuir*, **25**, 6300–6306.

137. Mikołajczyk, T., Rabiej, S., and Bogun, M. (2006) Analysis of the structural parameters of polyacrylonitrile fibers containing nanohydroxyapatite, *J. Appl. Polym. Sci.*, **101**, 760–765.

138. There are both nano-sized biomaterials and nanostructured biomaterials, which should be differentiated from each other. Nano-sized biomaterials refer to individual molecular level biomaterials such as single proteins (are not considered in this review), while nanostructured biomaterials refer to any biomaterials whose structure or morphology can be engineered to get features with nanometer-scale dimensions [139]. This review is limited to calcium orthophosphate-based nanostructured biomaterials only.

139. Thomas, V., Dean, D. R., and Vohra, Y. K. (2006) Nanostructured biomaterials for regenerative medicine, *Curr. Nanosci.*, **2**, 155–177.

140. LeGeros, R. Z. (1993) Biodegradation and bioresorption of calcium phosphate ceramics, *Clin. Mater.*, **14**, 65–88.

141. Wang, J., and Shaw, L. L. (2007) Morphology-enhanced low-temperature sintering of nanocrystalline hydroxyapatite, *Adv. Mater.*, **19**, 2364–2369.

142. Fomin, A. S., Barinov, S. M., Ievlev, V. M., Smirnov, V. V., Mikhailov, B. P., Belonogov, E. K., and Drozdova, N. A. (2008) Nanocrystalline hydroxyapatite ceramics produced by low-temperature sintering after high-pressure treatment, *Dokl. Chem.*, **418**, 22–25.

143. Drouet, C., Bosc, F., Banu, M., Largeot, C., Combes, C., Dechambre, G., Estournes, C., Raimbeaux, G., and Rey, C. (2009) Nanocrystalline apatites: from powders to biomaterials, *Powder Technol.*, **190**, 118–122.

144. Ramesh, S., Tan, C. Y., Bhaduri, S. B., Teng, W. D., and Sopyan, I. (2008) Densification behaviour of nanocrystalline hydroxyapatite bioceramics, *J. Mater. Process. Technol.*, **206**, 221–230.

145. Skorokhod, V. V., Solonin, S. M., Dubok, V. A., Kolomiets, L. L., Katashinskii, V. P., and Shinkaruk, A. V. (2008) Pressing and sintering of nanosized hydroxyapatite powders, *Powder Metall. Metal Ceram.*, **47**, 518–524.

146. Sung, Y. M., Lee, J. C., and Yang, J. W. (2004) Crystallization and sintering characteristics of chemically precipitated hydroxyapatite nanopowder, *J. Cryst. Growth*, **262**, 467–472.

147. Lin, K., Chang, J., Lu, J., Wu, W., and Zeng, Y. (2007) Properties of β-$Ca_3(PO_4)_2$ bioceramics prepared using nanosized powders, *Ceram. Int.*, **33**, 979–985.

148. Tanaka, Y., Hirata, Y., and Yoshinaka, R. (2003) Synthesis and characteristics of ultra-fine hydroxyapatite particles, *J. Ceram. Proc. Res.*, **4**, 197–201.

149. Wang, J., and Shaw, L. L. (2009) Nanocrystalline hydroxyapatite with simultaneous enhancements in hardness and toughness, *Biomaterials*, **30**, 6565–6572.

150. Stupp, S. I., and Ciegler, G. W. (1992) Organoapatites: materials for artificial bone. I. Synthesis and microstructure, *J. Biomed. Mater. Res.*, **26**, 169–183.

151. Webster, T. J., Ergun, C., Doremus, R. H., Siegel, R. W., and Bizios, R. (2001) Enhanced osteoclast-like cell functions on nanophase ceramics, *Biomaterials*, **22**, 1327–1333.

152. Huang, J., Best, S. M., Bonfield, W., Brooks, R. A., Rushton, N., Jayasinghe, S. N., and Edirisinghe, M. J. (2004) *In vitro* assessment of the biological response to nanosized hydroxyapatite, *J. Mater. Sci. Mater. Med.*, **15**, 441–445.

153. Kim, H. W., Kim, H. E., and Salih, V. (2005) Stimulation of osteoblast responses to biomimetic nanocomposites of gelatin-hydroxyapatite for tissue engineering scaffolds, *Biomaterials*, **26**, 5221–5230.

154. Webster, T. J., Siegel, R. W., and Bizios, R. (1999) Osteoblast adhesion on nanophase ceramics, *Biomaterials*, **20**, 1221–1227.

155. Webster, T. J., Ergun, C., Doremus, R. H., Siegel, R. W., and Bizios, R. (2000) Enhanced functions of osteoblast on nanophase ceramics, *Biomaterials*, **21**, 1803–1810.

156. Smith, I. O., McCabe, L. R., and Baumann, M. J. (2006) MC3T3-E1 osteoblast attachment and proliferation on porous hydroxyapatite scaffolds fabricated with nanophase powder, *Int. J. Nanomed*, **1**, 189–194.

157. Nelson, M., Balasundaram, G., and Webster, T. J. (2006) Increased osteoblast adhesion on nanoparticulate crystalline hydroxyapatite functionalized with KRSR, *Int. J. Nanomed.*, **1**, 339–349.

158. Liu, H., Yazici, H., Ergun, C., Webster, T. J., and Bermek, H. (2008) An *in vitro* evaluation of the Ca/P ratio for the cytocompatibility of nano-to-micron particulate calcium phosphates for bone regeneration, *Acta Biomater.*, **4**, 1472–1479.

159. Sato, M., Sambito, M. A., Aslani, A., Kalkhoran, N. M., Slamovich, E. B., and Webster, T. J. (2006) Increased osteoblast functions on undoped and yttrium-doped nanocrystalline hydroxyapatite coatings on titanium, *Biomaterials*, **27**, 2358–2369.

160. Thian, E. S., Huang, J., Best, S. M. Barber, Z. H., Brooks, R. A., Rushton, N., and Bonfield, W. (2006) The response of osteoblasts to nanocrystalline silicon-substituted hydroxyapatite thin films, *Biomaterials*, **27**, 2692–2698.

161. Palin, E., Liu, H., and Webster, T. J. (2005) Mimicking the nanofeatures of bone increases bone-forming cell adhesion and proliferation, *Nanotechnology*, **16**, 1828–1835.

162. Sun, W., Chu, C., Wang, J., and Zhao, H. (2007) Comparison of periodontal ligament cells responses to dense and nanophase hydroxyapatite, *J. Mater. Sci. Mater. Med.*, **18**, 677–683.

163. Ergun, C., Liu, H., Webster, T. J., Olcay, E., Yılmaz, Ş., and Sahin, F. C. (2008) Increased osteoblast adhesion on nanoparticulate calcium phosphates with higher Ca/P ratios, *J. Biomed. Mater. Res. A*, **85A**, 236–241.

164. Lewandrowski, K. U., Bondre, S. P., Wise, D. L., and Trantolo, D. J. (2003) Enhanced bioactivity of a poly(propylene fumarate) bone graft substitute by augmentation with nano-hydroxyapatite, *Biomed. Mater. Eng.*, **13**, 115–124.

165. Zhou, D. S., Zhao, K. B., Li, Y., Cui, F. Z., and Lee, I. S. (2006) Repair of segmental defects with nano-hydroxyapatite / collagen / PLA composite combined with mesenchymal stem cells, *J. Bioactive Compat. Polym.*, **21**, 373–384.

166. Xu, Z., Sun, J., Changsheng, L., and Jie, W. (2009) Effect of hydroxyapatite nanoparticles of different concentrations on rat osteoblast, *Mater. Sci. Forum*, 610–613, 1364–1369.

167. Krut'ko, V. K., Kulak, A. I., Lesnikovich, L. A., Trofimova, I. V., Musskaya, O. N., Zhavnerko, G. K., and Paribok, I. V. (2007) Influence of the dehydration procedure on the physicochemical properties of nanocrystalline hydroxylapatite xerogel, *Russ. J. General Chem.*, **77**, 336–342.

168. Severin, A. V., Komarov, V. F., Bozhevol'nov, V. E., and Melikhov, I. V. Morphological selection in suspensions of nanocrystalline hydroxylapatite leading to spheroidal aggregates, *Russ. J. Inorg. Chem.*, **50**, 72–77.

169. Biggemann, D., da Silva, M. H. P., Rossi, A. M., and Ramirez, A. J. High-resolution transmission electron microscopy study of nanostructured hydroxyapatite, *Microsc. Microanal.*, **14**, 433–438.

170. Kester, M., Heakal, Y., Fox, T., Sharma, A., Robertson, G. P., Morgan, T. T., Altinoğlu, E. I., Tabaković, A., Parette, M. R., Rouse, S. M., Ruiz-Velasco, V., and Adair, J. H. (2008) Calcium phosphate nanocomposite particles for *in vitro* imaging and encapsulated chemotherapeutic drug delivery to cancer cells, *Nano Lett.*, **8**, 4116–4121.

171. Welzel, T., Meyer-Zaika, W., and Epple, M. (2004) Continuous preparation of functionalised calcium phosphate nanoparticles with adjustable crystallinity, *Chem. Commun.*, 1204–1205.

172. Nichols, H. L., Zhang, N., Zhang, J., Shi, D., Bhaduri, S., and Wen, X. (2007) Coating nanothickness degradable films on nanocrystalline hydroxyapatite particles to improve the bonding strength between nanohydroxyapatite and degradable polymer matrix, *J. Biomed. Mater. Res. A*, **82A**, 373–382.

173. Rey, C., Hina, A., Tofighi, A., and Glimcher, M. J. (1995) Maturation of poorly crystalline apatites: chemical and structural aspects *in vivo* and *in vitro*, *Cell Mater.*, **5**, 345–356.

174. Dorozhkin, S. V. (2007) Calcium orthophosphates, *J. Mater. Sci.*, **42**, 1061–1095.

175. Dorozhkin, S. V. (2009) Calcium orthophosphates in nature, biology and medicine, *Materials*, **2**, 399–498.

176. Elliott, J. C. (1994) *Structure and Chemistry of the Apatites and Other Calcium Orthophosphates*, Elsevier, Amsterdam, p. 404.

177. Olszta, M. J., Cheng, X., Jee, S. S., Kumar, R., Kim, Y. Y., Kaufmane, M. J., Douglas, E. P., and Gower, L. B. (2007) Bone structure and formation: a new perspective, *Mater. Sci. Eng. R*, **58**, 77–116.

178. Cui, F. Z., Li, Y., and Ge, J. (2007) Self-assembly of mineralized collagen composites, *Mater. Sci. Eng. R*, **57**, 1–27.

179. Meyers, M. A., Chen, P. Y., Lin, A. Y. M., and Seki, Y. (2008) Biological materials: structure and mechanical properties, Prog, *Mater. Sci.*, **53**, 1–206.

180. Currey, J. D. (2005) Hierarchies in biomineral structures, *Science*, **309**, 253–254.

181. Rubin, M. A., Jasiuk, I., Taylor, J., Rubin, J., Ganey, T., and Apkarian, R. P. (2003) TEM analysis of the nanostructure of normal and osteoporotic human trabecular bone, *Bone*, **33**, 270–282.

182. Hartgerink, J. D., Beniash, E., and Stupp, S. I. (2001) Self-assembly and mineralization of peptide-amphiphile nanofibers, *Science*, **294**, 1684–1688.

183. Ji, B., and Gao, H. (2006) Elastic properties of nanocomposite structure of bone, *Compos. Sci. Technol.*, **66**, 1212–1218.

184. Wang, L., Nancollas, G. H., Henneman, Z. J., Klein, E., and Weiner, S. (2006) Nanosized particles in bone and dissolution insensitivity of bone mineral, *Biointerphases*, **1**, 106–111.

185. Gao, H., Ji, B., Jager, I. L., Arz, E., and Fratzl, P. (2003) Materials become insensitive to flaws at nanoscale: lessons from nature, *Proc. Natl. Acad. Sci. U. S. A.*, **100**, 5597–5660.

186. Gupta, H. S., Seto, J., Wagermaier, W., Zaslansky, P., Boesecke, P., and Fratzl, P. (2006) Cooperative deformation of mineral and collagen in bone at the nanoscale, *Proc. Natl. Acad. Sci. U. S. A.*, **103**, 17741–17746.

187. Currey, J. D. (2006) *Bones: Structure and Mechanics, Princeton University Press, Princeton*, p. 456.

188. Porter, A. E., Nalla, R. K., Minor, A., Jinschek, J. R., Kisielowski, C., Radmilovic, V., Kinney, J. H., Tomsia, A. P., and Ritchie, R. O. (2005) A transmission electron microscopy study of mineralization in age-induced transparent dentin, *Biomaterials*, **26**, 7650–7660.

189. Kirkham, J., Brookes, S. J., Shore, R. C., Wood, S. R., Smith, D. A., Zhang, J., Chen, H., and Robinson, C. (2002) Physico-chemical properties of crystal surfaces in matrix-mineral interactions during mammalian biomineralisation, *Curr. Opin. Colloid Interf. Sci.*, **7**, 124–132.

190. Daculsi, G., Mentanteau, J., Kerebel, L. M., and Mitre, D. (1984) Length and shape of enamel crystals, *Calcif. Tissue Int.*, **36**, 550–555.

191. Robinson, C., Connell, S., Kirkham, J., Shorea, R., and Smith, A. (2004) Dental enamel — a biological ceramic: regular substructures in enamel hydroxyapatite crystals revealed by atomic force microscopy, *J. Mater. Chem.*, **14**, 2242–2248.

192. Chen, H., Tang, Z., Liu, J., Sun, K., Chang, S. R., Peters, M. C., Mansfield, J. F., Czajka-Jakubowska, A., and Clarkson, B. H. (2006) Acellular synthesis of a human enamel-like microstructure, *Adv. Mater.*, **18**, 1846–1851.

193. Chen, H., Clarkson, B. H., Sun, K., and Mansfield, J. F. (2005) Self-assembly of synthetic hydroxyapatite nanorods into an enamel prism-like structure, *J. Colloid Interf. Sci.*, **288**, 97–103.

194. Robinson, C. (2007) Self-oriented assembly of nano-apatite particles: a subunit mechanism for building biological mineral crystals, *J. Dental Res.*, **86**, 677–679.

195. Cui, F. Z., and Ge, J. (2007) New observations of the hierarchical structure of human enamel, from nanoscale to microscale, *J. Tissue Eng. Regen. Med.*, **1**, 185–191.

196. He, L. H., and Swain, M. V. (2007) Enamel — a "metallic-like" deformable biocomposite, *J. Dent.*, **35**, 431–437.

197. Nelson, S. J. (2009) *Wheeler's Dental Anatomy, Physiology and Occlusion*, 9th edn., W. B. Saunders, Philadelphia, p. 368.

198. Suvorova E. I., and Buffat P. A. (1999) Electron diffraction from micro- and nanoparticles of hydroxyapatite, *J. Microscopy*, **196**, 46–58.

199. Panda, R. N., Hsieh, M. F., Chung, R. J., and Chin, T. S. (2001) X-ray diffractometry and X-ray photoelectron spectroscopy investigations of nanocrytalline hydroxyapatite synthesized by a hydroxide gel technique, *Jpn. J. Appl. Phys.*, **40**, 5030–5035.

200. Panda, R. N., Hsieh, M. F., Chung, R. J., and Chin, T. S. (2003) FTIR, XRD, SEM and solid state NMR investigations of carbonate-containing hydroxyapatite nano-particles synthesized by hydroxide-gel technique, *J. Phys. Chem. Solids*, **64**, 193–199.

201. Eichert, D. Sfihi, H., Combes, C., and Rey, C. (2004) Specific characteristics of wet nanocrystalline apatites. Consequences on biomaterials and bone tissue, *Key Eng. Mater.*, 254–256, 927–930.

202. Rey, C., Combes, C., Drouet, C., Sfihi, H., and Barroug, A. (2007) Physico-chemical properties of nanocrystalline apatites: implications for biominerals and biomaterials, *Mater. Sci. Eng. C*, **27**, 198–205.

203. Eichert, D., Drouet, C., Sfihi, H., Rey, C., and Combes, C. (2007) Nanocrystalline apatite-based biomaterials: synthesis, processing and characterization, in *Biomaterials Research Advances* (ed. Kendall, J. B.), Nova Science Publishers, Chapter 5, pp. 93–143.

204. Aronov, D., and Rosenman, G. (2007) Trap state spectroscopy studies and wettability modification of hydroxyapatite nanobioceramics, *J. Appl. Phys.*, **101**, 034701 (5 pages).

205. Jäger, C., Welzel, T., Meyer-Zaika, W., and Epple, M. (2006) A solid-state NMR investigation of the structure of nanocrystalline hydroxyapatite, *Magn. Reson. Chem.*, **44**, 573–580.

206. Isobe, T., Nakamura, S., Nemoto, R., Senna, M., and Sfihi, H. (2002) Solid-state double nuclear magnetic resonance of calcium phosphate nanoparticules synthesized by wet-mechanochemical reaction, *J. Phys. Chem. B*, **106**, 5169–5176.

207. Bertinetti, L., Tampieri, A., Landi, E., Ducati, C., Midgley, P. A., Coluccia, S., and Martra, G. (2007) Surface structure, hydration, and cationic sites of nanohydroxyapatite: UHR-TEM, IR, and microgravimetric studies, *J. Phys. Chem. C*, **111**, 4027–4035.

208. Bertinetti, L., Tampieri, A., Landi, E., Bolis, V., Busco, C., and Martra, G. (2008) Surface structure, hydration and cationic sites of nanohydroxyapatite, *Key Eng. Mater.*, pp. 361–363, 87–90.

209. Bertinetti, L., Drouet, C., Combes, C., Rey, C., Tampieri, A., Coluccia, S., and Martra, G. (2009) Surface characteristics of nanocrystalline apatites: effect of Mg surface enrichment on morphology, surface hydration species, and cationic environments, *Langmuir*, **25**, 5647–5654.

210. Gopi, D., Indira, J., Prakash, V. C. A., and Kavitha, L. (2009) Spectroscopic characterization of porous nanohydroxyapatite synthesized by a novel amino acid soft solution freezing method, *Spectrochim. Acta A*, **74A**, 282–284.

211. Rossi, A. M., da Silva, M. H. P., Ramirez, A. J., Biggemann, D., Caraballo, M. M., Mascarenhas, Y. P., Eon, J. G., and Moure, G. T. (2007) Structural properties of hydroxyapatite with particle size less than 10 nanometers, *Key Eng. Mater.*, 330–332, 255–258.

212. Ramirez, C. A. O., Costa, A. M., Bettini, J., Ramirez, A. J., da Silva, M. H. P., and Rossi, A. M. (2009) Structural properties of nanostructured carbonate apatites, *Key Eng. Mater.*, 396–398, 611–614.

213. Pasteris, J. D., Wopenka, B., Freeman, J. J., Rogers, K., Valsami-Jones, E., van der Houten, J. A. M., and Silva, M. J. (2004) Lack of OH in nanocrystalline apatite as a function of degree of atomic order: implications for bone and biomaterials, *Biomaterials*, **25**, 229–238.

214. Sakhno, Y., Bertinetti, L., Iafisco, M., Tampieri, A., Roveri, N., and Martra, G. (2010) Surface hydration and cationic sites of nanohydroxyapatites with amorphous or crystalline surfaces: a comparative study, *J. Phys. Chem. C*, **114**, 16640–16648.

215. Zyman, Z. Z., Epple, M., Rokhmistrov, D., and Glushko, V. (2009) On impurities and the internal structure in precipitates occurring during the precipitation of nanocrystalline calcium phosphate, *Mat. -wiss. u. Werkstofftech.*, **40**, 297–301.

216. Cazalbou, S., Combes, C., Eichert, D., and Rey, C. (2004) Adaptative physico-chemistry of bio-related calcium phosphates, *J. Mater. Chem.*, **14**, 2148–2153.

217. Eichert, D., Salomé, M., Banu, M., Susini, J., and Rey, C. (2005) Preliminary characterization of calcium chemical environment in apatitic and non-apatitic calcium phosphates of biological interest by X-ray absorption spectroscopy, *Spectrochim. Acta B*, **60B**, 850–858.

218. Rosenman, G., Aronov, D., Oster, L., Haddad, J., Mezinskis, G., Pavlovska, I., Chaikina, M., and Karlov, A. (2007) Photoluminescence and surface

photovoltage spectroscopy studies of hydroxyapatite nano-bio-ceramics, *J. Luminescence*, 122–123, 936–938.

219. Melikhov, I. V., Teterin, Y. A., Rudin, V. N., Teterin, A. Y., Maslakov, K. I., and Severin, A. V. (2009) An X-ray electron study of nanodisperse hydroxyapatite, *Russ. J. Phys. Chem. A*, **83**, 91–97.

220. Aronov, D., Rosenman, G., Karlov, A., and Shashkin, A. (2006) Wettability patterning of hydroxyapatite nanobioceramics induced by surface potential modification, *Appl. Phys. Lett.*, **88**, 163902 (3 pages).

221. Rau, J. V., Generosi, A., Ferro, D., Minozzi, F., Paci, B., Albertini, V. R., Dolci, G., and Barinov, S. M. (2009) *In situ* time-resolved X-ray diffraction study of evolution of nanohydroxyapatite particles in physiological solution, *Mater. Sci. Eng. C*, **29**, 1140–1143.

222. Arora, A. (2004) Ceramics in nanotech revolution, *Adv. Eng. Mater.*, **6**, 244–247.

223. Ioku, K., and Yoshimura, M. (1991) Stochiometric apatite fine single crystals by hydrothermal synthesis, *Phosphorus Res. Bull.*, **1**, 15–20.

224. Chen, J. D., Wang, Y. J., Wei, K., Zhang, S. H., and Shi, X. T. (2007) Self-organization of hydroxyapatite nanorods through oriented attachment, *Biomaterials*, **28**, 2275–2280.

225. Guo, X., Xiao, P., Liu, J., and Shen, Z. (2004) Fabrication of nanostructured hydroxyapatite via hydrothermal synthesis and spark plasma sintering, *J. Am. Ceram. Soc.*, **88**, 1026–1029.

226. Brown, P. W., and Constantz B. (eds.) (1994) *Hydroxyapatite and Related Materials*, CRC Press, Boca Raton, p. 343.

227. Amjad, Z. (ed.) (1997) *Calcium Phosphates in Biological and Industrial Systems*, Kluwer Academic Publishers, Boston, p. 529.

228. Hughes, J. M., Kohn, M., and Rakovan, J. (eds.) (2002) *Phosphates: Geochemical, Geobiological and Materials Importance*, Series: Reviews in Mineralogy and Geochemistry, Mineralogical Society of America, Washington, D.C., **48**, p. 742.

229. Chow, L. C., and Eanes, E. D. (eds.) (2001) *Octacalcium Phosphate*, Karger, Basel, p. 168.

230. Brès, E., and Hardouin, P. (eds.) (1998) Les matériaux en phosphate de calcium, *Aspects fondamentaux, / Calcium phosphate materials, Fundamentals,* Sauramps Medical, Montpellier, p. 176.

231. Komarov, V. F., and Kibalchitz, V. (1979) Precipitation of apatite through highly saturated solutions, *Moscow Univ. Bull. Chem. Dic.*, 2680–2685.

232. Prakash, K. H., Kumar, R., Ooi, C. P., Cheang, P., and Khor, K. A. (2006) Conductometric study of precursor compound formation during wet-

chemical synthesis of nanocrystalline hydroxyapatite, *J. Phys. Chem. B*, **110**, 24457–24462.

233. Tao, J., Pan, H., Wang, J., Wu, J., Wang, B., Xu, X., and Tang, R. (2008) Evolution of amorphous calcium phosphate to hydroxyapatite probed by gold nanoparticles, *J. Phys. Chem. C*, **112**, 14929–14933.

234. Chane-Ching, J. Y., Lebugle, A., Rousselot, I., Pourpoint, A., and Pelle, F. (2007) Colloidal synthesis and characterization of monocrystalline apatite nanophosphors, *J. Mater. Chem.*, **17**, 2904–2913.

235. Zyman, Z. Z., Rokhmistrov, D. V., and Glushko, V. I. (2010) Structural and compositional features of amorphous calcium phosphate at the early stage of precipitation, *J. Mater. Sci. Mater. Med.*, **21**, 123–130.

236. Wei, M., Ruys, A. J., Milthorpe, B. K., and Sorrell, C. C. (1999) Solution ripening of hydroxyapatite nanoparticles: effects on electrophoretic deposition, *J. Biomed. Mater. Res.*, **45**, 11–19.

237. Zhu, X., Eibl, O., Berthold, C., Scheideler, L., and Geis-Gerstorfer, J. (2006) Structural characterization of nanocrystalline hydroxyapatite and adhesion of pre-osteoblast cells, *Nanotechnology*, **17**, 2711–2721.

238. Rusu, V. M., Ng, C. H., Wilke, M., Tiersch, B., Fratzl, P., and Peter, M. G. (2005) Size-controlled hydroxyapatite nanoparticles as self-organized organic — inorganic composite materials, *Biomaterials*, **26**, 5414–5426.

239. Wang, Y. J., Lai, C., Wei, K., and Tang, S. Q. (2005) Influence of temperature, ripening time and cosurfactant on solvothermal synthesis of calcium phosphate nanobelts, *Mater. Lett.*, **59**, 1098–1104.

240. Li, Y. B., Li, D., and Weng, W. (2008) Preparation of nano carbonate-substituted hydroxyapatite from an amorphous precursor, *Int. J. Appl. Ceram. Technol.*, **5**, 442–448.

241. Zhang, S., and Gonsalves, K. E. (1997) Preparation and characterization of thermally stable nanohydroxyapatite, *J. Mater. Sci. Mater. Med.*, **8**, 25–28.

242. Ferraz, M. P., Monteiro, F. J., and Manuel, C. M. (2004) Hydroxyapatite nanoparticles: a review of preparation methodologies, *J. Appl. Biomater. Biomech.*, **2**, 74–80.

243. Ahn, E. S., Gleason, N. J., Nakahira, A., and Ying, J. Y. (2001) Nanostructure processing of hydroxyapatite-based bioceramics, *Nano Lett.*, **1**, 149–153.

244. Mazelsky, R., Hopkins, R. H., and Kramer, W. E. (1968) Czochralski-growth of calcium fluorophosphates, *J. Cryst. Growth*, 3–4, 260–264.

245. Loutts, G. B., and Chai, B. H. T. (1993) Growth of high-quality single crystals of FAP ($Ca_5(PO_4)_3F$) and its isomorphs, *Proc. SPIE Int. Soc. Opt. Eng.*, **1863**, 31–34.

246. Siegel, R. W. (1996) Creating nanophase materials, *Sci. Am.*, **275**, 42–47.

247. Hu, J., Odom, T. W., and Lieber, C. M. (1999) Chemistry and physics in one dimension: synthesis and properties of nanowires and nanotubes, *Acc. Chem. Res.*, **32**, 435–445.

248. Schmidt, H. K. (2000) Nanoparticles for ceramic and nanocomposite processing, *Mol. Cryst. Liq. Cryst.*, **353**, 165–179.

249. Cushing, B. L., Kolesnichenko, V. L., and O'Connor, C. J. (2004) Recent advances in the liquid-phase syntheses of inorganic nanoparticles, *Chem. Rev.*, **104**, 3893–3946.

250. Wang, X., Zhuang, J., Peng, Q., and Li, Y. (2005) A general strategy for nanocrystal synthesis, *Nature*, **437**, 121–124.

251. Yin, Y., and Alivisatos, A. P. (2005) Colloidal nanocrystal synthesis and the organic-inorganic interface, *Nature*, **437**, 664–670.

252. Mao, Y., Park, T. J., Zhang, F., Zhou, H., and Wong, S. S. (2007) Environmentally friendly methodologies of nanostructure synthesis, *Small*, **3**, 1122–1139.

253. Ma, M. G., and Zhu, J. F. (2010) Recent progress on fabrication of calcium-based inorganic biodegradable nanomaterials, *Recent Patents on Nanotechnology*, **4**, 164–170.

254. Takagi, S., Chow, L. C., and Ishikawa, K. (1998) Formation of hydroxyapatite in new calcium phosphate cements, *Biomaterials*, **19**, 1593–1599.

255. Meejoo, S., Maneeprakorn, W., and Winotai, P. (2006) Phase and thermal stability of nanocrystalline hydroxyapatite prepared via microwave heating, *Thermochim. Acta*, **447**, 115–120.

256. Kumta, P., Sfeir, C., Lee, D. H., Olton, D., and Choi, D. (2005) Nanostructured calcium phosphates for biomedical applications: novel synthesis and characterization, *Acta Biomater.*, **1**, 65–83.

257. Liou, S. C., Chen, S. Y., Lee, H. Y., and Bow, J. S. (2004) Structural characterization of nanosized calcium deficient apatite powders, *Biomaterials*, **25**, 189–196.

258. Mollazadeh, S., Javadpour, J., and Khavandi, A. (2007) *In situ* synthesis and characterization of nanosized hydroxyapatite in poly(vinyl alcohol) matrix, *Ceram. Int.*, **33**, 1579–1583.

259. Bigi, A., Boanini, E., Gazzano, M., Rubini, K., and Torricelli, P. (2004) Nanocrystalline hydroxyapatite — polyaspartate composites, *Biomed Mater. Eng.*, **14**, 573–579.

260. Chen, H., Sun, K., Tang, Z., Law, R. V., Mansfield, J. F., Czajka-Jakubowska, A., and Clarkson, B. H. (2006) Synthesis of fluorapatite nanorods and nanowires by direct precipitation from solution, *Cryst. Growth Des.*, **6**, 1504–1508.

261. Kong, L., Gao, Y., Cao, W., Gong, Y., Zhao, N., and Zhang, X. (2005) Preparation and characterization of nano-hydroxyapatite/chitosan composite scaffolds, *J. Biomed. Mater. Res. A*, **75A**, 275–282.

262. Kong, L., Gao, Y., Lu, G., Gong, Y., Zhao, N., and Zhang, X. (2006) A study on the bioactivity of chitosan/nano-hydroxyapatite composite scaffolds for bone tissue engineering, *Eur. Polym. J.*, **42**, 3171–3179.

263. Melikhov, I. V., Komarov, V. F., Severin, A. V., Bozhevol'nov, V. E., and Rudin, V. N. (2000) Two-dimensional crystalline hydroxyapatite, *Dokl. Phys. Chem.*, **373**, 355–358.

264. Zhao, Y., Zhang, Y., Ning, F., Guo, D., and Xu, Z. (2007) Synthesis and cellular biocompatibility of two kinds of HAP with different nanocrystal morphology, *J. Biomed. Mater. Res. B (Appl. Biomater.)*, **83B**, 121–126.

265. Ganesan, K., and Epple, M. (2008) Calcium phosphate nanoparticles as nuclei for the preparation of colloidal calcium phytate, *New J. Chem.*, **32**, 1326–1330.

266. Zhang, Y., and Lu, J. (2007) A simple method to tailor spherical nanocrystal hydroxyapatite at low temperature, *J. Nanopart. Res.*, **9**, 589–594.

267. Bouyer, E., Gitzhofer, F., and Boulos, M. I. (2000) Morphological study of hydroxyapatite nanocrystal suspension, *J. Mater. Sci. Mater. Med.*, **11**, 523–531.

268. Pang, Y. X., and Bao, X. (2003) Influence of temperature, ripening time and calcination on the morphology and crystallinity of hydroxyapatite nanoparticles, *J. Eur. Ceram. Soc.*, **23**, 1697–1704.

269. Kumar, R., Prakash, K. H., Cheang, P. and Khor, K. A. (2004) Temperature driven morphological changes of chemically precipitated hydroxyapatite nanoparticles, *Langmuir*, **20**, 5196–5200.

270. Li-yun, C., Chuan-bo, Z., and Jian-feng, H. (2005) Influence of temperature, [Ca^{2+}], Ca/P ratio and ultrasonic power on the crystallinity and morphology of hydroxyapatite nanoparticles prepared with a novel ultrasonic precipitation method, *Mater. Lett.*, **59**, 1902–1906.

271. Afshar, A., Ghorbani, M., Ehsani, N., Saeri, M. R., and Sorrell, C. C. (2003) Some important factors in the wet precipitation process of hydroxyapatite, *Mater. Des.*, **24**, 197–202.

272. Wei, M., Ruys, A. J., Milthorpe, B. K., and Sorrell, C. C. (2005) Precipitation of hydroxyapatite nanoparticles: effects of precipitation method on electrophoretic deposition, *J. Mater. Sci. Mater. Med.*, **16**, 319–324.

273. Liu, Y., Hou, D., and Wang, G. (2004) A simple wet chemical synthesis and characterization of hydroxyapatite nanorods, *Mater. Chem. Phys.*, **86**, 69–73.

274. Mobasherpour, I., Heshajin, M. S., Kazemzadeh, A., and Zakeri, M. (2007) Synthesis of nanocrystalline hydroxyapatite by using precipitation method, *J. Alloys Compd.*, **430**, 330–333.

275. Phillips, M. J., Darr, J. A., Luklinska, Z. B., and Rehman, I. (2003) Synthesis and characterization of nanobiomaterials with potential osteological applications, *J. Mater. Sci. Mater. Med.*, **14**, 875–882.

276. Lee, S. J., Yoon, Y. S., Lee, M. H., and Oh, N. S. (2007) Nanosized hydroxyapatite powder synthesized from eggshell and phosphoric acid, *J. Nanosci. Nanotechnol.*, **7**, 4061–4064.

277. Monmaturapoj, N. (2008) Nanosize hydroxyapatite powders preparation by wet-chemical precipitation route, *J. Metals Mater. Miner.*, **18**, 15–20.

278. Ramesh, S., Tan, C. Y., Sopyan, I., Hamdi, M., and Teng, W. D. (2007) Consolidation of nanocrystalline hydroxyapatite powder, *Sci. Technol. Adv. Mater.*, **8**, 124–130.

279. Zhou, W., Zhang, S. M., Hu, W., Qiu, Z. Y., and Liu, Y. H. (2007) Dialysis efficiency in rapid synthesis of nano-hydroxyapatite, *Key Eng. Mater.*, 330–332, 211–214.

280. Shi, H. B., Zhong, H., Liu, Y., Gu, J. Y., and Yang, C. S. (2007) Effect of precipitation method on stoichiometry and morphology of hydroxyapatite nanoparticles, *Key Eng. Mater.*, 330–332, 271–274.

281. Monkawa, A., Ikoma, T., Yunoki, S., Ohta, K., and Tanaka, J. (2006) Electrophoretic deposition of hydroxyapatite nanocrystal, *Key Eng. Mater.*, 309–311, 643–646.

282. Ong, H. T., Loo, J. S. C., Boey, F. Y. C., Russell, S. J., Ma, J., and Peng, K. W. (2008) Exploiting the high-affinity phosphonate — hydroxyapatite nanoparticle interaction for delivery of radiation and drugs, *J. Nanopart. Res.*, **10**, 141–150.

283. Silva, G. W. C., Ma, L., Hemmers, O., and Lindle, D. (2008) Micro-structural characterization of precipitation-synthesized fluorapatite

nano-material by transmission electron microscopy using different sample preparation techniques, *Micron*, **39**, 269–274.

284. Poinern, G. E., Brundavanam, R. K., Mondinos, N., and Jiang, Z. T. (2009) Synthesis and characterisation of nanohydroxyapatite using an ultrasound assisted method, *Ultrason. Sonochem.*, **16**, 469–474.

285. Doğan, Ö., and Öner, M. (2008) The influence of polymer architecture on nanosized hydroxyapatite precipitation, *J. Nanosci. Nanotechnol.*, **8**, 667–674.

286. Loo, S. C. J., Siew, Y. E., Ho, S., Boey, F. Y. C., and Ma, J. (2008) Synthesis and hydrothermal treatment of nanostructured hydroxyapatite of controllable sizes, *J. Mater. Sci. Mater. Med.*, **19**, 1389–1397.

287. Guo, X., Gough, J. E., Xiao, P., Liu, J., and Shen, Z. (2007) Fabrication of nanostructured hydroxyapatite and analysis of human osteoblastic cellular response, *J. Biomed. Mater. Res. A*, **82A**, 1022–1032.

288. Safronova, T. V., Putlyaev, V. I., Sergeeva, A. I., Kunenkov, E. V., and Tret'yakov, Y. D. (2009) Synthesis of nanocrystalline calcium hydroxyapatite from calcium saccharates and ammonium hydrogen phosphate, *Dokl. Chem.*, **426**, 118–123.

289. Iafisco, M., Palazzo, B., Marchetti, M., Margiotta, N., Ostuni, R., Natile, G., Morpurgo, M., Gandin, V., Marzano, C., and Roveri, N. (2009) Smart delivery of antitumoral platinum complexes from biomimetic hydroxyapatite nanocrystals, *J. Mater. Chem.*, **19**, 8385–8392.

290. Wang, P., Li, C., Gong, H., Jiang, X., Wang, H., and Li, K. (2010) Effects of synthesis conditions on the morphology of hydroxyapatite nanoparticles produced by wet chemical process, *Powder Technol.*, **203**, 315–321.

291. Leskiv, M., Lagoa, A. L. C., Urch, H., Schwiertz, J., da Piedade, M. E. M., and Epple, M. (2009) Energetics of calcium phosphate nanoparticle formation by the reaction of $Ca(NO_3)_2$ with $(NH_4)_2HPO_4$, *J. Phys. Chem. C*, **113**, 5478–5484.

292. Rodrigues, L. R., Motisuke, M., Zavaglia, C. A. C. (2009) Synthesis of nanostructured hydroxyapatite: a comparative study between sol-gel and aqueous solution precipitation, *Key Eng. Mater.*, 396–398, 623–626.

293. López-Macipe, A., Gómez-Morales, J., and Rodríguez-Clemente, R. (1998) Nanosized hydroxyapatite precipitation from homogeneous calcium/citrate/phosphate solutions using microwave and conventional heating, *Adv. Mater.*, **10**, 49–53.

294. Siddharthan, A., Seshadri, S. K., and Kumar, T. S. S. (2005) Rapid synthesis of calcium deficient hydroxyapatite nanoparticles by microwave irradiation, *Trends Biomater. Artif. Organs*, **18**, 110–113.

295. Ioku, K., Yamauchi, S., Fujimori, H., Goto, S., and Yoshimura, M. (2002) Hydrothermal preparation of fibrous apatite and apatite sheet, *Solid State Ionics*, **151**, 147–150.

296. Kasahara, H., Ogata, N., and Ogihara, T. (2004) Effect of starting solution on the formation of calcium phosphate nano particles by hydrothermal process, *J. Ceram. Soc. Jpn.*, **112**, 650–654.

297. Lemos, A. F., Rocha, J. H. G., Quaresma, S. S. F., Kannana, S., Oktar, F. N., Agathopoulos, S., and Ferreira, J. M. F. (2006) Hydroxyapatite nano-powders produced hydrothermally from nacreous material, *J. Eur. Ceram. Soc.*, **26**, 3639–3646.

298. Chaudhry, A. A., Haque, S., Kellici, S., Boldrin, P., Rehman, I., Khalid, F. A., and Darr, J. A. (2006) Instant nano-hydroxyapatite: a continuous and rapid hydrothermal synthesis, *Chem. Commun.*, 2286–2288.

299. Cao, M., Wang, Y., Guo, C., Qi, Y., and Hu, C. (2004) Preparation of ultrahigh-aspect-ratio hydroxyapatite nanofibers in reverse micelles under hydrothermal conditions, *Langmuir*, **20**, 4784–4786.

300. Jinlong, N. (2007) Hydrothermal synthesis of nano-crystalline hydroxyapatite, *Key Eng. Mater*, 330–332, 247–250.

301. Ryu, I. Y., Kim, D. J., Han, J. S., and Lee, M. H. (2008) Influence of two-step sintering variables on phase stability of hydrothermally prepared HAp nano powders, *Key Eng. Mater.*, 361–363, 91–94.

302. Han, J. K., Song, H. Y., Saito, F., and Lee, B. T. (2006) Synthesis of high purity nanosized hydroxyapatite powder by microwave-hydrothermal method, *Mater. Chem. Phys.*, **99**, 235–239.

303. Suchanek, W. L., Shuk, P., Byrappa, K., Riman, R. E., TenHuisen, K. S., and Janas, V. F. (2002) Mechanochemical-hydrothermal synthesis of carbonated apatite powders at room temperature, *Biomaterials*, **23**, 699–710.

304. Guo, X., and Xiao, P. (2006) Effects of solvents on properties of nanocrystalline hydroxyapatite produced from hydrothermal process, *J. Eur. Ceram. Soc.*, **26**, 3383–3391.

305. Xin, R., and Yu, K. (2009) Ultrastructure characterization of hydroxyapatite nanoparticles synthesized by EDTA-assisted hydrothermal method, *J. Mater. Sci.*, **44**, 4205–4209.

306. Zhang, C., Yang, J., Quan, Z., Yang, P., Li, C., Hou, Z., and Lin, J. (2009) Hydroxyapatite nano- and microcrystals with multiform morphologies:

controllable synthesis and luminescence properties, *Cryst. Growth Des.*, **9**, 2725–2733.

307. Zhang, H. B., Zhou, K. C., Li, Z. Y., and Huang, S. P. (2009) Plate-like hydroxyapatite nanoparticles synthesized by the hydrothermal method, *J. Phys. Chem. Solids*, **70**, 243–248.

308. Abdel-Aal, E. A., El-Midany, A. A., and El-Shall, H. (2008) Mechanochemical-hydrothermal preparation of nano-crystallite hydroxyapatite using statistical design, *Mater. Chem. Phys.*, **112**, 202–207.

309. Sun, Y., Guo, G., Tao, D., and Wang, Z. (2007) Reverse microemulsion-directed synthesis of hydroxyapatite nanoparticles under hydrothermal conditions, *J. Phys. Chem. Solids*, **68**, 373–377.

310. Du, X., Chu, Y., Xing, S., and Dong, L. (2009) Hydrothermal synthesis of calcium hydroxyapatite nanorods in the presence of PVP, *J. Mater. Sci.*, **44**, 6273–6279.

311. Xin, R., Ren, F., and Leng, Y. (2010) Synthesis and characterization of nano-crystalline calcium phosphates with EDTA-assisted hydrothermal method, *Mater. Des.*, **31**, 1691–1694.

312. Yan, L., Li, Y., Deng, Z., Zhuang, J., and Sun, X. (2001) Surfactant-assisted hydrothermal synthesis of hydroxyapatite nanorods, *Int. J. Inorg. Mater.*, **3**, 633–637.

313. Zhang, F., Zhou, Z., Yang, S., Mao, L., Chen, H., and Yu, X. (2005) Hydrothermal synthesis of hydroxyapatite nanorods in the presence of anionic starburst dendrimer, *Mater. Lett.*, **59**, 1422–1425.

314. Byrappa, K., and Haber, M. (2002) *Handbook of Hydrothermal Technology: A Technology for Crystal Growth and Materials Processing*, Noyes Publications, New Jersey, p. 893.

315. Chai, C. S., and Ben-Nissan, B. (1999) Bioactive nanocrystalline sol-gel hydroxyapatite coatings, *J. Mater. Sci. Mater. Med.*, **10**, 465–469.

316. Ben-Nissan, B., Green, D. D., Kannangara, G. S. K., Chai, C. S., and Milev, A. (2001) ^{31}P NMR studies of diethyl phosphite derived nanocrystalline hydroxyapatite, *J. Sol-Gel Sci. Technol.*, **21**, 27–37.

317. Ben-Nissan, B., and Choi, A. H. (2006) Sol-gel production of bioactive nanocoatings for medical applications, Part 1: an introduction, *Nanomedicine*, **1**, 311–319.

318. Choi, A. H., and Ben-Nissan, B. (2007) Sol-gel production of bioactive nanocoatings for medical applications, Part 2: current research and development, *Nanomedicine*, **2**, 51–61.

319. Kim, T. S., and Kumta, P. N. (2004) Sol-gel synthesis and characterization of nanostructured hydroxyapatite powder, *Mater. Sci. Eng. B*, **111**, 232–236.

320. Rajabi-Zamani, A. H., Behnamghader, A., and Kazemzadeh, A. (2008) Synthesis of nanocrystalline carbonated hydroxyapatite powder via nonalkoxide sol-gel method, *Mater. Sci. Eng. C*, **28**, 1326–1329.

321. Sopyan, I., Toibah, A. R., and Natasha, A. N. (2008) Nanosized bioceramic hydroxyapatite powders via sol-gel method, *Int. J. Mech. Mater. Eng.*, **3**, 133–138.

322. Yuan, Y., Liu, C., Zhang, Y., and Shan, X. (2008) Sol-gel auto-combustion synthesis of hydroxyapatite nanotubes array in porous alumina template, *Mater. Chem. Phys.*, **112**, 275–280.

323. Kuriakose, T. A., Kalkura, S. N., Palanichamy, M., Arivuoli, D., Dierks, K., Bocelli, G., and Betzel, C. (2004) Synthesis of stoichiometric nano crystalline hydroxyapatite by ethanol-based sol-gel technique at low temperature, *J. Cryst. Growth*, **263**, 517–523.

324. Jahandideh, R., Behnamghader, A., Rangie, M., Youzbashi, A., Joughehdoust, S., and Tolouei, R. (2009) Sol-gel synthesis of FHA nanoparticles and CDHA agglomerates from a mixture with a nonstochiometric Ca/P ratio, *Key Eng. Mater.*, 396–398, 607–610.

325. Sanosh, K. P., Chu, M. C., Balakrishnan, A., Lee, Y. J., Kim, T. N., and Cho, S. J. (2009) Synthesis of nano hydroxyapatite powder that simulate teeth particle morphology and composition, *Curr. Appl. Phys.*, **9**, 1459–1462.

326. Darroudi, M., Eshtiagh-Hosseini, H., Housaindokht, M. R., and Youssefi, A. (2010) Preparation and characterization of fluorohydroxyapatite nanopowders by nonalkoxide sol-gel method, *Digest J. Nanomater. Biostruct.*, **5**, 29–33.

327. Li, B., Wang, X. L., Guo, B., Xiao, Y. M., Fan, H. S., and Zhang, X. D. (2007) Preparation and characterization of nano hydroxyapatite, *Key Eng. Mater.*, 330–332, 235–238.

328. Tas, A. C. (2000) Synthesis of biomimetic Ca-hydroxyapatite powders at 37°C in synthetic body fluids, *Biomaterials*, **21**, 1429–1438.

329. Wu, Y. S., Lee, Y. H., and Chang, H. C. (2009) Preparation and characteristics of nanosized carbonated apatite by urea addition with coprecipitation method, *Mater. Sci. Eng. C*, **29**, 237–241.

330. Rameshbabu, N., Kumar, T. S. S., Murugan, R., and Rao, K. P. (2005) Mechanochemical synthesis of nanocrystalline fluorinated hydroxyapatite, *Int. J. Nanosci.*, **4**, 643–649.

331. Yeong, K. C. B., Wang, J., and Ng, S. C. (2001) Mechanochemical synthesis of nanocrystalline hydroxyapatite from CaO and $CaHPO_4$, *Biomaterials*, **22**, 2705–2712.

332. Coreno, J. A., Coreno, O. A., Cruz, R. J. J., and Rodriguez, C. C. (2005) Mechanochemical synthesis of nanocrystalline carbonate-substituted hydroxyapatite, *Opt. Mater.*, **27**, 1281–1285.

333. el Briak-Ben Abdeslam, H., Mochales, C., Ginebra, M. P., Nurit, J., Planell J. A., and Boudeville, P. (2003) Dry mechanochemical synthesis of hydroxyapatites from dicalcium phosphate dihydrate and calcium oxide: a kinetic study, *J. Biomed. Mater. Res, A*, **67A**, 927–937.

334. Nakamura, S., Isobe, T., and Senna, M. (2001) Hydroxyapatite nano sol prepared via a mechanochemical route, *J. Nanopart. Res.*, **3**, 57–61.

335. Nasiri-Tabrizi, B., Honarmandi, P., Ebrahimi-Kahrizsangi, R., and Honarmandi, P. (2009) Synthesis of nanosize single-crystal hydroxyapatite via mechanochemical method, *Mater. Lett.*, **63**, 543–546.

336. Fathi, M. H., and Zahrani, E. M. (2009) Fabrication and characterization of fluoridated hydroxyapatite nanopowders via mechanical alloying, *J. Alloys Compd.*, **475**, 408–414.

337. Fathi, M. H., and Zahrani, E. M. (2009) Mechanical alloying synthesis and bioactivity evaluation of nanocrystalline fluoridated hydroxyapatite, *J. Cryst. Growth*, **311**, 1392–1403.

338. Silva, C. C., Graça, M. P. F., Valente, M. A., and Sombra, A. S. B. (2007) Crystallite size study of nanocrystalline hydroxyapatite and ceramic system with titanium oxide obtained by dry ball milling, *J. Mater. Sci.*, **42**, 3851–3855.

339. Zahrani, E. M., and Fathi, M. H. (2009) The effect of high-energy ball milling parameters on the preparation and characterization of fluorapatite nanocrystalline powder, *Ceram. Int.*, **35**, 2311–2323.

340. Xu, J. L., Khor, K. A., Dong, Z. L., Gu, Y. W., Kumar, R., and Cheang, P. (2004) Preparation and characterization of nanosized hydroxyapatite powders produced in a radio frequency (rf) thermal plasma, *Mater. Sci. Eng. A*, **374**, 101–108.

341. Xu, J. L., Khor, K. A., Kumar, R., and Cheang, P. (2006) RF induction plasma synthesized calcium phosphate nanoparticles, *Key Eng. Mater.*, 309–311, 511–514.

342. Ruksudjarit, A., Pengpat, K., Rujijanagul, G., and Tunkasiri, T. (2008) Synthesis and characterization of nanocrystalline hydroxyapatite from natural bovine bone, *Curr. Appl. Phys.*, **8**, 270–272.

343. Cho, J. S., and Kang, Y. C. (2008) Nano-sized hydroxyapatite powders prepared by flame spray pyrolysis, *J. Alloys Compd.*, **464**, 282–287.

344. Wang, X., Zhuang, J., Peng, Q., and Li, Y. (2006) Liquid-solid-solution synthesis of biomedical hydroxyapatite nanorods, *Adv. Mater.*, **18**, 2031–2034.

345. Shirkhanzadeh, M. (1998) Direct formation of nanophase hydroxyapatite on cathodically polarized electrodes, *J. Mater. Sci. Mater. Med.*, **9**, 67–72.

346. Montalbert-Smith, R., Palma, C. A., Arias, J. D., and Montero, M. L. (2009) Formation of hydroxyapatite nanosized and other apatites by electrolysis process, *Key Eng. Mater.*, 396–398, 579–582.

347. Liu, J., Li, K., Wang, H., Zhu, M., Xu, H., and Yan, H. (2005) Self-assembly of hydroxyapatite nanostructures by microwave irradiation, *Nanotechnology*, **16**, 82–87.

348. Rameshbabu, N., Rao, K. P., and Kumar, T. S. S. (2005) Accelerated microwave processing of nanocrystalline hydroxyapatite, *J. Mater. Sci.*, **40**, 6319–6323.

349. Ran, X., Chen, J., Ran, J., Gou, L., and Zhang, X. (2007) Synthesis of nanosized carbonated hydroxyapatite under microwave irradiation, *Key Eng. Mater.*, 330–332, 303–306.

350. Siddharthan, A., Seshadri, S. K., and Kumar, T. S. S. (2004) Microwave accelerated synthesis of nanosized calcium deficient hydroxyapatite, *J. Mater. Sci. Mater. Med.*, **15**, 1279–1284.

351. Krishna, D. S. R., Siddharthan, A., Seshadri, S. K., and Kumar, T. S. S. (2007) A novel route for synthesis of nanocrystalline hydroxyapatite from eggshell waste, *J. Mater. Sci. Mater. Med.*, **18**, 1735–1743.

352. Seo, D. S., Hwang, K. H., and Lee, J. K. (2008) Nanostructured hydroxyapatite by microwave sintering, *J. Nanosci. Nanotechnol.*, **8**, 944–948.

353. Arami, H., Mohajerani, M., Mazloumi, M., Khalifehzadeh, R., Lak, A., and Sadrnezhaad, S. K. (2009) Rapid formation of hydroxyapatite nanostrips via microwave irradiation, *J. Alloys Compd.*, **469**, 391–394.

354. Rameshbabu, N., Kumar, T. S. S., and Rao, K. P. (2006) Synthesis of nanocrystalline fluorinated hydroxyapatite by microwave processing and its *in vitro* dissolution study, *Bull. Mater. Sci.*, **29**, 611–615.

355. Kumar, A. R., Kalainathan, S., and Saral, A. M. (2010) Microwave assisted synthesis of hydroxyapatite nano strips, *Cryst. Res. Technol.*, **45**, 776–778.

356. Shih, W. J., Chen, Y. F., Wang, M. C., and Hon, M. H. (2004) Crystal growth and morphology of the nanosized hydroxyapatite powders synthesized from $CaHPO_4{\cdot}2H_2O$ and $CaCO_3$ by hydrolysis method, *J. Cryst. Growth*, **270**, 211–218.

357. Zhang, Y., and Lu, J. (2008) The transformation of single-crystal calcium phosphate ribbon-like fibres to hydroxyapatite spheres assembled from nanorods, *Nanotechnology*, **19**, 155608 (10 pages).

358. Furuichi, K., Oaki, Y., and Imai, H. (2006) Preparation of nanotextured and nanofibrous hydroxyapatite through dicalcium phosphate with gelatin, *Chem. Mater.*, **18**, 229–234.

359. Yoruç, A. B. H., and Koca, Y. (2009) Double step stirring: a novel method for precipitation of nano-sized hydroxyapatite powder, *Digest J. Nanomater. Biostructures*, **4**, 73–81.

360. Jarudilokkul, S., Tanthapanichakoon, W., and Boonamnuayvittaya, V. (2007) Synthesis of hydroxyapatite nanoparticles using an emulsion liquid membrane system, *Colloids Surf. A*, **296**, 149–153.

361. Lim, G. K., Wang, J., Ng, S. C., and Gan, L. M. (1999) Nanosized hydroxyapatite powders from microemulsions and emulsions stabilized by a biodegradable surfactant, *J. Mater. Chem.*, **9**, 1635–1639.

362. Guo, G., Sun, Y., Wang, Z., and Guo, H. (2005) Preparation of hydroxyapatite nanoparticles by reverse microemulsion, *Ceram. Int.*, **31**, 869–872.

363. Lim, G. K., Wang, J., Ng, S. C., and Gan, L. M. (1999) Formation of nanocrystalline hydroxyapatite in nonionic surfactant emulsions, *Langmuir*, **15**, 7472–7477.

364. Sun, Y., Guo, G., Wang, Z., and Guo, H. (2006) Synthesis of single-crystal HAP nanorods, *Ceram. Int.*, **32**, 951–954.

365. Bose, S., and Saha, S. K. (2003) Synthesis and characterization of hydroxyapatite nanopowders by emulsion technique, *Chem. Mater.*, **15**, 4464–4469.

366. Lai, C., Tang, S. Q., Wang, Y. J., and Wei, K. (2005) Formation of calcium phosphate nanoparticles in reverse microemulsions, *Mater. Lett.*, **59**, 210–214.

367. Jiang, F. X., Lu, X. Y., Zhang, M. L., and Weng, J. (2008) Regulating size, morphology and dispersion of nano-crystallites of hydroxyapatite by pH value and temperature in microemulsion system, *Key Eng. Mater.*, 361–363, 195–198.

368. Sato, K., Hotta, Y., Nagaoka, T., Yasuoka, M., and Watari, K. (2006) Agglomeration control of hydroxyapatite nano-crystals grown in phase-separated microenvironments, *J. Mater. Sci.*, **41**, 5424–5428.

369. Li, H., Zhu, M. Y., Li, L. H., and Zhou, C. R. (2008) Processing of nanocrystalline hydroxyapatite particles via reverse microemulsions, *J. Mater. Sci.*, **43**, 384–389.

370. Koetz, J., Baier, J., and Kosmella, S. (2007) Formation of zinc sulfide and hydroxylapatite nanoparticles in polyelectrolyte-modified microemulsions, *Colloid Polym. Sci.*, **285**, 1719–1726.

371. Lim, H. N., Kassim, A., and Huang, N. M. (2010) Preparation and characterization of calcium phosphate nanorods using reverse microemulsion and hydrothermal processing routes, *Sains Malaysiana*, **39**, 267–273.

372. Furuzono, T., Walsh, D., Sato, K., Sonoda, K., and Tanaka, J. (2001) Effect of reaction temperature on the morphology and size of hydroxyapatite nanoparticles in an emulsion system, *J. Mater. Sci. Lett.*, **2**, 111–114.

373. Shen, S. C., Chia, L., Ng, W. K., Dong, Y. C., and Tan, R. B. H. (2010) Solid-phase steam-assisted synthesis of hydroxyapatite nanorods and nanoparticles, *J. Mater. Sci.*, **45**, 6059–6067.

374. Wang, Y. J., Lai, C., Wei, K., Chen, X., Ding, Y., and Wang, Z. L. (2006) Investigations on the formation mechanism of hydroxyapatite synthesized by the solvothermal method, *Nanotechnology*, **17**, 4405–4412.

375. Cao, L. Y., Zhang, C. B., and Huang, J. F. (2005) Synthesis of hydroxyapatite nanoparticles in ultrasonic precipitation, *Ceram. Int.*, **31**, 1041–1044.

376. Liu, J., Wu, Q., and Ding, Y. (2005) Self-assembly and fluorescent modification of hydroxyapatite nanoribbon spherulites, *Eur. J. Inorg. Chem.*, **20**, 4145–4149.

377. Huang, J., Jayasinghe, S. N., Su, X., Ahmad, Z., Best, S. M., Edirisinghe, M. J., Brooks, R. A., Rushton, N., and Bonfield, W. (2006) Electrostatic atomisation spraying: a novel deposition method for nano-sized hydroxyapatite, *Key Eng. Mater.*, 309–311, 635–638.

378. Hwang, K. S., and Kim, B. H. (2005) Preparation of calcium phosphate nano-powders prepared by sol-gel assisted-electrostatic spraying method, *J. Mater. Sci.*, **40**, 4665–4666.

379. Uota, M., Arakawa, H., Kitamura, N., Yoshimura, T., Tanaka, J., and Kijima, T. (2005) Synthesis of high surface area hydroxyapatite nanoparticles by mixed surfactant-mediated approach, *Langmuir*, **21**, 4724–4728.

380. Chu, M., and Liu, G. (2005) Preparation and characterization of hydroxyapatite/liposome core — shell nanocomposites, *Nanotechnology*, **16**, 1208–1212.

381. Ye, F., Guo, H., and Zhang, H. (2008) Biomimetic synthesis of oriented hydroxyapatite mediated by nonionic surfactants, *Nanotechnology*, **19**, 245605 (7 pages).

382. Han, Y., Wang, X., and Li, S. (2009) A simple route to prepare stable hydroxyapatite nanoparticles suspension, *J. Nanoparticle Res.*, **11**, 1235–1240.

383. Tseng, Y. H., Kuo, C. S., Li, Y. Y., and Huang, C. P. (2009) Polymer-assisted synthesis of hydroxyapatite nanoparticle, *Mater. Sci. Eng. C*, **29**, 819–822.

384. Klinkaewnarong, J., Swatsitang, E., and Maensiri, S. (2009) Nanocrystalline hydroxyapatite powders by a chitosan-polymer complex solution route: synthesis and characterization, *Solid State Sci.*, **11**, 1023–1027.

385. Li, Y., Li, D., and Xu, Z. (2009) Synthesis of hydroxyapatite nanorods assisted by Pluronics, *J. Mater. Sci.*, **44**, 1258–1263.

386. Nayar, S., Sinha, M. K., Basu, D., and Sinha, A. (2006) Synthesis and sintering of biomimetic hydroxyapatite nanoparticles for biomedical applications, *J. Mater. Sci. Mater. Med.*, **17**, 1063–1068.

387. Hong, Y., Fan, H., Li, B., Guo, B., Liu, M., and Zhang, X. (2010) Fabrication, biological effects, and medical applications of calcium phosphate nanoceramics, *Mater. Sci. Eng. R*, **70**, 225–242.

388. Tadic, D., Veresov, A., Putlayev, V. I., and Epple, M. (2003) *In-vitro* preparation of nanocrystalline calcium phosphates as bone substitution materials in surgery, *Mat.-Wiss. u. Werkstofftech.*, **34**, 1048–1051.

389. Qiu, Y., Xia, H., and Jiang, H. (2010) Fabrication of nano-hydroxyapatite using a novel ultrasonic atomization precipitation method, *J. Nanosci. Nanotechnol.*, **10**, 2213–2218.

390. Rouhani, P., Taghavinia, N., and Rouhani, S. (2010) Rapid growth of hydroxyapatite nanoparticles using ultrasonic irradiation, *Ultrasonics Sonochemistry*, **17**, 853–856.

391. Giardina, M. A., and Fanovich, M. A. (2010) Synthesis of nanocrystalline hydroxyapatite from $Ca(OH)_2$ and H_3PO_4 assisted by ultrasonic irradiation, *Ceram. Int.*, **36**, 1961–1969.

392. Sadjadi, M. S., Meskinfam, M., Sadeghi, B., Jazdarreh, H., and Zare, K. (2010) *In situ* biomimetic synthesis, characterization and *in vitro*

investigation of bone-like nanohydroxyapatite in starch matrix, *Mater. Chem. Phys.*, **124**, 217–222.

393. Mhin, S. W., Ryu, J. H., Kim, K. M., Park, G. S., Ryu, H. W., Shim, K. B., Sasaki, T., and Koshizaki, N. (2009) Simple synthetic route for hydroxyapatite colloidal nanoparticles via a Nd:YAG laser ablation in liquid medium, *Appl. Phys. A*, **96A**, 435–440.

394. Musaev, O. R., Dusevich, V., Wieliczka, D. M., Wrobel, J. M., and Kruger, M. B. (2008) Nanoparticle fabrication of hydroxyapatite by laser ablation in water, *J. Appl. Phys.*, **104**, 084316 (5 pages).

395. Boutinguiza, M., Lusquiños, F., Riveiro, A., Comesaña, R., and Pou, J. (2009) Hydroxylapatite nanoparticles obtained by fiber laser-induced fracture, *Appl. Surf. Sci.*, **255**, 5382–5385.

396. Zuo, Y., Li, Y. B., Wei, J., and Yan, Y. (2003) Influence of ethylene glycol on the formation of calcium phosphate nanocrystals, *J. Mater. Sci. Technol.*, **19**, 628–630.

397. Barinov, S. M., Belonogov, E. K., Ievlev, V. M., Kostyuchenko, A. V., Putlyaev, V. I., Tret'yakov, Y. D., Smirnov, V. V., and Fadeeva, I. V. (2007) Synthesis of dense nanocrystalline hydroxyapatite films, *Dokl. Phys. Chem.*, **412**, 15–18.

398. Mello, A., Mavropoulos, E., Hong, Z., Ketterson, J. B., and Rossi, A. M. (2009) Nanometer coatings of hydroxyapatite characterized by glancing-incidence X-ray diffraction, *Key Eng. Mater.*, pp. 396–398, 369–372.

399. Iafisco, M., Morales, J. G., Hernández-Hernández, M. A., García-Ruiz, J. M., and Roveri, N. (2010) Biomimetic carbonate-hydroxyapatite nanocrystals prepared by vapor diffusion, *Adv. Eng. Mater.*, **12**, B218–B223.

400. Luo, P., and Nieh, T. G. (1995) Synthesis of ultrafine hydroxyapatite particles by a spray dry method, *Mater. Sci. Eng. C*, **3**, 75–78.

401. Chen, F., Wang, Z. C., and Chang, J. L. (2002) Preparation and characterization of nanosized hydroxyapatite particles and hydroxyapatite / chitosan nano-composite for use in biomedical materials, *Mater. Lett.*, **57**, 858–861.

402. Sarig, S., and Kahana, F. (2002) Rapid formation of nanocrystalline apatite, *J. Cryst. Growth*, 237–239, 55–59.

403. Bose, S., and Saha, S. K. (2003) Synthesis of hydroxyapatite nanopowders via sucrose-templated sol-gel method, *J. Am. Ceram. Soc.*, **86**, 1055–1057.

404. Han, Y., Li, S., Wang, X., and Chen, X. (2004) Synthesis and sintering of nanocrystalline hydroxyapatite powders by citric acid sol-gel combustion method, *Mater. Res. Bull.*, **39**, 25–32.

405. Liu, D. M., Yang, Q., Troczynski, T., and Tseng, W. J. (2002) Structural evolution of sol-gel-derived hydroxyapatite, *Biomaterials*, **23**, 1679–1687.

406. Liu, D. M., Troczynski, T., and Tseng, W. J. (2001) Water-based sol-gel synthesis of hydroxyapatite: process development, *Biomaterials*, **22**, 1721–1730.

407. Wang, F., Li, M. S., Lu, Y. P., and Ge, S. S. (2005) Synthesis of nanocrystalline hydroxyapatite powders in stimulated body fluid, *J. Mater. Sci.*, **40**, 2073–2076.

408. Wang, J., and Shaw, L. L. (2009) Synthesis of high purity hydroxyapatite nanopowder via sol-gel combustion process, *J. Mater. Sci. Mater. Med.*, **20**, 1223–1227.

409. Varma, H. K., Kalkura, S. N., and Sivakumar, R. (1998) Polymeric precursor route for the preparation of calcium phosphate compounds, *Ceram. Int.*, **24**, 467–470.

410. Loher, S., Stark, W. J., Maciejewski, M., Baiker, A., Pratsinis, S. E., Reichardt, D., Maspero, F., Krumeich, F., and Günther, D. (2005) Fluoro-apatite and calcium phosphate nanoparticles by flame synthesis, *Chem. Mater.*, **17**, 36–42.

411. Trommer, R. M., Santos, L. A., and Bergmann, C. P. (2009) Nanostructured hydroxyapatite powders produced by a flame-based technique, *Mater. Sci. Eng. C*, **29**, 1770–1775.

412. Chow, L. C., Sun, L., and Hockey, B. (2004) Properties of nanostructured hydroxyapatite prepared by a spray drying technique, *J. Res. Natl. Inst. Stand. Technol.*, **109**, 543–551.

413. Li, J., Chen, Y. P., Yin, Y., Yao, F., and Yao, K. (2007) Modulation of nano-hydroxyapatite size via formation on chitosan-gelatin network film *in situ*, *Biomaterials*, **28**, 781–790.

414. Zhai, Y., Cui, F. Z., and Wang, Y. (2005) Formation of nano hydroxyapatite on recombinant human like collagen fibrils, *Curr. Appl. Phys.*, **5**, 429–432.

415. Liou, S. C., Chen, S. Y., and Liu, D. M. (2003) Synthesis and characterization of needlelike apatitic nanocomposite with controlled aspect ratios, *Biomaterials*, **24**, 3981–3988.

416. Liou, S. C., Chen, S. Y., and Liu, D. M. (2005) Manipulation of nanoneedle and nanosphere apatite/poly(acrylic acid) nanocomposites, *J. Biomed. Mater. Res. B (Appl. Biomater.)*, **73B**, 117–122.

417. Amjad, Z. (1995) Performance of polymeric additives as HA crystal growth inhibitors, *Phosphorus Res. Bull.*, **5**, 1–12.

418. Kamitahara, M., Kawashita, M., Kokubo, T., and Nakamura, T. (2001) Effect of polyacrylic acid on the apatite formation of a bioactive ceramic in a simulated body fluid: fundamental examination of the possibility of obtaining bioactive glass-ionomer cements for orthopedic use, *Biomaterials*, **22**, 3191–3196.

419. Wang, X., Li, Y., Wei, J., and de Groot, K. (2002) Development of biomimetic nano-hydroxyapatite/poly(hexamethylene adipamide) composites, *Biomaterials*, **23**, 4787–4791.

420. Sinha, A., Nayar, S., and Agrawak, A. C. (2003) Synthesis of nanosized and microporous precipitated hydroxyapatite in synthetic polymers and biopolymers, *J. Am. Ceram. Soc.*, **86**, 357–359.

421. Liao, S., Watari, F., Zhu, Y., Uo, M., Akasaka, T., Wang, W., Xu, G., and Cui, F. (2007) The degradation of the three layered nano-carbonated hydroxyapatite/collagen/PLGA composite membrane *in vitro*, *Dental Mater.*, **23**, 1120–1128.

422. Gonzalez-McQuire, R., Chane-Ching, J. Y., Vignaud, E., Lebugle, A., and Mann, S. Synthesis and characterization of amino acid-functionalized hydroxyapatite nanorods, *J. Mater. Chem.*, **14**, 2277–2281.

423. Rosseeva, E. V., Golovanova, O. A., and Frank-Kamenetskaya, O. V. (2007) The influence of amino acids on the formation of nanocrystalline hydroxyapatite, *Glass Phys. Chem.*, **33**, 283–286.

424. Zhan, J., Tseng, Y. H., Chan, J. C. C., and Mou, C. Y. (2005) Biomimetic formation of hydroxyapatite nanorods by a single-crystal-to-single-crystal transformation, *Adv. Funct. Mater.*, **15**, 2005–2010.

425. Xu, A. W., Ma, Y., and Cölfen, H. (2007) Biomimetic mineralization, *J. Mater. Chem.*, **17**, 415–449.

426. Pileni, M. (2003) The role of soft colloidal templates in controlling the size and shape of inorganic nanocrystals, *Nature Mater.*, **12**, 145–150.

427. Wu, Y., and Bose, S. (2005) Nanocrystalline hydroxyapatite: micelle templated synthesis and characterization, *Langmuir*, **21**, 3232–3234.

428. Wei, K., Lai, C., and Wang, Y. (2006) Solvothermal synthesis of calcium phosphate nanowires under different pH conditions, *J. Macromolec. Sci. A*, **43A**, 1531–1540.

429. Lai, C., Tang, S. Q., Wang, Y. J., Wei, K., and Zhang, S. Y. (2005) Insight into shape control mechanism of calcium phosphate nanopartiles in reverse micelles solution, *Synth. React. Inorg. Met. Met. Org. Nano-Metal Chem.*, **35**, 717–725.

430. Shchukin, D. G., Sukhorukov, G. B., and Möhwald, H. (2003) Biomimetic fabrication of nanoengineered hydroxyapatite/polyelectrolyte composite shell, *Chem. Mater.*, **15**, 3947–3950.

431. Mateus, A. Y. P., Ferraz, M. P., and Monteiro, F. J. (2007) Microspheres based on hydroxyapatite nanoparticles aggregates for bone regeneration, *Key Eng. Mater.*, 330–332, 243–246.

432. Cai, Y., Liu, Y., Yan, W., Hu, Q., Tao, J., Zhang, M., Shi, Z., and Tang, R. (2007) Role of hydroxyapatite nanoparticle size in bone cell proliferation, *J. Mater. Chem.*, **17**, 3780–3787.

433. Quantum dots, first developed in the early 1980's, are crystalline semi-conducting nanodimensional particles comprised of a metalloid crystalline core and a "cap" or "shell" that shields the core and renders the dots biologically compatible. They are used or being developed for use in electronics, biomedical imaging and surveillance.

434. Guo, Y., Shi, D., Lian, J., Dong, Z., Wang, W., Cho, H., Liu, G., Wang, L., and Ewing, R. C. (2008) Quantum dot conjugated hydroxylapatite nanoparticles for *in vivo* imaging, *Nanotechnology*, **19**, pp. 175102 (6 pages).

435. Liu, Q., de Wijn, J. R., de Groot, K., and van Blitterswijk, C. A. (1998) Surface modification of nano-apatite by grafting organic polymer, *Biomaterials*, **19**, 1067–1072.

436. Palazzo, B., Iafisco, M., Laforgia, M., Margiotta, N., Natile, G., Bianchi, C. L., Walsh, D., Mann, S., and Roveri, N. (2007) Biomimetic hydroxyapatite-drug nanocrystals as potential bone substitutes with antitumor drug delivery properties, *Adv. Funct. Mater.*, **17**, 2180–2188.

437. Lee, H. J., Choi, H. W., Kim, K. J., and Lee, S. C. (2006) Modification of hydroxyapatite nanosurfaces for enhanced colloidal stability and improved interfacial adhesion in nanocomposites, *Chem. Mater.*, **18**, 5111–5118.

438. Lee, S. C., Choi, H. W., Lee, H. J., Kim, K. J., Chang, J. H., Kim, S. Y., Choi, J., Oh, K. S., and Jeong, Y. K. (2007) *In-situ* synthesis of reactive hydroxyapatite nanocrystals for a novel approach of surface grafting polymerization, *J. Mater. Chem.*, **17**, 174–180.

439. Li, L., Liu, Y. K., Tao, J. H., Zhang, M., Pan, H. H., Xu, X. R., and Tang, R. K. (2008) Surface modification of hydroxyapatite nanocrystallite by a small amount of terbium provides a biocompatible fluorescent probe, *J. Phys. Chem. C*, **112**, 12219–12224.

440. Wang, W., Shi, D., Lian, J., Guo, Y., Liu, G., Wang, L., and Ewing, R. C. (2006) Luminescent hydroxylapatite nanoparticles by surface functionalization, *Appl. Phys. Lett.*, **89**, 183106 (3 pages).

441. Bow, J. S., Liou, S. C., and Chen, S. Y. (2004) Structural characterization of room-temperature synthesized nanosized β-tricalcium phosphate, *Biomaterials*, **25**, 3155–3161.

442. Brunner, T. J., Bohner, M., Dora, C., Gerber, C., and Stark, W. J. (2007) Comparison of amorphous TCP nanoparticles to micron-sized α-TCP as starting materials for calcium phosphate cements, *J. Biomed. Mater. Res. B (Appl. Biomater.)*, **83B**, 400–407.

443. Brunner, T. J., Grass, R. N., Bohner, M., and Stark, W. J. (2007) Effect of particle size, crystal phase and crystallinity on the reactivity of tricalcium phosphate cements for bone reconstruction, *J. Mater. Chem.*, **17**, 4072–4078.

444. Döbelin, N., Brunner, T. J., Stark, W. J., Eggimann, M., Fisch, M., and Bohner, M. (2009) Phase evolution of thermally treated amorphous tricalcium phosphate nanoparticles, *Key Eng. Mater.*, 396–398, 595–598.

445. Bohner, M., Brunner, T. J., Döbelin, N., Tang, R., and Stark, W. J. (2008) Effect of thermal treatments on the reactivity of nanosized tricalcium phosphate powders, *J. Mater. Chem.*, **18**, 4460–4467.

446. Liu, Y. H., Zhang, S. M., Liu, L., Zhou, W., Hu, W., Li, J., and Qiu, Z. Y. (2007) Rapid wet synthesis of nano-sized β-TCP by using dialysis, *Key Eng. Mater.*, 330–332, 199–202.

447. Abdel-Fattah, W. I., Reicha, F. M., and Elkhooly, T. A. (2008) Nano-beta-tricalcium phosphates synthesis and biodegradation: 1, Effect of microwave and SO_4^{2-} ions on β-TCP synthesis and its characterization, *Biomed. Mater.*, **3**, 034121 (13 pages).

448. Sanosh, K. P., Chu, M. C., Balakrishnan, A., Kim, T. N., and Cho, S. J. (2010) Sol-gel synthesis of pure nano sized β-tricalcium phosphate crystalline powders, *Curr. Appl. Phys.*, **10**, 68–71.

449. Dasgupta, S., Bandyopadhyay, A., and Bose, S. (2009) Reverse micelle-mediated synthesis of calcium phosphate nanocarriers for controlled release of bovine serum albumin, *Acta Biomater.*, **5**, 3112–3121.

450. Xia, C., Deng, X., Lin, Y. H., and Nan, C. W. (2010) Preparation and characterisation of nano-sized beta-tricalcium phosphate with a ps template method, *Int. J. Mater. Product Technol.*, **37**, 257–262.

451. Choi, D., and Kumta, P. N. (2007) Mechano-chemical synthesis and characterization of nanostructured β-TCP powder, *Mater. Sci. Eng. C*, **27**, 377–381.

452. Nikcevic, I., Maravic, D., Ignjatovic, N., Mitric, M., Makovec, D., and Uskokovic, D. (2006) The formation and characterization of

nanocrystalline phases by mechanical milling of biphasic calcium phosphate/poly-L-lactide biocomposite, *Mater. Transact.*, **47**, 2980–2986.

453. Cho, J. S., Jung, D. S., Han, J. M., and Kang, Y. C. (2009) Nano-sized α and β-TCP powders prepared by high temperature flame spray pyrolysis, *Mater. Sci. Eng. C*, **29**, 1288–1292.

454. Jalota, S., Tas, A. C., and Bhaduri, S. B. (2004) Microwave-assisted synthesis of calcium phosphate nanowhiskers, *J. Mater. Res.*, **19**, 1876–1881.

455. Rameshbabu, N., and Rao, K. P. (2009) Microwave synthesis, characterization and in-vitro evaluation of nanostructured biphasic calcium phosphates, *Curr. Appl. Phys.*, **9**, S29–S31.

456. Li, B., Chen, X., Guo, B., Wang, X., Fan, H., and Zhang, X. (2009) Fabrication and cellular biocompatibility of porous carbonated biphasic calcium phosphate ceramics with a nanostructure, *Acta Biomater.*, **5**, 134–143.

457. Guha, A. K., Singh, S., Kumaresan, R., Nayar, S., and Sinha, A. (2009) Mesenchymal cell response to nanosized biphasic calcium phosphate composites, *Colloids Surf. B*, **73**, 146–151.

458. Layrolle, P., and Lebugle, A. (1996) Synthesis in pure ethanol and characterization of nanosized calcium phosphate fluoroapatite, *Chem. Mater.*, **8**, 134–144.

459. Andres, C., Sinani, V., Lee, D., Gun'ko, Y., and Kotov, N. (2006) Anisotropic calcium phosphate nanoparticles coated with 2-carboxyethylphosphonic acid, *J. Mater. Chem.*, **16**, 3964–3968.

460. Lim, H., Kassim, A., Huang, N., Hashim, R., Radiman, S., Khiew, P., and Chiu, W. (2009) Fabrication and characterization of 1D brushite nanomaterials via sucrose ester reverse microemulsion, *Ceram. Int.*, **35**, 2891–2897.

461. Shirkhanzadeh, M., and Sims, S. (1997) Immobilization of calcium phosphate nano-clusters into alkoxy-derived porous TiO_2 coatings, *J. Mater. Sci. Mater. Med.*, **8**, 595–601.

462. Schmidt, H. T., and Ostafin, A. E. (2002) Liposome directed growth of calcium phosphate nanoshells, *Adv. Mater.*, **14**, 532–535.

463. Schmidt, H. T., Gray, B. L., Wingert, P. A., and Ostafin, A. E. (2004) Assembly of aqueous-cored calcium phosphate nanoparticles for drug delivery, *Chem. Mater.*, **16**, 4942–4947.

464. Xu, H. H. K., Sun, L., Weir, M. D., Antonucci, J. M., Takagi, S., Chow, L. C., and Peltz, M. (2006) Nano DCPA — whisker composites with high strength and Ca and PO_4 release, *J. Dent. Res.*, **85**, 722–727.

465. Xu, H. H. K., Weir, M. D., Sun, L., Takagi, S., and Chow, L. C. (2007) Effects of calcium phosphate nanoparticles on Ca-PO_4 composite, *J. Dent. Res.*, **86**, 378–383.

466. Xu, H. H. K., Weir, M. D., and Sun, L. (2007) Nanocomposites with Ca and PO_4 release: effects of reinforcement, dicalcium phosphate particle size and silanization, *Dental Mater.*, **23**, 1482–1491.

467. Singh, S., Bhardwaj, P., Singh, V., Aggarwal, S., and Mandal, U. K. (2008) Synthesis of nanocrystalline calcium phosphate in microemulsion — effect of nature of surfactants, *J. Colloid Interf. Sci.*, **319**, 322–329.

468. Wals, D., and Mann, S. (1996) Chemical synthesis of microskeletal calcium phosphate in bicontinuous microemulsions, *Chem. Mater.*, **8**, 1944–1953.

469. Urch, H., Vallet-Regí, M., Ruiz, L., Gonzalez-Calbet, J. M., and Epple, M. (2009) Calcium phosphate nanoparticles with adjustable dispersability and crystallinity, *J. Mater. Chem.*, **19**, 2166–2171.

470. Holt, C., Wahlgren, N. M., and Drakenberg, T. (1996) Ability of a β-casein phosphopeptide to modulate the precipitation of calcium phosphate by forming amorphous dicalcium phosphate nanoclusters, *Biochem. J.*, **314**, 1035–1039.

471. Holt, C., Timmins, P. A., Errington, N., and Leaver, J. (1998) A core-shell model of calcium phosphate nanoclusters stabilized by β-casein phosphopeptides, derived from sedimentation equilibrium and small-angle X-ray and neutron-scattering measurements, *Eur. J. Biochem.*, **252**, 73–78.

472. Duan, B., Wang, M., Zhou, W. Y., and Cheung, W. L. (2008) Synthesis of Ca-P nanoparticles and fabrication of Ca-P/PHBV nanocomposite microspheres for bone tissue engineering applications, *Appl. Surf. Sci.*, **255**, 529–533.

473. Hwang, K. S., Jeon, K. O., Jeon, Y. S., and Kim, B. H. (2006) Hydroxyapatite forming ability of electrostatic spray pyrolysis derived calcium phosphate nano powder, *J. Mater. Sci.*, **41**, 4159–4162.

474. Hwang, K. S., Jeon, K. O., Jeon, Y. S., and Kim, B. H. (2007) Hydroxyapatite forming ability of electrostatic spray pyrolysis derived calcium phosphate nano powder, *J. Mater. Sci. Mater. Med.*, **18**, 619–622.

475. Perkin, K. K., Turner, J. L., Wooley, K. L., and Mann, S. (2005) Fabrication of hybrid nanocapsules by calcium phosphate mineralization of shell cross-linked polymer micelles and nanocages, *Nano Lett.*, **5**, 1457–1461.

476. Tjandra, W., Ravi, P., Yao, J., and Tam, K. C. (2006) Synthesis of hollow spherical calcium phosphate nanoparticles using polymeric nanotemplates, *Nanotechnology*, **17**, 5988–5994.

477. Sadasivan, S., Khushalani, D., and Mann, S. (2005) Synthesis of calcium phosphate nanofilaments in reverse micelles, *Chem. Mater.*, **17**, 2765–2770.

478. Morgan, T. T., Muddana, H. S., Altinoglu, E. I., Rouse, S. M., Tabakovic, A., Tabouillot, T., Russin, T. J., Butler, P. J., Eklund, P., Yun, J. K., Kester, M., and Adair, J. H. (2008) Encapsulation of organic molecules in calcium phosphate nanocomposite particles for intracellular imaging and drug delivery, *Nano Lett.*, **8**, 4108–4115.

479. Lai, C., Wang, Y. J., and Wei, K. (2008) Nucleation kinetics of calcium phosphate nanoparticles in reverse micelle solution, *Colloids Surf. A*, **315**, 268–274.

480. Yang, X., Gao, X., Gan, Y., Gao, C., Zhang, X., Ting, K., Wu, B. M., and Gou, Z. (2010) Facile synthesis of octacalcium phosphate nanobelts: growth mechanism and surface adsorption properties, *J. Phys. Chem. C*, **114**, 6265–6271.

481. Socol, G., Torricelli, P., Bracci, B., Iliescu, M., Miroiu, F., Bigi, A., Werckmann, J., and Mihailescu, I. N. (2004) Biocompatible nanocrystalline octacalcium phosphate thin films obtained by pulsed laser deposition, *Biomaterials*, **25**, 2539–2545.

482. Cho, J. S., Ko, Y. N., Koo, H. Y., and Kang, Y. C. (2010) Synthesis of nano-sized biphasic calcium phosphate ceramics with spherical shape by flame spray pyrolysis, *J. Mater. Sci. Mater. Med.*, **21**, 1143–1149.

483. Farzadi, A., Solati-Hashjin, M., Tahmasebi-Birgani, Z., and Aminian, A. (2010) Microwave-assisted synthesis and characterization of biphasic calcium phosphate nanopowders. *Ceram. Transact.*, **218**, 59–65.

484. Pan, L., Li, Y., Zou, C., Weng, W., Cheng, K., Song, C., Du, P., Zhao, G., Shen, G., Wang, J., and Han, G. (2007) Surface modification of nanosized biphasic α-TCP/HA powders, *Key Eng. Mater.*, 330–332, 223–226.

485. Urch, H., Franzka, S., Dahlhaus, D., Hartmann, N., Hasselbrink, E., and Epple, M. (2006) Preparation of two-dimensionally patterned layers of functionalised calcium phosphate nanoparticles by laser direct writing, *J. Mater. Chem.*, **16**, 1798–1802.

486. Sokolova, V., Prymak, O., Meyer-Zaika, W., Cölfen, H., Rehage, H., Shukla, A., and Epple, M. (2006) Synthesis and characterization of DNA functionalized calcium phosphate nanoparticles, *Mat. -wiss. u. Werkstofftech.*, **37**, 441–445.

487. Muddana, H. S., Morgan, T. T., Adair, J. H., and Butler, P. J. (2009) Photophysics of Cy3-encapsulated calcium phosphate nanoparticles, *Nano Lett.*, **9**, 1559–1566.

488. Altinoğlu, E. I., Russin, T. J., Kaiser, J. M., Barth, B. M., Eklund, P. C., Kester, M., and Adair, J. H. (2008) Near-infrared emitting fluorophore-doped calcium phosphate nanoparticles for *in vivo* imaging of human breast cancer, *ACS Nano*, **2**, 2075–2084.

489. Schwiertz, J., Wiehe, A., Gräfe, S., Gitter, B., and Epple, M. (2009) Calcium phosphate nanoparticles as efficient carriers for photodynamic therapy against cells and bacteria, *Biomaterials*, **30**, 3324–3331.

490. Schwiertz, J., Meyer-Zaika, W., Ruiz-Gonzalez, L., González-Calbet, J. M., Vallet-Regí, M., and Epple, M. (2008) Calcium phosphate nanoparticles as templates for nanocapsules prepared by the layer-by-layer technique, *J. Mater. Chem.*, **18**, 3831–3834.

491. Hayakawa, S., Li, Y., Tsuru, K., Osaka, A., Fujii, E., and Kawabata, K. (2009) Preparation of nanometer-scale rod array of hydroxyapatite crystal, *Acta Biomater.*, **5**, 2152–2160.

492. Liao, S. S., Cui, F. Z., Zhang, W., and Feng, Q. L. (2004) Hierarchically biomimetic bone scaffold materials: nano-HA/collagen/PLA composite, *J. Biomed. Mater. Res. B (Appl. Biomater.)*, **69B**, 158–165.

493. Thomas, V., Dean, D. R., Jose, M. V., Mathew, B., Chowdhury, S., and Vohra, Y. K. (2007) Nanostructured biocomposite scaffolds based on collagen co-electrospun with nanohydroxyapatite, *Biomacromolecules*, **8**, 631–637.

494. de Yoreo, J. J., and Vekilov, P. G. (2003) Principles of crystal nucleation and growth, *Rev. Mineral. Geochem.*, **54**, 57–93.

495. Liao, S., Xu, G., Wang, W., Watari, F., Cui, F., Ramakrishna, S., and Chan, C. K. (2007) Self-assembly of nano-hydroxyapatite on multi-walled carbon nanotubes, *Acta Biomater.*, **3**, 669–675.

496. Penn, R. L., and Banfield, J. F. (1998) Imperfect oriented attachment: dislocation generation in defect-free nanocrystals, *Science*, **281**, 969–971.

497. Tao, J., Pan, H., Zeng, Y., Xu, X., and Tang, R. (2007) Roles of amorphous calcium phosphate and biological additives in the assembly of hydroxyapatite nanoparticles, *J. Phys. Chem. B*, **111**, 13410–13418.

498. Hing, K. A. (2004) Bone repair in the twenty-first century: biology, chemistry or engineering? *Philos. Trans. R. Soc. Lond. A*, **362**, 2821–2850.

499. Kokubo, T., Kim, H. M., and Kawashita, M. (2003) Novel bioactive materials with different mechanical properties, *Biomaterials*, **24**, 2161–2175.

500. Fu, J. M., Miao, B., Jia, L. H., and Lü, K. L. (2009) Nano-hydroxyapatite for repair of rabbit jaw bone defect: bone mineral density analysis, *J. Clin. Rehabil. Tissue Eng. Res.*, **13**, 2387–2390.

501. Barralet, J. E., Lilley, K. J., Grover, L. M., Farrar, D. F., Ansell, C., and Gbureck, U. (2004) Cements from nanocrystalline hydroxyapatite.,J. *Mater. Sci. Mater. Med.*, **15**, 407–411.

502. Lilley, K. J., Gbureck, U., Wright, A. J., Farrar, D. F., and Barralet, J. E. (2005) Cement from nanocrystalline hydroxyapatite: effect of calcium phosphate ratio, *J. Mater. Sci. Mater. Med*, **16**, 1185–1190.

503. Neira, I. S., Kolen'ko, Y. V., Lebedev, O. I., van Tendeloo, G., Gupta, H. S., Matsushita, N., Yoshimura, M., and Guitián, F. (2009) Rational synthesis of a nanocrystalline calcium phosphate cement exhibiting rapid conversion to hydroxyapatite, *Mater. Sci. Eng. C*, **29**, 2124–2132.

504. Dorozhkin, S. V. (2008) Calcium orthophosphate cements for biomedical application, *J. Mater. Sci.*, **43**, 3028–3057.

505. Dorozhkin, S. V. (2009) Calcium orthophosphate cements and concretes, *Materials* **2**, 221–291.

506. Fu, Q., Zhou, N., Huang, W., Wang, D., Zhang, L., and Li, H. (2005) Effects of nano HAP on biological and structural properties of glass bone cement, *J. Biomed. Mater. Res. A*, **74A**, 156–163.

507. Strnadova, M., Protivinsky, J., Strnad, J., and Vejsicka, Z. (2008) Preparation of porous synthetic nanostructured HA scaffold, *Key Eng. Mater.*, 361–363, 211–214.

508. Kim, J. Y., Lee, J. W., Lee, S. J., Park, E. K., S. Y. Kim, and Cho, D. W. (2007) Development of a bone scaffold using HA nanopowder and micro-stereolithography technology, *Microelectronic Engineering*, **84**, 1762–1765.

509. Severin, A. V., Komarov, V. F., Bozhevol'nov, V. E., and Melikhov, I. V. (2005) Morphological selection in suspensions of nanocrystalline hydroxylapatite leading to spheroidal aggregates, *Russ. J. Inorg. Chem.*, **50**, 72–77.

510. Krylova, I. V., Ivanov, L. N., Bozhevol'nov, V. E., and Severin, A. V. (2007) Self-organization processes and phase transitions in nanocrystalline hydroxyapatite according to exoemission data, *Russ. J. Phys. Chem. A*, **81**, 241–245.

511. Veljovic, D., Jokic, B., Jankovic-Castvan, I., Smiciklas, I., Petrovic, R., and Janackovic, D. (2007) Sintering behaviour of nanosized HAP powder, *Key Eng. Mater.*, 330–332, 259–262.

512. Zhang, F., Lin, K., Chang, J., Lu, J., and Ning, C. (2008) Spark plasma sintering of macroporous calcium phosphate scaffolds from nanocrystalline powders, *J. Eur. Ceram. Soc.*, **28**, 539–545.

513. Grossin, D., Banu, M., Sarda, S., Martinet-Rollin, S., Drouet, C., Estournès, C., Champion, E., Rossignol, F., Combes, C., and Rey, C. (2010) Low temperature consolidation of nanocrystalline apatites toward a new generation of calcium phosphate ceramics, *Ceram. Eng. Sci. Proc.*, **30**, 113–126.

514. Chaudhry, A. A., Yan, H., Gong, K., Inam, F., Viola, G., Reece, M. J., Goodall, J. B. M., ur Rehman, I., McNeil-Watson, F. K., Corbett, J. C. W., Knowles, J. C, and Darr, J. A. (2011) High-strength nanograined and translucent hydroxyapatite monoliths via continuous hydrothermal synthesis and optimized spark plasma sintering, *Acta Biomater.* **7**, 791–799.

515. Kutty, M. G., Loertscher, J., Bhaduri, S., Bhaduri, S. B., and Tinga, W. R. (2001) Microwave sintering of nanocrystalline hydroxyapatite, *Ceram. Eng. Sci. Proc.*, **22**, 3–10.

516. Vijayan, S., and Varma, H. (2002) Microwave sintering of nanosized hydroxyapatite powder compacts, *Mater. Lett.*, **56**, 827–831.

517. Ramesh, S., Tan, C. Y., Bhaduri, S. B., and Teng, W. D. (2007) Rapid densification of nanocrystalline hydroxyapatite for biomedical applications, *Ceram. Int.*, **33**, 1363–1367.

518. Okada, M., and Furuzono, T. (2006) Fabrication of high-dispersibility nanocrystals of calcined hydroxyapatite, *J. Mater. Sci.*, **41**, 6134–6137.

519. Okada, M., and Furuzono, T. (2007) Nanosized ceramic particles of hydroxyapatite calcined with an anti-sintering agent, *J. Nanosci. Nanotechnol.*, **7**, 848–851.

520. Okada, M., and Furuzono, T. (2007) Calcination of rod-like hydroxyapatite nanocrystals with an anti-sintering agent surrounding the crystals, *J. Nanopart. Res.*, **9**, 807–815.

521. Müller-Mai, C. M., Stupp, S. I., Voigt, C., and Gross, U. (1995) Nanoapatite and organoapatite implants in bone: histology and ultrastructure of the interface, *J. Biomed. Mater. Res.*, **29**, 9–18.

522. Du, C., Cui, F. Z., Feng, Q. L., Zhu, X. D., and de Groot, K. (1998) Tissue response to nano-hydroxyapatite/collagen composite implants in marrow cavity, *J. Biomed. Mater. Res.*, **42**, 540–548.

523. Du, C., Cui, F. Z., Zhu, X. D., and de Groot, K. (1999) Three-dimensional nano-HAp/collagen matrix loading with osteogenic cells in organ culture, *J. Biomed. Mater. Res.*, **44**, 407–415.

524. Paul, W., and Sharma, C. P. (2006) Nanoceramic matrices: biomedical applications, *Am. J. Biochem. Biotechnol.*, **2**, 41–48.

525. Huber, F. X., McArthur, N., Hillmeier, J., Kock, H. J., Baier, M., Diwo, M., Berger, I., and Meeder, P. J. (2006) Void filling of tibia compression fracture zones using a novel resorbable nanocrystalline hydroxyapatite paste in combination with a hydroxyapatite ceramic core: first clinical results, *Arch. Orthop. Trauma Surg.*, **126**, 533–540.

526. Smeets, R., Jelitte, G., Heiland, M., Kasaj, A., Grosjean, M., Riediger, D., Yildirim, M., Spiekermann, H., and Maciejewski, O. (2008) Hydroxylapatit-Knochenersatzmaterial (Ostim®) bei der Sinusbodenelevation, *Schweiz Monatsschr. Zahnmed.*, **118**, 203–208.

527. Gerlach, K. L., and Niehues, D. (2007) Die Behandlung der Kieferzysten mit einem neuartigen nanopartikulären Hydroxylapatit, *Mund Kiefer GesichtsChir.*, **11**, 131–137.

528. Schwarz, F., Bieling, K., Latz, T., Nuesry, E., and Becker, J. (2006) Healing of intrabony periimplantitis defects following application of a nanocristalline hydroxyapatite (Ostim™) or a bovine-derived xenograft (Bio-Oss™) in combination with a collagen membrane (Bio-Gide™), *A case series. J. Clin. Periodontol.*, **33**, 491–499.

529. Strietzel, F. P., Reichart, P. A., and Graf, H. L. (2007) Lateral alveolar ridge augmentation using a synthetic nano-crystalline hydroxyapatite bone substitution material (Ostim®), *Preliminary clinical and histological results. Clin. Oral Implants Res.*, **18**, 743–751.

530. Spies, C., Schnürer, S., Gotterbarm, T., and Breusch, S. (2008) Tierexperimentelle Untersuchung des Knochenersatzstoffs Ostim™ im knöchernen Lager des Göttinger Miniaturschweins, *Z. Orthop. Unfall.*, **146**, 64–69.

531. Thorwarth, M., Schultze-Mosgau, S., Kessler, P., Wiltfang, J., and Schlegel, K. A. (2005) Bone regeneration in osseous defects using a resorbable nanoparticular hydroxyapatite, *J. Oral Maxillofac. Surg.*, **63**, 1626–1633.

532. Brandt, J., Henning, S., Michler, G., Schulz, M., and Bernstein, A. (2008) Nanocrystalline hydroxyapatite for bone repair, *Key Eng. Mater.*, 361–363, 35–38.

533. Huber, F. X., Hillmeier, J., Herzog, L., McArthur, N., Kock, H. J., and Meeder, P. J. (2006) Open reduction and palmar plate-osteosynthesis

in combination with a nanocrystalline hydroxyapatite spacer in the treatment of comminuted fractures of the distal radius, *J. Hand Surg. (Brit.)*, **31B**, 298–303.

534. Huber, F. X., Hillmeier, J., McArthur, N., Kock, H. J., and Meeder, P. J. (2006) The use of nanocrystalline hydroxyapatite for the reconstruction of calcaneal fractures: preliminary results, *J. Foot Ankle Surg.*, **45**, 322–328.

535. Laschke, M. W., Witt, K., Pohlemann, T., and Menger, M. D. (2007) Injectable nanocrystalline hydroxyapatite paste for bone substitution: *in vivo* analysis of biocompatibility and vascularization, *J. Biomed. Mater. Res. B (Appl. Biomater.)*, **82B**, 494–505.

536. Spies, C. K. G., Schnürer, S., Gotterbarm, T., and Breusch, S. (2009) The efficacy of Biobon™ and Ostim™ within metaphyseal defects using the Göttinger Minipig, *Arch. Orthop. Trauma Surg.*, **129**, 979–988.

537. Huber, F. X., Belyaev, O., Hillmeier, J., Kock, H. J., Huber, C., Meeder, P. J., and Berger, I. (2006) First histological observations on the incorporation of a novel nanocrystalline hydroxyapatite paste OSTIM® in human cancellous bone, *BMC Musculoskelett. Disord.*, **7**, 50 (14 pages).

538. Huber, F. X., Berger, I., McArthur, N., Huber, C., Kock, H. P., Hillmeier, J., and Meeder, P. J. (2008) Evaluation of a novel nanocrystalline hydroxyapatite paste and a solid hydroxyapatite ceramic for the treatment of critical size bone defects (CSD) in rabbits, *J. Mater. Sci. Mater. Med.*, **19**, 33–38.

539. Arts, J. J. C., Verdonschot, N., Schreurs, B. W., and Buma, P. (2006) The use of a bioresorbable nano-crystalline hydroxyapatite paste in acetabular bone impaction grafting, *Biomaterials*, **27**, 1110–1118.

540. Zhang, W., Liao, S. S., and Cui, F. Z. (2003) Hierarchical self-assembly of nano-fibrils in mineralized collagen, *Chem. Mater.*, **15**, 3221–3226.

541. Li, X., Huang, J., and Edirisinghe, M. J. (2008) Development of nano-hydroxyapatite coating by electrohydrodynamic atomization spraying, *J. Mater. Sci. Mater. Med.*, **19**, 1545–1551.

542. Guo, L., and Li, H. (2004) Fabrication and characterization of thin nano-hydroxyapatite coatings on titanium, *Surf. Coat. Technol.*, **185**, 268–274.

543. Thian, E. S., Ahmad, Z., Huang, J., Edirisinghe, M. J., Jayasinghe, S. N., Ireland, D. C., Brooks, R. A., Rushton, N., Bonfield, W., and Best, S. M. (2008) Electrosprayed nanoapatite: a new generation of bioactive material, *Key Eng. Mater.*, 361–363, 597–600.

544. Han, Y., Xu, K., Montay, G., Fu, T., and Lu, J. (2002) Evaluation of nanostructured carbonated hydroxyapatite coatings formed by a hybrid process of plasma spraying and hydrothermal synthesis, *J. Biomed. Mater. Res.*, **60**, 511–516.

545. Li, P. (2003) Biomimetic nano-apatite coating capable of promoting bone ingrowth, *J. Biomed. Mater. Res. A*, **66A**, 79–85.

546. Mendes, V. C., Moineddin, R., and Davies, J. E. (2007) The effect of discrete calcium phosphate nanocrystals on bone-bonding to titanium surfaces, *Biomaterials*, **28**, 4748–4755.

547. Oh, S. H., Finõnes, R. R., Daraio, C., Chen, L. H., and Jin, S. (2005) Growth of nano-scale hydroxyapatite using chemically treated titanium oxide nanotubes, *Biomaterials*, **26**, 4938–4943.

548. Ma, J., Wong, H., Kong, L. B., and Peng, K. W. (2003) Biomimetic processing of nanocrystallite bioactive apatite coating on titanium, *Nanotechnology*, **14**, 619–623.

549. Gu, Y. W., Tay, B. Y., Lim, C. S., and Yong, M. S. (2006) Nanocrystallite apatite formation and its growth kinetics on chemically treated porous NiTi, *Nanotechnology*, **17**, 2212–2218.

550. Hu, R., Lin, C. J., and Shi, H. Y. (2007) A novel ordered nano hydroxyapatite coating electrochemically deposited on titanium substrate, *J. Biomed. Mater. Res. A*, **80A**, 687–692.

551. Bigi, A., Boanini, E., Bracci, B., Facchini, A., Panzavolta, S., Segatti, F., and Sturba, L. (2005) Nanocrystalline hydroxyapatite coatings on titanium: a new fast biomimetic method, *Biomaterials*, **26**, 4085–4089.

552. Narayanan, R., Seshadri, S. K., Kwon, T. Y., and Kim, K. H. (2007) Electrochemical nano-grained calcium phosphate coatings on Ti-6Al-4V for biomaterial applications, *Scripta Mater.*, **56**, 229–232.

553. Thian, E. S., Huang, J., Best, S. M., Barber, Z. H., and Bonfield, W. (2006) Nanostructured apatite coatings for rapid bone repair, *Key Eng. Mater.*, 309–311, 519–522.

554. Cai, X., Gong, P., Man, Y., Chen, Z., and He, G. (2007) The construction and characterization of nano-FHA bioceramic coating on titanium surface, *Key Eng. Mater.*, 330–332, 333–336.

555. Citterio, H., Jakani, S., Benmarouane, A., Millet, P., and Lodini, A. (2008) Nano-hydroxyapatite coatings on titanium substrates. Finite element analysis of process and experimental plasma thermal sprayed coatings, *Key Eng. Mater.*, 361–363, 745–748.

566. Lee, S. H., Kim, H. E., and Kim, H. W. (2007) Nanosized hydroxyapatite coatings on Ti substrate with TiO_2 buffer layer by e-beam deposition, *J. Am. Ceram. Soc.*, **90**, 50–56.

557. Nishimura, I., Huang, Y., Butz, F., Ogawa, T., Lin, A., and Wang, C. J. (2007) Discrete deposition of hydroxyapatite nanoparticles on a titanium implant with predisposing substrate microtopography accelerated osseointegration, *Nanotechnology*, **18**, 245101 (9 pages).

558. Narayanan, R., Kwon, T. Y., and Kim, K. H. (2008) Preparation and characteristics of nano-grained calcium phosphate coatings on titanium from ultrasonated bath at acidic pH, *J. Biomed. Mater. Res. B (Appl. Biomater.)*, **85B**, 231–239.

559. Hahn, B. D., Park, D. S., Choi, J. J., Ryu, J., Yoon, W. H., Kim, K. H., Park, C., and Kim, H. E. (2009) Dense nanostructured hydroxyapatite coating on titanium by aerosol deposition, *J. Am. Ceram. Soc.*, **92**, 683–687.

560. Narayanan, R., Kwon, T. Y., and Kim, K. H. (2008) Direct nanocrystalline hydroxyapatite formation on titanium from ultrasonated electrochemical bath at physiological pH, *Mater. Sci. Eng. C*, **28**, 1265–1270.

561. Yousefpour, M., Afshar, A., Yang, X., Li, X., Yang, B., Wu, Y., Chen, J., and Zhang, X. (2006) Nano-crystalline growth of electrochemically deposited apatite coating on pure titanium, *J. Electroanal. Chem.*, **589**, 96–105.

562. Mendes, V. C., Moineddin, R., and Davies, J. E. (2009) Discrete calcium phosphate nanocrystalline deposition enhances osteoconduction on titanium-based implant surfaces, *J. Biomed. Mater. Res. A*, **90A**, 577–585.

563. Yang, Y., Kim, K. H., and Ong, J. L. (2005) A review on calcium phosphate coatings produced using a sputtering process — an alternative to plasma spraying, *Biomaterials*, **26**, 327–337.

564. Nies, B., Rößler, S., and Reinstorf, A. (2007) Formation of nano hydroxyapatite — a straightforward way to bioactivate bone implant surfaces, *Int. J. Mat. Res. (formerly Z. Metallkd.)*, **98**, 630–636.

565. Jalota, S., Bhaduri, S. B., and Tas, A. C. (2006) Effect of carbonate content and buffer type on calcium phosphate formation in SBF solutions, *J. Mater. Sci. Mater. Med.*, **17**, 697–707.

566. Chen, F., Lam, W. M., Lin, C. J., Qiu, G. X., Wu, Z. H., Luk, K. D. K., and Lu, W. W. (2007) Biocompatibility of electrophoretical deposition of nanostructured hydroxyapatite coating on roughen titanium surface: *in vitro* evaluation using mesenchymal stem cells, *J. Biomed. Mater. Res. B (Appl. Biomater.)*, **82B**, 183–191.

567. Thian, E. S., Ahmad, Z., Huang, J., Edirisinghe, M. J., Jayasinghe, S. N., Ireland, D. C., Brooks, R. A., Rushton, N., Bonfield, W., and Best, S. M.

(2008) The role of electrosprayed nanoapatites in guiding osteoblast behaviour, *Biomaterials*, **29**, 1833–1843.

568. Bigi, A., Nicoli-Aldini, N., Bracci, B., Zavan, B., Boanini, E., Sbaiz, F., Panzavolta, S., Zorzato, G., Giardino, R., Facchini, A., Abatangelo, G., and Cortivo, R. (2007) *In vitro* culture of mesenchymal cells onto nanocrystalline hydroxyapatite coated Ti13Nb13Zr alloy, *J. Biomed. Mater. Res. A*, **82A**, 213–221.

569. Bigi, A., Fini, M., Bracci, B., Boanini, E., Torricelli, P., Giavaresi, G., Aldini, N. N., Facchini, A., Sbaiz, F., and Giardino, R. (2008) The response of bone to nanocrystalline hydroxyapatite-coated Ti13Nb11Zr alloy in an animal model, *Biomaterials*, **29**, 1730–1736.

570. Thian, E. S., Huang, J., Ahmad, Z., Edirisinghe, M. J., Jayasinghe, S. N., Ireland, D. C., Brooks, R. A., Rushton, N., Best, S. M., and Bonfield, W. (2008) Influence of nanohydroxyapatite patterns deposited by electrohydrodynamic spraying on osteoblast response, *J. Biomed. Mater. Res. A*, **85A**, 188–194.

571. Furuzono, T., Masuda, M., Okada, M., Yasuda, S., Kadono, H., Tanaka, R., and Miyatake, K. (2006) Increase in cell adhesiveness on a poly(ethylene terephthalate) fabric by sintered hydroxyapatite nanocrystal coating in the development of an artificial blood vessel, *ASAIO J.*, **52**, 315–320.

572. Yanagida, H., Okada, M., Masuda, M., Ueki, M., Narama, I., Kitao, S., Koyama, Y., Furuzono, T., and Takakuda, K. (2009) Cell adhesion and tissue response to hydroxyapatite nanocrystal-coated poly(L-lactic acid) fabric, *J. Biosci. Bioeng.*, **108**, 235–243.

573. Li, X., Huang, J., and Edirisinghe, M. J. (2008) Development of template-assisted electrohydrodynamic atomization spraying for nanoHA patterning, *Key Eng. Mater.*, 361–363, 585–588.

574. Shi, Z. L., Huang, X., Cai, Y. R., Tang, R. K., and Yang, D. S. (2009) Size effect of hydroxyapatite nanoparticles on proliferation and apoptosis of osteoblast-like cells, *Acta Biomater.*, **5**, 338–345.

575. Liu, Y., Wang, G., Cai, Y., Ji, H., Zhou, G., Zhao, X., Tang, R., and Zhang, M. (2009) *In vitro* effects of nanophase hydroxyapatite particles on proliferation and osteogenic differentiation of bone marrow-derived mesenchymal stem cells, *J. Biomed. Mater. Res. A*, **15**, 1083–1091.

576. Zhu, X., Eibl, O., Scheideler, L., and Geis-Gerstorfer, J. (2006) Characterization of nano hydroxyapatite/collagen surfaces and cellular behaviors, *J. Biomed. Mater. Res. A*, **79A**, 114–127.

577. Wang, H., Li, Y., Zuo, Y., Li, J., Ma, S., and Cheng, L. (2007) Biocompatibility and osteogenesis of biomimetic nano-hydroxyapatite/polyamide

composite scaffolds for bone tissue engineering, *Biomaterials*, **28**, 3338–3348.

578. Zhang, Y. F., Cheng, X. R., Chen, Y., Shi, B., Chen, X. H., Xu, D. X., and Ke, J. (2007) Three-dimensional nanohydroxyapatite / chitosan scaffolds as potential tissue engineered periodontal tissue, *J. Biomater. Appl.*, **21**, 333–349.

579. Huang, Y. X., Ren, J., Chen, C., Ren, T. B., and Zhou, X. Y. (2008) Preparation and properties of poly(lactide-co-glycolide) (PLGA) / nano-hydroxyapatite (NHA) scaffolds by thermally induced phase separation and rabbit mscs culture on scaffolds, *J. Biomater. Appl.*, **22**, 409–432.

580. Thian, E. S., Ahmad, Z., Huang, J., Edirisinghe, M. J., Jayasinghe, S. N., Ireland, D. C., Brooks, R. A., Rushton, N., Bonfield, W., and Best, S. M. (2007) Bioactivity of nanoapatite produced by electrohydrodynamic atomization, *J. Bionanosci.*, **1**, 60–63.

581. Pezzatini, S., Solito, R., Morbidelli, L., Lamponi, S., Boanini, E., Bigi, A., and Ziche, M. (2006) The effect of hydroxyapatite nanocrystals on microvascular endothelial cell viability and functions, *J. Biomed. Mater. Res. A*, **76A**, 656–663.

582. Pezzatini, S., Morbidelli, L., Solito, R., Paccagnini, E., Boanini, E., Bigi, A., and Ziche, M. (2007) Nanostructured HA crystals up-regulate FGF-2 expression and activity in microvascular endothelium promoting angiogenesis, *Bone*, **41**, 523–534.

583. Hu, Q., Tan, Z., Liu, Y., Tao, J., Cai, Y., Zhang, M., Pan, H., Xu, X., and Tang, R. (2007) Effect of crystallinity of calcium phosphate nanoparticles on adhesion, proliferation, and differentiation of bone marrow mesenchymal stem cells, *J. Mater. Chem.*, **17**, 4690–4698.

584. Balasundaram, G., Sato, M., and Webster, T. J. (2006) Using hydroxyapatite nanoparticles and decreased crystallinity to promote osteoblast adhesion similar to functionalizing with RGD, *Biomaterials*, **27**, 2798–2805.

585. Detsch, R., Hagmeyer, D., Neumann, M., Schaefer, S., Vortkamp, A., Wuelling, M., Ziegler, G., and Epple, M. (2010) The resorption of nanocrystalline calcium phosphates by osteoclast-like cells, *Acta Biomater.*, **6**, 3223–3233.

586. Stevens, M. M., and George, J. H. (2005) Exploring and engineering the cell surface interface, *Science*, **310**, 1135–1138.

587. Martínez, E., Engel, E., Planell, J. A., and Samitier, J. (2009) Effects of artificial micro- and nano-structured surfaces on cell behaviour, *Annals Anat.*, **191**, 126–135.

588. Lee, D. H., Han, J. S., Yang, J. H., and Lee, J. B. (2009) MC3T3-E1 cell response to pure titanium, zirconia and nano-hydroxyapatite, *Int. J. Modern Phys. B*, **23**, 1535–1540.

589. Onuma, K., Yamagishi, K., and Oyane, A. (2005) Nucleation and growth of hydroxyapatite nanocrystals for nondestructive repair of early caries lesions, *J. Cryst. Growth*, **282**, 199–207.

590. Roveri, N., Battistella, E., Bianchi, C. L., Foltran, I., Foresti, E., Iafisco, M., Lelli, M., Naldoni, A., Palazzo, B., and Rimondini, L. (2009) Surface enamel remineralization: biomimetic apatite nanocrystals and fluoride ions different effects, *J. Nanomater.*, 746383 (9 pages).

591. Lv, K., Zhang, J., Meng, X., and Li, X. (2007) Remineralization effect of the nano-HA toothpaste on artificial caries, *Key Eng. Mater.*, 330–332, 267–270.

592. Jeong, S. H., Jang, S. O., Kim, K. N., Kwon, H. K., Park, Y. D., and Kim, B. I. (2006) Remineralization potential of new toothpaste containing nano-hydroxyapatite, *Key Eng. Mater.*, 309–311, 537–540.

593. Kim, B. I., Jeong, S. H., Jang, S. O., Kim, K. N., Kwon, H. K., and Park, Y. D. (2006) Tooth whitening effect of toothpastes containing nano-hydroxyapatite, *Key Eng. Mater.*, 309–311, 541–544.

594. Kim, M. Y., Kwon, H. K., Choi, C. H., and Kim, B. I. (2007) Combined effects of nano-hydroxyapatite and NaF on remineralization of early caries lesion, *Key Eng. Mater.*, 330–332, 1347–1350.

595. Lee, H. J., Min, J. H., Choi, C. H., Kwon, H. G., and Kim, B. I. (2007) Remineralization potential of sports drink containing nano-sized hydroxyapatite, *Key Eng. Mater.*, 330–332, 275–278.

596. Hong, Y. W., Kim, J. H., Lee, B. H., Lee, Y. K., Choi, B. J., Lee, J. H., and Choi, H. J. (2008) The effect of nano-sized β-tricalcium phosphate on remineralization in glass ionomer dental luting cement, *Key Eng. Mater.*, 361–363, 861–864.

597. Li, L., Pan, H. H., Tao, J. H., Xu, X. R., Mao, C. Y., Gu, X. H., and Tang, R. K. (2008) Repair of enamel by using hydroxyapatite nanoparticles as the building blocks, *J. Mater. Chem.*, **18**, 4079–4084.

598. Meng, X., Lv, K., Zhang, J., and Qu, D. (2007) Caries inhibitory activity of the nano-HA *in vitro*, *Key Eng. Mater.*, 330–332, 251–254.

599. Li, B. G., Wang, J. P., Zhao, Z. Y., Sui, Y. F., and Zhang, Y. X. (2007) Mineralizing of nano-hydroxyapatite powders on artificial caries, *Rare Metal. Mat. Eng.*, **36**, 128–130.

600. Borum, L., and Wilson, O. C. (2003) Surface modification of hydroxyapatite, *Part II. Silica. Biomaterials*, **24**, 3681–3688.

601. Lee, H. J., Kim, S. E., Choi, H. W., Kim, C. W., Kim, K. J., and Lee, S. C. (2007) The effect of surface-modified nano-hydroxyapatite on biocompatibility of poly(ε-caprolactone)/hydroxyapatite nanocomposites, *Eur. Polym. J.*, **43**, 1602–1608.

602. Wilson, O. C., and Hull, J. R. (2008) Surface modification of nanophase hydroxyapatite with chitosan, *Mater. Sci. Eng. C*, **28**, 434–437.

603. Liao, J. G., Wang, X. J., Zuo, Y., Zhang, L., Wen, J. Q., and Li, Y. B. (2008) Surface modification of nano-hydroxyapatite with silane agent, *J. Inorg. Mater.*, **23**, 145–149.

604. Ramachandran, R., Paul, W., and Sharma, C. P. (2009) Synthesis and characterization of PEGylated calcium phosphate nanoparticles for oral insulin delivery, *J. Biomed. Mater. Res. B (Appl. Biomater.)*, **88B**, 41–48.

605. Loo, S. C., Moore, T., Banik, B., and Alexis, F. (2010) Biomedical applications of hydroxyapatite nanoparticles, *Curr. Pharm. Biotechnol.*, **11**, 333–342.

606. Fu, H., Hu, Y., McNelis, T., and Hollinger, J. O. (2005) A calcium phosphate-based gene delivery system, *J. Biomed. Mater. Res. A*, **74A**, 40–48.

607. Liu, T. Y., Chen, S. Y., Liu, D. M., and Liou, S. C. (2005) On the study of BSA-loaded calcium-deficient hydroxyapatite nano-carriers for controlled drug delivery, *J. Control. Release*, **107**, 112–121.

608. Barroug, A., Kuhn, L. T., Gerstenfeld, L. C., and Glimcher, M. J. (2004) Interactions of cisplatin with calcium phosphate nanoparticles: *in vitro* controlled adsorption and release, *J. Orthop. Res.*, **22**, 703–708.

609. Cheng, X. G., and Kuhn, L .T. (2007) Chemotherapy drug delivery from calcium phosphate nanoparticles, *Int. J. Nanomed.*, **2**, 667–674.

610. Maitra, A. (2005) Calcium phosphate nanoparticles: second-generation nonviral vectors in gene therapy, *Expert Rev. Mol. Diagn.*, **5**, 893–905.

611. Yang, X. C., Walboomers, X. F., van den Dolder, J., Yang, F., Bian, Z., Fan, M. W., and Jansen, J. A. (2008) Non-viral bone morphogenetic protein 2 transfection of rat dental pulp stem cells using calcium phosphate nanoparticles as carriers, *Tissue Eng. A*, **14**, 71–81.

612. Bauer, I. W., Li, S. P., Han, Y. C., Yuan, L., and Yin, M. Z. (2008) Internalization of hydroxyapatite nanoparticles in liver cancer cells, *J. Mater. Sci. Mater. Med.*, **19**, 1091–1095.

613. Altinoğlu, E. I., and Adair, J. H. (2009) Calcium phosphate nanocomposite particles: a safer and more effective alternative to conventional chemotherapy? *Future Oncology*, **5**, 279–281.

614. Joyappa, D. H., Kumar, C. A., Banumathi, N., Reddy, G. R., and Suryanarayana, V. V. S. (2009) Calcium phosphate nanoparticle prepared with foot and mouth disease virus P1-3CD gene construct protects mice and guinea pigs against the challenge virus, *Veter. Microbiol.*, **139**, 58–66.

615. Dreesen, I. A. J., Lüchinger, N. A., Stark, W. J., and Fussenegger, M. (2009) Tricalcium phosphate nanoparticles enable rapid purification, increase transduction kinetics, and modify the tropism of mammalian viruses, *Biotechnol. Bioeng.*, **102**, 1197–1208.

616. Chu, T. C., He, Q., and Potter, D. E. (2002) Biodegradable calcium phosphate nanoparticlesas a new vehicle for delivery of a potential ocular hypotensive agent, *J. Ocular Pharmacol. Therapeutics*, **18**, 507–514.

617. Paul, W., and Sharma, C. P. (2001) Porous hydroxyapatite nanoparticles for intestinal delivery of insulin, *Trends Biomater. Artif. Organs*, **14**, 37–38.

618. Victor, S. P., and Kumar, T. S. S. (2008) Tailoring calcium-deficient hydroxyapatite nanocarriers for enhanced release of antibiotics, *J. Biomed. Nanotechnol.*, **4**, 203–209.

619. Kilian, O., Alt, V., Heiss, C., Jonuleit, T., Dingeldein, E., Flesch, I., Fidorra, U., Wenisch, S., and Schnettler, R. (2005) New blood vessel formation and expression of VEGF receptors after implantation of platelet growth factor-enriched biodegradable nanocrystalline hydroxyapatite, *Growth Factors*, **23**, 125–133.

620. Ling, J. Y., Loo, S. C., Phung, N. T., Boey, F., and Ma, J. (2008) Controlled size and morphology of EDTMP-doped hydroxyapatite nanoparticles as model for ^{153}Samarium-EDTMP doping, *J. Mater. Sci. Mater. Med.*, **19**, 2993–3003.

621. Jordan, M., Schallhorn, A., and Wurm, F. M. (1996) Transfecting mammalian cells: optimization of critical parameters affecting calcium-phosphate precipitate formation, *Nucleic Acids Res.*, **24**, 596–601.

622. Sokolova, V. V., and Epple, M. (2008) Inorganic nanoparticles as carriers of nucleic acids into cells, *Angew. Chem. Int. Ed.*, **47**, 1382–1395.

623. Liu, T., Tang, A., Zhang, G. Y., Chen, Y. X., Zhang, J. Y., Peng, S. S., and Cai, Z. M. (2005) Calcium phosphate nanoparticles as a novel nonviral vector for efficient transfection of DNA in cancer gene therapy, *Cancer Biother. Radiopharm.*, **20**, 141–149.

624. Czupryna, J., and Tsourkas, A. (2006) Suicide gene delivery by calcium phosphate nanoparticles. A novel method of targeted therapy for gastric cancer, *Cancer Biol. Ther.*, **5**, 1691–1692.

625. Olton, D., Li, J., Wilson, M. E., Rogers, T., Close, J., Huang, L., Kumta, P. N., and Sfeir, C. (2007) Nanostructured calcium phosphates (NanoCaPs) for non-viral gene delivery: influence of the synthesis parameters on transfection efficiency, *Biomaterials*, **28**, 1267–1279.

626. Bisht, S., Bhakta, G., Mitra, S., and Maitra, A. (2005) pDNA loaded calcium phosphate nanoparticles: highly efficient non-viral vector for gene delivery, *Int. J. Pharm.*, **288**, 157–168.

627. Chowdhury, E. H., and Akaike, T. (2005) A bio-recognition device developed onto nano-crystals of carbonate apatite for cell-targeted gene delivery, *Biotechnol. Bioeng.*, **90**, 414–421.

628. Bisht, S., Chattopadhyay, D., and Maitra, A. (2006) Intraperitoneal administration of calcium phosphate nanoparticles encapsulating pSVβgal elicits immune response to encoded protein, *J. Biomed. Nanotechnol.*, **2**, 229–238.

629. Zhu, S. H., Huang, B. Y., Zhou, K. C., Huang, S. P., Liu, F., Li, Y. M., Xue, Z. G., and Long, Z.G. (2004) Hydroxyapatite nanoparticles as a novel gene carrier, *J. Nanopart. Res.*, **6**, 307–311.

630. Chowdhury, E. H., Kutsuzawa, K., andAkaike, T. (2005) Designing smart nano-apatite composites: the emerging era of non-viral gene delivery, *Gene Ther. Mol. Biol.*, **9**, 301–316.

631. Chowdhury, E. H., Maruyama, A., Kano, A., Nagaoka, M., Kotaka, M., Hirose, S., Kunou, M., and Akaike, T. (2006) pH-sensing nano-crystals of carbonate apatite: effects on intracellular delivery and release of DNA for efficient expression into mammalian cells, *Gene*, **376**, 87–94.

632. Chowdhury, E. H. (2007) pH-sensitive nano-crystals of carbonate apatite for smart and cell-specific transgene delivery, *Expert Opin. Drug Deliv.*, **4**, 193–196.

633. Chowdhury, E. H., and Akaike, T. (2007) High performance DNA nano-carriers of carbonate apatite: multiple factors in regulation of particle synthesis and transfection efficiency, *Int. J. Nanomed.*, **2**, 101–106.

634. Pedraza, C. E., Bassett, D. C., McKee, M. D., Nelea, V., Gbureck, U., Barralet, J. E. (2008) The importance of particle size and DNA condensation salt for calcium phosphate nanoparticle transfection, *Biomaterials*, **29**, 3384–3392.

635. Zhou, C., Yu, B., Yang, X., Huo, T., Lee, L. J., Barth, R. F., and Lee, R. J. (2010) Lipid-coated nano-calcium-phosphate (LNCP) for gene delivery, *Int. J. Pharm.*, **392**, 201–208.

636. Welzel, T., Radtke, I., Meyer-Zaika, W., Heumann, R., and Epple, M. (2004) Transfection of cells with custom-made calcium phosphate nanoparticles coated with DNA, *J. Mater. Chem.*, **14**, 2213–2217.

637. Sokolova, V. V., I. Radtke, R. Heumann, and Epple, M. (2006) Effective transfection of cells with multi-shell calcium phosphate-DNA nanoparticles, *Biomaterials*, **27**, 3147–3153.

638. Sokolova, V. V., Kovtun, A., Heumann, R., and Epple, M. (2007) Tracking the pathway of calcium phosphate/DNA nanoparticles during cell transfection by incorporation of red-fluorescing tetramethylrhodamine isothiocyanate-bovine serum albumin into these nanoparticles, *J. Biol. Inorg. Chem.*, **12**, 174–179.

639. Sokolova, V. V., Kovtun, A., Prymak, O., Meyer-Zaika, W., Kubareva, E. A., Romanova, E. A., Oretskaya, T. S., Heumann, R., and Epple, M. (2007) Functionalisation of calcium phosphate nanoparticles by oligonucleotides and their application for gene silencing, *J. Mater. Chem.*, **17**, 721–727.

640. Neumann, S., Kovtun, A., Dietzel, I. D., Epple, M., and Heumann, R. (2009) The use of size-defined DNA-functionalized calcium phosphate nanoparticles to minimise intracellular calcium disturbance during transfection, *Biomaterials*, **30**, 6794–6802.

641. Graham, F. L., and van der Eb, A. J. (1973) A new technique for the assay of infectivity of human adenovirus 5 DNA, *Virology*, **52**, 456–467.

642. Kovtun, A., Heumann, R., and Epple, M. (2009) Calcium phosphate nanoparticles for the transfection of cells, *Bio-Med. Mater. Eng.*, **19**, 241–247.

643. Roy, I., Mitra, S., Maitra, A., and Mozumdar, S. (2003) Calcium phosphate nanoparticles as novel non-viral vectors for targeted gene delivery, *Int. J. Pharm.*, **250**, 25–33.

644. Li, J., Chen, Y. C., Tseng, Y. C., Mozumdar, S., and Huang, L. (2010) Biodegradable calcium phosphate nanoparticle with lipid coating for systemic siRNA delivery, *J. Controlled Release*, **142**, 416–421.

645. Kakizawa, Y., and Kataoka, K. (2002) Block copolymer self-assembly into monodispersive nanoparticles with hybrid core of antisense DNA and calcium phosphate, *Langmuir*, **18**, 4539–4543.

646. Wang, K. W., Zhou, L. Z., Sun, Y., Wu, G. J., Gu, H. C., Duan, Y. R., Chen, F., and Zhu, Y. J. (2010) Calcium phosphate/PLGA-mPEG hybrid porous nanospheres: a promising vector with ultrahigh gene loading and transfection efficiency, *J. Mater. Chem.*, **20**, 1161–1166.

647. Epple, M., and Kovtun, A. (2010) Functionalized calcium phosphate nanoparticles for biomedical application, *Key Eng. Mater.*, **441**, 299–305.

648. Sokolova, V., Neumann, S., Kovtun, A., Chernousova, S., Heumann, R., and Epple, M. (2010) An outer shell of positively charged poly(ethyleneimine) strongly increases the transfection efficiency of calcium phosphate/DNA nanoparticles, *J. Mater. Sci.*, **45**, 4952–4957.

649. Epple, M., Ganesan, K., Heumann, R., Klesing, J., Kovtun, A., Neumann, S., and Sokolova, V. (2010) Application of calcium phosphate nanoparticles in biomedicine, *J. Mater. Chem.*, **20**, 18–23.

650. He, Q., Mitchell, A. R., Johnson, S. L., Wagner-Bartak, C., Morcol, T., and Bell, S. J. D. (2000) Calcium phosphate nanoparticle adjuvant, *Clin. Diagn. Lab. Immunol.*, **7**, 899–903.

651. He, Q., Mitchell, A. R., Morcol, T., Bell, S. J. D. (2002) Calcium phosphate nanoparticles induce mucosal immunity and protection against herpes simplex virus type 2, *Clin. Diagn. Lab. Immunol.*, **9**, 1021–1024.

652. Liu, Z .S., Tang, S. L., and Ai, Z. L. (2003) Effects of hydroxyapatite nanoparticles on proliferation and apoptosis of human hepatoma BEL-7402 cells, *World J. Gastroenterol.*, **9**, 1968–1971.

653. Sun, J., and Ding, T. (2009) P53 reaction to apoptosis induced by hydroxyapatite nanoparticles in rat macrophages, *J. Biomed. Mater. Res. A*, **88A**, 673–679.

654. Yuan, Y., Liu, C., Qian, J., Wang, J., and Zhang, Y. (2010) Size-mediated cytotoxicity and apoptosis of hydroxyapatite nanoparticles in human hepatoma HepG2 cells, *Biomaterials*, **31**, 730–740.

655. Allen, T. M., and Cullis, P. R. (2004) Drug delivery systems: entering the mainstream, *Science*, **303**, 1818–1822.

656. Schmidt, H. T., Kroczynski, M., Maddox, J., Chen, Y., Josephs, R., and Ostafin, A. E. J. (2006) Antibody-conjugated soybean oil-filled calcium phosphate nanoshells for targeted delivery of hydrophobic molecules, *Microencapsulation*, **23**, 769–781.

657. Ferraz, M. P., Mateus, A. Y., Sousa, J. C., and Monteiro, F. J. (2007) Nanohydroxyapatite microspheres as delivery system for antibiotics: release kinetics, antimicrobial activity, and interaction with osteoblasts, *J. Biomed. Mater. Res. A*, **81A**, 994–1004.

658. Cai, Y., Pan, H., Xu, X., Hu, Q., Li, L., and Tang, R. (2007) Ultrasonic controlled morphology transformation of hollow calcium phosphate nanospheres: a smart and biocompatible drug release system, *Chem. Mater.*, **19**, 3081–3083.

659. Zhou, W. Y., Wang, M., Cheung, W. L., Guo, B. C., and Jia, D. M. (2008) Synthesis of carbonated hydroxyapatite nanospheres through nanoemulsion, *J. Mater. Sci. Mater. Med.*, **19**, 103–110.

660. Wingert, P. A., Mizukami, H., and Ostafin, A. E. (2007) Enhanced chemiluminescent resonance energy transfer in hollow calcium phosphate nanoreactors and the detection of hydrogen peroxide, *Nanotechnology*, **18**, pp. 295707 (7 pages).

661. Yih, T. C., and Al-Fandi, M. (2006) Engineered nanoparticles as precise drug delivery systems, *J. Cell. Biochem.*, **97**, 1184–1190.

662. Celotti, G., Tampieri, A., Sprio, S., Landi, E., Bertinetti, L., Martra, G., and Ducati, C. (2006) Crystallinity in apatites: how can a truly disordered fraction be distinguished from nanosize crystalline domains? *J. Mater. Sci. Mater. Med.*, **17**, 1079–1087.

663. Christenson, E. M., Anseth, K. S., van den Beucken, J. J. J. P., Chan, C. K., Ercan, B., Jansen, J .A., Laurencin, C. T., Li, W. J., Murugan, R., Nair, L. S., Ramakrishna, S., Tuan, R. S., Webster, T. J., and Mikos, A. G. (2007) Nanobiomaterial applications in orthopedics, *J. Orthop. Res.*, **25**, 11–22.

664. Schmidt, S. M., Moran, K. A., Kent, A. M. T., Slosar, J. L., Webber, M. J., McCready, M. J., Deering, C., Veranth, J. M., and Ostafin, A. (2008) Uptake of calcium phosphate nanoshells by osteoblasts and their effect on growth and differentiation., *J. Biomed. Mater. Res. A*, **87A**, 418–428.

665. Powell, M. C., and Kanarek, M. S. (2006) Nanomaterials health effects — Part 1: background and current knowledge, *Wisconsin Med. J.*, **105**, 16–20.

666. Powell, M. C., and Kanarek, M. S. (2006) Nanomaterials health effects — Part 2: uncertainties and recommendations for the future, *Wisconsin Med. J.*, **105**, 18–23.

667. Motskin, M., Wright, D. M., Muller, K., Kyle, N., Gard, T. G., Porter, A. E., and Skepper, J. N. (2009) Hydroxyapatite nano and microparticles: correlation of particle properties with cytotoxicity and biostability, *Biomaterials*, **30**, 3307–3317.

668. Liu, X., Qin, D., Cui, Y., Chen, L., Li, H., Chen, Z., Gao, L., Li, Y., and Liu, J. (2010) The effect of calcium phosphate nanoparticles on hormone production and apoptosis in human granulosa cells, *Reproductive Biology and Endocrinology*, **8**, 32 (8 pages).

669. Li, S., and Huang, L. (2008) Pharmacokinetics and biodistribution of nanoparticles, *Mol. Pharm.*, **5**, 496–504.

670. Bionanotechnologies are modeled after biological substances and structures, or combine nanomaterials with biological substances. They include materials such as biochips, drug release systems, nanofibers, hybrid nanobiodevices, molecular electronics and biomimetics (synthetic genes, proteins and viruses) [671].

671. Moghimi, S. J., Hunter, A. C., and Murray, J. C. (2005) Nanomedicine: current status and future prospects, *FASEB J.*, **19**, 311–330.

672. Xu, H. H. K., Weir, M. D., Simon, C. G., Jr. (2008) Injectable and strong nano-apatite scaffolds for cell/growth factor delivery and bone regeneration, *Dental Mater.*, **24**, 1212–1222.

673. Watari, F., Abe, S., Tamura, K., Uo, M., Yokoyama, A., and Totsuka, Y. (2008) Internal diffusion of micro/nanoparticles inside body, *Key Eng. Mater.*, 361–363, 95–98.

674. Oberdorster, G., Oberdorster, E., and Oberdorster, J. (2005) Nanotoxicology: an emerging discipline evolving from studies of ultrafine particles, *Environ. Health Perspect.*, **113**, 823–839.

675. Nel, A., Xia, T., Mädler, L., and Li, N. (2006) Toxic potential of materials at the nanolevel, *Science*, **311**, 622–627.

676. Jahnen-Dechent, W., and Simon, U. (2008) Function follows form: shape complementarity and nanoparticle toxicity, *Nanomedicine*, **3**, 601–603.

677. Singh, N., Manshian, B., Jenkins, G. J. S., Griffiths, S. M., Williams, P. M., Maffeis, T. G. G., Wright, C. J., and Doak, S. H. (2009) NanoGenotoxicology: the DNA damaging potential of engineered nanomaterials, *Biomaterials*, **30**, 3891–3914.

678. Dhawan, A., Sharma, V., and Parmar, D. (2009) Nanomaterials: a challenge for toxicologists, *Nanotoxicology*, **3**, 1–9.

679. Dwivedi, P. D., Misra, A., Shanker, R., and Das, M. (2009) Are nanomaterials a threat to the immune system? *Nanotoxicology*, **3**, 19–26.

Chapter 4

Calcium Orthophosphates as Bioceramics

4.1 Introduction

One of the most exciting and rewarding research areas of the materials science involves various applications to health care. Examples are sutures, catheters, heart valves, pacemakers, breast implants, fracture fixation plates, nails and screws in orthopedics, dental filling materials, orthodontic wires, as well as total joint replacement prostheses. Furthermore, during the past decades, both an ageing population and a democratization of high-risk sports have led to a surge of bone-related diseases and bone fractures, which must be treated. However, in order to be accepted by the living body, all implantable items must be prepared from a special class of materials, called biomedical materials or biomaterials, in short.

Overall, all solids are divided into four major groups of materials: metals, polymers ceramics and composites thereof. Similarly, all biomaterials are also divided into the same major groups: biometals, biopolymers, bioceramics and biocomposites. All of them play very important roles in both replacement and regeneration of the human tissues. However, due to a great number of publications, this chapter is limited to bioceramics only. In general, the modern bioceramics comprise various polycrystalline materials, glasses, glass-ceramics, as well as ceramic-filled bioactive [1] composites. All of them might be manufactured in both porous and dense forms in bulk, as well as

Calcium Orthophosphates: Applications in Nature, Biology, and Medicine
Sergey Dorozhkin

ISBN 978-981-4316-62-0 (Hardcover), 978-981-4364-17-1 (eBook)
www.panstanford.com

in the forms of powders, granules and/or coatings. An expansion of bioceramics to the health care has been characterized by a significant increasing in the number of publications and patents in this field and an ever-increasing number of major international conferences and themed meetings [2–6].

Interestingly, but the chemical elements used to manufacture bioceramics form just a small set of the Periodic Table. Namely, bioceramics might be prepared from alumina, zirconia, magnesia, carbon, silica-contained and calcium-contained compounds, as well as some other chemicals [4]; however, this chapter is further limited to calcium orthophosphates only. Calcium orthophosphate-based biomaterials and bioceramics are now used in a number of different applications throughout the body, covering all areas of the skeleton. Applications include dental implants, percutaneous devices and use in periodontal treatment, healing of bone defects, fracture treatment, total joint replacement (bone augmentation), orthopedics, cranio-maxillofacial reconstruction, otolaryngology, ophthalmology and spinal surgery [3–7]. Depending upon the required properties, different calcium orthophosphates might be used. For example, Fig. 4.1 shows some randomly chosen samples of the commercially available calcium orthophosphate bioceramics for bone graft applications.

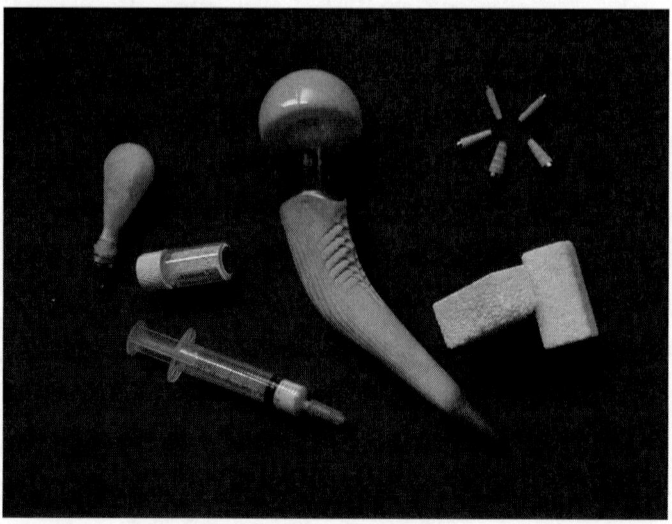

Figure 4.1 Several examples of the commercial calcium orthophosphate-based bioceramics.

In this chapter, the focus has been placed upon applications of calcium orthophosphate bioceramics as medical implants to repair and reconstruct the damaged or diseased hard tissues (usually, those of the musculo-skeletal system, such as bones or teeth) of the body and to describe some of the major developments in this field during the past ~40 years. To narrow the subject further, with a few important exceptions, bioceramics prepared from undoped and un-substituted calcium orthophosphates have been considered and discussed only. Furthermore, calcium orthophosphate bioceramics prepared from the biological resources, such as bones, teeth, corals, etc., is not considered either. The readers interested in these topics are advised to read the original papers [8–38].

4.2 General Knowledge on Biomaterials and Bioceramics

A number of definitions have been developed for the term "biomaterials." Until recently, the consensus developed by the experts in this field has been the following: biomaterials are defined as synthetic or natural materials to be used to replace parts of a living system or to function in intimate contact with living tissues [39]. However, in September 2009, a more advanced definition has been introduced: "A biomaterial is a substance that has been engineered to take a form which, alone or as part of a complex system, is used to direct, by control of interactions with components of living systems, the course of any therapeutic or diagnostic procedure, in human or veterinary medicine" [40]. These alterations in the definitions are accompanied by a shift in conceptual ideas on biomaterials and the expectations of their biological performance, which both have changed in time [41].

The biomaterials discipline is founded in the knowledge of the synergistic interaction of material science, biology, chemistry, medicine and mechanical science and requires the input of comprehension from all these areas so that implanted biomaterials perform adequately in a living body and interrupt normal body functions as little as possible [42]. As biomaterials mainly deal with all aspects of material synthesis and processing, the knowledge in chemistry, material science and engineering is essential. On the other hand, as clinical applications are the main purposes of biomaterials, biomedical sciences become the key part of the research. These include cell and molecular biology, anatomy and animal and human physiology. The final aim is to achieve the correct biological interaction of the implanted biomaterials with living tissues of a host. In order to achieve the goals, several stages have to be performed, namely: material synthesis, design and manufacturing of prostheses, followed by various types of tests. Furthermore, any potential biomaterial must also pass all regulatory requirements before its clinical application [43].

The major difference of biomaterials from other classes of materials is their ability to remain in a biological environment without damaging the surroundings and without being damaged in that process. Therefore, biomaterials in the form of implants (sutures, bone plates, joint replacements, ligaments, vascular grafts, etc.) and medical devices (pacemakers, biosensors, artificial hearts, blood tubes, etc.) are widely used to replace and/or restore the function of traumatized or degenerated tissues or organs, to assist in healing, to improve function, to correct abnormalities, and, thus, to improve the quality of life of the patients [44]. In any case, biomaterials are intended to interface with biological systems to evaluate, treat, augment or replace any tissue, organ or function of the body and are now used in a number of different applications throughout the body [5, 6, 45]. Thus, biomaterials are solely associated with the health care domain and must have an interface with tissues or tissue components. One should stress that any artificial materials those simply are in contact with skin, such as hearing aids and wearable artificial limbs, are not included in the definition of biomaterials since the skin acts as a protective barrier between the body and the external world.

Biomaterials must be distinguished from *biological materials* because the former are the materials that are accepted by living tissues and, therefore, they might be used for tissue replacements, while the latter are the materials being produced by various biological systems (wood, cotton, bones, chitin, etc.) [46]. In addition, there are *biomimetic materials*, which are not made by living organisms but have the composition, structure and properties similar to those of biological materials. Further, *bioceramics* (or biomedical ceramics) might be defined as biomaterials of the ceramic [47] origin. In general, bioceramics can have structural functions as joint or tissue replacements, be used as coatings to improve the biocompatibility

[48] of metal implants, as well as function as resorbable lattices, providing temporary structures and frameworks those are dissolved and/or replaced as the body rebuilds the damaged tissues [49–55]. Some types of bioceramics even feature a drug-delivery capability [56, 57].

A progressive deterioration of all tissues with age is the major contributor to the need for spare parts for the body. Bone is especially vulnerable to fracture in older people due to a loss of density and strength with age. This effect is especially severe in women due to the hormonal changes associated with menopause. A graphical representation of the effect of time on bone strength and density from the age of 30 years onward is available in literature [Ref. 51, Fig. 1]. Bone density decreases because bone-growing cells (osteoblasts) become progressively less productive in making new bone and repairing micro-fractures. The lower density greatly deteriorates the strength of bones and an unfortunate consequence is that many old people fracture their hips or have collapsed vertebrae and spinal problems [51].

Bioceramics are needed to alleviate pain and restore functions to diseased or damaged calcified tissues (bones and teeth) of the body. A great challenge facing its medical application is to replace old, deteriorating bone with a material that can function the remaining years of the patient's life and, ideally, be replaced by a new mature bone without transient loss of mechanical support [2]. Surface reactivity is one of the common characteristics of bioceramics. It contributes to their bone bonding ability and their enhancing effect on bone tissue formation. During implantation, various interactions occur at the material/tissue interfaces that lead to time-dependent changes in the surface characteristics of the implanted bioceramics and the surrounding tissues [58]. Because the average life span of humans is now 80+ years and the major need for spare parts begins at about 60 years of age, the implanted non-resorbable bioceramics need to last, at least, for 20+ years. This demanding requirement of survivability is under conditions of use that are especially harsh to implanted materials: corrosive saline solutions at 37°C under variable, multiaxial and cyclical mechanical loads. The excellent performance of the specially designed bioceramics that have survived these clinical conditions represents one of the most remarkable accomplishments of research, development, production and quality assurance during the past century [51].

4.3 Bioceramics of Calcium Orthophosphates

4.3.1 History

The performance of living tissues is the result of millions years of evolution, while the performance of acceptable artificial substitutions those humankind has designed to repair damaged tissues are only a few decades old. However, man's attempts to repair the human body with the use of implant materials are recorded in the early medical writings of the Hindu, Egyptian and Greek civilizations. The earliest successful implants were in the skeletal system. Historically, a selection of the materials was based on their availability and an ingenuity of the individual making and applying the prosthetic [59]. Archaeological findings exhibited in museums showed that materials used to replace missing human bones and teeth included animal or human (from corpses) bones and teeth, shells, corals, ivory (elephant tusk), wood, as well as some metals (gold or silver). For instance, the Etruscans learned to substitute missing teeth with bridges made from artificial teeth carved from the bones of oxen, while in ancient Phoenicia loose teeth were bound together with gold wires for tying artificial ones to neighboring teeth. In the seventeenth century, a piece of dog skull was successfully transplanted into the damaged skull of a Dutch duke. The Chinese recorded the first use of dental amalgam to repair decayed teeth in the year 659 AD, while pre-Columbian civilizations used gold sheets to heal cranial cavities following trepanation [60]. Furthermore, in 1970, Amadeo Bobbio discovered Mayan skulls, some of them more than ~4000 years old, in which missing teeth had been replaced by nacre substitutes [61]. Unfortunately, due to the practice of cremation in many societies, little is known about prehistoric materials used to replace bone lost to accident or disease.

Historically, plaster of paris was the first widely tested artificial bioceramics. However, in the past, many implantations failed due to infections, which tended to be exacerbated in the presence of implants, since they provided a region inaccessible to the body's immunologically competent cells. Thus, the use of biomaterials did not become practical until the advent of an aseptic surgical technique developed by Dr. J. Lister in the 1860s. Furthermore, there was a lack of knowledge about a toxicity of the selected materials.

In this frame, application of calcium orthophosphates appears to be logical due to their similarity with the mineral phases of bones and teeth [62–66]. Calcium orthophosphates are not toxic and do not cause cell death in the surrounding tissues. However, according to available literature, the first attempt to use them (it was TCP) as an artificial material to repair surgically created defects in rabbits was performed in 1920 [67]. Although this may be the first scientific study on use of a calcium orthophosphate for bone defects repair, it remains unclear whether the calcium orthophosphate was a precipitated or a ceramic material and whether it was in a powder or granular form. The second clinical report was only published 30 years later [68]. More than 20 years afterwards, the first dental application of a calcium orthophosphate (erroneously described as TCP) in surgically created periodontal defects [69] and the use of dense HA cylinders for immediate tooth root replacement [70] were reported. According to the available databases, the first paper with the term "bioceramics" in the abstract was published in 1971 [71], while with that in the title were published in 1972 [72, 73]. However, application of the ceramic materials as prostheses had been known before [74–77]. Further historical details are available in both Chapter 8 and other publications on the subject [78, 79]. On April 26, 1988, the first international symposium on bioceramics was held in Kyoto, Japan.

Commercialization of the dental and surgical applications of calcium orthophosphate (mainly, HA) bioceramics occurred in the 1980s, largely due to the pioneering efforts by Jarcho [80–83] in the United States, de Groot [84–86] in Europe and Aoki [87–90] in Japan. Shortly afterwards HA has become a bioceramic of reference in the field of calcium orthophosphates for biomedical applications. Preparation and biomedical applications of apatites derived from sea corals (coralline HA) [91–93] and bovine bone [94] have been reported at the same time [95].

4.3.2 Chemical Composition and Preparation

Currently, calcium orthophosphate bioceramics can be prepared from various sources [96–109]. Unfortunately, up to now, all attempts to synthesize bone replacement materials for clinical applications featuring the physiological tolerance, biocompatibility and a long-term stability have had only a relative success; it shows the superiority and a complexity of the natural structures [110].

In general, a characterization of calcium orthophosphate bioceramics should be observed from many viewpoints such as the chemical composition (stoichiometry and purity), homogeneity, phase distribution, morphology, grain sizes and shape, grain boundaries, crystallite size, crystallinity, pores, cracks, surface, etc. From the chemical point of view, the vast majority of calcium orthophosphate bioceramics is based on HA, β-TCP, α-TCP and/or biphasic calcium phosphate (BCP, which is an intimate mixture of either β-TCP + HA [111–126] or α-TCP + HA [8–12]) [127–136]. One should note, that recently the concept of BCP has been extended by preparation and characterization of biphasic TCP (BTCP), consisting of α-TCP and β-TCP phases [137–141]. The biphasic TCP is usually prepared by heating ACP precursors [139–141], in which the α-TCP/β-TCP ratio can be controlled by aging time and pH value during synthesis of the amorphous precursor [140]. Furthermore, a triphasic formulation, consisting of HA, α-TCP and β-TCP has been prepared [142]. Further details on this topic are available in section 1.3.14 Multiphasic and polyphasic calcium orthophosphate compositions.

The preparation techniques of various calcium orthophosphates have been extensively reviewed in literature [86, 143–151 and references therein] where the interested readers are referred to. Besides, some data are available in Chapter 1 of this book. When compared to both α- and β-TCP, HA is a more stable phase under the physiological conditions, as it has a lower solubility (Table 1.1) and, thus, a slower resorption kinetics [64, 127, 128]. Therefore, the BCP concept is determined by the optimum balance of a more stable phase of HA and a more soluble TCP. Due to a higher biodegradability of the α- or β-TCP component, the reactivity of BCP increases with the TCP/HA ratio increasing. Thus, *in vivo* bioresorbability of BCP can be controlled through the phase composition [120]. Similar conclusions are also valid for both the biphasic TCP (in which α-TCP is a more soluble phase) and the triphasic (HA, α-TCP and β-TCP) formulation.

As implants made of sintered HA are found in bone defects for many years after implantation (Fig. 4.2, bottom), bioceramics made of more soluble calcium orthophosphates [8–12, 111–126, 129–142, 152–154] is preferable for the biomedical purposes (Fig. 4.2, top). Furthermore, the experimental results showed that BCP had a higher ability to adsorb fibrinogen, insulin or type I collagen than HA [155]. Thus, according to both observed and measured bone formation parameters, calcium orthophosphates have been ranked as follows: low sintering temperature BCP (rough and smooth) ≈ medium sintering temperature BCP ≈ TCP > calcined low sintering temperature HA > non-calcined low sintering temperature HA > high sintering temperature BCP (rough and smooth) > high sintering temperature HA [156]. This sequence has been developed in year 2000 and, thus, neither biphasic TCP, nor triphasic (HA, α-TCP and β-TCP) formulations have been included. Recent developments in processing and surface modification of HA have been reviewed elsewhere [157].

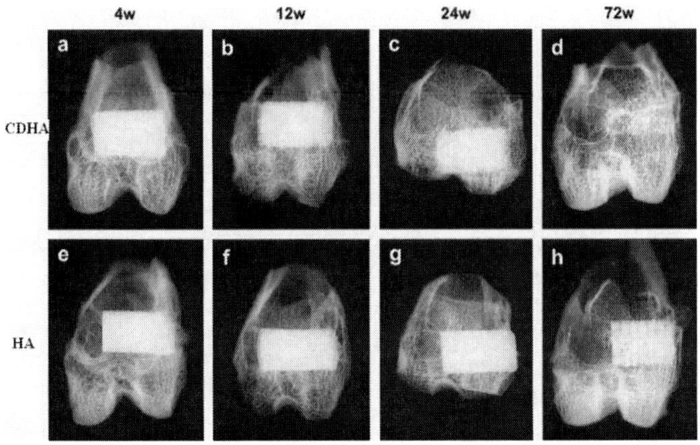

Figure 4.2 Soft X-ray photographs of the operated portion of the rabbit femur. Four weeks (a), 12 weeks (b), 24 weeks (c) and 72 weeks (d) after implantation of CDHA; 4 weeks (e), 12 weeks (f), 24 weeks (g) and 72 weeks (h) after implantation of sintered HA. Reprinted from Ref. [152] with permission.

4.3.3 Forming and Shaping

In order to fabricate bioceramics in progressively complex shapes, scientists are investigating the use of old and new manufacturing techniques. These techniques range from an adaptation of the age-old pottery techniques to the latest manufacturing methods for high-temperature ceramic parts for airplane engines. For example, reverse engineering and rapid prototyping technologies have revolutionized a generation of physical models, allowing an engineer to efficiently and accurately produce physical models and customized implants with high levels of geometric intricacy [158–160]. Combined with the computer-aided design and manufacturing (CAD/CAM), complex physical objects of the anatomical structure can be fabricated in a variety of shapes and sizes. In a typical application, an image of a bone defect in a patient can be taken and used to develop a three-dimensional (3D) CAD computer model [161–164]. A computer can then reduce the model to slices or layers. The 3D objects are constructed layer-by-layer using rapid prototyping techniques such as fused deposition modeling [165, 166], selective laser sintering [167–169], laser cladding [170], 3D printing [171–180], solid freeform fabrication [181–185] and/or stereo lithography [186–189]. Furthermore, a thermal printing process of melted calcium orthophosphates has been proposed as well [190]. A custom-made implant of actual dimensions would reduce the time it takes to perform the medical implantation procedure and subsequently lower the risk to the patient. Another advantage of a prefabricated, exact-fitting implant is that it can be used more effectively and applied directly to the damaged site rather than a replacement, which is formulated during surgery from a paste or granular material [182, 191, 192]. In some cases, laser processing might be applied as well [193].

The manufacturing technique depends greatly on the ultimate application of the bioceramic device, whether it is for a hard-tissue replacement or an integration of the device within the surrounding tissues. In general, three types of the processing technologies might be used: (1) employment of a lubricant and a liquid binder with ceramic powders for shaping and subsequent firing; (2) application of self-setting and self-hardening properties of water-wet molded powders (cementation—see Chapter 5); (3) materials are melted to form a liquid and are shaped during cooling and solidification [194–197]. Since calcium orthophosphates are either thermally unstable (MCPM, MCPA, DCPA, DCPD, OCP, ACP, CDHA) or have a melting point at temperatures exceeding ~1400°C with a partial decomposition (α-TCP, β-TCP, HA, FA, TTCP), only the first and the second consolidation approaches are used to prepare bulk bioceramics and scaffolds. The methods include uniaxial compaction [198, 199], isostatic pressing (cold or hot) [200–206], granulation [207, 208], loose packing [209], slip casting [210–213], gel casting [188, 189, 214–219], pressure mold forming [220], injection molding [221], polymer replication [222–225], extrusion [226–228], slurry dipping and spraying [229], as well as to form ceramic sheets from slurries tape casting [124, 216, 230, 231] doctor blade [232] and colander methods might be employed [194–197, 233]. A combination of several techniques is also possible [216, 234]. Furthermore, some of these processes might be performed under the magnetic field, which helps crystal aligning [235–238].

Powders are usually pressed damp in metal dies or dry in lubricated dies at pressures high enough to form sufficiently strong structures to hold together until they are sintered. An organic binder such as polyvinyl alcohol helps to bind the powder together [200]. Drying at about 100°C is a critical step in preparing damp-formed pieces for firing. Too much or too little water in the compacts can lead to blowing apart the ware on heating or crumbling, respectively. The binder is removed by heating in air to oxidize the organic phases to carbon dioxide and water [194–197].

Furthermore, forming and shaping of any ceramic products require a proper selection of the raw materials in terms of particle sizes and size distribution. Namely, tough and strong bioceramics consist of pure, fine and homogeneous microstructures. To attain this, pure powders with small average size and high surface area must be used as the starting sources. However, for maximum packing and least shrinkage after firing, mixing of ~70% coarse and ~30% fine powders have been suggested [197]. Mixing is usually carried out in a ball mill for uniformity of properties and reaction during subsequent firing. Mechanical die forming or sometimes extrusion through a die orifice can be used to produce a fixed cross-section. Drying involves removal of water and subsequent shrinkage of the product. However, due to local variations in water content, warping and even cracks may be developed during drying. Dry pressing and hydrostatic molding can minimize these problems [197]. Afterwards, the manufactured green samples are sintered.

Finally, to produce the accurate shaping, necessary for the fine design of bioceramics, machine finishing might be essential [163, 194, 239]. Unfortunately, cutting tools developed for metals are usually useless for bioceramics due to their fragility; therefore, grinding and polishing appear to be the convenient finishing techniques [163, 194]. Furthermore, the surface of bioceramics might be modified by various additional treatments [240].

4.3.4 Sintering and Firing

A sintering (or firing) procedure appears to be of a great importance to manufacture bulk bioceramics with the required mechanical properties. Usually, this stage is carried out according to controlled temperature programs of electric furnaces in adjusted ambience of air with necessary additional gasses; however, always at temperatures below the melting points of the materials. The firing step can include temporary holds at intermediate temperatures to burn out organic binders [194–197]. The heating rate, sintering temperature and holding time depend on the starting materials. For example, in the case of HA, these values are in the ranges of 0.5–3°C/min, 1000–1250°C and 2–5 h, respectively [241]. In the majority cases, sintering allows a structure to retain its shape. However, this process might be accompanied by a considerable degree of shrinkage [94, 242–244], which must be accommodated in the fabrication process. For instance, in the case of FA sintering, a linear shrinkage was found to occur at ~715°C and the material reached its final density at ~890°C. Above this value, grain growth became important and induced an intra-granular porosity, which was responsible for density decrease. At ~1180°C, a liquid phase was formed due to formation of a binary eutectic between FA and fluorite contained in the powder as impurity. This liquid phase further promoted the coarsening process and induced formation of large pores at high temperatures [245].

The sintering mechanism is controlled by both surface and volume diffusion at grain boundaries. In general, when solids are heated to high temperatures, the constituent ions or atoms are driven to move to fill up pores and open channels between the grains of powders, as well as to compensate for the surface energy differences among their convex and concave surfaces. At the initial stages, bottlenecks are formed and grow among the particles (Fig. 4.3). Existing vacancies tend to flow away from the surfaces of sharply curved necks; this is an equivalent of a material flow towards the necks, which grow as the voids shrink. Small contact areas among the particles expand and, at the same time, a density of the compact increases and the total void volume decreases. As the pores and open channels are closed during a heat treatment, the particles become tightly bonded together and density, strength and fatigue resistance of the sintered object improve greatly. Grain-boundary diffusion was identified as the dominant mechanism for densification [246]. Furthermore, strong chemical bonds are formed among the particles and loosely compacted green bodies are hardened to denser materials [194–197].

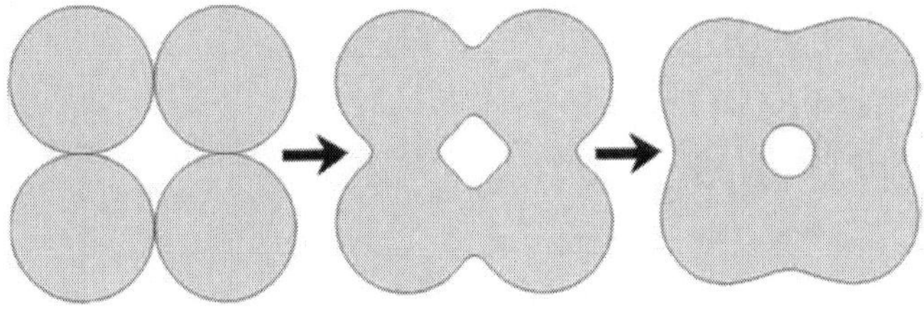

Figure 4.3 A schematic diagram representing the changes occurring with particles under sintering.

In the case of calcium orthophosphates, several specific processes occur during sintering. Firstly, moisture, carbonates and all other volatile chemicals remaining from the synthesis stage, such as ammonia, nitrates and any organic compounds, are removed as gaseous products. Secondly, unless powders are sintered, the removal of these gases facilitates production of denser ceramics with subsequent shrinkage of the samples (Fig. 4.4). Thirdly, all chemical changes are accompanied by a concurrent increase in crystal size and a decrease in the specific surface area. Fourthly, a chemical decomposition of all acidic orthophosphates and their transformation into other phosphates (e.g., $2HPO_4^{2-} \rightarrow P_2O_7^{4-} + H_2O\uparrow$) takes place. Besides, sintering causes toughening [247], densification [248, 249], partial dehydroxylation (in the case of HA) [249], as well as it increases the mechanical strength [250, 251]. The latter events are due to presence of air and other gases filling gaps among the particles of un-sintered powders. At sintering, the gases move towards the outside of powders and green bodies shrink owing to decrease of distances among the particles. For example, sintering of a biologically formed apatites was investigated [252, 253] and the obtained products were characterized [254, 255]. In all cases, the numerical value of Ca/P ratio in sintered apatites of biological origin was higher than that of the stoichiometric HA. One should mention that in the vast majority cases, calcium orthophosphates with Ca/P ratio < 1.5 (Table 1.1) are not sintered, since these compounds are thermally unstable, while sintering of CDHA and ACP leads to their transformation into various types of BCP (HA + β-TCP [121], HA + α-TCP [136], α-TCP + β-TCP [139–141]).

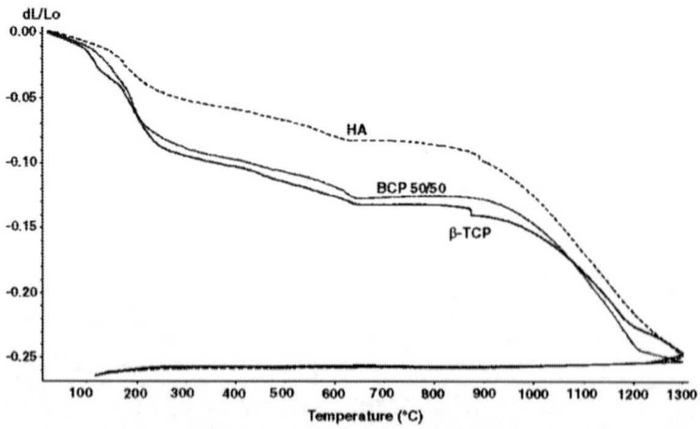

Figure 4.4 Linear shrinkage of the compacted ACP powders that were converted into β-TCP, BCP (50% HA + 50% β-TCP) and HA upon heating. According to the authors: "At 1300°C, the shrinkage reached a maximum of approximately ~25, ~30 and ~35% for the compacted ACP powders that converted into HA, BCP 50/50 and β-TCP, respectively" [243]. Reprinted from Ref. [243] with permission.

An extensive study on the effects of sintering temperature and time on the properties of HA bioceramics revealed a correlation between these parameters and density, porosity, grain size, chemical composition and strength of the scaffolds [256]. Namely, sintering below ~1000°C was found to result in initial particle coalescence, with little or no densification and a significant loss of the surface area and porosity. The degree of densification appeared to depend on the sintering temperature whereas the degree of ionic diffusion was governed by the period of sintering [256]. To enhance sinterability of calcium orthophosphates, a variety of sintering additives might be added [257–260].

HA powders can be pressurelessly sintered up to the theoretical density at 1000–1200°C. Processing at higher temperatures may lead to exaggerated grain growth and decomposition because HA becomes unstable at temperatures exceeding ~1300°C [64, 86, 145–151, 261]. The decomposition temperature of HA bioceramics is a function of the partial pressure of water vapor. Moreover, processing under vacuum leads to an earlier decomposition of HA, while processing under high partial pressure of water prevents from the decomposition. On the other hand, a presence of water in the sintering atmosphere was reported to inhibit densification of HA and accelerated grain growth [233, 262]. Unexpectedly, but application of a magnetic field during sintering was found to influence the growth of HA grains [263]. A definite correlation between hardness, density and a grain size in sintered HA bioceramics was found: despite exhibiting high bulk density, hardness started to decrease at a certain critical grain size limit [264–266].

Hot pressing [266–272], hot isostatic pressing [204, 205] or hot pressing with post-sintering [273, 274] processes make it possible to decrease a temperature of the densification process, diminish the grain size, as well as achieve higher densities. This leads to finer microstructures, higher thermal stability and subsequently better mechanical properties of calcium orthophosphate bioceramics. Microwave [108, 275–283] and spark plasma [284–292] sintering techniques are alternative methods to the conventional sintering, hot pressing and hot isostatic pressing. Both alternative methods were found to be time and energy efficient densification techniques. Further developments are still possible. For example, a hydrothermal hot pressing method has been developed to fabricate both OCP [293] and CDHA [294] bulk bioceramics with neither thermal dehydration nor thermal decomposition. Further details on the sintering and firing processes of calcium orthophosphate bioceramics are available in literature [50, 64, 146, 233, 295, 296].

To conclude this part, one should mention on an excellent review paper on various ceramic manufacturing techniques [297], where the interested readers are referred to extend their knowledge on the ceramic processing.

4.4 The Major Properties

4.4.1 Mechanical Properties

Ideally, a bone substitute should be replaced by a mature bone without transient loss of mechanical support. Unfortunately for material scientists, a human body provides one of the most inhospitable environments for the implanted materials. It is warm, wet and both chemically and biologically active. For example, a diversity of body fluids in various tissues might have a solution pH varying from 1 to 9. Furthermore, the body is capable of generating quite massive force concentrations and the variance in such characteristics among individuals might be enormous. Typically, bones are subjected to approximately 4 MPa, whereas tendons and ligaments experience peak stresses in the range 40–80 MPa. The hip joints are subjected to an average load up to three times body weight (3,000 N) and peak loads experienced during jumping can be as high as 10 times body weight. These stresses are repetitive and fluctuating depending on the nature of the activities, which can include standing, sitting, jogging, stretching and climbing. Therefore, all types of potential biomaterials and bioceramics must sustain attacks of a great variety of aggressive conditions [298]. Regrettably, there is presently no material fulfilling all these requirements.

On the other hand, any ceramics, when they fail, tend to do so in a dramatic manner. Namely, a brittle nature of calcium orthophosphate bioceramics is attributed to high-strength ionic bonds. Thus, it is not possible for plastic deformation to happen prior to failure, as slip cannot occur. Consequently, if a crack is initiated, its progress will not be hindered by the deformation of material ahead of the crack, as would be the case in a ductile material (e.g., a metal). In ceramics, the crack will continue to propagate, rapidly resulting in a catastrophic failure [195].

For dense bioceramics, the strength is a function of the grain sizes. It appears to be very sensitive to a slow crack growth [299]. Finer grain size bioceramics have smaller flaws at the grain boundaries and thus are stronger than one with larger grain sizes. In general, the mechanical properties decrease significantly with increasing content of an amorphous phase, microporosity and grain sizes, while a high crystallinity, a low porosity and small grain sizes tend to give a higher stiffness, a higher compressive and tensile strength and a greater fracture toughness. Accordingly, from the mechanical point of view, calcium orthophosphate bioceramics appear to be brittle polycrystalline materials for which the mechanical properties are governed by crystallinity, grain size, grain boundaries, porosity and composition [203]. Thus, it possesses poor mechanical properties (for instance, a low impact and fracture resistances) that do not allow calcium orthophosphate bioceramics to be used in load-bearing areas, such as artificial teeth or bones [49–55, 300]. For example, fracture toughness [301] of HA bioceramics does not exceed the value of ~1.2 MPa·$m^{1/2}$ [302] (human bone: 2–12 MPa·$m^{1/2}$). It decreases almost linearly with a porosity increasing [233]. Generally, fracture toughness increases with grain size decreasing. However, in some materials, especially non-cubic ceramics, fracture toughness reaches the maximum and rapidly drops with decreasing grain size. For example, Halouani et al. investigated a fracture toughness of pure hot pressed HA with grain sizes between 0.2 and 1.2 μm. They found two distinct trends, where fracture toughness decreased with increasing grain size above ~0.4 μm and subsequently decreased with decreasing grain size. The maximum fracture toughness measured was 1.20 ± 0.05 MPa·$m^{1/2}$ at ~0.4 μm [272]. Fracture energy of HA bioceramics is in the range of 2.3–20 J/m^2, while the Weibull modulus [303] is low (~5–12) in wet environments, which means that HA behaves as a typical brittle ceramics and indicates to a low reliability of HA implants [233]. Interestingly, but 3 peaks of internal friction were found at temperatures about −40, 80 and 130°C for HA but no internal friction peaks were obtained for FA in the measured temperature range; this effect was attributed to the differences of F^- and OH^- positions in FA and HA, respectively [304].

Bending, compressive and tensile strengths of dense HA bioceramics are in the ranges of 38–250 MPa, 120–900 MPa and 38–300 MPa, respectively. Similar values for porous HA bioceramics are in the ranges of 2–11 MPa, 2–100 MPa and ~3 MPa, respectively [233]. These wide variations in the properties are due to both structural variations (e.g., an influence of remaining microporosity, grain sizes, presence of impurities, etc.) and manufacturing processes, as well as they are caused by a statistical nature of the strength distribution. Strength [305] was found to increase with Ca/P ratio increasing, reaching the maximum value around Ca/P ~1.67 (stoichiometric HA) and decreases suddenly when Ca/P > 1.67 [233]. Furthermore, strength decreases almost exponentially with porosity increasing [112, 113]. However, by changing the pore geometry, it is possible to influence the strength of porous bioceramics. It is also worth mentioning that porous HA bioceramics is considerably less fatigue [306] resistant than dense ones. Both grain sizes and porosity are reported to influence the fracture path, which itself has a little effect on the fracture toughness of calcium orthophosphate bioceramics [203, 307]. Furthermore, no obvious decrease in mechanical properties was found after calcium orthophosphate bioceramics had been aged in the various solutions during the different time periods [308].

Young's (or elastic) modulus [309] of dense HA bioceramics is in the range of 35–120 GPa, which is more or less similar to those of the most resistant components of the natural calcified tissues (dental enamel: ~74 GPa, dentine: ~21 GPa, compact bone: ~18–22 GPa). Nevertheless, dense bulk compacts of HA have mechanical resistances of the order of 100 MPa versus ~300 MPa of human bones, diminishing drastically their resistances in the case of porous bulk compacts [310]. Young's modulus measured in bending is between 44 and 88 GPa. Recently, a considerable anisotropy in the stress–strain behavior of the perfect HA crystals was found by *ab initio* calculations [311]. The crystals appeared to be brittle for tension along the *z*-axis with the maximum stress of ~9.6 GPa at 10% strain. Furthermore, the structural analysis of the HA crystal under various stages of tensile strain revealed that the deformation behavior manifested itself mainly in the rotation of PO_4 tetrahedrons with concomitant movements of both the columnar and axial Ca ions [311]. Vickers hardness [312] of dense HA bioceramics is within 3–7 GPa, while the Poisson's ratio [313] for the synthetic HA is about 0.27, which is close to that of bones (~0.3). At temperatures within 1000–1100°C, dense HA bioceramics was found to exhibit superplasticity with a deformation mechanism based on grain boundary sliding. Furthermore, both wear resistance and friction coefficient of dense HA bioceramics are comparable to those of dental enamel [233].

Due to a high brittleness (associated to a low crack resistance), the biomedical applications of calcium orthophosphate bioceramics are focused on production of non-load-bearing implants, such as pieces for middle ear surgery, filling of bone defects in oral or orthopedic surgery, as well as coating of dental implants and metallic prosthesis (see below) [110, 314, 315]. In order to improve the reliability of calcium orthophosphate bioceramics, diverse reinforcements (ceramics, metals or polymers) have been applied to manufacture various biocomposites and hybrid biomaterials [316], but that is another story (see Chapter 4). However, successful hybrid formulations consisted of calcium orthophosphates only should be mentioned [317–322]. For example, bulk HA bioceramics might be reinforced by HA whiskers [318–321]. Furthermore, a superior superplasticity of HA/β-TCP biocomposites (i.e., BCP) to HA bioceramics has been detected [322].

Another approach to improve the mechanical properties of calcium orthophosphate bioceramics is to coat the items by a polymeric layer [323, 324]; however, this is still other story. Further details on the mechanical properties of calcium orthophosphate bioceramics are available elsewhere [233, 325], where the interested readers are referred to.

4.4.2 Electrical Properties

Occasionally, an interest to the electrical properties of calcium orthophosphate bioceramics is expressed. For example, a surface ionic conductivity of both porous and dense HA bioceramics was examined for humidity sensor applications, since the room temperature conductivity was influenced by relative humidity [326]. Namely, the ionic conductivity of HA has been a subject of research for its possible use as an alcohol [327], carbon dioxide [327] or carbon monoxide [328] gas sensors. Electrical measurements have also been used as a characterization tool to study the evolution of microstructure in HA bioceramics [329]. More to the point, Valdes et al. examined the dielectric properties of HA to understand its decomposition to β-TCP [330]. In the case of CDHA, the electrical properties, in terms of ionic conductivity, were found to increase after compression of the samples at 15 t/cm^2, which was attributed to establishment of some order within the apatitic network [331]. The conductivity mechanism of CDHA appeared to be multiple [332]. Furthermore, there was an attempt to develop CDHA whisker electrets for biomedical utilization [333].

Interestingly, but the electrical properties of calcium orthophosphate bioceramics appear to influence their biomedical applications. For example, there is an interest in polarization of HA bioceramics to generate a surface charge by the application of electric fields at elevated temperatures [334, 335]. The presence of surface charges on HA bioceramics was shown to have a significant effect on both *in vitro* and *in vivo* crystallization of biological apatite [336–340]. Furthermore, a growth of both biomimetic calcium orthophosphates and bones was found to be accelerated on negatively charged surfaces and decelerated at positively charged surfaces [340–349]. In addition, the electrical polarization of HA bioceramics was found to accelerate a cytoskeleton reorganization of osteoblast-like cells [350–352], extend bioactivity [353] and enhance bone ingrowth through the pores of porous HA implants [354]. There is an interesting study on the interaction of a blood coagulation factor on electrically polarized HA surfaces [355]. Further details on the electrical properties of calcium orthophosphate-based bioceramics are available in literature [279, 356–360].

4.4.3 Possible Transparency

Single crystals of all calcium orthophosphates are optically transparent for the visible light. As bioceramics of calcium orthophosphates have a polycrystalline nature with a random orientation of big amounts of small crystals, it is opaque and of white color, unless colored dopants have been added. However, in some cases, a transparency is convenient to provide some essential advantages (e.g., to enable direct viewing of living cells, their attachment, spreading, proliferation, and osteogenic differentiation cascade in a transmitted light). Thus, transparent calcium orthophosphate bioceramics (Fig. 4.5) [361] have been prepared and investigated [204, 206, 289, 292, 361–370]. They can exhibit an optical transmittance of ~66 % at a wavelength of 645 nm [370]. The preparation techniques, for example, include a hot isostatic pressing [204, 206], an ambient-pressure sintering [362], gel casting coupled with a low-temperature sintering [365, 368], a pulse electric current sintering [366], as well as a spark plasma sintering [289, 292]. Fully dense, transparent calcium orthophosphate bioceramics was obtained at temperatures above ~800°C. Depending on the preparation technique, the transparent bioceramics has a uniform grain sizes ranging from ~81 nm [370] to ~250 μm [365] and always is pore-free. Furthermore, a translucent calcium orthophosphate bioceramics is also known [371–373]. However, due to a lack of both porosity and the necessity to have see-through implants inside the body, the transparent and/ or translucent forms of calcium orthophosphate bioceramics will hardly be ever used in medicine with the specific eye implants as the only reasonable exception.

Figure 4.5 Transparent HA bioceramics prepared by spark plasma sintering at 900°C from nano-sized HA single crystals. Reprinted from Ref. [361] with permission.

4.4.4 Porosity

Porosity is defined as a percentage of void spaces in solids and it is a morphological property independent of the material. The surface area of porous bodies is much higher, which guarantees a good mechanical fixation in addition to providing sites on the surface that allow chemical bonding between the bioceramics and bones [374]. Furthermore, a porous material may have both closed (isolated) pores and open (interconnected) pores. The interconnected pores look like tunnels and are accessible by gases, liquids and particulate suspensions [375]. The open-cell nature of reticulated materials is a unique characteristic essential in many applications. Furthermore, dimensions of open pores are directly related to bone formation, since such pores grant both the surface and space for cell adhesion and bone ingrowth. On the other hand, pore interconnection provides the ways for cell distribution and migration, as well as it allows an efficient *in vivo* blood vessel formation suitable for sustaining bone tissue neo-formation and possibly remodeling [155, 376–385]. Namely, porous HA bioceramics can be colonized by bone tissues [381, 386–396]. Therefore, interconnecting macroporosity (pore size > 100 μm) [111, 374, 381, 397, 398], which is defined by its capacity to be colonized by cells, is intentionally introduced in solid bioceramics (Fig. 4.6). Macroporosity is usually formed due to a release of various volatile materials and, for that reason, incorporation of pore-creating additives (porogens) is the most popular technique to create macroporosity. The porogens are crystals or particles of either volatile (they evolve gases at elevated temperatures) or soluble substances, such as paraffin, naphthalene, sucrose, $NaHCO_3$, gelatin, polymethylmethacrylate or even hydrogen peroxide [112, 295, 400–407]. Obviously, the ideal porogen should be nontoxic and be removed at ambient temperature, thereby allowing the ceramic/porogen mixture to be injected directly into a defect site and allowing the scaffold to fit the defect [408]. Sintering particles, preferably spheres of equal size, is a similar way to generate porous 3D bioceramics of calcium orthophosphates (Fig. 4.7). However, pores resulting from this method are often irregular in size and shape and not fully interconnected with one another.

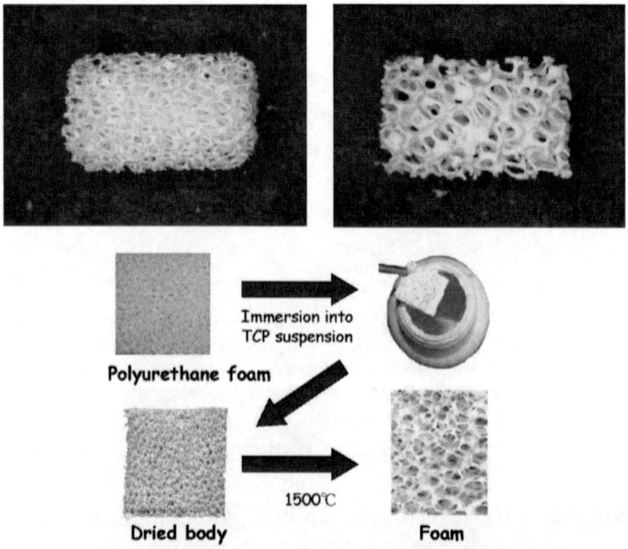

Figure 4.6 Photographs of a commercially available porous calcium orthophosphate bioceramics with different porosity (top) and a method of their production (bottom). For photos, the horizontal field width is 20 mm. The bottom picture is reprinted from Ref. [399] with permission.

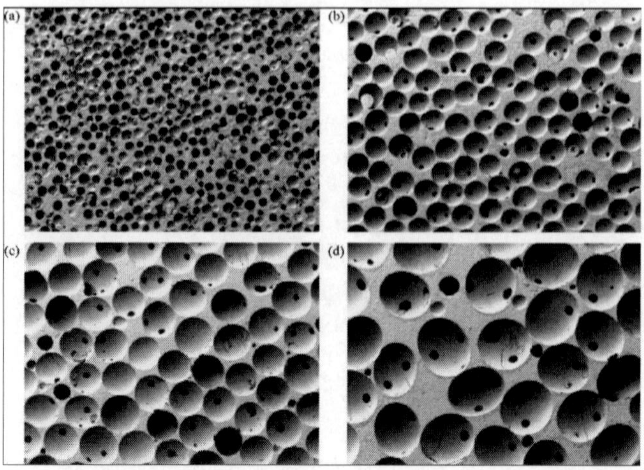

Figure 4.7 β-TCP porous ceramics with different pore sizes prepared using polymethylmethacrylate balls with diameter equal to: (a) 100–200, (b) 300–400, (c) 500–600 and (d) 700–800 μm. Horizontal field width is 45 mm. Reprinted from Ref. [411] with permission.

Several other techniques, such as replication of polymer foams by impregnation (Fig. 4.6), dual-phase mixing, particulate leaching, freeze casting, slip casting, stereo lithography, direct foaming of suspensions, as well as surfactant washing have been applied to fabricate porous calcium orthophosphate bioceramics [91, 195, 212, 216, 219, 222–224, 267, 268, 376–378, 397–451]. Some of them have been summarized in Table 4.1. [408]. Furthermore, natural porous materials, like coral skeletons made of $CaCO_3$, can be converted into porous HA under the hydrothermal conditions (250°C, 24–48 h) with the microstructure undamaged [91–93]. Porous HA bioceramics can also be obtained by hydrothermal hot pressing. This technique allows solidification of the HA powder at 100–300°C (30 MPa, 2 h) [415]. In another approach, bi-continuous water-filled microemulsions have been used as pre-organized systems for the fabrication of needle-like frameworks of crystalline HA (2°C, 3 weeks) [416, 417]. Porous HA bioceramics might be prepared by a combination of gel casting and foam burn out methods [219]. Lithography was used to print a polymeric material, followed by packing with HA and sintering [418]. Both hot pressing [267, 268] and ice templating [419] techniques might be applied as well. Besides, an HA suspension can be cast into a porous $CaCO_3$ skeleton, which is then dissolved, leaving a porous network [410]. Three-dimensional periodic macroporous frame of HA has been fabricated via a template-assisted colloidal processing technique [420, 421]. A superporous (~85% porosity) HA ceramics was developed as well [450, 451]. More to the point, porous HA bioceramics might be prepared by using different starting HA powders and sintering at various temperatures by a pressureless-sintering approach [427].

Porous bioceramics with an improved strength might be fabricated from calcium orthophosphate fibers or whiskers. In general, fibrous porous materials are known to exhibit an improved strength due to fiber interlocking, crack deflection and/or pullout [422]. Namely, porous bioceramics with well-controlled open pores was processed by sintering of fibrous HA particles [423]. In another approach, porosity was achieved by firing apatite-fiber compacts mixed with carbon beads and agar. By varying the compaction pressure, firing temperature and carbon/HA ratio, the total porosity was controlled in the ranges from ~40% to ~85% [412]. Additional examples are available in literature [397, 401, 408–451].

Table 4.1 The procedures used to manufacture porous calcium orthophosphate scaffolds for tissue engineering [408].

Year	Location	Process	Apatite from:	Sintering	Compressive strength	Pore size	Porosity	Method of porosity control
2006	Deville et al. Berkeley, CA	HA + ammonium methacrylate in PTFE mold, freeze dried and sintered	HA #30	Yes: 1300°C	16 MPa 65 MPa 145 MPa	Open unidirectional 50–150 μm	> 60% 56% 47%	Porosity control: slurry conc. structure controlled by physics of ice front formation.
2006	Saiz et al. Berkeley, CA	Polymer foams coated, compressed after infiltration, then calcined.	HA powder	Yes: 700–1300°C	—	100–200 μm	—	Porosity control: extent of compression, HA loading
2006	Murugan et al. Singapore + USA	Bovine bone cleaned, calcined	bovine bone	Yes: 500°C	—	Retention of nano-sized pores	—	Porosity control: native porosity of bovine bone
2006	Xu et al. Gaithersburg, MD	Directly injectable calcium orthophosphate cement, self hardens, mannitol as porogen	nanocrystalline HA	No	2.2–4.2 MPa (flexural)	0–50% macroporous	65–82%	Porosity control: mannitol mass fraction in mixture

2004	Landi et al. Italy + Indonesia	Sponge impregnation, isotactic pressing, sintering of HA in simulated body fluid	$CaO + H_3PO_4$	Yes: 1250°C for 1 h	23 ± 3.8 MPa	Closed 6% Open 60%	66%	Porosity control: possibly by controlling HA particle size. Not suggested by authors
2003	Charriere et al. EPFL, Switzerland	Thermoplastic negative porosity by Ink jet printing, slip casting process for HA	DCPA + calcite	No: 90°C for 1 day	12.5 ± 4.6 MPa	—	44%	Porosity control: negative printing
2003	Almirall et al. Barcelona, Spain	α-TCP foamed with hydrogen peroxide at different conc., liq. ratios, poured in PTFE molds	α-TCP + (10% and 20% H_2O_2)	No: 60°C for 2 h	1.41 ± 0.27 MPa 2.69 ± 0.91 MPa	35.7% macro 29.7% micro 26.8% macro 33.8% micro	65.5% 60.7%	Porosity control: different concentration, α-TCP particle sizes
2003	Ramay et al. Seattle, WA	Slurries of HA prepared: gel-casting + polymer sponge technique, sintered.	HA powder	Yes: 600°C for 1 h 1350°C for 2 h	0.5–5 MPa	200–400 μm	70–77%	Porosity control: replicate of polymer sponge template

(Contd.)

Table 4.1 (Contd.)

Year	Location	Process	Apatite from:	Sintering	Compressive strength	Pore size	Porosity	Method of porosity control
2003	Miao et al. Singapore	TTCP to calcium orthophosphate cement. Slurry cast on polymer foam, sintered.	TTCP	Yes: 1200°C for 2 h	—	1 mm macro 5 μm micro	~70%	Porosity control: Recoating time, polyurethane foam
2003	Uemura et al. China + Japan	Slurry of HA with polyoxyethylene-laurylether (cross-linked) and sintered	HA powders	Yes: 1200°C for 3 h	2.25 MPa (0 week) 4.92 MPa (12 weeks) 11.2 MPa (24 weeks)	500 μm 200 μm interconnects	~77%	Porosity control: polymer interconnects cross-linking
2003	Ma et al. Singapore + USA	Electrophoretic deposition of HA, sintering.	HA powders	Yes: 1200°C for 2 h	860 MPa	0.5 μm 130 μm	~20%	Porosity control: electrophoresis field
2002	Barralet et al. Birmingham, London, UK	Calcium orthophosphate cement + sodium phosphate ice, evaporated	$CaCO_3$ + DCPD	1st step: 1400°C for 1 day	0.6 ± 0.27 MPa	2 μm	62 ± 9%	Porosity control: porogen shape

In vivo response of calcium orthophosphate bioceramics of different porosity was investigated and a hardly any effect of macropore dimensions (~150, ~260, ~510 and ~1220 μm) was observed [452]. In another study, a greater differentiation of mesenchymal stem cells was observed when cultured on ~200 μm pore size HA scaffolds when compared to those on ~500 μm pore size HA [453]. The latter finding was attributed to the fact that a higher pore volume in ~500 μm macropore scaffolds might contribute to a lack of cell confluency leading to the cells proliferating before beginning differentiation. Besides, the authors hypothesized that bioceramics having a less than the optimal pore dimensions induced quiescence in differentiated osteoblasts due to reduced cell confluency [453]. Already in 1979, Holmes suggested that the optimal pore range was 200–400 μm with the average human osteon size of ~223 μm [92]. In 1997, Tsurga and coworkers implied that the optimal pore size of bioceramics that supported ectopic bone formation was 300–400 μm [454]. Thus, there is no need to create calcium orthophosphate bioceramics with very big pores; however, the pores must be interconnected [95, 384, 397, 398]. Interconnectivity governs a depth of cells or tissue penetration into the porous bioceramics, as well as it allows development of blood vessels required for new bone nourishing and wastes removal [455, 456].

Bioceramic microporosity (pore size < 10 μm), which is defined by its capacity to be impregnated by biological fluids [455], results from the sintering process, while the pore dimensions mainly depend on the material composition, thermal cycle and sintering time. The microporosity provides both a greater surface area for protein adsorption and increased ionic solubility. For example, embedded osteocytes distributed throughout microporous rods might form a mechanosensory network, which would not be possible in scaffolds without microporosity [457]. HA bioceramics with nanodimensional (<100 nm) pores might be fabricated as well [458]. Differences in porogens influence the macroporosity, while differences in sintering temperatures and conditions affect the percentage of microporosity. Usually, the higher the sintering temperature, the lower both the microporosity content and the specific surface area of bioceramics. Namely, HA bioceramics sintered at ~1200°C shows significantly less microporosity and a dramatic change in crystal sizes, if compared with that sintered at ~1050°C (Fig. 4.8). Furthermore, the average shape of pores was found to transform from strongly oblate to round at higher sintering temperatures [460]. The total porosity (macroporosity + microporosity) of calcium orthophosphate bioceramics was reported to be ~70% [461] or ~85% [450, 451] of the bioceramic volume. In the case of coralline HA or bovine-derived apatites, the porosity of the original biologic material (coral or bovine bone) is usually preserved during processing [94]. To conclude this topic, creation of the desired porosity in calcium orthophosphate bioceramics is a rather complicated engineering task and the interested readers are referred to other publications on this subject [112, 377, 402, 416, 462–489].

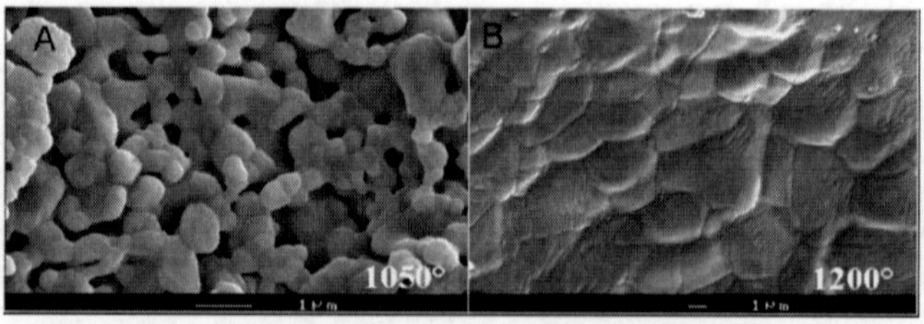

Figure 4.8 SEM pictures of HA bioceramics sintered at (A) 1050°C and (B) 1200°C. Note the presence of microporosity in A and not in B. Reprinted from Ref. [459] with permission.

Studies revealed that increasing of both the specific surface area and pore volume of bioceramics might greatly accelerate the *in vivo* process of apatite deposition and, therefore, enhance the bone-forming bioactivity. More importantly, a precise control over the porosity, pore dimensions and internal pore architecture of bioceramics on different length scales is essential for understanding of the structure-bioactivity relationship and the rational design of better bone-forming biomaterials [483, 490, 491]. Namely, in antibiotic charging experiments, a calcium orthophosphate bioceramics with nanodimensional (<100 nm) pores showed a much higher charging capacity (1621 μg/g) than that of commercially available calcium orthophosphate (100 μg/g), which did not contain nanodimensional porosity [478]. In other experiments, porous blocks of HA were found to be viable carriers with sustained release profiles for drugs [492] and antibiotics over 12 days [493] and 12 weeks [494], respectively. Unfortunately, porosity significantly decreases the strength of implants [233, 307, 325]. Thus, porous calcium orthophosphate implants cannot be loaded and are used to fill only small bone defects. However, their strength increases gradually when bones ingrow into the porous network of calcium orthophosphate implants [495–498]. For example, Martin et al. reported bending strengths of 40–60 MPa for a porous HA implant filled with 50–60% of cortical bone [495], while in another study an ingrown bone increased strength of porous HA bioceramics by a factor of 3–4 [497].

To conclude this topic, one should mention that filters for microbial filtration might be manufactured from porous HA [499].

4.5 Biomedical Applications

Since Levitt et al. described a method of preparing a FA bioceramics and suggested its possible use in medical applications in 1969 [500], calcium orthophosphate bioceramics have been widely tested for clinical applications. Namely, calcium orthophosphates in a number of forms and compositions (Table 4.2) currently are either in use or under a consideration in many areas of dentistry and orthopedics, with even more in development. For example, bulk materials, available in dense and porous forms, are used for alveolar ridge augmentation, immediate tooth replacement and maxillofacial reconstruction [64, 377]. Other examples include burr-hole buttons [505], orbital implants (Bio-Eye®) [506–509], other ophthalmic applications [510, 511], increment of the hearing ossicles, spine fusion and repair of bone defects [512, 513]. In order to permit growth of new bone into defects, a suitable bioresorbable material should fill these defects. Otherwise, ingrowth of fibrous tissue might prevent bone formation within the defects.

In spite of the aforementioned serious mechanical limitations, bioceramics of calcium orthophosphates is available in various physical forms: powders, particles, granules (or granulates [12]), dense blocks, porous scaffolds, self-setting formulations (Chapter 5), implant coatings and composite component (Chapter 6) of different origin (natural, biological or synthetic) often with the specific shapes, such as implants, prostheses or prosthetic devices (Table 4.3) [2, 95]. Furthermore, bone grafts are also proposed as non-hardening pastes (= "putty"). Generally, the latter materials consist of a mixture of calcium orthophosphate powders or granules and a "glue," typically a highly viscous hydrogel (see section 6.4.6 Injectable bone substitutes (IBS)) [2, 316]. More to the point, custom-designed shapes like wedges for tibial opening osteotomy, cones for spine and knee and inserts for vertebral cage fusion are also available [461]. Various trademarks of the commercially available types of calcium orthophosphate-based bioceramics and biomaterials have been summarized in Table 4.2.

Table 4.2 Various examples of the commercially available calcium orthophosphate-based bioceramics and biomaterials [13, 459, 461, 501–504]

Calcium orthophosphate	Trade name and producer
CDHA	Cementek (Teknimed, France)
	Osteogen (Impladent, NY, USA)
HA	Actifuse (ApaTech, UK)
	Apaceram (Pentax, Japan)
	ApaPore (ApaTech, UK)
	Bioroc (Depuy-Bioland, France)
	Bonefil (Pentax, Japan)
	Bonetite (Pentax, Japan)
	Boneceram (Sumitomo Osaka Cement, Japan)
	BoneSource (Stryker Orthopaedics, NJ, USA)
	Calcitite (Zimmer, IN, USA)
	Cerapatite (Ceraver, France)
	Neobone (Toshiba Ceramics, Japan)
	Ostegraf (Ceramed, CO, USA)
	Ostim (Heraeus Kulzer, Germany)
	Synatite (SBM, France)
HA/collagen	Bioimplant (Connectbiopharm, Russia)
	Bonject (Koken, Japan)
	Collagraft (Zimmer and Collagen Corporation, USA)
	CollapAn (Intermedapatite, Russia)
	HAPCOL (Polystom, Russia)
	LitAr (LitAr, Russia)

(Contd.)

Table 4.2 (Contd.)

Calcium orthophosphate	Trade name and producer
HA/sodium alginate	Bialgin (Biomed, Russia)
HA/poly-L-Lactic Acid	SuperFIXSORB30 (Takiron, Japan)
HA/polyethylene	HAPEX (Gyrus, TN, USA)
HA/$CaSO_4$	Hapset (LifeCore, MIN, USA)
Coralline HA	Interpore (Interpore, CA, USA)
	ProOsteon (Interpore, CA, USA)
Algae-derived HA	Algipore (Dentsply Friadent, Germany)
Bovine bone apatite (unsintered)	BioOss (Geitslich, Switzerland)
	Laddec (Ost-Developpement, France)
	Lubboc (Ost-Developpement, France)
	Oxbone (Bioland biomateriaux, France)
	Tutoplast (IOP, CA, USA)
Bovine bone apatite (sintered)	BonAP
	Cerabone (aap Implantate, Germany)
	Endobon (Merck, Germany)
	Osteograf (Ceramed, CO, USA)
	PepGen P-15 (Dentsply Friadent, Germany)
β-TCP	Bioresorb (Sybron Implant Solutions, Germany)
	Biosorb (SBM S.A., France)
	Calciresorb (Ceraver, France)
	CELLPLEX (Wright Medical Technology, TN, USA)
	Cerasorb (Curasan, Germany)
	Ceros (Thommen Medical, Switzerland)
	ChronOS (Synthes, PA, USA)
	Conduit (DePuy Spine, USA)
	JAX (Smith and Nephew Orthopaedics, USA)
	Osferion (Olympus Terumo Biomaterials, Japan)
	OsSatura TCP (Integra Orthobiologics, CA, USA)
	Vitoss (Orthovita, PA, USA)

(Contd.)

Table 4.2 (Contd.)

Calcium orthophosphate	Trade name and producer
BCP (HA + β-TCP)	4Bone (MIS, Israel)
	BCP (Medtronic, MN, USA)
	Biosel (Depuy Bioland, France)
	BoneCeramic (Straumann, Switzerland)
	BoneSave (Stryker Orthopaedics, NJ, USA)
	Calciresorb (Ceraver, France)
	CellCeram (Scaffdex, Finland)
	Ceraform (Teknimed, France)
	Ceratite (NGK Spark Plug, Japan)
	CuriOs (Progentix Orthobiology BV, Netherlands)
	Eurocer (FH Orthopedics, France)
	Graftys BCP (Graftys, France)
	Hatric (Arthrex, Naples, FL, USA)
	Indost (Polystom, Russia)
	Kainos (Signus, Germany)
	MasterGraft Granules (Medtronic Sofamor Danek, TN, USA)
	MBCP (Biomatlante, France)
	OpteMX (Exactech, FL, USA)
	OsSatura BCP (Integra Orthobiologics, CA, USA)
	Osteosynt (Einco, Brazil)
	SBS (Expanscience, France)
	TCH (Kasios, France)
	Triosite (Zimmer, IN, USA)
	Tribone (Stryker, Europe)
BCP (HA + α-TCP)	Skelite (Millennium Biologix, ON, Canada)
BCP/collagen	Allograft (Zimmer, IN, USA)
BCP/fibrin	TricOS (Baxter BioScience, France)
BCP/silicon	FlexHA (Xomed, FL, USA)
FA + BCP (HA + β-TCP)	FtAP (Polystom, Russia)
Carbonateapatite	Healos (Orquest, CA, USA)
	SRS (Norian, CA, USA)

Table 4.3 Specific features of the four most common forms of bone graft substitutes [2]

Form	Defect form	Mechanical stability	Resorption / bone formation	Handling
Granules (0.1–5 mm in diameter)	Open	Negligible	Throughout the defect	Fair (granule migration during and after surgery)
Macroporous blocks	Open and defined shape	Fair provided there is press-fitting into the defect	Throughout the defect	Very good (problems might arise to fit the block within the defect)
Cement paste	Closed	Fair	Peripheral	Fair to good (the paste might set too fast or might be poorly injectable)
Putty	Open or closed	Negligible	Peripheral or throughout the defect depending on the composition	Very good for pastes that have to be mixed in the operating room to excellent for ready-mixed pastes (the paste might be poorly injectable)

Note: The column "defect form" lists the types of defects that can be potentially filled with the listed bone graft substitute form. "Open" means that the defect has to be widely open, *e.g.* an open cancellous bone defect; "Defined shape" means that the defect has to have a well-defined shape, e.g. cylinder; "Closed" means that the material can be (potentially) injected into a closed defect, *e.g.* to reinforce an osteoporotic bone [2].

One should note, that among the existing calcium orthophosphates (Table 1.1), only certain compounds are useful for biomedical applications, because those having the Ca/P ionic ratio less than one are not suitable for implantation due to their high solubility and acidity. Furthermore, due to its basicity, TTCP alone is not suitable either. However, to be used in medicine, these "unsuitable" calcium orthophosphates might successfully be combined with either other calcium orthophosphates or other chemicals.

4.5.1 Cements and Concretes

The need of bioceramics for minimal invasive surgery has induced the development of a concept of self-setting formulations consisting of only calcium orthophosphates to be applied as injectable and/or mouldable bone substitutes [156, 400, 401, 418, 475, 514–522]. In addition, there are reinforced formulations, which, in a certain sense, might be defined as calcium orthophosphate concretes [516]. Furthermore, porous formulations of both the cements and the concretes are available [401, 418, 515–520].

Calcium orthophosphate cements and concretes belong to a low temperature bioceramics. They are divided into two major groups. The first one is a dry mixture of two different calcium orthophosphates (a basic one and an acidic one), in which, after being wetted, the setting reaction occurs according to an acid–base reaction. The second group of the cements contains only one calcium orthophosphate. Typical examples include ACP with Ca/P molar ratio within 1.50–1.67 and α-TCP: they form CDHA upon contact with an aqueous solution [156, 515, 516]. The setting reaction (= hardening, curing) of these materials is initiated by mixing the initial powder(s) with an aqueous solution. Chemically, hardening is due to the succession of dissolution and precipitation reactions. Mechanically, hardening results from crystal entanglement and intergrowth (Fig. 4.9) [2]. Setting of calcium orthophosphate cements and concretes occurs mostly within the initial ~6 h, yielding a ~80% conversion to the final products and a compressive strength of 40–60 MPa. A hardening rate is strongly influenced by a powder to liquid ratio, as well as by addition of other chemicals [156, 514–522]. Despite a large number of formulations, all calcium orthophosphate cements can only form two different end products: CDHA and DCPD [156, 515, 516].

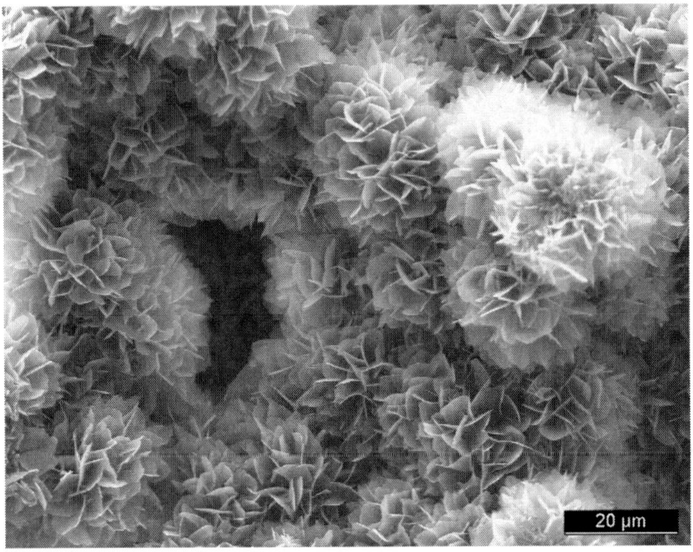

Figure 4.9 A typical microstructure of calcium orthophosphate cement after hardening. The mechanical stability is provided by the physical entanglement of crystals. Reprinted from Ref. [2] with permission.

All calcium orthophosphate cements and concretes are biocompatible, bioactive and bioresorbable. The first animal study of a calcium orthophosphate cement was performed in 1991. That time a cement consisting of TTCP and DCPA was investigated histologically by implanting disks made of this cement within the heads of nine cats [523, 524]. In 1996, that formulation received an approval by the US Food and Drug Administration (FDA), thus becoming the first commercially available calcium orthophosphate cement for use in humans [517]. As the structure and composition of the hardened cements is close to that of bone mineral, the material of the hardened cements can easily be used by bone remodeling cells for reconstruction of damaged parts of bones [156, 514–517]. A possibility to be injected (a minimally invasive technique), a low setting temperature, an adequate stiffness, an easy shaping and a good adaptation to the defect geometry are the major advantages of calcium orthophosphate cements and concretes, when compared to the prefabricated bulk bioceramics and porous scaffolds. Further details on this subject are available in Chapter 5 of this book.

4.5.2 Coatings

For many years, the clinical application of calcium orthophosphate-based bioceramics has been largely limited to non-load bearing parts of the skeleton due to their inferior mechanical properties. One of the major innovations in the last ~30 years has been to coat mechanically strong bioinert and/or biotolerant prostheses by calcium orthophosphates [525–527]. For example, metallic implants are encountered in endoprostheses (total hip joint replacements) and artificial teeth sockets. The requirement for a sufficient mechanical stability necessitates the use of a metallic body for such devices. As metals do not undergo bone bonding, i.e., do not form a mechanically stable link between the implant and bone tissue, ways have been sought to improve contacts at the interface. The major way is to coat metals with calcium orthophosphate bioceramics that exhibit a bone-bonding ability between the metal and bone [194, 205, 339, 525, 528–533]. Thickness of the coatings varies from submicron dimensions to several hundreds microns (Table 4.4) and this parameter appears to be very important. For example, if a calcium orthophosphate coating is too thick, it is easy to break. On the contrary, if the coating is too thin, it is easy to dissolve, because resorbability of HA, which is the second slowest to dissolve among calcium orthophosphates (Table 1.1), is about 15–30 μm per year [534]. One should stress, that calcium orthophosphate coatings are not limited to metals only; they can be applied on carbon, bioinert ceramics and polymers as well [535]. Furthermore, multi-walled carbon nanotubes can be functionalized by deposition of calcium orthophosphate coatings [536]. The list of most important coating techniques is comprised in Table 4.4, while the main advantages and drawbacks of each coating technique, as well as the major properties of the deposed calcium orthophosphates, are discussed in details elsewhere [194, 240, 295, 525, 526, 537–551]. Unfortunately, none of these methods is able to provide the perfect covering because each coating always contains cracks, pores, second phases and residual stresses that reduced their durability and might lead to a partial or complete disintegration of the coating in body fluids. The biomedical aspects of osteoconductive coatings for total joint arthroplasty have been reviewed elsewhere [552].

All coatings must adhere satisfactorily to the underlying substrate irrespective of their intended function. Specifically, a mechanical stability of calcium orthophosphate coatings should be high enough to maintain their bioactive functionality after a surgical implantation. Generally, tensile adhesion testing according to standards such as ASTM C633 is the most common procedure to determine the quantitative values for calcium orthophosphate coating adhesion to the underlying metallic substrates. Furthermore, fatigue [553, 554], scratch [555] and pullout [555] testing are among the most valuable techniques to provide additional information on the mechanical behavior of calcium orthophosphate coatings [194].

Already in the 1980s, de Groot et al. published a paper [556] on the development of plasma-sprayed HA coatings on metallic implants. A little bit later, Furlong and Osborn [557], two leading surgeons in the orthopedics field, began implanting plasma-sprayed HA stems in patients. Coated implants combine the surface biocompatibility and bioactivity of calcium orthophosphates with the core strength of strong substrates (Fig. 4.10). Moreover, calcium orthophosphate coatings decrease a release of potentially hazardous chemicals from the core implant and shield the substrate surface from environmental attack. In the case of porous implants, calcium orthophosphate coatings enhance bone ingrowth into the pores [233]. The clinical results for calcium orthophosphate-coated implants reveal that they have much longer life times after implantation than uncoated devices and they have been found to be particularly beneficial for younger patients. Studies concluded that there was significantly less pin loosening in the HA-coated groups [558]. HA coating as a system of fixation of hip implants was found to work well in the short to medium term (8 years [559], 10 to 15.5 years [560], 10 to 17 years [561], 15 years [562], 16 years [563], 17 years [564], 19 years [565] and 15 to 21 years [566]). In 2004, a special book summarizing the studies with HA-coated implants and the "state of the art" of HA coatings in orthopedics at the close of 2002 was published [567]. Similar data for HA-coated dental implants are also available [568, 569]. Even longer-term clinical results are awaited with a great interest.

Table 4.4. Various techniques to deposit bioresorbable coatings of calcium orthophosphates on metal implants [538, 539]

Technique	Thickness	Advantages	Disadvantages
Thermal spraying	30–200 μm	High deposition rates; low cost	Line of sight technique; high temperatures induce decomposition; rapid cooling produces amorphous coatings
Sputter coating	0.5–3 μm	Uniform coating thickness on flat substrates; dense coating	Line of sight technique; expensive; time consuming; produces amorphous coatings
Pulsed laser deposition	0.05–5 μm	Coating by crystalline and amorphous phases; dense and porous coating	Line of sight technique
Dynamic mixing method	0.05–1.3 μm	High adhesive strength	Line of sight technique; expensive; produces amorphous coatings
Dip coating	0.05–0.5 mm	Inexpensive; coatings applied quickly; can coat complex substrates	Requires high sintering temperatures; thermal expansion mismatch
Sol-gel technique	< 1 μm	Can coat complex shapes; low processing temperatures; relatively cheap as coatings are very thin	Some processes require controlled atmosphere processing; expensive raw materials

Electrophoretic deposition	0.1–2.0 mm	Uniform coating thickness; rapid deposition rates; can coat complex substrates	Difficult to produce crack-free coatings; requires high sintering temperatures
Biomimetic coating	< 30 μm	Low processing temperatures; can form bonelike apatite; can coat complex shapes; can incorporate bone growth stimulating factors	Time consuming; requires replenishment and a pH constancy of simulated body fluid
Hot isostatic pressing	0.2–2.0 μm	Produces dense coatings	Cannot coat complex substrates; high temperature required; thermal expansion mismatch; elastic property differences; expensive; removal/interaction of encapsulation material
Electrochemical deposition	0.05–0.5 mm	Uniform coating thickness; rapid deposition rates; can coat complex substrates; moderate temperature, low cost	The coating/substrate bonding is not strong enough

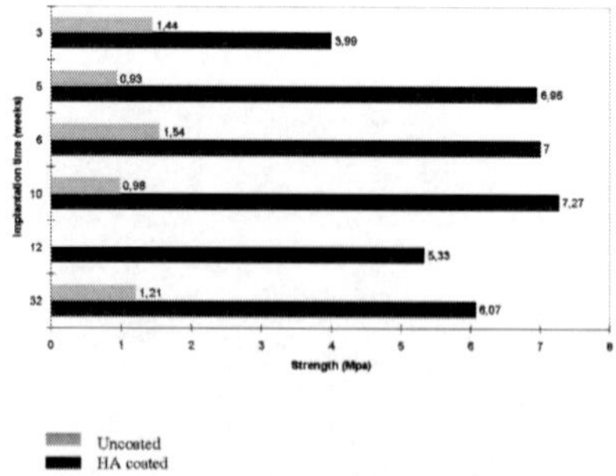

Figure 4.10 Shows how a plasma-sprayed HA coating on a porous titanium (dark bars) dependent on the implantation time will improve the interfacial bond strength compared to uncoated porous titanium (light bars). Reprinted from Ref. [49] with permission.

A number of factors appeared to influence the properties of calcium orthophosphate coatings. They include coating thickness (this will influence coating adhesion and fixation—the agreed optimum now seems to be within 50–100 μm), crystallinity (this affects the dissolution and biological behavior), phase and chemical purity, porosity and adhesion [526, 538]. Methods for the production of coatings and their properties are now largely standardized and, over recent years, calcium orthophosphate-coated implants have found highly successful clinical application, particularly in younger patients [520–572]. Further details on calcium orthophosphate coatings might be found in excellent reviews [573, 574].

4.5.3 Functionally Graded Bioceramics

In general, functionally gradient materials (FGMs) are defined as the materials, having gradient either compositional or structural changes from their surface to the interior of the materials. The idea of FGMs allows one device to possess two different properties. One of the most important combinations for the biomedical field is that of a mechanical strength and biocompatibility. Namely, only surface properties govern a biocompatibility of the entire device. In contrast, the strongest material determines the mechanical strength of the entire device. Although, this subject belongs to the coatings section (see above), in a certain sense, metallic implants covered by calcium orthophosphates might be considered as a FGM. The surface shows excellent biocompatibility because it consists of calcium orthophosphates, while the metallic core provides the excellent mechanical strength. The gradient change from calcium orthophosphates to the metal is important, for example, from the point of thermal expansion.

Functionally graded bioceramics consisting of calcium orthophosphates only [575] have been developed [411, 482, 484, 578–587]. For example, dense sintered bodies with gradual compositional changes from α-TCP to HA were prepared by sintering a diamond-coated HA compacts at 1280°C under a reduced pressure, followed by heating under the atmospheric conditions [578]. The content of α-TCP gradually decreased, while the content of HA increased with increasing depth from the surface. This functionally gradient bioceramics consisting of HA core and α-TCP surface showed a potential value as bone-substituting biomaterials [578]. Two types of functionally gradient FA/β-TCP biocomposites were prepared in another study [579]. As shown in Fig. 4.11, one of the graded biocomposites was in the shape of a disk and contained four different layers of about 1 mm thick. The other graded biocomposite was also in the shape of a disk but contained two sets of the four layers, each layer being 0.5 mm thick controlled by using a certain amount of the mixed powders. The final FA/β-TCP graded structures were formed at 100 MPa and sintered at 1300°C for 2 h [579].

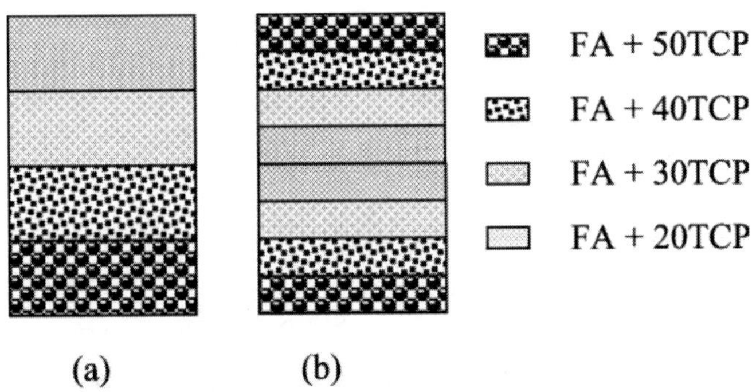

Figure 4.11 A schematic diagram showing the arrangement of the FA/β-TCP biocomposite layers. (a) A non-symmetric functionally gradient material (FGM); (b) symmetric FGM. Reprinted from Ref. [579] with permission.

Besides, it is well known that a bone cross-section from cancellous to cortical bone is non-uniform in porosity and pore dimensions. Thus, in various attempts to mimic the porous structure of bones, calcium orthophosphate bioceramics with graded porosity have been fabricated [375, 411, 482, 484, 578–585, 588]. For example, graded porous calcium orthophosphate bioceramics can be produced by means of tape casting and lamination (Fig. 4.12a). Other manufacturing techniques, such as a compression molding process (Fig. 4.12b) followed by impregnation and firing, are known as well [375]. In the first method, a HA slurry was mixed with a pore former. The mixed slurry was then cast into a tape. Using the same method, different tapes with different pore former sizes were prepared individually. The different tape layers were then laminated together. Firing was then done to remove the pore formers and sinter the HA particle compacts, resulting in graded porous bioceramics [582]. This method was also used to prepare graded porous HA with a dense part (core or layer) in order to improve the mechanical strength, as dense ceramics are much stronger than porous ceramics. However, as in the pressure infiltration of mixed particles, this multiple tape casting also has the problem of poor connectivity of pores, although the pore size and the porosity are relatively easy to control. Furthermore, the lamination step also introduces additional discontinuity of the porosity on the interfaces between the stacked layers.

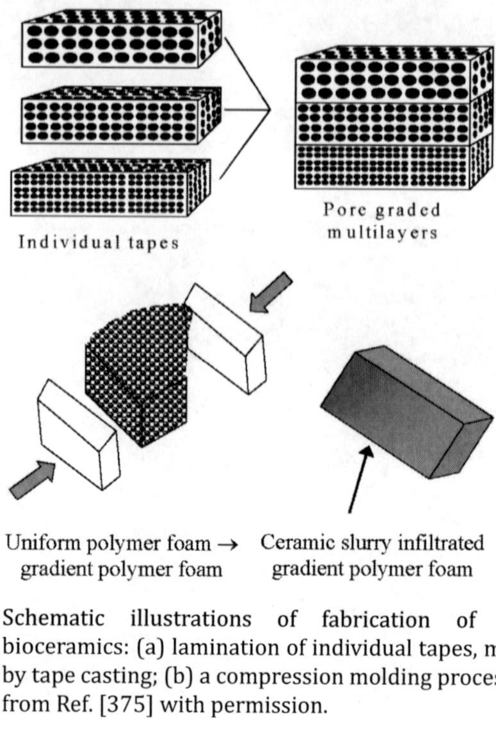

Figure 4.12 Schematic illustrations of fabrication of pore-graded bioceramics: (a) lamination of individual tapes, manufactured by tape casting; (b) a compression molding process. Reprinted from Ref. [375] with permission.

Since diverse biomedical applications require different configurations and shapes, the graded (or gradient) porous bioceramics can be grouped according to both the overall shape and the structural configuration [375]. The basic shapes include rectangular blocks and cylinders (or disks). For the cylindrical shape, there are configurations of dense core—porous layer, less porous core—more porous layer, dense layer—porous core and less porous layer—more porous core. For the rectangular shape, in the gradient direction, i.e., the direction with varying porosity, pore size or composition, there are configurations of porous top—dense bottom (same as porous bottom—dense top), porous top—dense center—porous bottom, dense top—porous center—dense bottom, etc. Concerning biomedical applications, a dense core—porous layer structure is suitable for implants of a high mechanical strength and with bone ingrowth for stabilization, whereas a less porous layer—more porous core configuration can be used for drug delivery systems. Furthermore, a porous top—dense bottom structure can be shaped into implants of articulate surfaces for wear resistance and with porous ends for bone ingrowth fixation; while a dense top—porous center—dense bottom arrangement mimics the structure of head skull. Further details on bioceramics with graded porosity might be found in literature [375].

4.6 Biological Properties and *in vivo* Behavior

The most important differences between bioactive bioceramics and all other implanted materials are the following: inclusion in the metabolic processes of the organism; adaptation of either surface or the entire material to the biomedium; integration of a bioactive implant with bone tissues at the molecular level or the complete replacement of a resorbable bioceramics by healthy bone tissues. All of the enumerated processes are related to the effect of an organism on the implant. Nevertheless, another aspect of implantation is also important—the effect of the implant on the organism. For example, using of bone implants from corpses or animals, even after they have been treated in various ways, provokes a substantially negative immune reactions in the organism, which substantially limits the application of such implants. In this connection, it is useful to dwell on the biological properties of bioceramic implants, particularly those of calcium orthophosphates, which in the course of time may be resorbed completely [589].

4.6.1 Interaction with Surrounding Tissues and the Host Responses

An interaction between an implant and surrounding tissues is a dynamic process. Water, dissolved ions, biomolecules and cells surround the implant surface during initial few seconds after the implantation. It has been accepted that no foreign material placed within a living body is completely compatible. The only substances that conform completely are those manufactured by the body itself (autogenous), while any other substance, which is recognized as foreign, initiates some types of reactions (a host-tissue response). The reactions occurring at the biomaterial/tissue interfaces lead to time-dependent changes in the surface characteristics of both the implanted biomaterials and the surrounding tissues [58, 590].

In order to develop new biomaterials, it is necessary to understand the *in vivo* host responses. Like any other species, biomaterials and bioceramics react chemically with their environment and, ideally, they should neither induce any changes nor provoke undesired reactions in the neighboring or distant tissues. In general, living organisms can treat artificial implants as biotoxic (or bioincompatible [54]), bioinert (or biostable [43]), biotolerant (or biocompatible [54]), bioactive and bioresorbable materials [4–7, 39, 40, 45, 49–54, 77, 531, 589–591]. Biotoxic (e.g., alloys containing cadmium, vanadium, lead and other toxic elements) materials release to the body substances in toxic concentrations and/or trigger the formation of antigens that may cause immune reactions ranging from simple allergies to inflammation to septic rejection with the associated severe health consequences. They cause atrophy, pathological change or rejection of living tissue near the material as a result of chemical, galvanic or other processes. Bioinert [592] (e.g., zirconia, alumina, carbon and titanium) and biotolerant (e.g., polymethylmethacrylate, titanium and Co–Cr alloy) materials do not release any toxic constituents but also do not show positive interaction with living tissue. They evoke a physiological response to form a fibrous capsule, thus, isolating the material from the body. In such cases, thickness of the layer of fibrous tissue separating the material from other tissues of an organism can serve as a measure of bioinertness. Generally, both bioactivity and bioresorbability phenomena are fine examples of chemical reactivity and calcium orthophosphates (both non-substituted and ion-substituted ones) fall into these two categories of bioceramics [4–7, 39, 40, 45, 49–54, 77, 531, 589–591]. A bioactive material will dissolve slightly but promote formation of a surface layer of biological apatite before interfacing directly with the tissue at the atomic level, that result in formation of a direct chemical bonds to bones. Such an implant provides a good stabilization for materials that are subject to mechanical loading. A bioresorbable material will dissolve over time (regardless of the mechanism leading to the material removal) and allow a newly formed tissue to grow into any surface irregularities but may not necessarily interface directly with the material. Consequently, the functions of bioresorbable materials are to participate in dynamic processes of formation and re-absorption occurring in bone tissues; thus, bioresorbable materials are used as scaffolds or filling spacers allowing to the tissues their infiltration and substitution [50, 194, 295, 593–595].

Interestingly, but a distinction between the bioactive and bioresorbable bioceramics might be associated with a structural factor only. For example, bioceramics made from non-porous, dense and highly crystalline HA behaves as a bioinert (but a bioactive) material and is retained in an organism for at least 5–7 years without noticeable changes (Fig. 4.2 bottom), while a highly porous bioceramics of the same composition can be resorbed approximately within a year. Furthermore, submicron-sized HA powders are biodegraded even faster than the highly porous HA scaffolds. Other examples of bioresorbable materials include porous bioceramic scaffolds made of BCP (which is an intimate mixture of either β-TCP + HA [111–126] or α-TCP + HA [8–12]) or bone grafts (dense or porous) made of CDHA [152], TCP [411, 596, 597] and/or ACP [462, 598]. One must stress that recently the concepts of bioactive and bioresorbable materials have converged and bioactive materials are made bioresorbable, while bioresorbable materials are made bioactive [599].

Although in certain *in vivo* experiments an inflammatory reaction was observed after implantation of calcium orthophosphate bioceramics [600–602], the general conclusion on using calcium orthophosphates with Ca/P ionic ratio within 1.0–1.7 is that all types of implants (bioceramics of various porosities and structures, powders or granules) are not only nontoxic but also induce neither inflammatory nor foreign-body reactions [603]. The biological response to implanted calcium orthophosphates follows a similar cascade observed in fracture healing. This cascade includes a hematoma formation, inflammation, neovascularization, osteoclastic resorption and a new bone formation. An intermediate layer of fibrous tissue between the implants and bones has been never detected. Furthermore, calcium orthophosphate implants display the ability to directly bond to bones [3–7, 39, 43, 49–56, 58, 525, 589]. For further details, the interested readers are referred to a good review on cellular perspectives of bioceramic scaffolds for bone tissue engineering [408].

One should note that the aforementioned rare cases of the inflammatory reactions to calcium orthophosphate bioceramics were caused by "other" reasons. For example, a high rate of wound inflammation occurred when highly porous HA was used. In that particular case, the inflammation was explained by sharp implant edges, which irritated surrounding soft tissues [601]. Another reason for inflammation produced by porous HA could be due to micro movements of the implants, leading to simultaneous disruption of a large number of micro-vessels, which grow into the pores of the bioceramics. This would immediately produce an inflammatory reaction. Additionally, problems could arise in clinical tests connected with migration of granules used for alveolar ridge augmentation, because it might be difficult to achieve a mechanical stability of implants at the implantation sites [601].

4.6.2 Osteoinduction

Before recently, it was generally considered, that alone, any type of synthetic bioceramics possessed neither osteogenic [604] nor osteoinductive [605] properties and demonstrated a minimal immediate structural support. When attached to healthy bones, an osteoid [606] is produced directly onto the surfaces of bioceramics in absence of a soft tissue interface. Consequently, the osteoid is mineralized and the resulting new bone undergoes remodeling [607]. However, several reports have already shown osteoinductive properties of certain types of calcium orthophosphate bioceramics [177, 459, 504, 608–621]. Namely, bone formation was found to occur in dog muscle inside porous calcium orthophosphates with surface microporosity, while bone was not observed on the surface of dense bioceramics [616]. Furthermore, implantation of porous β-TCP bioceramics appeared to induce bone formation in soft tissues of dogs, while no bone formation was detected in any α-TCP implants [613]. More to the point, titanium implants coated by a microporous layer of OCP were found to induce ectopic bone formation in goat muscles, while a smooth layer of carbonated apatite on the same implants was not able to induce bone formation there [619, 620].

Although the underlying mechanism(s) leading to bone induction by synthetic materials remains largely unknown, the dissolution/precipitation behavior of calcium orthophosphates [532], as well as their microporosity [621–623], chemistry [621], the specific surface area [623], as well as surface topography and geometry have been pointed out as the relevant parameters. A positive effect of increased microporosity on the ectopic bone formation could be direct and indirect. Firstly, an increased microporosity is directly related to the changes in surface topography, i.e. increases a surface roughness, which might affect the cellular differentiation. Secondly, an increased microporosity indirectly means a larger surface that is exposed to the body fluids leading to elevated dissolution/precipitation phenomena as compared to non-microporous surfaces. Furthermore, other hypotheses are available. Namely, Reddi explained the apparent osteoinductive properties as an ability of particular bioceramics to concentrate bone growth factors, which are circulating in biological fluids, and those growth factors induce bone formation [624]. Other researchers proposed a similar hypothesis that the intrinsic osteoinduction by calcium orthophosphate bioceramics is a result of adsorption of osteoinductive substances on their surface [618]. Moreover, Ripamonti [625] and Kuboki et al. [626] independently postulated that the geometry of calcium orthophosphate bioceramics is a critical parameter in bone induction. Specifically, bone induction by calcium orthophosphates was never observed on flat bioceramic surfaces. All osteoinductive cases were observed on either porous structures or structures contained well-defined concavities. What's more, bone formation was never observed on the peripheries of porous implants and was always found inside the pores or concavities, aligning the surface [194]. Some researchers speculated that a low oxygen tension in the central region of implants might provoke a dedifferentiation of pericytes from blood micro-vessels into osteoblasts [627]. Finally but yet importantly, both nano-structured rough surfaces and a surface charge on implants were found to cause an asymmetrical division of the stem cells into osteoblasts, which is important for osteoinduction [622].

4.6.3 Biodegradation

Shortly after implantation, a healing process is initiated by compositional changes of the surrounding bio-fluids and adsorption of biomolecules. Following this, various types of cells reach the bioceramic surface and the adsorbed layer dictates the ways the cells respond. Further, a biodegradation of the implanted bioceramics begins. This process can occur by either physicochemical dissolution with a possibility of phase transformations or cellular activity (so-called bioresorption), as well as by a combination of both processes. Dissolution is a physical chemistry process, which is controlled by some factors, such as solubility of the implant matrix (Table 1.1), surface area to volume ratio, local acidity, fluid convection and temperature. For HA, the dissolution process in acids has been described by a sequence of four successive Eqs. (7.1–7.4) [628, 629]. See section 7.2.5 Chemical model for details.

With few exceptions, dissolution rates of calcium orthophosphates are inversely proportional to the Ca/P ratio, phase purity and crystalline size, as well as it is directly related to the porosity and surface area. Phase transformations might occur with OCP, DCPA, DCPD, α-TCP, β-TCP and ACP because they are unstable in aqueous environment under the physiological conditions. Bioresorption is a biological process mediated by cells (mainly, osteoclasts and, in a lesser extent, macrophages) [630–632]. It depends on the response of cells to their environment. Osteoclasts attach firmly to the implant and dissolve calcium orthophosphates by secreting an enzyme carbonic anhydrase or any other acid, leading to a local pH drop to ~4–5 [633]. Furthermore, small particles of calcium orthophosphate can also be phagocyotosed by osteoclasts, i.e. they are incorporated into cytoplasm and thereafter dissolved by acid attack and/or enzymatic processes. In any case, biodegradation of calcium orthophosphates is a combination of various non-equilibrium processes, occurring simultaneously and/or in competition with each other.

Usually, an *in vitro* biodegradation of calcium orthophosphate bioceramics is estimated by suspending the material in a slightly acidic (pH ~5) buffer and monitoring the release of Ca^{2+} ions with time. The acidic buffer, to some extent, mimics the acidic environment during osteoclastic activity. In one study, an *in vivo* behavior of porous β-TCP bioceramics prepared from rod-shaped particles and that prepared from non-rod-shaped particles in the rabbit femur was compared. Although the porosities of both types of β-TCP bioceramics were almost the same, a more active osteogenesis was preserved in the region where rod-shaped bioceramics was implanted [634]. This result implied that the microstructure affected the activity of bone cells and subsequent bone replacement.

The experimental results demonstrated that both the dissolution kinetics and *in vivo* biodegradation of biologically relevant calcium orthophosphates proceed in the following decreasing order: β-TCP > bovine bone apatite (unsintered) > bovine bone apatite (sintered) > coralline HA > HA. In the case of BCP bioceramics, the biodegradation kinetics depends on the HA/TCP ratio: the higher the ratio, the lower the degradation rate. Similarly, *in vivo* degradation rate of biphasic TCP (α-TCP + β-TCP) bioceramics appeared to be lower than that of α-TCP and higher than that of β-TCP bioceramics, respectively [138]. Furthermore, incorporation of doping ions can either increase (e.g., CO_3^{2-}, Mg^{2+} or Sr^{2+}) or decrease (e.g., F^-) the solubility (therefore, biodegradability) of CDHA and HA. Contrarily to apatites, solubility of β-TCP is decreased by incorporation of either Mg^{2+} or Zn^{2+} ions [459]. Here, one should remind that ion-substituted calcium orthophosphates are not considered in this review; the interested readers are advised to read the original publications [8–38].

4.6.4 Bioactivity

Generally, bioactive materials interact with surrounding bone resulting in formation of a chemical bond to this tissue (bone bonding). The bioactivity phenomenon is determined by both chemical factors, such as crystal phases and molecular structures of a biomaterial, and physical factors, such as surface roughness and porosity. Currently, it is agreed that the newly formed bone bonds directly to biomaterials through a carbonated CDHA layer precipitating at the bone/biomaterial interface. Strange enough but a careful seeking in the literature resulted in just a few publications [459, 538, 635, 636], where the bioactivity mechanism of calcium orthophosphates was briefly described. For example, the chemical changes occurring after exposure of a synthetic HA bioceramics to both *in vivo* (implantation in human) and *in vitro* (cell culture) conditions were studied. A small amount of HA was phagocytozed but the major remaining part behaved as a secondary nucleator as evidenced by the appearance of a newly formed mineral [635]. *In vivo*, a cellular activity (e.g., of macrophages or osteoclasts) associated with an acidic environment were found to result in partial dissolution of calcium orthophosphates, causing liberation of calcium and orthophosphate ions to the microenvironment. The liberated ions increased a local supersaturation degree of the surrounding biologic fluids, causing precipitation of nano-sized crystals of biological apatite with simultaneous incorporating of various ions presented in the fluids. Infrared spectroscopic analyses demonstrated that these nano-sized crystals were intimately associated with bioorganic components (probably proteins), which might also have originated from the biologic fluids, such as serum [459].

Therefore, one should better rely on the bioactivity mechanism of other biomaterials, particularly of bioactive glasses—the concept introduced by Prof. Larry L. Hench [49–52]. The bonding mechanism of bioactive glasses to living tissues involves a sequence of 11 successive reaction steps. The initial 5 steps occurred on the surface of bioactive glasses are "chemistry" only, while the remaining 6 steps belong to "biology" because the latter include colonization by osteoblasts, followed by proliferation and differentiation of the cells to form a new bone that had a mechanically strong bond to the implant surface (Fig. 4.13). Therefore, in the case of bioactive glasses the border between "dead" and "alive" is postulated between stages 5 and 6. According to Hench, all bioactive materials "form a bone-like apatite layer on their surfaces in the living body and bond to bone through this apatite layer. The formation of bone-like apatite on artificial material is induced by functional groups, such as Si–OH (in the case of biological glasses), Ti–OH, Zr–OH, Nb–OH, Ta–OH, –COOH and $-H_2PO_4$ (in the case of other materials). These groups have specific structures revealing negatively charge and induce apatite formation via formations of an amorphous calcium compound, e.g., calcium silicate, calcium titanate and ACP" [49–52].

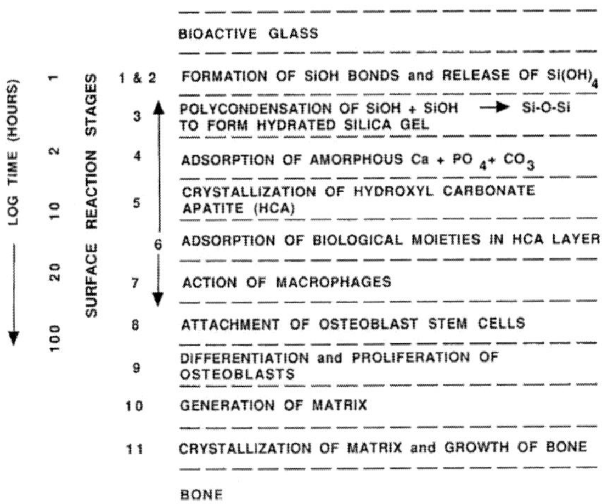

Figure 4.13 A sequence of interfacial reactions involved in forming a bond between tissue and bioactive ceramics. Reprinted from Refs. [49–52] with permission.

To extend the subject, it is important to refer to another set of 11 successive reaction steps for bonding mechanism of unspecified bioceramics, developed by Prof. Paul Ducheyne (Fig. 4.14) [58]. One can see that the Ducheyne's model is rather similar to that proposed by Hench; however, there are noticeable differences between them. For example, Ducheyne mentions on ion exchange and structural rearrangement at the bioceramic/tissue interface (stage 3), as well as on interdiffusion from the surface boundary layer into bioceramics (stage 4) and deposition with integration into the bioceramics (stage 7), which are absent in the Hench's model. On the other hand, Hench describes six biological stages (stages 6–11), while Ducheyne describes only four ones (stages 8–11). Both models have been developed almost 2 decades ago and, to the best of my knowledge, remain unchanged since then. Presumably, both approaches have *pro et contra* of their own and, obviously, should be updated and/or revised. Furthermore, in literature there are at least two other descriptions of the biological and cellular events occurring at the bone/implant interface [637, 638]; however, they include lesser number of stages. Recently, one more hypothesis has been proposed (Fig. 4.15). For the first time, it describes reasonable surface transformations, happening with calcium orthophosphate bioceramics (in this case, HA) shortly after the implantation [636].

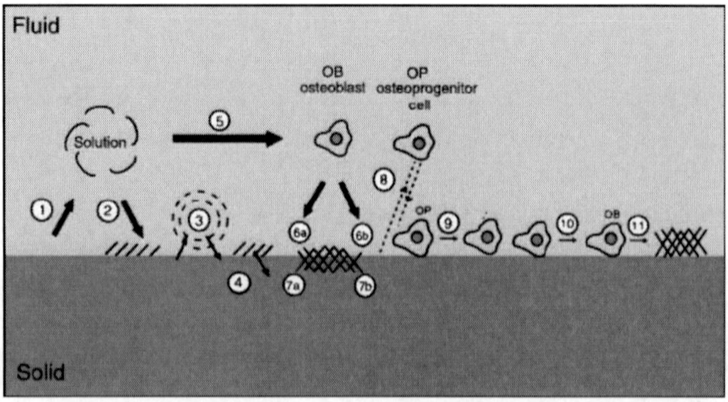

Figure 4.14 A schematic diagram representing the events, which take place at the interface between bioceramics and the surrounding biological environment: (1) dissolution of bioceramics; (2) precipitation from solution onto bioceramics; (3) ion exchange and structural rearrangement at the bioceramic/tissue interface; (4) interdiffusion from the surface boundary layer into the bioceramics; (5) solution-mediated effects on cellular activity; (6) deposition of either the mineral phase (a) or the organic phase (b) without integration into the bioceramic surface; (7) deposition with integration into the bioceramics; (8) chemotaxis to the bioceramic surface; (9) cell attachment and proliferation; (10) cell differentiation; (11) extracellular matrix formation. All phenomena, collectively, lead to the gradual incorporation of a bioceramic implant into developing bone tissue. Reprinted from Ref. [58] with permission.

An important study on formation of calcium orthophosphate precipitates on various types of bioceramic surfaces in both simulated body fluid (SBF) and rabbit muscle sites was performed [639]. The bioceramics were sintered porous solids, including bioglass, glass-ceramics, α-TCP, β-TCP and HA. An ability to induce calcium orthophosphate precipitation was compared among these types of bioceramics. The following conclusions were made: (1) OCP formation ubiquitously occurred on all types of bioceramic surfaces both *in vitro* and *in vivo*, except on β-TCP. (2) Apatite formation did not occur on every type of bioceramic surface; it was less likely to occur on the surfaces of HA and α-TCP. (3) Precipitation of calcium orthophosphates on the bioceramic surfaces was more difficult *in vivo* than *in vitro*. (4) Differences in calcium orthophosphate precipitation among the bioceramic surfaces were less noticeable *in vitro* than that *in vivo*. (5) β-TCP bioceramics showed a poor ability of calcium orthophosphate precipitation both *in vitro* and *in vivo*

[639]. These findings clearly revealed that apatite formation in the physiological environments could not be confirmed as the common feature of bioceramics. Nevertheless, for want of anything better, currently the bioactivity mechanism of calcium orthophosphate bioceramics could be described by a reasonable combination of Figs. 4.10–4.12, e.g., by updating the Ducheyne's and Hench's models by 3 initial stages taken from Fig. 4.15.

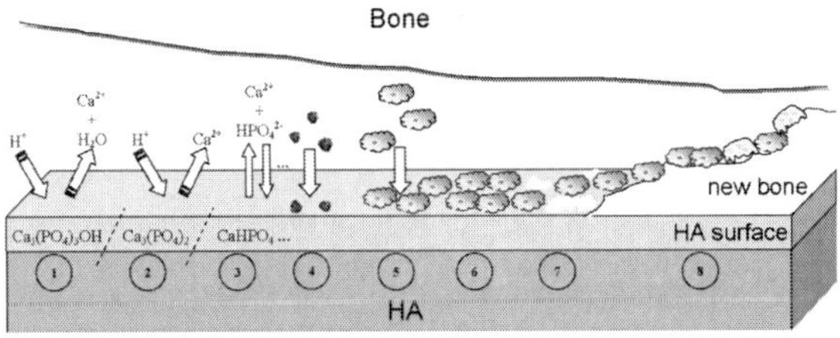

Figure 4.15 A schematic diagram representing the phenomena that occur on HA surface after implantation: (1) beginning of the implant procedure, where a solubilization of the HA surface starts; (2) continuation of the solubilization of the HA surface; (3) the equilibrium between the physiological solutions and the modified surface of HA has been achieved (changes in the surface composition of HA does not mean that a new phase of DCPA or DCPD forms on the surface); (4) adsorption of proteins and/or other bioorganic compounds; (5) cell adhesion; (6) cell proliferation; (7) beginning of a new bone formation; (8) new bone has been formed. Reprinted from Ref. [636] with permission.

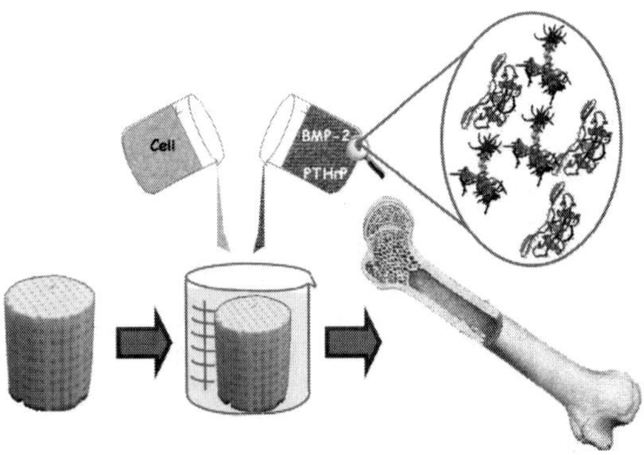

Figure 4.16 A schematic view of a third generation biomaterial, in which porous calcium orthophosphate bioceramics acts as a scaffold or a template for cells, growth factors, etc. Reprinted from Refs. [43, 56] with permission.

Interestingly, but bioactivity of HA bioceramics might be enhanced by a high-energy ion irradiation [640]. The effect was attributed to formation of a unique 3D macroporous apatite layer of decreased crystallinity and crystal size on the irradiated surfaces. To conclude this topic, the atomic and molecular phenomena occurring at the bioceramic surface in aqueous solutions and their effects on the relevant reaction pathways of cells and tissues must be elucidated in more details. However, further investigation of this topic requires a careful analysis of the available experimental data, which is beyond the scope of this review.

4.6.5 Cellular Response

Fixation of an implant in the human body is a dynamic process that remodels the interface zone between the implant and living tissues at all dimensional levels, from the molecular up to the cell and tissue morphology level, and at all time scales, from the first second up to several years after implantation. Immediately following the implantation, a space filled with bio-fluids appears next to the implant surface. With time, proteins will be adsorbed at the bioceramic surface that will give rise to osteoinduction by cell proliferation and their differentiation towards bone cells, revascularisation and eventual gap closing. Ideally, a strong bond will be formed between the implant and surrounding tissues [54]. A detailed study on interfacial interactions between calcined HA and substrates have been performed recently [641], where the interested readers are referred for further details.

Osteoblasts cultured on HA bioceramics are generally reported to be completely flattened and difficult to distinguish the cytoplasmic edge from the HA surfaces after ~2 h incubation [642]. These observations underscore an expected bioactivity of HA and make HA bioceramics well suited for bone reconstruction. Osteoblasts cultured on porous HA bioceramics appeared to exhibit a higher adhesion, an enhanced differentiation and suppressed proliferation rates when compared to the non-porous controls [643, 644]. Furthermore, formation of distinct resorption pits on HA [645] and β-TCP [634] surfaces in the presence of osteoclasts was observed. Moreover, a surface roughness of calcium orthophosphate bioceramics was reported to strongly influence the activation of mononuclear precursors to mature osteoclasts [645].

Cellular biodegradation of calcium orthophosphate bioceramics is known to depend on its phases. For example, a higher solubility of β-TCP was shown to prevent L-929 fibroblast cell adhesion, thereby leading to damage and rupture of the cells [646]. A mouse ectopic model study indicated the maximal bone growth for the 80 : 20 β-TCP : HA biphasic formulations preloaded with human mesenchymal stem cells when compared to other calcium orthophosphates [647]. The effects of substrate microstructure and crystallinity have been corroborated with an *in vivo* rabbit femur model, where rod-like crystalline β-TCP was reported to enhance osteogenesis when compared to non-rod like crystalline β-TCP [634]. Additionally, using a dog mandibular defect model, a higher bone formation on a scaffold surface coated by nanodimensional HA was observed when compared to that coated by a micro-dimensional HA [648]. Furthermore, recent studies revealed a stronger stress signaling response by osteoblast precursor cells in 3D scaffolds when compared to 2D surfaces [649]. Besides, variations in the physical properties of a bone substitute material (in that case, β-TCP was studied) were found to influence the inflammatory reaction and its consequences [650].

Mesenchymal stem cells are one of the most attractive cellular lines for application as bone grafts [651]. Early investigations by Okumura et al. indicated an adhesion, proliferation and differentiation, which ultimately became new bone and integrated with porous HA bioceramics [652]. Recently, Unger et al. showed a sustained co-culture of endothelial cells and osteoblasts on HA scaffolds for up to 6 weeks [653]. Furthermore, a release of factors by endothelial and osteoblast cells in co-culture supported proliferation and differentiation was suggested to ultimately result in microcapillary-like vessel formation and supported a neo-tissue growth within the scaffold [408]. More to the point, investigation of rat calvaria osteoblasts cultured on transparent HA bioceramics, as well as the analysis of osteogenic-induced human bone marrow stromal cells at different time points of culturing indicated to a good cytocompatibility of HA bioceramics and revealed favorable cell proliferation [368]. Positive results for other types of cells have been obtained in other studies [206, 363, 364, 367].

Interestingly, but HA scaffolds with marrow stromal cells in a perfused environment were reported to result in ~85% increase in mean core strength, a ~130% increase in failure energy and a ~355% increase in post-failure strength. The increase in mineral quantity and promotion of the uniform mineral distribution in that study was suggested to attribute to the perfusion effect [496]. Additionally, other investigators indicated to mechanical properties increasing for other calcium orthophosphate scaffolds after induced osteogenesis [495, 498].

Furthermore, the dimensions, extent and interconnectivity of pores in bioceramics are known to influence bone ingrowth, blood vessels formation and canaliculi networks [452, 453, 504]. Initial reports have estimated a minimum pore size of ~50 μm for blood vessel formation and a minimum pore size of ~200 μm for osteonal ingrowth [504]. Pore dimensions of ~100 μm and even ~50 μm [654] were reported in later studies to support bone ingrowth. Additionally, vascularization, cell migration and nutrient diffusion required for sustained cell viability and tissue function are possible if pores within the scaffolds are well interconnected. For example, an essential mean pore interconnection size of ~10 μm was found to be necessary to allow cell migration between the pores [655]. As such, both porosity and general architecture are critical in determining the rate of fluid transport through porous bioceramics, which, in turn, determines the rate and degree of bone ingrowth *in vivo* [115, 455, 456, 656].

4.7 Calcium Orthophosphate Bioceramics in Tissue Engineering

4.7.1 Tissue Engineering

All modern orthopedic implants lack three of the most critical abilities of living tissues: (i) self-repairing; (ii) maintaining of blood supply; (iii) self-modifying their structure and properties in response to external aspects such as a mechanical load [473]. Needless to mention, that bones not only possess all of these properties but, in addition, they are self-generating, hierarchical, multifunctional, nonlinear, composite and biodegradable; therefore, the ideal artificial bone grafts must possess similar properties [110].

The last decades have seen a surge in creative ideas and technologies developed to tackle the problem of repairing or replacing diseased and damaged tissues, leading to the emergence of a new field in healthcare technology now referred to as *tissue engineering*.

This is an interdisciplinary field that exploits a combination of living cells, engineering materials and suitable biochemical factors in a variety of ways to improve, replace, restore, maintain or enhance living tissues and whole organs [657, 658]. However, as two of three major components (namely, cells and biochemical factors) of the tissue engineering subject appear to be far beyond the scope of this book, the topic of tissue engineering is narrowed down to the engineering materials prepared from calcium orthophosphate bioceramics only.

Regeneration, rather than a repair, is the central goal of any tissue engineering strategy [659]. Thus, tissue engineering has a potential to create tissues and organs *de novo*. This field of science [661] started more than two decades ago [662, 663] and a famous review article by Langer and Vacanti [664] has greatly contributed to the promotion of tissue engineering research worldwide. The field of tissue engineering, particularly when applied to bone substitutes where tissues often function in a mechanically demanding environment [665], requires a collaboration of excellence in cell and molecular biology, biochemistry, material sciences, bioengineering and clinical research. For the success, it is necessary that researchers with expertise in one area have an appreciation of the knowledge and challenges of the other areas. However, since the technical, regulatory and commercial challenges might be substantial, the introduction of new products is likely to be slow [660].

Nowadays tissue engineering is at full research potential due to the following key advantages: (i) the solutions it provides are long-term, much safer than other options and cost-effective as well; (ii) the need for a donor tissue is minimal, which eliminates the immuno-suppression problems; (iii) the presence of residual foreign material is eliminated as well.

4.7.2 Scaffolds and Their Properties

It would be very convenient to both patients and physicians if devastated tissues or organs of patients can be regenerated by simple cell injections to the target sites but such cases are rare. The majority of large-sized tissues and organs with distinct 3D form require a support for their formation from cells. The support is called scaffold [666], template and/or artificial extracellular matrix [158, 159, 422, 662, 665, 667–670]. The major function of scaffolds is similar to that of the natural extracellular matrix that assists proliferation, differentiation and biosynthesis of cells. In addition, scaffolds placed at the regeneration sites will prevent disturbing cells from invasion into the sites of action [671, 672]. The role of scaffolds has been perfectly described by Andrés Segovia (1893–1987), a Spanish classical guitarist: "When one puts up a building one makes an elaborate scaffold to get everything into its proper place. But when one takes the scaffold down, the building must stand by itself with no trace of the means by which it was erected. That is how a musician should work."

The idea behind tissue engineering is to create or engineer autografts by either expanding autologous cells *in vitro* guided by a scaffold or implanting an acellular template *in vivo* and allowing the patient's cells to repair the tissue guided by the scaffold. The first phase is the *in vitro* formation of a tissue construct by placing the chosen cells and scaffolds in a metabolically and mechanically supportive environment with growth media (in a bioreactor), in which the cells proliferate and elaborate extracellular matrix. It is expected that cells infiltrate into the porous matrix and consequently proliferate and differentiate therein. In the second phase, the construct is implanted in the appropriate anatomic location, where remodeling *in vivo* is intended to recapitulate the normal functional architecture of an organ or a tissue [673, 674]. The key processes occurring during both *in vitro* and *in vivo* phases of the tissue formation and maturation are: (1) cell proliferation, sorting and differentiation, (2) extracellular matrix production and organization, (3) biodegradation of the scaffold, (4) remodeling and potentially growth of the tissue.

To achieve the goal of tissue reconstruction, the scaffolds must meet several specific requirements [158, 159, 667]. A reasonable surface roughness is necessary to facilitate cell seeding and fixation [675, 676]. A sufficient mechanical strength and stiffness are mandatory to oppose contraction forces and later for the remodeling of damaged tissues. A high porosity and an adequate pore dimensions (Tables 4.1 and 4.5) are very important to allow cell migration, vascularization, as well as a diffusion of nutrients [384]. Namely, scaffolds should have a network of interconnected pores where more than ~60 % of the pores should have a size ranging from ~150 μm to ~400 μm and at least ~20 % should be smaller than ~20 μm [12, 92, 384, 394, 395, 452–459, 504, 677–684]. Scaffolds must be manufactured from the materials with controlled biodegradability and/or bioresorbability, such as calcium orthophosphate bioceramics, so that a new bone will eventually replace the scaffold [685]. Furthermore, the resorption rate has to coincide as much as possible with the rate of bone formation (i.e., between a few months and about 2 years) [686]. This means that while cells are fabricating their own natural matrix structure around themselves, the scaffold is able to provide a structural integrity within the body and eventually it will break down leaving the newly formed tissue that will take over the mechanical load. Besides, scaffolds should be easily fabricated into a variety of shapes and sizes [687] and be malleable to fit irregularly shaped defects. In many cases, ease of processability, such as an easiness of conformation, and injectability, such as self-setting calcium orthophosphate formulations possess (Chapter 5), can determine the choice of a certain biomaterial. Finally, sterilization with no loss of properties is a crucial step in scaffold production at both a laboratory and an industrial level [665]. In conclusion, since ions of calcium and orthophosphate are known to regulate bone metabolism, calcium orthophosphates appear to be among the few bone graft substitute materials, which can be considered as a drug [2].

Table 4.5 A hierarchical pore size distribution that an ideal scaffold should exhibit [12]

Pore sizes of a 3D scaffold	A biochemical effect or function
< 1 μm	Interaction with proteins Responsible for bioactivity
1–20 μm	Type of cells attracted Cellular development Orientation and directionality of cellular ingrowth
100–1000 μm	Cellular growth Bone ingrowth Predominant function in the mechanical strength
>1000 μm	Implant functionality Implant shape Implant esthetics

Many fabrication techniques are available to produce porous calcium orthophosphate scaffolds (Table 4.1) with varying architectural features (for details, see sections 4.3.3 Forming and shaping and 4.4.4 Porosity). In order to achieve the desired properties at minimum expenses, the production process should be optimized [688]. With the advent of tissue engineering, the search is on for the ultimate option—a "tissue engineered bone substitute," consisting of a synthetic calcium orthophosphate scaffold impregnated with cells and growth factors. Figure 4.16 schematically depicts a possible fabrication process of such item that, afterwards, will be implanted into a living organism to induce bone regeneration [43, 56].

From the structural perspectives, a degree of scaffold porosity is responsible for regulating the bioactivity of bone graft substitutes as a function of its influence on structural permeability, which controls the initial rate of bone regeneration and the local mechanical environment, which mediates the equilibrium volume of new bone within the repair site. Parameters such as pore interconnectivity, pore geometry, strut topography and strut porosity all contribute to modulate this process of osteogenesis and act synergistically to promote or screen the osteoconductive or osteoinductive potential of bone graft substitutes [455, 689, 690]. However, since bones have very different structures depending on their functions and locations, the same pore sizes and shapes may not be ideal for all potential uses. Therefore, bioceramic scaffolds of various porosities are required.

4.7.3 Bioceramic Scaffolds from Calcium Orthophosphates

Philosophically, the increase in life expectancy requires biological solutions to orthopedic problems, which were previously managed with mechanical solutions. Therefore, since the end of 1990s, the biomaterials research focuses on tissue regeneration instead of tissue replacement [691]. The alternatives include use hierarchical bioactive scaffolds to engineer *in vitro* living cellular constructs for transplantation or use bioresorbable bioactive particulates or porous networks to activate *in vivo* the mechanisms of tissue regeneration [692, 693]. Thus, the aim of calcium orthophosphate bioceramics is to prepare artificial porous scaffolds able to provide the physical and chemical cues to guide cell seeding, differentiation and assembly into 3D tissues of a newly formed bone [648, 694–700]. Particle sizes, shape and surface roughness of the scaffolds are known to affect cellular adhesion, proliferation and phenotype. Additionally, the surface energy may play a role in attracting particular proteins to the bioceramic surface and, in turn, this will affect the cells affinity to the material. More to the point, cells are exceedingly sensitive to the chemical composition and their bone-forming functions can be dependent on grain morphology of the scaffolds. For example, osteoblast functions were found to increase on nanodimensional fibers if compared to nanodimensional spheres because the former more closely approximated the shape of biological apatite in bones [701]. Besides, a significantly higher osteoblast proliferation on HA bioceramics sintered at 1200°C as compared to that on HA bioceramics sintered at 800°C and 1000°C was reported [702]. Thus, to meet the tissue engineering requirements, much attention is devoted to further improvements of calcium orthophosphate bioceramics [703]. From the chemical point of view, the developments include synthesis of novel ion-substituted calcium orthophosphates [8–38]. From the material point of view, the major research topics include nanodimensional and nanocrystalline structures [704–708] and Chapter 3, organic–inorganic biocomposites and hybrid biomaterials (Chapter 6), fibers, whiskers and filaments [709–722], micro- and nano-sized spheres and beads [722–737], micro- and nano-sized tubes [738–740], porous 3D scaffolds made of ACP [462, 598], TCP [432–435], HA [176, 397, 398, 436, 438, 485, 688, 741–747] and various biphasic formulations [484, 700, 733, 748–752], structures with graded porosity [375, 411, 482, 484, 578–585, 588] and hierarchically organized ones [753, 754]. Furthermore, an addition of defects through an intensive milling [755, 756] or their removal by a thermal treatment [757] can be used to modify a chemical reactivity of calcium orthophosphates. Besides, more attention should be paid to a crystallographically aligned calcium orthophosphate bioceramics [758].

There are three principal therapeutic strategies for treating diseased or injured tissues in patients: (i) implantation of freshly isolated or cultured cells; (ii) implantation of tissues assembled *in vitro* from cells and scaffolds; (iii) *in situ* tissue regeneration. For cellular implantation, individual cells or small cellular aggregates from the patient or a donor are either injected into the damaged tissue directly or arc combined with a degradable scaffold *in vitro* and then implanted. For tissue implantation, a complete 3D tissue is grown *in vitro* using patient or donor cells and a bioresorbable scaffold and then is implanted into the patients to replace diseased or damaged tissues. For *in situ* regeneration, a scaffold implanted directly into the injured tissue stimulates the body's own cells to promote local tissue repair [657, 759]. In any case, simply trapping cells at the particular point on a surface is not enough: the cells must be encouraged to differentiate, which is impossible without the presence of suitable biochemical factors [760]. All previously mentioned clearly indicates that for the purposes of tissue engineering, calcium orthophosphate bioceramics plays an auxiliary role; namely, it acts as a suitable material to manufacture the appropriate 3D templates, substrates or scaffolds to be colonized by living cells before the successive implantation [761, 762]. The *in vitro* evaluation of potential calcium orthophosphate scaffolds for tissue engineering has been described elsewhere [763], while the data on the mechanical properties of calcium orthophosphate bioceramics for use in tissue engineering are also available [764, 765]. The effect of a HA-based biomaterial on gene expression in osteoblast-like cells was reported as well [766]. To conclude this part, the excellent biocompatibility of calcium orthophosphate bioceramics, its possible osteoinductivity [177, 459, 504, 608–621] and a high affinity for drugs [767], proteins and cells make them very functional for the tissue engineering applications. The feasible production of scaffolds with tailored structures and properties opens up a spectacular future for calcium orthophosphates [766–774].

4.7.4 A Clinical Experience

During the last decade, several groups have made steps towards a clinical application of cell-seeded calcium orthophosphate bioceramics for bone tissue engineering of humans. For example, Quarto et al. [775] were the first to report a treatment of large (4–7 cm) bone defects of the tibia, ulna and humerus in three patients from 16 to 41 years old, where the conventional surgical therapies had failed. The authors implanted a custom-made unresorbable porous HA scaffolds seeded with *in vitro* expanded autologous bone marrow stromal cells. In all three patients, radiographs and computed tomographic scans revealed abundant callus formation along the implants and good integration at the interfaces with the host bones by the second month after surgery [775]. In the same year, Vacanti et al. [776] reported the case of a man who had a traumatic avulsion of the distal phalanx of a thumb. The phalanx was replaced with a specially treated natural coral (porous HA; 500-pore ProOsteon (see Table 4.2)) implant that was previously seeded with *in vitro* expanded autologous periosteal cells. The procedure resulted in the functional restoration of a stable and biomechanically sound thumb of normal length, without the pain and complications that are usually associated with harvesting a bone graft.

Morishita et al. [777] treated a defect resulting from surgery of benign bone tumors in three patients using HA scaffolds seeded with *in vitro* expanded autologous bone marrow stromal cells after osteogenic differentiation of the cells. Two bone defects in a tibia and one defect in a femur were treated. Although ectopic implants in nude mice were mentioned to show the osteogenicity of the cells, details such as the percentage of the implants containing bone and at what quantities were not reported. Furthermore, cell-seeded calcium orthophosphate scaffolds were found to be superior to autograft, allograft or cell-seeded allograft in terms of bone formation at ectopic implantation sites [778]. Besides, it has been hypothesized that dental follicle cells combined with β-TCP bioceramics might become a novel therapeutic strategy to restore periodontal defects [779].

4.8 Conclusions and Outlook

The available chronology of seeking for a suitable bioceramics for bone substitutes is as follows: since the 1950s, the first aim was to use bioinert [592] bioceramics, which had no reaction with living tissues. They included inert and tolerant compounds, which were designed to withstand physiological stress without, however, stimulating any specific cellular responses. Later on, in the 1980s, the trend changed towards exactly the opposite: the idea was to implant bioceramics that reacted with the surrounding tissues by producing newly formed bone (a "responsive" bioceramics because it was able to elicit biological responses). These two stages have been referred to as the first and the second generations of bioceramics, respectively [780] and, currently, both of them are extensively commercialized. Thus, the majority of the marketable products listed in Table 4.2 belong to the first and the second generations of bone substitute biomaterials. However, the progress keeps going and, in current century, scientists search for the third generation of bioceramics [759], which will be able to "instruct" the physiological environment toward desired biological responses (i.e., bioceramics will be able to regenerate bone tissues by stimulating specific responses at the molecular level) [41, 43, 56]. One should note that these three generations should not be interpreted as the chronological but the conceptual, since each generation represents an evolution on the requirements and properties of the biomaterials involved. This means that at present, research and development is still devoted to biomaterials and bioceramics that, according to their properties, could be considered to be of the first or the second generations, because the second generation of bioceramics with added porosity is one of the initial approaches in developing of the third generation of bioceramics [781]. Furthermore, there is another classification of the history of biomaterials introduced by Prof. James M. Anderson. According to Anderson, within 1950–1975 the researchers studied bioMATERIALS, within 1975–2000 they studied BIOMATERIALS and since 2000 the time for BIOmaterials has been coming [782]. Here, the capital letters emphasis the major direction of the research efforts in the complex subject of biomaterials. As bioceramics are biomaterials of the ceramic origin (see section 4.2 General knowledge on biomaterials and bioceramics), the Anderson's historical classification appears to be applicable to the bioceramics field as well.

The widespread use of biomaterials, however, experiences two major difficulties. The first is an incomplete understanding of the physical and chemical functioning of biomaterials and of the human response to these materials. Recent advances in material characterization and computer science, as well as in cell and molecular biology are expected to play a significant role in studies of biomaterials. A second difficulty is that many biomaterials do not perform as desirably as we would like. This is not surprising, since many materials used in medicine were not designed for medical purposes. It needs to be mentioned here that biomaterials are expected to perform in our body's internal environment, which is very aggressive. For example, solution pH of body fluids in various tissues varies in the range from 1 to 9. During daily activities, bones are subjected to a stress of ~4 MPa, whereas the tendons and ligaments experience peak stresses in the range of 40–80 MPa. The mean load on a hip joint is up to three times body weight (3000 N) and peak load during jumping can be as high as ~10 times body weight. More importantly, these stresses are repetitive and fluctuating, depending on the activities, such as standing, sitting, jogging, stretching and climbing. All of these require careful designing of biomaterials in terms of composition, shape, physical and biocompatibility properties. Therefore, a significant challenge is the rational design of human biomaterials based on a systematic evaluation of desired biological, chemical and engineering requirements [44].

Nevertheless, the field of biomaterials is in the midst of a revolutionary change in which the life sciences are becoming equal in importance to materials science and engineering as the foundation of the field. Simultaneously, advances in engineering (for example nanotechnology) are greatly increasing the sophistication with which biomaterials are designed and have allowed fabrication of biomaterials with increasingly complex functions [59]. Specifically, during last ~40 years, calcium orthophosphate bioceramics has become an integral and vital segment of our modern health care delivery system. In the modern fields of the third generation bioceramics (Hench) or BIOceramics (Anderson), the full potential of calcium orthophosphates has only begun to be recognized. Namely, calcium orthophosphates, which were intended as osteoconductive bioceramics in the past, stand for materials to fabricate osteoinductive implants nowadays [177, 459, 504, 608–621]. The initial steps in this direction have been already made by both fabricating BCP-based scaffolds for bone tissue engineering through the design of controlled 3D-porous structures and increasing the biological activity through development of novel ion-substituted calcium orthophosphate bioceramics [12, 461]. In future, the composition, microstructure and molecular surface chemistry of various types of calcium orthophosphates will be tailored to match the specific biological and metabolic requirements of tissues or disease states. This new generation of calcium orthophosphate bioceramics should enhance the quality of life of millions of people, as they grow older [783].

In spite of the great progress, there is still a great potential for major advances to be made in the field of calcium orthophosphate bioceramics [5]. This includes requirements for

- Improvement of the mechanical performance of existing types of bioceramics.

- Enhanced bioactivity in terms of gene activation.
- Improvement in the performance of biomedical coatings in terms of their mechanical stability and ability to deliver biological agents.
- Development of smart biomaterials capable of combining sensing with bioactivity.
- Development of improved biomimetic composites.

Furthermore, still there are needs for a better understanding of the biological systems. For example, the bonding mechanism between the bone mineral and collagen remains unclear. It is also unclear whether a rapid repair that is elicited by the new generation of bioceramics is a result of the enhancement of mineralization *per se* or whether there is a more complex signaling process involving proteins in collagen. If we were able to understand the fundamentals of bone response to specific ions and the signals they activate, then we would be able to design better bioceramics for the future [5].

To finalize this chapter, it is completely obvious that the present status of research and development in the field of calcium orthophosphate bioceramics is still at the starting point for the solution of new problems at the confluence of materials science, biology and medicine, concerned with the restoration of damaged functions in the human organisms. A large increase in active elderly people has dramatically raised the need for load-bearing bone graft substitutes, for example, for bone reconstruction during revision arthroplasty or for the reinforcement of osteoporotic bones. Strategies applied in the last four decades towards this goal have failed. So new strategies, possibly based on self-assembling and/or nanofabrication, will have to be proposed and developed [784]. Furthermore, in future, it should be feasible to design a new generation of gene-activating calcium orthophosphate based scaffolds tailored for specific patients and disease states. Perhaps, sometime bioactive stimuli will be used to activate genes in a preventative treatment to maintain the health of aging tissues. Currently this concept seems impossible. However, we need to remember that only ~40 years ago the concept of a material that would not be rejected by living tissues also seemed impossible [599].

References and Notes

1. Bioactivity is defined as the property of the material to develop a direct, adherent, and strong bonding with bone.
2. Bohner, M. (2010) Resorbable biomaterials as bone graft substitutes, *Mater. Today*, **13**, 24–30.
3. Doremus, R. H. (1992) Bioceramics, *J. Mater. Sci.*, **27**, 285–297.
4. Vallet-Regí, M. (2001) Ceramics for medical applications, *J. Chem. Soc. Dalton Trans.*, 97–108.
5. Best, S. M., Porter, A. E., Thian, E. S., and Huang, J. (2008) Bioceramics: past, present and for the future, *J. Eur. Ceram. Soc.*, **28**, 1319–1327.
6. Vallet-Regí, M. (2010) Evolution of bioceramics within the field of biomaterials, *C. R. Chimie*, **13**, 174–185.
7. Rahaman, M. N., Yao, A., Bal, B. S., Garino, J. P., and Ries, M. D. (2007) Ceramics for prosthetic hip and knee joint replacement, *J. Am. Ceram. Soc.*, **90**, 1965–1988.
8. Langstaff, S. D., Sayer, M., Smith, T. J. N., Pugh, S. M., Hesp, S. A. M., and Thompson, W. T. (1999) Resorbable bioceramics based on stabilized calcium phosphates. Part I: Rational design, sample preparation and material characterization, *Biomaterials*, **20**, 1727–1741.
9. Langstaff, S. D., Sayer, M., Smith, T. J. N., and Pugh, S. M. (2001) Resorbable bioceramics based on stabilized calcium phosphates. Part II: Evaluation of biological response, *Biomaterials*, **22**, 135–150.
10. Sayer, M., Stratilatov, A. D., Reid, J. W., Calderin, L., Stott, M. J., Yin, X., MacKenzie, M., Smith, T. J. N., Hendry, J. A., and Langstaff, S. D. (2003) Structure and composition of silicon-stabilized tricalcium phosphate, *Biomaterials*, **24**, 369–382.
11. Reid, J. W., Pietak, A. M., Sayer, M., Dunfield, D., and Smith, T. J. N. (2005) Phase formation and evolution in the silicon substituted tricalcium phosphate/apatite system, *Biomaterials*, **26**, 2887–2897.
12. Sanchez-Sálcedo, S., Arcos, D., and Vallet-Regí, M. (2008) Upgrading calcium phosphate scaffolds for tissue engineering applications, *Key Eng. Mater.*, **377**, 19–42.
13. Ito, A., and LeGeros, R. Z. (2008) Magnesium- and zinc-substituted beta-tricalcium phosphates as potential bone substitute biomaterials, *Key Eng. Mater.*, **377**, 85–98.
14. Ergun, C., Webster, T. J., Bizios, R., and Doremus, R. H. (2002) Hydroxylapatite with substituted magnesium, zinc, cadmium, and

yttrium. I. Structure and microstructure, *J. Biomed. Mater. Res.*, **59**, 305–311.

15. Webster, T. J., Ergun, C., Doremus, R. H., and Bizios, R. (2002) Hydroxylapatite with substituted magnesium, zinc, cadmium, and yttrium. II. Mechanisms of osteoblast adhesion, *J. Biomed. Mater. Res.*, **59**, 312–317.
16. Kim, S. R., Lee, J. H., Kim, Y. T., Riu, D. H., Jung, S. J., Lee, Y. J., Chung, S. C., and Kim, Y. H. (2003) Synthesis of Si, Mg substituted hydroxyapatites and their sintering behaviors, *Biomaterials*, **24**, 1389–1398.
17. Patel, N., Best, S. M., and Bonfield, W. (2005) Characterisation of hydroxyapatite and substituted-hydroxyapatites for bone grafting, *J. Australasian Ceram. Soc.*, **41**, 1–22.
18. Vallet-Regí, M., and Arcos, D. (2005) Silicon substituted hydroxyapatites. A method to upgrade calcium phosphate based implants, *J. Mater. Chem.*, **15**, 1509–1516.
19. Landi, E., Celotti, G., Logroscino, G., and Tampieri, A. (2003) Carbonated hydroxyapatite as bone substitute, *J. Eur. Ceram. Soc.*, **23**, 2931–2937.
20. Gbureck, U., Thull, R., and Barralet, J. E. (2005) Alkali ion substituted calcium phosphate cement formation from mechanically activated reactants, *J. Mater. Sci. Mater. Med.*, **16**, 423–427.
21. Kannan, S., Ventura, J. M., and Ferreira, J. M. F. (2005) *In situ* formation and characterization of flourine-substituted biphasic calcium phosphate ceramics of varied F-HAP / β-TCP ratios, *Chem. Mater.*, **17**, 3065–3068.
22. Tas, A. C., Bhaduri S. B., and Jalota, S. (2007) Preparation of Zn-doped β-tricalcium phosphate (β-$Ca_3(PO_4)_2$) bioceramics, *Mater. Sci. Eng. C*, **27**, 394–401.
23. Gbureck, U., Knappe, O., Grover, L. M., and Barralet, J. E. (2005) Antimicrobial potency of alkali ion substituted calcium phosphate cements, *Biomaterials*, **26**, 6880–6886.
24. Tardei, C., Grigore, F., Pasuk, I., and Stoleriu, S. (2006) The study of Mg^{2+}/Ca^{2+} substitution of β-tricalcium phosphate, *J. Optoelectronics Adv. Mater.*, **8**, 568–571.
25. Pietak, A. M., Reid, J. W., Stott, M. J., and Sayer, M. (2007) Silicon substitution in the calcium phosphate bioceramics, *Biomaterials*, **28**, 4023–4032.
26. Yao, Z. P., Liu, W. G., and Ni, G. X. (2008) Biology characteristics and clinical application of strontium substituted hydroxyapatite bone cement, *J. Clin. Rehabil. Tissue Eng. Res.*, **12**, 7151–7154.

27. Landi, E., Tampieri, A., Mattioli-Belmonte, M., Celotti, G., Sandri, M., Gigante, A., Fava, P., and Biagini, G. (2006) Biomimetic Mg- and Mg,CO_3-substituted hydroxyapatites: synthesis characterization and *in vitro* behaviour, *J. Eur. Ceram. Soc.*, **26**, 2593-2601.

28. Landi, E., Tampieri, A., Celotti, G., Sprio, S., Sandri, M., and Logroscino, G. (2007) Sr-substituted hydroxyapatites for osteoporotic bone replacement, *Acta Biomater.*, **3**, 961–969.

29. Kannan, S., Ventura, J. M. G., and Ferreira, J. M. F. (2007) Synthesis and thermal stability of potassium substituted hydroxyapatites and hydroxyapatite/β-tricalcium phosphate mixtures, *Ceram. Int.*, **33**, 1489–1494.

30. Kannan, S., Rebelo, A., Lemos, A. F., Barba, A., and Ferreira, J. M. F. (2007) Synthesis and mechanical behaviour of chlorapatite and chlorapatite/β-TCP composites, *J. Eur. Ceram. Soc.*, **27**, 2287–2294.

31. Kannan, S., Goetz-Neunhoeffer, F., Neubauer, J., and Ferreira, J. M. F. (2008) Ionic substitutions in biphasic hydroxyapatite and β-tricalcium phosphate mixtures: structural analysis by Rietveld refinement, *J. Am. Ceram. Soc.*, **91**, 1–12.

32. Lafon, J. P., Champion, E., and Bernache-Assollant, D. (2008) Processing of AB-type carbonated hydroxyapatite $Ca_{10-x}(PO_4)_{6-x}(CO_3)_x(OH)_{2-x-2y}(CO_3)_y$ ceramics with controlled composition, *J. Eur. Ceram. Soc.*, **28**, 139–147.

33. Ren, F., Xin, R., Ge, X., and Leng, Y. (2008) An experimental and computational study on Zn-substituted hydroxyapatite, *Adv. Mater. Res.*, 47–50, Part 2, 1379–1382.

34. Meejoo, S., Pon-On, W., Charnchai, S., and Amornsakchai, T. (2008) Substitution of iron in preparation of enhanced thermal property and bioactivity of hydroxyapatite, *Adv. Mater. Res.*, 55–57, 689–692.

35. Kannan, S., Goetz-Neunhoeffer, F., Neubauer, J., and Ferreira, J. M. F. (2009) Synthesis and structure refinement of zinc-doped β-tricalcium phosphate powders, *J. Am. Ceram. Soc.*, **92**, 1592–1595.

36. Natasha, A. N., Sopyan, I., and Zuraida, A. (2008) Fourier transform infrared study on sol-gel derived manganese-doped hydroxyapatite, *Adv. Mater. Res.*, 47–50, Part 2, 1185–1188.

37. Matsumoto, N., Yoshida, K., Hashimoto, K., and Toda, Y. (2009) Thermal stability of β-tricalcium phosphate doped with monovalent metal ions, *Mater. Res. Bull.*, **44**, 1889–1894.

38. Boanini, E., Gazzano, M., and Bigi, A. (2010) Ionic substitutions in calcium phosphates synthesized at low temperature, *Acta Biomater.*, **6**, 1882–1894.

39. Williams, D. F. (1999) *The Williams Dictionary of Biomaterials*, Liverpool University Press, Liverpool, p. 368.

40. Williams, D. F. (2009) On the nature of biomaterials, *Biomaterials*, **30**, 5897–5909.

41. Bongio, M., van den Beucken, J. J. J. P., Leeuwenburgh, S. C. G., and Jansen, J. A. (2010) Development of bone substitute materials: from 'biocompatible' to 'instructive', *J. Mater. Chem.*, **20**, 8747–8759.

42. Mann, S. (ed.) (1996) *Biomimetic Materials Chemistry*, Wiley-VCH, UK, p. 400.

43. Vallet-Regí, M. (2008) Bioceramics: where do we come from and which are the future expectations, *Key Eng. Mater.*, **377**, 1–18.

44. Tripathi, G., Choudhury, P., and Basu, B. (2010) Development of polymer based biocomposites: a review, *Mater. Technol.*, **25**, 158–176.

45. Jandt, K. D. (2007) Evolutions, revolutions and trends in biomaterials science — a perspective, *Adv. Eng. Mater.*, **9**, 1035–1050.

46. Meyers, M. A., Chen, P. Y., Lin, A. Y. M., and Seki, Y. (2008) Biological materials: structure and mechanical properties, *Prog. Mater. Sci.*, **53**, 1–206.

47. According to Wikipedia, the free encyclopedia: "The word *ceramic* comes from the Greek word κεραμικός (keramikos) meaning pottery, which is said to derive from the Indo-European word ker, meaning heat. A ceramic is an inorganic, non-metallic solid prepared by the action of heat and subsequent cooling. Ceramic materials may have a crystalline or partly crystalline structure, or may be amorphous (e.g., a glass). Because most common ceramics are crystalline, the definition of ceramic is often restricted to inorganic crystalline materials, as opposed to the non-crystalline glasses. Ceramic may be used as an adjective describing a material, product or process; or as a singular noun, or, more commonly, as a plural noun, ceramics." http://en.wikipedia.org/wiki/Ceramics (accessed in November 2010).

48. Biocompatibility is the ability of a material to perform with an appropriate host response in a specific application [49–52].

49. Hench, L. L. (1991) Bioceramics: from concept to clinic, *J. Am. Ceram. Soc.*, **74**, 1487–1510.

50. Cao, W., and Hench, L. L. (1996) Bioactive materials, *Ceram. Int.*, **22**, 493–507.

51. Hench, L. L. (1998) Bioceramics, *J. Am. Ceram. Soc.*, **81**, 1705–1728.

52. Hench, L. L., Day, D. E., Höland, W., Rheinberger, and V. M. (2010) Glass and medicine, *Int. J. Appl. Glass Sci.*, **1**, 104–117.

53. Pinchuk, N. D., and Ivanchenko, L. A. (2003) Making calcium phosphate biomaterials, *Powder Metall. Metal Ceram.*, **42**, 357–371.

54. Heimann, R. B. (2002) Materials science of crystalline bioceramics: a review of basic properties and applications, *CMU J.*, **1**, 23–46.

55. Dorozhkin, S. V. (2007) Bioceramics based on calcium orthophosphates (Review), *Glass Ceram.*, **64**, 442–447.

56. Salinas, A. J., and Vallet-Regí, M. (2007) Evolution of ceramics with medical applications, *Z. Anorg. Allg. Chem.*, **633**, 1762–1773.

57. Tomoda, K., Ariizumi, H., Nakaji, T., and Makino, K. (2010) Hydroxyapatite particles as drug carriers for proteins, *Colloids Surf. B*, **76**, 226–235.

58. Ducheyne, P., and Qiu, Q. (1999) Bioactive ceramics: the effect of surface reactivity on bone formation and bone cell function, *Biomaterials*, **20**, 2287–2303.

59. Huebsch N., and Mooney, D. J. (2009) Inspiration and application in the evolution of biomaterials, *Nature*, **462**, 426–432.

60. Ring, M. E. (1992) Dentistry: An Illustrated History, Harry N. Abrams, New York, p. 320.

61. Bobbio, A. (1970) The first endosseous alloplastic implant in the history of man, *Bull. Hist. Dent.*, **20**, 1–6.

62. Lowenstam, H. A., and Weiner, S. (1989) *On Biomineralization*, Oxford University Press, London, p. 324.

63. Weiner, S., and Wagner, H. D. (1998) Material bone: structure-mechanical function relations, *Ann. Rev. Mater. Sci.*, **28**, 271–298.

64. LeGeros, R. Z. (1991) *Calcium Phosphates in Oral Biology and Medicine*, Karger, Basel, p. 201.

65. Weiner, S., and Dove, P. M. (2003) An overview of biomineralization processes and the problem of the vital effect, in *Biomineralization*, Series: *Reviews in Mineralogy and Geochemistry*, (ed. Dove, P. M., de Yoreo, J. J., and Weiner, S.), vol. 54, Mineralogical Society of America, Washington, D.C., pp. 1–29.

66. Weiner, S., Traub, W., and Wagner, H. D. (1999) Lamellar bone: structure-function relations, *J. Struct. Biol.*, **126**, 241–255.

67. Albee, F. H. (1920) Studies in bone growth: triple calcium phosphate as a stimulus to osteogenesis, *Ann. Surg.*, **71**, 32–39.

68. Ray, R. D., and Ward, A. A., Jr. (1951) A preliminary report on studies of basic calcium phosphate in bone replacement, *Surg. Form.*, **3**, 429–434.

69. Nery, E. B., Lynch, K. L., Hirthe W. M., and Mueller, K. H. (1975) Bioceramic implants in surgically produced infrabony defects, *J. Periodontol.*, **46**, 328–347.

70. Denissen, H. W., and de Groot, K. (1979) Immediate dental root implants from synthetic dense calcium hydroxylapatite, *J. Prosthet. Dent.*, **42**, 551–556.

71. Blakeslee, K. C., and Condrate, R. A., Sr. (1971) Vibrational spectra of hydrothermally prepared hydroxyapatites, *J. Am. Ceram. Soc.*, **54**, 559–563.

72. Garrington, G. E., and Lightbody, P. M. (1972) Bioceramics and dentistry, *J. Biomed. Mater. Res.*, **6**, 333–343.

73. Cini, L., Sandrolini, S., Paltrinieri, M., Pizzoferrato, A., and Trentani, C. (1972) Materiali bioceramici in funzione sostitutiva. Nota preventiva. (Bioceramic materials for replacement purposes. Preliminary note.), *La Chirurgia Degli Organi Di Movimento*, **60**, 423–430.

74. Rivault, M. A. (1966) Evolution, conception et technologie des travaux de prothèse fixe, réalisés en céramo-métallique. (Evolution, conception and technology of fixed prosthesis made of ceramic and metal), *Revue Francaise d'Odonto-Stomatologie*, **13**, 1367–1402.

75. Dumont, A., Appel, M., and Favard, E. (1968) Prothèses plurales en céramique sur métal. Soudage et artifices de jonction. (Multiple prostheses made of ceramics on metal. Soldering and artifacts of the junction), *Annales Odonto-Stomatologiques*, **25**, 231–240.

76. Hulbert, S. F., Young, F. A., Mathews, R. S., Klawitter, J. J., Talbert, C. D., and Stelling, F. H. (1970) Potential of ceramic materials as permanently implantable skeletal prostheses, *J. Biomed. Mater. Res.*, **4**, 433–456.

77. Hench, L. L., Splinter, R. J., Allen, W. C., and Greenlee, T. K. (1971) Bonding mechanisms at the interface of ceramic prosthetic materials, *J. Biomed. Mater. Res.*, **2**, 117–141.

78. Hulbert, S. F., Hench, L. L., Forbers, D., and Bowman, L. S. (1982) History of bioceramics, *Ceram. Int.*, **8**, 131–140.

79. Hulbert, S. F., Hench, L. L., Forbers, D., and Bowman, L. S. (1983) History of bioceramics, in *Ceramics in Surgery*, (ed. Vincenzini, P.), Elsevier, Amsterdam, pp. 3–29.

80. Jarcho, M., Bolen, C. H., Thomas, M. B., Bobick, J., Kay, J. F., and Doremus, R. H. (1976) Hydroxylapatite synthesis and characterization in dense polycrystalline form, *J. Mater. Sci.*, **11**, 2027–2035.

81. Jarcho, M., O'Connor, J. R., and Paris, D. A. (1977) Ceramic hydroxylapatite as a plaque growth and drug screening substrate, *J. Dental Res.*, **56**, 151–156.

82. Jarcho, M., Salsbury, R. L., Thomas, M. B., and Doremus, R. H. (1979) Synthesis and fabrication of β-tricalcium phosphate ceramics for potential prosthetic applications, *J. Mater. Sci.*, **14**, 142–150.

83. Jarcho, M. (1981) Calcium phosphate ceramics as hard tissue prosthetics, *Clin. Orthop. Rel. Res.*, **157**, 259–278.

84. Rejda, B. V., Peelen, J. G. J., and de Groot, K. (1977) Tricalcium phosphate as a bone substitute, *J. Bioeng.*, **1**, 93–97.

85. de Groot, K. (1980) Bioceramics consisting of calcium phosphate salts, *Biomaterials*, **1**, 47–50.

86. de Groot, K. (ed.) (1983) *Bioceramics of Calcium Phosphate*, CRC Press, Boca Raton, p. 146.

87. Aoki, H., Kato, K. M., Ogiso, M., and Tabata, T. (1977) Studies on the application of apatite to dental materials, *J. Dent. Eng.*, **18**, 86–89.

88. Kato, K., Aoki, H., Tabata, T., and Ogiso, M. (1979) Biocompatibility of apatite ceramics in mandibles, *Biomater. Med. Dev. Artif. Organs*, **7**, 291–297.

89. Akao, M., Aoki, H., and Kato, K. (1981) Mechanical properties of sintered hydroxyapatite for prosthetic applications, *J. Mater. Sci.*, **16**, 809–812.

90. Akao, M., Aoki, H., Kato, K., and Sato, A. (1982) Dense polycrystalline β-tricalcium phosphate for prosthetic applications, *J. Mater. Sci.*, **17**, 343–346.

91. Roy, D. M., and Linnehan, S. K. (1974) Hydroxyapatite formed from coral skeletal carbonate by hydrothermal exchange, *Nature*, **247**, 220–222.

92. Holmes, R. E. (1979) Bone regeneration within a coralline hydroxyapatite implant, *Plast. Reconstr. Surg.*, **63**, 626–633.

93. Elsinger, E. C., and Leal, L. (1996) Coralline hydroxyapatite bone graft substitutes, *J. Foot Ankle Surg.*, **35**, 396–399.

94. Shipman, P., Foster, G., and Schoeninger, M. (1984) Burnt bones and teeth: an experimental study of color, morphology, crystal structure and shrinkage, *J. Archaeological Sci.*, **11**, 307–325.

95. LeGeros, R. Z., and LeGeros, J. P. (2003) Calcium phosphate bioceramics: past, present, future, *Key Eng. Mater.*, 240–242, 3–10.

96. Oktar, F. N., Kesenci, K., and Pişkin, E. (1999) Characterization of processed tooth hydroxyapatite for potential biomedical implant applications, *Artif. Cells Blood Substit. Immobil. Biotechnol.*, **27**, 367–379.

97. Rivera, E. M., Araiza, M., Brostow, W., Castaño, V. M., Díaz-Estrada, J. R., Hernández, R., and Rodríguez, J. R. (1999) Synthesis of hydroxyapatite from eggshells, *Mater. Lett.*, **41**, 128–134.

98. Lee, S. J., and Oh, S. H. (2003) Fabrication of calcium phosphate bioceramics by using eggshell and phosphoric acid, *Mater. Lett.*, **57**, 4570–4574.

99. Balazsi, C., Weber, F., Kover, Z., Horvath, E., and Nemeth, C. (2007) Preparation of calcium-phosphate bioceramics from natural resources, *J. Eur. Ceram. Soc.*, **27**, 1601–1606.

100. Murugan, R., and Ramakrishna, S. (2005) Crystallographic study of hydroxyapatite bioceramics derived from various sources, *Cryst. Growth Des.*, **5**, 111–112.

101. Ruksudjarit, A., Pengpat, K., Rujijanagul, G., and Tunkasiri, T. (2008) Synthesis and characterization of nanocrystalline hydroxyapatite from natural bovine bone, *Curr. Appl. Phys.*, **8**, 270–272.

102. Ooi, C. Y., Hamdi, M., and Ramesh, S. (2007) Properties of hydroxyapatite produced by annealing of bovine bone, *Ceram. Int.*, **33**, 1171–1177.

103. Lee, S. J., Lee, Y. C., and Yoon, Y. S. (2007) Characteristics of calcium phosphate powders synthesized from cuttlefish bone and phosphoric acid, *J. Ceram. Process. Res.*, **8**, 427–430.

104. Oktar, F. N. (2007) Microstructure and mechanical properties of sintered enamel hydroxyapatite, *Ceram. Int.*, **33**, 1309–1314.

105. Figueiredo, M., Fernando, A., Martins, G., Freitas, J., Judas, F., and Figueiredo, H. (2010) Effect of the calcination temperature on the composition and microstructure of hydroxyapatite derived from human and animal bone, *Ceram. Int.*, **36**, 2383–2393.

106. Han, F., and Wu, L. (2010) Preparing and characterizing natural hydroxyapatite ceramics, *Ceram. Int.*, **220**, 281–285.

107. Gergely, G., Wéber, F., Lukács, I., Illés, L., Tóth, A. L., Horváth, Z. E., Mihály, J., and Balázsi, C. (2010) Nano-hydroxyapatite preparation from biogenic raw materials, *Central Eur. J. Chem.*, **8**, 375–381.

108. Harabi, A., Belamri, D., Karboua, N., and Mezahi, F. Z. (2011) Sintering of bioceramics using a modified domestic microwave oven — natural hydroxyapatite sintering, *J. Thermal Analysis Calorimetry*, **104**, 383–388.

109. Mondal, S., Mahata, S., Kundu, S., and Mondal, B. (2010) Processing of natural resourced hydroxyapatite ceramics from fish scale, *Adv. Appl. Ceram.*, **109**, 234–239.

110. Vallet-Regí, M., and González-Calbet, J. M. (2004) Calcium phosphates as substitution of bone tissues, *Progr. Solid State Chem.*, **32**, 1–31.

111. Lecomte, A., Gautier, H., Bouler, J. M., Gouyette, A., Pegon, Y., Daculsi, G., and Merle, C. (2008) Biphasic calcium phosphate: a comparative study of interconnected porosity in two ceramics, *J. Biomed. Mater. Res. B (Appl. Biomater.)*, **84B**, 1–6.

112. Tancret, F., Bouler, J. M., Chamousset, J., and Minois, L. M. (2006) Modelling the mechanical properties of microporous and macroporous biphasic calcium phosphate bioceramics, *J. Eur. Ceram. Soc.*, **26**, 3647–3656.

113. Bouler, J. M., Trecant, M., Delecrin, J., Royer, J., Passuti, N., and Daculsi, G. (1996) Macroporous biphasic calcium phosphate ceramics: influence of five synthesis parameters on compressive strength, *J. Biomed. Mater. Res.*, **32**, 603–609.

114. Kivrak, N., and Tas, A. C. (1998) Synthesis of calcium hydroxyapatite–tricalcium phosphate (HA–TCP) composite bioceramic powders and their sintering behavior, *J. Am. Ceram. Soc.*, **81**, 2245–2252.

115. Gauthier, O., Bouler, J. M., Aguado, E., Pilet, P., and Daculsi, G. (1998) Macroporous biphasic calcium phosphate ceramics: influence of macropore diameter and macroporosity percentage on bone ingrowth, *Biomaterials*, **19**, 133–139.

116. Wang, J., Chen, W., Li, Y., Fan, S., Weng, J., and Zhang, X. (1998) Biological evaluation of biphasic calcium phosphate ceramic vertebral laminae, *Biomaterials*, **19**, 1387–1392.

117. Daculsi, G. (1998) Biphasic calcium phosphate concept applied to artificial bone, implant coating and injectable bone substitute, *Biomaterials*, **19**, 1473–1478.

118. Daculsi, G., Weiss, P., Bouler, J. M., Gauthier, O., Millot, F., and Aguado, E. (1999) Biphasic calcium phosphate / hydrosoluble polymer composites: a new concept for bone and dental substitution biomaterials, *Bone*, **25**(2), 59S–61S.

119. LeGeros, R. Z. Lin, S., Rohanizadeh, R., Mijares, D., and LeGeros, J. P. (2003) Biphasic calcium phosphate bioceramics: preparation, properties and applications, *J. Mater. Sci. Mater. Med.*, **14**, 201–209.

120. Daculsi, G., Laboux, O., Malard, O., and Weiss, P. (2003) Current state of the art of biphasic calcium phosphate bioceramics, *J. Mater. Sci. Mater. Med.*, **14**, 195–200.

121. Dorozhkina, E. I., and Dorozhkin, S. V. (2002) Mechanism of the solid-state transformation of a calcium-deficient hydroxyapatite (CDHA)

into biphasic calcium phosphate (BCP) at elevated temperatures, *Chem. Mater.*, **14**, 4267–4272.

122. Alam, I., Asahina, I., Ohmamiuda, K., and Enomoto, S. (2001) Comparative study of biphasic calcium phosphate ceramics impregnated with rhBMP-2 as bone substitutes, *J. Biomed. Mater. Res.*, **54**, 129–138.

123. Daculsi, G. (2006) Biphasic calcium phosphate granules concept for injectable and mouldable bone substitute, *Adv. Sci. Technol.*, **49**, 9–13.

124. Tanimoto, Y., Shibata, Y., Murakami, A., Miyazaki, T., and Nishiyama, N. (2009) Effect of varying HAP/TCP ratios in tape-cast biphasic calcium phosphate ceramics on responcce *in vitro*, *J. Hard Tissue Biology*, **18**, 71–76.

125. Daculsi, G., Baroth, S., and LeGeros, R. Z. (2010) 20 years of biphasic calcium phosphate bioceramics development and applications, *Ceram. Eng. Sci. Proc.*, **30**, 45–58.

126. Lukić, M., Stojanović, Z., Škapin, S. D., Maček-Kržmanc, M., Mitrić, M., Marković, S., and Uskoković, D. (2011) Dense fine-grained biphasic calcium phosphate (BCP) bioceramics designed by two-step sintering, *J. Eur. Ceram. Soc.*, **31**, 19–27.

127. O'Neill, W. C. (2007) The fallacy of the calcium–phosphorus product, *Kidney Int.*, **72**, 792–796.

128. da Silva, R. V., Bertran, C. A., Kawachi, E. Y., and Camilli, J. A. (2007) Repair of cranial bone defects with calcium phosphate ceramic implant or autogenous bone graft, *J. Craniofac. Surg.*, **18**, 281–286.

129. Metsger, D. S., Driskell, T. D., and Paulsrud, J. R. (1982) Tricalcium phosphate ceramic — a resorbable bone implant: review and current status, *J. Am. Dent. Assoc.*, **105**, 1035–1038.

130. Reid, J. W., Tuck, L., Sayer, M., Fargo, K., and Hendry, J. A. (2006) Synthesis and characterization of single-phase silicon substituted α-tricalcium phosphate, *Biomaterials*, **27**, 2916–2925.

131. Yin, X., Stott, M. J., and Rubio, A. (2003) α- and β-tricalcium phosphate: a density functional study, *Phys. Rev. B*, **68**, 205205 (7 pages).

132. Yin, X., and Stott, M. J. (2005) Theoretical insights into bone grafting Si-stabilized α-tricalcium phosphate, *J. Chem. Phys.*, **122**, 024709 (9 pages).

133. Ruan, J. M., Zou, J. P., Zhou, J. N., and Hu, J. Z. (2006) Porous hydroxyapatite — tricalcium phosphate bioceramics, *Powder Metall.*, **49**, 66–69.

134. Yin, X., and Stott, M. J. (2006) Surface and adsorption properties of α-tricalcium phosphate, *J. Chem. Phys.*, **124**, 124701 (9 pages).

135. Kamitakahara, M., Ohtsuki, C., and Miyazaki, T. (2008) Review paper: behavior of ceramic biomaterials derived from tricalcium phosphate in physiological condition, *J. Biomater. Appl.*, **23**, 197–212.

136. Li, Y., Kong, F., and Weng, W. (2009) Preparation and characterization of novel biphasic calcium phosphate powders (α-TCP/HA) derived from carbonated amorphous calcium phosphates, *J. Biomed. Mater. Res. B* (*Appl. Biomater.*), **89B**, 508–517.

137. Oishi, M., Ohtsuki, C., Kitamura, M., Kamitakahara, M., Ogata, S., Miyazaki, T., and Tanihara, M. (2004) Fabrication and chemical durability of porous bodies consisting of biphasic tricalcium phosphates, *Phosphorus Res. Bull.*, **17**, 95–100.

138. Kamitakahara, M., Ohtsuki, C., Oishi, M., Ogata, S., Miyazaki, T., and Tanihara, M. (2005) Preparation of porous biphasic tricalcium phosphate and its *in vivo* behavior, *Key Eng. Mater.*, 284–286, 281–284.

139. Wang, R., Weng, W., Deng, X., Cheng, K., Liu, X., Du, P., Shen, G., and Han, G. (2006) Dissolution behavior of submicron biphasic tricalcium phosphate powders, *Key Eng. Mater.*, 309–311, 223–226.

140. Li, Y., Weng, W., and Tam, K. C. (2007) Novel highly biodegradable biphasic tricalcium phosphates composed of α-tricalcium phosphate and β-tricalcium phosphate, *Acta Biomater.*, **3**, 251–254.

141. Li, Y., Li, D., and Weng, W. (2008) *In vitro* dissolution behavior of biphasic tricalcium phosphate composite powders composed of α-tricalcium phosphate and β-tricalcium phosphate, *Key Eng. Mater.*, 368–372, 1206–1208.

142. Vani, R., Girija, E. K., Elayaraja, K., Prakash Parthiban, S., Kesavamoorthy, R., and Narayana Kalkura, S. (2009) Hydrothermal synthesis of porous triphasic hydroxyapatite/(α and β) tricalcium phosphate, *J. Mater. Sci. Mater. Med.*, **20**(1), 43–48.

143. Dorozhkin, S. V. (2007) Calcium orthophosphates, *J. Mater. Sci.*, **42**, 1061–1095.

144. Dorozhkin, S. V. (2009) Calcium orthophosphates in nature, biology and medicine, *Materials*, **2**, 399–498.

145. Narasaraju, T. S. B., and Phebe, D. E. (1996) Some physico-chemical aspects of hydroxylapatite, *J. Mater. Sci.*, **31**, 1–21.

146. Elliott, J. C. (1994) Structure and chemistry of the apatites and other calcium orthophosphates, *Studies in Inorganic Chemistry*, vol. 18, Elsevier, Amsterdam, p. 389.

147. Brown, P. W., and Constantz, B. (eds.) (1994), Hydroxyapatite and related materials, CRC Press, Boca Raton, p. 343.

148. Amjad, Z. (ed.) (1997), Calcium phosphates in biological and industrial systems, Kluwer Academic Publishers, Boston, p. 529.

149. Hughes, J. M., Kohn, M., and Rakovan, J. (eds.) (2002), Phosphates: geochemical, geobiological and materials importance. *Series: Reviews in Mineralogy and Geochemistry*, vol. 48, Mineralogical Society of America, Washington, DC, p. 742.

150. Chow, L. C., and Eanes, E. D. (eds.) (2001), Octacalcium phosphate, Monographs in Oral Science, vol. 18, S. Karger AG, Basel, p. 168.

151. Brès, E., and Hardouin, P. (eds.) (1998), Les matériaux en phosphate de calcium. Aspects fondamentaux./Calcium phosphate materials. Fundamentals, Sauramps Medical, Montpellier, p. 176.

152. Okuda, T., Ioku, K., Yonezawa, I., Minagi, H., Gonda, Y., Kawachi, G., Kamitakahara, M., Shibata, Y., Murayama, H., Kurosawa, H., and Ikeda, T. (2008) The slow resorption with replacement by bone of a hydrothermally synthesized pure calcium-deficient hydroxyapatite, *Biomaterials*, **29**, 2719–2728.

153. Daculsi, G., Bouler, J. M., and LeGeros, R. Z. (1997) Adaptive crystal formation in normal and pathological calcifications in synthetic calcium phosphate and related biomaterials, *Int. Rev. Cytology*, **172**, 129–191.

154. Astala, R., Calderin, L., Yin, X., and Stott, M. J. (2006) *Ab initio* simulation of Si-doped hydroxyapatite *Chem. Mater.*, **18**, 413–422.

155. Zhu, X. D., Zhang, H. J., Fan, H. S., Li, W., and Zhang, X. D. (2010) Effect of phase composition and microstructure of calcium phosphate ceramic particles on protein adsorption, *Acta Biomater.*, **6**, 1536–1541.

156. Bohner, M. (2000) Calcium orthophosphates in medicine: from ceramics to calcium phosphate cements, *Injury*, **31**(4), D37–D47.

157. Norton, J., Malik, K. R., Darr, J. A., and Rehman, I. (2006) Recent developments in processing and surface modification of hydroxyapatite, *Adv. Appl. Ceram.*, **105**, 113–139.

158. Yang, S., Leong, K. F., Du, Z., and Chua, C. K. (2002) The design of scaffolds for use in tissue engineering. Part II. Rapid prototyping techniques, *Tissue Eng.*, **8**, 1–11.

159. Yeong, W. Y., Chua, C. K., Leong, K. F., and Chandrasekaran, M. (2004) Rapid prototyping in tissue engineering: challenges and potential, *Trends Biotechnol.*, **22**, 643–652.

160. Hieu, L. C., Zlatov, N., Sloten, J. V., Bohez, E., Khanh, L., Binh, P. H., Oris, P., and Toshev, Y. (2005) Medical rapid prototyping applications and methods, *Assembly Automation*, **25**, 284–292.

161. Eufinger, H., Wehniöller, M., Machtens, E., Heuser, L., Harders, A., and Kruse, D. (1995) Reconstruction of craniofacial bone defects with individual alloplastic implants based on CAD/CAM-manipulated CT-data, *J. Cranio-Maxillofacial Surg.*, **23**, 175–181.

162. Klein, M., and Glatzer, C. (2006) Individual CAD/CAM fabricated glass-bioceramic implants in reconstructive surgery of the bony orbital floor, *Plastic Reconstruct. Surg.*, **117**, 565–570.

163. Yin, L., Song, X. F., Song, Y. L., Huang, T., and Li, J. (2006) An overview of *in vitro* abrasive finishing & CAD/CAM of bioceramics in restorative dentistry, *Int. J. Machine Tools Manufact.*, **46**, 1013–1026.

164. Li, J., Hsu, Y., Luo, E., Khadka, A., and Hu, J. (2011) Computer-aided design and manufacturing and rapid prototyped nanoscale hydroxyapatite/polyamide (n-HA/PA) construction for condylar defect caused by mandibular angle ostectomy, **35**, 636–640.

165. Yardimci, M. A., Guceri, S. I., and Danforth, S. C. (1996) Process modeling for fused deposition of ceramics, *Ceram. Eng. Sci. Proceed.*, **17**, 78–82.

166. Bellini, A., Shor, L., and Guceri, S. I. (2005) New developments in fused deposition modeling of ceramics, *Rapid Prototyping J.*, **11**, 214–220.

167. Tan, K. H., Chua, C. K., Leong, K. F., Cheah, C. M., Cheang, P., Abu Bakar, M. S., and Cha, S. W. (2003) Scaffold development using selective laser sintering of polyetheretherketone-hydroxyapatite biocomposite blends, *Biomaterials*, **24**, 3115–3123.

168. Wiria, F. E., Leong, K. F., Chua, C. K., and Liu, Y. (2007) Poly-ε-caprolactone/hydroxyapatite for tissue engineering scaffold fabrication via selective laser sintering, *Acta Biomater.*, **3**, 1–12.

169. Zhou, W. Y., Lee, S. H., Wang, M., Cheung, W. L., and Ip, W. Y. (2008) Selective laser sintering of porous tissue engineering scaffolds from poly(L-lactide)/carbonated hydroxyapatite nanocomposite microspheres, *J. Mater. Sci. Mater. Med.*, **19**, 2535–2540.

170. Comesaña, R., Lusquiños, F., del Val, J., Malot, T., López-Álvarez, M., Riveiro, A., Quintero, F., Boutinguiza, M., Aubry, P., de Carlos, A., and Pou, J. (2011) Calcium phosphate grafts produced by rapid prototyping based on laser cladding, *J. Eur. Ceram. Soc.*, **31**, 29–41.

171. Sachs, E., Cima, M., Williams, P., Brancazio, D., and Cornie, J. (1992) Three dimensional printing. Rapid tooling and prototypes directly from a CAD model, *J. Eng. Ind.*, **114**, 481–488.

172. Gbureck, U., Hölzel, T., Doillon, C. J., Müller, F. A., and Barralet, J. E. (2007) Direct printing of bioceramic implants with spatially localized angiogenic factors, *Adv. Mater.*, **19**, 795–800.

173. Gbureck, U., Hölzel, T., Klammert, U., Würzler, K., Müller, F. A., and Barralet, J. E. (2007) Resorbable dicalcium phosphate bone substitutes prepared by 3D powder printing, *Adv. Function. Mater.*, **17**, 3940–3945.

174. Vorndran, E., Klarner, M., Klammert, U., Grover, L. M., Patel, S., Barralet, J. E., and Gbureck, U. (2008) 3D powder printing of β-tricalcium phosphate ceramics using different strategies, *Adv. Eng. Mater.*, **10**, B67–B71.

175. Seitz, H., Rieder, W., Irsen, S., Leukers, B., and Tille, C. (2005) Three-dimensional printing of porous ceramic scaffolds for bone tissue engineering, *J. Biomed. Mater. Res. B*, **74B**, 782–788.

176. Leukers, B., Gülkan, H., Irsen, S. H., Milz, S., Tille, C., Schieker, M., and Seitz, H. (2005) Hydroxyapatite scaffolds for bone tissue engineering made by 3D printing, *J. Mater. Sci. Mater. Med.*, **16**, 1121–1124.

177. Habibovic, P., Gbureck, U., Doillon, C. J., Bassett, D. C., van Blitterswijk, C. A., and Barralet, J. E. (2008) Osteoconduction and osteoinduction of low-temperature 3D printed bioceramic implants, *Biomaterials*, **29**, 944–953.

178. Fierz, F. C., Beckmann, F., Huser, M., Irsen, S. H., Leukers, B., Witte, F., Degistirici, O., Andronache, A., Thie, M., and Müller, B. (2008) The morphology of anisotropic 3D-printed hydroxyapatite scaffolds, *Biomaterials*, **29**, 3799–3806.

179. Seitz, H., Deisinger, U., Leukers, B., Detsch, R., and Ziegler, G. (2009) Different calcium phosphate granules for 3-D printing of bone tissue engineering scaffolds, *Adv. Eng. Mater.*, **11**, B41–B46.

180. Bergmann, C., Lindner, M., Zhang, W., Koczur, K., Kirsten, A., Telle, R., and Fischer, H. (2010) 3D printing of bone substitute implants using calcium phosphate and bioactive glasses, *J. Eur. Ceram. Soc.*, **30**, 2563–2567.

181. Porter, N. L., Pilliar, R. M., and Grynpas, M. D. (2001) Fabrication of porous calcium polyphosphate implants by solid freeform fabrication: a study of processing parameters and in vitro degradation characteristics, *J. Biomed. Mater. Res.*, **56**, 504–515.

182. Leong, K. F., Cheah, C. M., and Chua, C. K. (2003) Solid freeform fabrication of three-dimensional scaffolds for engineering replacement tissues and organs, *Biomaterials*, **24**, 2363–2378.

183. Calvert, J. W., Brenner, K. A., Mooney, M. P., Kumta, P., and Weiss, L. E. (2004) Cellular fusion of hydroxyapatite layers: solid freeform fabrication of synthetic bone grafts, *Rivista Italiana di Chirurgia Plastica*, **36**, 145–150.

184. Jongpaiboonkit, L., Lin, C. Y., Krebsbach, P. H., Hollister, S. J., and Halloran, J. W. (2006) Mechanical behavior of complex 3D calcium phosphate cement scaffolds fabricated by indirect solid freeform fabrication *in vivo*, *Key Eng. Mater.*, 309–311, 957–960.

185. Shanjani, Y., de Croos, J. N. A., Pilliar, R. M., Kandel, R. A., and Toyserkani, E. (2010) Solid freeform fabrication and characterization of porous calcium polyphosphate structures for tissue engineering purposes, *J. Biomed. Mater. Res. B* (*Appl. Biomater.*), **93B**, 510–519.

186. Chu, T. M. G., Halloran, J. W., Hollister, S. J., and Feinberg, S. E. (2001) Hydroxyapatite implants with designed internal architecture, *J. Mater. Sci. Mater. Med.*, **12**, 471–478.

187. Li, X., Li, D., Lu, B., Tang, Y., Wang, L., and Wang, Z. (2005) Design and fabrication of CAP scaffolds by indirect solid free form fabrication, *Rapid Prototyping J.*, **11**, 312–318.

188. Woesz, A., Rumpler, M., Stampfl, J., Varga, F., Fratzl-Zelman, N., Roschger, P., Klaushofer, K., and Fratzl, P. (2005) Towards bone replacement materials from calcium phosphates via rapid prototyping and ceramic gel casting, *Mater. Sci. Eng. C*, **25**, 181–186.

189. Li, X., Li, D., Lu, B., and Wang, C. (2008) Fabrication of bioceramic scaffolds with pre-designed internal architecture by gel casting and indirect stereolithography techniques, *J. Porous Mater.*, **15**, 667–671.

190. Saber-Samandari, S., and Gross, K. A. (2010) The use of thermal printing to control the properties of calcium phosphate deposits, *Biomaterials*, **31**, 6386–6393.

191. Nather, A. (ed.) (2005), Bone grafts and bone substitutes: basic science and clinical applications, *World Scientific*, Singapore, p. 592.

192. Bártolo, P., and Bidanda, B. (eds.) (2008), *Bio-materials and Prototyping Applications in Medicine*, Springer, New York, p. 216.

193. Narayan, R. J., Jin, C., Doraiswamy, A., Mihailescu, I. N., Jelinek, M., Ovsianikov, A., Chichkov, B., and Chrisey, D. B. (2005) Laser processing of advanced bioceramics, *Adv. Eng. Mater.*, **7**, 1083–1098.

194. Kokubo, T. (ed.) (2008), *Bioceramics and Their Clinical Applications*, Woodhead Publishing, Cambridge, p. 784.

195. Narayan, R. (ed.) (2009), *Biomedical Materials*. Springer, New York, p. 566.

196. Raksujarit, A., Pengpat, K., Rujijanagul, G., and Tunkasiri, T. (2010) Processing and properties of nanoporous hydroxyapatite ceramics, *Mater. Design*, **31**, 1658–1660.

197. Park, J. (2008) *Bioceramics: Properties, Characterizations, and Applications*, Springer, New York, p. 364.

198. Rodríguez-Lorenzo, L. M., Vallet-Regí, M., and Ferreira, J. M. F. (2001) Fabrication of hydroxyapatite bodies by uniaxial pressing from a precipitated powder, *Biomaterials*, **22**, 583–588.

199. Nazarpak, M. H., Solati-Hashjin, M., and Moztarzadeh, F. (2009) Preparation of hydroxyapatite ceramics for biomedical applications, *J. Ceram. Process. Res.*, **10**, 54–57.

200. Itoh, H., Wakisaka, Y., Ohnuma, Y., and Kuboki, Y. (1994) A new porous hydroxyapatite ceramic prepared by cold isostatic pressing and sintering synthesized flaky powder, *Dental Mater.*, **13**, 25–35.

201. Gautier, H., Merle, C., Auget, J. L., and Daculsi, G. (2000) Isostatic compression, a new process for incorporating vancomycin into biphasic calcium phosphate: comparison with a classical method, *Biomaterials*, **21**, 243–249.

202. Tadic, D., and Epple, M. (2003) Mechanically stable implants of synthetic bone mineral by cold isostatic pressing, *Biomaterials*, **24**, 4565–4571.

203. Pecqueux, F., Tancret, F., Payraudeau, N., and Bouler, J. M. (2010) Influence of microporosity and macroporosity on the mechanical properties of biphasic calcium phosphate bioceramics: modelling and experiment, *J. Eur. Ceram. Soc.*, **30**, 819–829.

204. Uematsu, K., Takagi, M., Honda, T., Uchida, N., and Saito, K. (1989) Transparent hydroxyapatite prepared by hot isostatic pressing of filter cake, *J. Am. Ceram. Soc.*, **72**, 1476–1478.

205. Onoki, T., and Hashida, T. (2006) New method for hydroxyapatite coating of titanium by the hydrothermal hot isostatic pressing technique, *Surf. Coat. Technol.*, **200**, 6801–6807.

206. Takikawa, K., and Akao, M. (1996) Fabrication of transparent hydroxyapatite and application to bone marrow derived cell/ hydroxyapatite interaction observation *in-vivo*, *J. Mater. Sci. Mater. Med.*, **7**, 439–445.

207. Viana, M., Désiré, A., Chevalier, E., Champion, E., Chotard, R., and Chulia, D. (2009) Interest of high shear wet granulation to produce drug loaded porous calcium phosphate pellets for bone filling, *Key Eng. Mater.*, 396–398, 535–538.

208. Chevalier, E., Viana, M., Cazalbou, S., and Chulia, D. (2009) Comparison of low-shear and high-shear granulation processes: effect on implantable calcium phosphate granule properties, *Drug Dev. Ind. Pharm.*, **35**, 1255–1263.

209. Reikerås, O., Johansson, C. B., and Sundfeldt, M. (2006) Bone ingrowths to press-fit and loose-fit implants: comparisons between titanium and hydroxyapatite, *J. Long-Term Eff. Med. Implant.*, **16**, 157–164.

210. Rao, R. R., and Kannan, T. S. (2001) Dispersion and slip casting of hydroxyapatite, *J. Am. Ceram. Soc.*, **84**, 1710–1716.

211. Sakka, Y., Takahashi, K., Matsuda, N., and Suzuki, T. S. (2007) Effect of milling treatment on texture development of hydroxyapatite ceramics by slip casting in high magnetic field, *Mater. Transact.*, **48**, 2861–2866.

212. Zhang, Y., Yokogawa, Y., Feng, X., Tao, Y., and Li, Y. (2010) Preparation and properties of bimodal porous apatite ceramics through slip casting using different hydroxyapatite powders, *Ceram. Int.*, **36**, 107–113.

213. Liu, Y., Kim, J. H., Young, D., Kim, S., Nishimoto, S. K., and Yang, Y. (2010) Novel template-casting technique for fabricating β-tricalcium phosphate scaffolds with high interconnectivity and mechanical strength and *in vitro* cell responses, *J. Biomed. Mater. Res. A*, **92A**, 997–1006.

214. Sepulveda, P., Ortega, F. S., Innocentini, M. D. M., and Pandolfelli, V. C. (2000) Properties of highly porous hydroxyapatite obtained by the gel casting of foams, *J. Am. Ceram. Soc.*, **83**, 3021–3024.

215. Padilla, S., Vallet-Regí, M., Ginebra, M. P., and Gil, F. J. (2005) Processing and mechanical properties of hydroxyapatite pieces obtained by the gel-casting method, *J. Eur. Ceram. Soc.*, **25**, 375–383.

216. Sánchez-Salcedo, S., Werner, J., and Vallet-Regí, M. (2008) Hierarchical pore structure of calcium phosphate scaffolds by a combination of gel-casting and multiple tape-casting methods, *Acta Biomater.*, **4**, 913–922.

217. Chen, B., Zhang, T., Zhang, J., Lin, Q., and Jiang, D. (2008) Microstructure and mechanical properties of hydroxyapatite obtained by gel-casting process, *Ceram. Int.*, **34**, 359–364.

218. Marcassoli, P., Cabrini, M., Tirillò, J., Bartuli, C., Palmero, P., and Montanaro, L. (2010) Mechanical characterization of hydroxiapatite micro/macro-porous ceramics obtained by means of innovative gel-casting process, *Key Eng. Mater.*, 417–418, 565–568.

219. Padilla, S., Roman, J., and Vallet-Regí, M. (2002) Synthesis of porous hydroxyapatites by combination of gel casting and foams burn out methods, *J. Mater. Sci. Mater. Med.*, **13**, 1193–1197.

220. Fomin, A. S., Barinov, S. M., Ievlev, V. M., Smirnov, V. V., Mikhailov, B. P., Belonogov, E. K., and Drozdova, N. A. (2008) Nanocrystalline hydroxyapatite ceramics produced by low-temperature sintering after high-pressure treatment, *Dokl. Chem.*, **418**, 22–25.

221. Kankawa, Y., Kaneko, Y., and Saitou, K. (1991) Injection molding of highly-purified hydroxylapatite and TCP utilizing solid phase reaction method, *J. Ceram. Soc. Jpn.*, **99**, 438–442.

222. Kwon, S. H., Jun, Y. K., Hong, S. H., Lee, I. S., Kim, H. E., and Won, Y. Y. (2002) Calcium phosphate bioceramics with various porosities and dissolution rates, *J. Am. Ceram. Soc.*, **85**, 3129–3131.

223. Fooki, A. C. B. M., Aparecida, A. H., Fideles, T. B., Costa, R. C., and Fook, M. V. L. (2009) Porous hydroxyapatite scaffolds by polymer sponge method, *Key Eng. Mater.*, 396–398, 703–706.

224. Sopyan, I., and Kaur, J. (2009) Preparation and characterization of porous hydroxyapatite through polymeric sponge method, *Ceram. Int.*, **35**, 3161–3168.

225. Bellucci, D., Cannillo, V., and Sola, A. (2010) Shell scaffolds: a new approach towards high strength bioceramic scaffolds for bone regeneration, *Mater. Lett.*, **64**, 203–206.

226. Velayudhan, S., Ramesh, P., Sunny, M. C., and Varma, H. K. (2000) Extrusion of hydroxyapatite to clinically significant shapes, *Mater. Lett.*, **46**, 142–146.

227. Yang, H. Y., Thompson, I., Yang, S. F., Chi, X. P., Evans, J. R. G., and Cook, R. J. (2008) Dissolution characteristics of extrusion freeformed hydroxyapatite — tricalcium phosphate scaffolds, *J. Mater. Sci. Mater. Med.*, **19**, 3345–3353.

228. Yang, S., Yang, H., Chi, X., Evans, J. R. G., Thompson, I., Cook, R. J., and Robinson, P. (2008) Rapid prototyping of ceramic lattices for hard tissue scaffolds, *Mater. Design*, **29**, 1802–1809.

229. Muthutantri, A. I., Huang, J., Edirisinghe, M. J., Bretcanu, O., and Boccaccini, A. R. (2008) Dipping and electrospraying for the preparation of hydroxyapatite foams for bone tissue engineering, *Biomed. Mater.*, **3**, 25009 (14 pages).

230. Roncari, E., Galassi, C., and Pinasco, P. (2000) Tape casting of porous hydroxyapatite ceramics, *J. Mater. Sci. Lett.*, **19**, 33–35.

231. Tian, T., Jiang, D., Zhang, J., and Lin, Q. (2007) Aqueous tape casting process for hydroxyapatite, *J. Eur. Ceram. Soc.*, **27**, 2671–2677.

232. Suzuki, S., Itoh, K., Ohgaki, M., Ohtani, M., and Ozawa, M. (1999) Preparation of sintered filter for ion exchange by a doctor blade method with aqueous slurries of needlelike hydroxyapatite, *Ceram. Int.*, **25**, 287–291.

233. Suchanek, W. L., and Yoshimura, M. (1998) Processing and properties of hydroxyapatite-based biomaterials for use as hard tissue replacement implants, *J. Mater. Res.*, **13**, 94–117.

234. Yang, T. Y., Lee, J. M., Yoon, S. Y., and Park, H. C. (2010) Hydroxyapatite scaffolds processed using a TBA-based freeze-gel casting/polymer sponge technique, *J. Mater. Sci. Mater. Med.*, **21**, 1495–1502.

235. Inoue, K., Sassa, K., Yokogawa, Y., Sakka, Y., Okido, M., and Asai, S. (2003) Control of crystal orientation of hydroxyapatite by imposition of a high magnetic field, *Mater. Transact.*, **44**, 1133–1137.

236. Iwai, K., Akiyama, J., and Asai, S. (2007) Structure control of hydroxyapatite using a magnetic field, *Mater. Sci. Forum*, 561–565, Part 2, 1565–1568.

237. Iwai, K., Akiyama, J., Tanase, T., and Asai, S. (2007) Alignment of HAp crystal using a sample rotation in a static magnetic field, *Mater. Sci. Forum*, 539–543, Part 1, 716–719.

238. Sakka, Y., Takahashi, K., Suzuki, T. S., Ito, S., and Matsuda, N. (2008) Texture development of hydroxyapatite ceramics by colloidal processing in a high magnetic field followed by sintering, *Mater. Sci. Eng. A*, **475**, 27–33.

239. Kang, J., and Hadfield, M. (2001) Parameter optimization by Taguchi methods for finishing advanced ceramic balls using a novel eccentric lapping machine, *Proc. Inst. Mech. Eng. B*, **215**, 69–78.

240. Kurella, A., and Dahotre, N. B. (2005) Surface modification for bioimplants: the role of laser surface engineering, *J. Biomater. Appl.*, **20**, 5–50.

241. Oktar, F. N., Genc, Y., Goller, G., Erkmen, E. Z., Ozyegin, L. S., Toykan, D., Demirkiran, H., and Haybat, H. (2004) Sintering of synthetic hydroxyapatite compacts, *Key Eng. Mater.*, 264–268, 2087–2090.

242. Georgiou, G., Knowles, J. C., and Barralet, J. E. (2004) Dynamic shrinkage behavior of hydroxyapatite and glass-reinforced hydroxyapatites, *J. Mater. Sci.*, **39**, 2205–2208.

243. Fellah, B. H., and Layrolle, P. (2009) Sol-gel synthesis and characterization of macroporous calcium phosphate bioceramics containing microporosity, *Acta Biomater.*, **5**, 735–742.

244. Dudek, A., and Kolan, C. (2010) Assessments of shrinkage degree in bioceramic sinters HA+ZrO_2, *Diffusion Defect Data B Solid State Phenomena*, **165**, 25–30.

245. Ben Ayed, F., Bouaziz, J., and Bouzouita, K. (2000) Pressureless sintering of fluorapatite under oxygen atmosphere, *J. Eur. Ceram. Soc.*, **20**, 1069–1076.

246. He, Z., Ma, J., and Wang, C. (2005) Constitutive modeling of the densification and the grain growth of hydroxyapatite ceramics, *Biomaterials*, **26**, 1613–1621.

247. Khalil, K. A., Kim, S. W., Dharmaraj, N., Kim, K. W., and Kim, H. Y. (2007) Novel mechanism to improve toughness of the hydroxyapatite bioceramics using high-frequency induction heat sintering, *J. Mater. Process. Technol.*, 187–188, 417–420.

248. Landi, E., Tampieri, A., Celotti, G., and Sprio, S. (2000) Densification behaviour and mechanisms of synthetic hydroxyapatites, *J. Eur. Ceram. Soc.*, **20**, 2377–2387.

249. Laasri, S., Taha, M., Laghzizil, A., Hlil, E. K., and Chevalier, J. (2010) The affect of densification and dehydroxylation on the mechanical properties of stoichiometric hydroxyapatite bioceramics, *Mater. Res. Bull.*, **45**, 1433–1437.

250. Ruys, A. J., Wei, M., Sorrell, C. C., Dickson, M. R., Brandwood, A., and Milthorpe, B. K. (1995) Sintering effect on the strength of hydroxyapatite, *Biomaterials*, **16**, 409–415.

251. van Landuyt, P., Li, F., Keustermans, J. P., Streydio, J. M., Delannay, F., and Munting, E. (1995) The influence of high sintering temperatures on the mechanical properties of hydroxylapatite, *J. Mater. Sci. Mater. Med.*, **6**, 8–13.

252. Haberko, K., Bućko, M. M., Brzezińska-Miecznik, J., Haberko, M., Mozgawa, W., Panz, T., Pyda, A., and Zarebski, J. (2006) Natural hydroxyapatite — its behaviour during heat treatment, *J. Eur. Ceram. Soc.*, **26**, 537–542.

253. Haberko, K., Bućko, M. M., Mozgawa, W., Pyda, A., Brzezińska-Miecznik, J., and Carpentier, J. (2009) Behaviour of bone origin hydroxyapatite at elevated temperatures and in O_2 and CO_2 atmospheres, *Ceram. Int.*, **35**, 2537–2540.

254. Janus, A. M., Faryna, M., Haberko, K., Rakowska, A., and Panz, T. (2008) Chemical and microstructural characterization of natural hydroxyapatite derived from pig bones, *Mikrochim. Acta*, **161**, 349–353.

255. Bahrololoom, M. E., Javidi, M., Javadpour, S., and Ma, J. (2009) Characterisation of natural hydroxyapatite extracted from bovine cortical bone ash, *J. Ceram. Process. Res.*, **10**, 129–138.

256. Mostafa, N. Y. (2005) Characterization, thermal stability and sintering of hydroxyapatite powders prepared by different routes, *Mater. Chem. Phys.*, **94**, 333–341.

257. Suchanek, W., Yashima, M., Kakihana, M., and Yoshimura, M. (1997) Hydroxyapatite ceramics with selected sintering additives, *Biomaterials*, **18**, 923–933.

258. Kalita, S. J., Bose, S., Bandyopadhyay, A., and Hosick, H. L. (2003) Oxide based sintering additives for HAp ceramics, *Ceram. Transact.*, **147**, 63–72.

259. Kalita, S. J., Bose, S., Hosick, H. L., and Bandyopadhyay, A. (2004) $CaO-P_2O_5-Na_2O$-based sintering additives for hydroxyapatite (HAp) ceramics, *Biomaterials*, **25**, 2331–2339.

260. Safronova, T. V., Putlyaev, V. I., Shekhirev, M. A., Tretyakov, Y. D., Kuznetsov, A. V., and Belyakov, A. V. (2009) Densification additives for hydroxyapatite ceramics, *J. Eur. Ceram. Soc.*, **29**, 1925–1932.

261. Muralithran, G., and Ramesh, S. (2000) Effects of sintering temperature on the properties of hydroxyapatite, *Ceram. Int.*, **26**, 221–230.

262. Bernache-Assollant, D., Ababou, A., Champion, E., and Heughebaert, M. (2003) Sintering of calcium phosphate hydroxyapatite $Ca_{10}(PO_4)_6(OH)_2$ I. Calcination and particle growth, *J. Eur. Ceram. Soc.*, **23**, 229–241.

263. Chen, S., Wang, W., Kono, H., Sassa, K., and Asai, S. (2010) Abnormal grain growth of hydroxyapatite ceramic sintered in a high magnetic field, *J. Cryst. Growth*, **312**, 323–326.

264. Ramesh, S., Tan, C. Y., Bhaduri, S. B., Teng, W. D., and Sopyan, I. (2008) Densification behaviour of nanocrystalline hydroxyapatite bioceramics, *J. Mater. Process. Technol.*, **206**, 221–230.

265. Wang, J., Shaw, L. L. (2010) Grain-size dependence of the hardness of submicrometer and nanometer hydroxyapatite, *J. Am. Ceram. Soc.*, **93**, 601–604.

266. Kobayashi, S., Kawai, W., and Wakayama, S. (2006) The effect of pressure during sintering on the strength and the fracture toughness of hydroxyapatite ceramics, *J. Mater. Sci. Mater. Med.*, **17**, 1089–1093.

267. Kasuga, T., Ota, Y., Tsuji, K., and Abe, Y. (1996) Preparation of high-strength calcium phosphate ceramics with low modulus of elasticity containing β-$Ca(PO_3)_2$ fibers, *J. Am. Ceram. Soc.*, **79**, 1821–1824.

268. Suchanek, W. L., and Yoshimura, M. (1998) Preparation of fibrous, porous hydroxyapatite ceramics from hydroxyapatite whiskers, *J. Am. Ceram. Soc.*, **81**, 765–767.

269. Kim, Y., Kim, S. R., Song, H., and Yoon, H. (2005) Preparation of porous hydroxyapatite/TCP composite block using a hydrothermal hot pressing method, *Mater. Sci. Forum*, 486–487, 117–120.

270. Li, J. G., and Hashida, T. (2007) Preparation of hydroxyapatite ceramics by hydrothermal hot-pressing method at 300°C, *J. Mater. Sci.*, **42**, 5013–5019.

271. Li, J. G., and Hashida, T. (2006) *In situ* formation of hydroxyapatite-whisker ceramics by hydrothermal hot-pressing method, *J. Am. Ceram. Soc.*, **89**, 3544–3546.

272. Halouani, R., Bernache-Assolant, D., Champion, E., and Ababou, A. (1994) Microstructure and related mechanical properties of hot pressed hydroxyapatite ceramics, *J. Mater. Sci. Mater. Med.*, **5**, 563–568.

273. Nakahira, A., Murakami, T., Onoki, T., Hashida, T., and Hosoi, K. (2005) Fabrication of porous hydroxyapatite using hydrothermal hot pressing and post-sintering, *J. Am. Ceram. Soc.*, **88**, 1334–1336.

274. Auger, M. A., Savoini, B., Muñoz, A., Leguey, T., Monge, M. A., Pareja, R., and Victoria, J. (2009) Mechanical characteristics of porous hydroxyapatite/oxide composites produced by post-sintering hot isostatic pressing, *Ceram. Int.*, **35**, 2373–2380.

275. Fang, Y., Agrawal, D. K., Roy, D. M., and Roy, R. (1994) Microwave sintering of hydroxyapatite ceramics, *J. Mater. Res.*, **9**, 180–187.

276. Yang, Y., Ong, J. L., and Tian, J. (2002) Rapid sintering of hydroxyapatite by microwave processing, *J. Mater. Sci. Lett.*, **21**, 67–69.

277. Nath, S., Basu, B., and Sinha, A. (2006) A comparative study of conventional sintering with microwave sintering of hydroxyapatite synthesized by chemical route, *Trends Biomater. Artif. Organs*, **19**, 93–98.

278. Ramesh, S., Tan, C. Y., Bhaduri, S. B., and Teng, W. D. (2007) Rapid densification of nanocrystalline hydroxyapatite for biomedical applications, *Ceram. Int.*, **33**, 1363–1367.

279. Silva, C. C., Graça, M. P. F., Sombra, A. S. B., and Valente, M. A. (2009) Structural and electrical study of calcium phosphate obtained by a microwave radiation assisted procedure, *Phys. Rev. B Condens. Matter*, **404**, 1503–1508.

280. Chanda, A., Dasgupta, S., Bose, S., Bandyopadhyay, A. Microwave sintering of calcium phosphate ceramics. Mater. Sci. Eng. C 2009, 29, 1144-1149.

281. Veljović, D., Zalite, I., Palcevskis, E., Smiciklas, I., Petrović, R., and Janaćković, D. (2010) Microwave sintering of fine grained HAP and HAP/TCP bioceramics, *Ceram. Int.*, **36**, 595–603.

282. Kalita, S. J., and Verma, S. (2010) Nanocrystalline hydroxyapatite bioceramic using microwave radiation: synthesis and characterization, *Mater. Sci. Eng. C*, **30**, 295–303.

283. Veljović, D., Palcevskis, E., Dindune, A., Putić, S., Balać, I., Petrović, R., and Janaćković, D. (2010) Microwave sintering improves the mechanical properties of biphasic calcium phosphates from hydroxyapatite microspheres produced from hydrothermal processing, *J. Mater. Sci.*, **45**, 3175–3183.

284. Gu, Y. W., Loh, N. H., Khor, K. A., Tor, S. B., and Cheang, P. (2002) Spark plasma sintering of hydroxyapatite powders, *Biomaterials*, **23**, 37–43.

285. Guo, X., Xiao, P., Liu, J., and Shen, Z. (2005) Fabrication of nanostructured hydroxyapatite via hydrothermal synthesis and spark plasma sintering, *J. Am. Ceram. Soc.*, **88**, 1026–1029.

286. Drouet, C., Largeot, C., Raimbeaux, G., Estournès, C., Dechambre, G., Combes, C., and Rey, C. (2006) Bioceramics: spark plasma sintering (SPS) of calcium phosphates, *Adv. Sci. Tech.*, **49**, 45–50.

287. Li, H., Khor, K. A., Chow, V., and Cheang, P. (2007) Nanostructural characteristics, mechanical properties and osteoblast response of spark plasma sintered hydroxyapatite, *J. Biomed. Mater. Res. A*, **82A**, 296–303.

288. Nakamura, T., Fukuhara, T., and Izui, H. (2006) Mechanical properties of hydroxyapatites sintered by spark plasma sintering, *Ceram. Transact.*, **194**, 265–272.

289. Kawagoe, D., Ioku, K., Fujimori, H., and Goto, S. (2004) Transparent β-tricalcium phosphate ceramics prepared by spark plasma sintering, *J. Ceram. Soc. Jpn.*, **112**, 462–463.

290. Zhang, F., Lin, K., Chang, J., Lu, J., and Ning, C. (2008) Spark plasma sintering of macroporous calcium phosphate scaffolds from nanocrystalline powders, *J. Eur. Ceram. Soc.*, **28**, 539–545.

291. Grossin, D., Rollin-Martinet, S., Estournès, C., Rossignol, F., Champion, E., Combes, C., Rey, C., Geoffroy, C., and Drouet, C. (2010) Biomimetic apatite sintered at very low temperature by spark plasma sintering: physico-chemistry and microstructure aspects, *Acta Biomater.*, **6**, 577–585.

292. Chesnaud, A., Bogicevic, C., Karolak, F., Estournès, C., and Dezanneau, G. (2007) Preparation of transparent oxyapatite ceramics by combined use of freeze-drying and spark-plasma sintering, *Chem. Comm.*, 1550–1552.

293. Ishihara, S., Matsumoto, T., Onoki, T., Sohmura, T., and Nakahira, A. (2009) New concept bioceramics composed of octacalcium phosphate (OCP) and dicarboxylic acid-intercalated OCP via hydrothermal hot-pressing, *Mater. Sci. Eng. C*, **29**, 1885–1888.

294. Yanagisawa, K., Kim, J. H., Sakata, C., Onda, A., Sasabe, E., Yamamoto, T., Matamoros-Veloza, Z., and Rendón-Angeles, J. C. (2010) Hydrothermal sintering under mild temperature conditions: preparation of calcium-deficient hydroxyapatite compacts. Zeitschrift fur Naturforschung - Section B, *J. Chem. Sci.*, **65**, 1038–1044.

295. Gross, K. A., and Berndt C. C. (2002) Biomedical application of apatites. In: Phosphates: geochemical, geobiological and materials importance, *Series: Reviews in Mineralogy and Geochemistry*, vol. 48, (ed. Hughes, J. M., Kohn, M., and Rakovan, J.), Mineralogical Society of America, Washington, D.C., pp. 631–672.

296. LeGeros, R. Z., and LeGeros, J. P. (1993) Dense hydroxyapatite, in, *An Introduction to Bioceramics*, (ed. Hench, L. L., and Wilson, J.), World Scientific, London, pp. 139–180.

297. Evans, J. R. G. (2008) Seventy ways to make ceramics, *J. Eur. Ceram. Soc.*, **28**, 1421–1432.

298. Black, J. (2005) Biological Performance of Materials: Fundamentals of Biocompatibility, 4th edn., CRC Press, Boca Raton, p. 520.

299. Benaqqa, C., Chevalier, J., Saädaoui, M., and Fantozzi, G. (2005) Slow crack growth behaviour of hydroxyapatite ceramics, *Biomaterials*, **26**, 6106–6112.

300. Linhart, W., Briem, D., Amling, M., Rueger, J. M., and Windolf, J. (2004) Mechanical failure of porous hydroxyapatite ceramics 7.5 years after implantation in the proximal tibial, *Unfallchirurg.*, **107**, 154–157.

301. In materials science, fracture toughness is a property, which describes the ability of a material containing a crack to resist fracture and is one of the most important properties of any material for virtually all design applications.

302. Ramesh, S., Tan, C. Y., Sopyan, I., Hamdi, M., and Teng, W. D. (2007) Consolidation of nanocrystalline hydroxyapatite powder, *Sci. Technol. Adv. Mater.*, **8**, 124–130.

303. The nature of flaws in most ceramics is statistical. As such, the strength of ceramics is not one specific value, but a distribution of strengths. The Weibull modulus is a measure of the distribution of flaws, usually for a brittle material. The modulus is a dimensionless number corresponding to the variability in measured strength and reflects the distribution of flaws in the material.

304. Suzuki, S., Sakamura, M., Ichiyanagi, M., and Ozawa, M. (2004) Internal friction of hydroxyapatite and fluorapatite, *Ceram. Int.*, **30**, 625–627.

305. Strength can be defined as the maximum tensile stresses that a material can endure before fracture occurs. It might also be defined as force per unit area required to initiate and propagate a crack to the fracture point [194].

306. In materials science, fatigue is the progressive and localized structural damage that occurs when a material is subjected to cyclic loading.

307. le Huec, J. C., Schaeverbeke, T., Clement, D., Faber, J., and le Rebeller, A. (1995) Influence of porosity on the mechanical resistance of hydroxyapatite ceramics under compressive stress, *Biomaterials*, **16**, 113–118.

308. Hsu, Y. H., Turner, I. G., and Miles, A. W. (2009) Mechanical properties of three different compositions of calcium phosphate bioceramic following immersion in Ringer's solution and distilled water, *J. Mater. Sci. Mater. Med.*, **20**, 2367–2374.

309. The elastic modulus is the ratio of stress to strain below the proportional limit. It represents the magnitude of the interatomic stiffness of a material within the elastic range when tensile or compressive forces are applied. It is also an indication of the amount of reversible deformation that will occur in a structure when a load is applied [194].

310. de Aza, P. N., de Aza, A. H, and de Aza, S. (2005) Crystalline bioceramic materials, *Bol. Soc. Esp. Ceram. V.*, **44**, 135–145.

311. Ching, W. Y., Rulis, P., and Misra, A. (2009) *Ab initio* elastic properties and tensile strength of crystalline hydroxyapatite, *Acta Biomater.*, **5**, 3067–3075.

312. Hardness of a material usually is considered as a measure of the resistance to permanent indentation. A standardized force or weight is applied to an indenter pressing into the surface of a material to create a symmetrical shaped indentation [194].

313. Poisson's ratio (ν), named after Siméon Poisson, is the ratio, when a sample object is stretched, of the contraction or transverse strain

(perpendicular to the applied load), to the extension or axial strain (in the direction of the applied load).

314. Burger, E. L., and Patel, V. (2007) Calcium phosphates as bone graft extenders, *Orthopedics*, **30**, 939–942.

315. Rodriguez-Lorenzo, L. M., Vallet-Regí, M., Ferreira, J. M. F., Ginebra, M. P., Aparicio, C., and Planell, J. (2002) A hydroxyapatite ceramic bodies with tailored mechanical properties for different applications, *J. Biomed. Mater. Res.*, **60**, 159–166.

316. Dorozhkin, S. V. (2009) Calcium orthophosphate-based biocomposites and hybrid biomaterials, *J. Mater. Sci.*, **44**, 2343–2387.

317. Bouslama, N., Ben Ayed, F., and Bouaziz, J. (2009) Sintering and mechanical properties of tricalcium phosphate–fluorapatite composites, *Ceram. Int.*, **35**, 1909–1917.

318. Suchanek, W., Yashima, M., Kakihana, M., and Yoshimura, M. (1996) Processing and mechanical properties of hydroxyapatite reinforced with hydroxyapatite whiskers, *Biomaterials*, **17**, 1715–1723.

319. Suchanek, W., Yashima, M., Kakihana, M., and Yoshimura, M. (1997) Hydroxyapatite/hydroxyapatite-whisker composites without sintering additives: Mechanical properties and microstructural evolution, *J. Am. Ceram. Soc.*, **80**, 2805–2813.

320. Simsek, D., Ciftcioglu, R., Guden, M., Ciftcioglu, M., and Harsa, S. (2004) Mechanical properties of hydroxyapatite composites reinforced with hydroxyapatite whiskers, *Key Eng. Mater.*, 264–268, 1985–1988.

321. Bose, S., Banerjee, A., Dasgupta, S., and Bandyopadhyay, A. (2009) Synthesis, processing, mechanical, and biological property characterization of hydroxyapatite whisker-reinforced hydroxyapatite composites, *J. Am. Ceram. Soc.*, **92**, 323–330.

322. Shiota, T., Shibata, M., Yasuda, K., and Matsuo, Y. (2009) Influence of β-tricalcium phosphate dispersion on mechanical properties of hydroxyapatite ceramics, *J. Ceram. Soc. Jpn.*, **116**, 1002–1005.

323. Dorozhkin, S. V., and Ajaal, T. (2009) Toughening of porous bioceramic scaffolds by bioresorbable polymeric coatings, *Proc. Inst. Mech. Eng. H*, **223**, 459–470.

324. Dorozhkin, S. V., and Ajaal, T. (2009) Strengthening of dense bioceramic samples using bioresorbable polymers — a statistical approach, *J. Biomim. Biomater. Tissue. Eng.*, **4**, 27–39.

325. He, L. H., Standard, O. C., Huang, T. T., Latella, B. A., and Swain, M. V. (2008) Mechanical behaviour of porous hydroxyapatite, *Acta Biomater.*, **4**, 577–586.

326. Nagai, M., and Nishino, T. (1988) Surface conduction of porous hydroxyapatite ceramics at elevated temperatures, *Solid State Ionics*, 28–30, 1456–1461.

327. Yamashita, K., Owada, H., Umegaki, T., Kanazawa, T., and Futagamu, T. (1988) Ionic conduction in apatite solid solutions, *Solid State Ionics*, 28–30, 660–663.

328. Mahabole, M. P., Aiyer, R. C., Ramakrishna, C. V., Sreedhar, B., and Khairnar, R. S. (2005) Synthesis, characterization and gas sensing property of hydroxyapatite ceramic, *Bull. Mater. Sci.*, **28**, 535–545.

329. Fanovich, M. A., Castro, M. S., and Lopez, J. M. P. (1999) Analysis of the microstructural evolution in hydroxyapatite ceramics by electrical characterisation, *Ceram. Int.*, **25**, 517–522.

330. Valdes, J. J. P., Rodriguez, A. V., and Carrio, J. G. (1995) Dielectric properties and structure of hydroxyapatite ceramics sintered by different conditions, *J. Mater. Res.*, **10**, 2174–2177.

331. Bensaoud, A., Bouhaouss, A., and Ferhat, M. (2001) Electrical properties in compressed poorly crystalline apatite, *J. Solid State Electrochem.*, **5**, 362–365.

332. Tanaka, Y., Nakamura, M., Nagai, A., Toyama, T., and Yamashita, K. (2009) Ionic conduction mechanism in Ca-deficient hydroxyapatite whiskers, *Mater. Sci. Eng. B*, **161**, 115–119.

333. Tanaka, Y., Takata, S., Shimoe, K., Nakamura, M., Nagai, A., Toyama, T., and Yamashita, K. (2008) Conduction properties of non-stoichiometric hydroxyapatite whiskers for biomedical use, *J. Ceram. Soc. Jpn.*, **116**, 815–821.

334. Nakamura, S., Takeda, H., and Yamashita, K. (2001) Proton transport polarization and depolarization of hydroxyapatite ceramics, *J. Appl. Phys.*, **89**, 5386–5392.

335. Gittings, J. P., Bowen, C. R., Turner, I. G., Baxter, F. R., and Chaudhuri, J. B. (2008) Polarisation behaviour of calcium phosphate based ceramics, *Mater. Sci. Forum*, 587–588, 91–95.

336. Itoh, S., Nakamura, S., Kobayashi, T., Shinomiya, K., Yamashita, K., and Itoh, S. (2006) Effect of electrical polarization of hydroxyapatite ceramics on new bone formation, *Calcif. Tissue Int.*, **78**, 133–142.

337. Iwasaki, T., Tanaka, Y., Nakamura, M., Nagai, A., Hashimoto, K., Toda, Y., Katayama, K., and Yamashita, K. (2008) Rate of bonelike apatite formation accelerated on polarized porous hydroxyapatite, *J. Am. Ceram. Soc.*, **91**, 3943–3949.

338. Itoh, S., Nakamura, S., Kobayashi, T., Shinomiya, K., and Yamashita, K. (2006) Enhanced bone ingrowth into hydroxyapatite with interconnected pores by electrical polarization, *Biomaterials*, **27**, 5572–5579.

339. Kobayashi, T., Itoh, S., Nakamura, S., Nakamura, M., Shinomiya, K., and Yamashita, K. (2007) Enhanced bone bonding of hydroxyapatite-coated titanium implants by electrical polarization, *J. Biomed. Mater. Res. A*, **82A**, 145–151.

340. Bodhak, S., Bose, S., and Bandyopadhyay, A. (2009) Role of surface charge and wettability on early stage mineralization and bone cell-materials interactions of polarized hydroxyapatite, *Acta Biomater.*, **5**, 2178–2188.

341. Yamashita, K., Oikawa, N., and Umegaki, T. (1996) Acceleration and deceleration of bone-like crystal growth on ceramic hydroxyapatite by electric poling, *Chem. Mater.*, **8**, 2697–2700.

342. Teng, N. C., Nakamura, S., Takagi, Y., Yamashita, Y., Ohgaki, M., and Yamashita, K. (2001) A new approach to enhancement of bone formation by electrically polarized hydroxyapatite, *J. Dental Res.*, **80**, 1925–1929.

343. Kobayashi, T., Nakamura, S., and Yamashita, K. (2001) Enhanced osteobonding by negative surface charges of electrically polarized hydroxyapatite, *J. Biomed. Mater. Res.*, **57**, 477–484.

344. Park, Y. J., Hwang, K. S., Song, J. E., Ong, J. L., and Rawls, H. R. (2002) Growth of calcium phosphate on poling treated ferroelectric $BaTiO_3$ ceramics, *Biomaterials*, **23**, 3859–3864.

345. Hwang, K. S., Song, J. E., Jo, J. W., Yang, H. S., Park, Y. J., Ong, J. L., and Rawls, H. R. (2002) Effect of poling conditions on growth of calcium phosphate crystal in ferroelectric $BaTiO_3$ ceramics, *J. Mater. Sci. Mater. Med.*, **13**, 133–138.

346. Yamashita, K. (2003) Enhanced bioactivity of electrically poled hydroxyapatita ceramics and coatings, *Mater. Sci. Forum*, 426–432, 3237–3242.

347. Nakamura, S., Kobayashi, T., and Yamashita, K. (2005) Highly orientated calcification in newly formed bones on negatively charged hydroxyapatite electrets, *Key Eng. Mater.*, 284–286, 897–900.

348. Kato, R., Nakamura, S., Katayama, K., and Yamashita, K. (2005) Electrical polarization of plasma-spray-hydroxyapatite coatings for improvement of osteoconduction of implants, *J. Biomed. Mater. Res. A*, **74A**, 652–658.

349. Nakamura, S., Kobayashi, T., Nakamura, M., Itoh, S., and Yamashita, K. (2010) Electrostatic surface charge acceleration of bone ingrowth of porous hydroxyapatite/β-tricalcium phosphate ceramics, *J. Biomed. Mater. Res. A*, **92A**, 267–275.

350. Itoh, S., Nakamura, S., Nakamura, M., Shinomiya, K., and Yamashita, K. (2006) Enhanced bone regeneration by electrical polarization of hydroxyapatite, *Artif. Organs*, **30**, 863–869.

351. Nakamura, M., Nagai, A., Ohashi, N., Tanaka, Y., Sekilima, Y., and Nakamura, S. (2008) Regulation of osteoblast-like cell behaviors on hydroxyapatite by electrical polarization, *Key Eng. Mater.*, 361–363, 1055–1058.

352. Nakamura, M., Nagai, A., Tanaka, Y., Sekilima, Y., and Yamashita, K. (2010) Polarized hydroxyapatite promotes spread and motility of osteoblastic cells, *J. Biomed. Mater. Res. A*, **92A**, 783–790.

353. Nakamura, S., Kobayashi, T., and Yamashita, K. (2002) Extended bioactivity in the proximity of hydroxyapatite ceramic surfaces induced by polarization charges, *J. Biomed. Mater. Res.*, **61**, 593–599.

354. Wang, W., Itoh, S., Tanaka, Y., Nagai, A., and Yamashita, K. (2009) Comparison of enhancement of bone ingrowth into hydroxyapatite ceramics with highly and poorly interconnected pores by electrical polarization, *Acta Biomater.*, **5**, 3132–3140.

355. Nakamura, M., Niwa, K., Nakamura, S., Sekijima, Y., and Yamashita, K. (2007) Interaction of a blood coagulation factor on electrically polarized hydroxyapatite surfaces, *J. Biomed. Mater. Res. B (Appl. Biomater.)*, **82B**, 29–36.

356. Nagai, M., Shibuya, Y., Nishino, T., Saeki, T., Owada, H., Yamashita, K., and Umegaki, T. (1991) Electrical conductivity of calcium phosphate ceramics with various Ca/P ratios, *J. Mater. Sci.*, **26**, 2949–2953.

357. Laghzizil, A., Elherch, N., Bouhaouss, A., Lorente, G., Coradin, T., and Livage, J. (2001) Electrical behavior of hydroxyapatites $M_{10}(PO_4)_6(OH)_2$ (M = Ca, Pb, Ba), *Mater. Res. Bull.*, **36**, 953–962.

358. Louati, B., Guidara, K., and Gargouri, M. (2009) Dielectric and ac ionic conductivity investigations in the monetite, *J. Alloys Compd.*, **472**, 347–351.

359. Gittings, J. P., Bowen, C. R., Dent, A. C., Turner, I. G., Baxter, F. R., and Chaudhuri, J. B. (2009) Electrical characterization of hydroxyapatite-based bioceramics, *Acta Biomater.*, **5**, 743–754.

360. Tofail, S. A. M., Baldisserri, C., Haverty, D., McMonagle, J. B., and Erhart, J. (2009) Pyroelectric surface charge in hydroxyapatite ceramics, *J. Appl. Phys.*, **106**, 106104.

361. Ioku, K. (2010) Tailored bioceramics of calcium phosphates for regenerative medicine, *J. Ceram. Soc. Jpn.*, **118**, 775–783.

362. Fang, Y., Agrawal, D. K., Roy, D. M., and Roy, R. (1995) Fabrication of transparent hydroxyapatite ceramics by ambient-pressure sintering, *Mater. Lett.*, **23**, 147–151.

363. Kotobuki, N., Kawagoe, D., Fujimori, H., Goto, S., Loku, K., and Ohgushi, H. (2004) *In vitro* osteogenic activity of rat bone marrow derived mesenchymal stem cells cultured on transparent hydroxyapatite ceramics, *Key Eng. Mater.*, 254–256, 1055–1058.

364. Kotobuki, N., Ioku, K., Kawagoe, D., Nomura, D., Fujimori, H., Goto, S., and Ohgushi, H. (2005) *In vitro* osteogenic activity of rat mesenchymal cells cultured on transparent β-tricalcium phosphate ceramics, *Key Eng. Mater.*, 284–286, 663–666.

365. Varma, H., Vijayan, S. P., and Babu, S. S. (2002) Transparent hydroxyapatite ceramics through gel-casting and low-temperature sintering, *J. Am. Ceram. Soc.*, **85**, 493–495.

366. Watanabe, Y., Ikoma, T., Monkawa, A., Suetsugu, Y., Yamada, H., Tanaka, J., and Moriyoshi, Y. (2005) Fabrication of transparent hydroxyapatite sintered body with high crystal orientation by pulse electric current sintering, *J. Am. Ceram. Soc.*, **88**, 243–245.

367. Kotobuki, N., Ioku, K., Kawagoe, D., Fujimori, H., Goto, S., and Ohgushi, H. (2005) Observation of osteogenic differentiation cascade of living mesenchymal stem cells on transparent hydroxyapatite ceramics, *Biomaterials*, **26**, 779–785.

368. John, A., Varma, H. K., Vijayan, S., Bernhardt, A., Lode, A., Vogel, A., Burmeister, B., Hanke, T., Domaschke, H., and Gelinsky, M. (2009) *In vitro* investigations of bone remodeling on a transparent hydroxyapatite ceramic, *Biomed. Mater.*, **4**, 015007 (9 pages).

369. Kotobuki, N., Kawagoe, D., Nomura, D., Katou, Y., Muraki, K., Fujimori, H., Goto, S., Ioku, K., and Ohgushi, H. (2006) Observation and quantitative analysis of rat bone marrow stromal cells cultured *in vitro* on newly formed transparent β-tricalcium phosphate, *J. Mater. Sci. Mater. Med.*, **17**, 33–41.

370. Wang, J., and Shaw, L. L. (2010) Transparent nanocrystalline hydroxyapatite by pressure-assisted sintering, *Scripta Mater.*, **63**, 593–596.

371. Kobune, M., Mineshige, A., Fujii, S., and Iida, H. (1997) Preparation of translucent hydroxyapatite ceramics by HIP and their physical properties, *J. Ceram. Soc. Jpn.*, **105**, 210–213.

372. Barralet, J. E., Fleming, G. J. P., Campion, C., Harris, J. J., and Wright, A. J. (2003) Formation of translucent hydroxyapatite ceramics by sintering in carbon dioxide atmospheres, *J. Mater. Sci.*, **38**, 3979–3993.

373. Chaudhry, A. A., Yan, H., Gong, K., Inam, F., Viola, G., Reece, M. J., Goodall, J. B. M., ur Rehman, I., McNeil-Watson, F. K., Corbett, J. C. W., Knowles, J. C., and Darr, J. A. (2011) High-strength nanograined and translucent hydroxyapatite monoliths via continuous hydrothermal synthesis and optimized spark plasma sintering, *Acta Biomater.*, **7**, 791–799.

374. Tancred, D. C., McCormack, B. A., and Carr, A. J. (1998) A synthetic bone implant macroscopically identical to cancellous bone, *Biomaterials*, **19**, 2303–2311.

375. Miao, X., and Sun, D. (2010) Graded/gradient porous biomaterials, *Materials*, **3**, 26–47.

376. Gauthier, O., Bouler, J. M., Weiss, P., Bosco, J., Daculsi, G., and Aguado, E. (1999) Kinetic study of bone ingrowth and ceramic resorption associated with the implantation of different injectable calcium-phosphate bone substitutes, *J. Biomed. Mater. Res.*, **47**, 28–35.

377. Hing, K. A., Best, S. M., and Bonfield, W. (1999) Characterization of porous hydroxyapatite, *J. Mater. Sci. Mater. Med.*, **10**, 135–145.

378. Carotenuto, G., Spagnuolo, G., Ambrosio, L., and Nicolais, L. (1999) Macroporous hydroxyapatite as alloplastic material for dental applications, *J. Mater. Sci. Mater. Med.*, **10**, 671–676.

379. Bucholz, R. W., Carlton, A., and Holmes, R. (1989) Interporous hydroxyapatite as a bone graft substitute in tibial plateau fractures, *Clin. Orthop.*, **240**, 53–62.

380. Cavagna, R., Daculsi, G., and Bouler, J. M. (1999) Macroporous calcium phosphate ceramic: a prospective study of 106 cases in lumbar spinal fusion, *J. Long Term Eff. Med. Implants*, **9**, 403–412.

381. Lu, J. X., Flautre, B., Anselme, K., Hardouin, P., Gallur, A., Descamps, M., and Thierry, B. (1999) Role of interconnections in porous bioceramics on bone recolonization *in vitro* and *in vivo*, *J. Mater. Sci. Mater. Med.*, **10**, 111–120.

382. Ayers, R. A., Simske, S. J., Nunes, C. R., and Wolford, L. M. (1998) Long-term bone ingrowth and residual microhardness of porous block hydroxyapatite implants in humans, *J. Oral Maxillof. Surg.*, **56**(5), 1297–1302.

383. Jones, A. C., Arns, C. H., Sheppard, A. P., Hutmacher, D. W., Milthorpe, B. K., and Knackstedt, M. A. (2007) Assessment of bone ingrowth into porous biomaterials using MICRO-CT, *Biomaterials*, **28**, 2491–2504.

384. Karageorgiou, V., and Kaplan, D. (2005) Porosity of 3D biomaterial scaffolds and osteogenesis, *Biomaterials*, **26**, 5474–5491.

385. Tamai, N., Myoui, A., Kudawara, I., Ueda, T., and Yoshikawa, H. (2010) Novel fully interconnected porous hydroxyapatite ceramic in surgical treatment of benign bone tumor, *J. Orthop. Sci.*, **15**, 560–568.

386. Ohgushi, H., Goldberg, V. M., and Caplan, A. I. (1989) Heterotopic osteogenesis in porous ceramics induced by marrow cells, *J. Orthop. Res.*, **7**, 568–578.

387. Cheung, H. S., and Haak, M. H. (1989) Growth of osteoblasts on porous calcium phosphate ceramic: an *in vitro* model for biocompatibility study, *Biomaterials*, **10**, 63–67.

388. Zyman, Z., Ivanov, I., Glushko, V., Dedukh, N., and Malyshkina, S. (1998) Inorganic phase composition of remineralisation in porous CaP ceramics, *Biomaterials*, **19**, 1269–1273.

389. Yoshikawa, T., Ohgushi, H., and Tamai, S. (1996) Immediate bone forming capability of prefabricated osteogenic hydroxyapatite, *J. Biomed. Mater. Res.*, **32**, 481–492.

390. Chang, B. S., Lee, C. K., Hong, K. S., Youn, H. J., Ryu, H. S., Chung, S. S., and Park, K. (2000) Osteoconduction at porous hydroxyapatite with various pore configurations, *Biomaterials*, **21**, 1291–1298.

391. Flautre, B., Descamps, M., Delecourt, C., Blary, M. C., and Hardouin, P. (2001) Porous HA ceramic for bone replacement: role of the pores and interconnections–experimental study in the rabbits, *J. Mater. Sci. Mater. Med.*, **12**, 679–682.

392. McAfee, P. C., Cunningham, B. W., Orbegoso, D. O., Sefter, J. C., Dmitriev, A. E., and Fedder, I. L. (2003) Analysis of porous ingrowth in intervertebral disc prostheses, *Spine*, **28**, 332–340.

393. Tamai, N., Myoui, A., Tomita, T., Nakase, T., Tanaka, J., Ochi, T., and Yoshikawa, H. (2002) Novel hydroxyapatite ceramics with an interconnective porous structure exhibit superior osteoconduction *in vivo*, *J. Biomed. Mater. Res.*, **59**, 110–117.

394. Mastrogiacomo, M., Scaglione, S., Martinetti, R., Dolcini, L., Beltrame, F., Cancedda, R., and Quarto, R. (2006) Role of scaffold internal structure on *in vivo* bone formation in macroporous calcium phosphate bioceramics, *Biomaterials*, **27**, 3230–3237.

395. Okamoto, M., Dohi, Y., Ohgushi, H., Shimaoka, H., Ikeuchi, M., Matsushima, A., Yonemasu, K., and Hosoi, H. (2006) Influence of the porosity of hydroxyapatite ceramics on *in vitro* and *in vivo* bone formation by cultured rat bone marrow stromal cells, *J. Mater. Sci. Mater. Med.*, **17**, 327–336.

396. Zhang, L., Hanagata, N., Maeda, M., Minowa, T., Ikoma, T., Fan, H., and Zhang, X. (2009) Porous hydroxyapatite and biphasic calcium phosphate ceramics promote ectopic osteoblast differentiation from mesenchymal stem cells, *Sci. Technol. Adv. Mater.*, **10**, 025003 (9 pages).

397. Omae, H., Mochizuki, Y., Yokoya, S., Adachi, N., and Ochi, M. (2006) Effects of interconnecting porous structure of hydroxyapatite ceramics on interface between grafted tendon and ceramics, *J. Biomed. Mater. Res. A*, **79A**, 329–337.

398. Yoshikawa, H., Tamai, N., Murase, T., and Myoui, A. (2009) Interconnected porous hydroxyapatite ceramics for bone tissue engineering, *J. R. Soc. Interface*, **6**, S341–S348.

399. Ishikawa, K. (2010) Bone substitute fabrication based on dissolution-precipitation reactions, *Materials*, **3**, 1138–1155.

400. Takagi, S., and Chow, L. C. (2001) Formation of macropores in calcium phosphate cement implants, *J. Biomed. Mater. Res.*, **12**, 135–139.

401. Walsh, D., and Tanaka, J. (2001) Preparation of a bone-like apatite foam cement, *J. Mater. Sci. Mater. Med.*, **12**, 339–344.

402. Chevalier, E., Chulia, D., Pouget, C., and Viana, M. (2008) Fabrication of porous substrates: a review of processes using pore forming agents in the biomaterial field, *J. Pharm. Sci.*, **97**, 1135–1154.

403. Komlev, V. S., and Barinov, S. M. (2002) Porous hydroxyapatite ceramics of bi-modal pore size distribution, *J. Mater. Sci. Mater. Med.*, **13**, 295–299.

404. Sepulveda, P., Binner, J. G., Rogero, S. O., Higa, O. Z., and Bressiani, J. C. (2000) Production of porous hydroxyapatite by the gel-casting of foams and cytotoxic evaluation, *J. Biomed. Mater. Res.*, **50**, 27–34.

405. Hsu, Y. H., Turner, I. G., and Miles, A. W. (2007) Mechanical characterization of dense calcium phosphate bioceramics with interconnected porosity, *J. Mater. Sci. Mater. Med.*, **18**, 2319–2329.

406. Zhang, H. G., and Zhu, Q. (2007) Preparation of porous hydroxyapatite with interconnected pore architecture, *J. Mater. Sci. Mater. Med.*, **18**, 1825–1829.

407. le Ray, A. M., Gautier, H., Bouler, J. M., Weiss, P., and Merle, C. (2010) A new technological procedure using sucrose as porogen compound to manufacture porous biphasic calcium phosphate ceramics of appropriate micro- and macrostructure, *Ceram. Int.*, **36**, 93–101.

408. Guda, T., Appleford, M., Oh, S., and Ong, J. L. (2008) A cellular perspective to bioceramic scaffolds for bone tissue engineering: the state of the art, *Curr. Top. Medicinal Chem.*, **8**, 290–299.

409. Ota, Y., Kasuga, T., and Abe, Y. (1997) Preparation and compressive strength behaviour of porous ceramics with β-$Ca_3(PO_3)_2$ fiber skeletons, *J. Am. Ceram. Soc.*, **80**, 225–231.

410. White, E., and Shors, E. C. (1986) Biomaterial aspects of Interpore-200 porous hydroxyapatite, *Dent. Clin. North Am.*, **30**, 49–67.

411. Descamps, M., Duhoo, T., Monchau, F., Lu, J., Hardouin, P., Hornez, J. C., and Leriche, A. (2008) Manufacture of macroporous β-tricalcium phosphate bioceramics, *J. Eur. Ceram. Soc.*, **28**, 149–157.

412. Engin, N. O., and Tas, A. C. (1999) Manufacture of macroporous calcium hydroxyapatite bioceramics, *J. Eur. Ceram. Soc.*, **19**, 2569–2572.

413. Liu, D. M. (1997) Fabrication of hydroxyapatite ceramic with controlled porosity, *J. Mater. Sci. Mater. Med.*, **8**, 227–232.

414. Rivera-Munoz, E., Diaz, J. R., Rodriguez, J. R., Brostow, W., and Castano, V. M. (2001) Hydroxyapatite spheres with controlled porosity for eye ball prosthesis: processing and characterization, *J. Mater. Sci. Mater. Med.*, **12**, 305–311.

415. Hosoi, K., Hashida, T., Takahashi, H., Yamasaki, N., and Korenaga, T. (1996) New processing technique for hydroxyapatite ceramics by the hydrothermal hot-pressing method, *J. Am. Ceram. Soc.*, **79**, 2771–2774.

416. Walsh, D., Hopwood, J. D., and Mann, S. (1994) Crystal tectonics: construction of reticulated calcium phosphate frameworks in bicontinuous reverse microemulsions, *Science*, **264**, 1576–1578.

417. Walsh, D., and Mann, S. (1996) Chemical synthesis of microskeletal calcium phosphate in bicontinuous microemulsions, *Chem. Mater.*, **8**, 1944–1953.

418. Charriere, E., Lemaitre, J., and Zysset, P. (2003) Hydroxyapatite cement scaffolds with controlled macroporosity: fabrication protocol and mechanical properties, *Biomaterials*, **24**, 809–817.

419. Zhao, K., Tang, Y. F., Qin, Y. S., and Wei, J. Q. (2011) Porous hydroxyapatite ceramics by ice templating: freezing characteristics and mechanical properties, *Ceram. Int.*, **37**, 635–639.

420. Eichenseer, C., Will, J., Rampf, M., Wend, S., and Greil, P. (2010) Biomorphous porous hydroxyapatite-ceramics from rattan (*Calamus Rotang*), *J. Mater. Sci. Mater. Med.*, **21**, 131–137.

421. Zhou, L., Wang, D., Huang, W., Yao, A., Kamitakahara, M., and Ioku, K. (2009) Preparation and characterization of periodic porous frame of hydroxyapatite, *J. Ceram. Soc. Jpn.*, **117**, 521–524.

422. Moroni, L., de Wijn, J. R., and van Blitterswijk, C. A. (2008) Integrating novel technologies to fabricate smart scaffolds, *J. Biomater. Sci. Polymer*, **19**, 543–572.

423. Aizawa, M., Scott Howell, F., Itatani, K., Yokogawa, Y., Nishizawa, K., Toriyama, M., and Kameyama, T. (2000) Fabrication of porous ceramics with well-controlled open pores by sintering of fibrous hydroxyapatite particles, *J. Ceram. Soc. Jpn.*, **108**, 249–253.

424. Kawata, M., Uchida, H., Itatani, K., Okada, I., Koda, S., and Aizawa, M. (2004) Development of porous ceramics with well-controlled porosities and pore sizes from apatite fibers and their evaluations, *J. Mater. Sci. Mater. Med.*, **15**, 817–823.

425. Li, S. H., de Wijn, J. R., Layrolle, P., and de Groot, K. (2003) Novel method to manufacture porous hydroxyapatite by dual-phase mixing, *J. Am. Ceram. Soc.*, **86**, 65–72.

426. Koh, Y. H., Kim, H. W., Kim, H. E., and Halloran, J. W. (2002) Fabrication of macrochannelled-hydroxyapatite bioceramic by a coextrusion process, *J. Am. Ceram. Soc.*, **85**, 2578–2580.

427. Nakahira, A., Tamai, M., Sakamoto, K., and Yamaguchi, S. (2000) Sintering and microstructure of porous hydroxyapatite, *J. Ceram. Soc. Jpn.*, **108**, 99–104.

428. de Sousa, F. C. G., and Evans, J. R. G. (2005) Tubular hydroxyapatite scaffolds, *Adv. Appl. Ceram.*, **104**, 30–34.

429. Kitamura, M., Ohtsuki, C., Ogata, S. I., Kamitakahara, M., Tanihara, M., and Miyazaki, T. (2005) Mesoporous calcium phosphate via post-treatment of α-TCP, *J. Am. Ceram. Soc.*, **88**, 822–826.

430. Walsh, D., Boanini, E., Tanaka, J., and Mann, S. (2005) Synthesis of tri-calcium phosphate sponges by interfacial deposition and thermal transformation of self-supporting calcium phosphate films, *J. Mater. Chem.*, **15**, 1043–1048.

431. Gonzalez-McQuire, R., Green, D., Walsh, D., Hall, S., Chane-Ching, J. Y., Oreffo, R. O. C., and Mann, S. (2005) Fabrication of hydroxyapatite sponges by dextran sulphate/amino acid templating, *Biomaterials*, **26**, 6652–6656.

432. Ioku, K., Kawachi, G., Nakahara, K., Ishida, E. H., Minagi, H., Okuda, T., Yonezawa, I., Kurosawa, H., and Ikeda, T. (2006) Porous granules of β-tricalcium phosphate composed of rod-shaped particles, *Key Eng. Mater.*, 309–311, 1059–1062.

433. Kitamura, M., Ohtsuki, C., Ogata, S., Kamitakahara, M., and Tanihara, M. (2004) Microstructure and bioresorbable properties of α-TCP ceramic

porous body fabricated by direct casting method, *Mater. Transact.*, **45**, 983–988.

434. Sohier, J., Daculsi, G., Sourice, S., de Groot, K., and Layrolle, P. (2010) Porous beta tricalcium phosphate scaffolds used as a BMP-2 delivery system for bone tissue engineering, *J. Biomed. Mater. Res. A*, **92A**, 1105–1114.

435. Stähli, C., Bohner, M., Bashoor-Zadeh, M., Doebelin, N., and Baroud, G. (2010) Aqueous impregnation of porous β-tricalcium phosphate scaffolds, *Acta Biomater.*, **6**, 2760–2772.

436. Ramay, H. R., and Zhang, M. (2003) Preparation of porous hydroxyapatite scaffolds by combination of the gel-casting and polymer sponge methods, *Biomaterials*, **24**, 3293–3302.

437. Potoczek, M., Zima, A., Paszkiewicz, Z., and Ślósarczyk, A. (2009) Manufacturing of highly porous calcium phosphate bioceramics via gel-casting using agarose, *Ceram. Int.*, **35**, 2249–2254.

438. Xu, S., Li, D., Lu, B., Tang, Y., Wang, C., and Wang, Z. (2007) Fabrication of a calcium phosphate scaffold with a three dimensional channel network and its application to perfusion culture of stem cells, *Rapid Prototyping J.*, **13**, 99–106.

439. Li, S. H., de Wijn J. R., Layrolle, P., and de Groot, K. (2002) Synthesis of macroporous hydroxyapatite scaffolds for bone tissue engineering, *J. Biomed. Mater. Res.*, **61**, 109–120.

440. Almirall, A., Larrecq, G., Delgado, J. A., Martínez, S., Planell, J. A., and Ginebra, M. P. (2004) Fabrication of low temperature macroporous hydroxyapatite scaffolds by foaming and hydrolysis of an α-TCP paste, *Biomaterials*, **25**, 3671–3680.

441. Kamitakahara, M., Ohtsuki, C., Kawachi, G., Wang, D., and Ioku, K. (2008) Preparation of hydroxyapatite porous ceramics with different porous structures using a hydrothermal treatment with different aqueous solutions, *J. Ceram. Soc. Jpn.*, **116**, 6–9.

442. Peña, J., Román, J., Cabañas, M. V., and Vallet-Regí, M. (2010) An alternative technique to shape scaffolds with hierarchical porosity at physiological temperature, *Acta Biomater.*, **6**, 1288–1296.

443. Marcassoli, P., Cabrini, M., Tirillò, J., Bartuli, C., Palmero, P., and Montanaro, L. (2010) Mechanical characterization of hydroxyapatite micro/macro-porous ceramics obtained by means of innovative gel-casting process, *Key Eng. Mater.*, 417–418, 565–568.

444. Nakamura, S., and Nakahira, A. (2008) Synthesis and evaluation of porous hydroxyapatite prepared by hydrothermal treatment and subsequent sintering method, *J. Ceram. Soc. Jpn.*, **116**, 42–45.

445. Song, H. Y., Islam, S., and Lee, B. T. (2008) A novel method to fabricate unidirectional porous hydroxyapatite body using ethanol bubbles in a viscous slurry, *J. Am. Ceram. Soc.*, **91**, 3125–3127.

446. Impens, S., Schelstraete, R., Luyten, J., Mullens, S., Thijs, I., van Humbeeck, J., and Schrooten, J. (2009) Production and characterisation of porous calcium phosphate structures with controllable hydroxyapatite/β-tricalcium phosphate ratios, *Adv. Appl. Ceram.*, **108**, 494–500.

447. Kawachi, G., Misumi, H., Fujimori, H., Goto, S., Ohtsuki, C., Kamitakahara, M., and Ioku, K. (2010) Fabrication of porous blocks of calcium phosphate through hydrothermal processing under glycine coexistence, *J. Ceram. Soc. Jpn.*, **118**, 559–563.

448. Ng, S., Guo, J., Ma, J., and Loo, S. C. J. (2010) Synthesis of high surface area mesostructured calcium phosphate particles, *Acta Biomater.*, **6**, 3772–3781.

449. Zhao, K., Tang, Y. F., Qin, Y. S., and Luo, D. F. (2011) Polymer template fabrication of porous hydroxyapatite scaffolds with interconnected spherical pores, *J. Eur. Ceram. Soc.*, **31**, 225–229.

450. Sakamoto, M., Nakasu, M., Matsumoto, T., and Okihana, H. (2007) Development of superporous hydroxyapatites and their examination with a culture of primary rat osteoblasts, *J. Biomed. Mater. Res. A*, **82A**, 238–242.

451. Sakamoto, M. (2010) Development and evaluation of superporous hydroxyapatite ceramics with triple pore structure as bone tissue scaffold, *J. Ceram. Soc. Jpn.*, **118**, 753–757.

452. von Doernberg, M. C., von Rechenberg, B., Bohner, M., Grünenfelder, S., van Lenthe, G. H., Müller, R., Gasser, B., Mathys, R., Baroud, G., and Auer, J. (2006) *In vivo* behavior of calcium phosphate scaffolds with four different pore sizes, *Biomaterials*, **27**, 5186–5198.

453. Mygind, T., Stiehler, M., Baatrup, A., Li, H., Zou, X., Flyvbjerg, A., Kassem, M., and Bunger, C. (2007) Mesenchymal stem cell ingrowth and differentiation on coralline hydroxyapatite scaffolds, *Biomaterials*, **28**, 1036–1047.

454. Tsuruga, E., Takita, H., Wakisaka, Y., and Kuboki, Y. (1997) Pore size of porous hydoxyapatite as the cell-substratum controls BMP-induced osteogenesis, *J. Biochem.*, **121**, 317–324.

455. Hing, K., Annaz, B., Saeed, S., Revell, P., and Buckland, T. (2005) Microporosity enhances bioactivity of synthetic bone graft substitutes, *J. Mater. Sci. Mater. Med.*, **16**, 467–475.

456. Woodard, J. R., Hilldore, A. J., Lan, S. K., Park, C. J., Morgan, A. W., Eurell, J. A. C., Clark, S. G., Wheeler, M. B., Jamison, R. D., and Wagoner,

J. A. J. (2007) The mechanical properties and osteoconductivity of hydroxyapatite bone scaffolds with multi-scale porosity, *Biomaterials*, **28**, 45–54.

457. Lan Levengood, S. K., Polak, S. J., Wheeler, M. B., Maki, A. J., Clark, S. G., Jamison, R. D., and Wagoner Johnson, A. J. (2010) Multiscale osteointegration as a new paradigm for the design of calcium phosphate scaffolds for bone regeneration, *Biomaterials*, **31**, 3552–3563.

458. Ruksudjarit, A., Pengpat, K., Rujijanagul, G., and Tunkasiri, T. (2008) The fabrication of nanoporous hydroxyapatite ceramics, *Adv. Mater. Res.*, 47–50, Part 2, 797–800.

459. LeGeros, R. Z. (2008) Calcium phosphate-based osteoinductive materials, *Chem. Rev.*, **108**, 4742–4753.

460. Prokopiev, O., and Sevostianov, I. (2006) Dependence of the mechanical properties of sintered hydroxyapatite on the sintering temperature, *Mater. Sci. Eng. A*, **431**, 218–227.

461. Daculsi, G., Jegoux, F., and Layrolle, P. (2009) The micro macroporous biphasic calcium phosphate concept for bone reconstruction and tissue engineering, in *Advanced Biomaterials: Fundamentals, Processing and Applications* (ed. Basu, B., Katti, D. S., and Kumar, A.), Wiley-American Ceramic Society, p. 768.

462. Tadic, D., Beckmann, F., Schwarz, K., and Epple, M. (2004) A novel method to produce hydroxylapatite objects with interconnecting porosity that avoids sintering, *Biomaterials*, **25**, 3335–3340.

463. Woyansky, J. S., Scott, C. E., and Minnear, W. P. (1992) Processing of porous ceramics, *Am. Ceram. Soc. Bull.*, **71**, 1674–1682.

464. Vaz, L., Lopes, A. B., and Almeida, M. (1999) Porosity control of hydroxyapatite implants, *J. Mater. Sci. Mater. Med.*, **10**, 239–242.

465. Yao, X., Tan, S., and Jiang, D. (2005) Improving the properties of porous hydroxyapatite ceramics by fabricating methods, *J. Mater. Sci.*, **40**, 4939–4942.

466. Zhang, J., Fujiwara, M., Xu, Q., Zhu, Y., Iwasa, M., and Jiang, D. (2008) Synthesis of mesoporous calcium phosphate using hybrid templates, *Micropor. Mesopor. Mater.*, **111**, 411–416.

467. Simske, S. J., Ayers R. A., and Bateman T. A. (1997) Porous materials for bone engineering, *Mater. Sci. Forum*, **250**, 151–182.

468. Wang, H., Zhai, L., Li, Y., and Shi, T. (2008) Preparation of irregular mesoporous hydroxyapatite, *Mater. Res. Bull.*, **43**, 1607–1614.

469. Cyster, L. A., Grant, D. M., Howdle, S. M., Rose, F. R. A. J., Irvine, D. J., Freeman, D., Scotchford, C.A., and Shakesheff, K. M. (2005) The influence

of dispersant concentration on the pore morphology of hydroxyapatite ceramics for bone tissue engineering, *Biomaterials*, **26**, 697–702.

470. Tian, J. (2001) Preparation of porous hydroxyapatite, *J. Mater. Sci.*, **36**, 3061–3066.

471. Rodriguez-Lorenzo, L. M., Vallet-Regí, M., and Ferreira, J. M. F. (2002) Fabrication of porous hydroxyapatite bodies by a new direct consolidation method: starch consolidation, *J. Biomed. Mater. Res.*, **60**, 232–240.

472. Lee, E. J., Koh, Y. H., Yoon, B. H., Kim, H. E., and Kim, H. W. (2007) Highly porous hydroxyapatite bioceramics with interconnected pore channels using camphene-based freeze casting, *Mater. Lett.*, **61**, 2270–2273.

473. Jones, J. R., and Hench, L. L. (2003) Regeneration of trabecular bone using porous ceramics, *Curr. Opin. Solid State Mater. Sci.*, **7**, 301–307.

474. Ota, T., Eitsuka, T., Yoshida, H., and Adachi, N. (2006) Porous apatite ceramics derived from woods, *Adv. Mater. Res.*, 11–12, 247–250.

475. Tas, A. C. (2008) Preparation of porous apatite granules from calcium phosphate cement, *J. Mater. Sci. Mater. Med.*, **19**, 2231–2239.

476. Fu, Q., Rahaman, M. N., Dogan, F., and Bal, B. S. (2008) Freeze casting of porous hydroxyapatite scaffolds. I. Processing and general microstructure, *J. Biomed. Mater. Res. B (Appl. Biomater.)*, **86B**, 125–135.

477. Tang, Y. J., Tang, Y. F., Lv, C. T., and Zhou, Z. H. (2008) Preparation of uniform porous hydroxyapatite biomaterials by a new method, *Appl. Surf. Sci.*, **254**, 5359–5362.

478. Fan, J., Lei, J., Yu, C., Tu, B., and Zhao, D. (2007) Hard-templating synthesis of a novel rod-like nanoporous calcium phosphate bioceramics and their capacity as antibiotic carriers, *Mater. Chem. Phys.*, **103**, 489–493.

479. Sopyan, I., Mel, M., Ramesh, S., and Khalid, K. A. (2007) Porous hydroxyapatite for artificial bone applications, *Sci. Technol. Adv. Mater.*, **8**, 116–123.

480. Rice, R. W. (1998) *Porosity of Ceramics*, Marcel Dekker, New York, p. 560.

481. Fu, Y. C., Ho, M. L., Wu, S. C., Hsieh, H. S., and Wang, C. K. (2008) Porous bioceramic bead prepared by calcium phosphate with sodium alginate gel and PE powder, *Mater. Sci. Eng. C*, **28**, 1149–1158.

482. Hsu, Y. H., Turner, I. G., and Miles, A. W. (2007) Fabrication of porous bioceramics with porosity gradients similar to the bimodal structure of cortical and cancellous bone, *J. Mater. Sci. Mater. Med.*, **18**, 2251–2256.

483. Munch, E., Franco, J., Deville, S., Hunger, P., Saiz, E., and Tomsia, A. P. (2008) Porous ceramic scaffolds with complex architectures, *JOM*, **60**, 54–59.

484. Macchetta, A., Turner, I. G., and Bowen, C. R. (2009) Fabrication of HA/TCP scaffolds with a graded and porous structure using a camphene-based freeze-casting method, *Acta Biomater.*, **5**, 1319–1327.

485. Deville, S., Saiz, E., and Tomsia, A. P. (2006) Freeze casting of hydroxyapatite scaffolds for bone tissue engineering, *Biomaterials*, **27**, 5480–5489.

486. Zuo, K. H., Zeng, Y. P., and Jiang, D. (2010) Effect of polyvinyl alcohol additive on the pore structure and morphology of the freeze-cast hydroxyapatite ceramics, *Mater. Sci. Eng. C*, **30**, 283–287.

487. Ginebra, M. P., Espanol, M., Montufar, E. B., Perez, R. A., and Mestres, G. (2010) New processing approaches in calcium phosphate cements and their applications in regenerative medicine, *Acta Biomater.*, **6**, 2863–2873.

488. Yun, H. S. (2010) Design of hierarchically porous materials for bone tissue regeneration, *Key Eng. Mater.*, **441**, 139–153.

489. Deisinger, U. (2010) Generating porous ceramic scaffolds: processing and properties, *Key Eng. Mater.*, **441**, 155–179.

490. Yan, X., Yu, C., Zhou, X., Tang, J., and Zhao, D. (2004) Highly ordered mesoporous bioactive glasses with superior *in vitro* bone-forming bioactivities, *Angew. Chem. Int. Ed. Engl.*, **43**, 5980–5984.

491. Izquierdo-Barba, I., Ruiz-González, L., Doadrio, J. C., González-Calbet, J. M., and Vallet-Regí, M. (2005) Tissue regeneration: a new property of mesoporous materials, *Solid State Sci.*, **7**, 983–989.

492. Cosijns, A., Vervaet, C., Luyten, J., Mullens, S., Siepmann, F., van Hoorebeke, L., Masschaele, B., Cnudde, V., and Remon, J. P. (2007) Porous hydroxyapatite tablets as carriers for low-dosed drugs, *Eur. J. Pharm. Biopharm.*, **67**, 498–506.

493. Uchida, A., Shinto, Y., Araki, N., and Ono, K. (1992) Slow release of anticancer drugs from porous calcium hydroxyapatite ceramic, *J. Orthop. Res.*, **10**, 440–445.

494. Shinto, Y., Uchida, A., Korkusuz, F., Araki, N., and Ono, K. (1992) Calcium hydroxyapatite ceramic used as a delivery system for antibiotics, *J. Bone. Joint. Surg. Br.*, **74**, 600–604.

495. Martin, R. B., Chapman, M. W., Sharkey, N. A., Zissimos, S. L., Bay, B., and Shors, E. C. (1993) Bone ingrowth and mechanical properties of

coralline hydroxyapatite 1 yr after implantation, *Biomaterials*, **14**, 341–348.

496. Kazakia, G. J., Nauman, E. A., Ebenstein, D. M., Halloran, B. P., and Keaveny, T. M. (2006) Effects of *in vitro* bone formation on the mechanical properties of a trabeculated hydroxyapatite bone substitute, *J. Biomed. Mater. Res. A*, **77A**, 688–699.

497. Hing, K. A., Best, S. M., Tanner, K. E., Bonfield, W., and Revell, P. A. (2004) Mediation of bone ingrowth in porous hydroxyapatite bone graft substitutes, *J. Biomed. Mater. Res. A*, **68A**, 187–200.

498. Vuola, J., Taurio, R., Goransson, H., and Asko-Seljavaara, S. (1998) Compressive strength of calcium carbonate and hydroxyapatite implants after bone marrow induced osteogenesis, *Biomaterials*, **19**, 223–227.

499. Yang, L., Ning, X., Chen, K., and Zhou, H. (2007) Preparation and properties of hydroxyapatite filters for microbial filtration, *Ceram. Int.*, **33**, 483–489.

500. Levitt, G. E., Crayton, P. H., Monroe, E. A., and Condrate, R. A. (1969) Forming methods for apatite prosthesis, *J. Biomed. Mater. Res.*, **3**, 683–685.

501. Tadic, D., and Epple, M. (2004) A thorough physicochemical characterization of 14 calcium phosphate-based bone substitution materials in comparison to natural bone, *Biomaterials*, **25**, 987–994.

502. Bodde, E. W. H., Wolke, J. G. C., Kowalski, R. S. Z., and Jansen, J. A. (2007) Bone regeneration of porous β-tricalcium phosphate (Conduit™ TCP) and of biphasic calcium phosphate ceramic (Biosel®) in trabecular defects in sheep, *J. Biomed. Mater. Res. A*, **82A**, 711–722.

503. le Guéhennec, L., Layrolle, P., and Daculsi, G. (2004) A review of bioceramics and fibrin sealant, *Eur. Cell Mater.*, **8**, 1–11.

504. LeGeros, R. Z. (2002) Properties of osteoconductive biomaterials: calcium phosphates, *Clin. Orthop. Rel. Res.*, **395**, 81–98.

505. Kashimura, H., Ogasawara, K., Kubo, Y., Yoshida, K., Sugawara, A., and Ogawa, A. (2010) A newly designed hydroxyapatite ceramic burr-hole button, *Vasc. Health Risk Manag.*, **6**, 105–108.

506. Jordan, D. R., Gilberg, S., and Bawazeer, A. (2004) Coralline hydroxyapatite orbital implant (Bio-Eye): experience with 158 patients, *Ophthal. Plast. Reconstr. Surg.*, **20**, 69–74.

507. Yoon, J. S., Lew, H., Kim, S. J., and Lee, S. Y. (2008) Exposure rate of hydroxyapatite orbital implants. A 15-year experience of 802 cases, *Ophthalmology*, **115**, 566–572.e2.

508. Chai, G. R., and Chen, M. (2010) Clinical effect of hydroxyapatite orbital implantation, *Int. J. Ophthal.*, **10**, 999–1000.

509. Tabatabaee, Z., Mazloumi, M., Rajabi, T. M., Khalilzadeh, O., Kassaee, A., Moghimi, S., Eftekhar, H., and Goldberg, R. A. (2011) Comparison of the exposure rate of wrapped hydroxyapatite (Bio-Eye) versus unwrapped porous polyethylene (Medpor) orbital implants in enucleated patients, *Ophthal. Plast. Reconstr. Surg.*, **27**, 114–118.

510. Yoon, J. S., Lew, H., Kim, S. J., and Lee, S. Y. (2008) Exposure rate of hydroxyapatite orbital implants a 15-year experience of 802 cases, *Ophthalmology*, **115**, 566–572.

511. Liao, H. F., Xiao, W., and Chen, Q. J. (2008) Ophthalmic applications of hydroxyapatite and its polymer composites, *J. Clin. Rehabil. Tiss. Eng. Res.*, **12**, 8905–8908.

512. Schnettler, R., Stahl, J. P., Alt, V., Pavlidis, T., Dingeldein, E., and Wenisch, S. (2004) Calcium phosphate-based bone substitutes, *Eur. J. Trauma*, **4**, 219–229.

513. Zyman, Z. Z., Glushko, V., Dedukh, N., Malyshkina, S., and Ashukina, N. (2008) Porous calcium phosphate ceramic granules and their behaviour in differently loaded areas of skeleton, *J. Mater. Sci. Mater. Med.*, **19**, 2197–2205.

514. Constantz, B. R., Ison, I. C., Fulmer, M. T., Poser, R. D., Smith, S. T., Vanwagoner, M., Ross, J., Goldstein, S. A., Jupiter, J. B., and Rosenthal, D. I. (1995) Skeletal repair by *in situ* formation of the mineral phase of bone, *Science*, **267**, 1796–1799.

515. Dorozhkin, S. V. (2008) Calcium orthophosphate cements for biomedical application, *J. Mater. Sci.*, **43**, 3028–3057.

516. Dorozhkin, S. V. (2009) Calcium orthophosphate cements and concretes, *Materials*, **2**, 221–291.

517. Chow, L. C. (2009) Next generation calcium phosphate-based biomaterials, *Dent. Mater. J.*, **28**, 1–10.

518. del Real, R. P., Wolke, J. G. C., Vallet-Regí, M., and Jansen, J. A. (2002) A new method to produce macropores in calcium phosphate cements, *Biomaterials*, **23**, 3673–3680.

519. Victor, S. P., and Kumar, T. S. S. (2008) Processing and properties of injectable porous apatitic cements, *J. Ceram. Soc. Jpn.*, **116**, 105–107.

520. Hesaraki, S., Nemati, R., and Nosoudi, N. (2009) Preparation and characterisation of porous calcium phosphate bone cement as antibiotic carrier, *Adv. Appl. Ceram.*, **108**, 231–240.

521. Weiss, D. D., Sachs, M. A., and Woodard, C. R. (2003) Calcium phosphate bone cements: a comprehensive review, *J. Long Term Eff. Med. Implants*, **13**, 41–47.

522. Ambard, A. J., and Mueninghoff, L. (2006) Calcium phosphate cement: review of mechanical and biological properties, *J. Prosthodont.*, **15**, 321–328.

523. Costantino, P. D., Friedman, C. D., Jones, K., Chow, L. C., Pelzer, H. J., and Sisson, G. A. (1991) Hydroxyapatite cement. I. Basic chemistry and histologic properties, *Arch. Otolaryngol. Head Neck Surg.*, **117**, 379–384.

524. Costantino, P. D., Friedman, C. D., Jones, K., Chow, L. C., Pelzer, H. J., and Sisson, G. A. (1991) Hydroxyapatite cement. II. Obliteration and reconstruction of the cat frontal sinus, *Arch. Otolaryngol. Head Neck Surg.*, **117**, 385–389.

525. Ong, J. L., and Chan, D. C. N. (1991) Hydroxyapatite and their use as coatings in dental implants: a review, *Crit. Rev. Biomed. Eng.*, **28**, 667–707.

526. de Groot, K., Wolke, J. G. C., and Jansen, J. A. (1998) Calcium phosphate coatings for medical implants, *Proc. Inst. Mech. Eng. Part H: J. Eng. Med.*, **212**, 137–147.

527. Campbell, A. A. (2003) Bioceramics for implant coatings, *Mater. Today*, **6**, 26–30.

528. Epinette, J. A. M. D., and Geesink, R. G. T. (1995) *Hydroxyapatite Coated Hip and Knee Arthroplasty*, Elsevier, Amsterdam, p. 394.

529. Willmann, G. (1999) Coating of implants with hydroxyapatite — material connections between bone and metal, *Adv. Eng. Mater.*, **1**, 95–105.

530. Schliephake, H., Scharnweber, D., Roesseler, S., Dard, M., Sewing, A., and Aref, A. (2006) Biomimetic calcium phosphate composite coating of dental implants, *Int. J. Oral Max. Impl.*, **21**, 738–746.

531. Kokubo, T., Kim, H. M., and Kawashita, M. (2003) Novel bioactive materials with different mechanical properties, *Biomaterials*, **24**, 2161–2175.

532. Habibovic, P., Li, J., van der Valk, C. M., Meijer, G., Layrolle, P., van Blitterswijk, C. A., and de Groot, K. (2005) Biological performance of uncoated and octacalcium phosphate-coated Ti6Al4V, *Biomaterials*, **26**, 23–36.

533. Hahn, B. D., Park, D. S., Choi, J. J., Ryu, J., Yoon, W. H., Kim, K. H., Park, C., and Kim, H. E. (2009) Dense nanostructured hydroxyapatite coating on titanium by aerosol deposition, *J. Am. Ceram. Soc.*, **92**, 683–687.

534. Gineste, L., Gineste, M., Ranz, X., Ellefterion, A., Guilhem, A., Rouquet, N., and Frayssinet, P. (1999) Degradation of hydroxylapatite, fluorapatite, and fluorhydroxyapatite coatings of dental implants in dogs, *J. Biomed. Mater. Res.*, **48**, 224–234.

535. Okada, M., and Furuzono, T. (2010) Hydroxyapatite nanocrystal coating on biodegradable microspheres, *Mater. Sci. Eng. B*, **173**, 199–203.

536. Xiao, Y., Gong, T., and Zhou, S. (2010) The functionalization of multi-walled carbon nanotubes by in situ deposition of hydroxyapatite, *Biomaterials*, **31**, 5182–5190.

537. Habibovic, P., Barrère, F., van Blitterswijk, C. A., de Groot, K., and Layrolle, P. (2002) Biomimetic hydroxyapatite coating on metal implants, *J. Am. Ceram. Soc.*, **85**, 517–522.

538. Sun, L., Berndt, C. C., Gross, K. A., and Kucuk, A. (2001) Review: material fundamentals and clinical performance of plasma sprayed hydroxyapatite coatings, *J. Biomed. Mater. Res.* (*Appl. Biomater.*), **58**, 570–592.

539. Yang, Y., Kim, K. H., and Ong, J. L. (2005) A review on calcium phosphate coatings produced using a sputtering process — an alternative to plasma spraying, *Biomaterials*, **26**, 327–337.

540. Oliveira, A. L., Mano, J. F., and Reis, R. L. (2003) Nature-inspired calcium phosphate coatings: present status and novel advances in the science of mimicry, *Curr. Opin. Solid State Mater. Sci.*, **7**, 309–318.

541. Manso, M., Jiménez, C., Morant, C., Herrero, P., and Martínez-Duart, J. (2000) Electrodeposition of hydroxyapatite coatings in basic conditions, *Biomaterials*, **21**, 1755–1761.

542. Kuo, M. C., and Yen, S. K. (2002) The process of electrochemical deposited hydroxyapatite coatings on biomedical titanium at room temperature, *Mater. Sci. Eng. C*, **20**, 153–160.

543. Eliaz, N., Sridhar, T. M., Mudali, U. K., and Raj, B. (2005) Electrochemical and electrophoretic deposition of hydroxyapatite for orthopaedic applications, *Surf. Eng.*, **21**, 238–242.

544. Ma, M., Ye, W., and Wang, X. X. (2008) Effect of supersaturation on the morphology of hydroxyapatite crystals deposited by electrochemical deposition on titanium, *Mater. Lett.*, **62**, 3875–3877.

545. Ong, J. L., Appleford, M., Oh, S., Yang, Y., Chen, W. H., Bumgardner, J. D., and Haggard, W. O. (2006) The characterization and development of bioactive hydroxyapatite coatings, *JOM*, **58**, 67–69.

546. Qu, H., Wei, M. and The effect of temperature and initial pH on biomimetic apatite coating, *J. Biomed. Mater. Res. B* (*Appl. Biomater.*), **87B**, 204–212.

547. Esfehanian, M., Raissi-Dehkordi, B., and Moztarzadeh, F. (2005) Electrophoretic deposition of hydroxyapatite on metal implants, *InterCeram*, **54**, 92–98.

548. Narayanan, R., Seshadri, S. K., Kwon, T. Y., and Kim, K. H. (2008) Calcium phosphate-based coatings on titanium and its alloys, *J. Biomed. Mater. Res. B* (*Appl. Biomater.*), **85B**, 279–299.

549. Sugino, A., Tsuru, K., Hayakawa, S., Kikuta, K., Kawachi, G., Osaka, A., and Ohtsuki, C. (2009) Induced deposition of bone-like hydroxyapatite on thermally oxidized titanium substrates using a spatial gap in a solution that mimics a body fluid, *J. Ceram. Soc. Jpn.*, **117**, 515–520.

550. Ye, G., and Troczynski, T. (2008) Hydroxyapatite coatings by pulsed ultrasonic spray pyrolysis, *Ceram. Int.*, **34**, 511–516.

551. Zheng, M., Fan, D., Li, X. K., Zhang, J. B., and Liu, Q. B. (2010) Microstructure and *in vitro* bioactivity of laser-cladded bioceramic coating on titanium alloy in a simulated body fluid, *J. Alloys Compounds*, **489**, 211–214.

552. Geesink, R. G. T. (2002) Osteoconductive coating for total joint arthroplasty, *Clin. Orthop. Related Res.*, **395**, 53–65.

553. Mukherjee, D. P., Dorairaj, N. R., Mills, D. K., Graham, D., and Krauser, J. T. (2000) Fatigue properties of hydroxyapatite-coated dental implants after exposure to a periodontal pathogen, *J. Biomed. Mater. Res.*, **53**, 467–474.

554. Gledhill, H. C., Turner, I. G., and Doyle, C. (2001) *In vitro* fatigue behaviour of vacuum plasma and detonation gun sprayed hydroxyapatite coatings, *Biomaterials*, **22**, 1233–1240.

555. Cheng, K., Ren, C., Weng, W., Du, P., Shen, G., Han, G., and Zhang, S. (2009) Bonding strength of fluoridated hydroxyapatite coatings: a comparative study on pull-out and scratch analysis, *Thin Solid Films*, **517**, 5361–5364.

556. de Groot, K., Geesink, R. G. T., Klein, C. P. A. T., and Serekian, P. (1987) Plasma sprayed coatings of hydroxylapatite, *J. Biomed. Mater. Res.*, **21**, 1375–1381.

557. Furlong, R. J., and Osborn, J. F. (1991) Fixation of hip prostheses by hydroxyapatite ceramic coating, *J. Bone Joint Surg. B*, **73**, 741–745.

558. Saithna, A. (2010) The influence of hydroxyapatite coating of external fixator pins on pin loosening and pin track infection: a systematic review, *Injury*, **41**, 128–132.

559. Chang, J. K., Chen, C. H., Huang, K. Y., and Wang, G. J. (2006) Eight-year results of hydroxyapatite-coated hip arthroplasty, *J. Arthroplasty*, **21**, 541–546.

560. Matsumine, A., Myoui, A., Kusuzaki, K., Araki, N., Seto, M., Yoshikawa, H., and Uchida, A. (2004) Calcium hydroxyapatite ceramic implants in bone tumor surgery. A long-term follow-up study, *J. Bone Joint Surg. B*, **86**, 719–725.

561. Muirhead-Allwood, S. K., Sandiford, N., Skinner, J. A., Hua, J., Kabir, C., and Walker, P. S. (2010) Uncemented custom computer-assisted design and manufacture of hydroxyapatite-coated femoral components: survival at 10 to 17 years, *J. Bone Joint Surg. B*, **92**, 1079–1084.

562. Shetty, A. A., Slack, R., Tindall, A., James, K. D., and Rand, C. (2005) Results of a hydroxyapatite-coated (Furlong) total hip replacement. A 13- to 15-year follow-up, *J. Bone Joint Surg. B*, **87**, 1050–1054.

563. Buchanan, J. M. (2005) 16 year review of hydroxyapatite ceramic coated hip implants — a clinical and histological evaluation, *Key Eng. Mater.*, 284–286, 1049–1052.

564. Buchanan, J. M. (2006) 17 year review of hydroxyapatite ceramic coated hip implants — a clinical and histological evaluation, *Key Eng. Mater.*, 309–311, 1341–1344.

565. Buchanan, J. M., and Goodfellow, S. (2008) Nineteen years review of hydroxyapatite ceramic coated hip implants: a clinical and histological evaluation, *Key Eng. Mater.*, 361–363, 1315–1318.

566. Rajaratnam, S. S., Jack, C., Tavakkolizadeh, A., George, M. D., Fletcher, R. J., Hankins, M., and Shepperd, J. A. N. (2008) Long-term results of a hydroxyapatite-coated femoral component in total hip replacement: a 15- to 21-year follow-up study, *J. Bone Joint Surg. B*, **90**, 27–30.

567. Epinette, J. A., and Manley, M. T. (eds.) (2004) Fifteen years of clinical experience with hydroxyapatite coatings in joint arthroplasty, Springer, France, p. 452.

568. Binahmed, A., Stoykewych, A., Hussain, A., Love, B., and Pruthi, V. (2007) Long-term follow-up of hydroxyapatite-coated dental implants — a clinical trial, *Int. J. Oral Max. Impl.*, **22**, 963–968.

569. Iezzi, G., Scarano, A., Petrone, G., and Piattelli, A. (2007) Two human hydroxyapatite-coated dental implants retrieved after a 14-year loading period: a histologic and histomorphometric case report, *J. Periodontol.*, **78**, 940–947.

570. Bauer, T. W., Geesink, R. G. T., Zimmerman, R., and McMahon, J. T. (1991) Hydroxyapatite-coated femoral stems. Histological analysis

of components retrieved at autopsy, *J. Bone Joint Surg. A*, **73**, 1439–1452.

571. Buma, P., and Gardeniers, J. W. (1995) Tissue reactions around a hydroxyapatite-coated hip prostheses: case report of a retrieveal specimen, *J. Arthroplasty*, **10**, 389–395.

572. Capello, W. D., D'Antonio, J. A., Feinberg, J. R., and Manley, M. T. (1997) Hydroxyapatite-coated total hip femoral components in patients less than fifty years old. Clinical and radiographic results after five to eight years of follow-up, *J. Bone Joint Surg. A*, **79**, 1023–1029.

573. Paital, S. R., and Dahotre, N. B. (2009) Calcium phosphate coatings for bio-implant applications: materials, performance factors, and methodologies, *Mater. Sci. Eng. R*, **66**, 1–70.

574. León, B., and Jansen, J. A. (eds.) (2009), Thin Calcium Phosphate cCatings for Medical Implants, Springer, New York, p. 326.

575. To narrow the subject of this review, functionally graded bioceramics prepared from calcium orthophosphates with various dopants is not considered. The interested readers are advised to read the original papers [316, 576, 577].

576. Manjubala, I., and Kumar, T. S. S. (2000) Effect of TiO_2 — Ag_2O additives on the formation of calcium phosphate based functionally graded bioceramics, *Biomaterials*, **21**, 1995–2002.

577. Wei, T., Ruys, A., and Milthorpe, B. (2003) Hydroxyapatite-zirconia functionally graded bioceramics prepared by hot isostatic pressing, *Key Eng. Mater.*, 240–242, 591–594.

578. Kon, M., Ishikawa, K., Miyamoto, Y., and Asaoka, K. (1995) Development of calcium phosphate based functional gradient bioceramics, *Biomaterials*, **16**, 709–714.

579. Wong, L. H., Tio, B., and Miao, X. (2002) Functionally graded tricalcium phosphate/fluoroapatite composites, *Mater. Sci. Eng. C*, **20**, 111–115.

580. Tampieri, A., Celotti, G., Sprio, S., Delcogliano, A., and Franzese, S. (2001) Porosity-graded hydroxyapatite ceramics to replace natural bone, *Biomaterials*, **22**, 1365–1370.

581. Lu, W. W., Zhao, F., Luk, K. D. K., Yin, Y. J., Cheung, K. M. C., Cheng, G. X., Yao, K. D., and Leong, J. C. Y. (2003) Controllable porosity hydroxyapatite ceramics as spine cage: fabrication and properties evaluation, *J. Mater. Sci. Mater. Med.*, **14**, 1039–1046.

582. Werner, J., Linner-Krcmar, B., Friess, W., and Greil, P. (2002) Mechanical properties and *in vitro* cell compatibility of hydroxyapatite ceramics with graded pore structure, *Biomaterials*, **23**, 4285–4294.

583. Kaito, T., Mukai, Y., Nishikawa, M., Ando, W., Yoshikawa, H., and Myoui, A. (2006) Dual hydroxyapatite composite with porous and solid parts: experimental study using canine lumbar interbody fusion model, *J. Biomed. Mater. Res. B*, **78B**, 378–384.

584. Rodriguez-Lorenzo, L. M., and Ferreira, J. M. F. (2004) Development of porous ceramic bodies for applications in tissue engineering and drug delivery systems, *Mater. Res. Bull.*, **39**, 83–91.

585. Lu, W. W., Zhao, F., Luk, K. D. K., Yin, Y. J., Cheung, K. M. C., Cheng, G. X., Yao, K. D., and Leong, J. C. Y. (2003) Controllable porosity hydroxyapatite ceramics as spine cage: fabrication and properties evaluation, *J. Mater. Sci. Mater. Med.*, **14**, 1039–1046.

586. Watanabe, T., Fukuhara, T., Izui, H., Fukase, Y., and Okano, M. (2009) Properties of HAp/β-TCP functionally graded material by spark plasma sintering, *Transact. Jpn. Soc. Mech. Eng. A*, **75**, 612–618.

587. Roy, M., Balla, V. K., Bandyopadhyay, A., and Bose, S. (2011) Compositionally graded hydroxyapatite/tricalcium phosphate coating on Ti by laser and induction plasma, *Acta Biomater.*, **7**, 866–873.

588. Soon, Y. M., Shin, K. H., Koh, Y. H., Lee, J. H., Choi, W. Y., and Kim, H. E. (2011) Fabrication and compressive strength of porous hydroxyapatite scaffolds with a functionally graded core/shell structure, *J. Eur. Ceram. Soc.*, **31**, 13–18.

589. Dubok, V. A. (2000) Bioceramics — yesterday, today, tomorrow, *Powder Metallurgy Metal Ceramics*, **39**, 381–394.

590. Heness, G., and Ben-Nissan, B. (2004) Innovative bioceramics, *Mater. Forum*, **27**, 104–114.

591. Ohtsuki, C., Kamitakahara, M., and Miyazaki, T. (2009) Bioactive ceramic-based materials with designed reactivity for bone tissue regeneration, *J. R. Soc. Interface*, **6**, S349–S360.

592. The term "bioinert" should be used with care, since it is clear that any material introduced into the physiological environment will induce a response. However, for the purposes of biomedical implants, the term can be defined as a minimal level of response from the host tissue [5].

593. Greenspan, D. C. (1999) Bioactive ceramic implant materials, *Curr. Opin. Solid State Mater. Sci.*, **4**, 389–393.

594. Blokhuis, T. J., Termaat, M. F., den Boer, F. C., Patka, P., Bakker, F. C., and Haarman, H. J. T. M. (2000) Properties of calcium phosphate ceramics in relation to their *in vivo* behavior, *J. Trauma*, **48**, 179–189.

595. Kim, H. M. (2001) Bioactive ceramics: challenges and perspectives, *J. Ceram. Soc. Jpn.*, **109**, S49–S57.

596. Seeley, Z., Bandyopadhyay, A., and Bose, S. (2008) Tricalcium phosphate based resorbable ceramics: influence of NaF and CaO addition, *Mater. Sci. Eng. C*, **28**, 11–17.

597. Descamps, M., Richart, O., Hardouin, P., Hornez, J. C., and Leriche, A. (2008) Synthesis of macroporous β-tricalcium phosphate with controlled porous architectural, *Ceram. Int.*, **34**, 1131–1137.

598. Cushnie, E. K., Khan, Y. M., and Laurencin, C. T. (2008) Amorphous hydroxyapatite-sintered polymeric scaffolds for bone tissue regeneration: physical characterization studies, *J. Biomed. Mater. Res. A*, **84A**, 54–62.

599. Hench, L. L. (2005) Challenges for bioceramics in the 21st century, *Am. Ceram. Soc. Bull.*, **84**, 18–21.

600. Nagase, M., Baker, D. G., and Schumacher, H. R. (1988) Prolonged inflammatory reactions induced by artificial ceramics in the rat pouch model, *J. Rheumatology*, **15**, 1334–1338.

601. Rooney, T., Berman, S., and Indersano, A. T. (1988) Evaluation of porous block hydroxylapatite for augmentation of alveolar ridges, *J. Oral Maxillofac. Surg.*, **46**, 15–18.

602. Prudhommeaux, F., Schiltz, C., Lioté, F., Hina, A., Champy, R., Bucki, B., Ortiz-Bravo, E., Meunier, A., Rey, C., and Bardin, T. (1996) Variation in the inflammatory properties of basic calcium phosphate crystals according to crystal type, *Arthritis Rheum.*, **39**, 1319–1326.

603. Uchida, A., Araki, N., Shinto, Y., Yoshikawa, H., Kurisaki, E., and Ono, K. (1990) The use of calcium hydroxyapatite ceramic in bone tumour surgery, *J. Bone Joint Surg. B*, **72**, 298–302.

604. Osteogenesis is the process of laying down new bone material by osteoblasts.

605. Osteoinduction is the property of the material to induce bone formation *de novo* or ectopically (*i.e.*, in non-bone forming sites). It happens due to a stimulation and activation of host mesenchymal stem cells from the surrounding tissues, which differentiate into bone-forming osteoblasts.

606. Osteoid is a bioorganic portion of the matrix of bone tissue. Osteoblasts begin the process of forming bone tissue by secreting the osteoid as several specific proteins. When the osteoid becomes mineralized, it and the adjacent bone cells have developed into new bone tissues.

607. Giannoudis P. V., Dinopoulos, H., and Tsiridis, E. () Bone substitutes: an update, *Injury*, **36**(1), S20–S27.

608. Yamasaki, H., and Sakai, H. (1992) Osteogenic response to porous hydroxyapatite ceramics under the skin of dogs, *Biomaterials*, **13**, 308–312.

609. Klein, C., de Groot, K., Chen, W., Li, Y., and Zhang, X. (1994) Osseous substance formation induced in porous calcium phosphate ceramics in soft tissues, *Biomaterials*, **15**, 31–34.

610. Yuan, H., Yang, Z., Li, Y., Zhang, Z., de Bruijn, J. D., and de Groot, K. (1998) Osteoinduction by calcium phosphate biomaterials, *J. Mater. Sci. Mater. Med.*, **9**, 723–726.

611. Ripamonti, U. (1996) Osteoinduction in porous hydroxyapatite implanted in heterotopic sites of different animal models, *Biomaterials*, **17**, 31–35.

612. Ripamonti, U., Crooks, J., and Kirkbride, A. N. (1999) Sintered porous hydroxyapatites with intrinsic osteoinductive activity: geometric induction of bone formation, *South African J. Sci.*, **95**, 335–343.

613. Yuan, H. P., de Bruijn, J. D., Li, Y. B., Feng, J. Q., Yang, Z. J., de Groot, K., and Zhang, X. D. (2001) Bone formation induced by calcium phosphate ceramics in soft tissue of dogs: a comparative study between porous α-TCP and β-TCP, *J. Mater. Sci. Mater. Med.*, **12**, 7–13.

614. Matsushita, N., Terai, H., Okada, T., Nozaki, K., Inoue, H., Miyamoto, S., and Takaoka, K. (2004) A new bone-inducing biodegradable porous β-tricalcium phosphate, *J. Biomed. Mater. Res. A*, **70A**, 450–458.

615. le Nihouannen, D., Daculsi, G., Saffarzadeh, A., Gauthier, O., Delplace, S., Pilet, P., and Layrolle, P. (2005) Ectopic bone formation by microporous calcium phosphate ceramic particles in sheep muscles, *Bone*, **36**, 1086–1093.

616. Yuan, H., Kurashina, K., de Bruijn, D. J., Li, Y., de Groot, K., and Zhang, X. (1999) A preliminary study of osteoinduction of two kinds of calcium phosphate bioceramics, *Biomaterials*, **20**, 1799–1806.

617. Gosain, A. K., Song, L., Riordan, P., Amarante, M. T., Nagy, P. G., Wilson, C. R., Toth, J. M., and Ricci, J. L. (2002) A 1-year study of osteoinduction in hydroxyapatite-derived biomaterials in an adult sheep model: Part I. Plast, *Reconstr. Surg.*, **109**, 619–630.

618. Cheng, L., Ye, F., Yang, R., Lu, X., Shi, Y., Li, L., Fan, H., and Bu, H. (2010) Osteoinduction of hydroxyapatite/β-tricalcium phosphate bioceramics in mice with a fractured fibula, *Acta Biomater.*, **6**, 1569–1574.

619. Barrere, F., van der Valk, C. M., Dalmeijer, R. A., Meijer, G., van Blitterswijk, C. A., de Groot, K., and Layrolle, P. (2003) Osteogenecity of octacalcium phosphate coatings applied on porous titanium, *J. Biomed Mater. Res. A*, **66A**, 779–788.

620. Habibovic, P., van der Valk, C. M., van Blitterswijk, C. A., de Groot, K., and Meijer, G. (2004) Influence of octacalcium phosphate coating on osteoinductive properties of biomaterials, *J. Mater. Sci. Mater. Med.*, **15**, 373–380.

621. Yuan, H., Fernandes, H., Habibovic, P., de Boer J., Barradas, A. M. C., de Ruiter, A., Walsh, W. R., van Blitterswijk, C. A., and de Bruijn, J. D. (2010) Osteoinductive ceramics as a synthetic alternative to autologous bone grafting, *Proc. Natl. Acad. Sci. USA*, **107**, 13614–13619.

622. Habibovic, P., Yuan, H., van der Valk, C. M., Meijer, G., van Blitterswijk, C. A., and de Groot, K. (2005) 3D microenvironment as essential element for osteoinduction by biomaterials, *Biomaterials*, **26**, 3565–3575.

623. Habibovic, P., Sees, T. M., van den Doel, M. A., van Blitterswijk, C. A., and de Groot, K. (2006) Osteoinduction by biomaterials — physicochemical and structural influences, *J. Biomed. Mater. Res. A*, **77A**, 747–762.

624. Reddi, A. H. (2000) Morphogenesis and tissue engineering of bone and cartilage: Inductive signals, stem cells and biomimetic biomaterials, *Tissue Eng.*, **6**, 351–359.

625. Ripamonti, U. (1991) The morphogenesis of bone in replicas of porous hydroxyapatite obtained by conversion of calcium carbonate exoskeletons of coral, *J. Bone Joint Surg. A*, **73**, 692–703.

626. Kuboki, Y., Takita, H., and Kobayashi, D. (1998) BMP-induced osteogenesis on the surface of hydroxyapatite with geometrically feasible and nonfeasible structures: topology of osteogenesis, *J. Biomed. Mater. Res.*, **39**, 190–199.

627. Diaz-Flores, L., Gutierrez, R., Lopez-Alonso, A., Gonzalez, R., and Varela, H. (1992) Pericytes as a supplementary source of osteoblasts in periosteal osteogenesis, *Clin. Orthop. Relat. Res.*, **275**, 280–286.

628. Dorozhkin, S. V. (1999) Inorganic chemistry of the dissolution phenomenon, the dissolution mechanism of calcium apatites at the atomic (ionic) level, *Comment Inorg. Chem.*, **20**, 285–299.

629. Dorozhkin, S. V. (2002) A review on the dissolution models of calcium apatites, *Prog. Cryst. Growth Charact.*, **44**, 45–61.

630. Muller-Mai, C. M., Voigt, C., and Gross, U. (1990) Incorporation and degradation of hydroxyapatite implants of different surface roughness and surface structure in bone, *Scan. Microscopy*, **4**, 613–624.

631. Wenisch, S., Stahl, J. P., Horas, U., Heiss, C., Kilian, O., Trinkaus, K., Hild, A., and Schnettler, R. (2003) *In vivo* mechanisms of hydroxyapatite ceramic degradation by osteoclasts: fine structural microscopy, *J. Biomed. Mater. Res. A*, **67A**, 713–718.

632. Riihonen, R., Nielsen, S., Väänänen, H. K., Laitala-Leinonen, T., and Kwon, T. H. (2010) Degradation of hydroxyapatite *in vivo* and *in vitro* requires osteoclastic sodium-bicarbonate co-transporter NBCn1, *Matrix Biol.*, **29**, 287–294.

633. Teitelbaum, S. L. (2000) Bone resorption by osteoclasts, *Science*, **289**, 1504–1508.

634. Okuda, T., Ioku, K., Yonezawa, I., Minagi, H., Kawachi, G., Gonda, Y., Murayama, H., Shibata, Y., Minami, S., Kamihara, S., Kurosawa, H., and Ikeda, T. (2007) The effect of the microstructure of β-tricalcium phosphate on the metabolism of subsequently formed bone tissue, *Biomaterials*, **28**, 2612–2621.

635. Orly, I., Gregoire, M., Menanteau, J., Heughebaert, M., and Kerebel, B. (1989) Chemical changes in hydroxyapatite biomaterial under *in vivo* and *in vitro* biological conditions, *Calcif. Tissue Int.*, **45**, 20–26.

636. Bertazzo, S., Zambuzzi, W. F., Campos, D. D. P., Ogeda, T. L., Ferreira, C. V., and Bertran, C. A. (2010) Hydroxyapatite surface solubility and effect on cell adhesion, *Colloids Surf. B*, **78**, 177–184.

637. Schwartz, Z., and Boyan, B. D. (1994) Underlying mechanisms at the bone-biomaterial interface, *J. Cell. Biochem.*, **56**, 340–347.

638. Puleo, D. A., and Nanci, A. (1999) Understanding and controlling the bone-implant interface, *Biomaterials*, **20**, 2311–2321.

639. Xin, R., Leng, Y., Chen, J., and Zhang, Q. (2005) A comparative study of calcium phosphate formation on bioceramics *in vitro* and *in vivo*, *Biomaterials*, **26**, 6477–6486.

640. Girija, E. K., Parthiban, S. P., Suganthi, R. V., Elayaraja, K., Joshy, M. I. A., Vani, R., Kularia, P., Asokan, K., Kanjilal, D., Yokogawa, Y., and Narayana Kalkura, S. (2008) High energy irradiation — a tool for enhancing the bioactivity of hydroxyapatite, *J. Ceram. Soc. Jpn.*, **116**, 320–324.

641. Okada, M., Furukawa, K., Serizawa, T., Yanagisawa, Y., Tanaka, H., Kawai, T., and Furuzono, T. (2009) Interfacial interactions between calcined hydroxyapatite nanocrystals and substrates, *Langmuir*, **25**, 6300–6306.

642. Malik, M. A., Puleo, D. A., Bizios, R., and Doremus, R. H. (1992) Osteoblasts on hydroxyapatite, alumina and bone surfaces *in vitro*: morphology during the first 2 h of attachment, *Biomaterials*, **13**, 123–128.

643. Norman, M. E., Elgendy, H. M., Shors, E. C., El-Amin, S. F., and Laurencin, C. T. (1994) An in-vitro evaluation of coralline porous hydroxyapatite as a scaffold for osteoblast growth, *Clin. Mater.*, **17**, 85–91.

644. Shu, R., McMullen, R., Baumann, M. J., and McCabe, L. R. (2003) Hydroxyapatite accelerates differentiation and suppresses growth of MC3T3-E1 osteoblasts, *J. Biomed. Mater. Res. A*, **67A**, 1196–1204.

645. Gomi, K., Lowenberg, B., Shapiro, G., and Davies, J. E. (1992) Resorption of sintered synthetic hydroxyapatite by osteoclasts *in vitro*, *Biomaterials*, **20**, 91–96.

646. Suzuki, T., Ohashi, R., Yokogawa, Y., Nishizawa, K., Nagata, F., Kawamoto, Y., Kameyama, T., and Toriyama, M. (1999) Initial anchoring and proliferation of fibroblast L-929 cells on unstable surface of calcium phosphate ceramics, *J. Biosci. Bioeng.*, **87**, 320–327.

647. Arinzeh, T. L., Tran, T., McAlary, J., and Daculsi, G. (2005) A comparative study of biphasic calcium phosphate ceramics for human mesenchymal stem-cell-induced bone formation, *Biomaterials*, **26**, 3631–3638.

648. Oh, S., Oh, N., Appleford, M., and Ong, J. L. (2006) Bioceramics for tissue engineering applications — a review, *Am. J. Biochem. Biotechnol.*, **2**, 49–56.

649. Appleford, M., Oh, S., Cole, J. A., Carnes, D. L., Lee, M., Bumgardner, J. D., Haggard, W. O., and Ong, J. L. (2007) Effects of trabecular calcium phosphate scaffolds on stress signaling in osteoblast precursor cells, *Biomaterials*, **28**, 2747–2753.

650. Ghanaati, S., Barbeck, M., Orth, C., Willershausen, I., Thimm, B. W., Hoffmann, C., Rasic, A., Sader, R. A., Unger, R. E., Peters, F., and Kirkpatrick, C. J. (2010) Influence of β-tricalcium phosphate granule size and morphology on tissue reaction *in vivo*, *Acta Biomater.*, **6**, 4476–4487.

651. Ohgushi, H., and Caplan, A. I. (1999) Stem cell technology and bioceramics: From cell to gene engineering, *J. Biomed. Mater. Res.*, **48**, 913–927.

652. Okumura, M., Ohgushi, H., and Tamai, S. (1990) Bonding osteogenesis in coralline hydroxyapatite combined with bone marrow cells, *Biomaterials*, **12**, 28–37.

653. Unger, R. E., Sartoris, A., Peters, K., Motta, A., Migliaresi, C., Kunkel, M., Bulnheim, U., Rychly, J., and Kirkpatrick, C. J. (2007) Tissue-like self-assembly in cocultures of endothelial cells and osteoblasts and the formation of microcapillary like structures on three-dimensional porous biomaterials, *Biomaterials*, **28**, 3965–3976.

654. Saiz, E., Gremillard, L., Menendez, G., Miranda, P., Gryn, K., and Tomsia, A. P. (2007) Preparation of porous hydroxyapatite scaffolds, *Mater. Sci. Eng. C*, **27**, 546–550.

655. Tamai, N., Myoui, A., Hirao, M., Kaito, T., Ochi, T., Tanaka, J., Takaoka, K., and Yoshikawa, H. (2005) A new biotechnology for articular cartilage repair: subchondral implantation of a composite of interconnected porous hydroxyapatite, synthetic polymer (PLAPEG), and bone morphogenetic protein-2 (rhBMP-2), *Osteoarthritis Cartilage*, **13**, 405–417.

656. Bignon, A., Chouteau, J., Chevalier, J., Fantozzi, G., Carret, J. P., Chavassieux, P., Boivin, G., Melin, M., and Hartmann, D. (2003) Effect of micro- and macroporosity of bone substitutes on their mechanical properties and cellular response, *J. Mater. Sci. Mater. Med.*, **14**, 1089–1097.

657. Griffith, L. G., and Naughton, G. (2002) Tissue engineering — current challenges and expanding opportunities, *Science*, **295**, 1009–1014.

658. van Blitterswijk, C. A., Thomsen, P., Hubbell, J., Cancedda R., de Bruijn, J. D., Lindahl, A., Sohier, J., and Williams, D. F. (eds.) (2008) *Tissue Engineering*, Academic Press, Burlington, p. 760.

659. Repair is usually a rapid occurrence that is required for the survival of the individual but is not necessary for its optimal function. It usually results in a dense connective tissue scar that fills the space; however, it may not be responsive to the highly loaded mechanical environment required of musculoskeletal tissue. Regeneration is a slow process that is usually characterized by the recapitulation of aspects of embryonic development [660].

660. Goldberg, V. M., and Caplan, A. I. (2004) *Orthopedic Tissue Engineering Basic Science and Practice*, Marcel Dekker, New York, p. 338.

661. In 2003, the NSF published a report titled: "The emergence of tissue engineering as a research field", which provides a thorough description of the history of this field. See: http://www.nsf.gov/pubs/2004/nsf0450/start.htm (accessed in February 2010).

662. Ikada, Y. (2006) Challenges in tissue engineering, *J. R. Soc. Interface*, **3**, 589–601.

663. Cima, L. G., and Langer, R. (1993) Engineering human tissue, *Chem. Eng. Prog.*, **89**, 46–54.

664. Langer, R., and Vacanti, J. P. (1993) Tissue engineering, *Science*, **260**, 920–926.

665. El-Ghannam, A. (2005) Bone reconstruction: from bioceramics to tissue engineering, *Expert Rev. Med. Dev.*, **2**, 87–101.

666. In general, a scaffold is a temporary structure used to support people and material in the construction or repair of buildings and other large structures. It is usually a modular system of metal pipes (termed tubes in Britain), although it can be made out of other materials.

667. Yang, S., Leong, K. F., Du, Z., and Chua, C. K. (2001) The design of scaffolds for use in tissue engineering. Part I. Traditional factors, *Tissue Eng.*, **7**, 679–689.

668. Hutmacher, D. W. (2000) Scaffolds in tissue engineering bone and cartilage, *Biomaterials*, **21**, 2529–2543.

669. Ma, P. X. (2004) Scaffolds for tissue fabrication, *Mater. Today*, **7**, 30–40.

670. Yasuhiko, T. (2009) Biomaterial technology for tissue engineering applications, *J. R. Soc. Interface*, **6**, S311–S324.

671. Ma, P. X., and Elisseeff, J. (eds.) (2006) *Scaffolding in Tissue Engineering*, CRC Press, Boca Raton, p. 638.

672. Schieker, M., Seitz, H., Drosse, I., Seitz, S., and Mutschler, W. (2006) Biomaterials as scaffold for bone tissue engineering, *Eur. J. Trauma*, **32**, 114–124.

673. Okumura, M., Ohgushi, H., Dohi, Y., Katuda, T., Tamai, S., Koerten, H. K., and Tabata, S. (1997) Osteoblastic phenotype expression on the surface of hydroxyapatite ceramics, *J. Biomed. Mater. Res.*, **37**, 122–129.

674. Service, R. F. (2000) Tissue engineers build new bone, *Science*, **289**, 1498–1500.

675. Fini, M., Giardino, R., Borsari, V., Torricelli, P., Rimondini, L., Giavaresi, G., and Aldini, N. N. (2003) *In vitro* behaviour of osteoblasts cultured on orthopaedic biomaterials with different surface roughness, uncoated and fluorohydroxyapatite-coated, relative to the *in vivo* osteointegration rate, *Int. J. Artif. Organs*, **26**, 520–528.

676. Sato, M., and Webster, T. J. (2006) Designing orthopedic implant surfaces: harmonization of nanotopographical and chemical aspects, *Nanomed.*, **1**, 351–354.

677. Annaz, B., Hing, K. A., Kayser, M., Buckland, T., and di Silvio, L. (2004) Porosity in phase-pure hydroxyapatite, *J. Microscopy*, **216**, 97–109.

678. Li, S., de Wijn, J. R., Li, J., Layrolle, P., and de Groot, K. (2003) Macroporous biphasic calcium phosphate scaffold with high permeability/porosity ratio, *Tissue Eng.*, **9**, 535–548.

679. Ebaretonbofa, E., and Evans, J. R. (2002) High porosity hydroxyapatite foam scaffolds for bone substitute, *J. Porous Mater.*, **9**, 257–263.

680. Specchia, N., Pagnotta, A., Cappella, M., Tampieri, A., and Greco, F. (2002) Effect of hydroxyapatite porosity on growth and differentiation of human osteoblast-like cells, *J. Mater. Sci.*, **37**, 577–584.

681. Malmström, J., Adolfsson, E., Arvidsson, A., and Thomsen, P. (2007) Bone response inside free-form fabricated macroporous hydroxyapatite

scaffolds with and without an open microporosity, *Clin. Implant Dentistry Rel. Res.*, **9**, 79–88.

682. Teixeira, C. C., Nemelivsky, Y., Karkia, C., and LeGeros, R. Z. (2006) Biphasic calcium phosphate: a scaffold for growth plate chondrocyte maturation, *Tissue Eng.*, **12**, 2283–2289.

683. Teixeira, S., Oliveira, S., Ferraz, M. P., and Monteiro, F. J. (2008) Three dimensional macroporous calcium phosphate scaffolds for bone tissue engineering, *Key Eng. Mater.*, 361–363, 947–950.

684. Peng, Q., Jiang, F., Huang, P., Zhou, S., Weng, J., Bao, C., Zhang, C., and Yu, H. (2010) A novel porous bioceramics scaffold by accumulating hydroxyapatite spherules for large bone tissue engineering *in vivo*. I. Preparation and characterization of scaffold, *J. Biomed. Mater. Res. A*, **93A**, 920–929.

685. Stevens, M. M. (2008) Biomaterials for bone tissue engineering, *Mater. Today*, **11**, 18–25.

686. Artzi, Z., Weinreb, M., Givol, N., Rohrer, M. D., Nemcovsky, C. E., Prasad, H. S., and Tal, H. (2004) Biomaterial resorbability and healing site morphology of inorganic bovine bone and beta tricalcium phosphate in the canine: a 24-month longitudinal histologic study and morphometric analysis, *Int. J. Oral Max. Impl.*, **19**, 357–368.

687. Burg, K. J. L., Porter, S., and Kellam, J. F. (2000) Biomaterial developments for bone tissue engineering, *Biomaterials*, **21**, 2347–2359.

688. Ajaal, T. T., and Smith, R. W. (2009) Employing the Taguchi method in optimizing the scaffold production process for artificial bone grafts, *J. Mater. Process. Technol.*, **209**, 1521–1532.

689. Boyde, A., Corsi, A., Quarto, R., Cancedda, R., and Bianco, P. (1999) Osteoconduction in large macroporous hydroxyapatite ceramic implants: evidence for a complementary integration and disintegration mechanism, *Bone*, **24**, 579–589.

690. Hing, K. A. (2005) Bioceramic bone graft substitutes: influence of porosity and chemistry, *Int. J. Appl. Ceram. Technol.*, **2**, 184–199.

691. Peppas, N. A., and Langer, R. (1994) New challenges in biomaterials, *Science*, **263**, 1715–1720.

692. Hench, L. L. (1998) Biomaterials: a forecast for the future, *Biomaterials*, **19**, 1419–1423.

693. Barrère, F., Mahmood, T. A., de Groot, K., and van Blitterswijk, C. A. (2008) Advanced biomaterials for skeletal tissue regeneration: instructive and smart functions, *Mater. Sci. Eng. R*, **59**, 38–71.

694. Dellinger, J. G., Eurell, J. A. C., and Jamison, R. D. (2006) Bone response to 3D periodic hydroxyapatite scaffolds with and without tailored microporosity to deliver bone morphogenetic protein 2, *J. Biomed. Mater. Res. A*, **76A**, 366–376.

695. Pal, K., and Pal, S. (2006) Development of porous hydroxyapatite scaffolds, *Mater. Manuf. Process.*, **21**, 325–328.

696. Bae, C. J., Kim, H. W., Koh, Y. H., and Kim, H. E. (2006) Hydroxyapatite (HA) bone scaffolds with controlled macrochannel pores, *J. Mater. Sci. Mater. Med.*, **17**, 517–521.

697. Guo, H., Wei, J., Kong, H., Liu, C., and Pan, K. (2008) Biocompatibility and osteogenesis of calcium phosphate cement scaffolds for bone tissue engineering, *Adv. Mater. Res.*, 47–50, Part 2, 1383–1386.

698. Guo, H., Su, J., Wei, J., Kong, H., and Liu, C. (2009) Biocompatibility and osteogenicity of degradable Ca-deficient hydroxyapatite scaffolds from calcium phosphate cement for bone tissue engineering, *Acta Biomater.*, **5**, 268–278.

699. Wilson, C. E., van Blitterswijk, C. A., Verbout, A. J., Dhert, W. J. A., and de Bruijn, J. D. (2011) Scaffolds with a standardized macro-architecture fabricated from several calcium phosphate ceramics using an indirect rapid prototyping technique, *J. Mater. Sci. Mater. Med.*, **22**, 97–105.

700. Schumacher, M., Deisinger, U., Detsch, R., and Ziegler, G. (2010) Indirect rapid prototyping of biphasic calcium phosphate scaffolds as bone substitutes: influence of phase composition, macroporosity and pore geometry on mechanical properties, *J. Mater. Sci. Mater. Med.*, **21**, 3119–3127.

701. Liu, H., and Webster, T. J. (2007) Nanomedicine for implants: a review of studies and necessary experimental tools, *Biomaterials*, **28**, 354–369.

702. Wang, C., Duan, Y., Markovic, B., Barbara, J., Howlett, C. R., Zhang, X., and Zreiqat, H. (2004) Proliferation and bone-related gene expression of osteoblasts grown on hydroxyapatite ceramics sintered at different temperature, *Biomaterials*, **25**, 2949–2956.

703. Matsumoto, T., Okazaki, M., Nakahira, A., Sasaki, J., Egusa, H., and Sohmura, T. (2007) Modification of apatite materials for bone tissue engineering and drug delivery carriers, *Curr. Med. Chem.*, **14**, 2726–2733.

704. Traykova, T., Aparicio, C., Ginebra, M. P., and Planell, J. A. (2006) Bioceramics as nanomaterials, *Nanomedicine*, **1**, 91–106.

705. Kalita, S. J., Bhardwaj, A., and Bhatt, H. A. (2007) Nanocrystalline calcium phosphate ceramics in biomedical engineering, *Mater. Sci. Eng. C*, **27**, 441–449.

706. Dorozhkin, S. V. (2009) Nano-sized and nanocrystalline calcium orthophosphates in biomedical engineering, *J. Biomimetics Biomaterials Tiss. Eng.*, **3**, 59–92.

707. Dorozhkin, S. V. (2009) Nanodimensional and nanocrystalline apatites and other calcium orthophosphates in biomedical engineering, biology and medicine, *Materials*, **2**, 1975–2045.

708. Dorozhkin, S. V. (2010) Nanosized and nanocrystalline calcium orthophosphates, *Acta Biomater.*, **6**, 715–734.

709. Aizawa, M., Porter, A. E., Best, S. M., and Bonfield, W. (2005) Ultrastructural observation of single-crystal apatite fibres, *Biomaterials*, **26**, 3427–3433.

710. Park, Y. M., Ryu, S. C., Yoon, S. Y., Stevens, R., and Park, H. C. (2008) Preparation of whisker-shaped hydroxyapatite / β-tricalcium phosphate composite, *Mater. Chem. Phys.*, **109**, 440–447.

711. Morisue, H., Matsumoto, M., Chiba, K., Matsumoto, H., Toyama, Y., Aizawa, M., Kanzawa, N., Fujimi, T. J., Uchida, H., and Okada, I. (2006) A novel hydroxyapatite fiber mesh as a carrier for recombinant human bone morphogenetic protein-2 enhances bone union in rat posterolateral fusion model, *Spine*, **31**, 1194–1200.

712. Matsuda, A., Ikoma, T., Kobayashi, H., and Tanaka, J. (2004) Preparation and mechanical property of core-shell type chitosan / calcium phosphate composite fiber, *Mater. Sci. Eng. C*, **24**, 723–728.

713. Tas, A. C. (2001) Molten salt synthesis of calcium hydroxyapatite whiskers, *J. Am. Ceram. Soc.*, **84**, 295–300.

714. Aizawa, M., Ueno, H., Itatani, K., and Okada, I. (2006) Syntheses of calcium-deficient apatite fibres by a homogeneous precipitation method and their characterizations, *J. Eur. Ceram. Soc.*, **26**, 501–507.

715. Wu, Y., Hench, L. L., Du, J., Choy, K. L., and Guo, J. (2004) Preparation of hydroxyapatite fibers by electrospinning technique, *J. Am. Ceram. Soc.*, **87**, 1988–1991.

716. Ramanan, S. R., and Venkatesh, R. (2004) A study of hydroxyapatite fibers prepared via sol-gel route, *Mater. Lett.*, **58**, 3320–3323.

717. Seo, D. S., and Lee, J. K. (2008) Synthesis of hydroxyapatite whiskers through dissolution-reprecipitation process using EDTA, *J. Cryst. Growth*, **310**, 2162–2167.

718. Tas, A. C. (2007) Formation of calcium phosphate whiskers in hydrogen peroxide (H_2O_2) solutions at 90°C, *J. Am. Ceram. Soc.*, **90**, 2358–2362.

719. Neira, I. S., Guitián, F., Taniguchi, T., Watanabe, T., and Yoshimura, M. (2008) Hydrothermal synthesis of hydroxyapatite whiskers with sharp faceted hexagonal morphology, *J. Mater. Sci.*, **43**, 2171–2178.

720. Yang, H. Y., Yang, S. F., Chi, X. P., Evans, J. R. G., Thompson, I., Cook, R. J., and Robinson, P. (2008) Sintering behaviour of calcium phosphate filaments for use as hard tissue scaffolds, *J. Eur. Ceram. Soc.*, **28**, 159–167.

721. Cardoso, G. B. C., Ramos, S. L. F., Rodas, A. C. D., Higa, O. Z., Zavaglia, C. A. C., and Arruda, A. C. F. (2010) Scaffolds of poly (ε-caprolactone) with whiskers of hydroxyapatite, *J. Mater. Sci.*, **45**, 4990–4993.

722. Zhang, H., and Zhu, Q. (2009) Synthesis of nanospherical and ultralong fibrous hydroxyapatite and reinforcement of biodegradable chitosan/hydroxyapatite composite, *Modern Phys. Lett. B*, **23**, 3967–3976.

723. Ribeiro, C. C., Barrias, C. C., and Barbosa, M. A. (2006) Preparation and characterisation of calcium-phosphate porous microspheres with a uniform size for biomedical applications, *J. Mater. Sci. Mater. Med.*, **17**, 455–463.

724. Ribeiro, C. C., Barrias, C. C., and Barbosa, M. A. (2004) Calcium phosphate-alginate microspheres as enzyme delivery matrices, *Biomaterials*, **25**, 4363–4373.

725. Zhou, W. Y., Wang, M., Cheung, W. L., Guo, B. C., and Jia, D. M. (2008) Synthesis of carbonated hydroxyapatite nanospheres through nanoemulsion, *J. Mater. Sci. Mater. Med.*, **19**, 103–110.

726. Kim, H. W., Gu, H. J., and Lee, H. H. (2007) Microspheres of collagen-apatite nanocomposites with osteogenic potential for tissue engineering, *Tissue Eng.*, **13**, 965–973.

727. Kawai, T., Sekikawa, H., and Unuma, H. (2009) Preparation of hollow hydroxyapatite microspheres utilizing poly(divinylbenzene) as a template, *J. Ceram. Soc. Jpn.*, **117**, 340–343.

728. Descamps, M., Hornez, J. C., and Leriche, A. (2009) Manufacture of hydroxyapatite beads for medical applications, *J. Eur. Ceram. Soc.*, **29**, 369–375.

729. Cho, J. S., Jung, D. S., Han, J. M., and Kang, Y. C. (2008) Spherical shape hydroxyapatite powders prepared by flame spray pyrolysis, *J. Ceram. Process. Res.*, **9**, 348–352.

730. Kimura, I., Honma, T., and Riman, R. E. (2007) Preparation of hydroxyapatite microspheres by interfacial reaction in a multiple emulsion, *J. Ceram. Soc. Jpn.*, **115**, 888–893.

731. Yao, A., Ai, F., Liu, X., Wang, D., Huang, W., and Xu, W. (2010) Preparation of hollow hydroxyapatite microspheres by the conversion of borate glass at near room temperature, *Mater. Res. Bull.*, **45**, 25–28.

732. Jiang, D., Chen, M., Li, D., Zhu, J., Lü, X., and Xie, J. (2009) One-pot synthesis of hierarchical fluorapatite hollow microparticles, *Mater. Lett.*, **63**, 2639–2642.

733. Lim, J. H., Park, J. H., Park, E. K., Kim, H. J., Park, I. K., Shin, H. Y., and Shin, H. I. (2008) Fully interconnected globular porous biphasic calcium phosphate ceramic scaffold facilitates osteogenic repair, *Key Eng. Mater.*, 361–363, 119–122.

734. Cho, J. S., Ko, Y. N., Koo, H. Y., and Kang, Y. C. (2010) Synthesis of nano-sized biphasic calcium phosphate ceramics with spherical shape by flame spray pyrolysis, *J. Mater. Sci. Mater. Med.*, **21**, 1143–1149.

735. Ye, F., Guo, H., Zhang, H., and He, X. (2010) Polymeric micelle-templated synthesis of hydroxyapatite hollow nanoparticles for a drug delivery system, *Acta Biomater.*, **6**, 2212–2218.

736. He, W., Tao, J., Pan, H., Xu, R., and Tang, R. (2010) A size-controlled synthesis of hollow apatite nanospheres at water-oil interfaces, *Chem. Lett.*, **39**, 674–675.

737. Itatani, K., Tsugawa, T., Umeda, T., Musha, Y., Davies, I. J., and Koda, S. (2010) Preparation of submicrometer-sized porous spherical hydroxyapatite agglomerates by ultrasonic spray pyrolysis technique, *J. Ceram. Soc. Jpn.*, **118**, 462–466.

738. Kamitakahara, M., Murakami, S., Takahashi, H., Watanabe, N., and Ioku, K. (2010) Formation of hydroxyapatite microtubes assisted with anatase under hydrothermal conditions, *Chem. Lett.*, **39**, 854–855.

739. Chandanshive, B., Dyondi, D., Ajgaonkar, V. R., Banerjee, R., and Khushalani, D. (2010) Biocompatible calcium phosphate based tubes, *J. Mater. Chem.*, **20**, 6923–6928.

740. Kim, M. K., Kim, J. J., Shin, U. S., and Kim, H. W. (2010) Production of a biomimetic apatite nanotube mesh via biotemplating a polymer nanofiber mesh, *Mater. Lett.*, **64**, 2655–2658.

741. Simon, J. L., Michna, S., Lewis, J. A., Rekow, E. D., Thompson, V. P., Smay, J. E., Yampolsky, A., Parsons, J. R., and Ricci, J. L. (2007) *In vivo* bone response to 3D periodic hydroxyapatite scaffolds assembled by direct ink writing, *J. Biomed. Mater. Res. A*, **83A**, 747–758.

742. Chu, T. M., Orton, D. G., Hollister, S. J., Feinberg, S. E., and Halloran, J. W. (2002) Mechanical and *in vivo* performance of hydroxyapatite implants with controlled architectures, *Biomaterials*, **23**, 1283–1293.

743. Yoshikawa, H., and Myoui, A. (2005) Bone tissue engineering with porous hydroxyapatite ceramics, *J. Artif. Organs*, **8**, 131–136.

744. Detsch, R., Uhl, F., Deisinger, U., and Ziegler, G. (2008) *In vitro* studies of cell growth on three differently fabricated hydroxyapatite ceramic scaffolds for bone tissue engineering, *Key Eng. Mater.*, 361–363, 1181–1184.

745. Min, S. H., Jin, H. H., Park, H. Y., Park, I. M., Park, H. C., and Yoon, S. Y. (2006) Preparation of porous hydroxyapatite scaffolds for bone tissue engineering, *Mater. Sci. Forum*, 510–511, 754–757.

746. Deville, S., Saiz, E., Nalla, R. K., and Tomsia, A. P. (2006) Strong biomimetic hydroxyapatite scaffolds, *Adv. Sci. Technol.*, **49**, 148–152.

747. Buckley, C. T., and O'Kelly, K. U. (2010) Fabrication and characterization of a porous multidomain hydroxyapatite scaffold for bone tissue engineering investigations, *J. Biomed. Mater. Res. B (Appl. Biomater.)*, **93B**, 459–467.

748. Yang, W., Zhou, D., Yin, G., and Chen, H. (2004) Progress of biphasic calcium phosphate bioceramic as scaffold materials of bone tissue engineering, *J. Chinese Ceram. Soc.*, **32**, 1143–1149.

749. Ramay, H. R. R., and Zhang, M. (2004) Biphasic calcium phosphate nanocomposite porous scaffolds for load-bearing bone tissue engineering, *Biomaterials*, **25**, 5171–5180.

750. Yuan, H., van Blitterswijk, C. A., de Groot, K., and de Bruijn, J. D. (2006) Cross-species comparison of ectopic bone formation in biphasic calcium phosphate (BCP) and hydroxyapatite (HA) scaffolds, *Tissue Eng.*, **12**, 1607–1615.

751. Chen, G., Li, W., Zhao, B., and Sun, K. (2009) A novel biphasic bone scaffold: β-calcium phosphate and amorphous calcium polyphosphate, *J. Am. Ceram. Soc.*, **92**, 945–948.

752. Guo, D., Xu, K., and Han, Y. (2009) The *in situ* synthesis of biphasic calcium phosphate scaffolds with controllable compositions, structures, and adjustable properties, *J. Biomed. Mater. Res. A*, **88A**, 43–52.

753. Furuichi, K., Oaki, Y., Ichimiya, H., Komotori, J., and Imai, H. (2006) Preparation of hierarchically organized calcium phosphate-organic polymer composites by calcification of hydrogel, *Sci. Technol. Adv. Mater.*, **7**, 219–225.

754. Wei, J., Jia, J., Wu, F., Wei, S., Zhou, H., Zhang, H., Shin, J. W., and Liu, C. (2010) Hierarchically microporous/macroporous scaffold of magnesium-calcium phosphate for bone tissue regeneration, *Biomaterials*, **31**, 1260–1269.

755. Gbureck, U., Grolms, O., Barralet, J. E., Grover, L. M., and Thull, R. (2003) Mechanical activation and cement formation of β-tricalcium phosphate, *Biomaterials*, **24**, 4123–4131.

756. Gbureck, U., Barralet, J. E., Hofmann, M., and Thull, R. (2004) Mechanical activation of tetracalcium phosphate, *J. Am. Ceram. Soc.*, **87**, 311–313.

757. Bohner, M., Luginbühl, R., Reber, C., Doebelin, N., Baroud, G., and Conforto, E. (2009) A physical approach to modify the hydraulic reactivity of α-tricalcium phosphate powder, *Acta Biomater.*, **5**, 3524–3535.

758. Hagio, T., Tanase, T., Akiyama, J., Iwai, K., and Asai, S. (2008) Formation and biological affinity evaluation of crystallographically aligned hydroxyapatite, *J. Ceram. Soc. Jpn.*, **116**, 79–82.

759. Hench, L. L., and Polak, J. M. (2002) Third-generation biomedical materials, *Science*, **295**, 1014–1017.

760. Blawas, A. S., and Reichert, W. M. (1998) Protein patterning, *Biomaterials*, **19**, 595–609.

761. Kasai, T., Sato, K., Kanematsu, Y., Shikimori, M., Kanematsu, N., and Doi, Y. (2010) Bone tissue engineering using porous carbonate apatite and bone marrow cells, *J. Craniofac. Surg.*, **21**, 473–478.

762. Wang, L., Fan, H., Zhang, Z. Y., Lou, A. J., Pei, G. X., Jiang, S., Mu, T. W., Qin, J. J., Chen, S. Y., and Jin, D. (2010) Osteogenesis and angiogenesis of tissue-engineered bone constructed by prevascularized β-tricalcium phosphate scaffold and mesenchymal stem cells, *Biomaterials*, **31**, 9452–9461.

763. Sánchez-Salcedo, S., Izquierdo-Barba, I., Arcos, D., and Vallet-Regí, M. (2006) *In vitro* evaluation of potential calcium phosphate scaffolds for tissue engineering, *Tissue Eng.*, **12**, 279–290.

764. Meganck, J. A., Baumann, M. J., Case, E. D., McCabe, L. R., and Allar, J. N. (2005) Biaxial flexure testing of calcium phosphate bioceramics for use in tissue engineering, *J. Biomed. Mater. Res. A*, **72A**, 115–126.

765. Case, E. D., Smith, I. O., and Baumann, M. J. (2005) Microcracking and porosity in calcium phosphates and the implications for bone tissue engineering, *Mater. Sci. Eng. A*, **390**, 246–254.

766. Sibilla, P., Sereni, A., Aguiari, G., Banzi, M., Manzati, E., Mischiati, C., Trombelli, L., and del Senno, L. (2006) Effects of a hydroxyapatite-based biomaterial on gene expression in osteoblast-like cells, *J. Dent. Res.*, **85**, 354–358.

767. Verron, E., and Bouler, J. M. (2010) Calcium phosphate ceramics as bone drug-combined devices, *Key Eng. Mater.*, **441**, 181–201.

768. Ono, I., Ohura, T., Murata, M., Yamaguchi, H., Ohnuma, Y., and Kuboki, Y. (1992) A study on bone induction in hydroxyapatite combined with bone morphogenetic protein, *Plast. Reconstr. Surg.*, **90**, 870–879.

769. Ono, I., Yamashita, T., Jin, H. Y., Ito, Y., Hamada, H., Akasaka, Y., Nakasu, M., Ogawa, T., and Jimbow, K. (2004) Combination of porous hydroxyapatite and cationic liposomes as a vector for BMP-2 gene therapy, *Biomaterials*, **25**, 4709–4718.

770. Sawyer, A. A., Hennessy, K. M., and Bellis, S. L. (2007) The effect of adsorbed serum proteins, RGD and proteoglycan-binding peptides on the adhesion of mesenchymal stem cells to hydroxyapatite, *Biomaterials*, **28**, 383–392.

771. Mastrogiacomo, M., Muraglia, A., Komlev, V., Peyrin, F., Rustichelli, F., Crovace, A., and Cancedda, R. (2005) Tissue engineering of bone: search for a better scaffold, *Orthod. Craniofac. Res.*, **8**, 277–284.

772. Schek, R. M., Taboas, J. M., Hollister, S. J., and Krebsbach, P. H. (2005) Tissue engineering osteochondral implants for temporomandibular joint repair, *Orthod. Craniofac. Res.*, **8**, 313–319.

773. Krylova, E. A., Ivanov, A. A., Krylov, S. E., Plashchina, I. G., Grigorjan, A. S., Goldstein, D. V., Pulin, A. A., and Fatkhudinov, T. H. (2006) Hydroxyapatite — alginate structure as living cells supporting system, *Minerva Biotecnol.*, **18**, 17–22.

774. Nishikawa, M., Myoui, A., Ohgushi, H., Ikeuchi, M., Tamai, N., and Yoshikawa, H. (2004) Bone tissue engineering using novel interconnected porous hydroxyapatite ceramics combined with marrow mesenchymal cells: quantitative and three-dimensional image analysis, *Cell Transplant.*, **3**, 367–376.

775. Quarto, R., Mastrogiacomo, M., Cancedda, R., Kutepov, S. M., Mukhachev, V, Lavroukov, A., Kon, E., and Marcacci, M. (2001) Repair of large bone defects with the use of autologous bone marrow stromal cells, *N. Engl. J. Med.*, **344**, 385–386.

776. Vacanti, C. A., Bonassar, L. J., Vacanti, M. P., and Shufflebarger, J. (2001) Replacement of an avulsed phalanx with tissue-engineered bone, *N. Engl. J. Med.*, **344**, 1511–1514.

777. Morishita, T., Honoki, K., Ohgushi, H., Kotobuki, N., Matsushima, A., and Takakura, Y. (2006) Tissue engineering approach to the treatment of bone tumors: three cases of cultured bone grafts derived from patients' mesenchymal stem cells, *Artif. Organs*, **30**, 115–118.

778. Eniwumide, J. O., Yuan, H., Cartmell, S. H., Meijer, G. J., and de Bruijn, J. D. (2007) Ectopic bone formation in bone marrow stem cell seeded

calcium phosphate scaffolds as compared to autograft and (cell seeded) allograft, *Eur. Cell Mater.*, **14**, 30–39.

779. Zuolin, J., Hong, Q., and Jiali, T. (2010) Dental follicle cells combined with beta-tricalcium phosphate ceramic: a novel available therapeutic strategy to restore periodontal defects, *Med. Hypotheses*, **75**, 669–670.

780. Hench, L. L., and Wilson, J. (1984) Surface-active biomaterials, *Science*, **226**, 630–636.

781. Navarro, M., Michiardi, A., Castano, O., and Planell, J. A. (2008) Biomaterials in orthopaedics, *J. R. Soc. Interface*, **5**, 1137–1158.

782. Anderson, J. M. (2006) The future of biomedical materials, *J. Mater. Sci. Mater. Med.*, **17**, 1025–1028.

783. Chevalier, J., and Gremillard, L. (2009) Ceramics for medical applications: a picture for the next 20 years, *J. Eur. Ceram. Soc.*, **29**, 1245–1255.

784. Hartgerink, J. D., Beniash, E., and Stupp, S. I. (2001) Self-assembly and mineralization of peptide-amphiphile nanofibers, *Science*, **294**, 1684–1688.

Chapter 5

Self-Setting Calcium Orthophosphate Formulations: Cements, Concretes, Pastes, and Putties

5.1 Introduction

Calcium orthophosphates have been studied as bone-repair materials for the last 80 years. The first *in vivo* use of calcium orthophosphates was performed in 1920; that time the researchers implanted tricalcium phosphate (TCP) into animals to test its efficacy as a bone substitute [1]. In the following years, some other calcium orthophosphates were tested on animals to investigate their effect on the healing of nonunions [2]. However, it was 1951, when for the first time hydroxyapatite (HA) was implanted in rats and guinea pigs [3]. Those attempts might be characterized as initial medical trials with the first generation of bone-substituting biomaterials. However, it was already the 1970s, when other calcium orthophosphates were synthesized, characterized, investigated, and tried in medicine [4–10]. The list of known calcium orthophosphates, including their standard abbreviations and the major properties, is shown in Table 1.1 [11].

It is generally considered that the possibility to obtain a monolithic calcium orthophosphate bioceramics at ambient or body temperature via a cementation reaction was put forward by the scientists at the American Dental Association LeGeros et al. [12] and

Calcium Orthophosphates: Applications in Nature, Biology, and Medicine
Sergey Dorozhkin

ISBN 978-981-4316-62-0 (Hardcover), 978-981-4364-17-1 (eBook)
www.panstanford.com

Brown and Chow [13–16] in the early 1980s. However, there is an opinion [17] that self-setting calcium orthophosphate formulations for orthopedic and dental restorative applications have first been described in the early 1970s by Driskell et al. [18]. More to the point, there are researchers, who worked with similar reactions even earlier. Namely, in 1950, Kingery investigated chemical interactions among oxides and/or hydroxides of various metals (including CaO) with H_3PO_4 and discovered several self-hardening formulations [19]; thus, he appears to be the first (see Chapter 8). Leaving aside the priority topic, we further discuss the material subject, which currently is known as *calcium phosphate cements* (commonly referred to as CPC), and, due to their suitability for repair, augmentation and regeneration of bones, these biomaterials are also named as calcium phosphate *bone* cements (occasionally referred to as CPBC) [20]. In order to stress the fact that these cements consist either entirely or essentially from calcium *ortho*phosphates, this review is limited to consideration of calcium orthophosphate-based formulations only. The readers interested in formulations based on other types of calcium phosphates are requested to read the original publications [21].

Due to a good bioresorbability, all self-setting calcium orthophosphate formulations belong to the second generation of bone-substituting biomaterials [22]. These formulations are blends of amorphous and/or crystalline calcium orthophosphate powder(s) with an aqueous solution, which might be distilled water, phosphate-buffered saline (PBS), aqueous solutions of sodium orthophosphate (~ 0.25 M), orthophosphoric acid, citric acid (~ 0.5 M) [23], sodium silicate [24, 25], magnesium hydroorthophosphate [26], or even the revised simulated body fluid (rSBF) [27]. After the calcium orthophosphate powder(s) and the solution are mixed together, a viscous and moldable paste is formed that sets to a firm mass within a few minutes. When the paste becomes sufficiently stiff, it can be placed into a defect as a substitute for the damaged part of bone, where it hardens *in situ* within the operating theatre. The proportion of solid to liquid or the powder-to-liquid (P/L) ratio is a very important characteristic because it determines both bioresorbability and rheological properties. As the paste is set and hardened at room or body temperature, direct application in healing of bone defects became a new and innovative treatment modality by the end of the twentieth century. Moreover, calcium orthophosphate cements can be injected directly into the fractures and bone defects, where they intimately adapt to the bone cavity regardless its shape. More to the point, they were found to promote development of osteoconductive pathways, possess sufficient compressive strengths, be non-cytotoxic, create chemical bonds to the host bones, restore contour, and have both the chemical composition and X-ray diffraction patterns similar to those of bone [28]. Finally but yet importantly, they are osteotransductive, i.e., after implantation, calcium orthophosphate cements are replaced by a new bone tissue [29–31].

The aim of biomimetic bone cements is to disturb bone functions and properties as little as possible and, until a new bone has been grown, to behave temporary in a manner similar to that of bone. From a biological point of view, this term defines cements that can reproduce the composition, structure, morphology, and crystallinity of bone crystals [32, 33]. Therefore, the discovery of self-setting calcium orthophosphate formulations was a significant step forward in the field of bioceramics for bone regeneration, since it established good prospects for minimally invasive surgical techniques that were less aggressive than the classical surgical methods [34]. Such formulations provide surgeons with a unique ability of manufacturing, shaping, and implanting the bioactive bone-substitute materials on a patient-specific base in real time in the surgery room. Implanted bone tissues also take benefits from initial setting characteristics of the cements that give, in an acceptable clinical time, a suitable mechanical strength for a shorter tissue functional recovery. The major advantages of the self-setting calcium orthophosphate formulations include a fast setting time, excellent moldability, outstanding biocompatibility, and easy manipulation; therefore, they are more versatile in handling characteristics than prefabricated calcium orthophosphate granules or blocks. Besides, like any other type of calcium orthophosphate bioceramics, the self-setting formulations provide with the opportunity for bone grafting using alloplastic materials, which are unlimited in quantity and provide no risk of infectious diseases [35–37].

Since self-setting calcium orthophosphate formulations are intended for using as implanted biomaterials for parenteral application, for their chemical composition one might employ all ionic compounds of oligoelements occurring naturally in a human body. The list of possible additives includes (but is not limited to) the following cations: Na^+, K^+, Mg^{2+}, Ca^{2+}, Sr^{2+}, H^+ and anions: PO_4^{3-}, HPO_4^{2-}, $H_2PO_4^-$, $P_2O_7^{4-}$, CO_3^{2-}, HCO_3^-, SO_4^{2-}, HSO_4^-, Cl^-, OH^-, F^-, SiO_4^{4-} [29]. Therefore, mixed-type self-setting formulations consisting of calcium orthophosphates and other calcium salts (e.g., calcium sulfates [38–47], calcium pyrophosphate [48–50], calcium polyphosphates [51, 52], calcium carbonates [33, 53–55], calcium oxide [56–61], calcium hydroxide [62–64], calcium aluminate [26, 65, 66], calcium silicates [67–71], etc.), strontium orthophosphate [72–74], magnesium orthophosphate [74–78], magnesium oxide [79], Zn-containing compounds [80], as well as cements made of various ion-substituted calcium orthophosphates (e.g., $Ca_2KNa(PO_4)_2$, $NaCaPO_4$, $Na_3Ca_6(PO_4)_5$, magnesium substituted CDHA, strontium substituted CDHA, etc.) [81–90] are available. Furthermore, self-setting formulations might be prepared in the reaction-setting mixture of $Ca(OH)_2$–KH_2PO_4 system [91], as well as by treatment of calcium carbonates with orthophosphate solutions [92]. More to the point, self-setting formulations possessing magnetic properties due to incorporation of iron oxides have been developed as well [93, 94]. However, with a few important exceptions, such ion-substituted formulations have not been considered in this chapter.

The purpose of this chapter is to review the chemistry, physical and mechanical properties of the available self-setting calcium orthophosphate formulations with the specific reference to their biomedical applications in dentistry and surgery.

5.2 General Information and Data

According to Wikipedia, the free encyclopedia: "In the most general sense of the word, *cement* is a binder, a substance that sets and hardens independently and can bind other materials together. The name "cement" goes back to the Romans who used the term "*opus caementitium*" to describe masonry, which resembled concrete and was made from crushed rock with burnt lime as binder. The volcanic ash and pulverized brick additives, which were added to the burnt lime to obtain a hydraulic binder, were later referred to as cementum, cimentum, cäment, and cement" [95]. Thus, *calcium orthophosphate cement* appears to be a generic term to describe chemical formulations in the ternary system $Ca(OH)_2$–H_3PO_4–H_2O which can experience a transformation from a liquid or pasty state to a solid state and in which the end product of the chemical reactions is a calcium orthophosphate.

The first self-setting calcium orthophosphate cement formulation consists of the equimolar mixture of TTCP and dicalcium phosphate (DCPA or DCPD) [96] which is mixed with water at a P/L ratio of 4:1; the paste hardened in about 30 min and formed CDHA [97, 98]. This highly viscous, non-injectable paste can be molded and, therefore, is used mainly as a contouring material in craniofacial surgery. In 1990s, it was established that there were about 15 different binary combinations of calcium orthophosphates, which gave pastes upon mixing with water or aqueous solutions, so that the pastes set at room or body temperature into a solid cement. The list of these combinations is available in literature [99–101]. From these basic systems, secondary formulations could be derived containing additional or even non-reactive compounds but still setting like cements [29, 58, 99, 102–116]. Concerning their viscosity, both pasty cement formulations [117–120] and putties [121] of a very high viscosity [122–125] are known as well.

According to the classical solubility data of calcium orthophosphates (Fig. 1.6), depending upon the pH value of a cement paste, after setting all calcium orthophosphate cements can form only two major end-products: a precipitated poorly crystalline HA or CDHA [126] at pH > 4.2 and DCPD (also called "brushite" [127]) at pH < 4.2 [128]. However, the pH-border of 4.2 is shifted to a higher value of pH in the real cement formulations. Namely, DCPD might be formed at pH up to ~6, while CDHA normally is not formed at pH below 6.5–7 (Table 1.1). The results of the only study on an ACP cement demonstrated that this end product was rapidly converted into CDHA [113]; thus, it also belongs to apatite-forming formulations. Besides, there are some papers devoted to OCP-forming cements [129–132]; however, contrary to the reports of late 1980s [129] and early 1990s [130], in recent papers either simultaneous formation of OCP and CHDA has been detected [132] or no phase analysis has been performed [131]. Strong experimental evidences of the existence of a transient OCP phase during cement setting were found in still another study; however, after a few hours, the OCP phase disappeared giving rise to the final CDHA phase [25]. Thus, all existing formulations of calcium orthophosphate cements can be divided into two major groups: apatite cements and brushite cements [133]. The final hardened product of the cements is of the paramount importance because it determines the solubility and, therefore, *in vivo* bioresorbability. Since the chemical composition of mammalian bones is similar to an ion-substituted CDHA, apatite-forming cement formulations have been more extensively investigated. Nevertheless, many research papers on brushite cements have been published as well.

All self-setting calcium orthophosphate formulations are made of an aqueous solution and fine powders of one or several calcium orthophosphate(s). Here, dissolution of the initial calcium orthophosphate(s) (quickly or slowly depending on the chemical composition and solution pH) and mass transport appear to be the primary functions of an aqueous environment, in which the dissolved reactants form a supersaturated (very far away from the equilibrium) microenvironment with regard to precipitation of the final product(s) [135, 136]. The relative stability and solubility of various calcium orthophosphates (see Table 1.1) is the major driving force for setting reactions that occur in these cements. Therefore, mixing of a dry powder with an aqueous solution induces various chemical transformations, where crystals of the initial calcium orthophosphate(s) rapidly dissolve(s) and precipitate(s) into crystals of CDHA (precipitated HA) or DCPD with possible formation of intermediate precursor phases (e.g., ACP [113] and OCP [25, 129–132]). During precipitation, the newly formed crystals grow and form a web of intermingling microneedles or microplatelets of the final products, thus provide a mechanical rigidity to the hardened cements. In other words, entanglement of the newly formed crystals is the major reason of setting (Fig. 4.9). For the majority of apatite cements, water is not a reactant in the setting reaction. Therefore, the quantity of water, actually needed for setting of apatite cements, is very small [22, 135, 137]. However, for brushite cements, water always participates in the chemical transformations because it is necessary for DCPD formation. Due to this reason, brushite cements are always hydraulic, while usually this term is not associated with apatite cements.

Setting of calcium orthophosphate cements is a continuous process that always starts with dissolution of the initial compounds in an aqueous system. This process supplies ions of calcium and orthophosphate into the solution, where they chemically interact and precipitate in the form of either the end products or precursor phases, which causes the cement setting [138–140]. This was confirmed by Ishikawa and Asaoka, who showed that when TTCP and DCPA powders were mixed in double-distilled water, both powders were dissolved. The dissolved calcium and orthophosphate ions in the solution were then precipitated in the form of CDHA on the surface of the powders [141]. The precipitate can be either a gel or a conglomerate of crystals. Therefore, the hardening mechanism is either a sol-gel transition of ACP [113] or entanglement of the precipitated crystals of other calcium orthophosphates (Fig. 4.9) [29]. For example, for the classical Brown-Chow cement formulation, after the initial setting, petal- or needle-like crystals enlarge epitaxially and are responsible for the adherence and interlocking of the crystalline grains, which result in hardening. After ~ 2 h, the newly formed crystals become rod-like, resulting from higher crystallinity with the observation of more material at the inter-particle spaces. During this period, the cement setting reaction proceeded at a near-constant rate, suggesting that the reaction rate was limited by factors that are unrelated to the amounts of the starting materials and the reaction products present in the system. Such factors could be related to the surface area of DCPA or TTCP or to the diffusion distances over which the calcium and orthophosphate ions migrate in order to form CDHA [142–144]. At ~24 h, the crystals are completely formed, being very compacted in some areas of high density, and well separated in areas with more porosity [106, 111, 112].

The chemical reactions occurring during setting of calcium orthophosphate cements depend on their chemical composition. However, it can be stated that only two major chemical types of the setting reaction are possible. The first type occurs according to the classical rules of the acid–base interaction, i.e. a relatively acidic calcium orthophosphate reacts with a relatively basic one to produce a relatively neutral compound. The first cement by Brown and Chow is a typical example of this type because TTCP (basic) reacts with

DCPA (slightly acidic) in an aqueous suspension to form a poorly crystalline precipitated HA (slightly basic) [14, 15]:

$$2Ca_4(PO_4)_2O + 2CaHPO_4 \rightarrow Ca_{10}(PO_4)_6(OH)_2 \quad (5.1)$$

Earlier, it was believed that DCPA and TTCP reacted upon mixing with water to form the stoichiometric HA [13–16]. However, further investigations have shown that only the first nuclei consist of a nearly stoichiometric HA, whereas further growth of these nuclei occurs in the form of CDHA [145]. Besides, there is a study demonstrating that the initially formed stoichiometric HA further interacts with remaining DCPD to form CDHA [146].

According to Eq. (5.1), formation of HA releases neither acidic nor basic by-products. Thus, the liquid phase of the cement remains at a near constant pH of ~7.5 for the TTCP + DCPD and ~8.0 for the TTCP + DCPA formulations, respectively [142–144]. Various deviations from the stoichiometry of Eq. (5.1) were studied in details and various apatitic calcium orthophosphates with Ca/P ionic ratio within 1.5–1.67 were found as the end product [147]. The effect of mixing ratio and pH on the reaction between TTCP and DCPA is well described elsewhere [148]. Furthermore, the influence of Ca/P ionic ratio of TTCP on the properties of the TTCP + DCPD cement was studied as well [149].

A blend proposed by Lemaître et al. [150, 151] is another example of the acid–base interaction where β-TCP (almost neutral) reacts with MCPM (acidic) to form DCPD (slightly acidic):

$$\beta\text{-}Ca_3(PO_4)_2 + Ca(H_2PO_4)_2{\cdot}H_2O + 7H_2O \rightarrow 4CaHPO_4{\cdot}2H_2O \quad (5.2)$$

In Eq. (5.2) MCPM might easily be substituted by orthophosphoric acid [152–155] or MCPA, while β-TCP might be replaced by either α-TCP [156, 157], CDHA [158, 159], or HA [160]. For example:

$$Ca_9(HPO_4)(PO_4)_5(OH) + 3H_3PO_4 + 17H_2O \rightarrow 9CaHPO_4{\cdot}2H_2O \quad (5.3)$$

Furthermore, self-setting formulations based on mixtures of ACP + α-TCP [161], ACP + DCPD [162, 163], DCPA + α-TCP [157], OCP + TTCP [164], OCP + α-TCP [165, 166], and partially crystallized calcium orthophosphate + DCPA [167] as the initial reagents, are also available.

The second type of the setting reaction might be defined as hydrolysis of a metastable calcium orthophosphate in aqueous media. As the result, both the initial and final compounds have the same Ca/P ionic ratio. Due to the fact, that only one calcium orthophosphate is used; the solid part of such formulations might be called as a single-phase (or single-component) cement powder [168]. Cements made of ACP + an aqueous solution [169, 170], α-TCP + an aqueous solution [171–178], β-TCP + an aqueous solution [175, 179], DCPA + an aqueous solution [24], CDHA + an aqueous solution [25], TTCP + an aqueous solution [26, 180, 181], or γ-radiated TTCP + an aqueous solution [182–184] are the typical examples; the majority of them are re-crystallized to CDHA during setting:

$$Ca_xH_y(PO_4)_z{\cdot}nH_2O + H_2O \rightarrow Ca_{10-x}(HPO_4)_x(PO_4)_{6-x}(OH)_{2-x} + nH_2O \quad (5.4)$$

$$3(\alpha\text{- or }\beta\text{-})Ca_3(PO_4)_2 + H_2O \rightarrow Ca_9(HPO_4)(PO_4)_5(OH) \quad (5.5)$$

$$3Ca_4(PO_4)_2O + 3H_2O \rightarrow Ca_9(HPO_4)(PO_4)_5(OH) + 3Ca(OH)_2 \quad (5.6)$$

The experimental details on TTCP hydrolysis under a near-constant composition condition might be found elsewhere [185]. The details on α-TCP hydrolysis are also available. The results indicated that setting of α-TCP was initially controlled by surface dissolution; therefore, it depended on the surface area of the reactants [186–189]. Hydrolysis of DCPD to CDHA was studied as well [190]. Addition of ~2 wt.% of a precipitated poorly crystalline HA (i.e., CDHA) as a seed to α-TCP powder phase might be useful to accelerate the kinetics of reaction (5.5) [191].

Further, there is a single-phase cement powder consisting of K- and Na-containing CDHA (with the Ca/P ionic ratio of 1.64 ± 0.02) that sets and hardens after mixing with an aqueous solution of sodium citrate and sodium orthophosphate [17]. After setting, this formulation gives rise to formation of a weak cement (the compressive strength of 15 ± 3 MPa) consisting of the ion-substituted CDHA again (presumably, with another Ca/P ionic ratio), mimicking the bone mineral. Unfortunately, neither the setting reaction nor the setting mechanism of this cement has been disclosed in literature

[17]. What's more, self-setting cement might be prepared from the thermal decomposition product of HA [192].

The hydration process of calcium orthophosphate cements is slightly exothermic (which is beneficial for biomedical applications) and undergoes five periods: initiating period, induction period, accelerating period, decelerating period, and terminating period [193]. For the classical Brown-Chow cement formulation, the activation energy of the hydration reaction is 176 kJ/mol [194]. The rate of heat liberation during the solidification of calcium orthophosphate cements is low. The results of adiabatic experiments showed that the temperature rise arrived at the highest value of 37 °C 3 h later, which would cause no harm to surrounding tissues [193]. The results show that the hardening process of this cement is initially controlled by dissolution of the reactants in a 4 h period and subsequently by diffusion through the product layer of CDHA around the grains [112]. In general, setting of calcium orthophosphate cements occurs mostly within the initial ~6 h, yielding an ~80% conversion to the final products. The volume of the cements stays almost constant during setting. However, after hardening, calcium orthophosphate cements always form brittle ceramics with the tensile strength of 5–20 times lower than the compression strength [195, 196]. Since this material is weak under tensile forces, these cements can only be used either in combination with metal implants or in non-load bearing (e.g., craniofacial) applications [137, 197–199]. This is confirmed by the mechanical characterization of a bone-defect model filled with ceramic cements [200].

To conclude this part, one must stress, that chemical equations (5.1–5.6) of the cement setting are valid for the *in vitro* conditions only. There are evidences that samples of calcium orthophosphate cement retrieved 12 h after hardening *in vivo* already contained carbonateapatite, even though the initial cement mixture did not contain carbonate as one of the solid components [201]. The mass fraction of carbonate in the 12 h samples was about 1%. The results suggest that under the *in vivo* conditions, carbonate is readily available and this allows formation of carbonateapatite in favor of carbonate-free CDHA [201].

The United States Food and Drug Administration (FDA) has approved several cement formulations for clinical use [22, 202]. Some examples are given in Table 5.1. The same formulations have also received a Conformite Europene (CE) mark for certain maxillofacial indications and for use as a bone-void filler in the specific non-load-bearing orthopedic indications [137]. The major properties of these formulations are available in literature [22]. An extended list of the cement formulations is presented in Table 5.2 [125]. Other lists of the commercially available injectable bone cements with their chemical composition (when obtainable) might be found elsewhere [34, 144, 203, 204], while various types of bone cements and fillers are listed in still another review [198]. A general appearance of two randomly chosen commercial calcium orthophosphate cements is shown in Fig. 5.1. Besides, even more cement formulations are in experimental stages.

Table 5.1 Some self-setting calcium orthophosphate cement formulations having the 510(k) clearance from the FDA [20, 137, 202]. The technical data on these cements might be found in literature [22].

Product*	Manufacturer	Applications*
BoneSource™**	Striker Howmedica Osteonics (Rutherford, NJ)	Craniofacial
α-Bone Substitute Material (α-BSM®)***	Etex Corporation (Cambridge, MA)	Filling of bone defects and voids, dental, craniofacial
Skeletal Repair Systems (SRS®)	Norian Corporation (Cupertino, CA)	Skeletal distal radius fractures, craniofacial

* In Europe, other applications may apply, and the materials may be sold with a different commercial name.

** BoneSource™ is the original formulation of calcium orthophosphate cement developed by Brown and Chow.

*** In Europe, it is distributed by Biomet Merck (Zwijndrecht, The Netherlands) as Biobon® [137], while in North America it is marketed by Walter Lorenz Surgical (Jacksonville, FL) as Embarc® [22].

Table 5.2 A list of the commercial self-setting calcium orthophosphate formulations with the producer, product name, composition (when available) and main end product [125].

Producer	Commercial name	Composition	Product
Berkeley Advanced Biomaterials (US)	Cem-Ostetic™	Powder: calcium orthophosphates (details unknown); Solution: Sterile water	apatite
	Tri-Ostetic™	Powder: calcium orthophosphates (details unknown); Solution: Sterile water	apatite
Biomatlante (FR)	MCPC	Powder: mainly α-TCP, ACP, BCP (HA + β-TCP); Solution: phosphate buffered solution	apatite
Biomet (US) Interpore (US) Walter Lorenz Surgical (GER)	Calcibon®	Powder: α-TCP (61%), DCPA (26%), $CaCO_3$ (10%), CDHA (3%); Solution: H_2O, Na_2HPO_4	apatite
	Mimix™	Powder: TTCP, α-TCP, trisodium citrate; Solution: citric acid aqueous solution	apatite
	Quick Set Mimix™	Powder: Calcium orthophosphate powders, trisodium citrate; Solution: citric acid aqueous solution	apatite
Calcitec (US)	Osteofix	Powder: calcium orthophosphate and calcium oxide powders; Solution: phosphate buffer	apatite
ETEX (US)	α-BSM®; Embarc; Biobon	Powder: ACP (50%), DCPD (50%); Solution: Un-buffered aqueous saline solution	apatite

	β-BSM®	Composition: could not be found (it has apparently a higher compressive strength and better injectability than α-BSM®)	apatite
	γ-BSM®	Composition: could not be found (putty consistency)	apatite
	OssiPro	Composition: could not be found; the cement is claimed to be macroporous after hardening	apatite
	CarriGen	Composition: synthetic calcium orthophosphate, sodium carboxymethylcellulose, sodium bicarbonate and sodium carbonate	apatite
Graftys (FR)	Graftys® HBS	Powder: mainly β-TCP, ACP, BCP (HA + β-TCP); Solution: phosphate buffered solution	apatite
	Graftys® Quickset	Composition: calcium orthophosphate salts, hydroxypropylmethylcellulose and orthophosphate-based aqueous solution	apatite
Kasios (FR)	Jectos Eurobone®	Powder: β-TCP (98%), $Na_2P_2O_7$ (2%); Solution: H_2O, H_3PO_4 (3.0M), H_2SO_4 (0.1M)	brushite
	Jectos+	Composition: could not be found (likely to be close to that of Jectos)	brushite
Kyphon (US)	KyphOs™	Powder: β-TCP (77%), $Mg_3(PO_4)_2$ (14%), $MgHPO_4$ (4.8%), $SrCO_3$ (3.6%); Solution: H_2O, $(NH_4)_2HPO_4$ (3.5M)	apatite

(Contd.)

Table5.2 (Contd.)

Producer	Commercial name	Composition	Product
Mitsubishi Materials (J)	Biopex®	Powder: α-TCP (75%), TTCP (20–18%), DCPD (5%), HA (0–2%) Solution: H_2O, Na succinate (12–13%), Na chondroitin sulfate (5–5.4%)	apatite
	Biopex®-R	Powder: α-TCP, TTCP, DCPD, HA, $Mg_3(PO_4)_2$, $NaHSO_3$; Solution: H_2O, Na succinate, Na chondroitin sulfate	apatite
Produits Dentaires SA (CH) CalciphOs (CH)	VitalOs[4]	Solution 1: β-TCP (1.34 g), $Na_2H_2P_2O_7$ (0.025 g), H_2O, salts (0.05M PBS solution, pH 7.4); Solution 2: MCPM (0.78 g), $CaSO_4{\cdot}2H_2O$ (0.39 g), H_2O, H_3PO_4 (0.05M)	brushite
Shanghai Rebone Biomaterials Co (CN)	Rebone	Powder: TTCP, DCPA; Solution: H_2O	apatite
Skeletal Kinetics (US)	Callos™	Composition: α-TCP, $CaCO_3$, MCPM; Solution: sodium silicate	apatite
	Callos Inject™	Composition: α-TCP and unknown compounds (likely to be close to that of Callos™)	apatite
	OsteoVation EX Inject	Probably similar to Callos Inject™ (Product produced by S.K. but sold by OsteoMed)	apatite
Stryker (US) Leibinger (GER)	BoneSource™	Powder: TTCP (73%), DCPD (27%); Solution: H_2O, mixture of Na_2HPO_4 and NaH_2PO_4	apatite

Stryker (US)	HydroSet™	Powder: TTCP, DCPD, trisodium citrate; Solution: H_2O, polyvynilpyrrolidone, Na orthophosphate	apatite
Synthes (US)	Norian® SRS Norian® CRS	Powder: α-TCP (85%), $CaCO_3$ (12%), MCPM (3%); Solution: H_2O, Na_2HPO_4	apatite
	Norian® SRS Fast Set Putty Norian® CRS Fast Set Putty	Composition: could not be found (likely to be close to that of Norian SRS/CRS)	apatite
	Norian Drillable	Composition: calcium orthophosphate powder, bioresorbable fibers and Na hyaluronate solution	apatite
	chronOS™ Inject	Powder: β-TCP (73%), MCPM (21%), $MgHPO_4 \cdot 3H_2O$ (5%), $MgSO_4$ (<1%), $Na_2H_2P_2O_7$ (<1%); Solution: H_2O, Na hyaluronate (0.5%)	brushite
Teknimed (FR)	Cementek®	Powder: α-TCP, TTCP, Na glycerophosphate; Solution: H_2O, $Ca(OH)_2$, H_3PO_4	apatite
	Cementek® LV	Powder: α-TCP, TTCP, Na glycerophosphate, dimethylsiloxane; Solution: H_2O, $Ca(OH)_2$, H_3PO_4	apatite

Note: . The main endproduct of the reaction can be either an apatite (CDHA, carbonateapatite, etc...) or brushite (= DCPD) [125].

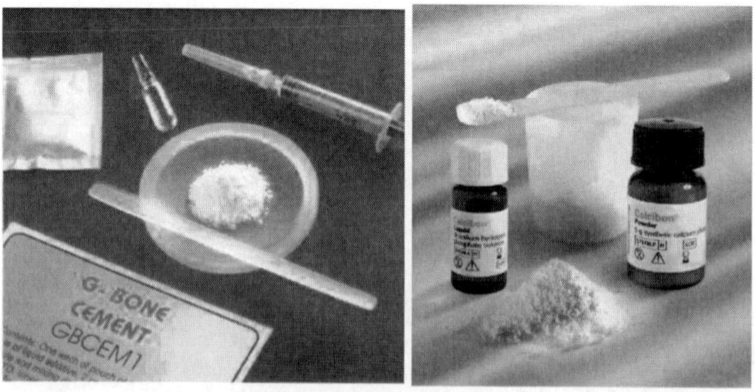

Figure 5.1 A presentation of two randomly chosen commercial calcium orthophosphate cements.

5.3 Two Major Types of Calcium Orthophosphate Cements

5.3.1 Apatite Cements

Typically, apatite cement formulations have a precipitated poorly crystalline HA and/or CDHA as the end product of the setting reaction (see chemical equations (5.1), (5.4–5.6)), although traces of the unreacted starting materials can be present [106]. A FA-forming cement is also known; it might be prepared by the same way but in the presence of F^--ions [205]. Due to the initial presence of carbonates, such commercial formulations as Norian SRS® and Biocement D® form a non-stoichiometric carbonatapatite or dahllite ($Ca_{8.8}(HPO_4)_{0.7}(PO_4)_{4.5}(CO_3)_{0.7}(OH)_{1.3}$) as the end product [53, 206]. As both CDHA and carbonateapatite are formed in an aqueous environment and have a low crystallinity, they appear to be similar to biological apatite of bones and teeth. These properties are believed to be responsible for their excellent *in vivo* resorption characteristics. Conventional apatite cements contain TCP and/or TTCP phases in their powder components [34], while a single component cement powder consisting of K- and Na-containing CDHA is also available [17]. The reactivity of TCP-based apatite cements varies as a function of TCP crystal phase, crystallinity and particle size [207, 208]. Generally, a higher reactivity is observed with a thermodynamically less stable phase (from β-TCP to α-TCP and further to ACP) and with a smaller particle size [175]. Nominally, it might be stated that formation of apatites through a cementation reaction is a sort of a biomimetic process because it occurs in physiological environment and at body temperature [37]; however, both the crystallization kinetics and a driving force are very far away from the biomimeticity. A unique feature of the hardened apatite cements is that the force linking the newly formed crystals (of both CDHA and carbonatapatite) is weak; therefore, crystals can be easily detached from the bulk of hardened cement, especially after dissolution has partly occurred. When this happens, osteoclasts and other cells can easily ingest the apatite crystals [209].

Immediately after implantation, any cement becomes exposed to blood and other tissue fluids that delays the setting time. Intrinsic setting time for apatite cements has been extensively studied and it appear to be rather long. For example, for the original formulation by Brown and Chow it ranges from 15 to 22 min [14, 15]. This may result in procedural complications. To remedy this, the amount of liquid might be reduced to a possible minimum. Therefore, all apatite cements look like viscous and easily moldable pastes, which tend to be difficult to inject. Besides playing with the P/L ratio, the setting time can also be reduced by using additives to the liquid phase (which is distilled water in the Brown-Chow formulation [14, 15]). The list of additives includes phosphoric acid, MCPM, and other soluble orthophosphates. These additives promote dissolution of the initial solids by lowering the solution pH. In such cases, a setting time in the range of 10–15 min can be obtained [169, 171–177, 210]. The influence of soluble orthophosphates (e.g., Na_2HPO_4 or NaH_2PO_4) on the setting time of apatite cements is explained by the fact that dissolution of DCPA and formation of CDHA during setting occur in a linear fashion, thus avoiding early formation of CDHA. This is important because too early formation of CDHA might engulf un-reacted DCPA, which slows down DCPA dissolution and thus the setting kinetics becomes slower, while the presence of sodium orthophosphates prevents DCPA particles from being isolated [211]. Particle size [191, 212, 213], temperature of the liquid phase, and initial presence of HA as a seed in the solid phase are other factors that influence the setting time [14, 15, 37, 207, 208]; however, *in vitro* studies demonstrated that these parameters did not affect significantly [106]. On the other hand, a reduction in particle size was found to result in a significant decrease in both initial and final setting times [191, 212, 213], an acceleration of the hardening rate [191], and hydration kinetics of the hardening cement [213]. Besides, the crystallite sizes of the final product can be strongly reduced by increasing the specific surface of the starting powder, which allows developing calcium orthophosphate cements with tailored structures at the micro- and nanoscale levels [191]. Unfortunately, an unclear correlation was found between the particle dimensions of the initial calcium orthophosphates and mechanical properties of the hardened cements: namely, a significant increase in compressive strength and storage modulus was reported for some formulations [212, 213] but a minor effect on compressive strength was discovered for other ones [191]. This inconsistence is not surprising because the manufacturing method used to produce test samples varies from one author to the other. Therefore, the only remaining fact is that calcium orthophosphate cements are brittle and hence worthless for load-bearing applications [197, 198].

Setting process of the most types of apatite cements occurs according to just one chemical reaction (see chemical equations (5.1), (5.4–5.6)) and at near the physiological pH. The latter may additionally contribute to the high biocompatibility observed for these materials [142–144]. For the classical formulation by Brown and Chow, the transmission electron microscopy results suggested the process for early-stage apatite formation as follows: when TTCP and DCPA powders were mixed in an orthophosphate-containing solution, TTCP powder quickly dissolved due to its higher solubility in acidic media. Then the dissolved ions of calcium and orthophosphate, along with ions already existing in the solution, were precipitated predominantly onto the surface of DCPA particles. Few apatite crystals were observed on the surface of TTCP powder. At a later stage of the reaction, an extensive growth of apatite crystals or whiskers effectively linked DCPA particles together and bridged the larger TTCP particles causing the cement setting [214].

However, Norian SRS® and Cementek® were found to set according to two chemical reactions: precipitation of DCPD, followed by precipitation of either CDHA or carbonatapatite:

$$\alpha\text{-}Ca_3(PO_4)_2 + Ca(H_2PO_4)_2{\cdot}H_2O + 7H_2O \rightarrow 4CaHPO_4{\cdot}2H_2O \quad (5.7)$$

$$5.2CaHPO_4{\cdot}2H_2O + 3.6CaCO_3 \rightarrow Ca_{8.8}(HPO_4)_{0.7}(PO_4)_{4.5}(CO_3)_{0.7}(OH)_{1.3} + 2.9CO_2 + 12H_2O \quad (5.8)$$

The initial chemical reaction (5.7) was very fast and provoked DCPD formation and setting of the cement pastes within seconds. The second step was slower: DCPD reacted completely within several hours with remaining α-$Ca_3(PO_4)_2$ and $CaCO_3$ forming carbonatapatite according to equation (5.8). The latter step caused the cement hardening. A similar two-step hardening mechanism was established for a cement consisting of MCPM and CaO: in the first step, during the mixing time, MCPM reacted with CaO immediately to give DCPD, which, in the second step, reacted more slowly with the remaining CaO to give CDHA [58].

In addition, the setting mechanism of an apatite cement was investigated in details for a three-component mixture of TTCP, β-TCP, and MCPM dry powders in convenient proportions and with the overall atomic Ca/P ratio equal to 1.67. Two liquid phases in a raw were used to damp the cement powder, initially it was water + ethanol (ethanol was added to slow down the hardening) and afterwards orthophosphoric acid and sodium glycerophosphate were added to water to prepare a reactive liquid [135]. At the very beginning, DCPD was found to form according to two chemical reactions:

$$Ca(H_2PO_4)_2{\cdot}H_2O + \beta\text{-}Ca_3(PO_4)_2 + 7H_2O \rightarrow 4CaHPO_4{\cdot}2H_2O \quad (5.9)$$

$$Ca_4(PO_4)_2O + 2H_3PO_4 + 7H_2O \rightarrow 4CaHPO_4{\cdot}2H_2O \quad (5.10)$$

The formation reactions of DCPD were fast and corresponded to the setting stage. Afterwards, TTCP reacted with the previously formed DCPD and with β-TCP to give CDHA according to the reactions:

$$2Ca_4(PO_4)_2O + 2CaHPO_4{\cdot}2H_2O \rightarrow Ca_{10-x}(HPO_4)_x(PO_4)_{6-x}(OH)_{2-x} + xCa(OH)_2 + (4-x)H_2O \quad (5.11)$$

$$2Ca_4(PO_4)_2O + 4\beta\text{-}Ca_3(PO_4)_2 + (2+2x)H_2O \rightarrow 2Ca_{10-x}(HPO_4)_x(PO_4)_{6-x}(OH)_{2-x} + 2xCa(OH)_2 \quad (5.12)$$

The formation reactions of the CDHA phase were quite slow and corresponded to the hardening stage. Although OCP was not detected in that study, its formation as an intermediate phase was postulated for this formulation [135]. A similar suggestion on the intermediate formation of OCP was made for the setting mechanism of Brown-Chow classical cement formulation [101, 106]; however, a reliable evidence for its presence is still lacking [172, 215]. In both cases, OCP was suggested to appear as an intermediate because it was a faster forming phase than CDHA. This hypothesis is based upon the classical studies performed by Prof. W. E. Brown et al. about the precursor phase formation during chemical crystallization of apatites in aqueous solutions [216–218].

Solubility of the hardened apatite cements in aqueous solutions is expected to be rather similar to that of bone mineral. This means that they are relatively insoluble at neutral pH and increasingly soluble as pH drops down; this is an important characteristic of normal bone mineral that facilitates controlled dissolution by osteoclasts [206].

To conclude this part, one should mention, that in 2000 the US bone-substitute market for Norian SRS® accounted for ~15% of the total sales, followed by BoneSource™ at ~13%, and α-BSM® at ~8.5% [137].

5.3.2 Brushite Cements

As indicated by its name, DCPD is the major end product of the setting reaction of brushite cements (chemical equations (5.2) and (5.3)). Mirtchi and Lemaître [150] and independently Bajpai et al. [151] introduced this type of the cements in 1987. Up to now, several formulations have been already proposed, e.g., β-TCP + MCPM [150, 152], β-TCP + H_3PO_4 [151, 153, 154], and TTCP + MCPM + CaO [219]. All brushite cements are set by the acid–base interaction only. As DCPD can only precipitate at the solution pH < 6, the paste of brushite cement is always acidic during setting [153, 220]. For example, during setting of a β-TCP + MCPM formulation, the cement pH varies from very acidic pH values of ~2.5, to almost neutral pH values of ~6.0 [153]. Replacing MCPM by orthophosphoric acid renders the cement paste very acidic for the initial ~30 s but then the pH profile follows that obtained with MCPM. It is important to notice that β-TCP + H_3PO_4 formulations have several advantages over β-TCP + MCPM formulations, namely: (i) easier and faster preparation, (ii) a better control of the chemical composition and reactivity, (iii) improved physico-chemical properties, such as longer setting times and larger tensile strengths due to a higher homogeneity. However, the use of orthophosphoric acid might impair the biocompatibility of the cement formulation, due to low pH values during setting [153]. If a cement formulation contains an excess of a basic phase, the equilibrium pH will be given by the intersection of the solubility isotherms of the basic phase with that of DCPD. For example, the equilibrium pH values of β-TCP + MCPM, HA + MCPM, and TTCP + MCPM mixtures are 5.9, 4.2, and 7.6, respectively [197, 198].

As the solubility of calcium orthophosphates decreases with increasing of their basicity (Table 1.1 and Fig. 1.6), the setting time of brushite cements much depends on the solubility of a basic phase: the higher its solubility, the faster the setting time. Therefore, the setting time of the cements made of MCPM + a basic calcium orthophosphate increases in the order: HA > β-TCP > α-TCP [197, 198]. For example, HA + MCPM mixtures have a setting time of several minutes: β-TCP + MCPM mixtures—of 30–60 s and α-TCP + MCPM mixtures—of a few seconds [150, 151]. Follow-up of the chemical composition by ^{31}P solid state NMR enabled to show that the chemical setting process for β-TCP + MCPM formulation appeared to reach an end after ~20 min [221]. Nevertheless, despite this initial high reactivity, the hardening reaction of brushite cements typically lasts one day until completion [207, 208]. Additives that inhibit the crystal growth of DCPD have been successfully used to increase the setting time of β-TCP + MCPM mixtures [222]. In contrast to apatite cements, the brushite cements can be initially liquid and still set within a short period of time [197, 198].

By itself, brushite is remarkably biocompatible and bioresorbable [220]. Due to both a better solubility of DCPD if compared to that of CDHA (Table 1.1 and Fig. 1.6) and metastability of DCPD under physiological conditions [223], after implantation brushite cements are faster degradable than apatite ones [224–226]. They are quickly resorbed *in vivo* and suffered from a rapid decrease in strength (although the mechanical properties of the healing bone increase as bone ingrowth occurs [35]). Short setting times, low mechanical strength, and limited injectability seem to prevent brushite cements from a broader clinical application. However, the major reason why brushite cements are not more widespread is probably not related to the mechanical issues but just to a later arrival on the market. Use of sodium citrate or citric acid as setting retardants is an option to get more workable and less viscous pastes of brushite cements [23, 227–230]. Similar effect might be achieved by addition of chondroitin 4-sulfate [231] and glycolic acid [232]. For the cement formulations with orthophosphoric acid as the initial reactant (see chemical equation (5.3)), acid-deficient formulations were also found to improve the workability. In this case, the setting reaction might be described by the following chemical equation [230]:

$$3.7\beta\text{-}Ca_3(PO_4)_2 + H_3PO_4 + 27.8H_2O \rightarrow 3CaHPO_4{\cdot}2H_2O + 2.7\beta\text{-}Ca_3(PO_4)_2 + 21H_2O \quad (5.13)$$

Although several studies revealed that too much of DCPD in a given volume was not detrimental to the biological properties of brushite cements [35, 206, 219], occasionally, when large quantities of brushite cements were used, a certain degree of tissue inflammation during the first weeks of *in vivo* implantation were reported [226, 230, 233]. Further investigations indicated that the inflammatory could be due to a partial transformation of DCPD into CDHA with release of orthophosphoric acid [234]:

$$(10-x)CaHPO_4{\cdot}2H_2O \rightarrow Ca_{10-x}(HPO_4)_x(PO_4)_{6-x}(OH)_{2-x} + (4-x)H_3PO_4 + (18-x)H_2O \quad (5.14)$$

Transformation of DCPD into CDHA occurs via two successive processes: dissolution and precipitation [235] and can be retarded by adding magnesium ions to the cement paste, thus reducing the possibility of inflammation [197, 198]. The aforementioned case of acid-deficient formulations of brushite cements (chemical equation (5.13)) is an alternative, because it reduces the amount of unreacted acid in the cement [230] with an option to consume liberating in chemical equation (5.14) orthophosphoric acid by the excess of β-TCP. Implantation of previously set brushite cement might be the third option, because a solid material was found to be better tolerated than paste implants. Besides, more bone was formed at the solid implant contact and the solid material degraded not so rapidly [236]. For brushite cements, a linear degradation rate of 0.25 mm per week was reported [237]. This rapid degradation rate might lead to formation of an immature bone. Adding β-TCP granules to the cement paste could solve this problem because β-TCP granules might act as bone anchors and encourage formation of a mature bone [237, 238].

5.4 Various Properties

5.4.1 Setting and Hardening

Generally, self-setting calcium orthophosphate formulations must set slowly enough to provide sufficient time to a surgeon to perform implantation but fast enough to prevent delaying the operation. Ideally, good mechanical properties should be reached within minutes after initial setting. Two main experimental approaches are used to study the cement setting process: a batch approach and a continuous approach. In the batch approach, the setting reaction is stopped at various times and the resulting samples are analyzed to determine, e.g., the composition and compressive strength of the samples [207, 208]. There are currently two standardized methods to apply this approach, namely, Gillmore needles method (ASTM C266-89) [239] and Vicat needle method (ASTM C191-92) [240]. The idea of both methods is to examine visually the surface of cement samples to decide whether the cement has already set, i.e. if no mark can be seen on the surface after indentation. Besides, the setting process might be monitored in real time by non-destructive methods (the continuous approach), e.g., pulse-echo ultrasound technique [241, 242], isothermal differential scanning calorimetry [174, 175, 243–248] and alternating current (AC) impedance spectroscopy [249]. For example, calorimetry measurements suggested that in Eq. (5.2) the endothermic MCPM dissolution and the highly exothermic β-TCP dissolution occurred simultaneously, followed by the exothermic crystallization of DCPD [247]. Moreover, acid–base reactions (5.1–5.3) can be and have been analyzed by measuring the pH evolution of a diluted cement paste [207]. Finally yet importantly, methods of Fourier-transform infrared spectroscopy [24–26, 248, 250], solid state NMR [221], X-ray diffraction [24, 26, 50, 156, 251], and energy dispersive X-ray diffraction [24–26, 252, 253] might be applied as well. The latter techniques proved to be powerful even though they have limitations such as the time required for each measurement (250 s for an X-ray diffraction scan is a problem for fast setting reactions); besides the analysis is located at the surface of the sample where evaporation and thermal effects can modify the reaction rate of the surface compared to that of the bulk. Furthermore, the continuous approach is an indirect one, which markedly complicates an interpretation of the collected data, particularly in complex cement formulations [207].

A way to assess the rate of a cement hardening is to measure its setting time, which means the time required to reach a certain compressive strength, generally close to 1 MPa. The most straightforward approach is to prepare cement samples with a well-controlled geometry (e.g., cylinders), incubating these samples for various times in the right environment (temperature, humidity) and assessing the composition and mechanical properties of the samples as a function of time [207]. One should stress that setting time for calcium orthophosphate cements often corresponds to an earlier stage in the overall setting reaction, typically 5–15% of the overall reaction, while the end of the cement setting is typically reached after several days [106, 172]. Gillmore needles have been used with success to measure the initial (*I*) and final (*F*) setting times of calcium orthophosphate cements [99]. Namely, a light and thick needle is used to measure the initial setting time *I*, while a heavy and thin needle for the final setting time *F* [134]. The clinical meaning is that the cement paste should be implanted before time *I* and that the wound can be closed after time *F* (Fig. 5.2).

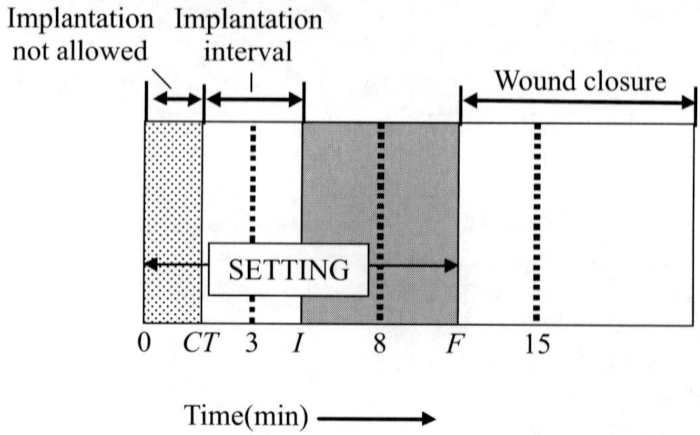

Figure 5.2 A diagram of the setting parameters relevant for a calcium orthophosphate cement: *CT*: cohesion time; *I*: initial setting time; *F*: final setting time. Adapted from Ref. [29] with permission.

The cement should not be deformed between times *I* and *F* because in that stage of the setting process any deformation could induce cracks [29]. The following handling requirements (in minutes) have been formulated for calcium orthophosphate cements, as a result [134, 254]:

$$3 \leq I < 8$$

$$I - CT \geq 1$$

$$F \leq 15$$

These parameters are represented schematically in Fig. 5.2. The second requirement means that the cohesion time (*CT*) must be at least 1 min before *I*, so that a clinician has at least 1 min to apply and to mold the material. *CT* is the time from which a cement no longer disintegrates when immersed in Ringer's solution [134]. As the mixing in a mortar is about 1 min, the shortest *CT* that can be allowed is about 2 min, so that a clinician has at least 1 min to collect the paste from the mortar and put it on a pallet knife or into a syringe with which it is to be transferred to the wound after *CT* and before *I* [134]. For dental applications, time *I* must be close to 3 min, whereas for orthopedic applications it must be close to 8 min. However, in no case it will be tolerable for the clinicians if time *F* becomes greater than 15 min [29, 134].

5.4.2 Mixing

In the clinical situation, self-setting calcium orthophosphate formulations can be either applied by the fingertips of a surgeon or injected from a syringe to the defect area of a bone. The first type of the clinical application requires formulation of high-viscosity cement pastes and putties, which can be applied manually as dough, while the second type requires formulation of low-viscosity compositions, which can be applied by injection from a syringe [134]. Currently, injection appears to be the preferred method between these two major options. Thus, a trade-off must be found between a high viscosity leading to too high injection forces and a low viscosity increasing the risk of cement extravasations. Viscosity values in the range of 100–2000 Pa·s are generally considered to be adequate [255].

In any case, before using a surgeon needs to have a cement powder and a liquid be mixed properly and thoroughly (to avoid the powder/liquid encapsulation) within the prescribed time. This process must be performed in a sterile environment. Therefore, a mixing procedure is very important because prior to be injected, a cement paste must be transferred from a mixing chamber into a syringe. Ideally, this should be done without trapping air bubbles by the cement paste [256]. Earlier, most calcium orthophosphate cements were manually mixed with aqueous solutions using a mortar and either a pestle or a spatula. That time, some concerns were raised about an insufficient and inhomogeneous mixing thus compromising the implant strength, as well as on inconsistencies between operators causing unpredictable variations in graft performance [257]. Mechanical mixing (e.g., by either an electrically powered mixing machine of Norian SRS/CRS® or Mini-malax® mixing system for Cementek® cement, produced by Teknimed S.A.) is the modern approach. It allows mixing the cement paste within 60–80 s and enables a rapid and reliable filling of the application syringe [34]. Besides, a cement powder and an aqueous solution might be placed into a syringe and mixed inside a shaker to produce a consistent cement paste of the desired viscosity [256]. A mechanical mixing was found to decrease both the mean viscosity of the curing cement paste and variability in the viscosity at a given time [258]. However, it did not improve the mechanical strength of the cement [197, 198].

Of the commercial formulations, listed in Table 5.1, Norian SRS® is sold as a reactant pack containing two components: a mixture of dry powders (MCPM + α-TCP + $CaCO_3$) and a liquid (aqueous solution of Na_2HPO_4). The components are mixed in the operating room. The paste that is formed is malleable and injectable for ~5 min; it hardens within ~10 min after injection [22, 206]. However, data are available that out of 4.5 ml Norian SRS® cement paste ~3 ml is injectable only, whereas up to 1.5 ml of the cement might remain uninjectable from the syringe [29]. This phenomenon is prescribed to the cement rheology and its interaction with the hydraulic forces of the syringe. α-Bone Substitute Material (α-BSM®) is also a two-component system; it is prepared from a mixture of ACP and DCPD powders and a saline solution [169]. Biopex® consists of four different calcium orthophosphates: 75 wt. % α-TCP, 18 wt. % TTCP, 5 wt. % DCPD and 2 wt. % HA. The aqueous solution contains 12 wt. % sodium succinate and 5 wt. % sodium chondroitin sulfate [259]. Effects of liquid phase on the basic properties of Biopex® were investigated. When mixed with neutral sodium hydrogen orthophosphate or succinic acid disodium salt solution, the initial setting times of the cement were 19.4 ± 0.55 and 11.8 ± 0.45 min, respectively. These setting times were much shorter than that of distilled water, 88.4 ± 0.55 min [260]. Biopex® is mixed with a spatula inside a syringe that can be opened from the front. After mixing, the front part is closed, a needle is inserted into this front part and the cement paste can be manually injected [197, 198].

Several systematic studies on the influence of composition and concentration of the liquids used in preparing of calcium orthophosphate cements were performed as well [23, 227]. Unfortunately, the results appeared to be rather unclear. For example, for several cements, mixing with sodium citrate or citric acid resulted in some effects on the initial setting time [23, 228], while for other cements the effect was insignificant [227]. Concentration increasing of sodium citrate solution resulted in initial setting time increasing [23, 227], although the injectability variations of the cement pastes were inconsistent [23, 228].

5.4.3 Rheological Properties of the Self-setting Formulations

Good injectability, adequate viscosity, and satisfactory cohesion are required for the successful biomedical applications of the self-setting calcium orthophosphate formulations [261, 262]. Injectability is an ability of a paste to be extruded through a small hole of a long needle (e.g., 2 mm diameter and 10 cm length) [263, 264] (other needles are also applied [265, 266]); and for certain applications, injectability is even a prerequisite. It is measured by the weight percentage of the cement paste that could be injected without demixing from a standard syringe by either a hand or a force of 100 N maximum (Fig. 5.3). Usually, injectability of a cement paste varies inversely with its viscosity, the P/L ratio, as well as the time after starting the mixing of liquid and powder [56, 264, 267]. Furthermore, injectability is improved with smaller particle sizes, with shorter and larger diameter cannula, as well as at smaller flow rates [263]. Moreover, particle shape of the powder is also expected to have effects on the injectability. Namely, powders with spherical shape or round particles may contribute good injectability. Spherical powder itself is easy to roll and thus good handling properties and injectability are found when a paste is prepared using spherical powder. Besides, it should be noted that the paste could become fluid with less amount of liquid phase since no captured liquid exists in the case of spherical powder [268].

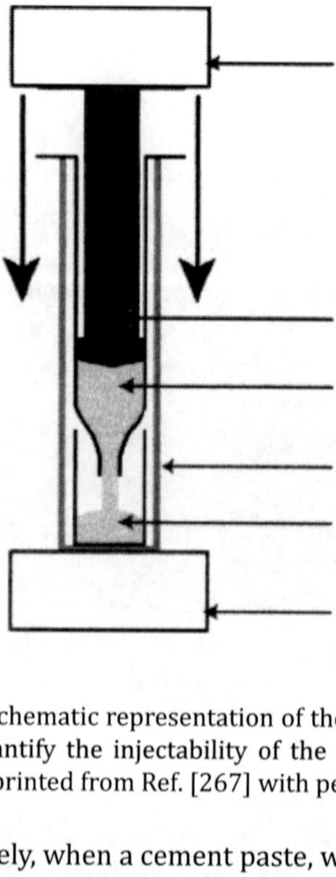

Figure 5.3 A schematic representation of the experimental setup used to quantify the injectability of the calcium phosphate cements. Reprinted from Ref. [267] with permission.

Unfortunately, when a cement paste, which is a biphasic mixture of a finely divided ceramic (powder, granules) and a liquid, is submitted to a pressure gradient, the liquid may flow faster than the solid, resulting in local changes of the paste composition. Specifically, the paste present in the region of the highest pressure (e.g., close to the plunger of a syringe) may become so depleted in liquid that the biphasic mixture in this zone is not longer a paste but a wet powder [263, 265]. Contrarily, the paste in the zone of the lowest pressure (e.g., at the cannula tip) is enriched in liquid. Since these effects are dynamic, the size of the zone depleted in liquid (wet powder) increases during injection, eventually reaching the tip of the injection device and plugging it. The phenomenon, in which the pressure applied to the cement paste provokes a phase separation after a certain injection time, is generally referred as filter pressing, phase separation, or phase migration [125] (see the aforementioned example for Norian SRS® [29], in which a thick mass remained inside a syringe).

Possible mechanisms underlying the limited injectability of hydraulic calcium orthophosphate cements have been discussed in literature [266, 269]. In the case of demixing, the exact composition of the extruded part of the paste becomes unknown. Moreover, due to a deviation from the initial P/L ratio, it becomes unclear whether the setting behavior and the mechanical and histological properties of the extruded part are still clinically acceptable. Therefore, a good cohesion of the paste is necessary in order to avoid these problems [270].

Cohesion (= cohesiveness, "non-decay") is the ability of a paste to keep its geometrical integrity in an aqueous solution [125]. For cements, a bad cohesion may prevent setting and may lead to negative *in vivo* reactions due to the release of microparticles [271]. Since a high cohesion is the result of strong attractive forces among the particles, factors enhancing van der Waals forces (attractive) and decreasing electrostatic forces (repulsive) can be used to improve cohesion [125]. For example, an appropriate cohesion was achieved when no disintegration of the cement paste was observed in the fluid [134, 270]. This can be accomplished by keeping a high viscosity for the cement paste [22] or using cohesion promoters (e.g., 1% aqueous solution of sodium alginate [176, 272, 273] and other chemicals [176, 274–276]). Some calcium orthophosphate cements fulfill both criteria, e.g., Norian SRS®, but others fulfill only one or even none of these requirements. For example, BoneSource™ [103] and Cementek® are not injectable and blood must be kept away from the implanting site until setting [197, 198]. A poor cohesion has been associated to a poor biocompatibility that might lead to inflammatory reactions [271]. Further details on the cohesion properties of various calcium orthophosphate pastes are available in literature [270].

Viscosity is a measure of the resistance of a fluid, which is being deformed by either shear stress or tensile stress. Generally, the cement viscosity in the range of 100–1000 Pa·s appears to be ideal [277] and, if possible, a cement should have a constant viscosity in the indicated range. Unfortunately, cement viscosity is not a constant value, which, after a decrease in the first seconds after mixing, increases considerably during curing, eventually leading to hardening. Furthermore, viscosity should be high enough to prevent extravasation; therefore, it is very important to define an adequate injection window [277].

5.4.4 Properties Improving

As written above, properties of the existing self-setting calcium orthophosphate formulations are not ideal. Several ways can be adopted to improve them. The first approach consists of injectability improvement. There are two options for this. Firstly, the injection device can be modified. For example, shorter cannulas with a larger diameter, as well as smaller injection rates favor a good injectability. The last option is not so straightforward: for example, Habib et al. have shown that large injection rates are not detrimental to injectability because of the shear-thinning behavior of many calcium orthophosphate cements [266]. Secondly, the cement composition can also be adapted. Namely, a decrease of the particle size, the P/L ratio and the plastic limit was found to contribute to a better injectability [263, 267]. For example, injectability was found to be unaffected by P/L ratio within the range of 3.85–4.50 g/ml but drops by nearly 100% between P/L ratio of 4.50 and 5.00 g/ml [23]. However, a decrease in P/L ratio leads to a decrease in the mechanical properties of the cements and cohesion might be destroyed. Furthermore, both the initial and final setting times decreased markedly with the P/L ratio increasing [227, 278]. Therefore, variations in the P/L ratio appear to be valid to a certain extent only. That is why the manufacturer of Biopex® suggests using a P/L ratio of 2.8 or 3.3 g/ml.

Particle size decreasing of calcium orthophosphate crystals is the second approach for the injectability improvement. For example, α-BSM® is well injectable because it consists of small particles. Even though small particles require a larger amount of mixing liquid to obtain a paste, injectability and cohesion of the cements are generally very good [197, 198]. An indirect approach is to add calcium orthophosphate particles those act as spacers between other particles. For example, DCPA is added to the formulation of Biocement D® to improve injectability [197, 198]. Similarly, there is an apatite cement containing spherical particles of TTCP to improve injectability [279].

Using various additives (this approach transforms cements into composites—see Chapter 6 of this book) is the second way to improve the physical properties of calcium orthophosphate cements [280]. For example, water demand of calcium orthophosphate cements can be reduced by ionically modifying the liquid component, e.g., by adding nontoxic sodium salts of α-hydroxy di- and tri-acids [281, 282]. A list of additives, that have been already studied, includes fluidificants, air-entraining agents, porogens, workability improvement agents, setting time controllers, and reinforcing additives [162, 203, 283]. Besides, various radiopacifiers might be used to simplify an uninvasive *in vivo* monitoring of the implanted cements [284–287]. The main role of fluidificants is to reduce a mixing time of the cement. Citric acid is an example of this reagent; it retards the dissolution–precipitation reactions in cements, decreases the compressive strength during initial setting, but increases its strength in the final stages of the cement hardening [228]. Furthermore, data are available that citric acid decreases the setting time and improves the mechanical properties of the hardened cements [288]. Adding of surfactants to calcium orthophosphate cements was found to have two different meanings: they might act as both air-entraining agents by lowering the surface tension [289, 290] and interaction modifiers by shifting the isoelectric point [291].

Besides, studies are available in which calcium orthophosphate cements have been modified by various bioorganic compounds in attempts to influence the bone-healing process [292–295]. For example, in a recent study, a calcium orthophosphate cement was set in the presence of cocarboxylase, glucuronic acid, tartaric acid, α-glucose-1-phosphate, L-arginine, L-aspartic acid, and L-lysine, respectively, with the aim to influence formation and growth of CDHA crystals through the functional groups of these biomolecules [295]. Except for glucuronic acid, all these modifications were found to result in the formation of smaller and more agglomerated CDHA particles, which had a positive impact on the biological performance indicated by first experiments with the human osteoblast cell line hFOB 1.19. Moreover, initial adhesion of human bone marrow-derived mesenchymal stem cells was improved on the cement modifications with cocarboxylase, arginine, and aspartic acid. Furthermore, cell proliferation was enhanced on the cement formulations modified with cocarboxylase and arginine whereas osteogenic differentiation remained unaffected. Besides, the cement modifications with arginine and aspartic acid, but not with cocarboxylase, led to a higher BMP-2 binding [295].

Porosity is a very important property to provide good *in vivo* bioresorption of implanted biomaterials. Thus, various air-entraining agents and porogens are commonly used to induce macroporosity inside calcium orthophosphate cements without affecting their normal setting. For example, crystals of mannitol, $CH_2OH(CHOH)_4CH_2OH$, were tested as an air-entraining agent; however, both loss of workability during the cement mixing and severe depreciation of mechanical properties were discovered simultaneously [296–301]. Other porogenic agents (e.g., oxygen peroxide [302] in the liquid phase and/or iced [303], sucrose granules, $NaHCO_3$ and Na_2HPO_4 crystals of 125–250 μm in size [304], poly(D,L-lactic-*co*-glycolic acid) microparticles with the average size of 66 ± 25 μm [305–310], pectin microspheres [311], calcium sulfate [46], calcite [219], NaCl crystals varying in size from 420 μm to 1 mm [312, 313], gelatin microspheres [314, 315], vesicants [316], cetyltrimethyl ammonium bromide [317], polytrimethylene carbonate [318], some immiscible liquids) have been also tested to create porosity. These additives could be applied on pre-set cements only, while the solubility degree of the particulates during the setting reaction influences both the content and dimensions of the macroporosity. After cement hardening, dissolution of the remaining soluble porogens in either water or body fluids produces macropores with the dimensions and shapes of the dissolved crystals. One important limitation that can be envisaged from this route is the need to add a large amount of porogenic agent to guarantee interconnectivity of the porosity, thus compromising not only the excellent biocompatibility and bioactivity of calcium orthophosphate cements but also the cements' injectability. Another shortcoming is the lack of strength of the resulting material, especially if particulates dissolve quickly, greatly limiting its applications. An innovative approach that aims at overcoming the lack of interconnectivity and initial strength consists in using resorbable fibers [319–326]. These fibers have the function of reinforcing the cement, providing the needed short-term strength and toughness, and gradually dissolving afterwards, leaving behind macropores suitable for bone ingrowth. One interesting advantage of long fibers over particulates and short fibers is the fact that once resorbed they can form interconnected pores inside the cement structure facilitating bone tissue regeneration [327].

One more approach consists in adding solid $NaHCO_3$ to the starting cement powder and using two different liquids: first, a basic liquid to form the paste and later an acid liquid to obtain CO_2 bubbles to create porosity [328]. Besides, pore forming CO_2 bubbles appear at hardening of an apatite cement, consisting of an acidic calcium orthophosphate, such as MCPM or DCPD, and either $CaCO_3$ [33, 53–55] or $NaHCO_3$ [329–331]. Furthermore, addition of an effervescent porogen formulation comprised from $NaHCO_3$ (54.52%) and citric acid monohydrate (45.48%) has been suggested [332]. More to the point, the liquid phase of a cement formulation might be foamed and subsequently mixed with the powder phase. Further, the setting reactions of a calcium orthophosphate cement transforms the liquid foam into a solid, which ideally maintains the geometry, size, and shape of the bubbles of the liquid foam. Thus, the liquid foam acts as a template for the macroporosity of the solid foam [333, 334]. In addition, several other porosity creation techniques in calcium orthophosphate cements are known and, for further details on the subject, the interested readers are referred to an excellent review [327].

The major examples of workability improvement agents, which are added to the cement powders, include water-soluble polymers. Specifically, polysaccharides [97, 108, 335–338], gelatin [278, 339–345], and polyacrylic acid [346–348] are of an interest due to their biocompatibility and good rheological properties. Only small amounts (a few weight %) are needed to dramatically increase the viscosity of the cement pastes. Besides, the cement paste becomes more cohesive and highly resistant to washout immediately after mixing. For example, a 5 wt. % sodium chondroitin sulfate solution is used as mixing liquid in Biopex® [197, 198]. In the case of gelatin, more than a 50% improvement of the compressive strength was detected [341]. The gelatin-cement after setting was found to exhibit reduced crystallinity, much smaller CDHA crystals and a more compact microstructure; all these phenomena might be accounted for the improved mechanical properties [342]. The presence of gelatin improved mechanical properties of the cements; in particular, calcium orthophosphate cements containing 2 wt. % gelatin were found to harden in an acceptable time and were recommended for clinical applications [345]. In general, the use of gelling agents widened a possible application of calcium orthophosphate cements because these cements can be used even when complete homeostasis is difficult. In some cases addition of a gelling agent might cause an increase in hardening time [349] but this was remedied by the use of a sodium orthophosphate solution as the cement liquid [143, 144]. Most polysaccharide solutions are thixotropic, i.e., the viscosity of the solution decreases as the shear rate increases. Certain polysaccharides, such as sodium alginate, pectize in contact with calcium ions. This property can be used to make putty-like cement pastes [22]. However, only few polysaccharides are accepted for parenteral use [197, 198].

Of two families of calcium orthophosphate cements, brushite cements react generally much faster than apatite ones. As a result, to satisfy the clinical requirements (Fig. 5.2), the setting time of brushite cements has to be prolonged, whereas that of apatite cements has to be shortened [197, 198]. In general, setting reactions of any calcium orthophosphate cements consist of three successive stages: (1) dissolution of reactants to saturate the mixing liquid in calcium and orthophosphate ions; (2) nucleation of crystals; and (3) growth of crystals. Therefore, experimental approaches to modify the setting reaction of calcium orthophosphate cements are to be targeted to these three stages. The available approaches have been summarized in Table 5.3 [207]. Furthermore, seven strategies have been described to control the setting time of calcium orthophosphate cements [208]. They are as follows: (i) mean particle size decreasing of the initial powders; (ii) the P/L ratio increasing; (iii) pH drop of the mixing liquid to increase calcium orthophosphate solubility and hence accelerate the chemical transformations; (iv) a nucleating phase addition, such as a nano-sized HA powder; (v) adding orthophosphate and/or calcium ions into the mixing liquid to accelerate the setting reaction according to the common-ion effect; (vi) solubility reducing of the reaction end-product, for example, by adding fluoride ions into the mixing liquid; (vii) solubility increasing of the starting material by amorphization, e.g., by prolonged milling. For further details on these strategies and approaches, as well as for application examples, the interested readers are referred to the original papers [207, 208].

Table 5.3 List of strategies and approaches to modify reactivity of calcium orthophosphate cements [207]

Strategy	Approach	Sub-approaches
1. Dissolution rate	1.1 Change contact area between reagent and mixing liquid	1.1.1 Change milling duration
		1.1.2 Use nano- or micron-sized powders
	1.2 Change solubility in the mixing liquid	1.2.1 Use more/less soluble phase
		1.2.2 Change of reaction pH
	1.3 Change saturation of the mixing liquid	
	1.4 Use dissolution inhibitors in the mixing liquid	
	1.5 Modify reagent surface	1.5.1 Chemical change (pre-reaction)
		1.5.2 Physical change (dissolution pits)
2. Nucleation rate	2.1 Use crystallization nuclei	
	2.2 Change the saturation of the reaction product in the mixing liquid	2.2.1 Change of saturation
		2.2.2 Change of end-product solubility
	2.3 Use nucleation inhibitors	
3. Growth rate	3.1 Change the saturation of the reaction product in the mixing liquid	3.1.1 Change of saturation
		3.1.2 Change of end-product solubility
	3.2 Use crystal growth inhibitors	

Various setting time controllers (accelerators and retardants) are used to influence the setting time. They include sodium hydrogen pyrophosphate ($Na_2H_2P_2O_7$) and magnesium sulfate (according to another study, ions of citrate, sulfate, and pyrophosphate are necessary [222]), which are added in amounts <1 wt. % [350].

Application of biocompatible α-hydroxylated organic acids (glycolic, lactic, malic, tartaric, and citric acids) and their calcium and sodium salts for modification of both rheological and setting properties of calcium orthophosphate cements is well described elsewhere [351, 352]. Besides, aqueous solutions of sodium orthophosphates [97, 211, 309, 353–356] and gelatinized starch [357] are also known as setting time accelerators. An extensive list of the compounds, which might be suitable as accelerators, retarders, additives, or reactants in calcium orthophosphate cement formulations, might be found in literature [99]. Interestingly, but in some cases a simple thermal treatment of the initial reagents (in that particular case, α-TCP powder) at ~ 500 °C could extend the initial part of the setting reaction from a few minutes to a few hours hence providing a potential approach to better control the setting process [358].

The subject of the reinforcing additives is discussed in detail below in section 5.7.

Concerning storage stability and shelf-life, the factors, significantly influencing these properties for the initial dry powders of calcium orthophosphate cements, were found to be temperature, humidity, and a mixing regime of the powders. Various storage conditions appeared to be effective in prolonging the stability of dry brushite cements. In the order of effectiveness, they were ranged: adding solid citric acid retardant > dry argon atmosphere ≈ gentle mixing (minimal mechanical energy input) >> low temperature [354]. Finally, the cement formulations must be sterilized before a clinical use. A detailed description of the sterilization techniques for calcium orthophosphate cements might be found elsewhere [359].

5.5 Bioresorption and Replacement of the Cements by Bones

Due to the excellent bioresorbability of DCPD and CDHA, a newly forming woven bone might substitute the hardened calcium orthophosphate cements. For example, the implants made of hardened BoneSource™ (an apatite cement) were partly resorbed and replaced by natural bone, depending upon the size of the cranial defect [103]. Replacement of BoneSource™ by bone with a minimal invasion of connective tissue was detected in another study, while

ChronOS™ Inject (a brushite cement) samples exhibited a higher rate of connective tissue formation and an insufficient osseointegration [360]. α-BSM® was evaluated in a canine femoral slot model. New bone was found to form in 3 weeks via an osteoconductive pathway. After 4 weeks, only 1.7% of the implanted material was observed. The hybrid bone possessed the strength of normal, unoperated bone after 12 weeks. In 26 weeks, the boundary between old and new bones was virtually indistinguishable, with only 0.36% of the implant recognizable [169]. Neither influence on general health, limb specific function and pain, nor associated complications with α-BSM® application were found past 2 years in another study [361]. Norian SRS® was evaluated in canine tibial and femoral metaphyseal defects. The cement appeared to be gradually remodeled over time, with blood vessels penetrating through it. However, some amounts of Norian SRS® were detected in the medullary area as long as 78 weeks after being implanted in dog femurs [32]. An interesting study on the *in vitro* resorption of three apatite cements (conventional, fast-setting, and anti-washout) by osteoclasts if compared with a similar resorption of sintered HA and a cortical bone revealed an intermediate behavior of the cements: they were resorbed slower than bone but faster than HA [362]. Furthermore, bone neo-formation was seen 7 seven days after implantation of an α-TCP cement [363]. The biodegradation rate of the cements might be influenced by ionic substitutions in calcium orthophosphates [364]. Evidences of the direct contact of bone and a calcium orthophosphate cement without soft tissue interposition might be found in literature [365, 366].

Different studies reported on both cement bioresorption and the progress of bone formation around calcium orthophosphate cements which in certain cases demonstrated both osteoconductive and osteoinductive properties [367]. However, there are studies in which the osteoinductive properties of calcium orthophosphate cements were not confirmed [368]. Besides, inflammatory reactions were noticed when the cement did not set [271]. As solubility of a non-stoichiometric CDHA is higher than that of stoichiometric HA, α- and β-TCP (Table 1.1) and the particle dimensions of a precipitated CDHA is smaller than those of sintered calcium orthophosphates, the biodegradability of apatite cements is always better than that of dense bioceramics made of the sintered stoichiometric calcium orthophosphates. For example, histologically, at 2 weeks, spicules of living bone with normal bone marrow and osteocytes in lacunae could be seen in an implanted cement. At 8 weeks, the cement was almost totally surrounded by mature bone. At this stage, no resorption of the cement was observed [369]. Only ~ 30% decrease of the implanted amount of Norian SRS® was reported after 24 months in a rabbit femur [370]. Moreover, several differences could be expected depending on the cement type. For example, as the end product of BoneSource™ and Cementek® is a crystalline CDHA, BoneSource™ and Cementek® are expected to resorb slower than other apatite cements. Indeed no resorption of BoneSource™ was observed after several years implantation; though some resorption of Biobone® was detected. However, porosity appears to be the main biodegradability factor at play: a more porous (for cells) hardened cement degrades faster than a less porous one [371]. For example, as Biobone® is more porous than BoneSource™, the discovered diversity could be due to the differences in the cement porosity [197, 198]. The latter conclusion is confirmed by the results of other studies: a positive influence of the cement porosity on the resorption rate was found [273]. The interested readers are referred to the study on the suitability of porous calcium orthophosphate cements as scaffold material for bone regeneration, using a rabbit model [372].

The resorption properties of bioceramics are generally believed to relate to the solubility of their constitutive phases. The implanted calcium orthophosphates might be resorbed by two possible mechanisms, namely: an active resorption, mediated by the cellular activity of macrophages, osteoclasts and other types of living cells (so-called phagocytosis or literally "cell-eating") [373–375] and a passive resorption due to either chemical dissolution [11] or chemical hydrolysis (brushite cements only) [230] in the body fluids. Unfortunately, the factors concerning the biodegradation of calcium orthophosphate biomaterials have not been completely elucidated yet. The chemical composition, physical characteristics and crystal structures certainly play an important role in the biological behavior of calcium orthophosphates. In addition, biodegradation may be influenced by the conditions, such as experimental models, implantation sites and animal species [374].

The data are available that macrophages and giant cells decompose quickly resorbed calcium orthophosphates (e.g., brushite cements) [226], while slowly (from months to years) resorbed apatite cements are decomposed by osteoclast-type cells [30, 209, 376]. Clearly, a fast resorption of brushite cements can only be achieved if the resorption occurs before conversion DCPD to CDHA according to Eq. (5.14) [48]. Both types of the resorption mechanisms (active + passive) might occur almost simultaneously, if a hardened cement consists of two different calcium orthophosphates, e.g., from DCPD and β-TCP. For example, the biphasic brushite cement ChronOS™ Inject was found to resorb by dissolution with cement disintegration and particle formation followed by the phagocytosis of the cement particles through macrophages [377]. A similar cement was found to be degraded through a dissolution process associated with a cellular process. The observations suggested that cell activities could be influenced by a small particle size, without close correlation between the particle size and the cell activities but with a correlation between particle concentration and the cell activities [374]. To get further details on this topic, the interested readers are referred to a very interesting review on the cellular mechanisms of calcium orthophosphate ceramic degradation [378].

The summary of studies on brushite cement implantation in various animal models and defect locations is available in literature [230]. Generally, in the same animal model, a degradation rate decreases with a sample size increases, as does DCPD to CDHA conversion time. The compositional changes of a brushite cement after implantation in sheep is well described elsewhere [350, 379].

The kinetics of passive resorption depends on porosity of the samples, ionic substitutions in calcium orthophosphates (when applicable), crystallinity and pH of the cement-tissue interface. The active resorption is due to cellular activity; however, it is also related to the passive one. Namely, the solution pH near macrophages and osteoclasts can drop to ~5 by excretion of lactic acid, which increases the solubility (Fig. 1.6), whereas near osteoblasts (bone forming cells) solution pH can become as high as 8.5 by excretion of ammonia [29]. Dissolution chemistry of CDHA (therefore, of hardened apatite cements) in acidic media (calcium orthophosphates are almost insoluble in alkaline solutions (Fig. 1.6)) might be described as a slightly modified sequence of four successive chemical equations [380, 381] (see section *7.2.5. Chemical model* for details):

$$Ca_{10-x}(HPO_4)_x(PO_4)_{6-x}(OH)_{2-x} + (2-x)H^+ \rightarrow Ca_{10-x}(HPO_4)_x(PO_4)_{6-x}(H_2O)_{2-x}{}^{(2-x)+} \quad (5.15)$$

$$Ca_{10-x}(HPO_4)_x(PO_4)_{6-x}(H_2O)_{2-x}{}^{(2-x)+} \rightarrow 3Ca_3(PO_4)_2 + (1-x)Ca^{2+} + (2-x)H_2O \quad (5.16)$$

$$Ca_3(PO_4)_2 + 2H^+ \rightarrow Ca^{2+} + 2CaHPO_4 \quad (5.17)$$

$$CaHPO_4 + H^+ \rightarrow Ca^{2+} + H_2PO_4^- \quad (5.18)$$

Obviously, the dissolution chemistry of DCPD (therefore, of hardened brushite cements) in acidic media is described by Eq. (5.18). One should stress that in Eq. (5.18) water is omitted for simplicity. Therefore, dissolution of DCPA is written instead.

The mechanism of bone healing caused by calcium orthophosphate cements is very multifactorial because the surface of the cements is rapidly colonized by cells. Several types of these cells degrade calcium orthophosphates by either phagocytotic mechanisms (fibroblasts, osteoblasts, monocytes/macrophages) or an acidic mechanism with a proton pump to reduce the pH of the microenvironment and resorb the hardened bioceramics (osteoclasts) [378, 382]. Various mesenchymal cells located at the implantation sites can induce solubilization of calcium orthophosphates. Upon the cells arrival, various active enzymes, such as acid phosphatase, are secreted that causes dissolution of the hardened cements [383–385]. Much more biology, than chemistry and material science altogether, is involved into this very complex process and many specific details still remain unknown. See section *4.6.4. Bioactivity* for further details.

It is well known that various polypeptides and growth factors present in bone matrix might be adsorbed onto HA [386–388] and modulate the local milieu of cells. This is supported by many purification protocols of growth factors and bone morphogenetic proteins/osteogenins involving HA chromatography [389, 390]. However, osteoblasts are not found in direct contact with calcium orthophosphates. A complex proteinaceous layer, usually osteoid, directly contacts the osteoblasts. After implantation of calcium orthophosphate cements, mitogenic events could occur either during the initial mesenchyma1 cell contact or after osteoid degradation by osteoblast collagenase. In a dense, mineralized material such as calcium orthophosphate cements, which provides a barrier to the free diffusion of circulating hormones, growth factors, and cytokines, it is questionable whether the local responses at the periphery of the material regulate osteoconduction [22]. The tissue response to injectable calcium orthophosphate cements is well described in literature [329, 362, 376, 391, 392]. Recent histological and mechanical evaluation of self-setting calcium orthophosphate cements in a sheep vertebral bone void model is available elsewhere [393]. The interested readers are also advised to get through a paper on the *in vitro* biodegradation of brushite cements by a macrophage cell-line [128].

To conclude this part, one should note that calcium orthophosphate cements are able to provide short-term biologically desirable properties and then be replaced by a new bone, which is very important [394]. The growth rate of a newly forming bone depends on age, sex and general metabolic health of the recipient as well as on the anatomic site, porosity, bulk site, crystallinity, chemical composition (brushite or apatite), particle sizes and P/L ratio of the cements. Considering all these factors, it might take from 3 to 36 months for different calcium orthophosphate cements to be completely resorbed and replaced by bones [202]. However, additional sound scientific data to determine the exact degree of biodegradability for calcium orthophosphate cements are still needed, viz. animal studies performed in a critical-size defect model. One must stress that the rate of cement resorption should be balanced with the rate of new bone formation to avoid collapse at the fracture site, which might occur if the resorption is too fast. Interestingly, but to advance calcium orthophosphate cements as bioabsorbable bone replaceable materials, it is essential to utilize the patient's own blood in combination with the cements [395].

5.6 The Mechanical Properties

As in most clinical applications self-setting calcium orthophosphate formulations are applied in direct contact with human trabecular bones, it may be stated as a mechanical requirement that the strength of the formulations must be at least as high as that of trabecular bones, which is close to 10 MPa [396]. Three-dimensional (3D) complex load is applied during the orthopedic and dental applications because of a combination of different forces that may include bending, torsion, tension and compression. Unfortunately, calcium orthophosphate cements are strong enough at compression only [195]. In theory, after setting, they can reach the mechanical properties comparable to those of calcium orthophosphate blocks with the same porosity.

However, in practice, the strength of the cements is lower than that of bones, teeth or sintered calcium orthophosphate bioceramics [144].

Having the ceramic origin, the set products of all calcium orthophosphate cements are brittle, have both a low impact resistance and a low tensile strength (within 1 to 10 MPa), whereas the compression strength varies within 10 to 100 MPa [141, 195, 196]. The latter value exceeds the maximum compression strength of human trabecular bones. Furthermore, at 12 weeks after implantation the compressive strength of these cements was found to be still significantly higher (60 to 70 MPa) than that of normal bone [35]. In general, hardened brushite cements are slightly weaker than hardened apatite cements. A tensile strength of ~ 10 MPa and a compressive strength of ~ 60 MPa were obtained for brushite cements [397]. In comparison, apatite cements can reach a tensile strength of ~ 16 MPa [398] and a compressive strength of ~ 83 MPa [399]. However, due to the inherent brittleness of ceramics, these values are close to be meaningless. Namely, the indication of a mean compressive strength of, say, 50 MPa measured on well-prepared (e.g., under vibrations and pressure) and perfectly shaped samples does not inform the readers with which probability this cement will fail *in situ* under a cyclic load of e.g., 10 MPa. Furthermore, a comparison of the compressive strength of hardened cements with that of cancellous bone is not very helpful either because cancellous bone is much less brittle than the ceramic cements [125].

Moreover, the mechanical properties of hardened calcium orthophosphate cements are not narrowly distributed around a mean value (as for metals), but widespread over a very large range of values, which strongly reduces their clinical application [400]. *In vivo*, the difference between the hardened apatite and brushite cements boosts: namely, the mechanical properties of apatite cements were found to increase [356], whereas those of brushite cements decreased [35]. This is attributed to a higher solubility of DCPD when compared with that of CDHA (Table 1.1). However, the mechanical properties of the hardened cements may vary with implantation time. For example, animal studies indicated that the mechanical properties of apatite cements tended to increase continually [356], in contrast to those of brushite cements, which initially decreased and again increased when bone was growing [35].

Furthermore, shear and tensile forces play a very important role. Thus, these parameters should also be considered, for example, using the Mohr circle approach [401]. Besides, it is difficult to compare the mechanical properties of different cement formulations. For example, the following numeric values of the compression strength and setting time were obtained: (i) Norian SRS® (~ 50% porosity): 33 ± 5 MPa and 8.5 ± 0.5 min, (ii) Cementek®: 8 ± 2 MPa and 17 ± 1 min, (iii) Biocement D® (~ 40% porosity): 83 ± 4 MPa and 6.5 ± 0.5 min, (iv) α-BSM® (~ 80% porosity): 4 ± 1 MPa and 19 ± 1 min, respectively [399]. Among them, Biocement D® has the highest compressive strength but the lowest porosity. A high compressive strength does not necessarily mean that Biocement D® is the least breakable implant [197]. Further details on the major properties of Norian SRS® are available elsewhere [203, 402]. Besides, the interested readers are suggested to get through the mechanical characterization of a bone-defect model filled with ceramic cements [200].

To improve the mechanical properties of the self-setting calcium orthophosphate formulations, addition of water-soluble polymers might be considered. For example, in early 1990s, Miyazaki et al. [403, 404] used a number of polymers, including polyacrylic acid and polyvinyl alcohol to improve the properties of a TTCP + DCPD cement. They noted marked increases (up to threefold) in mechanical properties but with an unacceptable reduction of workability and setting time. Later, another research group reported similar results using sodium alginate and sodium polyacrylate [405]. Afterwards, other researchers added several polyelectrolytes, polyethylene oxide, and a protein bovine serum albumin into α-BSM® cement pastes to create calcium orthophosphate—polymer biocomposites [406]. Biocomposites of α-BSM® with polycations (polyethylenimine and polyallylamine hydrochloride) exhibited compressive strengths up to six times greater than that of pure α-BSM® material. Biocomposites of α-BSM® with bovine serum albumin developed compressive strengths twice that of the original α-BSM® cement [406]. Similar strengthening effect was achieved by addition of some commercial superplasticizers [407]. The results showed that small additions, i.e. 0.5 vol. %, in the aqueous liquid phase improved the maximum compressive strength (35 MPa) of Biocement-H© by 71%, i.e. till ~ 60 MPa. Moreover, the addition of high amounts of superplasticizers, i.e. 50 vol. %, allowed for a significant increasing of the P/L ratio from 3.13 to 3.91 g/ml, without affecting the maximum strength and/or the workability of the cement [407]. This effect was explained by an inhibiting effect of the aforementioned additives on the crystal growth kinetics of newly forming crystals of calcium orthophosphates, which resulted in smaller crystallites and, hence, a denser and more interdigitated microstructure. However, the increased strength was attributed mainly to the polymer's capacity to bridge between multiple crystallites (thus forming a more cohesive composite) and to absorb energy through a plastic flow [406]. Other factors affecting strength are the materials used in the solid phase, particle sizes, incorporation of fillers (see section *5.7. Reinforced formulations and concretes* for details), the P/L ratio and various additives to the liquid phase [106].

As presence of pores simplifies for cracks to run throughout the ceramic mass, the mechanical properties of the hardened cements were found to decrease exponentially with the porosity increase [408]. In theory, calcium orthophosphate cements can be made with almost any porosity. However, for most commercial formulations, the pores are of 8–12 μm in diameter and, after the cement is set, porosity occupies about 40–50% of its volume [409]. Pressure can be applied to reduce the porosity of hardened cements [144, 410, 411]. Usually, the pore dimensions in hardened cements are too small to allow a fast bone ingrowth. Thus, there is a lack of macroporosity. Besides, unless special efforts have been performed, the available macroscopic pores are not interconnected. Due to these reasons, after injection, osteoclastic cells are able to degrade the hardened cements layer-by-layer only, starting at the bone-cement interface throughout its inner part (in other words, from the outside to the inside). This is the main drawback of the classical cement formulations when compared to calcium orthophosphate ceramic scaffolds with an open macroporosity [197, 198].

Since compression strength is reciprocally proportional to porosity [371], the former might be adjusted by varying the P/L ratio in the hardening mixture. Elevated compression strength would be applicable in cranioplasty for regions requiring significant soft-tissue support. For smaller bone defects, such as root canal fillings, low-compression cements might be used [137]. Concerning the tensile strength of calcium orthophosphate cements, as a rule of thumb, it appears to increase two-fold with each 10 vol. % decrease of the porosity, i.e. 5, 10, 20, 40 and 80 MPa for 80, 70, 60, 50 and 40% porosity, respectively [197, 198]. The effect of porosity on the compressive modulus of calcium orthophosphate cements is available as Fig. 4 in Ref. [411]. Ishikawa and Asaoka showed a linear relation ($R^2 = 0.94$) between ln diametral tensile strength and porosity of a calcium orthophosphate cement where porosity was controlled by compaction pressure (up to 173 MPa) [141]. Besides, an empirical relationship between strength, S, and porosity, P, has been introduced [412]:

$$S = S_0 e^{-bP},$$

where: S_0 is the theoretical strength at P = 0 (fully dense) and b is an empirical constant.

As the porosity is mainly due to an excess of water used in the cement compositions, attempts were made to reduce the amount of water. However, the amount of water determines the rheological properties of cement pastes: a decrease in water content leads to a large increase in viscosity, eventually leading to non-flowable pastes. As calcium orthophosphate cements set at an almost constant volume, the final porosity can be predicted from the initial composition [197, 198]. A shrinkage degree of ~1% causes no restrictions on clinical use [193]. Recent studies on the *in vivo* evaluation of an injectable macroporous calcium orthophosphate cements revealed a higher bioresorption rate due to both a higher surface contact with body fluids (which increases dissolution) and enhancing cellular activity due to particle degradation [273, 329].

Besides the addition of porogens [296–318], the porosity level of the self-setting calcium orthophosphate formulations might be controlled to a certain extent by adjusting particle sizes and the P/L ratio. When the P/L ratio is high, the porosity of the apatite cement is low [197, 198]. According to calculations, the tensile strength of the cements with zero porosity could be as high as 103 MPa [141]. However, a high density and a lack of pores decreases cement bioresorbability because a newly forming bone appears to be unable to grow into the implant; it might grow only simultaneously with dissolution of the cement. Thus, porosity of calcium orthophosphate cements is a very important factor for the cement biodegradability [197, 198].

The strength of the cement-prosthesis interface might be studied by a pullout test. The details are available elsewhere [65].

5.7 Reinforced Formulations and Concretes

Being aware on the excellent bioresorbability of DCPD and CDHA, researchers are focused on attempts to overcome the mechanical weakness of the self-setting calcium orthophosphate formulations by using different fillers, fibers and reinforcing additives that give rise to formation of various multiphasic biocomposites [104, 105, 109, 199, 259, 274, 409, 412–418]. Even carbon nanotubes have been successfully tested to reinforce the self-setting formulations [419]. Although the biomaterials community does not use this term (just 1 paper has been published [420]), a substantial amount of such formulations might be defined as calcium orthophosphate concretes [421]. The idea behind the concretes is simple: if a strong filler is present in the matrix, it might stop crack propagation. However, adding fillers always reduced the porosity that negatively influenced the ability of the concretes to allow bone ingrowth into the pores. Hence, a denser formulation has a slower resorption rate and thus a slower bone substitution [141]. Moreover, due to the presence of fillers, injectability and other rheological properties of calcium orthophosphate concretes frequently appear to be worse than those ones of calcium orthophosphate cements. Thus, it is difficult to increase strength of the cement formulations without having a negative influence on the other properties.

Calcium orthophosphate concretes can be prepared from both apatite and brushite cement formulations. For example, in an attempt to improve the mechanical properties of calcium orthophosphate cements, a group of investigators prepared concretes by adding human cadaveric femur bone chips in amounts of 25, 50 and 75% (w/w) to α-BSM® cement [414]. The mechanical tests revealed that the specimens of pure cement exhibited a relatively high stiffness but a low ductility. However, for the cement-bone concretes an increasing of bone content was found to result in the elastic modulus decreasing and the ductility increasing; however, the ultimate strength showed only small changes with no apparent trend [414]. A concrete of Biopex® cement with allografts taken from femurs and tibiae of rabbits is also available. Unfortunately, nothing is written on the mechanical properties improvement but, surprisingly, by the addition of allografts, the hydrolysis process of Biopex® was significantly changed [259]. By adding polymers, other researchers succeeded in improving the mechanical strength of the cement formulations up to ~ 30 MPa; however, the kinetics of CDHA formation and, thus, the bioactivity of the material were decreased [110, 422]. Xu et al. reported that incorporation of long carbon fibers at a volume fraction of 5.7% increased the flexural strength about 4 times and work of fracture ~100 times, if compared to un-reinforced formulations [423]. The reinforcement mechanisms were found to be crack bridging and fiber pullout, while fiber length and volume fraction were key microstructural parameters that determined the concrete properties [426]. Although addition of polypropylene, nylon and carbon fibers was found to reduce the compression strength of a double-setting calcium orthophosphate cement because of increased porosity, it strongly increased the cement's fracture toughness and tensile strength, relative to the values for the un-reinforced variant of this formulation [415]. A knitted two-dimensionally oriented polyglactin fiber-mesh was found to be effective in improving load-bearing behavior of a calcium orthophosphate cement for potential structural repair of bone defects [199]. To make the material stronger, fast setting and anti-washout, chitosan might be added [178, 344, 403, 424–434]. Furthermore, anti-washout properties might appear by adding sodium alginate [435]. Calcium orthophosphate concretes containing SiO_2 and TiO_2 particles showed a significant (~ 80–100 MPa) increase in the compressive strength, whilst no change in the mechanical behavior of the formulations was observed when ZrO_2 particles were added [416]. Additional examples of the properties improving comprise addition of calcium silicates [67], polypeptide copolymers [436] and collagen [437–443]. A strength improvement was found when DCPA and TiO_2 crystals were used as fillers for mechanically activated α-TCP cements [444]. Calcium orthophosphate concretes reinforced by whiskers of calcium carbonate [55] have been also developed.

Yet another team examined the effects of varying fiber type, fiber length and volume fraction of fiber-reinforced calcium orthophosphate concretes [424, 445]. Four fiber types were studied: aramid, carbon, E-glass and polyglactin. Fiber length ranged within 3–200 mm and fiber volume fraction ranged within 1.9–9.5%. The results indicated that a self-setting calcium orthophosphate cement was substantially strengthened via fiber reinforcement. Aramid contributed to the largest increase in the concrete strength, followed by carbon, E-glass and polyglactin. Fiber length, fiber volume fraction and fiber strength were found to be key microstructural parameters that controlled the mechanical properties of calcium orthophosphate concretes [424, 445]. Fiber reinforcement of porous formulations (mannitol was used as a porogen) was discovered as well [446]. Namely, reinforcement by aramid fibers (volume fraction of 6%) was found to improve the properties of a calcium orthophosphate concrete with the strength increasing threefold at 0% mannitol, sevenfold at 30% mannitol and nearly fourfold at 40% mannitol. Simultaneously, the work of fracture increased by nearly 200 times, however the modulus was not changed as a result of fiber reinforcement [446]. Addition of 20 wt. % of acrylamide and 1 wt. % ammonium polyacrylate to the liquid increased the compressive and tensile strength of α-TCP bone cement by 149 and 69% (55 and 21 MPa), respectively [447]. A positive influence of polyamide fibers [448] and bioactive glass [449] is also known. Interestingly, but reinforcement of calcium orthophosphate cements might be performed by infiltration of a preset cement formulation by a reactive polymer and then cross-linking the polymer *in situ* [450].

In the cases, when bioresorbable reinforcement fibers are used, strength augmentation is attained at the initial stages [319–326]. For example, the initial strength of a concrete was threefold higher than that of the unreinforced cement control [319]. The work of fracture (toughness) was found to increase by two orders of magnitude for other biocomposites of calcium orthophosphate with resorbable fibers (namely, Vicryl polyglactin 910 (Ethicon, Somerville, NJ) [320] and a mesh of copolymer of polyglycolic and polylactic acids [324]). When implanted *in vivo*, bioresorbable fibers would provide initial strength and then dissolve to form interconnecting macroscopic channels, which could facilitate bone ingrowth into implants [143, 144, 319, 324]. For example, interconnected macropores were formed in a hardened formulation at 84 days' immersion in a physiological solution [324]. One should note that, apart from the mechanical properties of the reinforcing materials, the structure of the incorporated fibers, regular or random, appears to be crucial for the resulting flexural strength and modulus of elasticity [322]. A higher strength might help extending the use of calcium orthophosphate cements to larger stress-bearing repairs, while the macropores might facilitate tissue ingrowth and integration of the hardened cement with an adjacent bone. To extend this idea further, several types of fibers with different rates of bioresorbability might be simultaneously incorporated into a cement formulation.

Besides the aforementioned, it is important to mention on concretes, after hardening consisting of calcium orthophosphates only [237, 238, 377, 451–454]. The first biphasic composition consisting of a hardened DCPD matrix filled with β-TCP granules was introduced in 1992 [452]. Further development of this formulation is well described in other papers [237, 377]; unfortunately, neither mechanical nor rheological properties of this concrete have been disclosed. At physiological pH, the *in vitro* solubility of DCPD is approximately 100 times higher than that of β-TCP (Table 1.1 and Fig. 1.6); roughly, the same order of magnitude applies for the *in vivo* resorption kinetics of these calcium orthophosphates. Thus, a new bone is formed in the space left after resorption of the DCPD matrix, while β-TCP granules act as guiding structures. This feature of the cement can be considered an inverse scaffolding effect [455]. Another group of investigators invented a formulation that incorporated as major powder components α-TCP, ACP and BCP (HA + β-TCP in various HA/β-TCP ratios) [413]. It was believed that after setting such a formulation could provide a porous ceramics *in vivo* due to preferential dissolution of a better soluble ACP component compared to the other calcium orthophosphates in the matrix. Further, this combination was extended to a multiphase concrete composition consisting of 70% w/w settable matrix (mixture of 45% α-TCP, 5% MCPM and 25% ACP [456]) with the average particle dimensions of 15 μm and 30% BCP granules (ranging between 80 and 200 μm) as a filler [451]. The role of BCP granules is quite interesting: after implantation of a formulation without BCP granules, the quality of newly formed bone was not identical to the host bone, while implantation of a concrete with BCP granules resulted in formation of a new bone identical to the host bone. The reason of this phenomenon is not clear yet; but, perhaps, it correlates with similar results for β-TCP granules, which act as bone anchors and encourage formation of a mature bone [237, 238]. Other ACP-containing cement formulations were elaborated as well [457].

Effects of added α-TCP and β-TCP were investigated to shed light on the setting reaction of apatite cement consisting of TTCP and DCPA [454]. Added β-TCP showed no reactivity, and thus resulted in extended setting time and decreased mechanical strength. In contrast, α-TCP dissolved to supply calcium and orthophosphate ions after initial apatite crystal formation by the chemical reaction (5.1). Although setting time was delayed because α-TCP was involved only in the latter reaction of apatite cement, larger apatite crystals were formed due to its addition. Because of larger apatite crystal formation, the mechanical strength of the α-TCP-added formulation increased by approximately 30%, as compared to α-TCP-free apatite cement [454]. Besides, calcium orthophosphate cements might be reinforced by calcium polyphosphate fibers [458, 459].

To conclude this part, one should briefly mention on the reverse situation: there are bone concretes made of various polymeric cements, reinforced by calcium orthophosphate powders or granules [460–468]. The calcium orthophosphates presented in these formulations act as fillers, which are necessary to both improve the mechanical properties and impart bioactivity; they do not participate in the hardening mechanisms. For example, the higher the amount of HA was in bioactive acrylic bone cements, the higher were the compressive and tensile moduli. Furthermore, as the percentage of HA increased to 20 wt. %, the heterogeneity of the material was higher [466]. Polymerization of monomers is primarily responsible for setting of such types of biocomposites and concretes. However, that is another story.

5.8 Clinical and Medical Applications

Injectable and self-setting calcium orthophosphate formulations have been introduced as an adjunct to internal fixation for treating selected fractures. Different studies have already shown that they are highly biocompatible and osteoconductive materials, which can stimulate tissue regeneration [22, 469]. The main purpose of calcium orthophosphate cements, concretes, pastes and putties is to fill voids in metaphyseal bone, thereby reducing the need for bone graft, although such formulations might also improve the holding strength around metal devices in osteoporotic bone. Bone augmentation (i.e., a reinforcement of osteoporotic bone through injection) appears to be a very promising application field of calcium orthophosphate cements. Such procedures ease the fixation of screws in mechanically poor bone (for example for osteosynthesis) and decrease pains associated with unstable vertebrae. The combination of a self-setting nature, biocompatibility, lack of any by-products and a great potential for replacement by bones make calcium orthophosphate cements, concretes, pastes and putties very promising materials for clinical and medical applications. In addition, they can easily be used by bone remodeling cells for reconstruction of damaged parts of bones [102, 103, 226, 392, 470–472]. The ability to be molded in place also is a very important property because these formulations can easily be delivered into the desired place and can be fitted perfectly with bone defects [103]. Besides, some formulations were found to possess an antimicrobial activity [56, 59, 61, 68, 473], as well as promote osteoblast cell adhesion and gene expression *in vitro* [474].

Recent studies reported optimistic results in relation to the clinical application of the self-setting calcium orthophosphate formulations. For example, the data on cytocompatibility and early osteogenic characteristics are available in literature [475]. The ratio of the cases determined to be "effective" or "better" among the 74 cases we found to be 97.3% [476]. Besides, the results of intra-articular degradation and resorption kinetics of these cements revealed no signs of pronounced acute or chronic inflammation [477]. Injected Norian SRS® cement was mainly found as a single particle, anterior to the cruciate ligaments. The cement became surrounded by synovial tissues within 4 weeks and showed signs of superficial resorption [477]. Unfortunately, disintegration or washout of calcium orthophosphate cements has been reported as a potential clinical problem [141, 210]. Perhaps, this problem could be solved by putting pressure on the paste during the setting period. In addition, sodium alginate might be added; however, the mechanical properties (strength) of this formulation are still poor [108].

According to the available information, the first animal study of calcium orthophosphate cements was performed in 1987 [152]. Afterwards, in 1991, a cement consisting of TTCP and DCPA was investigated histologically by implanting disks made of this cement within the heads of nine cats [478]. Simultaneously, another research group evaluated the tissue reaction to this cement in the teeth of monkeys [479]. Important examples of the most significant directions of current medical applications of the self-setting calcium orthophosphate formulations are given below.

5.8.1 Dental Applications

A group of investigators extracted all mandibular premolar teeth from beagles [480]. After one month of healing, alveolar bone was reduced to make space for a previously fabricated calcium orthophosphate cement block. One more month later, 8 mm HA implants were placed in such a manner that the apical half was embedded into alveolar bone and the coronal half in the calcium orthophosphate cement block. The investigators observed that the cement block was gradually replaced by bone and histopathologic features of the cement area were similar to that of natural bone. Moreover, the coronal half of the implants, previously surrounded by the calcium orthophosphate cement, was firmly attached by natural bone [480]. In another study, the same researchers used fluorescent labeling analysis and electron microanalysis to measure the extent of new bone formation and elemental (Ca, P, Mg) distribution [481]. The results indicated the presence of newly formed bone at ~ 1 month after surgery and similar elemental distributions in the calcium orthophosphate cement and natural bone areas at ~ 6 months after surgery [202]. Besides, calcium orthophosphate cements were tried as root canal fillers [59, 482, 483] and for pulp capping [484].

A hydraulic calcium orthophosphate cement was injected as a bone filler for gaps around oral implants placed on the medial femoral condyles of six goats and found excellent bone formation around the graft material. Unfortunately, the degradation rate of the cement appeared to be very slow and no resorption was observed [485]. In another study, a cement was placed on artificially created periodontal defects but no significant difference was found between the cement and control. However, the cement acted as a scaffold for bone formation and provided histocompatible healing of periodontal tissues [486]. Still other investigators used a cement for direct pulp capping [487, 488] and compared it to calcium hydroxide. Both materials were found to be equally capable of producing a secondary dentin at ~ 24 weeks [488]. Finally, injectable forms of calcium orthophosphate cements can be used as an adjunctive supportive agent for dental implants [489].

5.8.2 Craniofacial and Maxillofacial Applications

The use of self-setting calcium orthophosphate formulations for craniofacial applications seems logical, as there is little or no stress generated under these conditions. Moreover, the ability to mold the material at placement is an enormous advantage from a cosmetics standpoint [202]. For example, BoneSource^{TM} is indicated for the repair of neurosurgical burr holes, contiguous craniotomy cuts and other cranial defects with a surface area no larger than 25 cm^2 per a defect. In addition, it may be used in the sinus region for facial augmentation [103, 490] and the cement can be supported by metal hardware [103]. In dogs, BoneSource^{TM} was employed to supplement the supraorbital ridge and to augment skull base defects [491]. Another group performed trials to ascertain the inflammation around the site and the degree of loss of the implanted BoneSource^{TM}. The material was found to be osteoconductive with both periosteal and endosteal bone formation [492]. One more group presented excellent results using the material combined with an underlying resorbable mesh in calvarian defects of Yorkshire pigs. They found progressive bone ingrowths in all defects at 180 days, with nearly complete replacement by host bone [325]. Besides, excellent results for over 100 human patients were reported when a calcium orthophosphate cement was used in cranial defects. The success rate of the cement after 6 years was 97% [94]. The results of still other medical trials are available elsewhere [220, 493–505]. To conclude this part, one should stress that complications still occur: namely, two cases of apatite bone cement resorption and subsequent seroma formation have been recently reported for patients who had undergone retrosigmoid craniotomy [506].

5.8.3 Orthopedic Applications

Self-setting calcium orthophosphate formulations have successfully been used for treatment of the distal radius fracture [206, 507, 508]. Besides, other successful attempts have been made to use these formulations for calcaneal fractures [509], hip fractures [510, 511], augmentation of osteoporotic vertebral bodies [512], distal radius fractures [513], tibial plateau fractures [33, 513–517], restoration of pedicle screw fixation [518, 519], reinforcement of thoracolumbar burst fractures [520], cancellous bone screws [521, 522], in wrist arthrodesis [523] and for fixation of titanium implants [524]. A recent study on a cement augmentation of the femoral neck defect might be found elsewhere [525]. Considering their properties, calcium orthophosphate cements might potentially be applied to reinforce osteoporotic vertebral bodies [512, 526]. Further details on this topic are available elsewhere [527, 528]. Besides, calcium orthophosphate cements appear to be a reliable subchondral replacement material when the bone defect is adjacent to the articular cartilage [529].

5.8.4 Vertebroplasty and Kyphoplasty Applications

Vertebroplasty and kyphoplasty are two surgical procedures that recently have been introduced to medically manage of osteoporosis-induced vertebral compression fractures. Particularly, both procedures aim to augment the weakened vertebral body, stabilize it and/or restore it to as much of its normal height and functional state as possible. Both procedures involve injection of a self-setting paste of a calcium orthophosphate cement into the fractured vertebral body, which resulted in a faster healing [94, 203, 529–536]. Furthermore, prophylactic injections of calcium orthophosphate cements also have been performed.

5.8.5 Drug Delivery Applications

In general, a potential substrate to be used as a drug carrier must have the ability to incorporate a drug, retain it in a specific target site and deliver it progressively with time in the surrounding tissues. Additional advantages are provided if the material is injectable, biodegradable, sets at ambient temperature, has both near neutral pHs and a large surface area [36, 37]. These properties make self-setting calcium orthophosphate formulations to be very attractive candidates as drug carriers for therapeutic peptides [537], antibiotics [538–547], anticancer drugs [578], anti-inflammatory drugs [549, 550], cytokines [551], hormones [552] and bone morphogenetic proteins [432, 553–557]. For example, a "growth factor cement" has been reported [558]. In that study, a combination of bone morphogenetic protein-2 (BMP-2), transforming growth factor-beta (TGF-β1), platelet-derived growth factor and basic fibroblast growth factor (bFGF) was used in a calcium orthophosphate cement for treatment of peri-implant defects in a dog model. The findings indicated a significant effect of the "growth factor cement" on increased bone-to-implant contact and amount of bone per surface area if compared with both the cement-only and no-cement treatment groups [558]. Similar data were found for a combination of a calcium orthophosphate cement with an exogenous nerve growth factor [559]. Even more complicated combination of deproteinized osteoarticular allografts integrated with a calcium orthophosphate cement and recombinant human vascular endothelial cell growth factor plus recombinant human BMP-2 (rhBMP-2) has been studied as well [560]. Interestingly, but the drug delivery properties of calcium orthophosphate cements might be influenced by crystal morphology [561].

In the cement formulations, drugs might be incorporated into both a liquid and a powder phases. After setting, the drugs are slowly released through the cement pores [210, 543–546, 562, 563]. For example, a group of investigators added flomoxef sodium to a cement formulation and found that the release of antibiotic could be easily controlled *in vivo* by adjusting the content of sodium alginate in the formula [210]. *In vitro* elution of vancomycin from calcium orthophosphate cement has been studied as well [563]. The possibility of using self-setting calcium orthophosphate formulations as a drug-delivery system offers an attractive and efficient solution for the treatment of various bone diseases, e.g., tumours, osteoporosis and osteomyelitis, which normally require long and painful therapies.

The laboratory studies on drugs incorporation into self-setting calcium orthophosphate formulations cover different aspects. Firstly, it is necessary to verify that addition of a drug does not influence the setting reaction not only in terms of the setting and hardening mechanisms but also with respect to the rheological behavior and injectability. Secondly, it is necessary to determine the *in vitro* kinetics of drug release. Thirdly, the drug delivery properties of the formulation must be studied *in vivo*. Finally, but still importantly, the clinical performance of the drug delivery system must be evaluated as well [36, 37]. For example, recombinant human transforming growth factor β1 (rhTGF-β1) was added to a calcium orthophosphate cement [564–567]. This resulted in formation of a bioactivated cement that could be used as a bone filler and for the replacement of bone [564]. It appeared that after 8 weeks the addition of growth factors stimulated and increased bone formation (50% volume) and bone contact (65%) in comparison to control calvarian defects in an animal study. Besides, the growth factor group reduced the remaining volume of the cement by 20% [565]. Examples of rhBMP-2 release from a loaded porous calcium orthophosphate cement might be found elsewhere [567, 568], while an experimental study on calcium orthophosphate cement impregnated with dideoxy-kanamycin B is also available [569].

Although most materials currently used as drug carriers are polymers, in the specific field of the pharmacological treatment of skeletal disorders, self-setting calcium orthophosphate formulations have an added value due to their bioactive character and injectability. Further details and additional examples of the drug delivery application of calcium orthophosphate cements are well described elsewhere [29, 36, 37].

5.8.6 Brief Conclusions on the Medical Applications

To conclude this part, one should stress that despite several encouraging results, not every surgeon' expectation has been met yet [569]. First of all, self-setting calcium orthophosphate formulations are not superior to autografts, despite offering primary stability against compressive loading [570, 571]. One of the main concerns of clinicians is to reach higher rates of bioresorption, an improvement of bone reconstruction and to a lesser extent, higher mechanical resistance [33]. Besides, clinical application of the cements in comminuted fractures revealed penetration of the viscous paste into the joint space [572–574]. The interested readers are referred to a paper on cement leakage during vertebroplasty [575]. To date, cadaveric studies have already shown that using calcium orthophosphate cements with conventional metal fixation in certain fractures of the distal radius, tibial plateau, proximal femur and calcaneus can produce better stability, stiffness and strength than metal fixation alone. Early clinical results have revealed a reduced time to full load bearing when the cements were used for augmentation of tibial plateau and calcaneal fractures, more rapid gain of strength and range of motion when used in distal radius fractures and improved stability in certain hip fractures [471, 507]. However, surgeons reported on difficulties in filling the vertebral bodies (a bad injectability of present formulations) and other problems, such as filter pressing and cement decohesion, observed during vertebral body injection that resulted in bone instability due to low mechanical strength as well as long setting times of the cements [576]. This happens due to not only low mechanical properties of calcium orthophosphate cements but also some difficulties of filling vertebral bodies. In order to maintain a good cohesion and reduce filter pressing, calcium orthophosphate cements need to be more viscous (hence, less injectable) [197, 198]. For example, calcium orthophosphate cements might be modified by addition of polysaccharides [97, 108, 335–338] and/or gelatin [278, 339–344].

Another type of concerns has been raised that the use of self-setting calcium orthophosphate formulations for the augmentation of fractured and osteoporotic bones might aggravate cardiovascular deterioration in the event of pulmonary cement embolism by stimulating coagulation [577]. To investigate these potential problems, 2.0 ml of either calcium orthophosphate or polymethylmethacrylate (PMMA) cement were injected intravenously in 14 sheep. Intravenous injection of calcium orthophosphate cement resulted in a more severe increase in pulmonary arterial pressure and decrease in arterial blood pressure compared to the PMMA cement. Disintegration of the calcium orthophosphate cement seemed to be the reason for more severe reaction that represents a risk of cardiovascular complications. The authors concluded that further research efforts should aim at improving cohesion of calcium orthophosphate cements in an aqueous environment for future clinical applications such as vertebral body augmentation [577].

To conclude the biomedical part of this review, one should mention that, although the long-term outcomes are still poorly documented, currently there are no doubts concerning a very great potential of the clinical applications of self-setting calcium orthophosphate formulations for healing of bone and dental defects. For example, a bioresorbable calcium orthophosphate cement was once found to be a better choice, at least in terms of the prevention of subsidence, than autogenous iliac bone graft for the treatment of subarticular defects associated with unstable tibial plateau fractures [578]. Furthermore, BoneSource™ was found to be safe and effective when used to fill traumatic metaphyseal bone voids and appeared to be at least as good as autograft for treatment of these defects [579]. However, in other studies, autologous cancellous grafts were demonstrated to lead to a significantly better bone regeneration compared to the application of calcium orthophosphate granules produced from a self-setting calcium orthophosphate cement powder after 6 weeks [580]. As this text is intended to be read mainly by chemists and materials researchers, the biological, medical and clinical aspects of self-setting calcium orthophosphate formulations have not been discussed in many details. For additional biomedical details, the interested readers are referred to other papers and reviews [22, 29, 36, 37, 137, 471, 476, 570].

5.9 Future Developments

As self-setting calcium orthophosphate formulations represent an intriguing group of new materials for bone augmentation and reconstruction, there is a great potential for further improvement of their properties, in which the ideal characteristics (Table 5.4) should be approached by manipulations with the chemical composition, powder particle size and distribution, as well as by means of various additives. Several commercial cement formulations have been already approved for a clinical application (Tables 5.1 and 5.2). New formulations of both apatite and brushite cements are expected to appear in the market soon. The forthcoming commercial formulations will need to be improved in order to take the advantage of a variety of possibilities offered by calcium orthophosphate cements. New formulations will include: (i) injectable and open macroporous formulations to optimize their osteoconduction [278], (ii) formulations containing only one calcium orthophosphate (single-phase cement powders) [17] and (iii) drug-loaded and hormone-loaded cements for the treatment of bone diseases [29, 36, 37]. Furthermore, incorporation of autologous or allogenic osteo-progenitor cells in the cement formulations will be favorable [581–583]. Obviously, the former two directions deal with both chemistry and material science, while the last two directions are more related to tissue engineering and medicine.

Concerning the material point of view, an innovative approach of injectable self-setting formulations has been introduced relatively recently. The researches combined a water-reactive apatite cement such as a mixture of TTCP and DCPD powders with a nonaqueous but water-miscible liquid (e.g., glycerol, polyethylene glycol) + a gelling agent (e.g., hydroxypropylmethylcellulose, carboxymethylcellulose, chitosan) + a hardening accelerator (e.g., tartaric, malic, malonic, citric or glycolic acids) to form a stable paste that can be directly injected into a bone defect [584–589]. In literature, this type of cement pastes is called "premixed calcium phosphate cements" (occasionally referred to as PCPC) in which the paste remains stable during storage and hardens only after placement into the defect. The paste can be obtained of different consistencies, from low viscosity ones to putty-like plastic pastes [122–124]. Setting occurs *in vivo* upon a contact with body fluids or *in vitro* in a physiological solution and results in CDHA formation. This approach eliminates the powder-liquid mixing stage during surgery, which might improve the cement performance. Besides, it allows shortening the surgical time and the risk of operator-induced error is considerably reduced. Unfortunately, the setting reaction of the premixed formulations is difficult to control and the mechanical properties of the hardened cements are poor. Besides, such formulations must be protected from the environmental moisture during storage [590, 591]. Furthermore, little attention has been paid to the problem that the presence of water impurities in the non-aqueous liquid and/or the powdered solid can compromise the stability of the paste.

Earlier formulations of the premixed calcium orthophosphate cements had a setting time of longer than 1 h and a low mechanical strength [584]. Afterwards, improved formulations have been developed. They exhibit a rapid setting when immersed in a physiological solution, yielding a hardened ceramics with a higher mechanical strength, approached the reported strengths of sintered porous HA implants and cancellous bone [585–587]. Brushite-forming premixed cement formulations have been introduced as well [590, 592–594]; they are expected to have a shorter setting time then the aforementioned apatite-forming premixed cement formulations. Creation of premixed macroporous calcium orthophosphate cement scaffolds reinforced by slow-dissolving fibers (in other words, premixed macroporous concrete scaffolds) is the latest achievement of this approach [326].

Other researchers invented cements in the form of two injectable pastes that could be mixed together and injected at the time of implantation (e.g., by a static twin-chambered mixer incorporated in the injection cannula that allows injection of the paste immediately after mixing) [593, 594]. A wide range of possibilities appears by changing the calcium orthophosphate components. Furthermore, such formulations can also be modulated by adjoining different additives as setting retardants, polymeric adjuvants, visco-enhancing agents, suspension stabilizers, osteoinductive agents, radio-opaque fillers or macropore-forming agents [327]. Nevertheless, this approach is limited to acid–base cement formulations only [34].

Before recent publications [590, 592–594], no study had been available on a possibility of the premixed brushite formulations at ambient temperatures. Thus, researchers developed a way to overcome this problem at low temperatures by creation of pre-mixed frozen calcium orthophosphate "slabs" [155]. By freezing, the setting reaction of a cement was slowed down or even inhibited. When the cement had to be applied, the "slab" was softened at room temperature and molded by hands. Three different pre-mixed brushite formulations were prepared by freezing the cement pastes following combination of the powder and liquid components. When frozen and stored at t = –80°C or less, significant degradation in compression strength did not occur for the duration of the study (28 days). Interestingly, in the case of the brushite cement formed from a combination of β-TCP with 2 M H_3PO_4 solution, freezing the paste had the effect of increasing mean compressive strength fivefold (from 4 to 20 MPa), which was accompanied by a reduction in the setting rate of the cement. This strength improvement was attributed to a modification of crystal morphology and a reduction in damage caused to the cement matrix during manipulation [155].

A lack of macropores is a substantial disadvantage of many current self-setting calcium orthophosphate formulations [273]. As a result, biodegradation takes place layer-by-layer on the surface, from outside to inside. To solve this problem, various types of porogens are used [296–318]. Using a hydrophobic liquid instead of soluble particles could be an alternative. At the turn of the millennium, an open macroporous structure was obtained using a mixture of oil and a cement paste [595]; however, since than no research papers on this subject have been published. Besides, by means of surfactants, air bubbles might be incorporated into the bulk of the cements [290]. Unfortunately, the mechanical strength and porosity are conflicting requirements. As porosity of calcium orthophosphate cements appears to be of paramount importance to achieve the excellent bioresorbability, other experimental approaches have to be developed [596].

Recently, a layered structure was designed by combining a macroporous layer of calcium orthophosphate cement with a strong fiber-reinforced calcium orthophosphate cement layer. The rationale for such construction was for the macroporous layer to accept tissue ingrowth, while the fiber-reinforced strong layer would provide the needed early strength [597].

In the case of calcium orthophosphate concretes, future studies could combine in one formulation porogens and biodegradable fibers of different shapes and dissolution rates to form after *in vivo* hardening calcium orthophosphate scaffolds with sustained strength. In such a system, one porogen is quickly dissolved, which creates macropores to start a bone ingrowth process, while the second type of fibers provides the required strength to the implant. After significant bone ingrowth into the initial pores increased the implant strength, the second set of fibers would then be dissolved to create additional macropores for bone ingrowth [319]. Such complicated formulations have already been developed. For example, chitosan, sodium orthophosphate and hydroxypropylmethylcellulose were used to render calcium orthophosphate cements fast setting and resistant to washout, while absorbable fibers and mannitol porogen were incorporated for strength and macropores, respectively. Both strength and fracture resistance of this concrete were substantially increased and approached those values for sintered porous HA implants [598]. Turning on a bit of imagination, one might predict development of polymeric forms of drugs (already available [599, 600]), hormones, growth factors, etc. (e.g., prepared by either incorporation into or cross-linking with either water-soluble or bioresorbable polymers). Coupled with reinforcing biodegradable fibers and porogens, such types of "healing fibers" might be added to calcium orthophosphate concretes, which not only will accelerate the remedial process, but also will allow simultaneous improvement in both their strength and injectability.

Stability (insolubility) in normal physiological fluid environment and resorbability under acidic conditions produced by osteoclasts appears to be among the most important *in vivo* characteristics of modern types of calcium orthophosphate bioceramics. For some clinical applications, such as cranioplasty, a relatively slow resorption and replacement by bone is quite acceptable, whereas in other applications, such as periodontal bone defects repair, sinus lift, etc., the ability of the hardened cement to be replaced quickly by bone is crucial. Experimental results suggest that a number of parameters of the self-setting calcium orthophosphate formulations, such as the Ca/P ionic ratio, carbonate content, ionic substitution, crystallinity, etc. might affect the dissolution characteristics of the cements in slightly acidic solutions. This gives an opportunity to formulate compositions, possessing different resorption rates, which is suited for different biomedical applications [143, 144].

The discovery of self-setting calcium orthophosphate formulations has already opened up new perspectives in synthesis of bioceramic scaffolds, possessing sufficient mechanical properties [298, 302, 303, 412]. In the past, such scaffolds could only be manufactured by the sintering route at elevated temperatures [601]. Therefore, until recently it was impossible to produce resorbable preset low-temperature hydrated 3D bioceramics for various applications, e.g., scaffolds and granules, from low-temperature calcium orthophosphate phases, such as ACP, DCPA, DCPD, OCP and CDHA. Now, using the appropriate techniques, open macroporous 3D scaffolds consisting of the aforementioned low-temperature phases (currently, excluding ACP and OCP) can be produced via a cementitious reaction [297, 327, 333, 334, 412, 602–604], thus dramatically widening the application of these calcium orthophosphates as biomaterials and bioceramics. This type of biomaterials could be very promising for tissue engineering applications. Among them, CDHA is of a special interest due to its chemical similarity to bone material and a large specific surface area.

To conclude this part, one should stress, that the most promising direction of the future developments of self-setting calcium orthophosphate formulations is obviously seen in their functionalization by incorporation or impregnation of various hormones, growth factors, drugs, other bioorganic compounds, as well as incorporation of living cells and/or other tiny biological objects [605–616]. For example, silk fibroin can regular the mineralization process and bond with HA to form fibroin/HA nanodimensional biocomposites with increased gelation properties and, thus, it can be used as an additive to improve cohesion of calcium orthophosphate cements and decrease a risk of cardiovascular complications in its application in veterbroplasty and kyphoplasty [612].

While the simplicity in the processing of cements encourages the incorporation of cells, the principal difficulty remains to ensure cell survival. The harsh environment in terms of pH and high ionic strength together with the high stiffness achieved upon cement hardening can be thought as the principal threats for cell endurance. The initial attempts have already been performed but without a great success yet. For example, researchers have already found that unset calcium orthophosphate cements might have toxic effects when placed on cell monolayers, while the set formulations are biocompatible for the same type of cells (MC3T3-E1 osteoblast-like cells were tested). A gel encapsulation in alginate beads was found to be a possible solution to protect living cells for seeding into self-setting calcium orthophosphate pastes [582, 617]. *In vitro* cytotoxic effect of a calcium orthophosphate cement based on α-TCP was also observed [618]. In light of these results, the encapsulation approach [308] could potentially be used to seed a patient's *ex vivo* expanded stem cells into a cement to create an osteoinductive bone graft substitute that could be used to treat that patient. However, this becomes more related to tissue engineering and biology, rather than to chemistry and material science. A first possibility would be designing cement formulations that have setting reactions close to the physiological pH or by adding additives into the cement paste able to neutralize the acidic or basic ions released during the cementation reaction.

Finally, besides the aforementioned chemical, material and biomedical improvements of the self-setting calcium orthophosphate formulations, one should not forget on a better design of both the mixing equipment and delivery (injection) techniques. As an example, the interested readers are referred to a new cannula to ease cement injection during vertebroplasty [619]; however, this subject is beyond the scope of current review.

5.10 Conclusions

Thus, among the diverse range of bone replacing biomaterials, self-setting calcium orthophosphate formulations undoubtedly represent a distinct group because they are relatively simple materials formed by combining a calcium orthophosphate mixture with an aqueous solution. However, they symbolize an important breakthrough in the field of bone-repair biomaterials, since they offer the possibility of obtaining thermally unstable calcium orthophosphates in a monolithic form at room or body temperature by means of a cementation reaction. This particular fabrication technique implies that the self-setting formulations are moldable and therefore can adapt easily to the bone cavity providing a good fixation and the optimum tissue-biomaterial contact, necessary for stimulating bone ingrowth into them and their subsequent osteotransduction [29].

Unfortunately, the perfect grafting material does not exist. The self-setting calcium orthophosphate formulations are not an exception to this statement. While possessing excellent biological properties (osteoconduction and, occasionally, osteoinduction), adequate setting time, excellent moldability and the capability to deliver different bone-enhancing proteins/antibiotics at a local level, unfortunately, the material lacks adequate mechanical properties for applications other than non-loaded surgical sites (see Table 5.4 for other details). Nevertheless, even in its present state, the self-setting calcium orthophosphate formulations appear to be suitable for a number of applications. They can be injected into osteoporotic bone to reinforce it or can be used to make granules and blocks out of low-temperature calcium orthophosphates. Several types of the self-setting formulations are now on the market (Tables 5.1 and 5.2), while scaffolds made of low-temperature calcium orthophosphates are being tested. The use of slightly different chemical compositions and various dopants affects both the setting time and tensile strength that enables further improvements. In addition, new trials are conducted with the reinforced formulations and concretes, which represent additional attempts to improve the existing products.

Table 5.4 Major advantages and disadvantages of the calcium orthophosphate cements [36, 37, 202]

Advantages	Disadvantages
1. Self-setting ability *in vivo* 2. Good injectability that allows cement implantation by minimally invasive surgical techniques, which are less damageable than the traditional surgical techniques 3. Good osteoconductivity and occasional osteoinductivity: the initial biological properties of the hardened cements are similar to those of CDHA or brushite 4. Can be replaced by newly formed bone after a period of time (osteotransductivity) 5. Moldability: the perfect fit to the implant site, which assures good bone-material contact, even in geometrically complex defects. 6. Excellent biocompatibility and bioactivity 7. No toxicity 8. Low cost 9. Ease of preparation and handling 10. Setting at body temperature 11. Form chemical bonds to the host bone 12. Clinically safe materials in their powder components 13. Can be used to deliver antibiotics, anti-inflammatory drugs, growth factors, morphogenic proteins, etc., at local sites, which are able to stimulate certain biological responses*	1. Mechanical weakness: limited use due to potential collapse of material followed by soft tissue formation instead of bone formation (loaded areas). Until cements with adequate shear strength are available, most complex fractures that can be repaired with cement also will require metal supports 2. Can be washed out from surgical defect if excess of blood 3. Lack of macroporosity (especially interconnected pores), which prevents fast bone ingrowth and the cements degrade layer-by-layer from the outside to the inside only 4. The *in vivo* biodegradation of many formulations is slower than the growth rate of a newly forming bone

*Further studies are necessary.

It is anticipated that the use of self-setting calcium orthophosphate formulations will enable a faster and more aggressive rehabilitation, as the strength of the hardened cements makes it possible to allow full weight-bearing earlier than when bone graft is used. Although, preliminary clinical trials have already confirmed the great potential of this novel therapeutic product, the self-setting calcium orthophosphate formulations need to be improved further; in particular, their bioresorption needs to be accelerated as well as their injectability and mechanical properties need to get better. Besides, extra clinical studies are required to define the most appropriate indications and limitations of calcium orthophosphate cements for fracture repair.

In the author's humble opinion, mentioning the Prof. James M. Anderson's opinion on the history of biomaterials field would be the best way to conclude this subject. According to Prof. Anderson, within 1950–1975 researchers studied bioMATERIALS, within 1975–2000 they studied BIOMATERIALS and since 2000 the time for BIOmaterials has been coming [620]. Here, the capital letters emphasis the major direction of the research efforts in the complex subject of biomaterials. As the real history of self-setting calcium orthophosphate formulations started only in 1983, the aforementioned periods were shifted along the time scale. Certainly, the bioMATERIALS-epoch for the self-setting formulations is almost over (every possible combination of calcium orthophosphates has been already studied and tested), while the BIOmaterials-era (where cells are the key factor) either has not started yet or is just at the very beginning. Most likely, current state-of-the-art of the self-setting calcium orthophosphate formulations corresponds to BIOMATERIALS-phase with an approximately equal contribution of the biological and materials directions. Therefore, still there is much room for versatile ideas and approaches.

References and Notes

1. Albee, F., and Morrison, H. (1920) Studies in bone growth, *Ann. Surg.*, **71**, 32–38.
2. Haldeman, K., and Moore, J. (1934) Influence of a local excess of calcium and phosphorus on the healing of fractures *Arch. Surg.* **29**, 385–396.
3. Ray, R., Degge, J., Gloyd, P., and Mooney, G. (1952) Bone regeneration, *J. Bone Joint Surg. Am.* **34A**, 638–647.

4. Getter, L., Bhaskar, S., Cutright, D., Perez, B., Brady, J., Driskell, T., and O'Hara, M. (1972) Three biodegradable calcium orthophosphate slurry implants in bone, *J. Oral Surg.*, **30**, 263–268.
5. Roy, D., and Linnehan, S. (1974) Hydroxyapatite formed from coral skeletal carbonate by hydrothermal exchange, *Nature* **247**, 220–222.
6. Koster, K., Karbe, E., Kramer, H., Heide, H., and Konig, R. (1976) Experimenteller Knochenersatz durch resorbierbare Calciumphosphat-Keramik, *Langenbecks Arch. Chir.* **341**, 77–86.
7. Peelen, J., Rejda, B., Vermeiden, J., and de Groot, K. (1977) Sintered tricalcium orthophosphate as bioceramic, *Sci. Ceram.*, **9**, 226–236.
8. Jarcho, M., Kay, J., Gumaer, K., Doremus, R., and Drobeck, H. (1977) Tissue, cellular and subcellular events at a bone-ceramic hydroxyapatite interface, *J. Bioeng.*, **1**, 79–92.
9. Jarcho, M. (1981) Calcium phosphate ceramics as hard tissue prosthetics, *Clin. Orthop.*, **157**, 259–278.
10. LeGeros, R. Z. (1991) *Calcium Phosphates in Oral Biology and Medicine*, Karger, Basel, p. 210.
11. Dorozhkin, S. V. (2007) Calcium orthophosphates, *J. Mater. Sci.*, **42**, 1061–1095.
12. LeGeros, R. Z., Chohayeb, A., and Shulman, A. (1982) Apatitic calcium phosphates: possible dental restorative materials, *J. Dent. Res.*, **61**(Special issue), 343.
13. Brown, W. E., and Chow, L. C. (1983) A new calcium phosphate setting cement, *J. Dent. Res.*, **62**(Special issue), 672.
14. Brown, W. E., and Chow, L. C. (1986) A new calcium phosphate water setting cement, in *Cements Research Progress* (ed. Brown, P. W.), American Ceramic Society, Westerville, pp. 352–379.
15. Brown, W. E., and Chow, L. C. Dental restorative cement pastes. US Patent No. 4518430. May 21, 1985.
16. Gruninger, S. E., Siew, C., Chow, L. C., O'Young, A., Tsao, N. K., and Brown, W. E. (1984) Evaluation of the biocompatibility of a new calcium phosphate setting cement, *J. Dent. Res.*, **63**(Special issue), 200.
17. Tas, A. C. (2007) Porous, biphasic $CaCO_3$-calcium phosphate biomedical cement scaffolds from calcite ($CaCO_3$) powder, *Int. J. Appl. Ceram. Technol.*, **4**, 152–163.
18. Driskell, T. D., Heller, A. L., and Koenigs, J. F. Dental treatments. US Patent No. 3913229. October 21, 1975.
19. Kingery, W. D. II. (1950) Cold setting properties, *J. Am. Ceram. Soc.*, **33**, 242–246.

20. Driessens, F. C. M., Planell, J. A., and Gil, F. J. (1995) *Encyclopedic Handbook of Biomaterials and Bioengineering, Part B, Applications* (ed. Wise, D. L., Trantolo, D. J., Altobelli, D. E., Yaszemski, M. J., Gresser, J. D., and Schwarz, E. R.), vol. 2, Marcel Dekker, New York, pp. 855–877.

21. Bolarinwa, A., Gbureck, U., Purnell, P., Bold, M., and Grover, L. M. (2010) Cement casting of calcium pyrophosphate based bioceramics, *Adv. Appl. Ceram.*, **109**, 291–295.

22. Schmitz, J. P., Hollinger, J. O., and Milan, S. B. (1999) Reconstruction of bone using calcium phosphate bone cements: a critical review, *J. Oral Maxillofac. Surg.*, **57**, 1122–1126.

23. Gbureck, U., Barralet, J. E., Spatz, K., Grover, L. M., and Thull, R. (2004) Ionic modification of calcium phosphate cement viscosity. Part I: Hypodermic injection and strength improvement of apatite cement, *Biomaterials*, **25**, 2187–2195.

24. Generosi, A., Rau, J. V., Komlev, V. S., Albertini, V. R., Fedotov, A. Y., and Barinov, S. M. (2010) Anomalous hardening behavior of a calcium phosphate bone cement, *J. Phys. Chem. B*, **114**, 973–979.

25. Rau, J. V., Generosi, A., Komlev, V. S., Fosca, M., Barinov, S. M., and Albertini, V. R. (2010) Real-time monitoring of the mechanism of poorly crystalline apatite cement conversion in the presence of chitosan, simulated body fluid and human blood, *Dalton Trans.*, **21**, 11412–11423.

26. Smirnov, V. V., Rau, J. V., Generosi, A., Albertini, V. R., Ferro, D., and Barinov, S. M. (2010) Elucidation of real-time hardening mechanisms of two novel high-strength calcium phosphate bone cements, *J. Biomed. Mater. Res. B* (*Appl. Biomater.*), **93B**, 74–83.

27. Tagaya, M., Goto, H., Iinuma, M., Wakamatsu, N., Tamura, Y., and Doi, Y. (2005) Development of self-setting Te-Cp/alpha-TCP cement for pulpotomy, *Dent. Mater. J.*, **24**, 555–561.

28. Driessens, F. C. M., Boltong, M. G., Khairoun, I., de Maeyer, E. A. P., Ginebra, M. P., Wenz, R., Planell, J. A., and Verbeeck, R. M. H. (2000) Applied aspects of calcium phosphate bone cement, in *Biomaterials Engineering and Devices: Human Applications* (ed. Wise, D. L., Trantolo, D. J., Lewandrowski, K. U., Gresser, J. D., and Cattaneo, M. V.), vol. 2, *Orthopedic, Dental and Bone Graft Applications*, Humana Press, Totowa, pp. 253–260.

29. Driessens, F. C. M., Planell, J. A., Boltong, M. G., Khairoun, I., and Ginebra, M. P. (1998) Osteotransductive bone cements, *Proc. Inst. Mech. Eng. H: J. Eng. Med.*, **212**, 427–435.

30. Frankenburg, E. P., Goldstein, S. A., Bauer, T. W., Harris, S. A., and Poser, R. D. (1998) Biomechanical and histological evaluation of a calcium phosphate cement, *J. Bone Joint Surg. Am.*, **80A**, 1112–1124.

31. Frayssinet, P., Gineste, L., Conte, P., Fages, J., and Rouquet, N. (1998) Short-term implantation effects of a DCPD-based calcium phosphate cement, *Biomaterials*, **19**, 971–977.

32. Rey, C., Tofighi, A., Mounic, S., Combes, C., and Lee, D. (2002) Biomimetism and calcium phosphate cements, in *Actualités en Biomatériaux* (ed. Mainard, D., and Louis, J. P.), vol. 6, Editions Romillat, Paris, pp. 27–37.

33. Combes, C., Bareille, R., and Rey, C. (2006) Calcium carbonate-calcium phosphate mixed cement compositions for bone reconstruction, *J. Biomed. Mater. Res. A*, **79A**, 318–328.

34. Bohner, M., Gbureck, U., and Barralet, J. E. (2005) Technological issues for the development of more efficient calcium phosphate bone cements: a critical assessment, *Biomaterials*, **26**, 6423–6429.

35. Ikenaga, M., Hardouin, P., Lemaître, J., Andrianjatovo, H., and Flautre, B. (1998) Biomechanical characterization of a biodegradable calcium phosphate hydraulic cement: a comparison with porous biphasic calcium phosphate ceramics, *J. Biomed. Mater. Res.*, **40**, 139–144.

36. Ginebra, M. P., Traykova, T., and Planell, J. A. (2006) Calcium phosphate cements: competitive drug carriers for the musculoskeletal system? *Biomaterials*, **27**, 2171–2177.

37. Ginebra, M. P., Traykova, T., and Planell, J. A. (2006) Calcium phosphate cements as bone drug delivery systems: a review, *J. Control. Release*, **113**, 102–110.

38. It is interesting to note that calcium sulfate (gypsum or plaster of Paris) has been implanted into fracture gaps and voids of bones for more than 100 years [39–41].

39. Dreesmann, H. (1892) Knochenplombierung bei Hohlenforigen Defekten des Knochens, *Beitr. Klin. Chir.*, **9**, 804–810.

40. Peltier, L. F. (1961) The use of plaster of Paris to fill defects in bone, *Clin. Orthop.*, **21**, 1–29.

41. Kelly, C. M., Wilkins, R. M., Gitelis, S., Hartjen, C., Watson, J. T., and Kim, P. T. (2001) The use of a surgical grade calcium sulfate as a bone graft substitute: results of a multicenter trial, *Clin. Orthop.*, **382**, 42–50.

42. Bohner, M. (2004) New hydraulic cements based on α-tricalcium phosphate–calcium sulfate dihydrate mixtures, *Biomaterials*, **25**, 741–749.

43. Fernández, E., Vlad, M. D., Gel, M. M., Lopez, J., Torres, R., Cauich, J. V., and Bohner, M. (2005) Modulation of porosity in apatitic cements by the use of α-tricalcium phosphate–calcium sulphate dihydrate mixtures, *Biomaterials*, **26**, 3395–3404.

44. Hu, G., Xiao, L., Fu, H., Bi, D., Ma, H., and Tong, P. (2010) Degradable and bioactive scaffold of calcium phosphate and calcium sulphate from self-setting cement for bone regeneration, *J. Porous Mater.*, **17**, 605–613.

45. Hu, G., Xiao, L., Fu, H., Bi, D., Ma, H., and Tong, P. (2010) Study on injectable and degradable cement of calcium sulphate and calcium phosphate for bone repair, *J. Mater. Sci. Mater. Med.*, **21**, 627–634.

46. Nilsson, M., Fernández, E., Sarda, S., Lidgren, L., and Planell, J. A. (2002) Characterization of a novel calcium phosphate/sulphate bone cement, *J. Biomed. Mater. Res.*, **61**, 600–607.

47. Vlad, M. D., Şindilar, E. V., Mariñoso, M. L., Poeat, I., Torres, R., López, J., Barracó, M., and Fernández, E. (2010) Osteogenic biphasic calcium sulphate dihydrate/iron-modified α-tricalcium phosphate bone cement for spinal applications: *in vivo* study, *Acta Biomater.*, **6**, 607–616.

48. Grover, L. M., Gbureck, U., Wright, A. J., Tremaynec, M., and Barralet, J. E. (2006) Biologically mediated resorption of brushite cement *in vitro*, *Biomaterials*, **27**, 2178–2185.

49. Grover, L. M., Gbureck, U., Wright, A. J., and Barralet, J. E. (2005) Cement formulations in the calcium phosphate H_2O–H_3PO_4–$H_4P_2O_7$ system, *J. Am. Ceram. Soc.*, **88**, 3096–3103.

50. Grover, L. M., Gbureck, U., Young, A. M., Wright, A. J., and Barralet, J. E. (2005) Temperature dependent setting kinetics and mechanical properties of β-TCP–pyrophosphoric acid bone cement, *J. Mater. Chem.*, **46**, 4955–4962.

51. Oh, K. S., Jeong, Y. K., Yu, J. P., Chae, S. K., Kim, H. Y., Lee, H. Y., and Jeun, S. S. (2005) Preparation and *in vivo* studies of β-TCP based bone cement containing polyphosphate, *Key Eng. Mater.*, 284–286, 93–96.

52. Lilley, K. J., Gbureck, U., Wright, A. J., Knowles, J. C., Farrar, D. F., and Barralet, J. E. (2007) Brushite cements from polyphosphoric acid, calcium phosphate systems, *J. Am. Ceram. Soc.*, **90**, 1892–1898.

53. Fernández, E., Planell, J. A., and Best, S. M. (1999) Precipitation of carbonated apatite in the cement system α-$Ca_3(PO_4)_2$–$Ca(H_2PO_4)_2$–$CaCO_3$, *J. Biomed. Mater. Res.*, **47**, 466–471.

54. Calafiori, A. R., di Marco, G., Martino, G., and Marotta, M. (2007) Preparation and characterization of calcium phosphate biomaterials, *J. Mater. Sci. Mater. Med.*, **18**, 2331–2338.

55. Kon, M., Hirakata, L. M., Miyamoto, Y., Kasahara, H., and Asaoka, K. (2005) Strengthening of calcium phosphate cement by compounding calcium carbonate whiskers, *Dent. Mater. J.*, **24**, 104–110.

56. Serraj, S., Michailesco, P., Margerit, J., Bernard, B., and Boudeville, P. (2002) Study of a hydraulic calcium phosphate cement for dental applications, *J. Mater. Sci. Mater. Med.*, **13**, 125–131.

57. Nurit, L., Margerit, J., Terol, A., and Boudeville, P. (2002) pH-metric study of the setting reaction of monocalcium phosphate monohydrate/calcium oxide-based cements, *J. Mater. Sci. Mater. Med.*, **13**, 1007–1014.

58. Boudeville, P., Serraj, S., Leloup, J. M., Margerit, J., Pauvert, B., and Terol, A. (1999) Physical properties and self-setting mechanism of calcium phosphate cements from calcium bis-dihydrogenophosphate monohydrate and calcium oxide, *J. Mater. Sci. Mater. Med.*, **10**, 99–109.

59. Michaïlesco, P., Kouassi, M., Briak, H. E., Armynot, A., and Boudeville, P. (2005) Antimicrobial activity and tightness of a DCPD–CaO-based hydraulic calcium phosphate cement for root canal filling, *J. Biomed. Mater. Res. B* (*Appl. Biomater.*), **74B**, 760–767.

60. Briak, H. E., Durand, D., Nurit, J., Munier, S., Pauvert, B., and Boudeville, P. (2002) Study of a hydraulic dicalcium phosphate dihydrate/calcium oxide-based cement for dental applications, *J. Biomed. Mater. Res.* (*Appl. Biomater.*), **63**, 447–453.

61. Briak, H. E., Durand, D., and Boudeville, P. (2008) Study of a hydraulic DCPA/CaO-based cement for dental applications, *J. Mater. Sci. Mater. Med.*, **19**, 737–744.

62. Takagi, S., Chow, L. C., and Ishikawa, K. (1998) Formation of hydroxyapatite in new calcium phosphate cements, *Biomaterials*, **19**, 1593–1599.

63. Yang, Q., Troczynski, T., and Liu, D. M. (2002) Influence of apatite seeds on the synthesis of calcium phosphate cement, *Biomaterials*, **23**, 2751–2760.

64. Hsu, H. C., Chiu, C. Y., Tuan, W. H., and Lee, H. Y. (2008) Structural stability of calcium phosphate cement during aging in water, *Mater. Sci. Eng. C*, **28**, 429–433.

65. Roemhildt, M. L., McGee, T. D., and Wagner, S. D. (2003) Novel calcium phosphate composite bone cement, strength and bonding properties, *J. Mater. Sci. Mater. Med.*, **14**, 137–141.

66. Roemhildt, M. L., Wagner, S. D., and McGee, T. D. (2006) Characterization of a novel calcium phosphate composite bone cement: flow, setting, and aging properties, *J. Mater. Sci. Mater. Med.*, **17**, 1127–1132.

67. Wang, X., Ye, J., Wang, Y., and Chen, L. (2007) Self-setting properties of a β-dicalcium silicate reinforced calcium phosphate cement, *J. Biomed. Mater. Res. B* (*Appl. Biomater.*), **82B**, 93–99.

68. Huan, Z., and Chang, J. (2007) Novel tricalcium silicate/monocalcium phosphate monohydrate composite bone cement, *J. Biomed. Mater. Res. B* (*Appl. Biomater.*), **82B**, 352–359.

69. Huan, Z., and Chang, J. (2009) Calcium-phosphate-silicate composite bone cement, self-setting properties and *in vitro* bioactivity, *J. Mater. Sci. Mater. Med.*, **20**, 833–841.

70. Huan, Z., and Chang, J. (2009) Novel bioactive composite bone cements based on the β-tricalcium phosphate–monocalcium phosphate monohydrate composite cement system, *Acta Biomater.*, **5**, 1253–1264.

71. Shen, Q., Sun, J., Wu, J., Liu, C., and Chen, F. (2010) An *in vitro* investigation of the mechanical-chemical and biological properties of calcium phosphate/calcium silicate/bismutite cement for dental pulp capping, *J. Biomed. Mater. Res. B* (*Appl. Biomater.*), **94**, 141–148.

72. Guo, D., Xu, K., Zhao, X., and Han, Y. (2005) Development of a strontium-containing hydroxyapatite bone cement, *Biomaterials*, **26**, 4073–4083.

73. Wang, X., and Ye, J. (2008) Variation of crystal structure of hydroxyapatite in calcium phosphate cement by the substitution of strontium ions, *J. Mater. Sci. Mater. Med.*, **19**, 1183–1186.

74. Pina, S., Torres, P. M. C., and Ferreira, J. M. F. (2010) Injectability of brushite-forming Mg-substituted and Sr-substituted α-TCP bone cements, *J. Mater. Sci. Mater. Med.*, **21**, 431–438.

75. Wu, F., Su, J. C., Wei, J., Guo, H., and Liu, C. S. (2008) Injectable bioactive calcium-magnesium phosphate cement for bone regeneration, *Biomed. Mater.*, **3**, 044105 (7 pages).

76. Wu, F., Wei, J., Guo, H., Chen, F. P., Hong, H., and Liu, C. S. (2008) Self-setting bioactive calcium-magnesium phosphate cement with high strength and degradability for bone regeneration, *Acta Biomater.*, **4**, 1873–1884.

77. Pina, S., Olhero, S. M., Gheduzzi, S., Miles, A. W., and Ferreira, J. M. F. (2009) Influence of setting liquid composition and liquid-to-powder ratio on properties of a Mg-substituted calcium phosphate cement, *Acta Biomater.*, **5**, 1233–1240.

78. Klammert, U., Reuther, T., Blank, M., Reske, I., Barralet, J. E., Grover, L. M., Kübler, A. C., and Gbureck, U. (2010) Phase composition, mechanical performance and *in vitro* biocompatibility of hydraulic setting calcium magnesium phosphate cement, *Acta Biomater.*, **6**, 1529–1535.

79. Jia, J., Zhou, H., Wei, J., Jiang, X., Hua, H., Chen, F., Wei, S., Shin, J. W., and Liu, C. (2010) Development of magnesium calcium phosphate biocement for bone regeneration, *J. Royal Soc. Interf.*, **7**, 1171–1180.

80. Pina, S., Vieira, S. I., Torres, P. M. C., Goetz-Neunhoeffer, F., Neubauer, J., da Cruz E. Silva, O.A.B., da Cruz E. Silva, E. F., and Ferreira, J. M. F. (2010) *In vitro* performance assessment of new brushite-forming Zn- and ZnSr-substituted β-TCP bone cements, *J. Biomed. Mater. Res. B* (*Appl. Biomater.*), **94B**, 414–420.

81. Gbureck, U., Knappe, O., Grover, L. M., and Barralet, J. E. (2005) Antimicrobial potency of alkali ion substituted calcium phosphate cements, *Biomaterials*, **26**, 6880–6886.

82. Driessens, F. C. M., Boltong, M. G., de Mayer, E. A. P., Wenz, R., Nies, B., and Planell, J. A. (2002) The Ca/P range of nanoapatitic calcium phosphate cements, *Biomaterials*, **23**, 4011–4017.

83. Gbureck, U., Thull, R., and Barralet, J. E. (2005) Alkali ion substituted calcium phosphate cement formation from mechanically activated reactants, *J. Mater. Sci. Mater. Med.*, **16**, 423–427.

84. Doi, Y., Shimizu, Y., Moriwaki, Y., Aga, M., Iwanaga, H., Shibutani, T., Yamamoto, K., and Iwayama, Y. (2001) Development of a new calcium phosphate cement that contains sodium calcium phosphate, *Biomaterials*, **22**, 847–854.

85. Lilley, J., Gbureck, U., Knowles, J. C., Farrar, D. F., and Barralet, J. E. (2005) Cement from magnesium substituted hydroxyapatite, *J. Mater. Sci. Mater. Med.*, **16**, 455–460.

86. Ni, G. X., Lu, W. W., Tang, B., Ngan, A. H. W., Chiu, K. Y., Cheung, K. M. C., Li, Z. Y., and Luk, K. D. K. (2007) Effect of weight-bearing on bone-bonding behavior of strontium-containing hydroxyapatite bone cement, *J. Biomed. Mater. Res. A*, **83A**, 570–576.

87. Alkhraisat, M. H., Mariño, F. T., Rodríguez, C. R., Jerez, L. B., and Cabarcos, E. L. (2008) Combined effect of strontium and pyrophosphate on the properties of brushite cements, *Acta Biomater.*, **4**, 664–670.

88. Yao, Z. P., Liu, W. G., and Ni, G. X. (2008) Biology characteristics and clinical application of strontium substituted hydroxyapatite bone cement, *J. Clin. Rehabil. Tissue Eng. Res.*, **12**, 7151–7154.

89. Li, S., Liu, B., Cheng, J., and Hu, J. (2008) Composite cement of magnesium-bearing phosphoaluminate-hydroxyapatite reinforced by treated raw silk fiber, *Cement Concrete Composites*, **30**, 347–352.

90. Ni, G. X., Lin, J. H., Chiu, P. K. Y., Li, Z. Y., and Lu, W. W. (2010) Effect of strontium-containing hydroxyapatite bone cement on bone remodeling following hip replacement, *J. Mater. Sci. Mater. Med.*, **21**, 377–384.

91. Fadeeva, I. V., Barinov, S. M., Komlev, V. S., Fedotov, D. A., Durisin, J., and Medvecky, L. (2004) Apatite formation in the reaction-setting mixture of $Ca(OH)_2$–KH_2PO_4 system, *J. Biomed. Mater. Res. A*, **70A**, 303–308.

92. Tas, A. C. (2008) Use of vaterite and calcite in forming calcium phosphate cement scaffolds, *Ceram. Eng. Sci. Proc.*, **28**, 135–150.

93. Fernández, E., Vlad, M. D., Hamcerencu, M., Darie, A., Torres, R., and Lopez, J. (2005) Effect of iron on the setting properties of α-TCP bone cements, *J. Mater. Sci.*, **40**, 3677–3682.

94. Vlad, M. D., del Valle, L. J., Poeata, I., Barracó, M., López, J., Torres, R., and Fernández, E. (2008) Injectable iron-modified apatitic bone cement intended for kyphoplasty, cytocompatibility study, *J. Mater. Sci. Mater. Med.*, **19**, 3575–3583.

95. http://en.wikipedia.org/wiki/Cement (accessed in October 2010).

96. There are some differences between TTCP + DCPD and TTCP + DCPA cements. Due to a higher solubility of DCPD (Table 1.1), a TTCP + DCPD cement sets faster than a TTCP + DCPA cement. Besides, injectability of a TTCP + DCPD cement is better [97, 98].

97. Burguera, E. F., Xu, H. H. K., and Weir, M. D. (2006) Injectable and rapid-setting calcium phosphate bone cement with dicalcium phosphate dihydrate, *J. Biomed. Mater. Res. B* (*Appl. Biomater.*), **77B**, 126–134.

98. Burguera, E. F., Guitian, F., and Chow, L. C. (2004) A water setting tetracalcium phosphate–dicalcium phosphate dihydrate cement, *J. Biomed. Mater. Res. A*, **71A**, 275–282.

99. Driessens, F. C. M., Boltong, M. G., Bermudez, O., and Planell, J. A. (1993) Formulation and setting times of some calcium orthophosphate cements, a pilot study, *J. Mater. Sci. Mater. Med.*, **4**, 503–508.

100. Chow, L. C., Markovic, M., and Takagi, S. (1998) Calcium phosphate cements, in *Cements Research Progress*, (ed. Struble, L. J.), Chapter 7, American Ceramic Society, Westerville, pp. 215–238.

101. Driessens, F. C. M., Boltong, M. G., Bermudez, O., Planell, J. A., Ginebra, M. P., and Fernández, E. (1994) Effective formulations for the preparation of calcium phosphate bone cements, *J. Mater. Sci. Mater. Med.*, **5**, 164–170.

102. Kurashina, K., Hirano, M., Kotani, A., Klein, C. P. A. T., and de Groot, K. (1997) *In vivo* study of calcium phosphate cements, implantation of an α-tricalcium phosphate/dicalcium phosphate dibasic/tetracalcium phosphate monoxide cement paste, *Biomaterials*, **18**, 539–543.

103. Friedman, C. D., Costantino, P. D., Takagi, S., and Chow, L. C. (1998) BoneSource™ hydroxyapatite cement, a novel biomaterial for craniofacial skeletal tissue engineering and reconstruction, *J. Biomed. Mater. Res.*, **43**, 428–432.

104. Khairoun, I., Boltong, M. G., Driessens, F. C. M., and Planell, J. A. (1997) Effect of calcium carbonate on the compliance of apatitic calcium phosphate bone cement, *Biomaterials*, **18**, 1535–1539.

105. Fernández, E., Gil, F. J., Best, S. M., Ginebra, M. P., Driessens, F. C. M., and Planell, J. A. (1998) Improvement of the mechanical properties of new calcium phosphate bone cements in the $CaHPO_4$–α-$Ca_3(PO_4)_2$ system, compressive strength and microstructural development, *J. Biomed. Mater. Res.*, **41**, 560–567.

106. Fukase, Y., Eanes, E. D., Takagi, S., Chow, L. C., and Brown, W. E. (1990) Setting reactions and compressive strengths of calcium phosphate cements, *J. Dent. Res.*, **69**, 1852–1856.

107. Xie, L., and Monroe, E. A. (1991) Calcium phosphate dental cements, *Mat. Res. Soc. Symp. Proc.*, **179**, 25–39.

108. Ishikawa, K., Miyamoto, Y., Kon, M., Nagayama, M., and Asaoka, K. (1995) Non-decay type fast-setting calcium orthophosphate cement composite with sodium alginate, *Biomaterials*, **16**, 527–532.

109. Xu, H. H. K., Quinn, J. B., Takagi, S., and Chow, L. C. (2002) Processing and properties of strong and non-rigid calcium phosphate cement, *J. Dent. Res.*, **81**, 219–224.

110. Lee, Y. K., Lim, B. S., and Kim, C. W. (2003) Mechanical properties of calcium phosphate based dental filling and regeneration materials, *J. Oral Rehabil.*, **30**, 418–425.

111. Ginebra, M. P., Fernández, E., de Mayer, E. A. P., Verbeeck, R. M. H., Boltong, M. G., Ginebra, J., Driessens, F. C. M., and Planell, J. A. (1997) Setting reaction and hardening of an apatitic calcium phosphate cement, *J. Dent. Res.*, **76**, 905–912.

112. Liu, C., Shen, W., Gu, Y., and Hu, L. (1997) Mechanism of the hardening process for a hydroxyapatite cement, *J. Biomed. Mater. Res.*, **35**, 75–80.

113. Driessens, F. C. M., de Mayer, E. A. P., Fernández, E., Boltong, M. G., Berger, G., Verbeeck, R. M. H., Ginebra, M. P., and Planell, J. A. (1996) Amorphous calcium phosphate cements and their transformation into calcium deficient hydroxyapatite, *Bioceramics*, **9**, 231–234.

114. Driessens, F. C. M., Planell, J. A., and Gil, F. (1995) Calcium phosphate bone cements, in *Encyclopedic Handbook of Biomaterials and Bioengineering* (ed. Wise, D. L., Trantolo, D. J., Altobelli, D. E., Yaszemski, M. J., Cresser, J. D., Schwartz, E. R.), *Part B, Applications*, vol. 2, Marcel Dekker, New York, pp. 855–877.

115. Lemaître, J. (1995) Injectable calcium phosphate hydraulic cements: new developments and potential applications, *Inn. Tech. Biol. Med.*, **16**, 109–120.

116. Neira, I. S., Kolen'ko, Y. V., Lebedev, O. I., van Tendeloo, G., Gupta, H. S., Matsushita, N., Yoshimura, M., and Guitián, F. (2009) Rational synthesis of a nanocrystalline calcium phosphate cement exhibiting rapid conversion to hydroxyapatite, *Mater. Sci. Eng. C*, **29**, 2124–2132.

117. Tañag, M. A., Yano, K., and Hosokawa, K. (2004) Orbital floor reconstruction using calcium phosphate cement paste: an animal study, *Plast. Reconstr. Surg.*, **114**, 1826–1831.

118. Hatoko, M., Tada, H., Tanaka, A., Yurugi, S., Niitsuma, K., and Iioka, H. (2005) The use of calcium phosphate cement paste for the correction of the depressed nose deformity, *J. Craniofac. Surg.*, **16**, 327–331.

119. Tañag, M. A., Madura, T., Yano, K., and Hosokawa, K. (2006) Use of calcium phosphate cement paste in orbital volume augmentation, *Plast. Reconstr. Surg.*, **117**, 1186–1193.

120. Meng, D., Xie, Q. F., and Xiao, J. J. (2009) Effects of two calcium phosphate cement pastes on osteoblasts during solidification, *J. Clin. Rehabilit. Tiss. Eng. Res.*, **13**, 471–474.

121. According to Wikipedia, the free encyclopedia: "Putty is a generic term for a plastic material similar in texture to clay or dough typically used in domestic construction and repair as a sealant or filler." http://en.wikipedia.org/wiki/Putty (accessed in October 2010).

122. Ishikawa, K., Miyamoto, Y., Takechi, M., Toh, T., Kon, M., Nagayama, M., and Asaoka, K. (1997) Non-decay type fast-setting calcium phosphate cement: hydroxyapatite putty containing an increased amount of sodium alginate, *J. Biomed. Mater. Res.*, **36**, 393–399.

123. Ishikawa, K., Miyamoto, Y., Takechi, M., Ueyama, Y., Suzuki, K., Nagayama, M., and Matsumura, T. (1999) Effects of neutral sodium hydrogen phosphate on setting reaction and mechanical strength of hydroxyapatite putty, *J. Biomed. Mater. Res.*, **44**, 322–329.

124. Momota, Y., Miyamoto, Y., Ishikawa, K., Takechi, M., Yuasa, T., Tatehara, S., and Nagayama, M. (2004) Effects of neutral sodium hydrogen phosphate on the setting property and hemostatic ability of

hydroxyapatite putty as a local hemostatic agent for bone, *J. Biomed. Mater. Res. B (Appl. Biomater.)*, **69B**, 99–103.

125. Bohner, M. (2010) Design of ceramic-based cements and putties for bone graft substitution, *Eur. Cell Mater.*, **20**, 1–12.

126. In the vast majority cases, a precipitated poorly crystalline HA and CDHA are undistinguishable and might be considered as synonyms [11].

127. To honor Prof. George Jarvis Brush (1831–1912), an American mineralogist, Yale University, New Haven, Connecticut, USA.

128. Xia, Z., Grover, L. M., Huang, Y., Adamopoulos, I. E., Gbureck, U., Triffitt, J. T., Shelton, R. M., and Barralet, J. E. (2006) *In vitro* biodegradation of three brushite calcium phosphate cements by a macrophage cell-line, *Biomaterials*, **27**, 4557–4565.

129. Monma, H., Makishima, A., Mitomo, M., and Ikegami, T. (1988) Hydraulic properties of the tricalcium phosphate–dicalcium phosphate mixture, *J. Ceram. Soc. Jpn.*, **96**, 878–880.

130. Bermudez, O., Boltong, M. G., Driessens, F. C. M., and Planell, J. A. (1994) Development of an octacalcium phosphate cement, *J. Mater. Sci. Mater. Med.*, **5**, 144–146.

131. Sena, M., Yamashita, Y., Nakano, Y., Ohgaki, M., Nakamura, S., Yamashita, K., and Takagi, Y. (2004) Octacalcium phosphate-based cement as a pulp-capping agent in rats, *Oral Surg. Oral Med. Oral Pathol. Oral Radiol. Endod.*, **97**, 749–755.

132. Markovic, M., and Chow, L. C. (2010) An octacalcium phosphate forming cement, *J. Res. Natl. Inst. Stand. Technol.*, **115**, 257–265.

133. In early 1990s, depending on the type of calcium orthophosphate formed after the setting, five groups of the cement formulations were thought to exist: DCPD, CDHA, HA, ACP and OCP [101, 134], while currently only two cement groups remain.

134. Khairoun, I., Boltong, M. G., Driessens, F. C. M., and Planell, J. A. (1998) Limited compliance of some apatitic calcium phosphate bone cements with clinical requirements, *J. Mater. Sci. Mater. Med.*, **9**, 667–671.

135. Lacout, J., Mejdoubi, E., and Hamad, M. (1996) Crystallization mechanisms of calcium orthophosphate cement for biological uses, *J. Mater. Sci. Mater. Med.*, **7**, 371–374.

136. Song, Y., Feng, Z., and Wang, T. (2007) *In situ* study on the curing process of calcium phosphate bone cement, *J. Mater. Sci. Mater. Med.*, **18**, 1185–1193.

137. Weiss, D. D., Sachs, M. A., and Woodard, C. R. (2003) Calcium phosphate bone cements: a comprehensive review, *J. Long Term Eff. Med. Implants*, **13**, 41–47.

138. Fernández, E., Gil, F. J., Ginebra, M. P., Driessens, F. C. M., Planell, J. A., and Best, S. M. (1999) Calcium phosphate bone cements for clinical applications. Part I: Solution chemistry, *J. Mater. Sci. Mater. Med.*, **10**, 169–176.

139. Brown, W. E., and Chow, L. C. (1986) A new calcium phosphate water-setting cement, in *Cements Research Progress* (ed. Brown, P. W.), American Ceramic Society, Westerville, pp. 351–379.

140. Hatim, Z., Freche, M., Keribech, A., and Lacout, J. L. (1998) The setting mechanism of a phosphocalcium biological cement, *Ann. Chim. Sci. Mat.*, **23**, 65–68.

141. Ishikawa, K., and Asaoka, K. (1995) Estimation of ideal mechanical strength and critical porosity of calcium phosphate cement, *J. Biomed. Mater. Res.*, **29**, 1537–1543.

142. Chow, L. C. (1991) Development of self-setting calcium phosphate cements, *J. Ceram. Soc. Jpn.*, **99**, 954–964.

143. Chow, L. C. (2000) Calcium phosphate cements: chemistry, properties and applications, *Mat. Res. Soc. Symp. Proc.*, **599**, 27–37.

144. Chow, L. C. (2001) Calcium phosphate cements, in *Octacalcium Phosphate* (ed. Chow, L. C., and Eanes, E. D.), *Monographs in Oral Science*, vol. 18, Karger, Basel, pp. 148–163.

145. Brown, P. W., and Fulmer, M. T. (1991) Kinetics of hydroxyapatite formation at low temperature, *J. Am. Ceram. Soc.*, **74**, 934–940.

146. TenHuisen, K. S., and Brown, P. W. (1994) The formation of hydroxyapatite-ionomer cements at 38 °C, *J. Dent. Res.*, **3**, 598–606.

147. Ishikawa, K., Takagi, S., Chow, L. C., and Suzuki, K. (1999) Reaction of calcium phosphate cements with different amounts of tetracalcium phosphate and dicalcium phosphate anhydrous, *J. Biomed. Mater. Res.*, **46**, 504–510.

148. Matsuya, S., Takagi, S., and Chow, L. C. (2000) Effect of mixing ratio and pH on the reaction between $Ca_4(PO_4)_2O$ and $CaHPO_4$, *J. Mater. Sci. Mater. Med.*, **11**, 305–311.

149. Burguera, E. F., Guitian, F., and Chow, L. C. (2008) Effect of the calcium to phosphate ratio of tetracalcium phosphate on the properties of calcium phosphate bone cement, *J. Biomed. Mater. Res. A*, **85A**, 674–683.

150. Lemaître, J., Mirtchi, A. A., and Mortier, A. (1987) Calcium phosphate cements for medical use: state of the art and perspectives of development, *Silic. Ind.*, 9–10, 141–146.

151. Bajpai, P., Fuchs, C., and McCullum, D. (1987) Development of tricalcium orthophosphate ceramic cement, in *Quantitative Characterization and Performance of Porous Implants for Hard Tissue Applications* (ed. Lemons, J.), ASTM STP 953. Am. Soc. Test. Mater., Philadelphia, pp. 377–388.

152. Mirtchi, A. A., Lemaître, J., and Terao, N. (1989) Calcium phosphate cements: study of the β-tricalcium phosphate–monocalcium phosphate system, *Biomaterials*, **10**, 475–480.

153. Bohner, M., van Landuyt, P., Merkle, H. P., and Lemaître, J. (1997) Composition effects on the pH of a hydraulic calcium orthophosphate cement, *J. Mater. Sci. Mater. Med.*, **8**, 675–681.

154. Bohner, M., Lemaître, J., and Ring, T. A. (1996) Effects of sulfate, pyrophosphate and citrate ions on the physiochemical properties of cements made of β-tricalcium phosphate–phosphoric acid–water mixtures, *J. Am. Ceram. Soc.*, **79**, 1427–1434.

155. Grover, L. M., Hofmann, M. P., Gbureck, U., Kumarasami, B., and Barralet, J. E. (1923) Frozen delivery of brushite calcium phosphate cements, *Acta Biomater.*, **4**, 1916–1923.

156. Fernández, E., Gil, F. J., Best, S. M., Ginebra, M. P., Driessens, F. C. M., and Planell, J. A. (1998) The cement setting reaction in the $CaHPO_4$–α-$Ca_3(PO_4)_2$ system: an X-ray diffraction study, *J. Biomed. Mater. Res.*, **42**, 403–406.

157. Fernández, E., Gil, F. J., Ginebra, M. P., Driessens, F. C. M., Planell, J. A., and Best, S. M. (1999) Production and characterisation of new calcium phosphate bone cements in the $CaHPO_4$–α-$Ca_3(PO_4)_2$ system: pH, workability and setting times, *J. Mater. Sci. Mater. Med.*, **10**, 223–230.

158. Barralet, J. E., Lilley, K. J., Grover, L. M., Farrar, D. F., Ansell, C., and Gbureck, U. (2004) Cements from nanocrystalline hydroxyapatite, *J. Mater. Sci. Mater. Med.*, **15**, 407–411.

159. Lilley, K. J., Gbureck, U., Wright, A. J., Farrar, D. F., and Barralet, J. E. (2005) Cement from nanocrystalline hydroxyapatite: effect of calcium phosphate ratio, *J. Mater. Sci. Mater. Med.*, **16**, 1185–1190.

160. Alge, D. L., Cruz, G. S., Goebel, W. S., and Chu, T. M. G. (2009) Characterization of dicalcium phosphate dihydrate cements prepared using a novel hydroxyapatite-based formulation, *Biomed. Mater.*, **4**, 025016.

161. Wang, X., Ye, J., Wang, Y., Wu, X., and Bai, B. (2007) Control of crystallinity of hydrated products in a calcium phosphate bone cement, *J. Biomed. Mater. Res. A*, **81A**, 781–790.

162. Wang, X., Ye, J., and Wang, H. (2006) Effects of additives on the rheological properties and injectability of a calcium phosphate bone substitute material, *J. Biomed. Mater. Res. B* (*Appl. Biomater.*), **78B**, 259–264.

163. Tofighi, A., Schaffer, K., and Palazzolo, R. (2008) Calcium phosphate cement (CPC): a critical development path, *Key Eng. Mater.*, 361–363, 303–306.

164. de Maeyer, E. A. P., Verbeeck, R. M. H., and Vercruysse, C. W. J. (2000) Conversion of octacalcium phosphate in calcium phosphate cements, *J. Biomed. Mater. Res.*, **52**, 95–106.

165. Nakano, Y., Ohgaki, M., Nakamura, S., Takagi, Y., and Yamashita, K. (1999) *In vitro* and *in vivo* characterization and mechanical properties of α-TCP/OCP settings, *Bioceramics*, **12**, 315–318.

166. Nakano, Y. (2000) Preparation and characterization of porous octacalcium phosphate setting improved by α-tricalcium phosphate additive, *J. Dent. Mater.*, **19**, 65–76.

167. Wang, X., Ye, J., and Wang, Y. (2008) Hydration mechanism of a novel PCCP + DCPA cement system, *J. Mater. Sci. Mater. Med.*, **19**, 813–816.

168. Zoulgami, M., Lucas, A., Briard, P., and Gaudé, J. (2001) A self-setting single-component calcium phosphate cement, *Biomaterials*, **22**, 1933–1937.

169. Knaack, D., Goad, M. E., Aiolova, M., Rey, C., Tofighi, A., Chakravarthy, P., and Lee, D. D. (1998) Resorbable calcium phosphate bone substitute, *J. Biomed. Mater. Res.*, **43**, 399–409.

170. Tofighi, A., Mounic, S., Chakravarthy, P., Rey, C., and Lee, D. (2001) Setting reactions involved in injectable cements based on amorphous calcium phosphate, *Key Eng. Mater.*, 192–195, 769–772.

171. Monma, H., and Kanazawa, T. (2000) Hydration of α-tricalcium phosphate, *J. Ceram. Soc. Jpn.*, **108**, 575–580.

172. Fernández, E., Ginebra, M. P., Boltong, M. G., Driessens, F. C. M., Ginebra, J., de Maeyer, E. A. P., Verbeeck, R. M. H., and Planell, J. A. (1996) Kinetic study of the setting reaction of a calcium phosphate bone cement, *J. Biomed. Mater. Res.*, **32**, 367–374.

173. Gbureck, U., Barralet, J. E., Radu L., Klinger, H. G., and Thull, R. (2004) Amorphous α-tricalcium phosphate, preparation and aqueous setting reaction, *J. Am. Ceram. Soc.*, **87**, 1126–1132.

174. Bohner, M., Malsy, A. K., Camire, C. L., and Gbureck, U. (2006) Combining particle size distribution and isothermal calorimetry data to determine the reaction kinetics of α-tricalcium phosphate–water mixtures, *Acta Biomater.*, **2**, 343–348.

175. Brunner, T. J., Grass, R. N., Bohner, M., and Stark, W. J. (2007) Effect of particle size, crystal phase and crystallinity on the reactivity of tricalcium phosphate cements for bone reconstruction, *J. Mater. Chem.*, **38**, 4072–4078.

176. Alves, H. L. R., dos Santos, L. A., and Bergmann, C. P. (2008) Injectability evaluation of tricalcium phosphate bone cement, *J. Mater. Sci. Mater. Med.*, **19**, 2241–2246.

177. Jack, V., Buchanan, F. J., and Dunne, N. J. (2008) Particle attrition of α-tricalcium phosphate, effect on mechanical, handling, and injectability properties of calcium phosphate cements, *Proc. Inst. Mech. Eng. H: J. Eng. Med.*, **222**, 19–28.

178. Oh, S. A., Lee, G. S., Park, J. H., and Kim, H. W. (2010) Osteoclastic cell behaviors affected by the α-tricalcium phosphate based bone cements, *J. Mater. Sci. Mater. Med.*, **21**, 3019–3027.

179. Gbureck, U., Grolms, O., Barralet, J. E., Grover, L. M., and Thull, R. (2003) Mechanical activation and cement formation of β-tricalcium phosphate, *Biomaterials*, **24**, 4123–4131.

180. Gbureck, U., Barralet, J. E., Hofmann, M. P., and Thull, R. (2004) Nanocrystalline tetracalcium phosphate cement, *J. Dent. Res.*, **83**, 425–428.

181. Gbureck, U., Barralet, J. E., Hofmann, M. P., and Thull, R. (2004) Mechanical activation of tetracalcium phosphate, *J. Am. Ceram. Soc.*, **87**, 311–313.

182. Tsai, C. H., Ju, C. P., and Lin, J. H. C. (2008) Morphology and mechanical behavior of TTCP-derived calcium phosphate cement subcutaneously implanted in rats, *J. Mater. Sci. Mater. Med.*, **19**, 2407–2415.

183. Tsai, C. H., Lin, R. M., Ju, C. P., and Lin, J. H. C. (2008) Bioresorption behavior of tetracalcium phosphate-derived calcium phosphate cement implanted in femur of rabbits, *Biomaterials*, **29**, 984–993.

184. Tsai, C. H., Lin, J. H. C., and Ju, C. P. (2007) γ-radiation-induced changes in structure and properties of tetracalcium phosphate and its derived calcium phosphate cement, *J. Biomed. Mater. Res. B* (*Appl. Biomater.*), **80B**, 244–252.

185. Chow, L. C., Markovic, M., Frukhtbeyn, S. A., and Takagi, S. (2005) Hydrolysis of tetracalcium phosphate under a near-constant

composition condition — effects of pH and particle size, *Biomaterials*, **26**, 393–401.

186. TenHuisen, K. S., and Brown, P. W. (1998) Formation of calcium-deficient hydroxyapatite from α-tricalcium phosphate, *Biomaterials*, **19**, 2209–2217.

187. Ginebra, M. P., Fernández, E., Driessens, F. C. M., and Planell, J. A. (1999) Modeling of the hydrolysis of α-TCP, *J. Am. Ceram. Soc.*, **82**, 2808–2812.

188. Durucan, C., and Brown, P. W. (2000) α-tricalcium phosphate hydrolysis to hydroxyapatite at and near physiological temperature, *J. Mater. Sci. Mater. Med.*, **11**, 365–371.

189. Durucan, C., and Brown, P. W. (2002) Kinetic model for α-tricalcium phosphate hydrolysis, *J. Am. Ceram. Soc.*, **85**, 2013–2018.

190. Fulmer, M. T., and Brown, P. W. (1998) Hydrolysis of dicalcium phosphate dihydrate to hydroxyapatite, *J. Mater. Sci. Mater. Med.*, **9**, 197–202.

191. Ginebra, M. P., Driessens, F. C. M., and Planell, J. A. (2004) Effect of the particle size on the micro and nanostructural features of a calcium phosphate cement: a kinetic analysis, *Biomaterials*, **25**, 3453–3462.

192. Koshino, T., Kubota, W., and Morii, T. (1995) Bone formation as a reaction to hydraulic hydroxyapatite thermal decomposition product used as bone cement in rabbits, *Biomaterials*, **16**, 125–128.

193. Liu, C., Huang, Y., and Chen, J. (2004) The physicochemical properties of the solidification of calcium phosphate cement, *J. Biomed. Mater. Res. B (Appl. Biomater.)*, **69B**, 73–78.

194. Liu, C., Gai, W., Pan, S., and Liu, Z. (2003) The exothermal behavior in the hydration process of calcium phosphate cement, *Biomaterials*, **24**, 2995–3003.

195. Charrière, E., Terrazzoni, S., Pittet, C., Mordasini, P., Dutoit, M., Lemaître, J., and Zysset, P. (2001) Mechanical characterization of brushite and hydroxyapatite cements, *Biomaterials*, **22**, 2937–2945.

196. Morgan, H., and Dauskardt, R. H. (2003) Notch strength insensitivity of self-setting hydroxyapatite bone cements, *J. Mater. Sci. Mater. Med.*, **14**, 647–653.

197. Bohner, M. (2000) Calcium orthophosphates in medicine: from ceramics to calcium phosphate cements, *Injury*, **31**(4), S-D37–S-D47.

198. Bohner, M. (2001) Physical and chemical aspects of calcium phosphates used in spinal surgery, *Eur. Spine J.*, **10**, S114–S121.

199. von Gonten, A. S., Kelly, J. R., and Antonucci, J. M. (2000) Load-bearing behavior of a simulated craniofacial structure fabricated from a hydroxyapatite cement and bioresorbable fiber-mesh, *J. Mater. Sci. Mater. Med.*, **11**, 95–100.

200. Gisep, A., Kugler, S., Wahl, D., and Rahn, B. (2004) The mechanical characterization of a bone defect model filled with ceramic cements, *J. Mater. Sci. Mater. Med.*, **15**, 1065–1071.

201. Takagi, S., Chow, L. C., Markovic, M., Friedman, C. D., and Costantino, P. D. (2001) Morphological and phase characterizations of retrieved calcium phosphate cement implants, *J. Biomed. Mater. Res.* (*Appl. Biomater.*), **58**, 36–41.

202. Ambard, A. J., and Mueninghoff, L. (2006) Calcium phosphate cement: review of mechanical and biological properties, *J. Prosthodont.*, **15**, 321–328.

203. Lewis, G. (2006) Injectable bone cements for use in vertebroplasty and kyphoplasty, state-of-the-art review, *J. Biomed. Mater. Res. B* (*Appl. Biomater.*), **76B**, 456–468.

204. Kenny, S. M., and Buggy, M. (2003) Bone cements and fillers: a review, *J. Mater. Sci. Mater. Med.*, **14**, 923–938.

205. Takagi, S., Frukhtbeyn, S., Chow, L. C., Sugawara, A., Fujikawa, K., Ogata, H., Hayashi, M., and Ogiso, B. (2010) *In vitro* and *in vivo* characteristics of fluorapatite-forming calcium phosphate cements, *J. Res. Natl. Inst. Stand. Technol.*, **115**, 267–276.

206. Constantz, B. R., Ison, I. C., Fulmer, M. T., Poser, R. D., Smith, S. T., van Wagoner, M., Ross, J., Goldstein, S. A., Jupiter, J. B., and Rosenthal, D. I. (1995) Skeletal repair by *in situ* formation of the mineral phase of bone, *Science*, **267**, 1796–1799.

207. Bohner, M. (2007) Reactivity of calcium phosphate cements, *J. Mater. Chem.*, **38**, 3980–3986.

208. Bohner, M., Brunner, T. J., and Stark, W. J. (2008) Controlling the reactivity of calcium phosphate cements, *J. Mater. Chem.*, **18**, 5669–5675.

209. Yuan, H., Li, Y., de Bruijn, J. D., de Groot, K., and Zhang, X. (2000) Tissue responses of calcium phosphate cement, a study in dogs, *Biomaterials*, **21**, 1283–1290.

210. Takechi, M., Miyamoto, Y., Ishikawa, K., Toh, T., Yuasa, T., Nagayama, M., and Suzuki, K. (1998) Initial histological evaluation of anti-washout type fast-setting calcium phosphate cement following subcutaneous implantation, *Biomaterials*, **19**, 2057–2063.

211. Fulmer, M. T., and Brown, P. W. (1993) Effects of Na_2HPO_4 and NaH_2PO_4 on hydroxyapatite formation, *J. Biomed. Mater. Res.*, **27**, 1095–1102.

212. Otsuka, M., Matsuda, Y., Suwa, Y., Fox, J. L., and Higuchi, W. I. (1995) Effect of particle size of metastable calcium phosphates on mechanical strength of a novel self-setting bioactive calcium phosphate cement, *J. Biomed. Mater. Res.*, **29**, 25–32.

213. Liu, C., Shao, H., Chen, F., and Zheng, H. (2003) Effects of granularity of raw materials on the hydration and hardening process of calcium phosphate cement, *Biomaterials*, **24**, 4103–4113.

214. Chen, W. C., Lin, J. H. C., and Ju, C. P. (2003) Transmission electron microscopic study on setting mechanism of tetracalcium phosphate/dicalcium phosphate anhydrous-based calcium phosphate cement, *J. Biomed. Mater. Res.*, **64**, 664–671.

215. Fernández, E., Gil, F. J., Ginebra, M. P., Driessens, F. C. M., Planell, J. A., and Best, S. M. (1999) Calcium phosphate bone cements for clinical applications. Part II: Precipitate formation during setting reactions, *J. Mater. Sci. Mater. Med.*, **10**, 177–183.

216. Brown, W. E. (1966) Crystal growth of bone mineral, *Clin. Orthop. Rel. Res.*, **44**, 205–220.

217. Tung, M. S., and Brown, W. E. (1983) An intermediate state in hydrolysis of amorphous calcium phosphate, *Calcif. Tissue Int.*, **35**, 783–790.

218. Brown, W. E., Eidelman, N., and Tomazic, B. B. (1987) Octacalcium phosphate as a precursor in biomineral formation, *Adv. Dent. Res.*, **1**, 306–313.

219. Constantz, B. R., Barr, B. M., Ison, I. C., Fulmer, M. T., Baker, J., McKinney, L. A., Goodman, S. B., Gunasekaren, S., Delaney, D. C., Ross, J., and Poser, R. D. (1998) Histological, chemical and crystallographic analysis of four calcium phosphate cements in different rabbit osseous sites, *J. Biomed. Mater. Res.* (*Appl. Biomater.*), **43**, 451–461.

220. Elliott, J. C. (1994) Structure and chemistry of the apatites and other calcium orthophosphates, Elsevier, Amsterdam, p. 404.

221. Legrand, A. P., Sfihi, H., Lequeux, N., and Lemaître, J. (2009) ^{31}P solid-state NMR study of the chemical setting process of a dual-paste injectable brushite cements, *J. Biomed. Mater. Res. B* (*Appl. Biomater.*), **91B**, 46–54.

222. Bohner, M., Merkle, H. P., van Landuyt, P., Trophardy, G., and Lemaître, J. (2000) Effect of several additives and their admixtures on the physico-chemical properties of a calcium phosphate cement, *J. Mater. Sci. Mater. Med.*, **11**, 111–116.

223. Vereecke, G., and Lemaître, J. (1990) Calculation of the solubility diagrams in the system $Ca(OH)_2$–H_3PO_4–KOH–HNO_3–CO_2–H_2O, *J. Cryst. Growth*, **104**, 820–832.

224. Klein, C. P., de Groot, K., Driessen, A. A., and van der Lubbe, H. B. (1985) Interaction of biodegradable β-whitlockite ceramics with bone tissue, an *in vivo* study, *Biomaterials*, **6**, 189–192.

225. Liu, C., Shen, W., and Chen, J. (1999) Solution property of calcium phosphate cement hardening body, *Mater. Chem. Phys.*, **58**, 78–83.

226. Apelt, D., Theiss, F., El-Warrak, A. O., Zlinszky, K., Bettschart-Wolfisberger, R., Bohner, M., Matter, S., Auer, J. A., and von Rechenberg, B. (2004) *In vivo* behavior of three different injectable hydraulic calcium phosphate cements, *Biomaterials*, **25**, 1439–1451.

227. Barralet, J. E., Grover, L. M., and Gbureck, U. (2004) Ionic modification of calcium phosphate cement viscosity. Part II: Hypodermic injection and strength improvement of brushite cement, *Biomaterials*, **25**, 2197–2203.

228. Sarda, S., Fernández, E., Nilsson, M., Balcells, M., and Planell, J. A. (2002) Kinetic study of citric acid influence on calcium phosphate bone cements as water-reducing agent, *J. Biomed. Mater. Res.*, **61**, 653–659.

229. Qi, X., Ye, J., and Wang, Y. (2008) Improved injectability and *in vitro* degradation of a calcium phosphate cement containing poly(lactide-co-glycolide) microspheres, *Acta Biomater.*, **4**, 1837–1845.

230. Grover, L. M., Knowles, J. C., Fleming, G. J. P., and Barralet, J. E. (2003) *In vitro* ageing of brushite calcium phosphate cement, *Biomaterials*, **24**, 4133–4141.

231. Mariño, F. T., Mastio, J., Rueda, C., Blanco, L., and Cabarcos, E. L. (2007) Increase of the final setting time of brushite cements by using chondroitin 4-sulfate and silica gel, *J. Mater. Sci. Mater. Med.*, **18**, 1195–1201.

232. Mariño, F. T., Torres, J., Hamdan, M., Rodríguez, C. R., and Cabarcos, E. L. (2007) Advantages of using glycolic acid as a retardant in a brushite forming cement, *J. Biomed. Mater. Res. B* (*Appl. Biomater.*), **83B**, 571–579.

233. Flautre, B., Delecourt, C., Blary, M., van Landuyt, P., Lemaître, J., and Hardouin, P. (1999) Volume effect on biological properties of a calcium phosphate hydraulic cement, experimental study on sheep, *Bone*, **25**, S35–S39.

234. Bohner, M. (2001) pH variations of a solution after injecting brushite cements, *Key Eng. Mater.*, 192–195, 813–816.

235. Xie, J., Riley, C., and Chittur, K. (2001) Effect of albumin on brushite transformation to hydroxyapatite, *J. Biomed. Mater. Res.*, **57**, 357–365.

236. Frayssinet, P., Roudier, M., Lerch, A., Ceolin, J. L., Depres, E., and Rouquet, N. (2000) Tissue reaction against a self-setting calcium phosphate cement set in bone or outside the organism, *J. Mater. Sci. Mater. Med.*, **11**, 811–815.

237. Ohura, K., Bohner, M., Hardouin, P., Lemaître, J., Pasquier, G., and Flautre, B. (1996) Resorption of and bone formation from new β-tricalcium phosphate–monocalcium phosphate cements: an *in vivo* study, *J. Biomed. Mater. Res.*, **30**, 193–200.

238. Flautre, B., Maynou, C., Lemaître, J., van Landuyt, P., and Hardouin, P. (2002) Bone colonization of β-TCP granules incorporated in brushite cements, *J. Biomed. Mater. Res.* (*Appl. Biomater.*), **63**, 413–417.

239. Standard test method for time of setting of hydraulic cement paste by Gillmore needles. ASTM C266-89. In: Annual Book of ASTM Standards. Vol. 04.01: Cement, Lime, Gypsum. American Society for Testing and Materials: Philadelphia, USA, 1993; pp. 189–191.

240. Standard test method for time of setting of hydraulic cement paste by Vicat needle. ASTM C191-92. In: Annual Book of ASTM Standards. Vol. 04.01: Cement, Lime, Gypsum. American Society for Testing and Materials: Philadelphia, USA, 1993; pp. 158–160.

241. Nilsson, M., Carlson, J., Fernández, E., and Planell, J. A. (2002) Monitoring the setting of calcium-based bone cements using pulse-echo ultrasound, *J. Mater. Sci. Mater. Med.*, **13**, 1135–1141.

242. Carlson, J., Nilsson, M., Fernández, E., and Planell, J. A. (2003) An ultrasonic pulse-echo technique for monitoring the setting of $CaSO_4$-based bone cement, *Biomaterials*, **24**, 71–77.

243. Hofmann, M. P., Nazhat, S. N., Gbureck, U., and Barralet, J. E. (2006) Real-time monitoring of the setting reaction of brushite-forming cement using isothermal differential scanning calorimetry, *J. Biomed. Mater. Res. B* (*Appl. Biomater.*), **79B**, 360–364.

244. Martin, R. I., and Brown, P. W. (1997) The effects of magnesium on hydroxyapatite formation *in vitro* from $CaHPO_4$ and $Ca_4(PO_4)_2O$ at 37.4 °C, *Calcif. Tissue Int.*, **60**, 538–546.

245. Brunner, T. J., Bohner, M., Dora, C., Gerber, C., and Stark, W. J. (2007) Comparison of amorphous TCP nanoparticles to micron-sized α-TCP as starting materials for calcium phosphate cements, *J. Biomed. Mater. Res. B* (*Appl. Biomater.*), **83B**, 400–407.

246. Gao, W. Y., Wang, Y. W., Dong, L. M., and Yu, Z. W. (2006) Thermokinetic analysis of the hydration process of calcium phosphate cement, *J. Therm. Anal. Calorim.*, **85**, 785–789.

247. Bohner, M., and Gbureck, U. (2008) Thermal reactions of brushite cements, *J. Biomed. Mater. Res. B* (*Appl. Biomater.*), **84B**, 375–385.

248. Hofmann, M. P., Young, A. M., Nazhat, S. N., Gbureck, U., and Barralet, J. E. (2006) Setting kinetics observation of a brushite cement by FTIR and DSC, *Key Eng. Mater.*, 309–311, 837–840.

249. Liu, C., Huang, Y., and Zheng, H. (1999) Study of the hydration process of calcium phosphate cement by AC impedance spectroscopy, *J. Am. Ceram. Soc.*, **82**, 1052–1057.

250. Hofmann, M. P., Young, A. M., Gbureck, U., Nazhat, S. N., and Barralet, J. E. (2006) FTIR-monitoring of a fast setting brushite bone cement: effect of intermediate phases, *J. Mater. Chem.*, **16**, 3199–3206.

251. Hsu, H. C., Tuan, W. H., and Lee, H. Y. (2009) *In-situ* observation on the transformation of calcium phosphate cement into hydroxyapatite, *Mater. Sci. Eng. C*, **29**, 950–954.

252. Rau, J. V., Generosi, A., Smirnov, V. V., Ferro, D., Rossi, A. V., and Barinov, S. M. (2008) Energy dispersive X-ray diffraction study of phase development during hardening of calcium phosphate bone cements with addition of chitosan, *Acta Biomater.*, **4**, 1089–1094.

253. Generosi, A., Smirnov, V. V., Rau, J. V., Rossi, A. V., Ferro, D., and Barinov, S. M. (2008) Phase development in the hardening process of two calcium phosphate bone cements: an energy dispersive X-ray diffraction study, *Mater. Res. Bull.*, **43**, 561–571.

254. Ginebra, M. P., Fernández, E., Driessens, F. C. M., Boltong, M. G., Muntasell, J., Font, J., and Planell, J. A. (1995) The effects of temperature on the behaviour of an apatitic calcium phosphate cement, *J. Mater. Sci. Mater. Med.*, **6**, 857–860.

255. Baroud, G., Bohner, M., Heini, P., and Steffen, T. (2004) Injection biomechanics of bone cements used in vertebroplasty, *Biomed. Mater. Eng.*, **14**, 487–504.

256. Leung, K. S., Siu, W. S., Li, S. F., Qin, L., Cheung, W. H., Tam, K. F., Po, P., and Lui, Y. (2006) An *in vitro* optimized injectable calcium phosphate cement for augmenting screw fixation in osteopenic goats, *J. Biomed. Mater. Res. B* (*Appl. Biomater.*), **78B**, 153–160.

257. Eames, W. B., Monroe, S. D., Roan, J. D., and Oneal, S. J. (1977) Proportioning and mixing of cements — comparison of working times, *Oper. Dent.*, **2**, 97–104.

258. Baroud, G., Matsushita, C., Samara, M., Beckman, L., and Steffen, T. (2004) Influence of oscillatory mixing on the injectability of three acrylic and two calcium phosphate bone cements for vertebroplasty, *J. Biomed. Mater. Res. B* (*Appl. Biomater.*), **68B**, 105–111.

259. Nomoto, T., Haraguchi, K., Yamaguchi, S., Sugano, N., Nakayama, H., Sekino, T., and Niihara, K. (2006) Hydrolyses of calcium phosphates-allografts composite in physiological solutions, *J. Mater. Sci. Mater. Med.*, **17**, 379–385.

260. Oda, M., Takeuchi, A., Lin, X., Matsuya, S., and Ishikawa, K. (2008) Effects of liquid phase on basic properties of α-tricalcium phosphate-based apatite cement, *Dent. Mater. J.*, **27**, 672–677.

261. Sarda, S., Fernández, E., Llorens, J., Martinez, S., Nilsson, M., and Planell, J. A. (2001) Rheological properties of an apatitic bone cement during initial setting, *J. Mater. Sci. Mater. Med.*, **12**, 905–909.

262. Liu, C., Shao, H., Chen, F., and Zheng, H. (2006) Rheological properties of concentrated aqueous injectable calcium phosphate cement slurry, *Biomaterials*, **27**, 5003–5013.

263. Bohner, M., and Baroud, G. (2005) Injectability of calcium phosphate pastes, *Biomaterials*, **26**, 1553–1563.

264. Khairoun, I., Boltong, M. G., Driessens, F. C. M., and Planell, J. A. (1998) Some factors controlling the injectability of calcium phosphate bone cements, *J. Mater. Sci. Mater. Med.*, **9**, 425–428.

265. Burguera, E. F., Xu, H. H. K., and Sun, L. (2008) Injectable calcium phosphate cement: effects of powder-to-liquid ratio and needle size, *J. Biomed. Mater. Res. B* (*Appl. Biomater.*), **84B**, 493–502.

266. Habib, M., Baroud, G., Gitzhofer, F., and Bohner, M. (2008) Mechanisms underlying the limited injectability of hydraulic calcium phosphate paste, *Acta Biomater.*, **4**, 1465–1471.

267. Baroud, G., Cayer, E., and Bohner, M. (2005) Rheological characterization of concentrated aqueous beta-tricalcium phosphate suspensions: the effect of liquid-to-powder ratio, milling time and additives, *Acta Biomater.*, **1**, 357–363.

268. Ishikawa, K. (2003) Effects of spherical tetracalcium phosphate on injectability and basic properties of apatitic cement, *Key Eng. Mater.*, 240–242, 369–372.

269. Habib, M., Baroud, G., Gitzhofer, F., and Bohner, M. (2010) Mechanisms underlying the limited injectability of hydraulic calcium phosphate paste. Part II: Particle separation study, *Acta Biomater.*, **6**, 250–256.

270. Bohner, M., Doebelin, N., and Baroud, G. (2006) Theoretical and experimental approach to test the cohesion of calcium phosphate pastes, *Eur. Cell Mater.*, **12**, 26–35.

271. Miyamoto, Y., Ishikawa, K., Takechi, M., Toh, T., Yuasa, T., Nagayama, M., and Suzuki, K. (1999) Histological and compositional evaluations of three types of calcium phosphate cements when implanted in subcutaneous tissue immediately after mixing, *J. Biomed. Mater. Res. (Appl. Biomater.)*, **48**, 36–42.

272. Bermudez, O., Boltong, M. G., Driessens, F. C. M., and Planell, J. A. (1993) Compressive strength and diametral tensile strength of some calcium-orthophosphate cements, a pilot study, *J. Mater. Sci. Mater. Med.*, **4**, 389–393.

273. del Valle, S., Miño, N., Muñoz, F., González, A., Planell, J. A., and Ginebra, M. P. (2007) *In vivo* evaluation of an injectable macroporous calcium phosphate cement, *J. Mater. Sci. Mater. Med.*, **18**, 353–361.

274. Khairoun, I., Driessens, F. C. M., Boltong, M. G., Planell, J. A., and Wenz, R. (1999) Addition of cohesion promoters to calcium orthophosphate cements, *Biomaterials*, **20**, 393–398.

275. Alkhraisat, M. H., Rueda, C., Mariño, F. T., Torres, J., Jerez, L. B., Gbureck, U., and Cabarcos, E. L. (2009) The effect of hyaluronic acid on brushite cement cohesion, *Acta Biomater.*, **5**, 3150–3156.

276. Alkhraisat, M. H., Rueda, C., Jerez, L. B., Mariño, F. T., Torres, J., Gbureck, U., and Cabarcos, E. L. (2010) Effect of silica gel on the cohesion, properties and biological performance of brushite cement, *Acta Biomater.*, **6**, 257–265.

277. Low, K. L., Tan, S. H., Zein, S. H. S., Roether, J. A., Mouriño, V., and Boccaccini, A. R. (2010) Calcium phosphate-based composites as injectable bone substitute materials, *J. Biomed. Mater. Res. B (Appl. Biomater.)*, **94B**, 273–286.

278. Bigi, A., Bracci, B., and Panzavolta, S. (2004) Effect of added gelatin on the properties of calcium phosphate cement, *Biomaterials*, **25**, 2893–2899.

279. Ishikawa, K., Matsuya, S., Nakagawa, M., Udoh, K., and Suzuki, K. (2004) Basic properties of apatite cement containing spherical tetracalcium phosphate made with plasma melting method, *J. Mater. Sci. Mater. Med.*, **15**, 13–17.

280. Wang, X., Ye, J., and Wang, Y. (2008) Effect of additives on the morphology of the hydrated product and physical properties of a calcium phosphate cement, *J. Mater. Sci. Technol.*, **24**, 285–288.

281. Barralet, J. E., Hofmann, M., Grover, L. M., and Gbureck, U. (2005) High strength apatitic cement by modification with α-hydroxy acid salts, *Adv. Mater.*, **15**, 2091–2095.

282. Barralet, J. E., Duncan, C. O., Dover, M. S., Bassett, D. C., Nishikawa, H., Monaghan, A., and Gbureck, U. (2005) Cortical bone screw fixation in ionically modified apatite cements, *J. Biomed. Mater. Res. B* (*Appl. Biomater.*), **73B**, 238–243.

283. Ginebra, M. P., Boltong, M. G., Fernández, E., Planell, J. A., and Driessens, F. C. M. (1995) Effect of various additives and temperature on some properties of an apatitic calcium phosphate cement, *J. Mater. Sci. Mater. Med.*, **6**, 612–616.

284. Acarturk, O., Lehmicke, M., Aberman, H., Toms, D., Hollinger, J. O., and Fulmer, M. T. (2008) Bone healing response to an injectable calcium phosphate cement with enhanced radiopacity, *J. Biomed. Mater. Res. B* (*Appl. Biomater.*), **86B**, 56–62.

285. Wang, X., Ye, J., and Wang, Y. (2007) Influence of a novel radiopacifier on the properties of an injectable calcium phosphate cement, *Acta Biomater.*, **3**, 757–763.

286. Chen, F., Liu, C., and Mao, Y. (2010) Bismuth-doped injectable calcium phosphate cement with improved radiopacity and potent antimicrobial activity for root canal filling, *Acta Biomater.*, **6**, 3199–3207.

287. Romieu, G., Garric, X., Munier, S., Vert, M., and Boudeville, P. (2010) Calcium-strontium mixed phosphate as novel injectable and radio-opaque hydraulic cement, *Acta Biomater.*, **6**, 3208–3215.

288. Watanabe, M., Tanaka, M., Sakurai, M., and Maeda, M. (2006) Development of calcium phosphate cement, *J. Eur. Ceram. Soc.*, **26**, 549–552.

289. Bercier, A., Gonçalves, S., Lignon, O., and Fitremann, J. (2010) Calcium phosphate bone cements including sugar surfactants: part one — porosity, setting times and compressive strength, *Materials*, **3**, 4695–4709.

290. Sarda, S., Nilsson, M., Balcells, M., and Fernández, E. (2003) Influence of surfactant molecules as air-entraining agent for bone cement macroporosity, *J. Biomed. Mater. Res. A*, **65A**, 215–221.

291. Friberg, J., Fernández, E., Sarda, S., Nilsson, M., Ginebra, M. P., Martinez, S., and Planell, J. A. (2001) An experimental approach to the study of the rheology behavior of synthetic bone calcium phosphate cements, *Key Eng. Mater.*, 192–195, 777–780.

292. Reinstorf, A., Hempel, U., Olgemöller, F., Domaschke, H., Schneiders, W., Mai, R., Stadlinger, B, Rösen-Wolff, A., Rammelt, S., Gelinsky, M., and Pompe, W. (2006) *O*-phospho-L-serine modified calcium phosphate cements — material properties, *in vitro* and *in vivo* investigations, *Mat.-Wiss. u. Werkstofftech.*, **37**, 491–503.

293. Lode, A., Reinstorf, A., Bernhardt, A., Wolf-Brandstetter, C., König, U., and Gelinsky, M. (2008) Heparin modification of calcium phosphate bone cements for VEGF functionalization, *J. Biomed. Mater. Res. A*, **86A**, 749–759.

294. Mai, R., Lux, R., Proff, P., Lauer, G., Pradel, W., Leonhardt, H., Reinstorf, A., Gelinsky, M., Jung, R., Eckelt, U., Gedrange, T., and Stadlinger, B. (2008) *O*-phospho-L-serine: a modulator of bone healing in calcium-phosphate cements, *Biomed. Tech.*, **53**, 229–233.

295. Vater, C., Lode, A., Bernhardt, A., Reinstorf, A., Nies, B., and Gelinsky, M. (2010) Modifications of a calcium phosphate cement with biomolecules — influence on nanostructure, material, and biological properties, *J. Biomed. Mater. Res. A*, **95A**, 912–923.

296. Markovic, M., Takagi, S., and Chow, L. C. (2001) Formation of macropores in calcium phosphate cements through the use of mannitol crystals, *Key Eng. Mater.*, 192–195, 773–776.

297. Tajima, S., Kishi, Y., Oda, M., Maruta, M., Matsuya, S., and Ishikawa, K. (2006) Fabrication of biporous low-crystalline apatite based on mannitol dissolution from apatite cement, *Dent. Mater. J.*, **25**, 616–620.

298. Xu, H. H. K., Weir, M. D., Burguera, E. F., and Fraser, A. M. (2006) Injectable and macroporous calcium phosphate cement scaffold, *Biomaterials*, **27**, 4279–4287.

299. Cama, G., Barberis, F., Botter, R., Cirillo, P., Capurro, M., Quarto, R., Scaglione, S., Finocchio, E., Mussi, V., and Valbusa, U. (2009) Preparation and properties of macroporous brushite bone cements, *Acta Biomater.*, **5**, 2161–2168.

300. Vazquez, D., Takagi, S., Frukhtbeyn, S., and Chow, L. C. (2010) Effects of addition of mannitol crystals on the porosity and dissolution rates of a calcium phosphate cement, *J. Res. Natl. Inst. Stand. Technol.*, **115**, 225–232.

301. Shimogoryo, R., Eguro, T., Kimura, E., Maruta, M., Matsuya, S., and Ishikawa, K. (2009) Effects of added mannitol on the setting reaction and mechanical strength of apatite cement, *Dent. Mater. J.*, **28**, 627–633.

302. Almirall, A., Larrecq, G., Delgado, J. A., Martínez, S., Planell, J. A., and Ginebra, M. P. (2004) Fabrication of low temperature macroporous hydroxyapatite scaffolds by foaming and hydrolysis of an α-TCP paste, *Biomaterials*, **25**, 3671–3680.

303. Barralet, J. E., Grover, L., Gaunt, T., Wright, A. J., and Gibson, I. R. (2002) Preparation of macroporous calcium phosphate cement tissue engineering scaffold, *Biomaterials*, **23**, 3063–3072.

304. Takagi, S., and Chow, L. C. (2001) Formation of macropores in calcium phosphate cement implants, *J. Mater. Sci. Mater. Med.*, **12**, 135–139.

305. Simon, Jr. C. G., Khatri, C. A., Wight, S. A., and Wang, F. W. (2002) Preliminary report on the biocompatibility of a moldable, resorbable, composite bone graft consisting of calcium phosphate cement and poly(lactide-*co*-glycolide) microspheres, *J. Orthop. Res.*, **20**, 473–482.

306. Ruhe, P. Q., Hedberg, E. L., Padron, N. T., Spauwen, P. H. M., Jansen, J. A., and Mikos, A. G. (2005) Biocompatibility and degradation of poly(D,L-lactic-*co*-glycolic acid)/calcium phosphate cement composites, *J. Biomed. Mater. Res. A*, **74A**, 533–544.

307. Habraken, W. J. E. M., Wolke, J. G. C., Mikos, A. G., and Jansen, J. A. (2006) Injectable PLGA microsphere/calcium phosphate cements, physical properties and degradation characteristics, *J. Biomater. Sci. Polym. Ed.*, **17**, 1057–1074.

308. Habraken, W. J. E. M., Wolke, J. G. C., Mikos, A. G., and Jansen, J. A. (2008) PLGA microsphere/calcium phosphate cement composites for tissue engineering, *in vitro* release and degradation characteristics, *J. Biomater. Sci. Polym. Ed.*, **19**, 1171–1188.

309. Link, D. P., van den Dolder, J., van den Beucken, J. J. J. P., Cuijpers, V. M., Wolke, J. G. C., Mikos, A. G., and Jansen, J. A. (2008) Evaluation of the biocompatibility of calcium phosphate cement/PLGA microparticle composites, *J. Biomed. Mater. Res. A*, **87A**, 760–769.

310. Link, D. P., van den Dolder, J., Jurgens, W. J. F. M., Wolke, J. G. C., and Jansen, J. A. (2006) Mechanical evaluation of implanted calcium phosphate cement incorporated with PLGA microparticles, *Biomaterials*, **27**, 4941–4947.

311. Fullana, S. G., Ternet, H., Freche, M., Lacout, J. L., and Rodriguez, F. (2010) Controlled release properties and final macroporosity of a pectin microspheres-calcium phosphate composite bone cement, *Acta Biomater.*, **6**, 2294–2300.

312. Tas, A. C. (2008) Preparation of porous apatite granules from calcium phosphate cement, *J. Mater. Sci. Mater. Med.*, **19**, 2231–2239.

313. Tas, A. C. (2006) Preparation of self-setting cement-based micro- and macroporous granules of carbonated apatitic calcium phosphate, *Ceram. Eng. Sci. Proc.*, **27**, 49–60.

314. Li, M., Liu, X., Liu, X., Ge, B. and Chen, K. (2009) Creation of macroporous calcium phosphate cements as bone substitutes by using genipin — crosslinked gelatin microspheres, *J. Mater. Sci. Mater. Med.*, **20**, 925–934.

315. Habraken, W. J. E. M., de Jonge, L. T., Wolke, J. G. C., Yubao, L., Mikos, A. G., and Jansen, J. A. (2008) Introduction of gelatin microspheres into an injectable calcium phosphate cement, *J. Biomed. Mater. Res. A*, **87A**, 643–655.

316. Tang, P. F., Li, G., Wang, J. F., Zheng, Q. J., and Wang, Y. (2009) Development, characterization, and validation of porous carbonated hydroxyapatite bone cement, *J. Biomed. Mater. Res. B (Appl. Biomater.)*, **90B**, 886–893.

317. Wang X. P., Ye, J. D., Li, X., and Dong, H. (2008) Production of *in-situ* macropores in an injectable calcium phosphate cement by introduction of cetyltrimethyl ammonium bromide, *J. Mater. Sci. Mater. Med.*, **19**, 3221–3225.

318. Habraken, W. J. E. M., Zhang, Z., Wolke, J. G. C., Grijpma, D. W., Mikos, A. G., Feijen, J., and Jansen, J. A. (2008) Introduction of enzymatically degradable poly(trimethylene carbonate) microspheres into an injectable calcium phosphate cement, *Biomaterials*, **29**, 2464–2476.

319. Xu, H. H. K., and Simon, Jr., C. G. (2004) Self-hardening calcium phosphate composite scaffold for bone tissue engineering, *J. Orthop. Res.*, **22**, 535–543.

320. Burguera, E. F., Xu, H. H. K., Takagi, S., and Chow, L. C. (2005) High early strength calcium phosphate bone cement: effects of dicalcium phosphate dihydrate and absorbable fibers, *J. Biomed. Mater. Res. A*, **75A**, 966–975.

321. Xu, H. H. K., and Quinn, J. B. (2002) Calcium phosphate cement containing resorbable fibers for short-term reinforcement and macroporosity, *Biomaterials*, **23**, 193–202.

322. Gorst, N. J. S., Perrie, Y., Gbureck, U., Hutton, A. L., Hofmann, M. P., Grover, L. M., and Barralet, J. E. (2006) Effects of fiber reinforcement on the mechanical properties of brushite cement, *Acta Biomater.*, **2**, 95–102.

323. Zuo, Y., Yang, F., Wolke, J. G. C., Li, Y., and Jansen, J. A. (2010) Incorporation of biodegradable electrospun fibers into calcium phosphate cement for bone regeneration, *Acta Biomater.*, **6**, 1238–1247.

324. Xu, H. H. K., and Simon, Jr., C. G. (2004) Self-hardening calcium phosphate cement-mesh composite: reinforcement, macropores, and cell response, *J. Biomed. Mater. Res. A*, **69A**, 267–278.

325. Losee, J. E., Karmacharya, J., Gannon, F. H., Slemp, A. E., Ong, G., Hunenko, O., Gorden, A. D., Bartlett, S. P., and Kirschner, R. E. (2003) Reconstruction of the immature craniofacial skeleton with a carbonated calcium phosphate bone cement, interaction with bioresorbable mesh, *J. Craniofacial Surg.*, **14**, 117–124.

326. Xu, H. H. K., Carey, L. E., Simon, Jr., C. G. (2007) Premixed macroporous calcium phosphate cement scaffold, *J. Mater. Sci. Mater. Med.*, **18**, 1345–1353.

327. Ginebra, M. P., Espanol, M., Montufar, E. B., Perez, R. A., and Mestres, G. (2010) New processing approaches in calcium phosphate cements and their applications in regenerative medicine, *Acta Biomater.*, **6**, 2863–2873.

328. del Real, R. P., Wolke, J. G. C., Vallet-Regi, M., and Jansen, J. A. (2002) A new method to produce macropores in calcium phosphate cements, *Biomaterials*, **23**, 3673–3680.

329. del Real, R. P., Ooms, E., Wolke, J. G. C., Vallet-Regi, M., and Jansen, J. A. (2003) *In vivo* bone response to porous calcium phosphate cement, *J. Biomed. Mater. Res. A*, **65A**, 30–36.

330. Hesaraki, S., Moztarzadeh, F., and Sharifi, D. (2007) Formation of interconnected macropores in apatitic calcium phosphate bone cement with the use of an effervescent additive, *J. Biomed. Mater. Res. A*, **83A**, 80–87.

331. Hesaraki, S., Zamanian, A., and Moztarzadeh, F. (2008) The influence of the acidic component of the gas-foaming porogen used in preparing an injectable porous calcium phosphate cement on its properties, acetic acid versus citric acid, *J. Biomed. Mater. Res. B* (*Appl. Biomater.*), **86B**, 208–216.

332. Hesaraki, S., Moztarzadeh, F., and Solati-Hashjin, M. (2006) Phase evaluation of an effervescent-added apatitic calcium phosphate bone cement, *J. Biomed. Mater. Res. B* (*Appl. Biomater.*), **79B**, 203–209.

333. Ginebra, M. P., Delgado, J. A., Harr, I., Almirall, A., del Valle, S., and Planell, J. A. (2007) Factors affecting the structure and properties of an injectable self-setting calcium phosphate foam, *J. Biomed. Mater. Res. A*, **80A**, 351–361.

334. Montufar, E. B., Aguirre, A., Gil, C., Engel, E., Traykova, T., Planell, J. A., and Ginebra, M. P. (2010) Foamed surfactant solution as a template for

self-setting injectable hydroxyapatite scaffolds for bone regeneration, *Acta Biomater.*, **6**, 876–885.

335. Andrianjatovo, H., and Lemaître, J. (1995) Effects of polysaccharides on the cement properties in the monocalcium phosphate/β-tricalcium phosphate system, *Innov. Tech. Biol. Med.*, **16**, 140–147.

336. Cherng, A., Takagi, S., and Chow, L. C. (1997) Effects of hydroxypropylmethylcellulose and other gelling agents on the handling properties of calcium phosphate cement, *J. Biomed. Mater. Res.*, **35**, 273–277.

337. Yokoyama, A., Matsuno, H., Yamamoto, S., Kawasaki, T., Kohgo, T., Uo, M., Watari, F., and Nakasu, M. (2003) Tissue response to a newly developed calcium phosphate cement containing succinic acid and carboxymethyl-chitin, *J. Biomed. Mater. Res. A*, **64A**, 491–501.

338. Jyoti, M. A., Thai, V. V., Min, Y. K., Lee, B. T., and Song, H. Y. (2010) *In vitro* bioactivity and biocompatibility of calcium phosphate cements using hydroxy-propyl-methyl-cellulose (HPMC), *Appl. Surf. Sci.*, **257**, 1533–1539.

339. Bigi, A., Torricelli, P., Fini, M., Bracci, B., Panzavolta, S., Sturba, L., and Giardino, R. (2004) A biomimetic gelatin-calcium phosphate bone cement, *Int. J. Artif. Organs*, **27**, 664–673.

340. Bigi, A., Panzavolta, S., Sturba, L., Torricelli, P., Fini, M., and Giardino, R. (2006) Normal and osteopenic bone-derived osteoblast response to a biomimetic gelatin–calcium phosphate bone cement, *J. Biomed. Mater. Res. A*, **78A**, 739–745.

341. Fujishiro, Y., Takahashi, K., and Sato, T. (2001) Preparation and compressive strength of α-tricalcium phosphate/gelatin gel composite cement, *J. Biomed. Mater. Res.*, **54**, 525–530.

342. Bigi, A., Panzavolta, S., and Rubini, K. (2004) Setting mechanism of a biomimetic bone cement, *Chem. Mater.*, **16**, 3740–3745.

343. Panzavolta, S., Torricelli, P., Sturba, L., Bracci, B., Giardino, R., and Bigi, A. (2008) Setting properties and *in vitro* bioactivity of strontium-enriched gelatin-calcium phosphate bone cements, *J. Biomed. Mater. Res. A*, **84A**, 965–972.

344. Xu, L. X., Shi, X. T., Wang, Y. P., and Shi, Z. L. (2008) Performance of calcium phosphate bone cement using chitosan and gelatin as well as citric acid as hardening liquid, *J. Clin. Rehabil. Tissue Eng. Res.*, **12**, 6381–6384.

345. Shie, M. Y., Chen, D. C. H., Wang, C. Y., Chiang, T. Y., and Ding, S. J. (2008) Immersion behavior of gelatin-containing calcium phosphate cement, *Acta Biomater.*, **4**, 646–655.

346. Majekodunmi, A. O., Deb, S., and Nicholson, J. W. (2003) Effect of molecular weight and concentration of poly(acrylic acid) on the formation of a polymeric calcium phosphate cement, *J. Mater. Sci. Mater. Med.*, **14**, 747–752.

347. Majekodunmi, A. O., and Deb, S. (2007) Poly(acrylic acid) modified calcium phosphate cements, the effect of the composition of the cement powder and of the molecular weight and concentration of the polymeric acid, *J. Mater. Sci. Mater. Med.*, **18**, 1883–1888.

348. Chen, W. C., Ju, C. P., Wang, J. C., Hung, C. C., and Lin, J. H. C. (2008) Brittle and ductile adjustable cement derived from calcium phosphate cement/polyacrylic acid composites, *Dent. Mater.*, **24**, 1616–1622.

349. Komath, M., and Varma, H. K. (2003) Development of a fully injectable calcium phosphate cement for orthopedic and dental applications, *Bull. Mater. Sci.*, **26**, 415–422.

350. Bohner, M., Theiss, F., Apelt, D., Hirsiger, W., Houriet, R., Rizzoli, G., Gnos, E., Frei, C., Auer, J. A., and von Rechenberg, B. (2003) Compositional changes of a dicalcium phosphate dihydrate cement after implantation in sheep, *Biomaterials*, **24**, 3463–3474.

351. Leroux, L., Hatim, Z., Freche, M., and Lacout, J. L. (1999) Effects of various adjuvants (lactic acid, glycerol and chitosan) on the injectability of a calcium phosphate cement, *Bone*, **25**, S31–S34.

352. Barralet, J. E., Tremayne, M. J., Lilley, K. J., and Gbureck, U. (2005) Chemical modification of calcium phosphate cements with α-hydroxy acids and their salts, *Chem. Mater.*, **17**, 1313–1319.

353. Driessens, F. C. M., Boltong, M. G., de Maeyer, E. A. P., Verbeeck, R. M. H., and Wenz, R. (2000) Effect of temperature and immersion on the setting of some calcium phosphate cements, *J. Mater. Sci. Mater. Med.*, **11**, 453–457.

354. Gbureck, U., Dembski, S., Thull, R., and Barralet, J. E. (2005) Factors influencing calcium phosphate cement shelf life, *Biomaterials*, **26**, 3691–3697.

355. Ishikawa, K., Takagi, S., Chow, L. C., and Ishikawa, Y. (1995) Properties and mechanisms of fast-setting calcium phosphate cements, *J. Mater. Sci. Mater. Med.*, **6**, 528–533.

356. Miyamoto, Y., Ishikawa, K., Fukao, K., Sawada, M., Nagayama, M., Kon, M., and Asaoka, K. (1995) *In vivo* setting behavior of fast-setting calcium phosphate cement, *Biomaterials*, **16**, 855–860.

357. Kawai, T., Fujisawa, N., Suzuki, I., Ohtsuki, C., Matsushima, Y., and Unuma, H. (2010) Control of setting behavior of calcium phosphate paste using gelatinized starch, *J. Ceram. Soc. Jpn.*, **118**, 421–424.

358. Bohner, M., Luginbühl, R., Reber, C., Doebelin, N., Baroud, G., and Conforto, E. (2009) A physical approach to modify the hydraulic reactivity of α-tricalcium phosphate powder, *Acta Biomater.*, **5**, 3524–3535.

359. Takechi, M., Miyamoto, Y., Momota, Y., Yuasa, T., Tatehara, S., Nagayama, M., and Ishikawa, K. (2004) Effects of various sterilization methods on the setting and mechanical properties of apatite cement, *J. Biomed. Mater. Res. B* (*Appl. Biomater.*), **69B**, 58–63.

360. Schneider, G., Blechschmidt, K., Linde, D., Litschko, P., Körbs, T., and Beleites, E. (2010) Bone regeneration with glass ceramic implants and calcium phosphate cements in a rabbit cranial defect model, *J. Mater. Sci. Mater. Med.*, **21**, 2853–2859.

361. Johal, H. S., Buckley, R. E., Le, I. L. D., and Leighton, R. K. (2009) A prospective randomized controlled trial of a bioresorbable calcium phosphate paste (α-BSM) in treatment of displaced intra-articular calcaneal fractures, *J. Trauma Inj. Infect. Crit. Care*, **67**, 875–882.

362. Yuasa, T., Miyamoto, Y., Ishikawa, K., Takechi, M., Nagayama, M., and Suzuki, K. (2001) *In vitro* resorption of three apatite cements with osteoclasts, *J. Biomed. Mater. Res.*, **54**, 344–350.

363. Puricelli, E., Corsetti, A., Ponzoni, D., Martins, G. L., Leite, M. G., and Santos, L. A. (2010) Characterization of bone repair in rat femur after treatment with calcium phosphate cement and autogenous bone graft, *Head Face Med.*, **6**, art. no. 10.

364. Zhaoa, X., Lib, F., and Lic, S. (2010) Degradation characteristic of strontium-containing calcium phosphate cement *in vivo*, *Adv. Mater. Res.*, 105–106, 553–556.

365. Khairoun, I., Magne, D., Gauthier, O. Bouler, J. M., Aguado, E., Daculsi, G., and Weiss, P. (2002) *In vitro* characterization and *in vivo* properties of a carbonated apatite bone cement, *J. Biomed. Mater. Res.*, **60**, 633–642.

366. Mao, K., Yang, Y., Li, J., Hao, L., Tang, P., Wang, Z., Wen, N., Du, M., Wang, J., and Wang, Y. (2009) Investigation of the histology and interfacial bonding between carbonated hydroxyapatite cement and bone, *Biomed. Mater.*, **4**, 045003.

367. Sanzana, E. S., Navarro, M., Macule, F., Suso, S., Planell, J. A., and Ginebra, M. P. (2008) Of the *in vivo* behavior of calcium phosphate cements and glasses as bone substitutes, *Acta Biomater*, **4**, 1924–1933.

368. Bodde, E. W. H., Cammaert, C. T. R., Wolke, J. G. C., Spauwen, P. H. M., and Jansen, J. A. (2007) Investigation as to the osteoinductivity of macroporous calcium phosphate cement in goats, *J. Biomed. Mater. Res. B* (*Appl. Biomater.*), **83B**, 161–168.

369. Miyamoto, Y., Ishikawa, K., Takeshi, M., Toh, T., Yoshida, Y., Nagayama, M., Kon, M., and Asaoka, K. (1997) Tissue response to fast-setting calcium phosphate cement in bone, *J. Biomed. Mater. Res.*, **37**, 457–464.

370. Young, S., Holde, M., Gunasekaran, S., Poser, R., and Constantz, B. R. (1998) The correlation of radiographic, MRI and histological evaluations over two years of a carbonated apatite cement in a rabbit model, in *Proceedings of the 44th Annual Meeting*, Orthopedic Research Society, New Orleans, p. 846.

371. Feng, B., Guolin, M., Yuan, Y., Changshen, L., Zhen, W., and Jian, L. (2010) Role of macropore size in the mechanical properties and *in vitro* degradation of porous calcium phosphate cements, *Mater. Lett.*, **64**, 2028–2031.

372. Kroese-Deutman, H. C., Wolke, J. G. C., Spauwen, P. H. M., and Jansen, J. A. (2006) Closing capacity of cranial bone defects using porous calcium phosphate cement implants in a rabbit animal model, *J. Biomed. Mater. Res. A*, **79A**, 503–511.

373. Bourgeois, B., Laboux, O., Obadia, L., Gauthier, O., Betti, E., Aguado, E., Daculsi, G., and Bouler, J. M. (2003) Calcium-deficient apatite: a first *in vivo* study concerning bone in growth, *J. Biomed. Mater. Res. A*, **65A**, 402–408.

374. Lu, J., Descamps, M., Dejou, J., Koubi, G., Hardouin, P., Lemaître, J., and Proust, J. P. (2002) The biodegradation mechanism of calcium phosphate biomaterials in bone, *J. Biomed. Mater. Res.* (*Appl. Biomater.*), **63**, 408–412.

375. Wenisch, S., Stahl, J. P., Horas, U., Heiss, C., Kilian, O., Trinkaus, K., Hild, A., and Schnettler, R. (2003) *In vivo* mechanisms of hydroxyapatite ceramic degradation by osteoclasts, fine structural microscopy, *J. Biomed. Mater. Res. A*, **67A**, 713–718.

376. Ooms, E. M., Wolke, J. G. C., van der Waerden, J. P., and Jansen, J. A. (2002) Trabecular bone response to injectable calcium phosphate (Ca-P) cement, *J. Biomed. Mater. Res.*, **61**, 9–18.

377. Theiss, F., Apelt, D., Brand, B., Kutter, A., Zlinszky, K., Bohner, M., Matter, S., Frei, C., Auer, J. A., and von Rechenberg, B. (2005) Biocompatibility and resorption of a brushite calcium phosphate cement, *Biomaterials*, **26**, 4383–4394.

378. Heymann, D., Pradal, G., and Benahmad, M. (1999) Cellular mechanisms of calcium phosphate degradation, *Histol. Histopathol.*, **14**, 871–877.

379. Penel, G., Leroy, N., van Landuyt, P., Flautre, B., Hardouin, P., Lemaître, J., and Leroy, G. (1999) Raman microspectrometry studies of brushite cement: *in vivo* evolution in a sheep model, *Bone*, **25**(2), 81S–84S.

380. Dorozhkin, S. V. (1999) Inorganic chemistry of the dissolution phenomenon, the dissolution mechanism of calcium apatites at the atomic (ionic) level, *Comment Inorg. Chem.*, **20**, 285–299.

381. Dorozhkin, S. V. (2002) A review on the dissolution models of calcium apatites, *Prog. Cryst. Growth Charact.*, **44**, 45–61.

382. Knabe, C., Driessens, F. C. M., Planell, J. A., Gildenhaar, R., Berger, G., Reif, D., Fitzner, R., Radlanski, R. J., and Gross, U. (2000) Evaluation of calcium phosphates and experimental calcium phosphate bone cements using osteogenic cultures, *J. Biomed. Mater. Res.*, **52**, 498–508.

383. Teitelbaum, S. L. (2000) Bone resorption by osteoclasts, *Science*, **289**, 1504–1508.

384. Mostov, K., and Werb, Z. (1997) Journey across the osteoclast, *Science*, **276**, 219–220.

385. Rodan, G. A., and Martin, T. J. (2000) Therapeutic approaches to bone diseases, *Science*, **289**, 1508–1514.

386. Midy, V., Hollande, E., Rey, C., Dard, M., and Plouët, J. (2001) Adsorption of vascular endothelial growth factor to two different apatitic materials and its release, *J. Mater. Sci. Mater. Med.*, **12**, 293–298.

387. Hossain, M., Irwin, R., Baumann, M. J., and McCabe, L. R. (2005) Hepatocyte growth factor (HGF) adsorption kinetics and enhancement of osteoblast differentiation on hydroxyapatite surfaces, *Biomaterials*, **26**, 2595–2602.

388. Sun, L., Berndt, C. C., Gross, K. A., and Kucuk, A. (2001) Material fundamentals and clinical performance of plasma-sprayed hydroxyapatite coatings, a review, *J. Biomed. Mater. Res. B* (*Appl. Biomater.*), **58**, 570–592.

389. Renault, F., Chabriere, E., Andrieu, J. P., Dublet, B., Masson, P., and Rochu, D. (2006) Tandem purification of two HDL-associated partner proteins in human plasma., paraoxonase (PON1) and phosphate binding protein (HPBP) using hydroxyapatite chromatography, *J. Chromatogr. B*, **836**, 15–21.

390. Yoshitake, T., Kobayashi, S., Ogawa, T., and Okuyama, T. (2006) Hydroxyapatite chromatography of guanidine denatured proteins: 1. guanidine containing phosphate buffer system, *Chromatography*, **27**, 19–26.

391. Ooms, E. M., Egglezos, E. A., Wolke, J. G. C., and Jansen, J. A. (2003) Soft-tissue response to injectable calcium phosphate cements, *Biomaterials*, **24**, 749–757.

392. Ooms, E. M., Wolke, J. G. C., van de Heuvel, M. T., Jeschke, B., and Jansen, J. A. (2003) Histological evaluation of the bone response to calcium phosphate cement implanted in cortical bone, *Biomaterials*, **24**, 989–1000.

393. Kobayashi, N., Ong, K., Villarraga, M., Schwardt, J., Wenz, R., Togawa, D., Fujishiro, T., Turner, A. S., Seim, III H. B., and Bauer, T. W. (2007) Histological and mechanical evaluation of self-setting calcium phosphate cements in a sheep vertebral bone void model, *J. Biomed. Mater. Res. A*, **81A**, 838–846.

394. Wen, C. Y., Qin, L., Lee, K. M., and Chan, K. M. (2009) The use of brushite calcium phosphate cement for enhancement of bone-tendon integration in an anterior cruciate ligament reconstruction rabbit model, *J. Biomed. Mater. Res. B* (*Appl. Biomater.*), **89B**, 466–474.

395. Musha, Y., Umeda, T., Yoshizawa, S., Shigemitsu, T., Mizutani, K., and Itatani, K. (2010) Effects of blood on bone cement made of calcium phosphate: problems and advantages, *J J. Biomed. Mater. Res. B* (*Appl. Biomater.*), **92B**, 95–101.

396. Fernández, E., Ginebra, M. P., Bermudez, O., Boltong, M. G., Driessens, F. C. M., and Planell, J. A. (1995) Dimensional and thermal behaviour of calcium phosphate cements during setting compared to PMMA bone cements, *J. Mater. Sci. Lett.*, **14**, 4–5.

397. Andrianjatovo, H., Jose, F., and Lemaître, J. (1996) Effect of β-TCP granulometry on setting time and strength of calcium orthophosphate hydraulic cements, *J. Mater. Sci. Mater. Med.*, **7**, 34–39.

398. Ishikawa, K., Takagi, S., Chow, L. C., Ishikawa Y., Eanes, E. D., and Asaoka, K. (1994) Behavior of a calcium orthophosphate cement in simulated blood plasma *in vitro*, *Dent. Mater.*, **10**, 26–32.

399. Driessens, F. C. M. (1999) Chemistry and applied aspects of calcium orthophosphate bone cements, in *Concepts and Clinical Applications of Ionic Cements, 15th European Conference on Biomaterials*, Arcachon, Bordeaux, Sept. 8, 1999.

400. Yamamoto, H., Niwa, S., Hori, M., Hattori, T., Sawai, K., Aoki, S., Hirano, M., and Takeuchi, H. (1998) Mechanical strength of calcium phosphate cement *in vivo* and *in vitro*, *Biomaterials*, **19**, 1587–1591.

401. Pittet, C., and Lemaître, J. (2000) Mechanical characterization of brushite cements: a Mohr circles approach, *J. Biomed. Mater. Res.* (*Appl. Biomater.*), **53**, 769–780.

402. Morgan, E. F., Yetkinler, D. N., Constantz, B. R., and Dauskardt, R. H. (1997) Mechanical properties of carbonated apatite bone mineral

substitute: strength, fracture and fatigue behaviour, *J. Mater. Sci. Mater. Med.*, **8**, 559–570.

403. Miyazaki, K., Horibe, T., Antonucci, J. M., Takagi, S., and Chow, L. C. (1993) Polymeric calcium phosphate cements, analysis of reaction products and properties, *Dent. Mater.*, **9**, 41–45.

404. Miyazaki, K., Horibe, T., Antonucci, J. M., Takagi, S., and Chow, L. C. (1993) Polymeric calcium phosphate cements: setting reaction modifiers, *Dent. Mater.*, **9**, 46–50.

405. dos Santos, L. A., de Oliveira, L. C., Rigo, E. C. S., Carrodeguas, R. G., Boschi, A. O., and de Arruda, A. C. F. (1999) Influence of polymeric additives on the mechanical properties of α-tricalcium phosphate cement, *Bone*, **25**, 99S–102S.

406. Mickiewicz, R. A., Mayes, A. M., and Knaack, D. (2002) Polymer–calcium phosphate cement composites for bone substitutes, *J. Biomed. Mater. Res.*, **61**, 581–592.

407. Fernández, E., Sarda, S., Hamcerencu, M., Vlad, M. D., Gel, M., Valls, S., Torres, R., and López, J. (2005) High-strength apatitic cement by modification with superplasticizers, *Biomaterials*, **26**, 2289–2296.

408. Takahashi, T., Yamamoto, M., Ioku, K., and Goto, S. (1997) Relationship between compressive strength and pore structure of hardened cement pastes, *Adv. Cement Res.*, **9**, 25–30.

409. Costantino, P. D., Friedman, C. D., Jones, K., Chow, L. C., and Sisson, G. A. (1992) Experimental hydroxyapatite cement cranioplasty, *Plast. Reconstr. Surg.*, **90**, 174–185.

410. Chow, L. C., Hirayama, S., Takagi, S., and Parry, E. (2000) Diametral tensile strength and compressive strength of a calcium phosphate cement, effect of applied pressure, *J. Biomed. Mater. Res.* (*Appl. Biomater.*), **53**, 511–517.

411. Barralet, J. E., Gaunt, T., Wright, A. J., Gibson I. R., and Knowles, J. C. (2002) Effect of porosity reduction by compaction on compressive strength and microstructure of calcium phosphate cement, *J. Biomed. Mater. Res.* (*Appl. Biomater.*), **63**, 1–9.

412. Zhang, Y., Xu, H. H. K., Takagi, S., and Chow, L. C. (2006) *In situ* hardening hydroxyapatite-based scaffold for bone repair, *J. Mater. Sci. Mater. Med.*, **17**, 437–445.

413. Khairoun, I., LeGeros, R. Z., Daculsi, G., Bouler, J. M., Guicheux, J., and Gauthier, O. (2008) Macroporous, resorbable and injectable calcium phosphate-based cements (MCPC) for bone repair: augmentation, regeneration and osteoporosis treatment. US patent No. 7351280. April 1, 2008.

414. Speirs, A. D., Oxland, T. R., Masri, B. A., Poursartip, A., and Duncan, C. P. (2005) Calcium phosphate cement composites in revision hip arthroplasty, *Biomaterials*, **26**, 7310–7318.

415. dos Santos, L. A., Carrodeguas, R. G., Boschi, A. O., and de Arruda, A. C. F. (2003) Fiber-enriched double-setting calcium phosphate bone cement, *J. Biomed. Mater. Res. A*, **65A**, 244–250.

416. Gbureck, U., Spatz, K., and Thull, R. (2003) Improvement of mechanical properties of self-setting calcium phosphate bone cements mixed with different metal oxides, *Mat.-Wiss. u. Werkstofftech.*, **34**, 1036–1040.

417. Zhang, Y., and Xu, H. H. K. (2005) Effects of synergistic reinforcement and absorbable fiber strength on hydroxyapatite bone cement, *J. Biomed. Mater. Res. A*, **75A**, 832–840.

418. Buchanan, F., Gallagher, L., Jack, V., and Dunne, N. (2007) Short-fibre reinforcement of calcium phosphate bone cement, *Proc. Inst. Mech. Eng. H: J. Eng. Med.*, **221**, 203–212.

419. Wang, X., Ye, J., Wang, Y., and Chen, L. (2007) Reinforcement of calcium phosphate cement by bio-mineralized carbon nanotube, *J. Am. Ceram. Soc.*, **90**, 962–964.

420. Vélez, D., Arita, I. H., García-Garduño, M. V., and Castaño, V. M. (1994) Synthesis and characterization of a hydroxyapatite-zinc oxide-polyacrylic acid concrete, *Mater. Lett.*, **19**, 309–315.

421. According to Wikipedia, the free encyclopedia: "*Concrete* is a construction material that consists of a cement (commonly Portland cement), aggregates (generally gravel and sand) and water. It solidifies and hardens after mixing and placement due to a chemical process known as hydration. The water reacts with the cement, which bonds the other components together, eventually creating a stone-like material." http://en.wikipedia.org/wiki/Concrete (accessed in November 2010).

422. Dickens-Venz, S. H., Takagi, S., Chow, L. C., Bowen, R. L., Johnston, A. D., and Dickens, B. (1994) Physical and chemical properties of resin-reinforced calcium phosphate cements, *Dent. Mater.*, **10**, 100–106.

423. Xu, H. H. K., Eichmiller, F. C., and Barndt, P. R. (2001) Effects of fiber length and volume fraction on the reinforcement of calcium phosphate cement, *J. Mater. Sci. Mater. Med.*, **12**, 57–65.

424. Xu, H. H. K., Quinn, J. B., Takagi, S., and Chow, L. C. (2004) Synergistic reinforcement of *in situ* hardening calcium phosphate composite scaffold for bone tissue engineering, *Biomaterials*, **25**, 1029–1037.

425. Yokoyama, A., Yamamoto, S., Kawasaki, T., Kohgo, T., and Nakasu, M. (2002) Development of calcium phosphate cement using chitosan and citric acid for bone substitute materials, *Biomaterials*, **23**, 1091–1101.

426. Takagi, S., Chow, L. C., Hirayama, S., and Eichmiller, F. C. (2003) Properties of elastomeric calcium phosphate cement-chitosan composites, *Dent. Mater.*, **19**, 797–804.

427. Xu, H. H. K., and Simon Jr., C. G. (2005) Fast setting calcium phosphate-chitosan scaffold: mechanical properties and biocompatibility, *Biomaterials*, **26**, 1337–1348.

428. Sun, L., Xu, H. H. K., Takagi, S., and Chow, L. C. (2007) Fast setting calcium phosphate cement–chitosan composite, mechanical properties and dissolution rates, *J. Biomater. Appl.*, **21**, 299–316.

429. Pan, Z. H., Jiang, P. P., Fan, Q. Y., Ma, B., and Cai, H. P. (2007) Mechanical and biocompatible influences of chitosan fiber and gelatin on calcium phosphate cement, *J. Biomed. Mater. Res. B* (*Appl. Biomater.*), **82B**, 246–252.

430. Liu, H., Li, H., Cheng, W., Yang, Y., Zhu, M., and Zhou, C. (2006) Novel injectable calcium phosphate/chitosan composites for bone substitute materials, *Acta Biomater.*, **2**, 557–565.

431. Pan, Z. H., Cai, H. P., Jiang, P. P., and Fan, Q. Y. (2006) Properties of a calcium phosphate cement synergistically reinforced by chitosan fiber and gelatin, *J. Polymer Res.*, **13**, 323–327.

432. Weir, M. D., and Xu, H. H. K. (2008) High-strength, *in situ*-setting calcium phosphate composite with protein release, *J. Biomed. Mater. Res. A*, **85A**, 388–396.

433. Lian, Q., Li, D. C., He, J. K., and Wang, Z. (2008) Mechanical properties and *in-vivo* performance of calcium phosphate cement–chitosan fibre composite, *Proc. Inst. Mech. Eng. H: J. Eng. Med.*, **222**, 347–353.

434. Wang, X., Chen, L., Xiang, H., and Ye, J. (2007) Influence of anti-washout agents on the rheological properties and injectability of a calcium phosphate cement, *J. Biomed. Mater. Res. B* (*Appl. Biomater.*), **81B**, 410–418.

435. Tanaka, S., Kishi, T., Shimogoryo, R., Matsuya, S., and Ishikawa, K. (2003) Biopex acquires anti-washout properties by adding sodium alginate into its liquid phase, *Dent. Mater. J.*, **22**, 301–312.

436. Lin, J., Zhang, S., Chen, T., Liu, C., Lin, S., and Tian, X. (2006) Calcium phosphate cement reinforced by polypeptide copolymers, *J. Biomed. Mater. Res. B* (*Appl. Biomater.*), **76B**, 432–439.

437. Miyamoto, Y., Ishikawa, K., Takechi, M., Toh, T., Yuasa, T., Nagayama, M., and Suzuki, K. (1998) Basic properties of calcium phosphate cement containing atelocollagen in its liquid or powder phases, *Biomaterials*, **19**, 707–715.

438. Knepper-Nicolai, B., Reinstorf, A., Hofinger, I., Flade, K., Wenz, R., and Pompe, W. (2002) Influence of osteocalcin and collagen I on the mechanical and biological properties of Biocement D®, *Biomol. Eng.*, **19**, 227–231.

439. Hempel, U., Reinstorf, A., Poppe, M., Fischer, U., Gelinsky, M., Pompe, W., and Wenzel, K. W. (2004) Proliferation and differentiation of osteoblasts on Biocement D® modified with collagen type I and citric acid, *J. Biomed. Mater. Res. B* (*Appl. Biomater.*), **71B**, 130–143.

440. Reinstorf, A., Ruhnow, M., Gelinsky, M., Pompe, W., Hempel, U., Wenzel, K. W., and Simon, P. (2004) Phosphoserine — a convenient compound for modification of calcium phosphate bone cement collagen composites, *J. Mater. Sci. Mater. Med.*, **15**, 451–455.

441. Otsuka, M., Kuninaga, T., Otsuka, K., and Higuchi, W. I. (2006) Effect of nanostructure on biodegradation behaviors of self-setting apatite/collagen composite cements containing vitamin K_2 in rats, *J. Biomed. Mater. Res. B* (*Appl. Biomater.*), **79B**, 176–184.

442. Moreau, J. L., Weir, M. D., and Xu, H. H. K. (2009) Self-setting collagen–calcium phosphate bone cement: mechanical and cellular properties, *J. Biomed. Mater. Res. A*, **91A**, 605–613.

443. Otsuka, M., Nakagawa, H., Ito, A., and Higuchi, W. I. (2010) Effect of geometrical structure on drug release rate of a three-dimensionally perforated porous apatite/collagen composite cement, *J. Pharm. Sci.*, **99**, 286–292.

444. Gbureck, U., Spatz, K., Thull, R., and Barralet, J. E. (2005) Rheological enhancement of mechanically activated α-tricalcium phosphate cements, *J. Biomed. Mater. Res. B* (*Appl. Biomater.*), **73B**, 1–6.

445. Xu, H. H. K., Eichmiller, F. C., and Giuseppetti, A. A. (2000) Reinforcement of a self-setting calcium phosphate cement with different fibers, *J. Biomed. Mater. Res.*, **52**, 107–114.

446. Xu, H. H. K., Quinn, J. B., Takagi, S., Chow, L. C., and Eichmiller, F. C. (2001) Strong and macroporous calcium phosphate cement: effects of porosity and fiber reinforcement on mechanical properties, *J. Biomed. Mater. Res.*, **57**, 457–466.

447. dos Santos, L. A., Carrodeguas, R. G., Boschi, A. O., and de Arruda, A. C. F. (2003) Dual-setting calcium phosphate cement modified with ammonium polyacrylate, *Artif. Organs*, **27**, 412–418.

448. dos Santos, L. A., de Oliveira, L. C., da Silva Rigo, E. C., Carrodéguas, R. G., Boschi, A. O., and de Arruda, A. C. F. (2000) Fiber reinforced calcium phosphate cement, *Artif. Organs*, **24**, 212–216.

449. Liu, C. S., Chen, C. W., and Ducheyne, P. (2008) *In vitro* surface reaction layer formation and dissolution of calcium phosphate cement–bioactive glass composites,*Biomed. Mater.*, **3**, 034111 (11 pages).

450. Alge, D. L., and Chu, T. M. G. (2010) Calcium phosphate cement reinforcement by polymer infiltration and *in situ* curing: a method for 3D scaffold reinforcement, *J. Biomed. Mater. Res. A*, **94A**, 547–555.

451. Julien, M., Khairoun, I., LeGeros, R. Z., Delplace, S., Pilet, P., Weiss, P., Daculsi, G., Bouler, J. M., and Guicheux, J. (2007) Physico-chemical-mechanical and *in vitro* biological properties of calcium phosphate cements with doped amorphous calcium phosphates, *Biomaterials*, **28**, 956–965.

452. Lemaître, J., Munting, E., and Mirtchi, A. A. (1992) Setting, hardening and resorption of calcium phosphate hydraulic cements, *Rev. Stomatol. Chir. Maxillofac.*, **93**, 163–165.

453. Müller, F. A., Gbureck, U., Kasuga, T., Mizutani, Y., Barralet, J. E., and Lohbauer, U. (2007) Whisker-reinforced calcium phosphate cements, *J. Am. Ceram. Soc.*, **90**, 3694–3697.

454. Nakagawa, A., Matsuya, S., Takeuchi, A., and Ishikawa, K. (2007) Comparison of the effects of added α- and β-tricalcium phosphate on the basic properties of apatite cement, *Dent. Mater. J.*, **26**, 342–347.

455. Gisep, A., Wieling, R., Bohner, M., Matter, S., Schneider, E., and Rahn, B. (2003) Resorption patterns of calcium-phosphate cements in bone, *J. Biomed. Mater. Res. A*, **66A**, 532–540.

456. Ion-substituted ACPs contained other ions (carbonates, Mg, Zn, F) were used in that study [451].

457. van den Vreken, N. M. F., Pieters, I. Y., Declercq, H. A., Cornelissen, M. J., and Verbeeck, R. M. H. (2010) Characterization of calcium phosphate cements modified by addition of amorphous calcium phosphate, *Acta Biomater.*, **6**, 617–625.

458. Zhou, L., Yan, J. L., and Hu, C. J. (2007) Degradation of bone repairing composite of calcium polyphosphate fiber, calcium phospate cement and micromorselized bone *in vitro*, *J. Clin. Rehabilit. Tiss. Eng. Res.*, **11**, 33–36.

459. Xu, L. X., Shi, X. T., Wang, Y. P., and Shi, Z. L. (2009) Mechanical effect of calcium polyphosphate fiber on reinforcing calcium phosphate bone cement composites, *J. Clin. Rehabilit. Tiss. Eng. Res.*, **13**, 7474–7476.

460. Castaldini, A., and Cavallini, A. (1985) Setting properties of bone cement with added synthetic hydroxyapatite, *Biomaterials*, **6**, 55–60.

461. Sogal, A., and Hulbert, S. F. (1992) Mechanical properties of a composite bone cement, polymethylmethacrylate and hydroxyapatite, in *Proceedings of the 5th International Symposium on Ceramics in Medicine*: *Vol. 5. Bioceramics*, pp. 213–224.

462. Harper, E. J., Behiri, J. C., and Bonfield, W. (1995) Flexural and fatigue properties of a bone cement based upon polyethylmethacrylate and hydroxyapatite, *J. Mater. Sci. Mater. Med.*, **6**, 799–803.

463. Harper, E. J. (1998) Bioactive bone cements, *Proc. Inst. Mech. Eng. H: J. Eng. Med.*, **212**, 113–120.

464. Shinzato, S., Kobayashi, M., Mousa, W. F., Kamimura, M., Neo, M., Kitamura, Y., Kokubo, T., and Nakamura, T. (2000) Bioactive polymethylmethacrylate-based bone cement: comparison of glass beads, apatite- and wollastonite-containing glass-ceramic, and hydroxyapatite fillers on mechanical and biological properties, *J. Biomed. Mater. Res.*, **51**, 258–272.

465. Harper, E. J., Braden, M., and Bonfield, W. (2000) Mechanical properties of hydroxyapatite reinforced poly(ethylmethacrylate) bone cement after immersion in a physiological solution, Influence of a silane coupling agent, *J. Mater. Sci. Mater. Med.*, **11**, 491–497.

466. Espigares, I., Elvira, C., Mano, J. F., Vázquez, B., san Román, J., and Reis, R. L. (2002) New partially degradable and bioactive acrylic bone cements based on starch blends and ceramic fillers, *Biomaterials*, **23**, 1883–1895.

467. Pek, Y. S., Kurisawa, M., Gao, S., Chung, J. E., and Ying, J. Y. (2009) The development of a nanocrystalline apatite reinforced crosslinked hyaluronic acid-tyramine composite as an injectable bone cement, *Biomaterials*, **30**, 822–828.

468. Xu, H. H. K., and Quinn, J. B. (2001) Whisker-reinforced bioactive composites containing calcium phosphate cement fillers, Effects of filler ratio and surface treatments on mechanical properties, *J. Biomed. Mater. Res.*, **57**, 165–174.

469. Claes L., Höllen, I., and Ignatius, A. (1997) Resorbable bone cements, *Orthopäde*, **26**, 459–462.

470. Jansen, J. A., de Ruijter, J. E., Schaeken, H. G., van der Waerden, J. P. C., Planell, J. A., and Driessens, F. C. M. (1995) Evaluation of tricalciumphosphate/hydroxyapatite cement for tooth replacement, an experimental animal study, *J. Mater. Sci. Mater. Med.*, **6**, 653–657.

471. Larsson, S., and Bauer, T. W. (2002) Use of injectable calcium phosphate cement for fracture fixation: a review, *Clin. Orthop. Relat. Res.*, **395**, 23–32.

472. Oshtory, R., Lindsey, D. P., Giori, N. J., and Mirza, F. M. (2010) Bioabsorbable tricalcium phosphate bone cement strengthens fixation of suture anchors, *Clin. Orthop. Rel. Res.*, **468**, 3406–3412.

473. Gbureck, U., Knappe, O., Hofmann, N., and Barralet, J. E. (2007) Antimicrobial properties of nanocrystalline tetracalcium phosphate cements, *J. Biomed. Mater. Res. B*, (*Appl. Biomater.*), **83B**, 132–137.

474. Sethuraman, S., Nair, L. S., El-Amin, S., Nguyen, M. T. N., Greish, Y. E., Bender, J. D., Brown, P. W., Allcock, H. R., and Laurencin, C. T. (2007) Novel low temperature setting nanocrystalline calcium phosphate cements for bone repair: osteoblast cellular response and gene expression studies, *J. Biomed. Mater. Res. A*, **82A**, 884–891.

475. Link, D. P., van den Dolder, J., Wolke, J. G. C., and Jansen, J. A. (2007) The cytocompatibility and early osteogenic characteristics of an injectable calcium phosphate cement, *Tissue Eng.*, **13**, 493–500.

476. Oda, H., Nakamura, K., Matsushita, T., Yamamoto, S., Ishibashi, H., Yamazaki, T., and Morimoto, S. (2006) Clinical use of a newly developed calcium phosphate cement (XSB-671D), *J. Orthop. Sci.*, **11**, 167–174.

477. Braun, C., Rahn, B., Fulmer, M. T., Steiner, A., and Gisep, A. (2006) Intra-articular calcium Phosphate cement, its fate and impact on joint tissues in a rabbit model, *J. Biomed. Mater. Res. B* (*Appl. Biomater.*), **79B**, 151–158.

478. Costantino, P., Friedman, C., Jones, K., Chow, L. C., Pelzer, H., and Sisson, G. (1991) Hydroxyapatite cement. I. Basic chemistry and histologic properties, *Arch. Otolaryngol. Head Neck Surg.*, **117**, 379–384.

479. Hong, Y. C., Wang, J. T., Hong, C. Y., Brown, W. E., and Chow, C. Y. (1991) The periapical tissue reactions to a calcium phosphate cement in the teeth of monkeys, *J. Biomed. Mater. Res.*, **25**, 485–498.

480. Sugawara, A., Fujikawa, K., Kusama, K., Nishiyama, M., Murai, S., Takagi, S., and Chow, L. C. (2002) Histopathologic reaction of calcium phosphate cement for alveolar ridge augmentation, *J. Biomed. Mater. Res.*, **61**, 47–52.

481. Fujikawa, K., Sugawara, A., Kusama, K., Nishiyama, M., Murai, S., Takagi, S., and Chow, L. C. (2002) Fluorescent labeling analysis and electron probe microanalysis for alveolar ridge augmentation using calcium phosphate cement, *Dent. Mater. J.*, **21**, 296–305.

482. Sugawara, A., Chow, L. C., Takagi, S., and Chohayeb, H. (1990) *In vitro* evaluation of the sealing ability of a calcium phosphate cement when used as a root canal sealer-filler, *J. Endod.*, **16**, 162–165.

483. Noetzel, J., Özer, K., Reisshauer, B. H., Anil, A., Rössler, R., Neumann, K., and Kielbassa, A. M. (2006) Tissue responses to an experimental calcium phosphate cement and mineral trioxide aggregate as materials for furcation perforation repair, a histological study in dogs, *Clin. Oral Invest.*, **10**, 77–83.

484. Zhang, W., Walboomers, X. F., and Jansen, J. A. (2008) The formation of tertiary dentin after pulp capping with a calcium phosphate cement, loaded with PLGA microparticles containing TGF-β1, *J. Biomed. Mater. Res. A*, **85A**, 439–444.

485. Comuzzi, L., Ooms, E., Jansen, and J. A. (2002) Injectable calcium phosphate cement as a filler for bone defects around oral implants, an experimental study in goats, *Clin. Oral Implants Res.*, **13**, 304–311.

486. Shirakata, Y., Oda, S., Kinoshita, A., Kikuchi, S., Tsuchioka, H., and Ishikawa, I. (2002) Histocompatible healing of periodontal defects after application of injectable calcium phosphate bone cement. A preliminary study in dogs, *J. Periodontol.*, **73**, 1043–1053.

487. Lee, S. K., Lee, S. K., Lee, S. I., Park, J. H., Jang, J. H., Kim, H. W., and Kim, E. C. (2010) Effect of calcium phosphate cements on growth and odontoblastic differentiation in human dental pulp cells, *J. Endod.*, **36**, 1537–1542.

488. Chaung, H. M., Hong, C. H., Chiang, C. P., Lin, S. K., Kuo, Y. S., Lan, W. H., and Hsieh, C. C. (1996) Comparison of calcium phosphate cement mixture and pure calcium hydroxide as direct pulp-capping agents, *J. Formos Med. Assoc.*, **95**, 545–550.

489. Arisan, V., Anil, A., Wolke, J. G., and Özer, K. (2010) The effect of injectable calcium phosphate cement on bone anchorage of titanium implants: an experimental feasibility study in dogs, *Int. J. Oral Maxillofac. Surg.*, **39**, 463–468.

490. Aral, A., Yalçn, S., Karabuda, Z. C., Anıl, A., Jansen, J. A., and Mutlu, Z. (2008) Injectable calcium phosphate cement as a graft material for maxillary sinus augmentation: an experimental pilot study, *Clin. Oral Implants Res.*, **19**, 612–617.

491. Sliindo, M. L., Costantino, P. D., Friedman, C. D., and Chow, L. C. (1993) Facial skeletal augmentation using hydroxyapatite cement cranioplasty, *Arch. Otolaryngol. Head Neck Surg.*, **119**, 185–190.

492. Bifano, C. A., Edgin, W. A., Colleton, C., Bifano, S. L., and Constantino, P. D. (1998) Preliminary evaluation of hydroxyapatite cement as an augmentation device in the edentulous atrophic canine mandible, *Oral Surg.*, **85**, 512–516.

493. Friedman, C. D., Constantino, P. D., Jones, K., Chow, L. C., Pelzer, H., and Sisson, G. (1991) Hydroxyapatite cement. II. Obliteration and reconstruction of the cat frontal sinus, *Arch. Otolaryngol. Head Neck Surg.*, **117**, 385–389.

494. Sinikovic, B., Kramer, F. J., Swennen, G., Lubbers, H. T., and Dempf, R. (2007) Reconstruction of orbital wall defects with calcium phosphate cement: clinical and histological findings in a sheep model, *Int. J. Oral Maxillofac. Surg.*, **36**, 54–61.

495. Smartt, J. M., Karmacharya, J., Gannon, F. H., Ong, G., Jackson, O., Bartlett, S. P., Poser, R. D., and Kirschner, R. E. (2005) Repair of the immature and mature craniofacial skeleton with a carbonated calcium phosphate cement: assessment of biocompatibility, osteoconductivity and remodeling capacity, *Plast. Reconstr. Surg.*, **115**, 1642–1650.

496. Reddi, S. P., Stevens, M. R., Kline, S. N., and Villanueva, P. (1999) Hydroxyapatite cement in craniofacial trauma surgery, indications and early experience, *J. Cran. Maxillofac. Trauma*, **5**, 7–12.

497. Friedman, C. D., Costantino, P. D., Synderman, C. H., Chow, L. C., and Takagi, S. (2000) Reconstruction of the frontal sinus and frontofacial skeleton with hydroxyapatite cement, *Arch. Facial Plast. Surg.*, **2**, 124–129.

498. Kuemmerle, J. M., Oberle, A., Oechslin, C., Bohner, M., Frei, C., Boecken, I., and von Rechenberg, B. (2005) Assessment of the suitability of a new brushite calcium phosphate cement for cranioplasty — an experimental study in sheep, *J. Cran. Maxillofac. Surg.*, **33**, 37–44.

499. Luaces-Rey, R., García-Rozado, A., Crespo-Escudero, J. L., Seijas, B. P., Arenaz-Búa, J., and López-Cedrún, J. L. (2009) Use of carbonated calcium phosphate bone cement and resorbable plates for the treatment of frontal sinus fractures: two case reports, *J. Plastic Reconstr. Aesthetic Surg.*, **62**, 272–273.

500. Tamimi, F., Torres, J., Cabarcos, E. L., Bassett, D. C., Habibovic, P., Luceron, E., and Barralet, J. E. (2009) Minimally invasive maxillofacial vertical bone augmentation using brushite based cements, *Biomaterials*, **30**, 208–216.

501. Lee, D. W., Kim, J. Y., and Lew, D. H. (2010) Use of rapidly hardening hydroxyapatite cement for facial contouring surgery, *J. Craniofac. Surg.*, **21**, 1084–1088.

502. Singh, K. A., Burstein, F. D., and Williams, J. K. (2010) Use of hydroxyapatite cement in pediatric craniofacial reconstructive surgery: strategies for avoiding complications, *J. Craniofac. Surg.*, **21**, 1130–1135.

503. Bambakidis, N. C., Munyon, C., Ko, A., Selman, W. R., and Megerian, C. A. (2010) A novel method of translabyrinthine cranioplasty using hydroxyapatite cement and titanium mesh: a technical report, *Skull Base*, **20**, 157–161.

504. Abe, T., Anan, M., Kamida, T., and Fujiki, M. (2009) Surgical technique for anterior skull base reconstruction using hydroxyapatite cement and titanium mesh, *Acta Neurochirurgica*, **151**, 1337–1338.

505. Sanada, Y., Fujinaka, T., Yoshimine, T., and Kato, A. (2011) Optimal reconstruction of the bony defect after frontotemporal craniotomy with hydroxyapatite cement, *J. Clin. Neurosci.*, **18**, 280–282.

506. Benson, A. G., and Djalilian, H. R. (2009) Complications of hydroxyapatite bone cement reconstruction of retrosigmoid craniotomy: two cases, *Ear Nose Throat J.*, **88**, E1.

507. Liverneaux, P. (2005) Osteoporotic distal radius curettage-filling with an injectable calcium phosphate cement. A cadaveric study, *Eur. J. Orthop. Surg. Traumatol.*, **15**, 1–6.

508. Liverneaux, P., Vernet, P., Robert, C., and Diacono, P. (2006) Cement pinning of osteoporotic distal radius fractures with an injectable calcium phosphate bone substitute, report of 6 cases, *Eur. J. Orthop. Surg. Traumatol.*, **16**, 10–16.

509. Thordarson, D., Hedman, T., Yetkinler, D., Eskander, E., Lawrence, T., and Poser, R. (1999) Superior compressive strength of a calcaneal fracture construct augmented with remodelable cancellous bone cement, *J. Bone Joint Surg. Am.*, **81A**, 239–246.

510. Stankewich, C. J., Swiontkowski, M. F., Tencer, A. F., Yetkinler, D. N., and Poser, R. D. (1996) Augmentation of femoral neck fracture fixation with an injectable calcium-phosphate bone mineral cement, *J. Orthopaedic Res.*, **14**, 786–793.

511. Goodman, S., Bauer, T., Carter, D., Casteleyn, P. P., Goldstein, S. A., Kyle, R. F., Larsson, S., Stankewich, C. J., Swiontkowski, M. F., Tencer, A. F., Yetkinler, D. N., and Poser, R. D. (1998) Norian SRS® cement augmentation in hip fracture treatment, *Clin. Orthop. Rel. Res.*, **348**, 42–50.

512. Bai, B., Jazrawi, L., Kummer, F., and Spivak, J. (1999) The use of an injectable, biodegradable calcium orthophosphate bone substitute

for the prophylactic augmentation of osteoporotic vertebrae and the management of vertebral compression fractures, *Spine*, **24**, 1521–1526.

513. Ryf, C., Goldhahn, S., Radziejowski, M., Blauth, M., and Hanson, B. (2009) A new injectable brushite cement: first results in distal radius and proximal tibia fractures, *Eur. J. Trauma Emerg. Surg.*, **35**, 389–396.

514. Horstmann, W. G., Verheyen, C. C. P. M., and Leemans, R. (2003) An injectable calcium phosphate cement as a bone-graft substitute in the treatment of displaced lateral tibial plateau fractures, *Injury*, **34**, 141–144.

515. Simpson, D., and Keating, J. F. (2004) Outcome of tibial plateau fractures managed with calcium phosphate cement, *Injury*, **35**, 913–918.

516. Welch, R. D., Zhang, H., and Bronson, D. G. (2003) Experimental tibial plateau fractures augmented with calcium phosphate cement or autologous bone graft, *J. Bone Joint Surg. Am.*, **85A**, 222–231.

517. Keating, J. F., Hajducka, C. L., and Harper, J. (2003) Minimal internal fixation and calcium-phosphate cement in the treatment of fractures of the tibial plateau, *J. Bone Joint Surg. Br.*, **85B**, 68–73.

518. Moore, D., Maitra, R., Farjo, L., Graziano, G., and Goldstein, S. (1997) Restoration of pedicle screw fixation with an *in situ* setting calcium orthophosphate cement, *Spine*, **22**, 1696–1705.

519. Cho, W., Wu, C., Erkan, S., Kang, M. M., Mehbod, A. A., and Transfeldt, E. E. (2011) The effect on the pullout strength by the timing of pedicle screw insertion after calcium phosphate cement injection, *J. Spinal Disord. Tech.*, **24**, 116–120.

520. Mermelstein, L. E., McLain, R. F., and Yerby, S. A. (1998) Reinforcement of thoracolumbar burst fractures with calcium phosphate cement, *Spine*, **23**, 664–671.

521. Mermelstein, L. E., Chow, L. C., Friedman, C., and Crisco, J. (1996) The reinforcement of cancellous bone screws with calcium orthophosphate cement, *J. Orthop. Trauma*, **10**, 15–20.

522. Stadelmann, V. A., Bretton, E., Terrier, A., Procter, P., and Pioletti, D. P. (2010) Calcium phosphate cement augmentation of cancellous bone screws can compensate for the absence of cortical fixation, *J. Biomech.*, **43**, 2869–2874.

523. Liverneaux, P., and Khallouk, R. (2006) Calcium phosphate cement in wrist arthrodesis: three cases, *J. Orthop. Sci.*, **11**, 289–293.

524. Ooms, E. M., Wolke, J. G. C., van der Waerden, J. P. C. M., and Jansen, J. A. (2003) Use of injectable calcium phosphate cement for the fixation

of titanium implants: an experimental study in goats, *J. Biomed. Mater. Res. B* (*Appl. Biomater.*), **66B**, 447–456.

525. Strauss, E. J., Pahk, B., Kummer, F. J., and Egol, K. (2007) Calcium phosphate cement augmentation of the femoral neck defect created after dynamic hip screw removal, *J. Orthop. Trauma*, **21**, 295–300.

526. Schildhauer, T. A., Bennett, A. P., Wright, T. M., Lane, J. M., and O'Leary, P. F. (1999) Intravertebral body reconstruction with an injectable *in situ*-setting carbonated apatite: biomechanical evaluation of a minimally invasive technique, *J. Orthop. Res.*, **17**, 67–72.

527. Jansen, J. A., Ooms, E., Verdonschot, N., and Wolke, J. G. C. (2005) Injectable calcium phosphate cement for bone repair and implant fixation, *Orthop. Clin. North Am.*, **36**, 89–95.

528. Maestretti, G., Cremer, C., Otten, P., and Jakob, R. P. (2007) Prospective study of standalone balloon kyphoplasty with calcium phosphate cement augmentation in traumatic fractures, *Eur. Spine J.*, **16**, 601–610.

529. Hisatome, T., Yasunaga, Y., Ikuta, Y., and Fujimoto, Y. (2002) Effects on articular cartilage of subchondral replacement with polymethylmethacrylate and calcium phosphate cement, *J. Biomed. Mater. Res.*, **59**, 490–498.

530. Lim, T. H., Brebach, G. T., Renner, S. M., Kim, W. J., Kim, J. G., Lee, R. E., Andersson, G. B., and An, H. S. (2002) Biomechanical evaluation of an injectable calcium phosphate cement for vertebroplasty, *Spine*, **27**, 1297–1302.

531. Belkoff, S. M., Mathis, J. M., Jasper, L. E., and Deramond, H. (2001) An *ex vivo* biomechanical evaluation of a hydroxyapatite cement for use with vertebroplasty, *Spine*, **26**, 1542–1546.

532. Heini, P. F., Berlemann, U., Kaufmann, M., Lippuner, K., Fankhauser, C., and van Landuyt, P. (2001) Augmentation of mechanical properties in osteoporotic vertebral bones — a biomechanical investigation of vertebroplasty efficacy with different bone cements, *Eur. Spine J.*, **10**, 164–171.

533. Tomita, S., Kin, A., Yazu, M., and Abe, M. (2003) Biomechanical evaluation of kyphoplasty and vertebroplasty with calcium phosphate cement in a simulated osteoporotic compression fracture, *J. Orthop. Sci.*, **8**, 192–197.

534. Libicher, M., Hillmeier, J., Liegibel, U., Sommer, U., Pyerin, W., Vetter, M., Meinzer, H. P., Grafe, I., Meeder, P., Nöldge, G., Nawroth, P., and Kasperk, C. (2006) Osseous integration of calcium phosphate in osteoporotic

vertebral fractures after kyphoplasty: initial results from a clinical and experimental pilot study, *Osteoporos. Int.*, **17**, 1208–1215.

535. Khanna, A. J., Lee, S., Villarraga, M., Gimbel, J., Steffey, D., and Schwardt, J. (2008) Biomechanical evaluation of kyphoplasty with calcium phosphate cement in a 2-functional spinal unit vertebral compression fracture model, *Spine J.*, **8**, 770–777.

536. Zhu, X. S., Zhang, Z. M., Mao, H. Q., Geng, D. C., Wang, G. L., Gan, M. F., and Yang, H. L. (2008) Biomechanics of calcium phosphate cement in vertebroplasty, *J. Clin. Rehabil. Tissue Eng. Res.*, **12**, 8071–8074.

537. Otsuka, M., Matsuda, Y., Suwa, Y., Fox, J. L., and Higuchi, W. I. (1994) A novel skeletal drug-delivery system using a self-setting calcium orthophosphate cement. 3. Physicochemical properties and drug-release rate of bovine insulin and bovine albumin, *J. Pharm. Sci.*, **83**, 255–258.

538. Yu, D., Wong, J., Matsuda, Y., Fox, J. L., Higuchi, W. I., and Otsuka, M. (1992) Self-setting hydroxyapatite cement: a novel skeletal drug-delivery system for antibiotics, *J. Pharm. Sci.*, **81**, 529–531.

539. Bohner, M., Lemaître, J., van Landuyt, P., Zambelli, P., Merkle, H. P., and Gander, B. (1997) Gentamicin-loaded hydraulic calcium orthophosphate bone cement as antibiotic delivery system, *J. Pharm. Sci.*, **86**, 565–572.

540. Bohner, M., Lemaître, J., Merkle, H. P., and Gander, B. (2000) Control of gentamicin release from a calcium phosphate cement by admixed poly(acrylic acid), *J. Pharm. Sci.*, **89**, 1262–1270.

541. Ratier, A., Freche, M., Locout, J. L., and Rodriguez, F. (2004) Behaviour of an injectable calcium phosphate cement with added tetracycline, *Int. J. Pharm.*, **274**, 261–268.

542. Kisanuki, O., Yajima, H., Umeda, T., and Takakura, Y. (2007) Experimental study of calcium phosphate cement impregnated with dideoxy-kanamycin B, *J. Orthop. Sci.*, **12**, 281–288.

543. McNally, A., Sly, K., Lin, S., Bourges, X., and Daculsi, G. (2008) Release of antibiotics from macroporous injectable calcium phosphate cement, *Key Eng. Mater.*, 361–363, 359–362.

544. Hofmann, M. P., Mohammed, A. R., Perrie, Y., Gbureck, U., and Barralet, J. E. (2009) High-strength resorbable brushite bone cement with controlled drug-releasing capabilities, *Acta Biomater.*, **5**, 43–49.

545. Tamimi, F., Torres, J., Bettini, R., Ruggera, F., Rueda, C., López-Ponce, M., and Cabarcos, E. L. (2008) Doxycycline sustained release from brushite cements for the treatment of periodontal diseases, *J. Biomed. Mater. Res. A*, **85A**, 707–714.

546. Young, A. M., Ng, P. Y. J., Gbureck, U., Nazhat, S. N., Barralet, J. E., and Hofmann, M. P. (2008) Characterization of chlorhexidine-releasing, fast-setting, brushite bone cements, *Acta Biomater.*, **4**, 1081–1088.

547. Hesaraki, S., and Nemati, R. (2009) Cephalexin-loaded injectable macroporous calcium phosphate bone cement, *J. Biomed. Mater. Res. B* (*Appl. Biomater.*), **89B**, 342–352.

548. Otsuka, M., Matsuda, Y., Suwa, Y., Fox, J. L., and Higuchi, W. I. (1994) A novel skeletal drug delivery system using a self-setting calcium orthophosphate cement. 5. Drug release behavior from a heterogeneous drug-loaded cement containing an anticancer drug, *J. Pharm. Sci.*, **83**, 1565–1568.

549. Otsuka, M., Matsuda, Y., Suwa, Y., Fox, J. L., and Higuchi, W. I. (1994) A novel skeletal drug delivery system using a self-setting calcium orthophosphate cement. 2. Physicochemical properties and drug release rate of the cement-containing indomethacin, *J. Pharm. Sci.*, **83**, 611–615.

550. Panzavolta, S., Torricelli, P., Bracci, B., Fini, M., and Bigi, A. (2009) Alendronate and pamidronate calcium phosphate bone cements, setting properties and *in vitro* response of osteoblast and osteoclast cells, *J. Inorg. Biochem.*, **103**, 101–106.

551. le Nihouannen, D., Hacking, S. A., Gbureck, U., Komarova, S. V., and Barralet, J. E. (2008) The use of RANKL-coated brushite cement to stimulate bone remodeling, *Biomaterials*, **29**, 3253–3259.

552. Li, D. X., Fan, H. S., Zhu, X. D., Tan, Y. F., Xiao, W. Q., Lu, J., Xiao, Y. M., Chen, J. Y., and Zhang, X. D. (2007) Controllable release of salmon-calcitonin in injectable calcium phosphate cement modified by chitosan oligosaccharide and collagen polypeptide, *J. Mater. Sci. Mater. Med.*, **18**, 2225–2231.

553. Kamegai, A., Shimamura, N., Naitou, K., Nagahara, K., Kanematsu, N., and Mori, M. (1994) Bone formation under the influence of bone morphogenetic protein/self-setting apatite cement composite as delivery system, *Biomed. Mater. Eng.*, **4**, 291–307.

554. Fei, Z., Hu, Y., Wu, D., Wu, H., Lu, R., Bai, J., and Song, H. (2008) Preparation and property of a novel bone graft composite consisting of rhBMP-2 loaded PLGA microspheres and calcium phosphate cement, *J. Mater. Sci. Mater. Med.*, **19**, 1109–1116.

555. Ruhé, P. Q., Kroese-Deutman, H. C., Wolke, J. G. C., Spauwen, P. H. M., and Jansen, J. A. (2004) Bone inductive properties of rhBMP-2 loaded porous calcium phosphate cement implants in cranial defects in rabbits, *Biomaterials*, **25**, 2123–2132.

556. Bodde, E. W. H., Boerman, O. C., Russel, F. G. M., Mikos, A. G., Spauwen, P. H. M., and Jansen, J. A. (2008) The kinetic and biological activity of different loaded rhBMP-2 calcium phosphate cement implants in rats, *J. Biomed. Mater. Res. A*, **87A**, 780–791.

557. Perrier, M., Lu, Y., Nemke, B., Kobayashi, H., Peterson, A., and Markel, M. (2008) Acceleration of second and fourth metatarsal fracture healing with recombinant human bone morphogenetic protein-2/calcium phosphate cement in horses, *Vet. Surg.*, **37**, 648–655.

558. Meraw, S. J., Reeve, C. M., Lohse, C. M., and Sioussat, T. M. (2000) Treatment of perimplant defects with combination growth factor cement, *J. Periodont.*, **71**, 8–13.

559. Liu, H., Zang, X. F., Zhao, Z. P., Wang, J. L., and Mi, L. (2008) Co-transplantation of exogenous nerve growth factor and calcium phosphate cement composite for repairing rabbit radial bone defects, *J. Clin. Rehabil. Tissue Eng. Res.*, **12**, 8037–8041.

560. Qu, X. Y., Jiang, D. M., Li, M., Zhang, D. W., Qin, J. Q., and Liu, C. K. (2008) Deproteinized osteoarticular allografts integrated with calcium phosphate cement and recombinant human vascular endothelial cell growth factor plus recombinant human bone morphogenetic protein-2, an immunological study, *J. Clin. Rehabil. Tissue Eng. Res.*, **12**, 8067–8070.

561. Yu, T., Ye, J., Gao, C., Yu, L., and Wang, Y. (2010) Synthesis and drug delivery property of calcium phosphate cement with special crystal morphology, *J. Am. Ceram. Soc.*, **93**, 1241–1244.

562. Stallmann, H. P., de Roo, R., Faber, C., Amerongen, A. V. N., and Wuisman, P. I. J. M. (2008) *In vivo* release of the antimicrobial peptide hLFi-11 from calcium phosphate cement, *J. Orthop. Res.*, **26**, 531–538.

563. Sasaki, T., Ishibashi, Y., Katano, H., Nagumo, A., and Toh, S. (2005) *In vitro* elution of vancomycin from calcium phosphate cement, *J. Arthroplasty*, **20**, 1055–1059.

564. Blom, E. J., Klein-Nulend, J., Wolke, J. G. C., van Waas, M. A. J., Driessens, F. C. M., and Burger, E. H. (2002) Transforming growth factor-β1 incorporation in a calcium phosphate bone cement, Material properties and release characteristics, *J. Biomed. Mater. Res.*, **59**, 265–272.

565. Blom, E. J., Klein-Nulend, J., Yin, L., van Waas, M. A. J., and Burger, E. H. (2001) Transforming growth factor-β1 incorporated in calcium phosphate cement stimulates osteotransductivity in rat calvarial bone defects, *Clin. Oral Implants Res.*, **12**, 609–616.

566. Link, D. P., van den Dolder, J., van den Beucken, J. J., Wolke, J. G. C., Mikos, A. G., and Jansen, J. A. (2008) Bone response and mechanical strength

of rabbit femoral defects filled with injectable CaP cements containing TGF-β1 loaded gelatin microparticles, *Biomaterials*, **29**, 675–682.

567. Habraken, W. J. E. M., Boerman, O. C., Wolke, J. G. C., Mikos, A. G., and Jansen, J. A. (2009) *In vitro* growth factor release from injectable calcium phosphate cements containing gelatin microspheres, *J. Biomed. Mater. Res. A*, **91A**, 614–622.

568. Ruhé, P. Q., Boerman, O. C., Russel, F. G. M., Mikos, A. G., Spauwen, P. H. M., and Jansen, J. A. (2006) *In vivo* release of rhBMP-2 loaded porous calcium phosphate cement pretreated with albumin, *J. Mater. Sci. Mater. Med.*, **17**, 919–927.

569. Naito, K., Obayashi, O., Mogami, A., Itoi, A., and Kaneko, K. (2008) Fracture of the calcium phosphate bone cement which used to enchondroma of the hand, a case report, *Eur. J. Orthop. Surg. Traumatol.*, **18**, 405–408.

570. Blattert, T. R., Delling, G., and Weckbach, A. (2003) Evaluation of an injectable calcium phosphate cement as an autograft substitute for transpedicular lumbar interbody fusion: a controlled, prospective study in the sheep model, *Eur. Spine J.*, **12**, 216–223.

571. Cavalcanti, S. C., Santos, S. C., Pereira, C. L., Mazzonetto, R., de Moraes, M., and Moreira, R. W. F. (2008) Histological and histomorphometric analyses of calcium phosphate cement in rabbit calvaria, *J. Cran. Maxillofac. Surg.*, **36**, 354–359.

572. Sanchez-Sotelo, J., Munuera, L., and Madero, R. (2000) Treatment of fractures of the distal radius with a remodellable bone cement: a prospective, randomised study using Norian SRS®, *J. Bone Joint Surg. Br.*, **82B**, 856–863.

573. Lobenhoffer, P., Gerich, T., Witte, F., and Tscherne, H. (2006) Use of an injectable calcium phosphate bone cement in the treatment of tibial plateau fractures: a prospective study of twenty-six cases with twenty-month mean follow-up, *J. Orthop. Trauma*, **16**, 143–149.

574. Cassidy, C., Jupiter, J. B., Cohen, M., Delli-Santi, M., Fennell, C., Leinberry, C., Husband, J., Ladd, A., Seitz, W. R., and Constantz, B. R. (2003) Norian SRS® cement compared with conventional fixation in distal radial fractures, a randomized study, *J. Bone Joint Surg. Am.*, **85A**, 2127–2137.

575. Schmidt, R., Cakir, B., Mattes, T., Wegener, M., Puhl, W., and Richter, M. (2005) Cement leakage during vertebroplasty, an underestimated problem? *Eur. Spine J.*, **14**, 466–473.

576. Vlad, M. D., Torres, R., López, J., Barracó, M., Moreno, J. A., and Fernández, E. (2007) Does mixing affect the setting of injectable bone cement? An ultrasound study, *J. Mater. Sci. Mater. Med.*, **18**, 347–352.

577. Krebs, J., Aebli, N., Goss, B. G., Sugiyama, S., Bardyn, T., Boecken, I., Leamy, P. J., and Ferguson, S. J. (2007) Cardiovascular changes after pulmonary embolism from injecting calcium phosphate cement, *J. Biomed. Mater. Res. B* (*Appl. Biomater.*), **82B**, 526–532.

578. Russell, T. A., and Leighton, R. K. (2008) Comparison of autogenous bone graft and endothermic calcium phosphate cement for defect augmentation in tibial plateau fractures. A multicenter, prospective, randomized study, *J. Bone Joint Surg. Am.*, **90A**, 2057–2061.

579. Dickson, K. F., Friedman, J., Buchholz, J. G., and Flandry, F. D. (2002) The use of BoneSource^{TM} hydroxyapatite cement for traumatic metaphyseal bone void filling, *J. Trauma*, **53**, 1103–1108.

580. Jungbluth, P., Hakimi, M., Grassmann, J. P., Schneppendahl, J., Kessner, A., Sager, M., Hakimi, A. R., Becker, J., Windolf, J., and Wild, M. (2010) The progress of early phase bone healing using porous granules produced from calcium phosphate cement, *Eur. J. Med. Res.*, **15**, 196–203.

581. Yoshikawa, T., Suwa, Y., Ohgushi, H., Tamai, S., and Ichijima, K. (1996) Self-setting hydroxyapatite cement as a carrier for bone-forming cells, *Biomed. Mater. Eng.*, **6**, 345–351.

582. Simon, Jr. C. G., Guthrie, W. F., and Wang, F. W. (2004) Cell seeding into calcium phosphate cement, *J. Biomed. Mater. Res. A*, **68A**, 628–639.

583. Xu, H. H. K., Weir, M. D., and Simon, Jr. C. G. (2008) Injectable and strong nano-apatite scaffolds for cell/growth factor delivery and bone regeneration, *Dent. Mater.*, **24**, 1212–1222.

584. Takagi, S., Chow, L. C., Hirayama, S., and Sugawara, A. (2003) Premixed calcium phosphate cement pastes, *J. Biomed. Mater. Res. B* (*Appl. Biomater.*), **67B**, 689–696.

585. Carey, L. E., Xu, H. H. K., Simon, Jr. C. G., Takagi, S., and Chow, L. C. (2005) Premixed rapid-setting calcium phosphate composites for bone repair, *Biomaterials*, **26**, 5002–5014.

586. Xu, H. H. K., Carey, L. E., Simon, Jr. C. G., Takagi, S., and Chow, L. C. (2007) Premixed calcium phosphate cements: synthesis, physical properties, and cell cytotoxicity, *Dent. Mater.*, **23**, 433–441.

587. Shimada, Y., Chow, L. C., Takagi, S., and Tagami, J. (2010) Properties of injectable apatite-forming premixed cements, *J. Res. Natl. Inst. Stand. Technol.*, **115**, 233–241.

588. Sugawara, A., Fujikawa, K., Hirayama, S., Takagi, S., and Chow, L. C. (2010) *In vivo* characteristics of premixed calcium phosphate cements when implanted in subcutaneous tissues and periodontal bone defects, *J. Res. Natl. Inst. Stand. Technol.*, **115**, 277–290.

589. Rajzer, I., Castaño, O., Engel, E., and Planell, J. A. (2010) Injectable and fast resorbable calcium phosphate cement for body-setting bone grafts, *J. Mater. Sci. Mater. Med.*, **21**, 2049–2056.

590. Han, B., Ma, P. W., Zhang, L. L., Yin, Y. J., Yao, K. D., Zhang, F. J., Zhang, Y. D., Li, X. L., and Nie, W. (2009) β-TCP/MCPM-based premixed calcium phosphate cements, *Acta Biomater.*, **5**, 3165–3177.

591. Chow, and L. C., and Takagi, S. (2006) Premixed self-hardening bone graft pastes. US Patent Application No. 20060263443, November 23, 2006.

592. Aberg, J., Brisby, H., Henriksson, H. B., Lindahl, A., Thomsen, P., and Engqvist, H. (2010) Premixed acidic calcium phosphate cement: characterization of strength and microstructure, *J. Biomed. Mater. Res. B (Appl. Biomater.)*, **93**, 436–441.

593. Lemaître, J., Pittet, C., and Brendlen, D. (2008) Pasty or liquid multiple constituent compositions for injectable calcium phosphate cements. US Patent No. 7407542, May 8, 2008.

594. Chow, L. C., and Takagi, S. (2007) Dual-phase cement precursor systems for bone repair. US Patent Application No. 20070092580, April 26, 2007.

595. Bohner, M. (2001) Calcium phosphate emulsions: possible applications, *Key Eng. Mater.*, 192–195, 765–768.

596. Troczynski, T. (2004) A concrete solution, *Nature Mater.*, **3**, 13–14.

597. Xu, H. H. K., Burguera, E. F., and Carey, L. E. (2007) Strong, macroporous and *in situ*-setting calcium phosphate cement-layered structures, *Biomaterials*, **28**, 3786–3796.

598. Xu, H. H. K., Takagi, S., Quinn, J. B., and Chow, L. C. (2004) Fast-setting calcium phosphate scaffolds with tailored macropore formation rates for bone regeneration, *J. Biomed. Mater. Res. A*, **68A**, 725–734.

599. Ginebra, M. P., Rilliard, A., Fernández, E., Elvira, C., san Roman, J., and Planell, J. A. (2001) Mechanical and rheological improvement of a calcium phosphate cement by the addition of a polymeric drug, *J. Biomed. Mater. Res.*, **57**, 113–118.

600. García-Fernández, L., Halstenberg, S., Unger, R. E., Aguilar, M. R., Kirkpatrick, C. J., and san Román, J. (2010) Anti-angiogenic activity of heparin-like polysulfonated polymeric drugs in 3D human cell culture, *Biomaterials*, **31**, 7863–7872.

601. Andriotis, O., Katsamenis, O. L., Mouzakis, D. E., and Bouropoulos, N. (2010) Preparation and characterization of bioceramics produced from calcium phosphate cements, *Cryst. Res. Technol.*, **45**, 239–243.

602. Steffen, T., Stoll, T., Arvinte, T., and Schenk, R. K. (2001) Porous tricalcium phosphate and transforming growth factor used for anterior spine surgery, *Eur. Spine J.*, **10**, S132–S140.

603. Guo, H., Su, J., Wei, J., Kong, H., and Liu, C. (2009) Biocompatibility and osteogenicity of degradable Ca-deficient hydroxyapatite scaffolds from calcium phosphate cement for bone tissue engineering, *Acta Biomater.*, **5**, 268–278.

604. Guo, H., Wei, J., Kong, H., Liu, C., and Pan, K. (2008) Biocompatibility and osteogenesis of calcium phosphate cement scaffolds for bone tissue engineering, *Adv. Mater. Res.*, 47–50, 1383–1386.

605. Weir, M. D., Xu, H. H. K., and Simon, Jr. C. G. (2006) Strong calcium phosphate cement-chitosan-mesh construct containing cell-encapsulating hydrogel beads for bone tissue engineering, *J. Biomed. Mater. Res. A*, **77A**, 487–496.

606. Xu, J. H., Tan, W. Q., and Lin, J. (2007) Repair of madibular bone defect by combining calcium phosphate cement with bone morphogenetic protein composite as a bone graft material, *Chinese J. Biomed. Eng.*, **26**, 153–156.

607. Niikura, T., Tsujimoto, K., Yoshiya, S., Tadokoro, K., Kurosaka, M., and Shiba, R. (2007) Vancomycin-impregnated calcium phosphate cement for methicillin-resistant staphylococcus aureus femoral osteomyelitis, *Orthopedics*, **30**, 320–321.

608. Lode, A., Wolf-Brandstetter, C., Reinstorf, A., Bernhardt, A., König, U., Pompe, W., and Gelinsky, M. (2007) Calcium phosphate bone cements, functionalized with VEGF: release kinetics and biological activity, *J. Biomed. Mater. Res. A*, **81A**, 474–483.

609. Yoshikawa, M., and Toda, T. (2004) *In vivo* estimation of periapical bone reconstruction by chondroitin sulfate in calcium phosphate cement, *J. Eur. Ceram. Soc.*, **24**, 521–531.

610. Wang, J. L., Mi, L., Hou, G. H., and Zheng, Z. (2008) Repair of radial defects using calcium phosphate cements/poly lactic-co-glycolic acid materials combined with mesenchymal stem cells in rabbits, *J. Clin. Rehabil. Tissue Eng. Res.*, **12**, 8001–8005.

611. Zhao, L., Weir, M. D., and Xu, H. H. K. (2010) Human umbilical cord stem cell encapsulation in calcium phosphate scaffolds for bone engineering, *Biomaterials*, **31**, 3848–3857.

612. Ding, T., Yang, H., Maltenfort, M., and Xie, R. (2010) Silk fibroin added to calcium phosphate cement to prevent severe cardiovascular complications, *Case Rep. Clin. Pract. Rev.*, **16**, 23–26.

613. Panzavolta, S., Torricelli, P., Bracci, B., Fini, M., and Bigi, A. (2010) Functionalization of biomimetic calcium phosphate bone cements with alendronate, *J. Inorg. Biochem.*, **104**, 1099–1106.

614. Xu, H. H. K., Zhao, L., Detamore, M. S., Takagi, S., and Chow, L. C. (2010) Umbilical cord stem cell seeding on fast-resorbable calcium phosphate bone cement, *Tiss. Eng. A*, **16**, 2743–2753.

615. Li, M., Liu, X., Liu, X., and Ge, B. (2010) Calcium phosphate cement with BMP-2-loaded gelatin microspheres enhances bone healing in osteoporosis: a pilot study, *Clin. Orthop. Rel. Res.*, **468**, 1978–1985.

616. Weir, M. D., and Xu, H. H. K. (2010) Human bone marrow stem cell-encapsulating calcium phosphate scaffolds for bone repair, *Acta Biomater.*, **6**, 4118–4126.

617. Zhao, L., Weir, M. D., and Xu, H. H. K. (2010) An injectable calcium phosphate-alginate hydrogel-umbilical cord mesenchymal stem cell paste for bone tissue engineering, *Biomaterials*, **31**, 6502–6510.

618. dos Santos, L. A., Carrodéguas, R. G., Rogero, S. O., Higa, O. Z., Boschi, A. O., and de Arruda A. C. (2002) Alpha-tricalcium phosphate cement: "*in vitro*" cytotoxicity, *Biomaterials*, **23**, 2035–2042.

619. Baroud, G., and Steffen, T. (2005) A new cannula to ease cement injection during vertebroplasty, *Eur. Spine J.*, **14**, 474–479.

620. Anderson, J. M. (2006) The future of biomedical materials, *J. Mater. Sci. Mater. Med.*, **17**, 1025–1028.

Chapter 6

Calcium Orthophosphate-Based Biocomposites and Hybrid Biomaterials

Nomenclature

A-W	apatite-wollastonite
BMP	bone morphogenetic protein
BSA	bovine serum albumin
EVOH	a copolymer of ethylene and vinyl alcohol
IBS	injectable bone substitute
HDPE	high-density polyethylene
HIPS	high impact polystyrene
HPMC	hydroxypropylmethylcellulose
PAA	polyacrylic acid
PBT	polybutyleneterephthalate
PCL	poly(ε-caprolactone)
PDLLA	poly(D,L-lactic acid)
PE	polyethylene
PEEK	polyetheretherketone
PEG	polyethylene glycol
PGA	polyglycolic acid
PHB	polyhydroxybutyrate

Calcium Orthophosphates: Applications in Nature, Biology, and Medicine
Sergey Dorozhkin

ISBN 978-981-4316-62-0 (Hardcover), 978-981-4364-17-1 (eBook)
www.panstanford.com

PHBHV	poly(hydroxybutyrate-*co*-hydroxyvalerate)
PHEMA	polyhydroxyethyl methacrylate
PHV	polyhydroxyvalerate
PLA	polylactic acid
PLGA	poly(lactic-*co*-glycolic) acid
PLGC	*co*-polyester lactide-*co*-glycolide-*co*-ε-caprolactone
PLLA	poly(L-lactic acid)
PMMA	polymethylmethacrylate
PP	polypropylene
PPF	poly(propylene-*co*-fumarate)
PS	polysulfone
PSZ	partially stabilized zirconia
PTFE	polytetrafluoroethylene
PVA	polyvinyl alcohol
PVAP	polyvinyl alcohol phosphate
SEVA-C	a blend of EVOH with starch
UHMWPE	ultrahigh molecular weight polyethylene

6.1 Introduction

The fracture of bones due to various traumas or natural aging is a typical type of a tissue failure. An operative treatment frequently requires implantation of a temporary or a permanent prosthesis, which still is a challenge for orthopedic surgeons, especially in the cases of large bone defects. A fast aging of the population and serious drawbacks of natural bone grafts make the situation even worse; therefore, there is a high clinical demand for bone substitutes. Unfortunately, a medical application of xenografts (e.g., bovine bone) is generally associated with potential viral infections. In addition, xenografts have a low osteogenicity, an increased immunogenicity and, usually, resorb more rapidly than autogenous bone. Similar limitations are also valid for human allografts (i.e., tissue transplantation between individuals of the same species but of non-identical genetic composition), where the concerns about potential risks of transmitting tumor cells, a variety of bacterial and viral infections, as well as immunological and blood group incompatibility are even stronger [1–3]. Moreover, harvesting and conservation of allografts (exogenous bones) are additional limiting factors. Autografts (endogenous bones) are still the "golden standard" among any substitution materials because they are osteogenic, osteoinductive, osteoconductive, completely biocompatible, non-toxic and do not cause any immunological problems (non-allergic). They contain viable osteogenic cells, bone matrix proteins and support bone growth. Usually, autografts are well accepted by the body and rapidly integrated into the surrounding bone tissues. Due to these reasons, they are used routinely for a long period with good clinical results [3–6]; however, it is fair to say on complication cases, those frequently happened in the past [7, 8]. Unfortunately, a limited number of donor sites restrict the quantity of autografts harvested from the iliac crest or other locations of the patient's own body. In addition, their medical application is always associated with additional traumas and scars resulting from the extraction of a donor tissue during a superfluous surgical operation, which requires further healing at the donation site and can involve long-term postoperative pain [1, 8–11]. Thus, any types of biologically derived transplants appear to be imperfect solutions, mainly due to a restricted quantity of donor tissues, donor site morbidity, as well as potential risks of an immunological incompatibility and disease transfer [9, 11, 12]. In this light, manmade materials (alloplastic or synthetic bone grafts) stand out as a reasonable option because they are easily available, might be processed and modified to suit the specific needs of a given application [13–15]. What's more, there are no concerns about potential infections, immunological incompatibility, sterility and donor site morbidity. Therefore, investigations on artificial materials for bone tissue repair appear to be one of the key subjects in the field of biomaterials research for clinical applications [16].

Currently, there are several classes of synthetic bone grafting biomaterials for *in vivo* applications [17–21]. The examples include natural coral, coral-derived materials, bovine porous demineralized bone, human demineralized bone matrix, bioactive glasses, glass-ceramics and calcium orthophosphates [11]. All of these biomaterials are biocompatible and osteoconductive, guiding bone tissue from the edges toward the center of the defect, and aim to provide a scaffold of interconnected pores with pore dimensions ranging from 200 μm [22, 23] to 2 mm [24], to facilitate tissue and vessel ingrowths. Among them, porous bioceramics made of calcium orthophosphates appear to be very prominent due to both the excellent biocompatibility and bonding ability to living bone in the body. This is directly related to the fact that the inorganic material of mammalian calcified tissues, i.e. of bone and teeth, consists of calcium orthophosphates [25–27]. Due to this reason, other artificial materials are normally encapsulated by fibrous tissue, when implanted in body defects, while calcium orthophosphates are not [28]. Many types of calcium orthophosphate-based bioceramics with different chemical composition are already on the market (Tables 4.2 and 5.2). Unfortunately, as for any ceramic material, calcium orthophosphate bioceramics by itself lack the mechanical and elastic properties of the calcified tissues. Namely, scaffolds made of calcium orthophosphates only suffer from a low elasticity, a high brittleness, a poor tensile strength, a low mechanical reliability and fracture toughness, which leads to various concerns about their mechanical performance after implantation [29–31]. Besides, in many cases, it is difficult to form calcium orthophosphate bioceramics into the desired shapes.

The superior strength and partial elasticity of biological calcified tissues (e.g., bones) are due to the presence of bioorganic polymers (mainly, collagen type I fibers [32]) rather than to a natural ceramic (mainly, a poorly crystalline ion-substituted CDHA, often referred to as "biological apatite") phase [34, 35]. The elastic collagen fibers are aligned in bone along the main stress directions. The biochemical composition of bones is given in Table 6.1 [36]. A decalcified bone becomes very flexible being easily twisted, whereas a bone without collagen is very brittle; thus, the inorganic nano-sized crystals of biological apatite provide with the hardness and stiffness, while the bioorganic fibers are responsible for the elasticity and toughness [26, 37]. In bones, both types of materials integrate each other into a nanometric scale in such a way that the crystallite size, fibers orientation, short-range order between the components, etc., determine its nanostructure and therefore the function and mechanical properties of the entire composite [33, 38–42]. From the mechanical point of view, bone is a tough material at low strain rates but fractures more like a brittle material at high strain rates; generally, it is rather weak in tension and shear, particularly along the longitudinal plane. Besides, bone is an anisotropic material because its properties are directionally dependent [25, 26, 31]. For further details, see section *1.4* of this book.

Table 6.1 The biochemical composition* of bones [36].

Inorganic phases	Wt.%	Bioorganic phases	Wt. %
Calcium orthophosphates (biological apatite)	~ 60	Collagen type I	~ 20
Water	~ 9	Non-collagenous proteins: osteocalcin, osteonectin, osteopontin, thrombospondin, morphogenetic proteins, sialoprotein, serum proteins	~ 3
Carbonates	~ 4	Other traces: polysaccharides, lipids, cytokines	balance
Citrates	~ 0.9	Primary bone cells: osteoblasts, osteocytes, osteoclasts	balance
Sodium	~ 0.7		
Magnesium	~ 0.5		
Other traces: Cl^-, F^-, K^+ Sr^{2+}, Pb^{2+}, Zn^{2+}, Cu^{2+}, Fe^{2+}	Balance		

* The composition varies from species to species and from bone to bone.

It remains a great challenge to design the ideal bone graft that emulates nature's own structures or functions. Certainly, the successful design requires an appreciation of the structure of bone. According to expectations, the ideal bone graft should be benign, available in a variety of forms and sizes, all with sufficient mechanical properties for use in load-bearing sites, form a chemical bond at the bone/implant interface, as well as be osteogenic, osteoinductive, osteoconductive, biocompatible, completely biodegradable at the expense of bone growth and moldable to fill and restore bone defects [29, 40, 43]. Further, it should resemble the chemical composition of bones (thus, the presence of calcium orthophosphates is mandatory), exhibit contiguous porosity to encourage invasion by the live host tissue, as well as possess both viscoelastic and semi-brittle behavior, as bones do [44–47]. Moreover, the degradation kinetics of the ideal implant should be adjusted to the healing rate of the human tissue with absence of any chemical or biological irritation and/or toxicity caused by substances, which are released due to corrosion or degradation. Ideally, the combined mechanical strength of the implant and the ingrowing bone should remain constant throughout the regenerative process. Furthermore, the substitution implant material should not disturb significantly the stress environment of the surrounding living tissue [48]. Finally, there is an opinion, that in the case of a serious trauma, bone should fracture rather than the implant [29]. A good sterilizability, storability and processability, as well as a relatively low cost are also of a great importance to permit a clinical application. Unfortunately, no artificial biomaterial is yet available, which embodies all these requirements and unlikely it will appear in the nearest future. To date, most of the available biomaterials appear to be either predominantly osteogenic or osteoinductive or else purely osteoconductive [2].

Careful consideration of the bone type and mechanical properties are needed to design bone substitutes. Indeed, in high load-bearing bones such as the femur, the stiffness of the implant needs to be adequate, not too stiff to result in strain shielding, but rigid enough to present stability. However, in relatively low load-bearing applications such as cranial bone repairs, it is more important to have stability and the correct three-dimensional shapes for aesthetic reasons. One of the most promising alternatives is to apply materials with similar composition and nanostructure to that of bone tissue [40]. Mimicking the structure of calcified tissues and addressing the limitations of the individual materials, development of organic–inorganic hybrid biomaterials provides excellent possibilities for improving the conventional bone implants. In this sense, suitable biocomposites of tailored physical, biological and mechanical properties with the predictable degradation behavior can be prepared combining biologically relevant calcium orthophosphates with bioresorbable polymers [49, 50]. As a rule, the general behavior of these bioorganic/calcium orthophosphate biocomposites is dependent on nature, structure and relative contents of the constitutive components, although other parameters such as the preparation conditions also determine the properties of the final materials. Currently, biocomposites with calcium orthophosphates incorporated as either a filler or a coating (or both) either into or onto a biodegradable polymer matrix, in the form of particles or fibers, are increasingly considered for using as bone tissue engineering scaffolds due to their improved physical, biologic and mechanical properties [51–57]. In addition, such biocomposites could fulfill general requirements to the next generation of biomaterials, those should combine the bioactive and bioresorbable properties to activate *in vivo* mechanisms of tissue regeneration, stimulating the body to heal itself and leading to replacement of the implants by the regenerating tissue [50, 58, 59]. Thus, through the successful combinations of ductile polymer matrixes with hard and bioactive particulate bioceramic fillers, optimal materials can be designed and, ideally, this approach could lead to a superior construction to be used as either implants or posterior dental restorative material [60].

A lint-reinforced plaster was the first composite used in clinical orthopedics as an external immobilizer (bandage) in the treatment of bone fracture by Mathijsen in 1852 [61], followed by Dreesman in 1892 [62]. A great progress in the clinical application of various types of composite materials has been achieved since then. Based on the past experience and newly gained knowledge, various composite materials with tailored mechanical and biological performance can be manufactured and used to meet various clinical requirements [63]. However, this chapter presents only a brief history and advances in the field of calcium orthophosphate-based biocomposites and hybrid biomaterials suitable for biomedical application. The majority of the reviewed literature is restricted to the recent publications; a limited number of papers published in the twentieth century have been cited. Various aspects of the material constituents, fabrication technologies, structural and bioactive properties, as well as phase interaction have been considered and discussed in details. Finally, several critical issues and scientific challenges that are needed for further advancement are outlined.

6.2 General Information on Composites and Biocomposites

According to Wikipedia, the free encyclopedia, "*composite materials* (or *composites* for short) are engineered materials made from two or more constituent materials with significantly different physical or chemical properties and which remain separate and distinct on a macroscopic level within the finished structure" [64]. Thus, composites are always heterogeneous. Furthermore, the phases of any composite retain their identities and properties, and are bonded, which is why an interface is maintained between them. This provides improved specific or synergistic characteristics that are not obtainable by any of the original phases alone [65]. Following the point of view of some predecessors, we also consider that "for the purpose of this review, composites are defined as those having a distinct phase distributed through their bulk, as opposed to modular or coated components" [66, p. 1329]. For this reason, with a few important exceptions, the structures obtained by soaking of various materials in supersaturated solutions containing ions of calcium and orthophosphate (e.g., Refs. [67–73]), those obtained by coating of various materials by calcium orthophosphates (e.g., Refs. [74–82]), as well as calcium orthophosphates coated by other compounds [83–87] have not been considered; however, composite coatings have been considered. Occasionally, porous calcium orthophosphate scaffolds filled by cells inside the pores [88–91], as well as calcium orthophosphates impregnated by biologically active substances [92, 93] are also defined as composites and/or hybrids; nevertheless, such structures have not been considered in this review either.

In any composite, there are two major categories of constituent materials: a matrix (or a continuous phase) and (a) dispersed phase(s). To create a composite, at least one portion of each type is required. General information on the major fabrication and processing techniques might be found elsewhere [66, 94]. The continuous phase is responsible for filling the volume, as well as it surrounds and supports the dispersed material(s) by maintaining their relative positions. The dispersed phase(s) is(are) usually responsible for enhancing one or more properties of the matrix. Most of the composites target an enhancement of mechanical properties of the matrix, such as stiffness and strength; however, other properties, such as erosion stability, transport properties (electrical or thermal), radiopacity, density or biocompatibility might also be of a great interest. This synergism produces the properties, which are unavailable from the individual constituent materials [94, 95]. What's more, by controlling the volume fractions and local and global arrangement of the dispersed phase, the properties and design of composites can be varied and tailored to suit the necessary conditions. For example, in the case of ceramics, the dispersed phase serves to impede crack growth. In this case, it acts as reinforcement. A number of methods, including deflecting crack tips, forming bridges across crack faces, absorbing energy during pullout and causing a redistribution of stresses in regions, adjacent to crack tips, can be used to accomplish this [96]. Other factors to be considered in composites are the volume fraction of (a) dispersed phase(s), its(their) orientation and homogeneity of the overall composite. For example, higher volume fractions of reinforcement phases tend to improve the mechanical properties of the composites, while continuous and aligned fibers best prevent crack propagation with the added property of anisotropic behavior. Furthermore, the uniform distribution of the dispersed phase is also desirable, as it imparts consistent properties to the composite [64, 94, 95].

In general, composites might be simple, complex, graded and hierarchical. The term "a simple composite" is referred to the composites those result from the homogeneous dispersion of one dispersed phase throughout a matrix. The term "a complex composite" is referred to the composites those result from the homogeneous dispersion of several dispersed phases throughout one matrix. The term "a graded composite" is referred to the composites those result from the intentionally structurally inhomogeneous dispersion of one or several dispersed phases throughout one matrix. The term "a hierarchical composite" is referred to the cases, when fine entities of either a simple or a complex composite is somehow aggregated to form coarser ones (e.g., granules or particles) which afterwards are dispersed inside another matrix to produce the second hierarchical scale of the composite structure. There is another set of four types of composites: (i) fibrous composites, where the fibers are in a matrix; (ii) laminar composites, in which the phases are in layers; (iii) particulate composites, where the particles or flakes are in a matrix; and (iv) hybrid composites, which are combinations of any of the above. Still other classification type of the available composites is based on the matrix materials (metals, ceramics and polymers) [63].

In most cases, three interdependent factors must be considered in designing of any composite: (i) a selection of the suitable matrix and dispersed materials, (ii) a choice of appropriate fabrication and processing methods, (iii) both internal and external design of the device itself [66]. Furthermore, any composite must be formed to shape. To do this, the matrix material can be added before or after the dispersed material has been placed into a mold cavity or onto the mold surface. The matrix material experiences a melding event that, depending upon the nature of the matrix material, can occur in various ways such as chemical polymerization, setting, curing or solidification from a melted state. Due to a general inhomogeneity, the physical properties of many composite materials are not isotropic but rather orthotropic (i.e., there are different properties or strengths in different orthogonal directions) [64, 94, 95].

In order to prepare any type of a composite, at least two different materials must be mixed. Thus, a phase miscibility phenomenon appears to be of the paramount importance [97, 98]. Furthermore, the interfacial strength among the phases is a very important factor because a lack of adhesion among the phases will result in an early failure at the interface and thus in a decrease in the mechanical properties, especially the tensile strength. From a chemical point of view, we can distinguish several types of the interaction among the composite components: materials with strong (covalent, coordination, ionic) interactions; those with weak interactions (van der Waals forces, hydrogen bonds, hydrophilic–hydrophobic balance) or without chemical interactions among the components [99]. Wetting is also important in bonding or adherence of the materials. It depends on the hydrophilicity or polarity of the filler(s) and the available polar groups of the matrix.

Biocomposites are defined as nontoxic composites able to interact well with the human body *in vivo* and, ideally, contain one or more component(s) that stimulate(s) the healing process and uptake of the implant [100]. Thus, for biocomposites the biological compatibility appears to be more important than any other type of compatibility [63, 101, 102]. Interestingly, but, according to the databases, the first paper with the term "biocomposite" in the title was published in 1987 [103] and the one containing a combination of terms "biocomposite" and HA in the title was published in 1991 [104]. Thus, this subject appears to be quite new. The most common properties from the bioorganic and inorganic domains to be combined in biocomposites have been summarized in Table 6.2 [40]. For general advantages of the modern calcium orthophosphate-based biocomposites over calcium orthophosphate bioceramics and bioresorbable polymers individually, the interested readers are advised to get through "Composite materials strategy" section of Ref. [50].

Table 6.2 General respective properties from the bioorganic and inorganic domains, to be combined in various composites and hybrid materials [40].

Inorganic	Bioorganic
Hardness, brittleness	Elasticity, plasticity
High density	Low density
Thermal stability	Permeability
Hydrophilicity	Hydrophobicity
High refractive index	Selective complexation
Mixed valence slate (red-ox)	Chemical reactivity
Strength	Bioactivity

6.3 The Major Constituents of Biocomposites and Hybrid Biomaterials for Bone Grafting

6.3.1 Calcium Orthophosphates

The main driving force behind the use of calcium orthophosphates as bone substitute materials is their chemical similarity to the mineral component of mammalian bones and teeth [25–27]. As a result, in addition to being non-toxic, they are biocompatible, not recognized as foreign materials in the body and, most importantly, both exhibit bioactive behavior and integrate into living tissue by the same processes active in remodeling healthy bone. This leads to an intimate physicochemical bond between the implants and bone, termed osteointegration [105]. More to the point, calcium orthophosphates are also known to support osteoblast adhesion and proliferation [106, 107]. Even so, the major limitations to use calcium orthophosphates as load-bearing biomaterials are their mechanical properties; namely, they are brittle with poor fatigue resistance [29–31]. The poor mechanical behavior is even more evident for highly porous ceramics and scaffolds because porosity greater than 100 μm is considered as the requirement for proper vascularization and bone cell colonization [108–110]. That is why, in biomedical applications calcium orthophosphates are used primarily as fillers and coatings [27].

The complete list of known calcium orthophosphates, including their standard abbreviations and the major properties, is given in Table 1.1, while the detailed information on calcium orthophosphates, their synthesis, structure, chemistry, other properties and biomedical application has been comprehensively reviewed in the previous chapters of this book. Even more thorough information on calcium orthophosphates might be found in other books and monographs [111–117], as well as in other chapters of this book.

6.3.2 Polymers

Polymers are a class of materials consisting of large molecules, often containing many thousands of small units, or monomers, joined together chemically to form one giant chain, thus creating very ductile materials. In this respect, polymers are comparable with major functional components of the biological environment: lipids, proteins and polysaccharides. They differ from each other in chemical composition, molecular weight, polydispersity, crystallinity, hydrophobicity, solubility and thermal transitions. Besides, their properties can be fine-tuned over a wide range by varying the type of polymer, chain length, as well as by copolymerization or blending of two or more polymers [118–120]. Opposite to ceramics, polymers exhibit substantial viscoelastic properties and easily can be fabricated into complex structures, such as sponge-like sheets, gels or complex structures with intricate porous networks and channels [121]. Being X-ray transparent and non-magnetic polymeric materials are fully compatible with the modern diagnostic methods such as computed tomography and magnetic resonance imaging. Unfortunately, most of them are unable to meet the strict demands of the *in vivo* physiological environment. Namely, the main requirements to polymers suitable for biomedical applications are that they must be biocompatible, not eliciting an excessive or chronic inflammatory response upon implantation and, for those that degrade, that they breakdown into non-toxic products only. Unfortunately, polymers, for the most part, lack rigidity, ductility and ultimate mechanical properties required in load bearing applications. Thus, despite their good biocompatibility, many of the polymeric materials are mainly used for soft tissue replacements (such as skin, blood vessel, cartilage, ligament replacement, etc.). Moreover, the sterilization processes (autoclave, ethylene oxide and ^{60}Co irradiation) may affect the polymer properties [122].

There is a variety of biocompatible polymers suitable for biomedical applications [123, 124]. For example, polyacrylates, poly(acrylonitrile-*co*-vinylchloride) and polylysine have been investigated for cell encapsulation and immunoisolation [125, 126]. Polyorthoesters and PCL have been investigated as drug delivery devices, the latter for long-term sustained release because of their slow degradation rates [127]. PCL is a hydrolytic polyester having appropriate resorption period and releases nontoxic byproducts upon degradation [128]. Other polyesters and PTFE are used for vascular tissue replacement. Polyurethanes are in use as coatings for pacemaker lead insulation and have been investigated for reconstruction of the meniscus [129, 130]. Polymers considered for orthopedic purposes include polyanhydrides, which have also been investigated as delivery devices (due to their rapid and well-defined surface erosion), for bone augmentation or replacement since they can be photopolymerized *in situ* [127, 131, 132]. To overcome their poor mechanical properties, they have been co-polymerized with imides or formulated to be cross-linkable *in situ* [132]. Other polymers, such as polyphosphazenes, can have their properties (e.g., degradation rate) easily modified by varying the nature of their side groups and have been shown to support osteoblast adhesion, which makes them candidate materials for skeletal tissue regeneration [132]. PPF has emerged as a good bone replacement material, exhibiting good mechanical properties (comparable to trabecular bone), possessing the capability to cross-link *in vivo* through the C=C bond and being hydrolytically degradable. It has also been examined as a material for drug delivery devices [127, 131–134]. Polycarbonates have been suggested as suitable materials to make scaffolds for bone replacement and have been modified with tyrosine-derived amino acids to render them biodegradable [127]. Polydioxanone has been also tested for biomedical applications [135]. PMMA is widely used in orthopedics, as a bone cement for implant fixation, as well as to repair certain fractures and bone defects, for example, osteoporotic vertebral bodies [136, 137]. However, PMMA sets by a polymerization of toxic monomers, which also evolves significant amounts of heat that damages tissues. Moreover, it is neither degradable nor bioactive, does not bond chemically to bones, and might generate particulate debris leading to an inflammatory foreign body response [131, 138].

A number of other non-degradable polymers applied in orthopedic surgery include PE in its different modifications such as low density PE, HDPE, and UHMWPE (used as the articular surface of total hip replacement implants [139, 140]), polyethylene terepthalate, PP, and PTFE which are applied to repair knee ligaments [141]. Polyactive™, a block copolymer of PEG and PBT, was also considered for biomedical application [142–147]. Cellulose [148, 149] and its esters [150, 151] are also popular. Finally yet importantly, polyethylene oxide, PHB and blends thereof have also been tested for biomedical applications [50].

Nonetheless, the most popular synthetic polymers used in medicine are the linear aliphatic poly(α-hydroxyesters) such as PLA, PGA and their copolymers—PLGA (Table 6.3). These materials have been extensively studied; they appear to be the only synthetic and biodegradable polymers with an extensive FDA approval history [50, 132, 152–156]. They are biocompatible, mostly non-inflammatory, as well as degrade *in vivo* through hydrolysis and possible enzymatic action into products that are removed from the body by regular metabolic pathways [49, 127, 132, 156–161]. Besides, they might be used for drug delivery purposes [162]. Poly(α-hydroxyesters) have been investigated as scaffolds for replacement and regeneration of a variety of tissues, cell carriers, controlled delivery devices for drugs or proteins (e.g., growth factors), membranes or films, screws, pins, and plates for orthopedic applications [127, 132, 153, 154, 156, 163–165]. Additionally, the degradation rate of PLGA can be adjusted by varying the amounts of the two component monomers (Table 6.3), which in orthopedic applications can be exploited to create materials that degrade in concert with bone ingrowth [160, 166]. Furthermore, PLGA is known to support osteoblast migration and proliferation [59, 132, 157, 167], which is a necessity for bone tissue regeneration. Unfortunately, such polymers on their own, though they reduce the effect of stress-shielding, are too weak to be used in load bearing situations and are only recommended in certain clinical indications, such as ankle and elbow fractures [156, 161]. In addition, they exhibit bulk degradation, leading to both a loss in mechanical properties and lowering of the local solution pH that accelerates further degradation in an autocatalytic manner. As the body is unable to cope with the vast amounts of implant degradation products, this might lead to an inflammatory foreign body response [132, 156, 163]. Finally, poly(α-hydroxyesters) do not possess the bioactive and osteoconductive properties of calcium orthophosphates [153, 168].

Table 6.3 Major properties of several FDA-approved biodegradable polymers [152].

Polymer	Thermal properties* (°C)	Tensile modulus (GPa)	Degradation time (months)
Polyglycolic acid (PGA)	t_g = 35–40 t_m = 225–230	7.06	6–12 (strength loss within 3 weeks)
L-polylactic acid (LPLA)	t_g = 60–65 t_m = 173–178	2.7	> 24
D,L-polylactic acid (DLPLA)	t_g = 55–60 amorphous	1.9	12–16
85/15 D,L-polylactic-*co*-glycolic acid (85/15 DLPLGA)	t_g = 50–55 amorphous	2.0	5–6
75/25 D,L-polylactic-*co*-glycolic acid (75/25 DLPLGA)	t_g = 50–55 amorphous	2.0	4–5
65/35 D,L-polylactic-*co*-glycolic acid (65/35 DLPLGA)	t_g = 45–50 amorphous	2.0	3–4
50/50 D,L-polylactic-*co*-glycolic acid (50/50 DLPLGA)	t_g = 45–50 amorphous	2.0	1–2
Poly(ε-caprolactone) (PCL)	t_g = (– 60) – (– 65) t_m = 58–63	0.4	> 24

*t_g: glass transition temperature; t_m: melting point.

Several classifications of the biomedically relevant polymers are possible. For example, some authors distinguish between synthetic polymers such as PLA, PGA or their copolymers with PCL and polymers of biological origin like polysaccharides (starch, alginate, chitin/chitosan [169–171], gelatin, cellulose, hyaluronic acid derivatives), proteins (soy, collagen, fibrin [11], silk) and a variety of biofibers, such as lignocellulosic natural fibers [10, 172, 173]. Natural polymers often posses highly organized structures and may contain an extracellular substance, called ligand, which is necessary to bind with cell receptors. However, they always contain various impurities, which should be getting rid off prior use. As synthetic polymers can be produced under the controlled conditions, they in general exhibit predictable and reproducible mechanical and physical properties such as tensile strength, elastic modulus and degradation rate. Control of impurities is a further advantage of synthetic polymers. Other authors differentiate between resorbable or biodegradable (e.g., poly(α-hydroxyesters), polysaccharides and proteins) and non-resorbable (e.g., PE, PP, PMMA and cellulose) polymers [60, 173]. Furthermore, polymeric materials can be broadly classified as thermoplastics and thermosets. For example, HDPE and PEEK are the examples of thermoplastics, while polydimethylsiloxane and PMMA are the examples of thermosets [122]. The list of synthetic biodegradable polymers used for biomedical application as scaffold materials is available as Table 1 in Ref. [173], while further details on polymers suitable for biomedical applications are available in literature [122, 165, 174–183] where the interested readers are referred. Good reviews on the synthesis of different biodegradable polymers [184], as well as on the experimental trends in polymer composites [185] are available elsewhere.

6.3.3 Inorganic Materials and Compounds

6.3.3.1 Metals

Titanium (Ti) is one of the best biocompatible metals and used most widely as implant [16, 186, 187]. Besides, there are other metallic implants made of pure Zr, Hf, V, Nb, Ta, Re [186], Ni, Fe, Cu [188–190], Ag, stainless steels and various alloys [190] suitable for biomedical application. Recent studies revealed even a greater biomedical potential of porous metals [191–194]. The metallic implants provide the necessary strength and toughness that are required in load-bearing parts of the body and, due to these advantages, metals will continue to play an important role as orthopedic biomaterials in the future, even though there are concerns with regard to the release of certain ions from and corrosion products of metallic implants. Of course, neither metals nor alloys are biomimetic (the term biomimetic can be defined as a processing technique that either mimics or inspires the biological mechanism, in part or whole [195]) in terms of chemical composition because there are no elemental metals in the human body. In addition, even biocompatible metals are bioinert: while not rejected by the human body, any metallic implants cannot actively interact with the surrounding tissues. Nevertheless, in some cases (especially when they are coated by calcium orthophosphates; however, that is another story) the metallic implants can show a reasonable biocompatibility [196]. Only permanent implants are made of metals and alloys, in which degradation or corrosion is not desirable. However, during recent years a number of magnesium implants have been proposed which are aimed to degrade in the body in order to make room for ingrowing bones [193, 197, 198].

6.3.3.2 Glasses and glass-ceramics

Special types of glasses and glass-ceramics are also suitable materials for biomedical applications [199–201] and a special Na_2O – CaO – SiO_2 – P_2O_5 glass named Bioglass® [13, 28, 30, 31, 202, 203] is the most popular among them. They are produced via standard glass production techniques and require pure raw materials. Bioglass® is a biocompatible and osteoconductive biomaterial. It bonds to bone without an intervening fibrous connective tissue interface and, due to these properties, it has been widely used for filling bone defects [204]. The primary shortcoming of Bioglass® is mechanical weakness and low fracture toughness due to an amorphous two-dimensional glass network. The bending strength of most Bioglass® compositions is in the range of 40–60 MPa, which is not suitable for major load-bearing applications. Making porosity in Bioglass®-based scaffolds is beneficial for even better resorption and bioactivity [205].

By heat treatment, a suitable glass can be converted into glass-crystal composites containing crystalline phase(s) of controlled sizes and contents. The resultant glass-ceramics can have superior mechanical properties to the parent glass as well as to sintered crystalline ceramics. The bioactive A-W glass-ceramics is made from the parent glass in the pseudoternary system $3CaO{\cdot}P_2O_5$ – $CaO{\cdot}SiO_2$ – $MgO{\cdot}CaO{\cdot}2SiO_2$, which is produced by a conventional melt-quenching method. The bioactivity of A-W glass-ceramics is much higher than that of sintered HA. It possesses excellent mechanical properties and has therefore been used clinically for iliac and vertebrae prostheses and as intervertebral spacers [16, 206–208].

6.3.3.3 Ceramics

Metal oxide ceramics, such as alumina (Al_2O_3, high purity, polycrystalline, fine grained), zirconia (ZrO_2), and some other oxides (e.g., TiO_2, SiO_2) have been widely studied due to their bioinertness, excellent tribological properties, high wear resistance, fracture toughness and strength, as well as a relatively low friction [16, 209]. Unfortunately, due to transformation from the tetragonal to the monoclinic phase, a volume change occurs when pure zirconia is cooled down, which causes cracking of the zirconia ceramics. Therefore, additives such as calcia (CaO), magnesia (MgO), and yttria (Y_2O_3) must be mixed with zirconia to stabilize the material in either the tetragonal or the cubic phase. Such material is called PSZ [210–212]. However, the brittle nature of any ceramics has limited their scope of clinical applications and hence more research needs to be conducted to improve their properties.

6.3.3.4 Carbon

Due to its bioinertness, excellent tribological properties, fracture toughness and strength, as well as a low friction, elemental carbon has been used as a biomaterial, at least, since 1972 [213]. Applications include orthopedic prostheses, vitreous carbon roots for replacement teeth, structural skeletal extensions, bone bridges and hip prostheses. Biomedical properties of amorphous carbon were studied as well [214]. However, current trends represent investigations on biomedical applications of carbon nanotubes [215, 216].

Carbon nanotubes with their small dimensions, a high aspect (length to diameter) ratio as well as the exceptional mechanical properties, including extreme flexibility and strength, significant resistance to bending, high resilience and the ability to reverse any buckling of the tube, have the excellent potential to accomplish necessary mechanical properties [217]. Recent studies have even suggested that they may possess some bioactivity [218–221]. However, non-functionalized carbon nanotubes tend to agglomerate and form bundles. Besides, they are soluble in neither water nor organic solvents. Luckily, chemical functionalization [82, 222] allows carbon nanotubes to be dispersed more easily, which can improve interfacial bonding with other components of the composites. Furthermore, functionalization of carbon nanotubes with carboxylic groups was found to confer a capacity to induce calcification similar to woven bones [223]. Interestingly, but carbon nanotubes might be functionalized by *in situ* deposition of CDHA on their surface [224].

6.4 Calcium Orthophosphate-Based Biocomposites and Hybrid Biomaterials

Generally, the use of calcium orthophosphate-based biocomposites and hybrid biomaterials for clinical applications includes several (partly overlapping) broad areas:

- biocomposites with polymers,
- cement-based biocomposites and concretes,
- biocomposites based on nanodimensional calcium orthophosphates and nanodimensional biocomposites,
- biocomposites with collagen,
- biocomposites with other bioorganic compounds and/or biological macromolecules,
- injectable bone substitutes (IBS),
- biocomposites with glasses, inorganic compounds, carbon and metals,
- functionally graded biocomposites,
- biosensors.

The details on each subject are given below.

6.4.1 Biocomposites with Polymers

Typically, the polymeric components of biocomposites and hybrid biomaterials comprise polymers that both have shown a good biocompatibility and are routinely used in surgical applications. In general, since polymers have a low modulus (2–7 GPa, as the maximum) as compared to that of bone (3–30 GPa), calcium orthophosphate bioceramics need to be loaded at a high weight% ratio. Besides, general knowledge on composite mechanics suggests that any high aspect ratio particles, such as whiskers or fibers, significantly improve the modulus at a lower loading [179]. Thus, some attempts have been already performed to prepare biocomposites containing whisker-like [225–229] or needle-like [230–232] calcium orthophosphates, as well as calcium orthophosphate fibers [49, 233].

The history of implantable polymer-calcium orthophosphate biocomposites and hybrid biomaterials started in 1981 [234] from the pioneering study by Prof. William Bonfield and colleagues performed on HA/PE formulations [236, 237]. That initial study introduced a bone-analogue concept, when proposed biocomposites comprised a polymer ductile matrix of PE and a ceramic stiff phase of HA, and was substantially extended and developed in further investigations by that research group [102, 238–254]. More recent studies included investigations on the influence of surface topography of HA/PE composites on cell proliferation and attachment [255–261]. The material is composed of a particular combination of HA particles at a volume loading of ~40% uniformly dispensed in a HDPE matrix. In addition, PP might be used instead of PE [262–264]. The idea was to mimic bones by using a polymeric matrix that can develop a considerable anisotropic character through adequate orientation techniques reinforced with a bone-like bioceramics that assures both a mechanical reinforcement and a bioactive character of the composite. Following FDA approval in 1994, in 1995 this material has become commercially available under the trade-name HAPEX™ (Smith and Nephew, Richards, USA), and to date it has been implanted in over 300,000 patients with the successful results. It remains the only clinically successful bioactive composite that appeared to be a major step in the implant field [31, 265]. The major production stages of HAPEX™ include blending, compounding and centrifugal milling. A bulk material or device is then created from this powder by compression and injection molding [63]. Besides, HA/HDPE biocomposites might be prepared by a hot rolling technique that facilitated uniform dispersion and blending of the reinforcements in the matrix [266].

A mechanical interlock between the both phases of HAPEX™ is formed by shrinkage of HDPE onto the HA particles during cooling [102, 267]. Both HA particle size and their distribution in the HDPE matrix were recognized as important parameters affecting the mechanical behavior of HAPEX™ [247]. Namely, smaller HA particles were found to lead to stiffer composites due to general increasing of interfaces between the polymer and the ceramics; furthermore, rigidity of HAPEX™ was found to be proportional to HA volume fraction [239]. Furthermore, coupling agents, e.g. 3-trimethoxysiyl propylmethacrylate for HA and acrylic acid for HDPE might be used to improve bonding (by both chemical adhesion and mechanical coupling) between HA and HDPE [268, 269]. Obviously, other calcium orthophosphates might be used instead of HA in biocomposites with PE [270]. Furthermore, attempts were performed to improve the mechanical properties of HAPEX™ by incorporating other ceramic phases into the polymer matrix, such as PSZ [271] and alumina [272]. For example, a partial replacement of HA filler particles by PSZ particles was found to lead to an increase in the strength and fracture toughness of HA/HDPE biocomposites. The compressive stress, set up by the volume expansion associated with tetragonal to monoclinic phase transformation of PSZ, inhibits or retards the crack propagation within the composite. This results in an enhanced fracture toughness of the HA/ZrO_2/HDPE biocomposite [271].

Various studies revealed that HAPEX™ attached directly to bones by chemical bonding (a bioactive fixation), rather than forming fibrous encapsulation (a morphological fixation). Initial clinical applications of HAPEX™ came in orbital reconstruction [273] but since 1995, the main uses of this composite have been in the shafts of middle ear implants for the treatment of conductive hearing loss [274, 275]. In both applications, HAPEX™ offers the advantage of *in situ* shaping, so a surgeon can make final alterations to optimize the fit of the prosthesis to the bone of a patient and subsequent activity requires only limited mechanical loading with virtually no risk of failure from insufficient tensile strength [102, 202]. As compared to cortical bones, HA/PE composites have a superior fracture toughness for HA concentrations below ~40% and similar fracture toughness in the 45–50% range. Their Young's modulus is in the range of 1–8 GPa, which is quite close to that of bone. The examination of the fracture surfaces revealed that only mechanical bond occurs between HA and PE. Unfortunately, the HA/PE composites are not biodegradable, the available surface area of HA is low and the presence of bioinert PE decreases the ability to bond to bones. Furthermore, HAPEX™ has been designed with a maximized density to increase its strength but the resulting lack of porosity limits the ingrowth of osteoblasts when the implant is placed into the body [29, 203]. Further details on HAPEX™ are available elsewhere [102]. Except of HAPEX™, other types of HA/PE biocomposites are also known [276–282].

Both linear and branched PE was used as a matrix and the biocomposites with the former were found to give a higher modulus

[277]. The reinforcing mechanisms in calcium orthophosphate/polymer biocomposites have yet to be convincingly disclosed. Generally, if a poor filler choice is made, the polymeric matrix might be affected by the filler through reduction of molecular weight during composite processing, formation of an immobilized shell of polymer around the particles (transcrystallization, surface-induced crystallization or epitaxial growth), and changes in conformation of the polymer due to particle surfaces and inter-particle spacing [102]. On the other hand, the reinforcing effect of calcium orthophosphate particles might depend on the molding technique employed: a higher orientation of the polymeric matrix was found to result in a higher mechanical performance of the composite [282, 283].

Many other blends of calcium orthophosphates with various polymers are possible, including rather unusual formulations with dendrimers [284]. Even light-curable polymer/calcium orthophosphate formulations are known [285]. The list of the appropriate calcium orthophosphates is shown in Table 1.1 (except of MCPM and MCPA—both are too acidic and, therefore, are not biocompatible [27]; however, to overcome this drawback, they might be mixed with basic compounds, such as HA, TTCP, $CaCO_3$, CaO, etc.) many biomedically suitable polymers have been listed above. The combination of calcium orthophosphates and polymers into biocomposites has a twofold purpose. The desirable mechanical properties of polymers compensate for a poor mechanical behavior of calcium orthophosphate bioceramics, while in turn the desirable bioactive properties of calcium orthophosphates improve those of polymers, expanding the possible uses of each material within the body [158–160, 286–290]. Namely, polymers have been added to calcium orthophosphates in order to improve their mechanical strength [158, 286] and calcium orthophosphate fillers have been blended with polymers to improve their compressive strength and modulus, in addition to increasing their osteoconductive properties [52, 160, 168, 291–295]. Furthermore, biocompatibility of such biocomposites is enhanced because calcium orthophosphate fillers induce an increased initial flash spread of serum proteins compared to the more hydrophobic polymer surfaces [296]. What's more, experimental results of these biocomposites indicate favorable cell–material interactions with increased cell activities as compared to each polymer alone [288]. As a rule, with increasing of calcium orthophosphate content, both Young's modulus and bioactivity of the biocomposites increase, while the ductility decreases [29, 291]. Furthermore, such formulations can provide a sustained release of calcium and orthophosphate ions into the milieus, which is important for mineralized tissue regeneration [287]. Indeed, a combination of two different materials draws on the advantages of each one to create a superior biocomposite with respect to the materials on their own.

It is logical to assume that the proper biocomposite of a calcium orthophosphate (for instance, CDHA) with a bioorganic polymer (for instance, collagen) would yield the physical, chemical, and mechanical properties similar to those of human bones. Different ways have been already realized to bring these two components together into biocomposites, such as mechanical blending, compounding, ball milling, dispersion of ceramic fillers into a polymer-solvent solution, a melt extrusion of a ceramic/polymer powder mixture, co-precipitation, and electrochemical co-deposition [36, 63, 297–299]. Three methods for preparing a homogeneous blend of HA with PLLA were compared [297]. A dry process, consisting in mixing ceramic powder and polymer pellets before a compression-molding step, was used. The second technique was based on the dispersion of ceramic fillers into a polymer-solvent solution. The third method was a melt extrusion of a ceramic/polymer powder mixture. Mixing dry powders led to a ceramic particle network around the polymer pellets, whereas the solvent and melt methods also produced a homogeneous dispersion of HA in the matrix. The main drawback of the solvent casting method is the risk of potentially toxic organic solvent residues. The melt extrusion method was shown to be a good way to prepare homogeneous ceramic/polymer blends [297].

Besides, there is *in situ* formation, which involves either synthesizing the reinforcement inside a preformed matrix material or synthesizing the matrix material around the reinforcement [63, 300, 301]. This is one of the most attractive routes, since it avoids extensive particle agglomeration. For example, several papers have reported *in situ* formation technique to produce various composites of apatites with carbon nanotubes [302–308]. Other examples comprise using amino acid-capped gold nano-sized particles as scaffolds to grow CDHA [309] and preparation of nano-sized HA/

polyamide biocomposites [310, 311]. In certain cases, a mechano-chemical route [312], emulsions [313–316], freeze-drying [317] and freeze-thawing techniques [318], flame-sprayed technique [319], or gel-templated mineralization [320] might be applied to produce calcium othophosphates-based biocomposites. Various fabrication procedures are well described elsewhere [36, 63, 297], where the interested readers are referred.

The interfacial bonding between a calcium orthophosphate and a polymer is an important issue of any biocomposite. Four types of mutual arrangements of nanodimensional particles to polymer chains have been classified by Kickelbick (Fig. 6.1): (1) inorganic particles embedded in inorganic polymer, (2) incorporation of particles by bonding to the polymer backbone, (3) an interpenetrating network with chemical bonds, (4) an inorganic–organic hybrid polymer [321]. If adhesion among the phases is poor, the mechanical properties of a biocomposite suffer. To solve the problem, various approaches have been already introduced. For example, a diisocyanate coupling agent was used to bind PEG/PBT (Polyactive^{TM}) block copolymers to HA filler particles. Using surface-modified HA particles as a filler in a PEG/PBT matrix significantly improved the elastic modulus and strength of the polymer as compared to the polymers filled with ungrafted HA [293, 322]. Another group used processing conditions to achieve a better adhesion of the filler to the matrix. Ignjatovic et al. prepared PLLA/HA composites by pressing blends of varying PLLA and HA content at different temperatures and pressures [158, 159, 323]. They found that maximum compressive strength was achieved at ~15 wt.% of PLLA. By using blends with 20 wt.% of PLLA, the authors also established that increasing the pressing temperature and pressure improved the mechanical properties. The former was explained by decrease in viscosity of the PLLA associated with a temperature increase, hence leading to improved wettability of HA particles. The latter was explained by increased compaction and penetration of pores at higher pressure, in conjunction with a greater fluidity of the polymer at higher temperatures. The combination of high pressures and temperatures was found to decrease porosity and guarantee a close apposition of a polymer to the particles, thereby improving the compressive strength [286] and fracture energy [324] of the biocomposites. The PLLA/HA biocomposites scaffolds were found to improve cell survival over plain PLLA scaffolds [325].

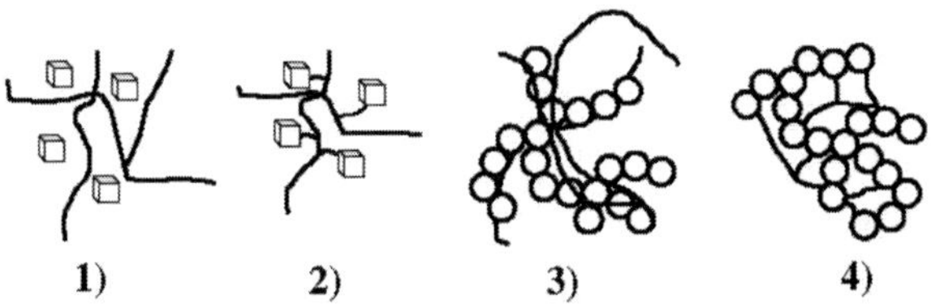

Figure 6.1 Four types of mutual arrangements of nano-sized particles to a polymer chain: (1) inorganic particles embedded in an inorganic polymer, (2) incorporation of particles by bonding to the polymer backbone, (3) interpenetrating network with chemical bonds, (4) inorganic-organic hybrid polymer. Reprinted from Ref. [321] with permission.

It is also possible to introduce porosity into calcium orthophosphate-based biocomposites, which is advantageous for most applications as bone substitution material. The porosity facilitates migration of osteoblasts from surrounding bones to the implant site [160, 326, 327]. Various material-processing strategies to prepare composite scaffolds with interconnected porosity comprise thermally induced phase separation, solvent casting and particle leaching, solid freeform fabrication techniques, microsphere sintering and coating [173, 328–330]. A supercritical gas foaming technique might be used as well [297, 331, 332].

6.4.1.1 Apatite-based biocomposites

A biological apatite is known to be the major inorganic phase of mammalian calcified tissues [25, 26]. Consequently, CDHA, HA, carbonateapatite (both with and without dopants) and, occasionally, FA have been applied to prepare biocomposites with other compounds, usually with the aim to improve the bioactivity. For example, PS composed with HA can be used as a starting material for long-term implants [333–335]. Retrieved *in vivo*, HA/PS biocomposite coated samples from rabbit distal femurs demonstrated direct bone apposition to the coatings, as compared to the fibrous encapsulation that occurred when uncoated samples were used [333]. The resorption time of such biocomposites is a very important factor, which depends on polymer's microstructure and the presence of modifying phases [334].

Various apatite-containing biocomposites with PVA [318, 336–344], PVAP [345], and several other polymeric components [346–

358] have been already developed. Namely, PVA/CDHA biocomposite blocks were prepared by precipitation of CDHA in aqueous solutions of PVA [318]. An artificial cornea consisted of a porous nano-sized HA/PVA hydrogel skirt and a transparent center of PVA hydrogel has been prepared as well. The results displayed a good biocompatibility and interlocking between artificial cornea and host tissues [340, 341]. PVAP has been chosen as a polymer matrix, because its phosphate groups can act as a coupling/anchoring agent, which has a higher affinity toward the HA surface [345]. Greish and Brown developed HA/Ca poly(vinyl phosphonate) biocomposites [349–351]. A template-driven nucleation and mineral growth process for the high-affinity integration of CDHA with PHEMA hydrogel scaffold has been developed as well [358].

PEEK [225, 227, 359–365, 367] and HIPS [366] were applied to create biocomposites with HA having a potential for clinical use in load bearing applications. The study on reinforcing PEEK with thermally sprayed HA particles revealed that the mechanical properties increased monotonically with the reinforcement concentration, with a maximum value in the study of ~40% volume fraction of HA particles [361–363]. The reported ranges of stiffness within 2.8–16.0 GPa and strength within 45.5–69 MPa exceeded the lower values for human bone (7–30 GPa and 50–150 MPa, respectively) [362]. Modeling of the mechanical behavior of HA/PEEK biocomposites is available elsewhere [364].

Biodegradable poly(α-hydroxyesters) are well established in clinical medicine. Currently, they provide with a good choice when a suitable polymeric filler material is sought. For example, HA/PLGA composites were developed which appeared to possess a cellular-compatibility suitable for bone tissue regeneration [368–376]. Zhang and Ma seeded highly porous PLLA foams with HA particles in order to improve the osteoconductivity of polymer scaffolds for bone tissue engineering [52, 292]. They pointed out that hydration of the foams prior to incubation in simulated body fluid increased the amount of carbonated CDHA material due to an increase of COOH and OH groups on the polymer surface, which apparently acted as nucleation sites for apatite. The following values of Young's modulus, compressive, bending, and tensile strengths for PLLA/HA composites have been achieved: 5–12 GPa, 78–137 MPa, 44–280 MPa, and 10–30 MPa, respectively [377]. However, these data do not appear to be in a good agreement with HA/PLLA biocomposite unit cell model predictions [378].

On their own, PGA and PLA are known to degrade to acidic products (glycolic and lactic acids, respectively) that both catalyze polymer degradation and cause inflammatory reactions of the surrounding tissues [379]. Thus, in biocomposites of poly(α-hydroxyesters) with calcium orthophosphates, the presence of slightly basic compounds (HA, TTCP) to some extent neutralizes the acid molecules, provides with a weak pH-buffering effect at the polymer surface and, therefore, more or less compensates these drawbacks [168, 380–382]. However, additives of even more basic chemicals (e.g., CaO, $CaCO_3$) might be necessary [173, 381, 383, 384]. Extensive cell culture experiments on pH-stabilized composites of PGA and carbonateapatite were reported, which afterwards were supported by extensive *in vitro* pH-studies [385]. A consequent development of this approach has led to designing of functionally graded composite skull implants consisting of polylactides, carbonateapatite, and $CaCO_3$ [386, 387]. Besides the pH-buffering effect, inclusion of calcium orthophosphates was found to modify both surface and bulk properties of the biodegradable poly(α-hydroxyesters) by increasing the hydrophilicity and water absorption of the polymer matrix, thus altering the scaffold degradation kinetics. For example, polymer biocomposites filled with HA particles were found to hydrolyze homogeneously due to water penetrating into interfacial regions [388].

Biocomposites of poly(α-hydroxyesters) with calcium orthophosphates are prepared mainly by incorporating the inorganic phase into a polymeric solution, followed by drying under vacuum. The resulting solid biocomposites might be shaped using different processing techniques. One can also prepare these biocomposites by mixing HA particles with L-lactide prior the polymerization [380] or by a combination of slip-casting technique and hot pressing [389]. Addition of a surfactant (surface active agent) might be useful to keep the suspension homogeneity [390]. Furthermore, HA/PLA [314, 315] and HA/PLGA [316] microspheres might be prepared by a microemulsion technique. More complex carbonated-FA/PLA [391] and PLGA/carbon nanotubes/HA [392] porous biocomposite scaffolds are also known. An interesting list of references, assigned to the different ways of preparing HA/poly(α-hydroxyesters) biodegradable composites, might be found in publications by

Durucan and Brown [53, 393, 394]. The authors prepared CDHA/PLA and CDHA/PLGA biocomposites by solvent casting technique with a subsequent hydrolysis of α-TCP to CDHA in aqueous solutions. The presence of both polymers was found to inhibit α-TCP hydrolysis, if compared with that of single-phase α-TCP; what is more, the inhibiting effect of PLA exceeded that of PLGA [53, 393, 394]. The physical interactions between calcium orthophosphates and poly(α-hydroxyesters) might be easily seen in Fig. 6.2 [53]. Another set of good pictures might be found in Ref. [87]. Nevertheless, it should not be forgotten that typically non-melt based routes lead to development of composites with lower mechanical performance and many times require the use of toxic solvents and intensive hand labor [178].

Figure 6.2 SEM micrographs of (a) α-TCP compact; (b) α-TCP-PLGA biocomposite (bars = 5 μm). Reprinted from Ref. [53] with permission.

The mechanical properties of poly(α-hydroxyesters) could be substantially improved by addition of calcium orthophosphates [395, 396]. Shikinami and Okuno developed CDHA/PLLA composites of very high mechanical properties [168]; mini-screws and mini-plates made of these composites have been manufactured and tested [388]. They have shown easy handling and shaping according to the implant site geometry, total resorbability, good ability to bond directly to the bone tissue without interposed fibrous tissue, osteoconductivity, biocompatibility, and high stiffness retainable for the period necessary to achieve bone union [388]. The initial bending strength of ~280 MPa exceeded that of cortical bone (120–210 MPa), while the modulus was as high as 12 GPa [168]. The strength could be maintained above 200 MPa up to 25 weeks in phosphate-buffered saline solution. Such biocomposites were obtained from precipitation of a PLLA/dichloromethane solution, where small granules of uniformly distributed CDHA microparticles (average size of 3 μm) could be prepared [167]. Porous scaffolds of PDLLA and HA have been manufactured as well [332, 397, 398]. Upon implantation into rabbit femora, a newly formed bone was observed and biodegradation was significantly enhanced if compared to single-phase HA bioceramics. This might be due to a local release of lactic acid, which in turn dissolves HA. In other studies, PLA and PGA fibers were combined with porous HA scaffolds. Such reinforcement did not hinder bone ingrowth into the implants, which supported further development of such biocomposites as bone graft substitutes [50, 51, 377, 399, 400].

Blends (named as SEVA-C) of EVOH with starch filled with 10–30 wt.% HA have been fabricated to yield biocomposites with modulus up to ~7 GPa with a 30% HA loading [401–406]. The incorporation of bioactive fillers such as HA into SEVA-C aimed to assure the bioactive behavior of the composite and to provide the necessary stiffness within the typical range of human cortical bone properties. These biocomposites exhibited a strong *in vitro* bioactivity that was supported by the polymer's water-uptake capability [407]. However, the reinforcement of SEVA-C by HA particles was found to affect the rheological behavior of the blend. A degradation model of these biocomposites has been developed [408].

Higher homologues poly(3-hydroxybutyrate), 3-PHB, and poly(3-hydroxyvalerate), 3-PHV, show almost no biodegradation. Nevertheless, biocomposites of these polymers with calcium orthophosphates showed a good biocompatibility both *in vitro* and *in vivo* [102, 409–415]. Both bioactivity and mechanical properties of these biocomposites can be tailored by varying the volume percentage of calcium orthophosphates. Similarly, biocomposites of PHBHV with both HA and amorphous carbonated apatite (almost ACP) appeared to have a promising potential for repair and replacement of damaged bones [416–419].

Along this line, PCL is used as a slowly biodegradable but well-biocompatible polymer. PCL/HA and PCL/CDHA biocomposites have been already discussed as suitable materials for substitution, regeneration, and repair of bone tissues [328, 420–433]. For example, biocomposites were obtained by infiltration of ε-caprolactone monomer into porous apatite blocks and *in situ* polymerization [423]. The composites were found to be biodegradable and might be applied as cancellous or trabecular bone replacement material or for a cartilage regeneration. Both the mechanical performance and biocompatibility in osteoblast cell culture of PCL were shown to be strongly increased when HA was added [434]. Several preparation techniques of PCL/HA biocomposites are known. For example, to make biocomposite fibers of PCL with nanodimensional HA, the desired amount of nanodimensional HA powder was dispersed in a solvent using magnetic stirrer followed by ultrasonication for 30 min. Then, PCL was dissolved in this suspension, followed by the solvent evaporation [435]. The opposite preparation order is also possible: PCL was initially dissolved in chloroform at room temperature (7–10% weight/volume), then HA (~10 μm particle size) was suspended in the solution, sonicated for 60 s, followed by the solvent evaporation [160] or salt-leaching [436]. The mechanical properties obtained by this technique were about one-third that of trabecular bone. In a comparative study, PCL and biological apatite were mixed in the ratio 19:1 in an extruder [437]. At the end of the preparation, the mixture was cooled in an atmosphere of nitrogen. The authors observed that the presence of biological apatite improved the modulus while concurrently increasing the hydrophilicity of the polymeric substrate. Besides, an increase in apatite concentration was found to increase both the modulus and yield stress of the composite, which indicated to good interfacial interactions between the biological apatite and PCL. It was also observed that the presence of biological apatite stimulated osteoblasts attachment to the biomaterial and cell proliferation [437]. In another study, a PCL/HA biocomposite was prepared by blending in melt form at 120 ºC until the torque reached equilibrium in the rheometer that was attached to the blender [438]. Then the sample was compression molded and cut into specimens of appropriate size for testing. It was observed that the composite containing 20 wt.% HA had the highest strength [438]. However, a direct grafting of PCL on the surface of HA particles seems to be the most interesting preparation technique [420]. In another study, HA porous scaffolds were coated by a PCL/HA composite coating [54]. In this system, PCL, as a coating component, was able to improve the brittleness and low strength of the HA scaffolds, while the particles in the coating were to improve the osteoconductivity and bioactivity of the coating layer. More complex formulations, such as PDLLA/PCL/HA [439], PLLA/PCL/HA [440], and supramolecular PCL/functionalized HA [441] biocomposites, have been prepared as well. Further details on both the PCL/HA biocomposites and the processing methodologies thereof might be found elsewhere [328].

A spread of attached human osteoblasts onto PLA and PCL films reinforced with CDHA and sintered HA was shown to be higher than that for the polymers alone [184]. Moreover, biochemical assays relating cell activity to DNA content allowed concluding that cell activity was more intense for the composite films [184]. Kim et al. coated porous HA blocks with PCL from dichloromethane solution and performed drug release studies. The antibiotic tetracycline hydrochloride was added into this layer, yielding a bioactive implant with drug release for longer than a week [54].

Yoon et al. investigated the highest mechanical and chemical stability of FA by preparing FA/collagen biocomposites and studied their effect in osteoblast-like cell culture [442]. The researchers found an increased cellular activity in FA composites compared to HA composites. This finding was confirmed in another study by means of variations in the fluoride content for FA-HA/PCL composites [443]. An interesting phenomenon of fractal growth of FA/gelatin composite crystals (Fig. 1.8) was achieved by diffusion of calcium- and orthophosphate + fluoride-solutions from the opposite sides into a tube filled with a gelatin gel [444–453]. The reasons of this phenomenon are not quite clear yet; besides, up to now nothing has yet been reported on a possible biomedical application of such very unusual structural composites.

6.4.1.2 TCP-based biocomposites

Both α-TCP and β-TCP have a higher solubility than HA (Table 1.1). Besides, they are faster resorbed *in vivo* [454]. Therefore, these calcium orthophosphates were widely used instead of apatites to prepare completely biodegradable biocomposites [456–479]. For example, a biodegradable and osteoconductive biocomposite made of β-TCP particles and gelatin was proposed [466]. This material was tested *in vivo* with good results. It was found to be biocompatible, osteoconductive, and biodegradable with no need for a second surgical operation to remove the device after healing occurred. Both herbal extracts [467] and K_2HPO_4 [468] might be added to this formulation. Another research group prepared biocomposites of cross-linked gelatin with β-TCP and both a good biocompatibility and bone formation upon subcutaneous implantation in rats were found [469]. Yang et al. [474] extended this to porous (porosity ~75%) β-TCP/gelatin biocomposites those also contained BMP-4. Furthermore, cell-compatible and possessive some osteoinductive properties porous β-TCP/alginate-gelatin hybrid scaffolds were prepared and successfully tested *in vitro* [471]. More to the point, biocomposites of β-TCP with PLLA [462–464] and PLGC [465] were prepared. Although β-TCP was able to counter the acidic degradation of the polyester to some extent, it did not prevent a pH drop down to ~6. Nevertheless, implantation of this biocomposite in beagles' mandibular bones was successful [465]. α-TCP/gelatin formulations are known as well [477].

Based on a self-reinforcement concept, biocomposites of TCP with polylactides were prepared and studied using conventional mechanical testing [480]. Resorbable scaffolds were fabricated from such biocomposites [481]. Chitosan was also used as the matrix for the incorporation of β-TCP by a solid/liquid phase separation of the polymer solution and subsequent sublimation of the solvent. Due to complexation of the functional groups of chitosan with calcium ions of β-TCP, these biocomposites had high compressive modulus and strength [482]. PCL/β-TCP biocomposites were developed in other studies [483–486] and their *in vitro* degradation behavior was systematically monitored by immersion in simulated body fluid at 37 °C [485]. To extend this topic further, PCL/β-TCP biocomposites might be loaded by drugs [486].

Cell culture tests on β-TCP/PLLA biocomposites were reported; the biocomposites showed no cytotoxicity and evidenced good cell attachment to its surface [456]. An *in vitro* study with primary rat calvarial osteoblasts showed an increased cellular activity in the BMP-loaded samples [474]. Other researchers investigated BMP-2-loaded porous β-TCP/gelatin biocomposites (porosity ~95%, average pore size 180–200 μm) [487] and confirmed the precious study. Biocomposites of β-TCP and glutaraldehyde cross-linked gelatin were manufactured and tested *in vitro* to measure the material cytotoxicity [470]. The experimental results revealed that the amount of glutaraldehyde cross-linking agent should be less than 8% to decrease the toxicity on the osteoblasts and to avoid inhibition of cellular growth caused by the release of residual or uncross-linked glutaraldehyde. A long-term implantation study of PDLLA/α-TCP composites in a loaded sheep implant model showed good results after 12 months but a strong osteolytic reaction after 24 months. This was ascribed to the almost complete dissolution of α-TCP to this time and an adverse reaction of the remaining PDLLA [488].

More complex calcium orthophosphate-based formulations are known as well. For example, there is a biocomposite consisting of three interpenetrating networks: TCP, CDHA and PLGA [489]. Firstly, a porous TCP network was produced by coating a polyurethane foam by hydrolysable α-TCP slurry. Then, a CDHA network was derived from a calcium orthophosphate cement filled in the porous TCP network. Finally, the remaining open pore network in the CDHA/α-TCP structures was infiltrated with PLGA. This biocomposite consists of three phases with different degradation behavior. It was postulated that bone would grow on the fastest degrading network of PLGA, while the remaining calcium orthophosphate phases would remain intact thus maintaining their geometry and load bearing capability [489].

6.4.1.3 Other calcium orthophosphate-based biocomposites

The number of research publications devoted to biocomposites based on other calcium orthophosphates is substantially lesser than those devoted to apatites and TCP. BCP, which is a solid composite of HA and β-TCP; however, similar composites of HA and α-TCP are possible, as well (see section *1.3.14* for details) appears to be most popular among the remaining calcium orthophosphates. For example, collagen-coated BCP ceramics was studied and the biocompatibility towards osteoblasts was found to increase upon coating with collagen [490]. Another research group created porous PDLLA/BCP scaffolds and coated them with a hydrophilic PEG/vancomycin composite for both drug delivery purposes and surface modification [491]. More to the point, both PLGA/BCP [492, 493] and PLLA/BCP [494] biocomposites were fabricated and their cytotoxicity and fibroblast properties were found to be acceptable for natural bone tissue reparation, filling and augmentation [495, 496]. Besides, PCL/

BCP [497] and gelatin/BCP [498, 499] biocomposites are known as well.

A choice of DCPD-based biocomposites of DCPD, albumin and duplex DNA was prepared by water/oil/water interfacial reaction method [313]. Core-shell type DCPD/chitosan biocomposite fibers were prepared by a wet spinning method in another study [500]. The energy-dispersive X-ray spectroscopy analysis indicated that Ca and P atoms were mainly distributed on the outer layer of the composite fibers; however, a little amount of P atoms remained inside the fibers. This indicated that the composite fibers formed a unique core-shell structure with shell of calcium orthophosphate and core of chitosan [500]. A similar formulation was prepared for further applications in bone cement biocomposites [501]. DCPA/BSA biocomposites were synthesized through the co-precipitation of BSA on the nanodimensional particles of DCPA performed in ethanol [502]. Nanodimensional DCPA was synthesized and incorporated into dental resins to form dental biocomposites [503–505]. Although, this is not to the point, it is interesting to mention that some DCPD/polymer composites could be used as proton conductors in battery devices [506, 507]. Nothing has been reported on their biocompatibility but, perhaps, sometime the improved formulations will be used to fabricate biocompatible batteries for implantable electronic devices.

Various ACP-based biocomposites and hybrid formulations for dental applications have been developed [508–511]. Besides, several ACP-based formulations were investigated as potential biocomposites for bone grafting [419, 512–514] and drug delivery [515]. Namely, ACP/PPF biocomposites were prepared by *in situ* precipitation [513], while PHB/carbonated ACP and PHBHV/carbonated ACP biocomposites appeared to be well suited as slowly biodegradable bone substitution material [419]. Another example comprises hybrid nanodimensional capsules of ~50–70 nm in diameter which were fabricated by ACP mineralization of shell cross-linked polymer micelles and nano-sized cages [514]. These nano-sized capsules consisted of a continuous ultrathin inorganic surface layer that infiltrated the outer cross-linked polymeric domains. They might be used as structurally robust, pH-responsive biocompatible hybrid nanostructures for drug delivery, bioimaging and therapeutic applications [514]. Additional examples of the biomedical applications of ACP-containing biocomposites and hybrid formulations are available in section *2.5*.

6.4.2 Calcium Orthophosphate Cement-Based Biocomposites and Concretes

Inorganic self-setting calcium orthophosphate cements, which harden in the body, were introduced by LeGeros et al. [516] and Brown and Chow [517, 518] in the early 1980s [519]. Since then, these cements have been broadly studied and many formulations have been proposed (see Chapter 5 for the details). The cements set and harden due to various chemical interactions among calcium orthophosphates that finally lead to formation of a monolithic body consisting of either CHDA or DCPD with possible admixtures of other phases. Unfortunately, having the ceramic nature, calcium orthophosphate cements are brittle after hardening and the setting time is sometimes unsuitable for clinical procedures [519]. Therefore, various attempts have been performed to transform the cements into biocomposites e.g., by adding hydroxylcarboxylic acids, to control the setting time [520], gelatin to improve both the mechanical properties and the setting time [473, 521–523] or osteocalcin/collagen to increase the bioactivity [524]. More to the point, various reinforcement additives of different shapes and nature are widely used to improve the mechanical properties of calcium orthophosphate cements (see section *5.7* for details). Even carbon nanotubes were used for this purpose [525]! Although the biomaterials community does not use this term, a substantial amount of the reinforced cement formulations might be defined as calcium orthophosphate-based concretes [526]. The idea behind the concretes is simple: if a strong filler is present in the matrix, it might stop crack propagation.

Various apatite-containing biocomposite formulations based on PMMA [527–540] and PEMA [102, 541, 542] have been already developed. Such biocomposites might be prepared by dispersion of apatite powder into a PMMA viscous fluid [543] and used for drug delivery purposes [544]. When the mechanical properties of the concretes composed of PMMA matrix and HA particles of various sizes were tested, the tensile results showed that strength was independent on particle sizes. In addition, up to 40% by weight HA could be added without impairing the mechanical properties [530,

531]. After immersion into Ringer's solution, the tensile strength was not altered whereas the fatigue properties were significantly reduced. The biocompatibility of PMMA/HA biocomposites was tested *in vivo* and enhanced osteogenic properties of the implants compared to single-phase PMMA were observed [528, 532–535]. It was shown that not only the mechanical properties of PMMA were improved but the osteoblast response of PMMA was also enhanced with addition of HA [532]. Thereby, by adding of calcium orthophosphates, a non-biodegradable PMMA was made more bioactive and osteoconductive, yielding a well-processible biocomposite concrete. As a drawback, the PMMA/HA formulations possess a low flexural, compressive, and tensile strength.

A biocomposite made from HA granules and *bis*-phenol-α-glycidylmethacrylate-based resin appeared to possess comparable mechanical and biological properties to typical PMMA cement, leading to potential uses for implant fixation [545]. To improve the mechanical properties of calcium orthophosphate cements and stabilize them at the implant site, various researchers have resorted to formulations that set *in situ*, primarily through cross-linking reactions of the polymeric matrix. For example, TTCP was reacted with PAA, forming a cross-linked CDHA/calcium polyacrylate biocomposite [546]. In aqueous solutions, TTCP hydrolyzes to CDHA [27] and the liberated calcium cations react with PAA, forming the cross-linked network [546]. Reed et al. synthesized a dicarboxy polyphosphazene that can be cross-linked by calcium cations and cement-based (TTCP + DCPD) CDHA/polyphosphazene biocomposites with a compressive strength ~10 MPa and of ~65% porosity were prepared as a result [547]. To mimic PMMA cements, PFF/β-TCP biocomposites were prepared with addition of vinyl monomer to cross-link PPF. As a result, quick setting and degradable biocomposite cements with a low heat output and compressive strengths in the range of 1–12 MPa were prepared by varying the molecular weight of PPF, as well as the contents of the monomer, β-TCP, initiator, and NaCl, as a porogen [548, 549]. An acrylic cement with Sr-containing HA as a filler [138], an injectable polydimethylsiloxane/HA cement [550], biocomposites consisting of PLGA microspheres and a calcium orthophosphate cement [551, 552], as well as a hybrid cement formulation of chitosan oligosaccharide/gelatin/calcium orthophosphate [553] were prepared as well.

In order to improve the mechanical properties of calcium orthophosphate cements, numerous researchers blended various polymers with the cements. For example, gelatin might be added to calcium orthophosphate cement formulations, primarily to stabilize the paste in aqueous solution before it develops adequate rigidity and, secondly, to improve the compressive strength [473, 521, 554]. Adding rod-like fillers to the cement formulations also caused an improvement in the mechanical properties [554]. For example, PAA and PVA were successfully used to improve the mechanical properties of a TTCP + DCPD cement but, unfortunately, with an inevitable and unacceptable reduction of both workability and setting time [555, 556]. Similar findings were reported in the presence of sodium alginate and sodium polyacrylate [557]. Other polymers, such as polyphosphazene, might be used as well [558–560]. Other examples of polymer/calcium orthophosphate cement formulations might be found elsewhere [561, 562].

Porous calcium orthophosphate scaffolds with interconnected macropores (~1 mm), micropores (~5 μm), and of high porosity (~80%) were prepared by coating polyurethane foams with a TTCP + DCPA cement, followed by firing at 1200 °C. In order to improve the mechanical properties of the scaffolds, the open micropores of the struts were then infiltrated by a PLGA solution to achieve an interpenetrating bioactive ceramic/biodegradable polymer composite structure. The PLGA filled struts were further coated with a 58S bioactive glass/PLGA composite coating. The obtained complex porous biocomposites could be used as tissue engineering scaffolds for low-load bearing applications [563]. A more complicated construction, in which the PLGA macroporous phase has been reinforced with a bioresorbable TTCP + DCPA cement, followed by surface coating of the entire construct by a non-stoichiomentic CDHA layer, has been designed as well [564]. The latter approach has culminated in a unique, three-phase biocomposite that is simple to fabricate, osteoconductive and completely biodegradable.

A porosity level of 42–80% was introduced into calcium orthophosphate cement/chitosan biocomposites by addition of the water-soluble mannitol [565]. Chitosan significantly improved the mechanical strength of the entire biocomposite [566]. A similar approach was used by other researchers who studied the effect of the addition of PLGA microparticles [567–570] (which can also be loaded with drugs or growth factors [571–573]) to calcium orthophosphate cements. These biocomposites were implanted into cranial defects of rats and a content of ~30 wt.% of the microparticles was found to give the best results [567], while the addition of a growth factor to the biocomposites significantly increased bone contact at 2 weeks and enhanced new bone formation at 8 weeks [573]. The *in vivo* rabbit femur implant tests showed that PLGA/calcium orthophosphate cement formulations exhibited outstanding biocompatibility and bioactivity, as well as a better osteoconduction and degradability than pure calcium orthophosphate cements [568]. Further details on this subject are available in section *5.7*.

6.4.3 Biocomposites Based on Nanodimensional Calcium Orthophosphates and Nanodimensional Biocomposites

Nanodimensional and nanophasic materials are the materials that have particles or grain sizes under ~100 nm, respectively (see section *3.2* for details). Thus, one should clearly differentiate between nanodimensional composites and composites based on nanodimensional compounds. The former might be any type of composites but disintegrated to particles with dimensions < 100 nm, while the latter consist of two or more materials, in which at least one of the materials is of a nanometer-scale.

Nanodimensional and nanophasic materials have different mechanical and optical properties if compared to the large grained materials of the same chemical composition. Namely, they possess the unique surface properties, such as an increased number of atoms, grain boundaries and defects at the surface, huge surface area and altered electronic structure, if compared to the conventional micron-sized materials. For example, nanodimensional HA (size ~ 67 nm) has a higher surface roughness of 17 nm if compared to 10 nm for the conventional submicron size HA (~ 180 nm), while the contact angles (a quantitative measure of the wetting of a solid by a liquid) are significantly lower for nanodimensional HA (6.1) if compared to the conventional HA (11.51). Additionally, the diameter of individual pores in nanodimensional HA compacts is five times smaller (pore diameter ~ 6.6 Å) than that in the conventional grain-sized HA compacts (pore diameter within 19.8–31.0 Å) [574–576]. Besides, nanodimensional HA promotes osteoblast cells adhesion, differentiation and proliferation, osteointegration and deposition of calcium containing minerals on its surface better than microcrystalline HA; thus enhancing formation of a new bone tissue within a short period [574–576]. More to the point, nanodimensional HA was found to cause apoptosis of the leukemia P388 cells [577].

Natural bones and teeth are hierarchical biocomposites of biological origin based on nanodimensional compounds because they consists of nano-sized blade-like crystals of biological apatite grown in intimate contact with the organic matrix rich in bioorganic fibers and organized in complicated hierarchical structures (see section *1.4* for details). Given the fact that the major bioorganic phase of bones is collagen, i.e. a natural polymer (Table 6.1), it is obvious that a composite of a nanodimensional calcium orthophosphate with a biodegradable polymer should be advantageous as bone-substitution material. The inorganic nanodimensional phase would be responsible for the mechanical strength (hardness) and bioactivity, while the polymeric phase would provide the elasticity. In addition, the solubility of calcium orthophosphates depends on their crystallite size (smaller crystals have a higher solubility) and on their carbonate content (higher carbonate content increases the solubility) [578]. To the author's best knowledge, among calcium orthophosphates listed in Table 1.1, before very recently only apatites (CDHA, HA and, perhaps, FA) have been available in nanodimensional state. However, recently, nano-sized DCPA [503–505] and nano-sized MCPM [579] have been synthesized and applied to prepare biocomposites with strong ionic release to combat tooth caries. Presumably, all calcium orthophosphates from Table 1.1 might be manufactured in nanodimensional and/or nanocrystalline state; however not all of them have been prepared yet (see Chapter 3 for details).

A number of investigations have been conducted recently to determine the mineralization, biocompatibility, and mechanical properties of the biocomposites based on various (bio)polymers and nanodimensional HA. Unfortunately, in the majority of the already published papers it often remained unclear whether "nanodimensional HA", in fact, represented the nanodimensional stoichiometric HA or a nanodimensional non-stoichiometric CDHA. These studies covered biocomposites with PLA [332, 580–589] and its copolymer with PGA [590–593], collagen [594–607], collagen + PLA [607–615], collagen + PVA [616], collagen + alginate [617, 618], gelatin [619–624], PPF [625–627], polyamide [310, 311, 628–639], PVA [340, 341, 640–642], PVAP [345], poly(ethylene-*co*-acrylic) acid [643, 644], chitosan [645–651] and its derivatives [652], konjac glucomannan + chitosan [653], PHEMA + PCL [654], PCL [390, 435, 655, 656], cellulose [70, 71, 657–659], Ti [660–662], PCL semi-interpenetrating biocomposites [663] and many other biocompatible hybrid formulations [279, 320, 335, 417, 664–683]. Furthermore, each from the aforementioned formulations might be covered by a layer of nanodimensional calcium orthophosphate, as it was done by Zandi et al. [684], who coated a biocomposite of nano-sized rods HA with gelatin by nano-sized HA. Several nanodimensional biocomposites were found to be applicable as carriers for delivery of drugs and growth factors [38, 685–687], as well as promising vectors with ultrahigh gene loading and transfection efficiency [688]. Data are available on the excellent biocompatibility of such biocomposites [605]. The dispersion state of nano-sized particles appears to be the critical parameter in controlling the mechanical properties of nanodimensional biocomposites, as nano-sized particles always tend to aggregate owing to their high surface energy [417]. A comparison was made of the mechanical properties of biocomposites with nano-sized and micron-sized HA with a polyamide. The results showed that the bending and tensile strengths of the biocomposite increased with increasing content of nanodimensional HA but decreased with increasing micron-sized HA content [310]. A SEM image of the mineralized collagen fibrils, demonstrating homogeneity of the nanodimensional biocomposite and the close interaction between the mineral phase and the reconstituted collagen fibrils, is shown in Fig. 6.3 [689].

Porous (porosity ~85%) biocomposites of nano-sized HA with collagen and PLA have been prepared by precipitation and freeze-drying; these biocomposites did not show a pH drop upon *in vitro* degradation [608–610]. They were implanted in the radius of rabbits and showed a high biocompatibility and partial resorption after 12 weeks. Nano-sized HA/chitosan biocomposites with improved mechanical stability were prepared from HA/chitosan nano-sized rods [690]. Nano-sized HA/PLLA biocomposites of high porosity (~90%) were prepared using thermally induced phase separation [691]. Besides, nanodimensional HA was used to prepare biocomposites with PAA and the nanostructure of the resulting nano-sized crystals exhibited a core-shell configuration [692, 693].

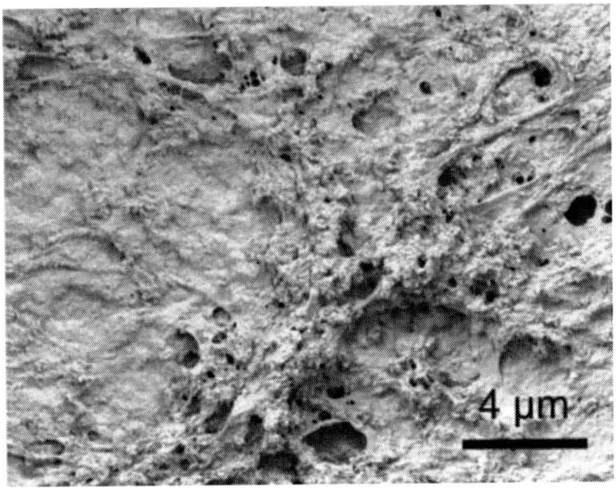

Figure 6.3 Scanning electron microscopy image of reconstituted mineralized collagen I fibrils. An example of an organic–inorganic nanostructural composite, mimicking the extracellular matrix of bone tissue on the nanometer scale. Reprinted from Ref. [689] with permission.

Nanodimensional crystals of HA appeared to be suitable for intra-osseous implantation and offered a potential to formulate enhanced biocomposites for clinical applications [694]. Thus, the biocompatibility of chitosan in osteoblast cell culture was significantly improved by addition of nano-sized HA [695]. Similar finding is valid for nanodimensional HA/polyamide biocomposites [630]. Further details on nanodimensional biocomposites might be found in an excellent review [36]. More to the point, a more general review on applications of nanodimensional biomaterials in orthopedics is also available [696], where the interested readers are referred.

6.4.4 Biocomposites with Collagen

The main constituent of the bioorganic matrix of bones is type I collagen (Table 6.1) with molecules about 300 nm in length. The structural and biochemical properties of collagens have been widely investigated and over 25 collagen subtypes have been identified [697, 698]. This protein is conducive to crystal formation in the associated inorganic matrix. It is easily degraded and resorbed by the body and allows good attachment to cells. Collagen alone is not effective as an osteoinductive material but it becomes osteoconductive in combination with calcium orthophosphates [699]. Both collagen type I and HA were found to enhance osteoblast differentiation [700] but combined together, they were shown to accelerate osteogenesis. However, this tendency is not so straightforward: the data are available that implanted HA/collagen biocomposites enhanced regeneration of calvaria bone defects in young rats but postponed the regeneration of calvaria bone in aged rats [701]. Finally, addition of calcium orthophosphates to collagen sheets was found to give a higher stability and an increased resistance to 3D swelling compared to the collagen reference [702]. Therefore, a bone analogue based on these two constituents should possess the remarkable properties. Furthermore, addition of bone marrow constituents gives osteogenic and osteoinductive properties to calcium orthophosphate/collagen biocomposites [1].

The unique characteristics of bones are the spatial orientation between the nanodimensional crystals of biological apatite and collagen macromolecules at the nano-scale [39], where the crystals (about 50 nm length) are aligned parallel to the collagen fibrils [25, 26, 35, 42], which is believed to be the source of the mechanical strength of bones. The collagen molecules and the crystals of biological apatite assembled into mineralized fibrils are approximately 6 nm in diameter and 300 nm long [35, 39, 42, 609, 703]. Although the complete mechanisms involved in the bone building strategy are still unclear, the strengthening effect of nanodimensional crystals of biological apatite in calcified tissues might be explained by the fact that the collagen matrix is a load transfer medium and thus transfers the load to the intrinsically rigid inorganic crystals. Furthermore, the crystals of biological apatite located in between tangled fibrils cross-link the fibers either through a mechanical interlocking or by forming calcium ion bridges, thus increasing deformation resistance of the collagenous fiber network [704].

When calcium orthophosphates are combined with collagen in a laboratory, the prepared biocomposites appear to be substantially different from natural bone tissue due to a lack of real interaction between the two components, i.e. the interactions that are able to modify the intrinsic characteristics of the singular components themselves. The main characteristics of the route, by which the mineralized hard tissues are formed *in vivo*, is that the organic matrix is laid down first and the inorganic reinforcing phase grows within this organic matrix [25, 26, 35, 42]. Although to date, neither the elegance of the biomineral assembly mechanisms nor the intricate composite nano-sized architectures have been duplicated by non-biological methods, the best way to mimic bone is to copy the way it is formed, namely by nucleation and growth of CDHA nano-sized crystals from a supersaturated solution both onto and within the collagen fibrils [705–707]. Such syntheses were denoted as "biologically inspired" which means they reproduce an ordered pattern and an environment very similar to natural ones [708–710]. The biologically inspired biocomposites of collagen and calcium orthophosphates (mainly, apatites) for bone substitute have a long history [33, 442, 597, 711–730] and started from the pioneering study by Mittelmeier and Nizard [731], who mixed calcium orthophosphate granules with a collagen web. Such combinations were found to be bioactive, osteoconductive, osteoinductive [33, 699, 732–734] and, in general, artificial grafts manufactured from this type of the biocomposites are likely to behave similarly to bones and be of more use in surgery than those prepared from any other materials. Indeed, data are available on the superiority of calcium orthophosphate/collagen biocomposite scaffolds over the artificial polymeric and calcium orthophosphate bioceramic scaffolds individually [735].

It has been found that calcium orthophosphates may be successfully precipitated onto a collagen substrate of whatever form or source [33, 40, 597, 736, 737]. However, adherence of calcium orthophosphate crystals to collagen did depend on how much the collagen had been denatured: the more fibrillar the collagen, the greater attachment. Clarke et al. first reported the production of a biocomposite produced by precipitation of DCPD onto a collagen matrix with the aid of phosphorylated amino acids commonly associated with fracture sites [716]. Apatite cements (DCPD + TTCP) have been mixed with a collagen suspension, hydrated, and allowed to set. CDHA crystals were found to nucleate on the collagen fibril network, giving a material with the mechanical properties weaker than those reported for bone. More to the point, the prepared biocomposites were without the nanostructure similar to that of bone [713, 738]. The oriented growth of OCP crystals on collagen was achieved by an experimental device in which Ca^{2+} and PO_4^{3-} ions diffused into a collagen disc from the opposite directions [737, 739, 740]. Unfortunately, these experiments were designed to simulate the mechanism of *in vivo* precipitation of biological apatite only; due to this reason, the mechanical properties of the biocomposites were not tested [741].

Conventionally, collagen/calcium orthophosphate biocomposites can be prepared by blending or mixing of collagen and calcium orthophosphates, as well as by biomimetic methods [33, 36, 38, 41, 594, 597, 609, 686, 703, 708–710, 713, 736, 745–757]. For example, Tampieri et al. [710] produced and compared artificial bone like tissue apatite/collagen biocomposites prepared by using two different methodologies: (1) dispersion of apatite in a collagen aqueous suspension and then freeze dried and (2) direct nucleation of an apatitic phase on assembling collagen fibrils. Biocomposites obtained using first way were similar to uncalcified natural collagen. The crystallite sizes were not uniform and were often aggregated and randomly distributed into the matrix, proving that there was no real interaction between apatite and collagen fibers. However, the second method allowed the direct nucleation of nano-sized crystals of apatite on self-assembled collagen fibers. In this case, the two components (CDHA and collagen) exhibited strong interactions, highlighted by several analysis techniques, which showed a complete analogy of the composite with calcified natural tissue [710]. Other production techniques are also possible. For example, using a polymer-induced liquid-precursor process, collagen/apatite biocomposites mimicking the nanostructure of bones, wherein nano-sized crystals of apatite were embedded within the collagen fibrils, were prepared [757]. More complicated formulations, such as a magnetite enriched HA/collagen [758] and HA/collagen/PVA [759] biocomposites, have been developed as well.

Furthermore, collagen might be incorporated into various calcium orthophosphate cements [713, 738, 760–764]. Typically, a type I collagen sponge is presoaked in PO_4^{3-}-containing a highly basic aqueous solution and then is immersed into a Ca^{2+}-containing solution to allow mineral deposition. Also, collagen I fibers might be dissolved in acetic acid and then this solution is added to phosphoric acid, followed by a neutralization synthesis (performed at 25 ºC and solution pH within 9–10) between an aqueous suspension of $Ca(OH)_2$ and the H_3PO_4/collagen solution [708, 709]. To ensure the quality of the final product, it is necessary to control the Ca/P ionic ratio in the reaction solution. One way to do this is to dissolve a commercial calcium orthophosphate in an acid; another one is to add

Ca^{2+} and PO_4^{3-} ions in a certain ratio to the solution and after that induce the reaction [39]. Biomimetically, one can achieve an oriented growth of CDHA crystals onto dissolved collagen fibrils in aqueous solutions via a self-organization mechanism [747]. A number of authors produced calcium orthophosphate/collagen biocomposites by mixing preformed ceramic particles with a collagen suspension [765–767]. However, in all blended composites, the crystallite sizes of calcium orthophosphates were not uniform and the crystals were often aggregated and randomly distributed within a fibrous matrix of collagen. Therefore, no structural similarity to natural bone was obtained and only a compositional similarity to that of natural bone was achieved. Besides, CDHA crystallization from aqueous solutions might be performed in the presence of a previously dispersed collagen [33, 597]. More to the point, collagen might be first dispersed in an acidic solution, followed by addition of calcium and orthophosphate ions and then co-precipitation of collagen and CDHA might be induced by either increasing the solution pH or adding mixing agents [41]. Although it resulted in biocomposites with poor mechanical properties, pressing of the apatite/collagen mixtures at 40 ºC under 200 MPa for several days is also known [768]. Attempts have been performed for a computer simulation of apatite/collagen composite formation process [769]. It is interesting to note, that such biocomposites were found to possess some piezoelectric properties [770].

As the majority of the collagen/HA biocomposites are conventionally processed by anchoring micron-sized HA particles into collagen matrix, it makes quite difficult to obtain a uniform and homogeneous composite graft. Besides, such biocomposites have inadequate mechanical properties; over and above, the proper pore sizes have not been achieved either. Further, microcrystalline HA, which is in contrast to nanocrystalline bone apatite, might take a longer time to be remodeled into a new bone tissue upon the implantation. In addition, some of the biocomposites exhibited very poor mechanical properties, probably due to a lack of strong interfacial bonding between the constituents. The aforementioned data clearly demonstrate that the chemical composition similar to bone is insufficient for manufacturing the proper grafts; both the mechanical properties and mimetic of the bone nanostructure are necessary to function as bone in recipient sites. There is a chance for improving osteointegration by reducing the grain size of HA crystals by activating of ultrafine apatite growth into the matrix. This may lead to enhance the mechanical properties and osteointegration with improved biological and biochemical affinity to the host bone. Besides, the porosity was found to have a positive influence on the ingrowth of the surrounding tissues into the pores of collagen/HA biocomposites [771, 772].

Bovine collagen might be mixed with calcium orthophosphates and such biocomposites are marketed commercially as bone-graft substitutes (Table 4.2). Those further can be combined with bone marrow aspirated from the iliac crest of the site of the fracture. Collagraft®, BioOss®, and Healos® are several examples of the commercially available grafts for clinical use [36]. Application of these materials was compared with autografts for the management of acute fractures of long bones with defects, which had been stabilized by internal or external fixation [773, 774]. These biocomposites are osteogenic, osteoinductive, and osteoconductive; however, they lack the structural strength and require a harvest of the patient's bone marrow. Although no transmission of diseases has been recorded yet, the use of bovine collagen might be a source of concern [2].

Collagen sponges with an open porosity (30–100 μm) were prepared by a freeze-drying technique and then their surface was coated by a 10 μm layer of biomimetic apatite precipitated from simulated body fluid [775]. The researchers found a good *in vitro* performance with fibroblast cell culture. Other preparation techniques are also possible [776]. Collagen/HA microspheres or gel beads have been prepared in the intention of making injectable bone fillers [777, 778]. Liao et al. succeeded in mimicking the bone structure by blending carbonateapatite with collagen [779]. A similar material (mineralized collagen) was implanted into femur of rats and excellent clinical results were observed after 12 weeks [780]. Collagen/HA biocomposites were prepared and their mechanical performance was increased by cross-linking the collagen fibers with glutaraldehyde [598, 600, 601]. These biocomposites were tested in rabbits and showed a good biological performance, osteoconductivity, and biodegradation. A similar approach was selected to prepare HA/collagen microspheres (diameter ~5 μm) by a water–oil emulsion technique in which the surface was also cross-linked by glutaraldehyde [778]. That material showed a good *in vitro* performance with osteoblast cell culture. A porous bone graft substitute was formed from a nano-sized HA/collagen biocomposite combined with PLA by a freeze-drying method; the resulting material was found to mimic natural bones at several hierarchical levels [609]. Subsequent *in vitro* experiments confirmed a good adhesion, proliferation, and migration of osteoblasts into this composite [608]. A further increase in biocompatibility might be achieved by addition of various dopants. For example, to enhance bone substitution, Si-substituted HA/collagen composites have been developed with silicon located preferentially in the collagen phase [599]. Porous (porosity level ~95% with interconnected pores of 50–100 μm) biocomposites of collagen (cross-linked with glutaraldehyde) and β-TCP have been prepared by a freeze-drying technique, followed by sublimation of the solvent; the biocomposites showed a good biocompatibility upon implantation in the rabbit jaw [781].

Biocomposites of calcium orthophosphates with collagen were found to be useful for drug delivery purposes [618, 721, 764, 782–784]. Namely, an HA/collagen—alginate (20 μl) with the rh-BMP2 (100 μg/ml, 15 μl) showed bone formation throughout the implant 5 weeks after implantation without obvious deformation of the material [618]. Gotterbarm et al. developed a two-layered collagen/β-TCP implant augmented with chondral inductive growth factors for repair of osteochondral defects in the trochlear groove of minipigs. This approach might be a new promising option for the treatment of deep osteochondral defects in joint surgery [783].

To conclude this part, one should note that biocomposites of apatites with collagen are a very hot topic of the research and up to now, just a few papers are devoted to biocomposites of other calcium orthophosphates with collagen [742–744, 783, 785–788]. These biomaterials mimic natural bones to some extent, while their subsequent biological evaluation suggests that they are readily incorporated into the bone metabolism in a way similar to bone remodeling, instead of acting as permanent implant [609, 731]. However, the performance of these biocomposites depends on the source of collagen from which it was processed. Several attempts have been made to simulate the collagen-HA interfacial behavior in real bone by means of cross-linking agents such as glutaraldehyde [598, 600, 601, 736, 778, 781] with the purpose to improve the mechanical properties of these biocomposites. Unfortunately, a further progress in this direction is restricted by a high cost, difficulty to control cross-infection, a poor definition of commercial sources of collagens, as well as by a lack of an appropriate technology to fabricate bone-resembling microstructures. Further details on calcium orthophosphate/collagen composites might be found elsewhere [36, 725].

6.4.5 Biocomposites with Other Bioorganic Compounds and/or Biological Macromolecules

The biggest practical problems with collagen type I are its cost and the poor definition of commercial sources of this material, which makes it difficult to follow up on well-controlled processing. Therefore, collagen type I can be replaced by other compounds. One should notice that, besides collagen, both human and mammalian bodies contain dozens types of various bioorganic compounds, proteins, and biological macromolecules. The substantial amount of them potentially might be used to prepare biocomposites with calcium orthophosphates. For example, a biologically strong adhesion (to prevent invasion of bacteria) between teeth and the surrounding epithelial tissues is attributed to a cell-adhesive protein, laminin [789]. In order to mimic the nature, a laminin/apatite biocomposite layer was successfully created on the surface of both titanium [790] and EVOH [791, 792] using the biomimetic approach. A more complicated laminin/DNA/apatite biocomposite layer was found to be an efficient gene transfer system [793].

Calcium orthophosphate/gelatin biocomposites are widely investigated as potential bone replacement biomaterials [317, 336–338, 444–452, 466–474, 487, 521–523, 554, 619–624, 794–807]. For example, gelatin foams were successfully mechanically reinforced by HA and then cross-linked by a carbodiimide derivative [317]. Such foams were shown to be a good carrier for antibiotic tetracycline [798]. Several biocomposites of calcium orthophosphates with alginates have been prepared [471, 617, 618, 622, 709, 808, 809]. For example, porous HA/alginate composites based on hydrogels were prepared both biomimetically [709] and by using a freeze-drying technique [808]. Another research group succeeded in preparation of biphasic but monolithic scaffolds using a similar preparation route [810]. Their biocompatibility in cell culture experiments and *in vitro* biodegradability were high; however, a mechanical strength could be better.

Various biocomposites of calcium orthophosphates with chitosan [298, 482, 500, 512, 527, 565, 645–653, 674, 675, 690, 695, 799, 807,

811–828] and chitin [232, 476, 612, 829–833] are also very popular. For example, a solution-based method was developed to combine HA powders with chitin, in which the ceramic particles were uniformly dispersed [829, 830]. Unfortunately, it was difficult to obtain the uniform dispersions. The mechanical properties of the final biocomposites were not very good; due to a poor adhesion between the filler and the matrix both the tensile strength and modulus were found to decrease with increase of the HA amount. Microscopic examination revealed that HA particles were intervened between the polymer chains, weakening their interactions and decreasing the entire strength [829, 830]. Other manufacturing techniques might be found in the aforementioned references; I just would like to mention on an interesting approach, in which a HA/chitosan biocomposite was produced by a hydrothermal process from natural $CaCO_3$/chitosan biocomposite of crab shells [827]. Biocomposites of natural HA with chitosan were found to possess both a good hard tissue biocompatibility and an excellent osteoconductivity, which is suitable for artificial bone implants and frame materials of tissue engineering [823]. Data are available that addition of HA into chitosan improved cell attachment, provided a higher cell proliferation and well-spread morphology when compared to chitosan alone [650]. More complex formulations, such as silk fibers reinforced HA/chitosan [834] and HA/collagen/chitosan [835] biocomposites, have been studied as well. Interestingly, but hybrid biocomposites of nano-sized HA with chitin/chitosan might be used for removal of Fe(III) [836] and fluorides [837, 838] from aqueous solutions. Further details on the biocomposites and hybrid biomaterials of calcium orthophosphates with chitosan are available in literature [836].

Biocomposites of CDHA with water-soluble proteins, such as BSA, might be prepared by a precipitation method [561, 839–842]. In such biocomposites, BSA is not strongly fixed to solid CDHA, which is useful for a sustained release. However, this is not the case if a water/oil/water interfacial reaction route has been used [313]. To extend this subject, inclusion of DNA into CDHA/BSA biocomposites was claimed [313, 843–845]. Furthermore, nanodimensional biocomposites of an unspecified calcium orthophosphate with DNA [846], as well as biocomposites of nano-sized crystals of biomimetic apatite with C_{60} and Au-DNA nano-sized particles [847] were prepared as well.

Akashi and co-workers developed a procedure to prepare calcium orthophosphate-based biocomposites by soaking hydrogels in supersaturated by Ca^{2+} and PO_4^{3-} ions solutions in order to precipitate CDHA in the hydrogels (up to 70% by weight of CDHA could be added to these biocomposites) [848]. This procedure was applied to chitosan; the 3D shape of the resulting biocomposite was controlled by the shape of the starting chitosan hydrogel [849]. Another research group developed biocomposites based on *in situ* calcium orthophosphate mineralization of self-assembled supramolecular hydrogels [850]. Other experimental approaches are also possible [851].

Various biocomposites of CDHA with glutamic and aspartic amino acids, as well as poly-glutamic and poly-aspartic amino acids, have been prepared and investigated by Bigi et al. [346, 347, 852–855]. These (poly)amino acids were quantitatively incorporated into CDHA crystals, provoking a reduction of the coherent length of the crystalline domains and decreasing the crystal sizes. The relative amounts of the (poly)amino acid content in the solid phase, determined through HPLC analysis, increased with their concentration in solution up to a maximum of about 7.8 wt.% for CDHA/aspartic acid and 4.3 wt.% for CDHA/glutamic acid biocomposites. The small crystal dimensions, which implied a great surface area, and the presence of (poly)amino acids were suggested to be relevant for possible application of these biocomposites for hard tissues replacement [346, 347, 852–855]. A schematic description of a biocomposite formation from amino acids and ACP is shown in Fig. 6.4 [856].

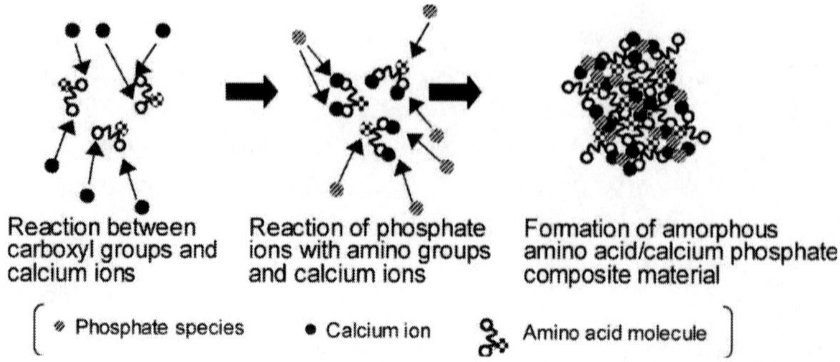

Figure 6.4 A proposed mechanism for the formation of ACP/amino acid biocomposites in aqueous solutions. Reprinted from Ref. [856] with permission.

Furthermore, BCP (HA + β-TCP)/agarose macroporous scaffolds with controlled and complete interconnection, high porosity, thoroughly open pores and tailored pore size were prepared for tissue engineering application [857, 858]. Agarose, a biodegradable polymer, was selected as the organic matrix because it was a biocompatible hydrogel, which acted as gelling agent leading to strong gels and fast room temperature polymerization. Porous scaffolds with the designed architecture were manufactured by combining a low temperature shaping method with stereo-lithography and two drying techniques. The biocompatibility of this BCP/agarose system was tested with mouse L929 fibroblast and human SAOS-2 osteoblast during different colonization times [859].

Fibrin sealants are non-cytotoxic, fully resorbable, biological matrices that simulate the last stages of a natural coagulation cascade, forming a structured fibrin clot similar to a physiological clot [860]. Biocomposites of calcium orthophosphates with fibrin sealants might develop the clinical applications of bone substitutes. The 3D mesh of fibrin sealant interpenetrates the macro- and micro-porous structure of calcium orthophosphate ceramics [11]. The physical, chemical, and biological properties of calcium orthophosphate bioceramics and the fibrin glue might be cumulated in biocomposites, suitable for preparation of advanced bone grafts [861–873].

Furthermore, there are biocomposites of calcium orthophosphates with bisphosphonates [874], silk fibroin (that is a hard protein extracted from silk cocoon) [312, 670–672, 677, 678, 875–881], chitosan + silk fibroin [882], fibronectin [883], chondroitin sulfate [299, 733, 884], casein phosphopeptides [885], and vitamins [886]. Besides, the reader's attention is pointed out to an interesting approach to crystallize CDHA inside poly(allylamine)/poly(styrene sulfonate) polyelectrolyte capsules resulting in empty biocomposite spheres of micron size [887]. Depending on the amount of precipitated CDHA, the thickness of the shell of biocomposite spheres can be varied between 25 and 150 nm. These biocomposite capsules might find application as medical agents for bone repairing and catalytic microreactors [887].

6.4.6 Injectable Bone Substitutes (IBS)

With the development of minimally invasive surgical methods, for example percutaneous surgery, directly injectable biomaterials are

Table 6.4 A list of some commercial non-setting calcium orthophosphate IBS and pastes with indication of producer, product name, composition (when available) and form [910].

Producer	Product name	Composition	Form
ApaTech (UK)	Actifuse™	HA, polymer and aqueous solution	Pre-mixed
	Actifuse™ Shape Actifuse™ ABX	Si-substituted calcium orthophosphate and a polymer	Pre-mixed
Baxter (US)	TricOs T TricOs	BCP (60% HA, 40% β-TCP) granules and Tissucol (fibrin glue)	To be mixed
Berkeley Advanced Biomaterials	Bi-Ostetic Putty	not disclosed	Not disclosed
BioForm (US)	Calcium hydroxylapatite implant	HA powder embedded in a mixture of glycerine, water and carboxymethylcellulose	Pre-mixed
Biomatlante (FR)	MBCP Gel®	BCP granules (60% HA, 40% β-TCP; 0.08–0.2 mm) and 2% HPMC	Pre-mixed
	Hydr'Os	BCP granules (60% HA, 40% β-TCP; micro- and nano-sized particles) and saline solution	Pre-mixed
Degradable solutions (CH)	Easy graft™	β-TCP or BCP granules (0.45–l.0 mm) coated with 10 μm PLGA, N-methyl-2-pyrrolydone	To be mixed
Dentsply (US)	PepgenP-15® flow	HA (0.25–0.42 mm), P-15 peptide and aqueous Na hyaluronate solutionT	To be mixed

DePuy Spine (US)	Healos® Fx	HA (20–30%) and collagen	To be mixed
Fluidinova (P)	nanoXIM TCP	β-TCP (5 or 15%) and water	Pre-mixed
	nanoXIM HA	HA (5, 15, 30 or 40%) and water	Pre-mixed
Integra LifeSciences (US)	Mozaik Osteoconductive Scaffold	β-TCP (80%) and type 1 collagen (20%)	To be mixed
Mathys Ltd (CH)	Ceros® Putty / cyclOS® Putty	β-TCP granules (0.125–0.71 mm; 94%) and recombinant Na hyaluronate powder (6%)	To be mixed
Medtronic (US)	Mastergraft®	BCP (85% HA, 15% β-TCP) and bovine collagen	To be mixed
Osartis / AAP (GER)	Ostim®	Nanocrystalline HA (35%) and water (65%)	Pre-mixed
Smith & Nephew (US)	JAXTCP	β-TCP granules and an aqueous solution of 1.75% carboxymethylcellulose and 10% glycerol	To be mixed
Stryker (US)	Calstrux™	β-TCP granules and carboxymethylcellulose	To be mixed
Teknimed (FR)	Nanogel	HA (100–200 nm) (30%) and water (70%)	Pre-Pixed
Therics (US)	Therigraft™ Putty	β-TCP granules and polymer	Pre-mixed
Zimmer (US)	Collagraft	BCP granules (65% HA, 35% β-TCP; 0.5–1.0 mm), bovine collagen and bone marrow aspirate	To be mixed

needed. The challenge is to place a biomaterial at the site of surgery by the least possible invasive method. In this regard, IBS appear to be a convenient alternative to solid bone-filling materials. They represent ready-to-use suspensions of calcium orthophosphate microspheres [888, 889], nano-sized rods [890] or powder(s) in a liquid carrier phase. However, addition of other phases, such as calcium sulfate [891], is possible. They look like opaque viscous pastes with the rheological properties, sufficient to inject them into bone defects by means of surgical syringes and needles. Besides, IBS could be easily produced in a sterile stage. Their stable composition and mechanical properties are suitable for reproducibility of the biological response [892]. All types of IBS are divided into 2 major groups: self-setting formulations and those, which does not set. The former ones belong to cements and concretes (see section *6.4.2* above and Chapter 5 for details), while the latter ones are described here.

IBS requires suitable rheological properties to ensure bonding of the mineral phase *in situ* with good cell permeability. Usually, the necessary level of viscosity is created by addition of water-soluble polymers [131, 893, 894]. Therefore, the majority of calcium orthophosphate-based IBS formulations might be considered as a subgroup of calcium orthophosphate/polymer biocomposites. For example, an IBS was described that involved a silanized hydroxyethylcellulose carrier with BCP (HA + β-TCP) [895]. The suspension is liquid at pH within 10–12, but gels quickly at pH < 9. Injectable composites can be formed with β-TCP to improve mechanical integrity [548]. Similarly, Bennett et al. showed that a polydioxanone-*co*-glycolide-based biocomposite reinforced with HA or β-TCP can be used as an injectable or moldable putty [896]. During the cross-linking reaction following injection, carbon dioxide is released allowing the formation of interconnected pores. Furthermore, HA/poly(L-lactide-*co*-ε-caprolactone) biocomposite microparticles were fabricated as an injectable scaffold via the Pickering emulsion route in the absence of any molecular surfactants. A stable injectable oil-in-water emulsion was obtained using water dispersed HA nano-sized crystals as the particulate emulsifier and a dichloromethane solution of poly(L-lactide-*co*-ε-caprolactone) as an oil phase [897].

Daculsi et al. developed viscous IBS biocomposites based on BCP (60% HA + 40% β-TCP) and 2% aqueous solution of HPMC that was said to be perfectly biocompatible, resorbable and easily fitted bone defects (due to an initial plasticity) [108, 894, 898–905]. The best ratio BCP/HPMC aqueous solution was found to be at ~ 65/35 w/w. To extend this subject further, IBS might be loaded by cells [906, 907], radiopaque elements [908] or microparticles [909], as well as functionalized by nucleic acids [890]. Self-hardening formulations, based on Si-HPMC hydrogel, are known as well [906]. The list of the commercially available calcium orthophosphate-based IBS formulations is presented in Table 6.4 [910].

The advanced characteristics of IBS come from their good rheological properties and biocompatibility and the ease of tissue regeneration. Although the fabrication of IBS biocomposites in most cases improved the mechanical properties of the system and provided the material with resistance to fluids penetration, these achievements were limited by the amount of polymer that can be added to the paste. For instance, Mickiewicz et al. reported that after a critical concentration (that depended on the type and molecular weight of the polymer, but was always around 10%), the polymer started forming a thick coating on the crystal clusters, preventing them from interlocking, originating plastic flow and, as a consequence, decreasing mechanical properties [561]. More to the point, Fujishiro et al. reported a decrease in mechanical properties with higher amounts of gel, which was attributed to formation of pores due to leaching of gelatin in solution [554]. Therefore, it seems that mechanical properties, although improved by the addition of polymers, are still a limitation for the application of calcium orthophosphate-based IBS formulations in load-bearing sites [178]. Further details on IBS might be found in a recent review [892].

6.4.7 Biocomposites with Glasses, Inorganic Materials, Carbon and Metals

To overcome the problem of poor mechanical properties of calcium orthophosphate bioceramics, suitable biocomposites of calcium orthophosphates reinforced by various inorganic materials, glasses, and metals have been developed. Such biocomposites are mainly prepared by the common ceramic processing techniques such as thermal treatment after kneading [911–913], powder slurry coating [914] and metal-sol mixing [915]. For example, HA was combined with Bioglass® (Novabone Products, Alachua, FL) [916, 917] and with other glasses [918] to form glass-ceramics biocomposites. Other reinforcement materials for calcium orthophosphates are differentiated by either shape of the fillers, namely, particles [919, 920], platelets [921, 922], whiskers [579, 923–925], fibers [926–930], or their chemical composition: zirconia and/or PSZ [313, 911–914, 923, 931–966], alumina [313, 919, 922, 965, 967–996], other oxides [925, 997–1004], silica and/or glasses [1005–1014], wollastonite [206, 1015–1025], mullite [1026, 1027], various metals and alloys [540, 928, 967, 997, 1028–1045], calcium sulfate [1046–1049], calcium carbonate [1050, 1051], silicon carbide [683, 924], barium titanate [1052], zeolite [1053], boron nitride [1054] and several other materials [335, 1055–1057]. More complicated formulations, such as HA/aluminum oxide/carbon nanotubes [1058], have been developed as well. All these materials have been added to calcium orthophosphate bioceramics to improve its reliability. Unfortunately, significant amounts of the reinforcing phases are needed to achieve the desired properties and, as these materials are either bioinert, significantly less bioactive than calcium orthophosphates or not bioresorbable, the ability of the biocomposites to form a stable interface with bone is poorer if compared with calcium orthophosphate bioceramics alone. Due to the presence of bioinert compounds, such formulations might be called bioinert/bioactive composites [1005]. The ideal reinforcement material would impart mechanical integrity to a biocomposite at low loadings, without diminishing its bioactivity.

There are several types of HA/glass biocomposites. The first one is also called bioactive glass-ceramics. A dense and homogeneous biocomposite was obtained after a heat treatment of the parent glass, which comprised ~38 wt.% oxy-FAP ($Ca_{10}(PO_4)_6(O,F)_2$) and ~34 wt.% β-wollastonite ($CaO{\cdot}SiO_2$) crystals, 50–100 nm in size in a MgO-CaO-SiO_2 glassy matrix [206, 1015–1025]. A-W glass-ceramics is an assembly of small apatite particles effectively reinforced by wollastonite. The bending strength, fracture toughness and Young's modulus of A-W glass-ceramics are the highest among bioactive glass and glass-ceramics, enabling it to be used in some major compression load-bearing applications, such as vertebral prostheses and iliac crest replacement. It combines a high bioactivity with the suitable mechanical properties [1059]. β-TCP/wollastonite biocomposites are also known [1060–1062]. More complicated formulations have been developed as well. For example, (A-W)/HDPE composite (AWPEX) biomaterials have been designed to match the mechanical strength of human cortical bone and to provide favorable bioactivity, with potential use in many orthopedic applications [1063–1066]. Other examples comprise wollastonite-reinforced HA/Ca polycarboxylate [1067], glass-reinforced HA/polyacrylate [1068], as well as collagen- [1069] and gelatin- [1070] calcium phosphate silicate/wollastonite biocomposites.

HA/glass biocomposites can be prepared by simple sintering of appropriate HA/glass powder mixtures [1071–1074]. If sintering is carried out below 1000 ºC, HA does not react with the bioactive glass [1072, 1073] or this reaction is limited [1074]. Besides, reaction between HA and glasses depends on the glass composition. In another approach, small quantities of bioactive glass have been added to HA bioceramics in order to improve densification and/or mechanical properties [29]. In addition, biocomposites might be sintered from HA and silica [1005]. In general, bioactive glass-ceramics maintain a high strength for a longer time than HA bioceramics under both the *in vitro* and *in vivo* conditions [1012, 1019].

Due to a huge difference in shapes, it is a challenge to prepare homogeneous mixtures of calcium orthophosphates and carbon nanotubes: "one can imagine something similar to achieving a homogeneous mixture of peas and spaghetti" [217, page 7]. Nevertheless, different strategies might be employed to prepare calcium orthophosphate/carbon nanotube biocomposites. For example, apatites might be chemically synthesized by using carboxyl functionalized carbon nanotubes as a matrix [302–308]. Physico-chemical characterization of these biocomposites showed that nucleation of CDHA initiates through the carboxyl group [302]. Hot pressing [1075], plasma spraying [1076], laser surface alloying [1077–1079], spark plasma sintering [1080], and precipitation [1081] techniques might be applied as well. Due to carbon oxidation at elevated temperature, sintering of calcium orthophosphate/carbon nanotube biocomposites must be performed in a deoxidizing atmosphere [1082]. The research on calcium orthophosphate (up to now, only apatites)/carbon nanotube biocomposites is in its early stages, with the first papers published in 2004 [307, 525]. Due to this reason, the mechanical property data for such biocomposites have been reported only in few papers; however, these results are encouraging. For example, Chen et al. performed nano-indentation tests on biocomposite coatings to give hardness and Young's modulus values [1079]. They found that the higher the loading of the nanotubes, the better the mechanical properties. Namely, at 20 wt.% loading, hardness was increased by ~43% and Young's modulus by ~21% over a single-phase HA coating [1079]. Scratching test results indicated that as alloyed HA biocomposite coatings exhibited improved wear resistance and lower friction coefficient with increasing the amount of carbon nanotubes in the precursor material powders [1078]. Additionally, measurements of the elastic modulus and hardness of the biocomposite coatings indicated that the mechanical properties were also affected by the amount of carbon nanotubes [1077]. Another research group performed compression tests on bulk HA/carbon nanotubes biocomposites and found an increase in strength over single-phase HA [307]. However, the highest compressive strength they achieved for any material was only 102 MPa, which is similar to that of cortical bone but much lower than the typical values for dense HA [217]. More complex formulations, such as poly-L-lysine/HA/carbon nanotube hybrid biocomposites, have been also developed [1083]. Furthermore, calcium orthophosphate/carbon nanotube biocomposites might be immobilized by hemoglobin [1084]. Unfortunately, carbon nanotubes are very stable substances; they are neither bioresorbable nor biodegradable. Therefore, during *in vivo* bioresorption, the nanotubes will get into the human body from the biocomposite matrix and might cause uncertain health problems. Certainly, this problem must be solved. To conclude the carbon subject, one should mention an application of carbon fibers of microscopic dimensions [1085–1087], nanodimensional diamonds [1088] and C_{60} [847] to reinforce HA bioceramics.

As clearly seen from the amount of the references, apatite/zirconia biocomposites are most popular ones among the researchers. The main disadvantage of HA reinforced by PSZ is degradation of zirconia in wet environments [923, 932, 933, 955]. Transformation of the tetragonal ZrO_2 to the monoclinic phase on the surface results in formation of microcracks and consequently lowers the strength of the implant [1089, 1090]. Interestingly, but Fe_3O_4/HA composites possess photocatalytic properties [1003, 1004].

Various biocomposites of calcium orthophosphates with metals and alloys have been fabricated as well [540, 928, 967, 997, 1028–1045]. For example, an HA-based biocomposite reinforced with 20 vol.% of Ti particles was fabricated by hot pressing [1030]. Besides, calcium orthophosphate/Ti biocomposites might be prepared by powder metallurgy processing [1032–1034]. At high temperatures, the presence of Ti metal phase was found to promote dehydration and decomposition of HA into β-TCP and TTCP [1030, 1032] or partial formation of β-TCP and calcium titanate instead of HA [661, 1033, 1034]. Comparing with pure HA bioceramics manufactured under the same conditions, the HA/Ti biocomposites possessed a higher fracture toughness, bending strength, work of fracture, porosity and lower elastic modulus, which is more suitable for biomedical applications. However, the mechanical properties appeared to be not high enough to use HA/Ti biocomposites in load-bearing applications. Luckily, the histological evaluations revealed that HA/Ti biocomposites could be partially integrated with newborn bone tissues after 3 weeks and fully osteointegrated at 12 weeks *in vivo* [1030]. Similar findings had been earlier made for HA bioceramics reinforced by addition of silver particulates (5–30 vol.%) and subsequent sintering of the HA/Ag powder compacts [1028, 1029]. Besides, addition of silver imparts an antimicrobial activity [1042]. Other studies on calcium orthophosphate/Ti biocomposites are available elsewhere [1035–1038].

To conclude this part, biocomposites consisting of calcium orthophosphates only should be briefly described. First of all, all multiphasic and polyphasic calcium orthophosphates (section *1.3.14*) should be mentioned. For example, in 1980s, BCP was called as "TCP ceramics complexed with HA" [1091]. Even nowadays BCP is occasionally called as a "nanocomposite" [1092]. Furthermore, FHA or HFA (see section *1.3.11. FA (or FAp)* for details) might mentioned as composites [1093]; however, an applicability of the "composite" term for such systems is doubtful. One should better consider 70% HA-powder + 30% HA-whisker biocomposites, which were fabricated by pressureless sintering, hot pressing and hot isostatic pressing. These biocomposites were found to exhibit an improved toughness, attaining the lower fracture-toughness limit of bone without a decrease of bioactivity and biocompatibility [1094, 1095]. Besides, a dual HA biocomposite that combined two HA materials with different porosities: HA with 75% porosity, for bone ingrowth and HA with 0% porosity, for load bearing was manufactured. This dual HA biocomposite appeared to be suitable for use as an implant material for spinal interbody fusion as a substitute for iliac bone grafts, which could eliminate the disadvantages associated with autograft harvesting [1096]. A biodegradable biocomposite porous scaffold comprising a β-TCP matrix and nano-sized fibers of HA was developed and studied for load-bearing bone tissue engineering. The nano-sized fibers of HA were prepared by a biomimetic precipitation method, the inclusion of which significantly enhanced the mechanical property of the scaffold, attaining a compressive strength of 9.87 MPa, comparable to the high-end value (2–10 MPa) of cancellous bone [1097]. Finally, it is interesting to mention on a successful reinforcement of carbonateapatite porous blocks by newly prepared carbonateapatite crystals (i.e., by the same compound; thus, a biocomposite of two different carbonateapatites was obtained) [1098]. First, a calcium salt was introduced to micropores of carbonateapatite blocks. Then, the calcium salt was carbonated to form calcite inside the micropores of the carbonateapatite blocks by exposing the blocks to carbon dioxide at the second step. On the third step, the blocks were immersed in a Na_2HPO_4 aqueous solution. In this process, calcite inside the micropores of the carbonateapatite blocks was transformed to carbonateapatite and the newly formed crystals of carbonateapatite entangled on those of the existing carbonateapatite blocks. Due to bonding between the newly formed carbonateapatite crystals and the existing ones in the carbonateapatite blocks, a mechanical strength of the blocks became ~1.5 times higher when compared to that before the treatment [1098].

6.4.8 Functionally Graded Biocomposites

Although, in most cases, the homogeneous distribution of filler(s) inside a matrix is required [426], there are composites, where this is not the case. For example, functionally graded materials (commonly referred to as FGM) might be characterized by the intentional variations in composition and/or structure gradually over volume, resulting in corresponding changes in the properties of the composite. The main feature of such materials is the almost continuously graded composition that results in two different properties at the two ends of the structure. Such composites can be designed for specific function and applications. Various approaches based on the bulk (particulate processing), preform processing, layer processing and melt processing are used to fabricate the functionally graded materials.

Bone is a biologically formed composite with variable density ranging from very dense and stiff (cortical bone) to a soft and foamed structure (trabecular bone). Normally the outer part of long bones consists of cortical bone with the density decreasing towards the core, where the trabecular bone is found. The trabecular bone is porous and the porosity is filled with osseous medulla [25, 26]. This brief description clearly indicates that bones are natural functionally graded composites.

The concept of FGM has been increasingly used for biomaterial design and currently it remains to be an important area of the research. For example, many studies have been performed to fabricate porosity-graded calcium orthophosphate bioceramics in attempts to mimic the porous structure of bones [1099–1102]. This is a structural approach to fabricate FGM. Besides, there is a compositional approach. For example, powder metallurgy methods have been used to fabricate HA/Ti functionally graded biocomposite dental implants offering the biocompatible HA on the tissue side and titanium on the outer side for mechanical strength [1103–1105]. The graded structure in the longitudinal direction contains more Ti in the upper section and more HA in the lower section. Actually, in the upper section the occlusal force is directly applied and Ti offers the required mechanical performance; in the lower part, which is implanted inside the bone, the HA confers the bioactive and osteoconductive properties to the material [1103]. Since the optimum conditions of sintering for Ti and HA are very different, HA/Ti functionally graded biocomposites are difficult to fabricate and the sintering conditions for their mixtures are obliged to compromise. The expected properties of this implant are shown in Fig. 6.5 [1104]. Such biocomposites might be both symmetrical

[1106] and asymmetrical [1107]. Furthermore, functionally graded HA/Ti biocomposite coatings might be prepared by rf-plasma spraying [1108]. More to the point, Ti alloy substrate has been combined with HA granules spread over the surface [1109].

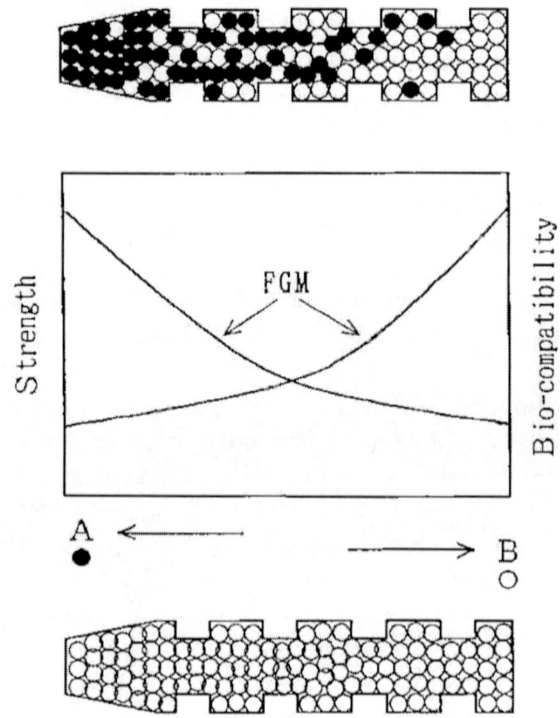

Figure 6.5 Expected properties of functionally graded biocomposite dental implant. For comparison, the upper drawing shows a functionally graded implant and the lower one shows a conventional uniform implant. The properties are shown in the middle. The implant with the composition changed from a biocompatible metal (Ti) at one end (left in the figure), increasing the concentration of bioceramics (HA) toward 100% HA at the other end (right in the figure), could control both mechanical properties and biocompatibility without an abrupt change due to the formation of discrete boundary. This FGM biocomposite was designed to provide more titanium for the upper part where occlusal force is directly applied and more HA for the lower part, which is implanted inside the jawbone. Reprinted from Ref. [1104] with permission.

A series of functionally graded HA coatings incorporated with various percentages of silver were deposited on titanium substrates using ion beam-assisted deposition. The analysis of the coating's cross-section revealed a decreased crystallinity as well as a distribution of nano-sized (10–50 nm) silver particles from the coating/substrate interface to top surface [1110]. A functionally graded HA/PMMA biocomposite was developed based on sedimentary HA distributions in a PMMA viscous fluid, using a centrifuge to avoid stress convergence on the interface. The stress-strain curves of this biocomposite showed a sufficient strength for biomedical applications along with the relaxation of brittleness and fragility [543]. A compositionally graded collagen/nanodimensional HA biocomposite scaffold might be prepared by an *in situ* diffusion method [1111]. Chemical and microstructural analysis revealed a gradient of the Ca to P ratio across the width of the scaffold template, resulting in the formation of a Ca-rich side and a Ca-depleted side of scaffold. The Ca-rich side featured low porosity and agglomerates of the nanodimensional HA crystallites, while the Ca-depleted side featured higher porosity and nanodimensional HA crystallites integrated with collagen fibrils to form a porous network structure [1111]. A three-layered graded biocomposite membrane, with one face of 8% nanodimensional carbonateapatite/collagen/PLGA porous membrane, the opposite face of pure PLGA non-porous membrane and the middle layer of 4% nanodimensional carbonateapatite/collagen/PLGA as the transition was prepared through the layer-by-layer casting method [611]. Functionally graded non-woven meshes of PCL incorporated by nano-sized particles of β-TCP were prepared using a hybrid twin-screw extrusion/electrospinning process [1112]. A functionally graded HA/silk fibroin biocomposite was prepared by pulse electric current sintering [1113]. HA/glass FGM layers were coated on titanium alloy (Ti – 6Al – 4V) substrates. The design of these layers and the use of the glass were for achieving a strong bonding between the FGM layered coatings and the substrates [1114, 1115].

Functionally graded β-TCP/FA biocomposites combine the biostability of FA with the bioresorbable properties of β-TCP [1116]. An interesting multilayered (each layer of 1 mm thick) structure consisting of β-TCP/FA biocomposites with different molar ratios has been prepared, giving rise to formation of a FGM (Fig. 4.11).

After implantation, the preferential dissolution of β-TCP phase would result in functionally gradient porosity for bone ingrowth [1116]. Functionally graded fluoridated HA (FHA) with a gradient of fluoride [1117] and carbonated HA with a gradient of carbonate [1118] were synthesized as well. HA/zirconia graded biocomposites were fabricated to enhance the mechanical properties of HA while retaining its bone bonding property [963]. TiO_2 and HA were found to be a good combination for FGM providing both a gradient of bioactivity and a good mechanical strength [1119]. Besides, graded HA/$CaCO_3$ biocomposite structures for bone ingrowth were developed [1120]. Functionally graded composite skull implants consisting of polylactides, carbonateapatite, and $CaCO_3$ are known as well [386, 387]. Thus, the research in this field is quite promising but currently the mechanical properties of the available biocomposites do not match the similar properties of bones [179].

6.4.9 Biosensors

A biosensor is a device for detection of an analyte that combines a biological component with a physicochemical detector component. Very briefly, it consists of three parts: a sensitive biological element; a transducer or a detector element that transforms the signal resulting from the interaction of the analyte with the biological element into another signal; and associated electronics that is primarily responsible for the display of the results in a user-friendly way [1121].

The surface of biologically relevant calcium orthophosphates (CDHA, HA, α-TCP, β-TCP, DCPD, DCPA) has an excellent ability of adsorption for functional biomolecules such as proteins, albumins, DNA, as well as some other types of chemicals. Therefore, several calcium orthophosphate-based biocomposites and hybrid biomaterials were found to be applicable for biosensor manufacturing [354, 642, 1041, 1083, 1122–1127]. For example, formation of poly-L-lysine/HA/carbon nanotube hybrid nanodimensional particles was described and a general design strategy for an immunosensing platform was proposed based on adsorption of antibodies onto this biocomposite [1083]. In another paper, a hybrid material formed by assembling of nanodimensional particles of gold onto nano-sized HA was employed for the interface design of piezoelectric immunosensor, on which the antibodies were bound. The developed sensing interface appeared to possess some advantages, such as activation-free immobilization and high antigen-binding activities of antibodies, over using nano-sized either HA or gold alone [1041]. A novel tyrosinase biosensor based on nano-sized HA/chitosan composite has been developed for the detection of phenolic compounds [1125]. Further details on the subject are available in the aforementioned references.

Up to date, not many papers have been published on the biosensor application of calcium orthophosphate-based biocomposites and hybrid biomaterials. Presumably, this subject will be further developed in future and, perhaps, sometime implantable biosensors will be designed to perform the continuous concentration monitoring of the important biological macromolecules *in vivo*. Possibly, those implantable biocencors will be able to use an electric power, generated by DCPD/polymer composite-based battery devices [506, 507].

6.5 Interaction Among the Phases in Calcium Orthophosphate-Based Biocomposites

An important aspect that should be addressed in details is a mutual interaction among calcium orthophosphates and other phases in biocomposites and hybrid biomaterials. In general, an interaction among the phases in any composite can be either mechanical, when it results from radial compression forces exerted by the matrix on the filler particles (for example, developed during cooling due to thermal contraction), or chemical, when the reactivity of the filler towards the matrix has an important role. In the latter case, it is important to distinguish a physical interaction from chemical bonding [282]. According to Wypych [1128], physical interaction is more or less temporary, implicating hydrogen bonding or van der Waals forces, whereas chemical bonding is stronger and more permanent, involving covalent bond formation. Thus, a chemical interfacial bond among the phases is preferred to achieve a higher strength of a composite. The magnitude of the interfacial bond among the phases determines how well a weak matrix transmits stress to the strong fibers. However, while a bond among the matrix and reinforcements must exist for the purpose of stress transfer, it should not be so strong that it prevents toughening mechanisms, such as debonding and fiber pullout [217].

There is still doubt as to the exact bonding mechanism among bone minerals (biological apatite) and bioorganics (collagen), which undoubtedly plays a critical role in determining the mechanical properties of bones. Namely, bone minerals are not directly bonded to collagen but through non-collagenous proteins that make up ~3% of bones (Table 6.1) and provide with active sites for biomineralization and for cellular attachment [36]. In bones, the interfacial bonding forces are mainly ionic bonds, hydrogen bonds and hydrophobic interactions, which give the bones the unique composite behavior [53]. There is an opinion that, opposite to bones, there is no sign of chemical bonding among the phases in conventional calcium orthophosphate/collagen biocomposites, probably due to a lack of suitable interfacial bonding during mixing [39]. However, this is not the case for phosphorylated collagens [753]. The interested readers are forwarded to a density functional theory study of the interaction of collagen peptides with hydroxyapatite surfaces [1129].

Anyway, Fourier-transformed infrared (FTIR) spectra of some calcium orthophosphate-based composites and collagen films were measured and transformed into absorption spectra using the Kramers-Kronig equation to demonstrate energy shifts of residues on the apatite/collagen interface. After comparing FTIR spectra of biocomposites and collagen films in detail, red shifts of the absorption bands for C–O bonds were observed in the spectra of the biocomposites. These red shifts were described as a decrease of bonding energies of C–O bonds and assumed to be caused by an interaction to Ca^{2+} ions located on the surfaces of apatite nano-sized crystals, as shown in Fig. 6.6 [747]. Another proof of a chemical interaction between apatite and collagen was also evaluated in FTIR spectra of CDHA/collagen biocomposites, in which a shift of the band corresponding to $-COO^-$ stretching from 1340 to 1337 cm^{-1} was observed [708, 709]. More to the point, nucleation of apatite crystals onto collagen through a chemical interaction with carboxylate groups of collagen macromolecules has been reported [1130–1132].

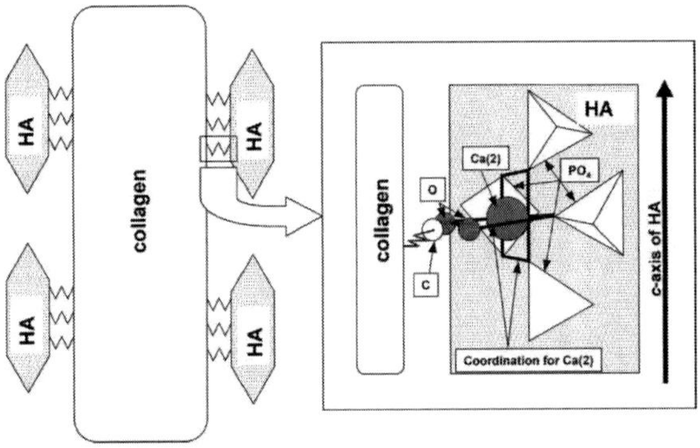

Figure 6.6 A schematic drawing of the relation between self-organization (directional deposition of HA on collagen) and interfacial interaction in biocomposites. Direction of interaction between HA and collagen is restricted by covalent bond between COO and Ca(2) to maintain regular coordination number of 7. Reprinted from Ref. [747] with permission.

FTIR spectroscopy seems to be the major tool to study a possible chemical bonding among the phases in calcium orthophosphate-based biocomposites and hybrid biomaterials [276, 310, 345, 353, 355, 462, 513, 600, 616, 629, 635, 638, 641, 644, 653, 664, 673, 678, 709, 753, 802, 803, 847, 882, 1133–1136]. For example, the characteristic bands at 2918, 2850 and 1472 cm^{-1} for the hydrocarbon backbone of PE appeared to have zero shift in an HA/PE biocomposite. However, in the case of polyamide, several FTIR bands indicated that the polar groups shifted apparently: the bands at 3304, 1273 and 692 cm^{-1} derived from stretching of N–H, stretching of C–N–H and vibrating of N–H moved to 3306, 1275 and 690 cm^{-1}, respectively, in the HA/polyamide biocomposites. Furthermore, both stretching (3568 cm^{-1}) and vibrating (692 cm^{-1}) modes of hydroxide in HA moved to 3570 and 690 cm^{-1} in the HA/polyamide biocomposites, respectively, indicating the formation of hydrogen bonds. Besides, bands at 1094 and 1031 cm^{-1} of PO_4 modes also shifted to 1093 and 1033 cm^{-1} in the HA/polyamide biocomposite. The bands shift in a fingerprint area indicated that the hydroxide and orthophosphate on the surface of HA might interact with plentiful carboxyl and amino groups of polyamide through nucleophilic addition [276]. Comparable conclusions were made for HA/PVA [641], CDHA/alginate [709], ACP/PPF [513], HA/maleic anhydride [355] and β-TCP/PLLA [462] biocomposites, in which weak chemical bonds were considered to form between Ca^{2+} ions located on the HA, CDHA, ACP or β-TCP surface, respectively, and slightly polarized O atoms of C=O bonds in the surrounding bioorganic compounds. The data obtained suggest that crystallization of calcium orthophosphates in chitosan-containing solutions is substantially modulated by a chemical interaction of the components; apparently, a part of calcium is captured by chitosan and does not participate in the formation of the main mineral phase [1136]. Schematically, this type of the chemical interaction is shown in Fig. 6.7 [709].

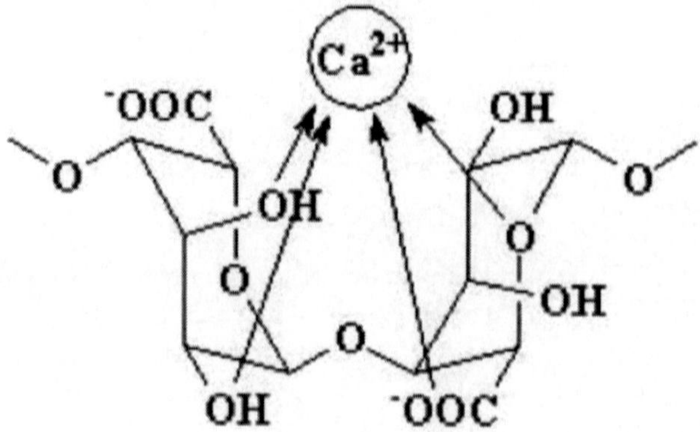

Figure 6.7 A schematic diagram of Ca^{2+} ion binding with alginate chains. Reprinted from Ref. [709] with permission.

Except of FTIR spectroscopy, other measurement techniques are also able to show some evidences of a chemical interaction among the phases in calcium orthophosphate-based biocomposites and hybrid biomaterials [345, 462, 635, 638, 641, 1134–1138]. For example, for nano-sized crystals of CDHA/alendronate such evidences were observed by thermogravimetric analysis: DTG plots of the crystals appeared to be quite different from those obtained from mechanical mixtures of CDHA and calcium alendronate with similar compositions [1137]. Analogous DTG results were obtained for nano-sized HA/PVA biocomposites [641]. In the case of biocomposites of nano-sized HA with polyamide, a hydrogen bonding among the phases was detected by differential scanning calorimetry technique [635]. Another example comprises application of the dynamic mechanical analysis to investigate softening mechanism of β-TCP/PLLA biocomposites

[462]. As to biocomposites of nano-sized HA with PVAP, some indirect evidences of a chemical bonding among the phases were found by X-ray diffraction and thermogravimetric analysis [345]. A strong structural correlation between the orientation of FA crystallites and gelatin within the FA/gelatin composite spheres was discovered that indicated to a substantial reorganization of the macromolecular matrix within the area of a growing aggregate [444]. Recently, chemical interactions between HA and organic molecules have been elucidated using *ab initio* calculation methods [1139].

By means of the X-ray photo-electronic spectroscopy (XPS) technique, binding energies of Ca, P, and O atoms were found to have some differences between nano-sized HA (Ca: 350.5 and 345.5; O: 530.2; P: 132.5 eV) and nano-sized HA/konjac glucomannan/chitosan biocomposite (Ca: 352.1 and 347.4; O: 531.2; P: 133.4 eV), respectively [653]. Further measurements by FTIR and X-ray diffraction revealed that nano-sized HA was mainly linked with konjac glucomannan and chitosan by hydrogen bonding among OH^- and PO_4^{3-} ions of HA and –C=O and –NH groups of konjac glucomannan and chitosan copolymer and there was a stable interface formed among the three phases in the biocomposite. Meanwhile, coordinate bonding might be formed between Ca^{2+} and –NH. Stable interfaces have been formed among the three phases in a biocomposite [653]. In HA/collagen biocomposites, a covalent bond formation between Ca^{2+} ions of HA and $RCOO^-$ groups of collagen molecules was found by XPS [601]. Similar XPS observations were also made for several other calcium orthophosphate-based biocomposites and hybrid biomaterials [629, 664, 673].

The interaction and adhesion between calcium orthophosphate fillers and respective matrixes have a significant effect on the properties of particulate filled reinforced materials, being essential to transfer the load among the phases and thus improve the mechanical performance of the biocomposites [353]. However, for a substantial amount of the aforementioned formulations, the interaction among the phases is mechanical in nature. This is because the matrix often consists of compounds with no functional groups or unsaturated bonds, which can form ionic complexes with the constituents of calcium orthophosphates. Obviously, less coupling exists between non-polar polymers and calcium orthophosphate ceramic particles. Therefore, polymers with functional groups pendant to the polymer backbone, which can act as sites for bridging to calcium orthophosphates, are more promising in this respect [53].

In order to improve the situation, various supplementary reagents are applied. Namely, if the primary effect of a processing additive is to increase the interaction between the phases, such additives can be regarded as coupling agents [1140]. These agents establish chemical bridges between the matrix and the fillers, promoting adhesion among the phases. In many cases, their effect is not unique; for example, it also might influence rheology of the composites [282]. In the case of calcium orthophosphates, a hexamethylene diisocyanate coupling agent was used to bind PEG/PBT (Polyactive™) block copolymers [293] and other polymers [1133] to HA filler particles. Thermogravimetric and infrared analysis demonstrated that the polymers were chemically bonded to the HA particles through the isocyanate groups, making it a suitable approach to improve the adhesion [1133]. Other researchers used glutaraldehyde as a cross-linked reagent [470, 474, 598, 600, 601, 619, 624, 699, 736, 778, 781, 1141]. Besides, the interfacial bonding among calcium orthophosphates and other components might be induced by silanes [242, 268, 269, 293, 406, 640, 1142–1145], zirconates [282, 406, 408, 1146, 1147], titanates [282, 406, 1146], phosphoric acid [643], alkaline pretreatment [877, 880], polyacids [143, 144, 293] and some other chemicals. Furthermore, some polymers might be grafted onto the surface of calcium orthophosphate particles [656]. Structural modifications of the polymeric matrices, for instance, with introduction of acrylic acid [245, 268, 269, 293], have also proved to be effective methods. For example, application of polyacids as a bonding agent for HA/Polyactive™ composites caused the surface modified HA particles to maintain better contact with the polymer at fracture and improved mechanical properties [143, 144, 293]. The use of titanate and zirconate coupling agents appeared to be very dependent on the molding technique employed [282]. Silane-coupled HA powders were tested before applying them as fillers in biodegradable composites [1143–1145]. This treatment allowed HA withstanding the attack of water without impairing overall bioactivity. Besides, a chemically modified reinforcement phase-matrix interface was found to improve the mechanical properties of the biocomposites. The examples include chemically coupled HA/PE [268, 269], chemically formed HA/Ca poly(vinylphosphonate) [349] and PLA/HA fibers [233]. These biocomposites are able to consume a large amount of energy in the fracture.

The action of some coupling agents was found to combine two distinct mechanisms: (i) cross-linking of the polymeric matrix (valid for zirconate and titanate coupling agents) and (ii) improvement of the interfacial interactions among the major phases of the biocomposites. This interfacial adhesion improvement appeared to be much dependent on the chemical nature (pH and type of metallic centre) of the coupling agents [406]. Several works claimed that silanes do interact with HA [242, 268, 269, 1143–1145]. It was shown that a silicon-containing inter-phase existed between HA and PE, which promoted the chemical adhesion between the HA particles and the polymer. A silane-coupling agent also facilitated penetration of PE into cavities of individual HA particles, which resulted in enhanced mechanical interlocking at the matrix-reinforcement interface [268, 269].

Thus, the optimization of biocomposite properties by coupling agents is currently an important area of the research. The control and development of molecular-level associations of polymers with calcium orthophosphates is suggested to be significant for the resulting mechanical responses in biocomposites. It appears that a fundamental molecular understanding of the interfacial behavior in biocomposites is an area not sufficiently addressed in literature. Various experimental characterization techniques using electron microscopy, vibrational spectroscopy, X-ray diffraction, scanning probe microscopy and others are used routinely to characterize these materials besides mechanical property characterization. In addition, atomic scale models for simulating the phase interaction and predicting responses in the novel material systems, where nanostructures and nano-interfaces are included, are important to understand and predict the load deformation behavior [179].

In addition to the aforementioned, the surface of calcium orthophosphates might be modified as well [144, 509, 510, 656, 1147–1154]. An interesting approach for HA surface modification was described by Lee et al. [1154]. First, *in situ* synthesis of surface thiol-functionalized HA (HA-SH) was realized by adding 3-mercaptopropionic acid during hydrothermal synthesis of HA, (Fig. 6.8A). This was followed by grafting polymerization of ethylene glycol methacrylate phosphate by radical chain transfer generating the sulfur-centered radicals on the HA surfaces (Fig. 6.8B), which initiated the surface grafting polymerization of ethylene glycol methacrylate phosphate (Fig. 6.8C) [1154]. Other examples might be found in literature [144, 509, 510, 656, 1147–1153]. In general, the purpose of surface modifying is not only to guarantee the even distribution of calcium orthophosphate particles at a high loading level in the matrix but also to prevent or delay the debonding process of calcium orthophosphate particles from the matrix. Obviously, all surface modifiers must satisfy several biomedical requirements, such as no toxicity, good biocompatibility and no changes in the biological or physico-chemical properties of the fillers.

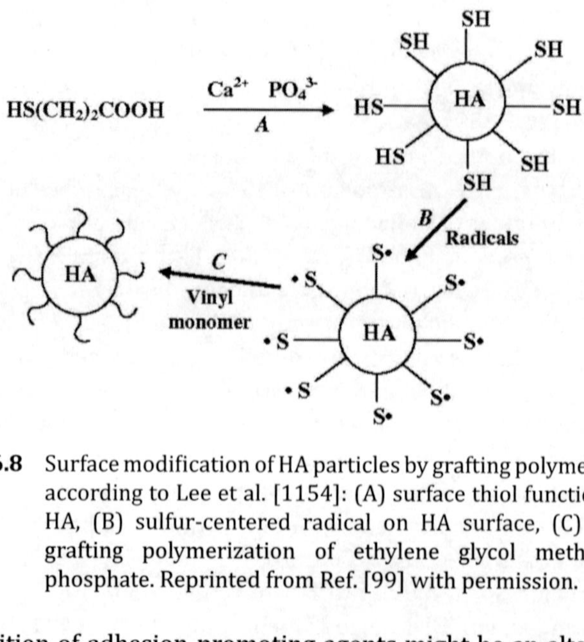

Figure 6.8 Surface modification of HA particles by grafting polymerization according to Lee et al. [1154]: (A) surface thiol functionalized HA, (B) sulfur-centered radical on HA surface, (C) surface grafting polymerization of ethylene glycol methacrylate phosphate. Reprinted from Ref. [99] with permission.

Addition of adhesion-promoting agents might be an alternative to improve the interaction between the fillers and the matrix. For example, Morita et al. incorporated 4-methacryloyloxyethyl trimellitate anhydride to promote adhesion of the polymer to HA [1155]. In another study, phosphoric ester was added to the liquid component of the formulation [1156]. Both the strength and the affinity index of biocomposites were found to increase, probably due to the effects of co-polymerization.

Possible interactions between BCP and HPMC have been investigated in IBS composites [900, 901, 1157]. After mixing, there was a decrease in the mean diameter of BCP granules and this influenced the viscosity of the paste. Dissolution of grain boundaries of β-TCP crystals and precipitation of CDHA on HA crystal surface were found during the interaction. Both phenomena were responsible for the observed granulometric changes [900, 901]; however, within the sensitivity of the employed measurement techniques, no chemical bonding between BCP and HPMC was detected [1157].

A co-precipitation technique was used to prepare CDHA/chitosan biocomposites [811]. Growth of CDHA crystals was inhibited by organic acids with more than two carboxyl groups, which strongly bind to CDHA surfaces via COO–Ca bonds. Transmission electron microscopy images revealed that CDHA formed elliptic aggregates with chemical interactions (probably coordination bond) between Ca on its surface and amino groups of chitosan; the nano-sized crystals of CDHA were found to align along the chitosan molecules, with the amino groups working as the nucleation sites [811]. Formation of calcium cross-linked polymer carboxylate salts was suggested during setting of calcium orthophosphate cement (TTCP + DCPA)/polyphosphazane biocomposites; a chemical involvement of the polymer in the cement setting was concluded based on the results of pH monitoring [558–560].

A chemical bond between the phases was presumed in PCL/HA composites, prepared by the grafting technique [420]; unfortunately, no strong experimental evidences were provided. In another study, CDHA/poly(α-hydroxyester) composites were prepared by a low temperature chemical route [393]. In that study, pre-composite structures were prepared by combining α-TCP with PLA, PLGA and copolymers thereof. The final biocomposite was achieved by *in situ* hydrolysis of α-TCP to CDHA performed at 56 ºC either in solvent cast or pressed pre-composites. That transformation occurred without any chemical reaction between the polymer and calcium orthophosphates, as it was determined by FTIR spectroscopy [393].

In nearly every study on HA/carbon nanotubes biocomposites, the nanotubes were functionalized before combining them with HA. Most researchers did this by oxidation [303–307], although non-covalent functionalizing with sodium dodecylsulfate [307] and coating the nanotubes by a polymer [1158] before combining them with HA were also reported. Several studies by transmission electron microscopy revealed evidences that the functionalization enhanced interaction between carbon nanotubes and HA [306, 307, 1159].

For calcium orthophosphate-based biocomposites able to sustain a high-temperature sintering (valid for the formulations consisting of inorganic components only), an inter-diffusion of chemical elements might take place among the phases. Such effect was detected by energy-dispersive X-ray spectroscopy in HA/TiO_2 biocomposite particles with partial formation of calcium titanates; this process was found to be favorable to enhancing the cohesive strength of particles in the composite coating [997]. A similar high-temperature interaction between HA and zirconia [911, 940], as well as between HA and Ti [661, 1030, 1032–1034] was also detected. Namely, lower Ti content composites sintered at 1200 ºC showed main crystalline phases as $CaTiO_3$, CaO and Ti_xP_y, while an increase in Ti content to 50 vol.% revealed Ti_2O and residual α-Ti as additional phases. Thus, the chemical reactions between HA and Ti were expressed by the following unbalanced illustrative equation [1032]:

$$Ti + Ca_{10}(PO_4)_6(OH)_2 \rightarrow CaTiO_3 + CaO + Ti_xP_y + (Ti_2O) + (Ca_4P_2O_9) + H_2O \quad (6.1)$$

Besides, partial decomposition of HA and formation of different calcium aluminates were detected in HA/Al_2O_3 biocomposites after sintering at 1200–1300 °C. This has been attributed to the diffusion of Ca^{2+} from HA into the alumina matrix and the depletion of Ca^{2+} from HA leads to the decomposition of HA into β-TCP [968, 974–976]. Presumably, all these processes influence the mechanical strength of the biocomposites.

6.6 Bioactivity and Biodegradation of Calcium Orthophosphate-Based Biocomposites

The continuous degradation of an implant causes a gradual load transfer to the healing tissue, preventing stress-shielding atrophy and stimulates the healing and remodeling of bones. Some requirements must be fulfilled by the ideal prosthetic biodegradable materials, such as biocompatibility, adequate initial strength and stiffness, retention of mechanical properties throughout sufficient time to assure its biofunctionality and non-toxicity of the degradation by-products [178]. Generally speaking, bioactivity (i.e., ability of bonding to bones) of biologically relevant calcium orthophosphates reinforced by other materials is usually lower than that of pure calcium orthophosphates [30, 31, 1160].

In general, both bioactivity and biodegradability of any biocomposite and/or hybrid biomaterial are determined by the same properties of the constituents. Both processes are very multi-factorial because, during implantation, the surface of any graft contacts with biological fluids and, shortly afterwards, is colonized by cells. Much more biology, than chemistry and material science altogether, is involved into these very complex processes and many specific details still remain unknown. To simplify the task, the biodegradability of the biologically relevant calcium orthophosphates might be described by a chemical dissolution in slightly acidic media (calcium orthophosphates are almost insoluble in alkaline solutions [111–117]), which, in the case of CDHA, might be described as a sequence of four successive chemical Eqs. (5.15–5.18) [519, 1161, 1162] (see sections *5.5* and *4.6.3*, as well as Chapter 7 for details). Further details on bioactivity and biodegradability of calcium orthophosphates are available in section *4.6.4*.

Biodegradability of polymers generally depends on the following factors: (1) chemical stability of the polymer backbone, (2) hydrophobicity of the monomer, (3) morphology of the polymer, (4) initial molecular weight, (5) fabrication processes, (6) geometry of the implant, (7) properties of the scaffold such as porosity and pore diameter [328]. A summary on degradation of PLA and PGA, as well as that of SEVA-C is available in literature [Ref. 178, p. 798 and p. 803, respectively], where the interested readers are referred to. Biodegradation of HA/PLLA and CDHA/PLLA biocomposite rods in subcutis and medullary cavities of rabbits were investigated mechanically and histologically; the degradation was found to be faster for the case of using uncalcinated CDHA instead of calcinated HA [1163]. In a more detailed study, new bone formation was detected at 2 weeks after implantation, especially for formulations with a high HA content [1164]. More to the point, a direct contact between bones and these composites without intervening fibrous tissue was detected in this case [1164, 1165]. Both SEVA-C and SEVA-C/HA biocomposite were found to exhibit a non-cytotoxic behavior [1166, 1167] inducing a satisfactory tissue response when implanted as shown by *in vivo* studies [1167]. Furthermore, SEVA-C/HA biocomposites induce a positive response on osteoblast-like cells to what concerns cell adhesion and proliferation [1166]. An *in vivo* study on biodegradation of microspheres (PLGA, gelatin, and poly(trimethylene carbonate) were used)/calcium orthophosphate cement biocomposites revealed that they exhibited microsphere degradation after 12 weeks of subcutaneous implantation, which was accompanied by compression strength decreasing [1168]. Interestingly, but the amount of calcium orthophosphates in biocomposites was found to have a greater effect on the early stages of osteoblast behavior (cell attachment and proliferation) rather than the immediate and late stages (proliferation and differentiation) [1169].

Both *in vitro* (the samples were immersed into 1% trypsin/phosphate-buffered saline solution at 37 °C) and *in vivo* (implantation of samples into the posterolateral lumbar spine of rabbits) biodegradation have been investigated for nano-sized HA/collagen/PLA biocomposites [610]. The results demonstrated that weight loss increased continuously *in vitro* with a reduction in mass of ~20% after 4 weeks. During the experimental period *in vitro*, a relative rate of reduction of the three components in this material was shown to differ greatly: collagen decreased the fastest, from 40% by weight to ~20% in the composite; HA content increased from 45 to ~60%; while PLA changed little. *In vivo*, the collagen/HA ratio appeared to be slightly higher near the transverse process than in the central part of the intertransverse process [610]. Hasegawa et al. [1170] performed *in vivo* study, spanning over a period of 5–7 years, on high strength HA/PLLA biocomposite rods for the internal fixation of bone fractures. In that work, both uncalcined CDHA and calcined HA were used as reinforcing phases in PLLA matrix. Those composites were implanted in the femur of 25 rabbits. It was found that the implanted materials were resorbed after 6 years of implantation. The presence of remodeled bone and trabecular bone bonding was the significant outcome. These data clearly demonstrate a biodegradation independence of various components of biocomposites.

6.7 Some Challenges and Critical Issues

The scientific information summarized in this review represents the recent developments of calcium orthophosphate-based biocomposites and hybrid biomaterials from a variety of approaches, starting from conventional ones to tissue engineering. Such formulations combined with osteoconductive, osteoinductive factors, and/or osteogenic cells have gained much interest as a new and versatile class of biomaterials and are perceived to be beneficial in many aspects as bone grafts [36, 1171]. However, current applications of these biomaterials in medicine and surgery are still remarkably less than might be expected. In many biomedical applications, research and testing of such formulations have been introduced and highly developed but only in a very few cases an industrial production and commercial distribution of medical devices partially or entirely made of biocomposites have started. The medical application of biocomposites and hybrid biomaterials requires a better understanding of the objectives and limitations involved. Recently, the main critical issues have been summarized as follows [265]:

- There are not enough reliable experimental and clinical data supporting the long-term performance of biocomposites with respect to monolithic traditional materials.
- The design of biocomposites and hybrid biomaterials is far more complex than that of conventional monolithic materials because of the large number of additional design variables that must be considered.
- The available fabrication methods may limit the possible reinforcement configurations, may be time consuming, expensive, highly skilled and may require special cleaning and sterilization processes.
- There are no satisfactory standards yet for biocompatibility testing of the biocomposite implants because the ways in which the different components of any biocomposite interact to living tissues are not completely understood.
- There are no adequate standards for the assessment of biocomposite fatigue performance because the fatigue behavior of such materials is far more complex and difficult to predict than that of traditional materials [265].

On the other hand, in spite of an enormous progress in biocomposite processing, to achieve the desired characteristics researchers still need to develop more advanced technologies to fabricate a bone-resembling hierarchical organization over several length scales. Development of novel grafting materials depends on the progress in research into the structure of natural bones. The key issues are not only to understand the fundamentals of biomineralization but also to translate such knowledge into practical synthetic pathways to produce better bone grafts. Unfortunately, when it comes to the fabrication of biocomposites mimicing natural bones from the nanometer to the micrometer dimensions, there are many key issues, including control of morphology, incorporation of foreign ions, interaction with biomolecules and assembly of the organic and inorganic phases, which are still not well understood. A processing gap between the lower-level building units and the higher-order architecture could severely limit the practical application of current calcium orthophosphate-based biocomposites and hybrid biomaterials. Therefore, further substantial research efforts have been outlined to address the following key challenges [36, 41]:

- Optimizing biocomposite processing conditions.
- Optimization of interfacial bonding and strength equivalent to natural bone.
- Optimization of the surface properties and pore size to maximize bone growth.
- Maintaining the adequate volume of the construct *in vivo* to allow bone formation to take place.
- Withstanding the load-bearing conditions.
- Matching the bioresorbability of the grafts and their biomechanical properties while forming new bone.
- Understanding the molecular mechanisms by which the cells and the biocomposite matrix interact with each other *in vivo* to promote bone regeneration.
- Supporting angiogenesis and vascularization for the growth of healthy bone cells and subsequent tissue formation and remodeling [36, 41].

The aforementioned critical issues have to be solved before a widespread commercial use of calcium orthophosphate-based biocomposites and hybrid biomaterials can be made in surgery and medicine.

6.8 Conclusions

All types of calcified tissues of humans and mammals appear to possess a complex hierarchical biocomposite structure. Their mechanical properties are outstanding (considering weak constituents from which they are assembled) and far beyond those, that can be achieved using the same synthetic materials with present technologies. This is because biological organisms produce biocomposites that are organized in terms of both composition and structure, containing both brittle calcium orthophosphates and ductile bioorganic components in very complex structures, hierarchically organized at the nano-, micro-, and meso-levels. Additionally, the calcified tissues are always multifunctional: for example, bone provides structural support for the body plus blood cell formation. The third defining characteristic of biological systems, in contrast with current synthetic systems, is their self-healing ability, which is nearly universal in nature. These complex structures, which have risen from millions of years of evolution, inspire materials scientists in the design of novel biomaterials [1172].

Obviously, no single-phase biomaterial is able to provide all the essential features of bones and/or other calcified tissues and therefore, there is a great need to engineer multi-phase biomaterials (biocomposites) with a structure and composition mimicking those of natural bones. The studies summarized in this review have shown that the proper combination of a ductile matrix with a brittle, hard, and bioactive calcium orthophosphate filler offers many advantages for biomedical applications. Namely, the desirable properties of some components can compensate for a poor mechanical behavior of calcium orthophosphate bioceramics, while in turn the desirable bioactive properties of calcium orthophosphates improve those of other phases, thus expanding the possible application of each material within the body [102]. However, the reviewed literature clearly indicates that among possible types of calcium orthophosphate-based biocomposites and hybrid biomaterials only simple, complex and graded ones, as well as fibrous, laminar and particulate ones (see classification types of the composites in section *6.2*) have been investigated. Presumably, a future progress in this subject will require concentrating efforts on elaboration and development of both hierarchical and hybrid biocomposites. Furthermore, following the modern tendency of tissue engineering, a novel generation of calcium orthophosphate-based biocomposites and hybrid biomaterials should also contain a biological living part.

To conclude, the future of the calcium orthophosphate-based biocomposites and hybrid biomaterials is now directly dependent on the formation of multidisciplinary teams composed of experts but primarily experts ready to collaborate in close collaboration with others and thus be able to deal efficiently with the complexity of the human organism. The physical chemistries of solids, solid surfaces, polymer dispersion and solutions, as well as material–cell interactions are among the phenomena to be tackled. Furthermore, much work remains to be done on a long way from a laboratory to clinics and the success depends on the effective co-operation of clinicians, chemists, biologists, bioengineers and materials scientists.

References and Notes

1. Keating, J. F., and McQueen, M. M. (2001) Substitutes for autologous bone graft in orthopaedic trauma, *J. Bone Joint Surg. (Br.)*, **82B**, 3–8.
2. Chau, A. M. T., and Mobbs, R. J. (2009) Bone graft substitutes in anterior cervical discectomy and fusion, *Eur. Spine J.*, **18**, 449–464.
3. Kaveh, K., Ibrahim, R., Bakar, M. Z. A., and Ibrahim, T. A. (2010) Bone grafting and bone graft substitutes, *J. Animal and Veterinary Advances*, **9**, 1055–1067.
4. Murugan, R., and Ramakrishna, S. (2005) Bioactive nanomaterials in bone grafting and tissue engineering, in: *Handbook of Nanostructured Biomaterials and their Applications in Nanobiotechnology*, (ed. Nalwa, H. S.), American Scientific Publishers, Stevenson Ranch, vol. 2, pp. 141–148.
5. Tazaki, J., Murata, M., Yuasa, T., Akazawa, T., Ito, K., Hino, J., Nida, A., Arisue, M., and Shibata, T. (2010) Autograft of human tooth and demineralized dentin matrices for bone augmentation, *J. Ceram. Soc. Jpn.*, **118**, 442–445.
6. Conway, J. D. (2010) Autograft and nonunions: morbidity with intramedullary bone graft versus iliac crest bone graft, *Orthopedic Clinics of North America*, **41**, 75–84.
7. Keller, E. E., and Triplett, W. W. (1987) Iliac crest bone grafting: review of 160 consecutive cases, *J. Oral Maxillofac. Surg.*, **45**, 11–14.
8. Laurie, S. W., Kaban, L. B., Mulliken, J. B., and Murraym, J. E. (1984) Donor-site morbidity after harvesting rib and iliac bone, *Plast. Reconstr. Surg.*, **73**, 933–938.
9. Younger, E. M., and Chapman, M. W. (1989) Morbidity at bone graft sites, *J. Orthop. Trauma*, **3**, 192–195.
10. Neumann, M., and Epple, M. (2006) Composites of calcium phosphate and polymers as bone substitution materials, *Eur. J. Trauma*, **32**, 125–131.

11. le Guéhennec, L., Layrolle, P., and Daculsi, G. (2004) A review of bioceramics and fibrin sealant, *Eur. Cell Mater.*, **8**, 1–11.

12. Fuchs, J. R., Nasseri, B. A., and Vacanti, J. P. (2001) Tissue engineering: a 21st century solution to surgical reconstruction, *Ann. Thorac. Surg.*, **72**, 557–591.

13. Hench, L. L., and Wilson, J. (1984) Surface-active biomaterials, *Science*, **226**, 630–636.

14. Rose, F. R. A. J., and Oreffo, R. O. C. (2002) Breakthroughs and views bone tissue engineering: hope vs. hype, *Biochem. Biophys. Res.*, **292**, 1–7.

15. Beaman, F. D., Bancroft, L. W., Peterson, J. J., and Kransdorf, M. J. (2006) Bone graft materials and synthetic substitutes, *Radiologic Clinics of North America*, **44**, 451–461.

16. Kokubo, T., Kim, H. M., and Kawashita, M. (2003) Novel bioactive materials with different mechanical properties, *Biomaterials*, **24**, 2161–2175.

17. Rueger, J. M. (1998) Bone replacement materials – state of the art and the way ahead, *Orthopäde*, **27**, 72–79.

18. Greenwald, A. S., Boden, S. D., Goldberg, V. M., Khan, Y., Laurencin, C. T., and Rosier, R. N. (2001) Bone graft substitutes: facts, fictions and applications, *J. Bone Joint Surg. (Am.)*, **83**, 98–103.

19. Finkemeier, C. G. (2002) Bone grafting and bone graft substitutes, *J. Bone Joint Surg. (Am.)*, **84**, 454–464.

20. Giannoudis, P. V., Dinopoulos, H., and Tsiridis, E. (2005) Bone substitutes: an update, *Injury*, **36**(3), , S20–S27.

21. Bohner, M. (2010) Resorbable biomaterials as bone graft substitutes, *Mater. Today*, **13**, 24–30

22. Yang, S., Leong, K. F., Du, Z., and Chua, C. K. (2001) The design of scaffolds for use in tissue engineering, Part I. Traditional factors, *Tissue Eng.*, **7**, 679–689.

23. Burg, K. J. L., Porter, S., and Kellam, J. F. (2000) Biomaterial developments for bone tissue engineering, *Biomaterials*, **21**, 2347–2359.

24. Holy, C. E., Shoichet, M. S., and Davies, J. E. (2000) Engineering three-dimensional bone tissue *in vitro* using biodegradable scaffolds: investigating initial cell-seeding density and culture period, *J. Biomed. Mater. Res.*, **51**, 376–382.

25. Lowenstam, H. A., and Weiner, S. (1989) *On Biomineralization*, Oxford University Press, New York.

26. Weiner, S., and Wagner, H. D. (1998) The material bone: structure-mechanical function relations, *Ann. Rev. Mater. Sci.*, **28**, 271–298.

27. Dorozhkin, S. V. (2009) Calcium orthophosphates in nature, biology and medicine, *Materials*, **2**, 399–498.

28. Hench, L. L., and Wilson, J. (1993) *An Introduction to Bioceramics* (ed. Hench, L. L., and Wilson, J), Advanced Series in Ceramics, vol. 1, World Scientific, Singapore, p. 1.

29. Suchanek, W., and Yoshimura, M. (1998) Processing and properties of hydroxyapatite-based biomaterials for use as hard tissue replacement implants, *J. Mater. Res.*, **13**, 94–117.

30. Hench, L. L. (1991) Bioceramics: from a concept to clinics, *J. Am. Ceram. Soc.*, **74**, 1487–1510.

31. Hench, L. L. (1998) Bioceramics, *J. Am. Ceram. Soc.*, **81**, 1705–1728.

32. One molecule of collagen type I is a triple helix with 338 repetitions of amino acid residues and is about 300 nm in length [33]. Additionally, bone contains small quantities of other bioorganic materials, such as proteins, polysaccharides and lipids, as well as bone contains cells and blood vessels.

33. Itoh, S., Kikuchi, M., Koyama, Y., Takakuda, K., Shinomiya, K., and Tanaka, J. (2002) Development of an artificial vertebral body using a novel biomaterial, hydroxyapatite/collagen composite, *Biomaterials*, **23**, 3919–3926.

34. Thompson, J. B., Kindt, J. H., Drake, B., Hansma, H. G., Morse, D. E., and Hansma, P. K. (2001) Bone indentation recovery time correlates with bond reforming time, *Nature*, **414**, 773–776.

35. Fratzl, P., Gupta, H. S., Paschalis, E. P., and Roschger, P. (2004) Structure and mechanical quality of the collagen-mineral nano-composite in bone, *J. Mater. Chem.*, **14**, 2115–2123.

36. Murugan, R., and Ramakrishna, S. (2005) Development of nanocomposites for bone grafting, *Compos. Sci. Technol.*, **65**, 2385–2406.

37. Burr, D. B. (2002) The contribution of the organic matrix to bone's material properties, *Bone*, **31**, 8–11.

38. Itoh, S., Kikuchi, M., Koyama, Y., Matumoto, H. N., Takakuda, K., Shinomiya, K., and Tanaka, J. (2005) Development of a novel biomaterial., hydroxyapatite/collagen composite for medical use, *Biomed. Mater. Eng.*, **15**, 29–41.

39. Cui, F. Z., Li, Y., and Ge, J. (2007) Self-assembly of mineralized collagen composites, *Mater. Sci. Eng. R*, **57**, 1–27.

40. Vallet-Regi, M., and Arcos, D. (2006) Nanostructured hybrid materials for bone tissue regeneration, *Curr. Nanosci.*, **2**, 179–189.

41. Chan, C. K., Kumar, T. S. S., Liao, S., Murugan, R., Ngiam, M., and Ramakrishna, S. (2006) Biomimetic nanocomposites for bone graft applications, *Nanomedicine*, **1**, 177–188.

42. Olszta, M. J., Cheng, X. G., Jee, S. S., Kumar, B. R., Kim, Y. Y., Kaufman, M. J., Douglas, E. P., and Gower, L. B. (2007) Bone structure and formation: a new perspective, *Mater. Sci. Eng. R*, **58**, 77–116.

43. Bauer, T., and Muschler, G. (2000) Bone grafts materials, an overview of the basic science, *Clin. Orthop. Rel. Res.*, **371**, 10–27.

44. Athanasiou, K. A., Zhu, C. F., Lanctot, D. R., Agrawal, C. M., and Wang, X. (2000) Fundamentals of biomechanics in tissue engineering of bone, *Tissue Engineering*, **6**, 361–381.

45. Zioupos, P. (1998) Recent developments in the study of solid biomaterials and bone: "fracture" and "pre-fracture" toughness, *Mater. Sci. Eng. C*, **6**, 33–40.

46. Doblaré, M., Garcia, J. M., and Gómez, M. J. (2004) Modelling bone tissue fracture and healing: a review, *Eng. Fract. Mech.*, **71**, 1809–1840.

47. Vallet-Regi, M. (2006) Revisiting ceramics for medical applications, *Dalton Trans.*, pp. 5211–5220.

48. Huiskes, R., Ruimerman, R., van Lenthe, H. G., and Janssen, J. D. (2000) Effects of mechanical forces on maintenance and adaptation of form in trabecular bone, *Nature*, **405**, 704–706.

49. Thomson, R. C., Yaszemski, M. J., Powers, J. M., and Mikos, A. G. (1998) Hydroxyapatite fiber reinforced poly(α-hydroxy ester) foams for bone regeneration, *Biomaterials*, **19**, 1935–1943.

50. Boccaccini, A. R., and Blaker, J. J. (2005) Bioactive composite materials for tissue engineering scaffolds, *Expert Rev. Med. Devices*, **2**, 303–317.

51. Verheyen, C. C. P. M., de Wijn, J. R., van Blitterswijk. C. A., de Groot, K., and Rozing, P. M. (1993) Hydroxyapatite/poly(L-lactide) composites: an animal study on push-out strengths and interface histology, *J. Biomed. Mater. Res.*, **27**, 433–444.

52. Zhang, R. Y., and Ma, P. X. (1999) Poly(α-hydroxyl acids) hydroxyapatite porous composites for bone-tissue engineering, I. Preparation and Morphology, *J. Biomed. Mater. Res.*, **44**, 446–455.

53. Durucan, C., and Brown, P. W. (2001) Biodegradable hydroxyapatite-polymer composites, *Adv. Eng. Mater.*, **3**, 227–231.

54. Kim, H. W., Knowles, J. C., and Kim, H. E. (2004) Hydroxyapatite/poly(ε-caprolactone) composite coatings on hydroxyapatite porous bone scaffold for drug delivery, *Biomaterials*, **25**, 1279–1287.

55. Hutmacher, D. W., Schantz, J. T., Lam, C. X. F., Tan, K. C., and Lim, T. C. (2007) State of the art and future directions of scaffold-based bone engineering from a biomaterials perspective, *J. Tissue Eng. Regen. Med.*, **1**, 245–260.

56. Guarino, V., Causa, F., and Ambrosio, L. (2007) Bioactive scaffolds for bone and ligament tissue, *Expert Review of Medical Devices*, **4**, 405–418.

57. Yunos, D. M., Bretcanu, O., and Boccaccini, A. R. (2008) Polymer-bioceramic composites for tissue engineering scaffolds, *J. Mater. Sci.*, **43**, 4433–4442.

58. Hench, L. L., and Polak, J. M. (2002) Third-generation biomedical materials, *Science*, **295**, 1014–1017.

59. Crane, G. M., Ishaug, S. L., and Mikos, A. G. (1995) Bone tissue engineering, *Nature Medicine*, **1**, 1322–1324.

60. LeGeros, R. Z. (1988) Calcium phosphate materials in restorative dentistry: a review, *Adv. Dent. Res.*, **2**, 164–180.

61. Mathijsen, A. (1852) *Nieuwe Wijze van Aanwending van het Gips-Verband bij Beenbreuken*, J. B. van Loghem, Haarlem.

62. Dreesman, H. (1892) Über Knochenplombierung, *Beitr. Klin. Chir.*, **9**, 804–810.

63. Wang, M. (2003) Developing bioactive composite materials for tissue replacement, *Biomaterials*, **24**, 2133–2151.

64. http://en.wikipedia.org/wiki/Composite_material (assessed in November 2010).

65. Gibson, R. F. (2010) A review of recent research on mechanics of multifunctional composite materials and structures, *Composite Structures*, **92**, 2793–2810.

66. Evans, S. L., and Gregson, P. J. (1998) Composite technology in load-bearing orthopaedic implants, *Biomaterials*, **19**, 1329–1342.

67. Habibovic, P., Barrère, F., van Blitterswijk, C. A., de Groot, K., and Layrolle, P. (2002) Biomimetic hydroxyapatite coating on metal implants, *J. Am. Ceram. Soc.*, **85**, 517–522.

68. Zhang, R. Y., and Ma, P. X. (2004) Biomimetic polymer/apatite composite scaffolds for mineralized tissue engineering, *Macromol. Biosci.*, **4**, 100–111.

69. Oliveira, A. L., Mano, J. F., and Reis, R. L. (2003) Nature-inspired calcium phosphate coatings: present status and novel advances in the science of mimicry, *Curr. Opin. Solid State Mater. Sci.*, **7**, 309–318.

70. Wan, Y. Z., Hong, L., Jia, S. R., Huang, Y., Zhu, Y., Wang, Y. L., and Jiang, H. J. (2006) Synthesis and characterization of hydroxyapatite-bacterial cellulose nanocomposites, *Compos. Sci. Technol.*, **66**, 1825–1832.

71. Wan, Y. Z., Huang, Y., Yuan, C. D., Raman, S., Zhu, Y., Jiang, H. J., He, F., and Gao, C. (2007) Biomimetic synthesis of hydroxyapatite/bacterial cellulose nanocomposites for biomedical applications, *Mater. Sci. Eng. C*, **27**, 855–864.

72. Ohtsuki, C., Kamitakahara, M., and Miyazaki, T. (2007) Coating bone-like apatite onto organic substrates using solutions mimicking body fluid, *J. Tissue Eng. Regen. Med.*, **1**, 33–38.

73. Oyane, A. (2010) Development of apatite-based composites by a biomimetic process for biomedical applications, *J. Ceram. Soc. Jpn.*, **118**, 77–81.

74. de Groot, K., Geesink, R. G. T., Klein, C. P. A. T., and Serekian, P. (1987) Plasma sprayed coatings of hydroxylapatite, *J. Biomed. Mater. Res.*, **21**, 1375–1381.

75. de Groot, K., Wolke, J. G. C., and Jansen, J. A. (1998) Calcium phosphate coatings for medical implants, *Proc. Instn. Mech. Engrs. Part H*, **212**, 137–147.

76. Ignjatovic, N. L., Liu, C. Z., Czernuszka, J. T., and Uskokovic, D. P. (2007) Micro- and nano-injectable composite biomaterials containing calcium phosphate coated with poly(D,L-lactide-*co*-glycolide), *Acta Biomater.*, **3**, 927–935.

77. Manso M., Langlet M., Fernandez M., Vasquez L., and Martinez-Duart J. M. (2003) Surface and interface analysis of hydroxyapatite/TiO_2 biocompatible structures, *Mater. Sci. Eng. C*, **23**, 451–454.

78. Sun, L., Berndt, C. C., Gross, K. A., and Kucuk, A. (2001) Material fundamentals and clinical performance of plasma-sprayed hydroxyapatite coatings: a review, *J. Biomed. Mater. Res. (Appl. Biomater.)*, **58**, 570–592.

79. Song, J., Malathong, V., and Bertozzi, C. R. (2005) Mineralization of synthetic polymer scaffolds: a bottom-up approach for the development of artificial bone, *J. Am. Chem. Soc.*, **127**, 3366–3372.

80. Yoshida, K., Hashimoto, K., Toda, Y., Udagawa, S., and Kanazawa, T. (2006) Fabrication of structure-controlled hydroxyapatite/zirconia composite, *J. Eur. Ceram. Soc.*, **26**, 515–518.

81. Madhumathi, K., Binulal, N. S., Nagahama, H., Tamura, H., Shalumon, K. T., Selvamurugan, N., Nair, S. V., and Jayakumar, R. (2009) Preparation and characterization of novel β-chitin-hydroxyapatite composite membranes for tissue engineering applications, *Int. J. Biol. Macromolecules*, **44**, 1–5.

82. Lee, M., Ku, S. H., Ryu, J., and Park, C. B. (2010) Mussel-inspired functionalization of carbon nanotubes for hydroxyapatite mineralization, *J. Mater. Chem.*, **20**, 8848–8853.

83. Planeix, J. M., Jaunky, W., Duhoo, T., Czernuszka, J. T., Hosseini, M. W., and Brès, E. F. (2003) A molecular tectonics-crystal engineering approach for building organic-inorganic composites, potential application to the growth control of hydroxyapatite crystals *J. Mater. Chem.*, **13**, 2521–2524.

84. Zhao, J., Guo, L. Y., Yang, X. B., and Weng, J. (2008) Preparation of bioactive porous HA/PCL composite scaffolds, *Appl. Surf. Sci.*, **255**, 2942–2946.

85. Dorozhkin, S., and Ajaal, T. (2009) Toughening of porous bioceramic scaffolds by bioresorbable polymeric coatings, *Proc. Inst. Mech. Eng. H*, **223**, 459–470.

86. Woo, A. S., Jang, J. L., Liberman, R. F., and Weinzweig, J. (2010) Creation of a vascularized composite graft with a cellular dermal matrix and hydroxyapatite, *Plastic and Reconstructive Surgery*, **125**, 1661–1669.

87. Zhao, J., Duan, K., Zhang, J. W., Lu, X., and Weng, J. (2010) The influence of polymer concentrations on the structure and mechanical properties of porous polycaprolactone-coated hydroxyapatite scaffolds, *Appl. Surf. Sci.*, **256**, 4586–4590.

88. Dong, J., Uemura, T., Kojima, H., Kikuchi, M., Tanaka, J., and Tateishi, T. (2001) Application of low-pressure system to sustain *in vivo* bone formation in osteoblast/porous hydroxyapatite composite, *Mater. Sci. Eng. C*, **17**, 37–43.

89. Zerbo, I. R., Bronckers, A. L. J. J., de Lange, G., and Burger, E. H. (2005) Localisation of osteogenic and osteoclastic cells in porous β-tricalcium phosphate particles used for human maxillary sinus floor elevation, *Biomaterials*, **26**, 1445–1451.

90. Mikán, J., Villamil, M., Montes, T., Carretero, C., Bernal, C., Torres, M. L., and Zakaria, F. A. (2009) Porcine model for hybrid material of carbonated apatite and osteoprogenitor cells, *Mater. Res. Innovations*, **13**, 323–326.

91. Oe, K., Miwa, M., Nagamune, K., Sakai, Y., Lee, S. Y., Niikura, T., Iwakura, T., Hasegawa, T., Shibanuma, N., Hata, Y., Kuroda, R., and Kurosaka, M. (2010) Nondestructive evaluation of cell numbers in bone marrow stromal cell/β-tricalcium phosphate composites using ultrasound, *Tiss. Eng. C*, **16**, 347–353.

92. Krout, A., Wen, H. B., Hippensteel, E., and Li, P. (2005) A hybrid coating of biomimetic apatite and osteocalcin, *J. Biomed. Mater. Res. A*, **73A**, 377–387.

93. Kundu, B., Soundrapandian, C., Nandi, S. K., Mukherjee, P., Dandapat, N., Roy, S., Datta, B. K., Mandal, T. K., Basu, D., and Bhattacharya, R. N. (2010) Development of new localized drug delivery system based on ceftriaxone-sulbactam composite drug impregnated porous hydroxyapatite: a systematic approach for *in vitro* and *in vivo* animal trial, *Pharmaceutical Res.*, **27**, 1659–1676.

94. Kickelbick, G. (ed.) (2007) *Hybrid Materials. Synthesis, Characterization, and Applications*, Wiley-VCH Verlag GmbH, Weinheim, p. 498.

95. Matthews, F. L., and Rawlings, R. D. (2000) *Composite Materials: Engineering and Science*, CRC Press, Boca Raton, p. 480.

96. Xia, Z., Riester, L., Curtin, W. A., Li, H., Sheldon, B. W., Liang, J., Chang, B., and Xu, J. M. (2004) Direct observation of toughening mechanisms in carbon nanotube ceramic matrix composites, *Acta Mater.*, **52**, 931–944.

97. Tavares, M. I. B., Ferreira, O., Preto, M., Miguez, E., Soares, I. L., and da Silva, E. P. (2007) Evaluation of composites miscibility by low field NMR, *Int. J. Polym. Mater.*, **56**, 1113–1118.

98. Kiran, E. (2009) Polymer miscibility, phase separation, morphological modifications and polymorphic transformations in dense fluids, *J. Supercritical Fluids*, **47**, 466–483.

99. Šupová, M. (2009) Problem of hydroxyapatite dispersion in polymer matrices: a review, *J. Mater. Sci. Mater. Med.*, **20**, 1201–1213.

100. Böstman, O., and Pihlajamäki, H. (2000) Clinical biocompatibility of biodegradable orthopaedic implants for internal fixation: a review, *Biomaterials*, **21**, 2615–2621.

101. John, M. J., and Thomas, S. (2008) Biofibres and biocomposites, *Carbohydrate Polymers*, **71**, 343–364.

102. Rea, S. M., and Bonfield, W. (2004) Biocomposites for medical applications, *J. Aust. Ceram. Soc.*, **40**, 43–57.

103. Gravitis, Ya. A., Tééyaér, R. E., Kallavus, U. L., Andersons, B. A., Ozol'-Kalnin, V. G., Kokorevich, A. G., Érin'sh, P. P., and Veveris, G. P.

(1987) Biocomposite structure of wood cell membranes and their destruction by explosive autohydrolysis, *Mechanics Composite Mater.*, **22**, 721–725.

104. Bernard, S. L., and Picha, G. J. (1991) The use of coralline hydroxyapatite in a 'biocomposite' free flap, *Plastic and Reconstructive Surgery*, **87**, 96–107.

105. Ong, J. L., and Chan, D. C. N. (1999) Hydroxyapatite and their use as coatings in dental implants: a review, *Crit. Rev. Biomed. Eng.*, **28**, 667–707.

106. Davies, J. E. (1996) *In vitro* modeling of the bone/implant interface, *Anat. Record*, **245**, 426–445.

107. Anselme, K. (2000) Osteoblast adhesion on biomaterials, *Biomaterials*, **21**, 667–681.

108. Gauthier, O., Bouler, J. M., Weiss, P., Bosco, J., Daculsi, G., and Aguado, E. (1999) Kinetic study of bone ingrowth and ceramic resorption associated with the implantation of different injectable calcium-phosphate bone substitutes, *J. Biomed. Mater. Res.*, **47**, 28–35.

109. Hing, K. A., Best, S. M., and Bonfield, W. (1999) Characterization of porous hydroxyapatite, *J. Mater. Sci. Mater. Med.*, **10**, 135–145.

110. Carotenuto, G., Spagnuolo, G., Ambrosio, L., and Nicolais, L. (1999) Macroporous hydroxyapatite as alloplastic material for dental applications, *J. Mater. Sci. Mater. Med.*, **10**, 671–676.

111. LeGeros, R. Z. (1991) Calcium phosphates in oral biology and medicine, *Monographs in Oral Science*, vol. 15, (ed. Myers H. M.), Karger, Basel, p. 201.

112. Elliott, J. C. (1994) Structure and chemistry of the apatites and other calcium orthophosphates, *Studies in Inorganic Chemistry*, vol. 18, Elsevier, Amsterdam, p. 389.

113. Brown, P. W., and Constantz, B. (eds.) (1994) *Hydroxyapatite and Related Materials*, CRC Press, Boca Raton, FL, p. 343.

114. Amjad, Z. (ed.) (1997) *Calcium Phosphates in Biological and Industrial Systems*, Kluwer Academic Publishers, Boston, p. 529.

115. Hughes, J. M., Kohn, M., and Rakovan, J. (eds.) (2002) Phosphates: geochemical., geobiological and materials importance. *Series: Reviews in Mineralogy and Geochemistry*, vol. 48, Mineralogical Society of America, Washington.

116. Chow, L. C., and Eanes, E. D. (eds.) (2001) Octacalcium phosphate, *Monographs in Oral Science*, vol. 18, S. Karger A. G., Basel, p.

168.117. Brès, E., and Hardouin, P. (eds.) (1998) Les matériaux en phosphate de calcium. Aspects fondamentaux. / Calcium phosphate materials, *Fundamentals. Sauramps Medical*, Montpellier, p. 176.

118. Walton, D. J., and Lorimer, J. P. (2001) *Polymers*, Oxford University Press, USA,p. 160.

119. Carraher, C. E. Jr. (2010) *Introduction to Polymer Chemistry*, 2nd edn. CRC Press, Boca Raton, p. 534.

120. Langer, R. (2000) Biomaterials in drug delivery and tissue engineering: one laboratory's experience, *Acc. Chem. Res.*, **33**, 94–101.

121. Thomson, R. C., Ak, S., Yaszemski, M. J., and Mikos, A. G. (2000) Polymer scaffold processing, in *Principles of Tissue Engineering*, Academic Press, New York, pp. 251–262.

122. Ramakrishna, S., Mayer, J., Wintermantel, E., and Leong, K. W. (2001) Biomedical applications of polymer-composite materials: a review, *Compos. Sci. Technol.*, **61**, 1189–1224.

123. Shastri, V. P. (2003) Non-degradable biocompatible polymers in medicine: past, present and future, *Curr. Pharmaceutical Biotechnol*, **4**, 331–337.

124. Chen, H., Yuan, L., Song, W., Wu, Z., and Li, D. (2008) Biocompatible polymer materials: role of protein-surface interactions, *Prog. Polym. Sci.*, **33**, 1059–1087.

125. Langer, R., and Vacanti, J. P. (1993) Tissue engineering, *Science*, **260**, 920–925.

126. Lanza, R. P., Hayes, J. L., and Chick, W. L. (1996) Encapsulated cell technology, *Nature Biotechnology*, **14**, 1107–1111.

127. Agrawal, C. M., and Ray, R. B. (2001) Biodegradable polymeric scaffolds for musculoskeletal tissue engineering, *J. Biomed. Mater. Res.*, **55**, 141–150.

128. Kweon, H., Yoo, M., Park, I., Kim, T., Lee, H., Lee, S., Oh, J., Akaike, T., and Cho, C. (2003) A novel degradable polycaprolactone network for tissue engineering, *Biomaterials*, **24**, 801–808.

129. de Groot, J. H., de Vrijer, R., Pennings, A. J., Klompmaker, J., Veth, R. P. H., and Jansen, H. W. B. (1996) Use of porous polyurethanes for meniscal reconstruction and meniscal prostheses, *Biomaterials*, **17**, 163–173.

130. Resiak, I., and Rokicki, G. (2000) Modified polyurethanes for biomedical applications, *Polimery*, **45**, 592–602.

131. Temenoff, J. S., and Mikos, A. G. (2000) Injectable biodegradable materials for orthopedic tissue engineering, *Biomaterials*, **21**, 2405–2412.

132. Behravesh, E., Yasko, A. W., Engel, P. S., and Mikos, A. G. (1999) Synthetic biodegradable polymers for orthopaedic applications, *Clin. Orthop. Rel. Res.*, **367S**, S118–S125.

133. Lewandrowski, K. U., Gresser, J. D., Wise, D. L., White, R. L., and Trantolo, D. J. (2000) Osteoconductivity of an injectable and bioresorbable poly(propyleneglycol-*co*-fumaric acid) bone cement, *Biomaterials*, **21**, 293–298.

134. Peter, S. J., Miller, M. J., Yaszemski, M. J., and Mikos, A. G. (1997) Poly(propylene fumarate), in *Handbook of Biodegradable Polymers* (ed. Domb, A. J., Kost, J., and Wiseman, D. M.), Harwood Academic, Amsterdam, pp. 87–97.

135. Boland, E. D., Coleman, B. D., Barnes, C. P., Simpson, D. G., Wnek, G. E., and Bowlin, G. L. (2005) Electrospinning polydioxanone for biomedical applications, *Acta Biomater.*, **1**, 115–123.

136. Gilbert, J. L. (2001) Acrylics in biomedical engineering, in *Encyclopedia of Materials: Science and Technology*, Elsevier, Amsterdam, pp. 11–18.

137. Frazer, R. Q., Byron, R. T., Osborne, P. B., and West, K. P. (2005) PMMA: an essential material in medicine and dentistry, *J. Long-Term Eff. Med. Implants*, **15**, 629–639.

138. Li, Y. W., Leong, J. C, Y., Lu, W. W., Luk, K. D. K., Cheung, K. M. C., Chiu, K. Y., and Chow, S. P. (2000) A novel injectable bioactive bone cement for spinal surgery: a developmental and preclinical study, *J. Biomed. Mater. Res.*, **52**, 164–170.

139. Mckellop, H., Shen, F., Lu, B., Campbell, P., and Salovey, R. (1999) Development of an extremely wear resistant UHMW polyethylene for total hip replacements, *J. Orthop. Res.*, **17**, 157–167.

140. Kurtz, S. M., Muratoglu, O. K., Evans, M., and Edidin, A. A. (1999) Advances in the processing, sterilization and crosslinking of ultra-high molecular weight polyethylene for total joint arthroplasty, *Biomaterials*, **20**, 1659–1688.

141. Laurencin, C. T., Ambrosio, M. A., Borden, M. D., and Cooper, J. A., Jr. (1999) Tissue engineering: orthopedic applications, *Ann. Rev. Biomed. Eng.*, **1**, 19–46.

142. Jansen, J. A., de Ruijter, J. E., Janssen, P. T., and Paquay, Y. G. (1995) Histological evaluation of biodegradable Polyactive™/hydroxyapatite membrane, *Biomaterials*, **16**, 819–827.

143. Liu, Q., de Wijn, J. R., Bakker, D., and van Blitterswijk, C. A. (1996) Surface modification of hydroxyapatite to introduce interfacial bonding with Polyactive™ 70/30 in a biodegradable composite, *J. Mater. Sci. Mater. Med.*, **7**, 551–557.

144. Liu, Q., de Wijn, J. R., Bakker, D., van Toledo, M., and van Blitterswijk, C. A. (1998) Polyacids as bonding agents in hydroxyapatite polyesterether (Polyactive™ 30/70) composite, *J. Mater. Sci. Mater. Med.*, **9**, 23–30.

145. Meijer, G. J., Cune, M. S., van Dooren, M., de Putter, C., and van Blitterswijk, C. A. (1997) A comparative study of flexible (Polyactive™) versus rigid (hydroxylapatite) permucosal dental implants, *I. Clinical aspects. J. Oral Rehabilitation*, **24**, 85–92.

146. Meijer, G. J., Dalmeijer, R. A., de Putter, C., and van Blitterswijk, C. A. (1997) A comparative study of flexible (Polyactive™) versus rigid (hydroxylapatite) permucosal dental implants, *II. Histological aspects. J. Oral Rehabilitation*, **24**, 93–101.

147. Waris, E., Ashammakhi, N., Lehtimäki, M., Tulamo, R. M., Törmälä, P., Kellomäki, M., and Konttinen, Y. T. (2008) Long-term bone tissue reaction to polyethylene oxide/polybutylene terephthalate copolymer (Polyactive®) in metacarpophalangeal joint reconstruction, *Biomaterials*, **29**, 2509–2515.

148. Svensson, A., Nicklasson, E., Harrah, T., Panilaitis, B., Kaplan, D. L., Brittberg, M., and Gatenholm, P. (2005) Bacterial cellulose as a potential scaffold for tissue engineering of cartilage, *Biomaterials*, **26**, 419–431.

149. Rampinelli, G., di Landro, L., Fujii, T. (2010) Characterization of biomaterials based on microfibrillated cellulose with different modifications, *J. Reinforced Plastics and Composites*, **29**, 1793–1803.

150. Granja, P. L., Barbosa, M. A., Pouysége, L., de Jéso, B., Rouais, F., and Baquuey, C. (2001) Cellulose phosphates as biomaterials, Mineralization of chemically modified regenerated cellulose hydrogels, *J. Mater. Sci.*, **36**, 2163–2172.

151. Granja, P. L., Jéso, B. D., Bareille, R., Rouais, F., Baquey, C., and Barbosa, M. A. (2006) Cellulose phosphates as biomaterials, *In vitro* biocompatibility studies, *Reactive and Functional Polymers*, **66**, 728–739.

152. Thomas, V., Dean, D. R., and Vohra, Y. K. (2006) Nanostructured biomaterials for regenerative medicine, *Curr. Nanosci.*, **2**, 155–177.

153. Li, S. M., Garreau, H., and Vert, M. (1990) Structure-property relationships in the case of the degradation of massive aliphatic

poly(α-hydroxyacids) in aqueous media, Part 1: poly(D,L-lactic acid), *J. Mater. Sci. Mater. Med.*, **1**, 123–130.

154. Daniels, A. U., Adriano, K. P., Smuts, W. P., Chang, M. K. O., and Keller, J. (1994) Evaluation of absorbable poly(orthoesters) for use in surgical implants, *J. Biomed. Mater. Res. (Appl. Biomater.)*, **5**, 51–64.

155. Adriano, K. P., Pohjonen, T., and Törmällä, P. (1994) Processing and characterization of absorbable polylactide polymers for use in surgical implants, *J. Biomed. Mater. Res. (Appl. Biomater.)*, **5**, 133–140.

156. Athanasiou, K. A., Niederauer, G. G., and Agrawal, C. M. (1996) Sterilization, toxicity, biocompatibility and clinical applications of polylactic acid/polyglycolic acid copolymers, *Biomaterials*, **17**, 93–102.

157. Dee, K. C., and Bizios, R. (1996) Mini-review: proactive biomaterials and bone tissue engineering, *Biotechnology and Bioengineering*, **50**, 438–442.

158. Ignjatovic, N., Tomic, S., Dakic, M., Miljkovic, M., Plavsic, M., and Uskokovic, D. (1999) Synthesis and properties of hydroxyapatite/poly-L-lactide composite biomaterials, *Biomaterials*, **20**, 809–816.

159. Ignjatovic, N., Savic, V., Najman, S., Plavsic, M., and Uskokovic, D. (2001) A study of HAp/PLLA composite as a substitute for bone powder using FT-IR spectroscopy, *Biomaterials*, **22**, 571–575.

160. Marra, K. G., Szem, J. W., Kumta, P. N., DiMilla, P. A., and Weiss, L. E. (1999) *In vitro* analysis of biodegradable polymer blend/hydroxyapatite composites for bone tissue engineering, *J. Biomed. Mater. Res.*, **47**, 324–335.

161. Ashammakhi, N., and Rokkanen, P. (1997) Absorbable polyglycolide devices in trauma and bone surgery, *Biomaterials*, **18**, 3–9.

162. Boyan, B., Lohmann, C., Somers, A., Neiderauer, G., Wozney, J., Dean, D., Carnes, D., and Schwartz, Z. (1999) Potential of porous poly-D,L-lactide-*co*-glycolide particles as a carrier for recombinant human bone morphogenetic protein-2 during osteoinduction *in vivo*, *J. Biomed. Mater. Res.*, **46**, 51–59.

163. Hofmann, G. O. (1995) Biodegradable implants in traumatology: a review of the state-of-the-art, *Arch. Orthopedic Trauma Surgery*, **114**, 123–132.

164. Hollinger, J. O., and Leong, K. (1996) Poly(α-hydroxyacids): carriers for bone morphogenetic proteins, *Biomaterials*, **17**, 187–194.

165. Griffith, L. G. (2000) Polymeric biomaterials, *Acta Materialia*, **48**, 263–277.

166. Peter, S. J., Miller, M. J., Yasko, A. W., Yaszemski, M. J., and Mikos, A. G. (1998) Polymer concepts in tissue engineering, *J. Biomed. Mater. Res.*, **43**, 422–427.

167. Ishuang, S. L., Payne, R. G., Yaszemski, M. J., Aufdemorte, T. B., Bizios, R., and Mikos, A. G. (1996) Osteoblast migration on poly(α-hydroxy esters), *Biothechnology and Bioengineering*, **50**, 443–451.

168. Shikinami, Y., and Okuno, M. (1999) Bioresorbable devices made of forged composites of hydroxyapatite (HA) particles and poly-L-lactide (PLLA): Part I. Basic characteristics, *Biomaterials*, **20**, 859–877.

169. Khor, E., and Lim, L. Y. (2003) Implantable applications of chitin and chitosan, *Biomaterials*, **24**, 2339–2349.

170. Ishihara, M., Nakanishi, K., Ono, K., Sato, M., and Kikuchi, M. (2002) Photocrosslinkable chitosan as a dressing for wound occlusion and accelerator in healing process, *Biomaterials*, **23**, 833–840.

171. Di Martino, A., Sittinger, M., and Risbud, M. V. (2005) Chitosan: a versatile biopolymer for orthopaedic tissue-engineering, *Biomaterials*, **26**, 5983–5990.

172. Piskin, E., Bölgen, N., Egri, S., and Isoglu, I. A. (2007) Electrospun matrices made of poly(α-hydroxy acids) for medical use, *Nanomedicine*, **2**, 441–457.

173. Rezwana, K., Chena, Q. Z., Blakera, J. J., and Boccaccini, A. R. (2006) Biodegradable and bioactive porous polymer/inorganic composite scaffolds for bone tissue engineering, *Biomaterials*, **27**, 3413–3431.

174. Mohanty, A. K., Misra, M., and Hinrichsen, G. (2000) Biofibres, biodegradable polymers and biocomposites: an overview, *Macromol. Mater. Eng.*, **276/277**, 1–24.

175. Kohane, D. S., and Langer, R. (2008) Polymeric biomaterials in tissue engineering, *Pediatric Res.*, **63**, 487–491.

176. Seal, B. L., Otero, T. C., and Panitch, A. (2001) Polymeric biomaterials for tissue and organ regeneration, *Mater. Sci. Eng. R*, **34**, 147–230.

177. An, Y. H., Woolf, S. K., and Friedman, R. J. (2000) Pre-clinical *in vivo* evaluation of orthopaedic bioabsorbable devices, *Biomaterials*, **21**, 2635–2652.

178. Mano, J. F., Sousa, R. A., Boesel, L. F., Neves, N. M., and Reis, R. L. (2004) Bioinert, biodegradable and injectable polymeric matrix

composites for hard tissue replacement: state of the art and recent developments, *Compos. Sci. Technol.*, **64**, 789–817.

179. Katti, K. S. (1994) Biomaterials in total joint replacement, *Colloids Surf. B*, **39**, 133–142.

180. Hayashi, T. (1994) Biodegradable polymers for biomedical uses, *Prog. Polym. Sci.*, **19**, 663–702.

181. Middleton, J., and Tipton, A. (2000) Synthetic biodegradable polymers as orthopedic devices, *Biomaterials*, **21**, 2335–2346.

182. Ma, P. X. (2008) Biomimetic materials for tissue engineering, *Adv. Drug Delivery Rev.*, **60**, 184–198.

183. Coombes, A. G., and Meikle, M. C. (2004) Resorbable synthetic polymers as replacements for bone graft, *Clin. Mater.*, **17**, 35–67.

184. Okada, M. (2002) Chemical syntheses of biodegradable polymers, *Prog. Polym. Sci.*, **27**, 87–133.

185. Jordan, J., Jacob, K. I., Tannenbaum, R., Sharaf, M. A., and Jasiuk, I. (2005) Experimental trends in polymer nanocomposites – a review, *Mater. Sci. Eng. A*, **393**, 1–11.

186. Matsuno, H., Yokoyama, A., Watari, F., Uo, M., and Kawasaki, T. (2001) Biocompatibility and osteogenesis of refractory metal implants, titanium, hafnium, niobium, tantalum and rhenium, *Biomaterials*, **22**, 1253–1262.

187. Hanawa, T. (2010) Biofunctionalization of titanium for dental implant, *Japanese Dental Sci. Review*, **46**, 93–101.

188. Uo, M., Watari, F., Yokoyama, A., Matsuno, H., and Kawasaki, T. (1999) Dissolution of nickel and tissue response observed by X-ray analytical microscopy, *Biomaterials*, **20**, 747–755.

189. Uo, M., Watari, F., Yokoyama, A., Matsuno, H., and Kawasaki, T. Visualization and detectability of rarely contained elements in soft tissue by X-ray scanning analytical microscopy and electron probe micro analysis, *Biomaterials*, **22**, 1787–1794.

190. Uo, M., Watari, F., Yokoyama, A., Matsuno, H., and Kawasaki, T. (2001) Tissue reaction around metal implants observed by X-ray scanning analytical microscopy, *Biomaterials*, **21**, 677–685.

191. Ryan, G., Pandit, A., and Apatsidis, D. P. (2006) Fabrication methods of porous metals for use in orthopaedic applications, *Biomaterials*, **27**, 2651–2670.

192. Shimko, D. A., Shimko, V. F., Sander, E. A., Dickson, K. F., and Nauman, E. A. (2005) Effect of porosity on the fluid flow characteristics and

mechanical properties of tantalum scaffolds, *J. Biomed. Mater. Res. B (Appl. Biomater.)*, **73B**, 315–324.

193. Wen, C. E., Yamada, Y., Shimojima, K., Chino, Y., Hosokawa, H., and Mabuchi, M. (2004) Compressibility of porous magnesium foam: dependency on porosity and pore size, *Mater. Lett.*, **58**, 357–360.

194. Levine, B. (2008) A new era in porous metals: applications in orthopaedics, *Adv. Eng. Mater.*, **10**, 788–792.

195. Green, D., Walsh, D., Mann, S., and Oreffo, R. O. C. (2002) The potentials of biomimesis in bone tissue engineering: lessons from the design and synthesis of invertebrate skeletons, *Bone*, **30**, 810–815.

196. Kokubo, T. (1998) Apatite formation on surfaces of ceramics, metals and polymers in body environment, *Acta Mater.*, **46**, 2519–2527.

197. Witte, F., Reifenrath, J., Müller, P. P., Crostack, H. A., Nellesen, J., Bach, F. W., Bormann, D., and Rudert, M. (2006) Cartilage repair on magnesium scaffolds used as a subchondral bone replacement, *Mat.-Wiss. u. Werkstofflech.*, **37**, 504–508.

198. Gu, X. N., Zhou, W. R., Zheng, Y. F., Liu, Y., and Li, Y. X. (2010) Degradation and cytotoxicity of lotus-type porous pure magnesium as potential tissue engineering scaffold material, *Mater. Lett.*, **64**, 1871–1874.

199. Uo, M., Mizuno, M., Kuboki, Y., Makishima, A., and Watari, F. (1998) Properties and cytotoxicity of water soluble Na_2O-CaO-P_2O_5 glasses, *Biomaterials*, **19**, 2277–2284.

200. Imai, T., Watari, F., Yamagata, S., Kobayashi, M., Nagayama, K., and Nakamura, S. (1998) Mechanical properties and estheticity of FRP orthodontic wire fabricated by hot drawing, *Biomaterial*, **19**, 2195–2200.

201. Watari, F., Yamagata, S., Imai, T., Nakamura, S., and Kobayashi, M. (1998) The fabrication and properties of aesthetic FRP wires for use in orthodontics, *J. Mater. Sci.*, **33**, 5661–5664.

202. Hench, L. L. (2006) The story of Bioglass®, *J. Mater. Sci. Mater. Med.*, **17**, 967–978.

203. Cao, W., Hench, L. L. (1996) Bioactive materials, *Ceram. Int.*, **22**, 493–507.

204. Vogel, M., Voigt, C., Gross, U., and Müller-Mai, C. (2001) *In vivo* comparison of bioactive glass particles in rabbits, *Biomaterials*, **22**, 357–362.

205. de Aza, P., Luklinska, Z., Santos, C., Guitian, F., and de Aza, S. (2003) Mechanism of bone-like formation on a bioactive implant *in vivo*, *Biomaterials*, **24**, 1437–1445.

206. Ikeda, N., Kawanabe, K., and Nakamura, T. (1999) Quantitative comparison of osteoconduction of porous, dense A-W glass-ceramic and hydroxyapatite granules (effects of granule and pore sizes), *Biomaterials*, **20**, 1087–1095.

207. Weizhong, Y., Dali, Z., and Guangfu, Y. (2003) Research and development of A-W bioactive glass ceramic, *J. Biomed. Eng.*, **20**, 541–545.

208. Encinas-Romero, M. A., Aguayo-Salinas, S., Valenzuela-García, J. L., Payán, S. R., and Castillón-Barraza, F. F. (2010) Mechanical and bioactive behavior of hydroxyapatite-wollastonite sintered composites, *Int. J. Appl. Ceram. Technol.*, **7**, 164–177.

209. Piconi, C., and Maccauro, G. (1999) Zirconia as a ceramic biomaterial, *Biomaterials*, **20**, 1–25.

210. Christel, P., Meunier, A., Heller, M., Torre, J. P., and Peille, C. N. (1989) Mechanical properties and short-term *in vivo* evaluation of yttrium-oxide-partially-stabilized-zirconia, *J. Biomed. Mater. Res.*, **23**, 45–61.

211. Garvie, R. C., Urban, D., Kennedy, D. R., and McMeuer, J. C. (1984) Biocompatibility of magnesium partially stabilized zirconia (Mg-PSZ ceramics), *J. Mater. Sci.*, **19**, 3224–3228.

212. Burger, W., Richter, H. G., Piconi, C., Vatteroni, R., Cittadini, A., and Boccalari, M. (1997) New Y-TZP powders for medical grade zirconia, *J. Mater. Sci. Mater. Med.*, **8**, 113–118.

213. Benson, J. (1972) Elemental carbon as a biomaterial, *J. Biomed. Mater. Res.*, **5**, 41–47.

214. Olborska, A., Swider, M., Wolowiec, R., Niedzielski, P., Rylski, A., and Mitura, S. (1994) Amorphous carbon – biomaterial for implant coatings, *Diamond and Related Materials*, **3**, 899–901.

215. Saito, N., Usui, Y., Aoki, K., Narita, N., Shimizu, M., Hara, K., Ogiwara, N., Nakamura, K., Ishigaki, N., Kato, H., Taruta, S, and Endo, M. (2009) Carbon nanotubes: biomaterial applications, *Chem. Soc. Rev.*, **38**, 1897–1903.

216. Vila, M., Manzano, M., and Vallet-Regí, M. (2010) Carbon nanotubes: a solution for processing smart biomaterials, *Key Eng. Mater.*, **441**, 3–29.

217. White, A. A., Best, S. M., and Kinloch, I. A. (2007) Hydroxyapatite-carbon nanotube composites for biomedical applications: a review, *Int. J. Appl. Ceram. Technol.*, **4**, 1–13.

218. Chlopek, J., Czajkowska, B., Szaraniec, B., Frackowiak, E., Szostak, K., and Beguin, F. (2006) *In vitro* studies of carbon nanotubes biocompatibility, *Carbon*, **44**, 1106–1111.

219. Price, R. L., Waid, M. C., Haberstroh, K. M., and Webster, T. J. (2003) Selective bone cell adhesion on formulations containing carbon nanofibers, *Biomaterials*, **24**, 1877–1887.

220. Zanello, L. P., Zhao, B., Hu, H., and Haddon, R. C. (2006) Bone cell proliferation on carbon nanotubes, *Nano Lett.*, **6**, 562–567.

221. Saito, N., Usui, Y., Aoki, K., Narita, N., Shimizu, M., Ogiwara, N., Nakamura, K., Ishigaki, N., Kato, H., and Taruta, S. (2008) Carbon nanotubes for biomaterials in contact with bone, *Curr. Medicinal Chem.*, **15**, 523–527.

222. Banerjee, S., Kahn, M. G. C., and Wong, S. S. (2003) Rational chemical strategies for carbon nanotube functionalization, *Chemistry - A European J.*, **9**, 1898–1908.

223. Beuvelot, J., Bergeret, C., Mallet, R., Fernandez, V., Cousseau, J., Baslé, M. F., and Chappard, D. (2010) *In vitro* calcification of chemically functionalized carbon nanotubes, *Acta Biomater.*, **6**, 4110–4117.

224. Xiao, Y., Gong, T., and Zhou, S. (2010) The functionalization of multi-walled carbon nanotubes by in situ deposition of hydroxyapatite, *Biomaterials*, **31**, 5182–5190.

225. Converse, G. L., Yue, W., and Roeder, R. K. (2007) Processing and tensile properties of hydroxyapatite-whisker-reinforced polyetheretherketone, *Biomaterials*, **28**, 927–935.

226. Yue, W., Roeder, and R. K. (2006) Micromechanical model for hydroxyapatite whisker reinforced polymer biocomposites, *J. Mater. Res.*, **21**, 2136–2145.

227. Converse, G. L., and Roeder, R. K. (2005) Tensile properties of hydroxyapatite whisker reinforced polyetheretherketone, *Mater. Res. Soc. Symp. Proc.*, **898**, 44–49.

228. Mizutani, Y., Hattori, M., Okuyama, M., Kasuga, T., and Nogami, M. (2006) Preparation of porous composites with a porous framework using hydroxyapatite whiskers and poly(L-lactic acid) short fibers, *Key Eng. Mater.*, pp. 309–311, 1079–1082.

229. Choi, W. Y., Kim, H. E., Kim, M. J., Kim, U. C., Kim, J. H., and Koh, Y. H. (2010) Production and characterization of calcium phosphate (CaP) whisker-reinforced poly(ε-caprolactone) composites as bone regenerative, *Mater. Sci. Eng. C*, **30**, 1280–1284.

230. Watanabe, T., Ban, S., Ito, T., Tsuruta, S., Kawai, T., and Nakamura, H. (2004) Biocompatibility of composite membrane consisting of oriented needle-like apatite and biodegradable copolymer with soft and hard tissues in rats, *Dental Mater. J.*, **23**, 609–612.

231. Li, H., Chen, Y., and Xie, Y. (2003) Photo-crosslinking polymerization to prepare polyanhydride/needle-like hydroxyapatite biodegradable nanocomposite for orthopedic application, *Mater. Lett.*, **57**, 2848–2854.

232. Peng, Q., Weng, J., Li, X., and Gu, Z. (2005) Manufacturing porous blocks of nano-composite of needle-like hydroxyapatite crystallites and chitin for tissue engineering, *Key Eng. Mater.*, pp. 288–289, 199–202.

233. Kasuga, T., Ota, Y., Nogami, M., and Abe, Y. (2000) Preparation and mechanical properties of polylactic acid composites containing hydroxyapatite fibers, *Biomaterials*, **22**, 19–23.

234. However, a more general topic "ceramic-plastic material as a bone substitute" is, at least, 18 years older [235].

235. Smith, L. (1963) Ceramic-plastic material as a bone substitute, *Arch. Surg.*, **87**, 653–661.

236. Bonfield, W., Grynpas, M. D., Tully, A. E., Bowman, J., and Abram, J. (1981) Hydroxyapatite reinforced polyethylene – a mechanically compatible implant material for bone replacement, *Biomaterials*, **2**, 185–189.

237. Bonfield, W., Bowman, J., and Grynpas, M. D. (1981) Composite material for use in orthopaedics, *UK Patent 8032647*.

238. Bonfield, W. (1988) Composites for bone replacement, *J. Biomed. Eng.*, **10**, 522–526.

239. Wang, M., Porter, D., and Bonfield, W. (1994) Processing, characterisation and evaluation of hydroxyapatite reinforced polyethylene composites, *British Ceram. Trans.*, **93**, 91–95.

240. Guild, F. J., and Bonfield, W. (1993) Predictive character of hydroxyapatite-polyethelene HAPEX™ composite, *Biomaterials*, **14**, 985–993.

241. Huang, J., di Silvio, L., Wang, M., Tanner, K. E., and Bonfield, W. (1997) *In vitro* mechanical and biological assessment of hydroxyapatite-reinforced polyethylene composite, *J. Mater. Sci. Mater. Med.*, **8**, 775–779.

242. Deb, S., Wang, M., Tanner, K. E., and Bonfield, W. (1996) Hydroxyapatite-polyethylene composites: effect of grafting and surface treatment of hydroxyapatite, *J. Mater. Sci. Mater. Med.*, **7**, 191–193.

243. Wang, M., Joseph, R., and Bonfield, W. (1998) Hydroxyapatite-polyethylene composites for bone substitution: effect of ceramic particle size and morphology, *Biomaterials*, **19**, 2357–2366.

244. Suwanprateeb, J., Tanner, K. E., Turner, S., and Bonfield, W. (1997) Influence of Ringer's solution on creep resistance of hydroxyapatite reinforced polyethylene composites, *J. Mater. Sci. Mater. Med.*, **8**, 469–472.

245. Ladizesky, N. H., Ward, I. M., and Bonfield, W. (1997) Hydroxyapatite/high-performance polyethylene fiber composites for high load bearing bone replacement materials, *J. Appl. Polym. Sci.*, **65**, 1865–1882.

246. Ladizesky, N. H., Pirhonen, E. M., Appleyard, D. B., Ward, I. M., and Bonfield, W. (1998) Fibre reinforcement of ceramic/polymer composites for a major load-bearing bone substitute material, *Compos. Sci. Technol.*, **58**, 419–434.

247. Nazhat, S. N., Joseph, R., Wang, M., Smith, R., Tanner, K. E., and Bonfield, W. (2000) Dynamic mechanical characterisation of hydroxyapatite reinforced polyethylene: effect of particle size, *J. Mater. Sci. Mater. Med.*, **11**, 621–628.

248. Ladizesky, N. H., Ward, I. M., and Bonfield, W. (1996) Hydrostatic extrusion of polyethylene filled with hydroxyapatite, *Polym. Adv. Technol.*, **8**, 496–504.

249. Guild, F. J., and Bonfield, W. (1998) Predictive modelling of the mechanical properties and failure processes of hydroxyapatite-polyethylene (HAPEX™) composite, *J. Mater. Sci. Mater. Med.*, **9**, 497–502.

250. di Silvio, L., Dalby M., and Bonfield, W. (1998) *In vitro* response of osteoblasts to hydroxyapatite/reinforced polyethylene composites, *J. Mater. Sci. Mater. Med.*, **9**, 845–848.

251. Wang M., Ladizesky N. H., Tanner, K. E., Ward I. M., and Bonfield, W. (2000) Hydrostatically extruded HAPEX™, *J. Mater. Sci.*, **35**, 1023–1030.

252. That P. T., Tanner, K. E., and Bonfield, W. (2000) Fatigue characterization of a hydroxyapatite-reinforced polyethylene composite, *I. Uniaxial fatigue. J. Biomed. Mater. Res.*, **51**, 453–460.

253. That P. T., Tanner, K. E., and Bonfield, W. (2000) Fatigue characterization of a hydroxyapatite-reinforced polyethylene composite, *II. Biaxial fatigue. J. Biomed. Mater. Res.*, **51**, 461–468.

254. Bonner M., Saunders L. S., Ward I. M., Davies G. W., Wang M., Tanner, K. E., and Bonfield, W. (2002) Anisotropic mechanical properties of oriented HAPEX™, *J. Mater. Sci.*, **37**, 325–334.

255. Dalby M. J., di Silvio, L., Davies G. W., and Bonfield, W. (2000) Surface topography and HA filler volume effect on primary human osteoblasts *in vitro*, *J. Mater. Sci. Mater. Med.*, **12**, 805–810.

256. di Silvio, L., Dalby, M. J., and Bonfield, W. (2002) Osteoblast behaviour on HA/PE composite surfaces with different HA volumes, *Biomaterials*, **23**, 101–107.

257. Dalby, M. J., Kayser, M. V., Bonfield, W., and di Silvio, L. (2002) Initial attachment of osteoblasts to an optimised HAPEX™ topography, *Biomaterials*, **23**, 681–690.

258. Dalby, M. J., di Silvio, L., Gurav, N., Annaz, B., Kayser, M. V., and Bonfield, W. (2002) Optimizing HAPEX™ topography influences osteoblast response, *Tissue Eng.*, **8**, 453–467.

259. Zhang, Y., Tanner, K. E., Gurav, N., and di Silvio, L. (2007) *In vitro* osteoblastic response to 30 vol% hydroxyapatite-polyethylene composite, *J. Biomed. Mater. Res. A*, **81A**, 409–417.

260. Rea, S. M., Best, S. M., and Bonfield, W. (2004) Bioactivity of ceramic-polymer composites with varied composition and surface topography, *J. Mater. Sci. Mater. Med.*, **15**, 997–1005.

261. Rea, S. M., Brooks, R. A., Schneider, A., Best, S. M., and Bonfield, W. (2004) Osteoblast-like cell response to bioactive composites-surface-topography and composition effects, *J Biomed Mater Res B Appl Biomater.*, **70**, 250–261.

262. Bonner M., Ward I. M., McGregor W., Tanner, K. E., and Bonfield, W. (2001) Hydroxyapatite/polypropylene composite: a novel bone substitute material, *J. Mater. Sci. Lett.*, **20**, 2049–2052.

263. Suppakarn, N., Sanmaung, S., Ruksakulpiwa, Y., and Sutapun, W. (2008) Effect of surface modification on properties of natural hydroxyapatite/polypropylene composites, *Key Eng. Mater.*, pp. 361–363, 511–514.

264. Younesi, M., and Bahrololoom, M. E. (2010) Formulating the effects of applied temperature and pressure of hot pressing process on the mechanical properties of polypropylene-hydroxyapatite bio-composites by response surface methodology, *Mater. Des.*, **31**, 4621–4630.

265. Salernitano, E., and Migliaresi, C. (2003) Composite materials for biomedical applications: a review, *J. Appl. Biomater. Biomech.*, **1**, 3–18.

266. Pandey, A., Jan, E., and Aswath, P. B. (2006) Physical and mechanical behavior of hot rolled HDPE/HA composites, *J. Mater. Sci.*, **41**, 3369–3376.

267. Sousa, R. A., Reis, R. L., Cunha, A. M., and Bevis, M. J. (2003) Processing and properties of bone-analogue biodegradable and bioinert polymeric composites, *Compos. Sci. Technol.*, **63**, 389–402.

268. Wang, M., Deb, S., and Bonfield, W. (2000) Chemically coupled hydroxyapatite-polyethylene composites: processing and characterisation, *Mater. Lett.*, **44**, 119–124.

269. Wang, M., and Bonfield, W. (2001) Chemically coupled hydroxyapatite-polyethylene composites: structure and properties, *Biomaterials*, **22**, 1311–1320.

270. Homaeigohar, S. S., Shokrgozar, M. A., Khavandi, A., and Sadi, A. Y. (2008) *In vitro* biological evaluation of β-TCP/HDPE – a novel orthopedic composite: a survey using human osteoblast and fibroblast bone cells, *J. Biomed. Mater. Res. A*, **84A**, 491–499.

271. Sadi, A. Y., Homaeigohar, S. Sh., Khavandi, A. R., and Javadpour, J. (2004) The effect of partially stabilized zirconia on the mechanical properties of the hydroxyapatite-polyethylene composites, *J. Mater. Sci. Mater. Med.*, **15**, 853–858.

272. Nath, S., Bodhak, S., and Basu, B. (2009) HDPE-Al_2O_3-HAp composites for biomedical applications: processing and characterizations, *J. Biomed. Mater. Res. B (Appl. Biomater.)*, **88B**, 1–11.

273. Downes, R. N., Vardy, S., Tanner, K. E., and Bonfield, W. (1991) Hydroxyapatite-polyethylene composite in orbital surgery, *Bioceramics 4.*, pp. 239–246.

274. Dornhoffer, H. L. (1998) Hearing results with the dornhoffer ossicular replacement prostheses, *Laryngoscope*, **108**, 531–536.

275. Swain, R. E., Wang, M., Beale, B., and Bonfield, W. (1999) HAPEX™ for otologic applications, *Biomed. Eng. Appl. Basis Commun.*, **11**, 315–320.

276. Yi, Z., Li, Y., Jidong, L., Xiang, Z., Hongbing, L., Yuanyuan, W., and Weihu, Y. (2007) Novel bio-composite of hydroxyapatite reinforced polyamide and polyethylene: composition and properties, *Mater. Sci. Eng. A*, pp. 452–453, 512–517.

277. Unwin, A. P., Ward, I. M., Ukleja, P., and Weng, J. (2001) The role of pressure annealing in improving the stiffness of polyethylene/hydroxyapatite composites, *J. Mater. Sci.*, **36**, 3165–3177.

278. Fang, L. M., Leng, Y., and Gao, P. (2006) Processing and mechanical properties of HA/UHMWPE nanocomposites, *Biomaterials*, **27**, 3701–3707.

279. Fang, L. M., Gao, P., and Leng, Y. (2007) High strength and bioactive hydroxyapatite nano-particles reinforced ultrahigh molecular weight polyethylene, *Composites B*, **38**, 345–351.

280. Fang, L. M., Leng, Y., and Gao, P. (2005) Processing of hydroxyapatite reinforced ultrahigh molecular weight polyethylene for biomedical applications, *Biomaterials*, **26**, 3471–3478.

281. Selvin, T. P., Seno, J., Murukan, B., Santhosh, A. A., Sabu, T., Weimin, Y., and Sri, B. (2010) Poly(ethylene-*co*-vinyl acetate)/calcium phosphate nanocomposites: thermo mechanical and gas permeability measurements, *Polymer Composites*, **31**, 1011–1019.

282. Sousa, R. A., Reis, R. L., Cunha, A. M., and Bevis, M. J. (2002) Structure development and interfacial interactions in high-density polyethylene/hydroxyapatite (HDPE/HA) composites molded with preferred orientation, *J. Appl. Polym. Sci.*, **86**, 2873–2886.

283. Reis, R. L., Cunha, A. M., Oliveira, M. J., Campos, A. R., and Bevis, M. J. (2001) Relationship between processing and mechanical properties of injection molded high molecular mass polyethylene + hydroxyapatite composites, *Mat. Res. Innovat.*, **4**, 263–272.

284. Donners, J. J. J. M., Nolte, R. J. M., Sommerdijk, N. A. J. M. (2003) Dendrimer-based hydroxyapatite composites with remarkable materials properties, *Adv. Mater.*, **15**, 313–316.

285. Schneider, O. D., Stepuk, A., Mohn, D., Luechinger, N. A., Feldman, K., and Stark, W. J. (2010) Light-curable polymer/calcium phosphate nanocomposite glue for bone defect treatment, *Acta Biomater.*, **6**, 2704–2710.

286. Ignjatovic, N. L., Plavsic, M., Miljkovic, M. S., Zivkovic, L. M., and Uskokovic, D. P. (1999) Microstructural characteristics of calcium hydroxyapatite/poly-L-lactide based composites, *J. Microscopy Oxford*, **196**, 243–248.

287. Skrtic, D., Antonucci, J. M., and Eanes, E. D. (2003) Amorphous calcium phosphate-based bioactive polymeric composites for mineralized tissue regeneration, *J. Res. Natl. Inst. Stand. Technol.*, **108**, 167–182.

288. Rizzi, S. C., Heath, D. J., Coombes, A. G. A., Bock, N., Textor, M., and Downes, S. (2001) Biodegradable polymer/hydroxyapatite composites: surface analysis and initial attachment of human osteoblasts, *J. Biomed. Mater. Res.*, **55**, 475–486.

289. Kato, K., Eika, Y., and Ikada, Y. (1997) *In situ* hydroxyapatite crystallization for the formation of hydroxyapatite/polymer composites, *J. Mater. Sci.*, **32**, 5533–5543.

290. Navarro, M., and Planell, J. A. (2010) Bioactive composites based on calcium phosphates for bone regeneration, *Key Eng. Mater.*, **441**, 203–233.

291. Damien, C. J., and Parsons, J. R. (1991) Bone-graft and bone-graft substitutes a review of current technology and applications, *J. Biomed. Mater. Res. (Appl. Biomater.)*, **2**, 187–208.

292. Zhang, R. Y., and Ma, P. X. (1999) Porous poly(L-lactic acid)/apatite composites created by biomimetic process, *J. Biomed. Mater. Res.*, **45**, 285–293.

293. Liu, Q., de Wijn, J. R., and van Blitterswijk, C. A. (1998) Composite biomaterials with chemical bonding between hydroxyapatite filler particles and PEG/PBT copolymer matrix, *J. Biomed. Mater. Res.*, **40**, 490–497.

294. Cerrai, P., Guerra, G. D., Tricoli, M., Krajewski, A., Ravaglioli, A., Martinetti, R., Dolcini, L., Fini, M., Scarano, A., and Piattelli, A. (1999) Periodontal membranes from composites of hydroxyapatite and bioresorbable block copolymers, *J. Mater. Sci. Mater. Med.*, **10**, 677–682.

295. Roeder, R. K., Sproul, M. M., and Turner, C. H. (2003) Hydroxyapatite whiskers provide improved mechanical properties in reinforced polymer composites, *J. Biomed. Mater. Res. A*, **67A**, 801–812.

296. Hutmacher, D. W. (2000) Scaffolds in tissue engineering bone and cartilage, *Biomaterials*, **21**, 2529–2543.

297. Mathieu, L. M., Bourban, P. E., and Manson, J. A. E. (2006) Processing of homogeneous ceramic/polymer blends for bioresorbable composites, *Compos. Sci. Technol.*, **66**, 1606–1614.

298. Redepenning, J., Venkataraman, G., Chen, J., and Stafford, N. (2003) Electrochemical preparation of chitosan/hydroxyapatite composite coatings on titanium substrates, *J. Biomed. Mater. Res. A*, **66A**, 411–416.

299. Rhee, S. H., and Tanaka, J. (2001) Synthesis of a hydroxyapatite/collagen/chondroitin sulfate nanocomposite by a novel precipitation method, *J. Am. Ceram. Soc.*, **84**, 459–461.

300. Pezzotti, G., and Asmus, S. M. F. (2001) Fracture behavior of hydroxyapatite/polymer interpenetrating network composites prepared by *in situ* polymerization process, *Mater. Sci. Eng. A*, **316**, 231–237.

301. Weickmann, H., Gurr, M., Meincke, O., Thomann, R., and Mülhaupt, R. (2010) A versatile solvent-free "one-pot" route to polymer

nanocomposites and the *in situ* formation of calcium phosphate/layered silicate hybrid nanoparticles, *Adv. Funct. Mater.*, **20**, 1778–1786.

302. Aryal, S., Bhattarai, S. R., Bahadur, K. C. R., Khil, M. S., Lee, D. R., and Kim, H. Y. (2006) Carbon nanotubes assisted biomimetic synthesis of hydroxyapatite from simulated body fluid, *Mater. Sci. Eng. A*, **426**, 202–207.

303. Kealley, C., Ben-Nissan, B., van Riessen, A., and Elcombe, M. (2006) Development of carbon nanotube reinforced hydroxyapatite bioceramics, *Key Eng. Mater.*, pp. 309–311, 597–600.

304. Kealley, C., Elcombe, M., van Riessen, A., and Ben-Nissan, B. (2006) Development of carbon nanotube reinforced hydroxyapatite bioceramics, *Physica B*, pp. 385–386, 496–498.

305. Aryal, S., Bahadur, K. C. R., Dharmaraj, N., Kim, K. W., and Kim, H. Y. (2006) Synthesis and characterization of hydroxyapatite using carbon nanotubes as a nano-matrix, *Scripta Mater.*, **54**, 131–135.

306. Wei, Q., Yang, X. P., Chen, G. Q., Tang, J. T., and Deng, X. L. (2005) The ultrasonic assisted synthesis of nano-hydroxyapatite and MWNT/hydroxyapatite composites, *New Carbon Mater.*, **20**, 164–170.

307. Zhao, L. P., and Gao, L. (2004) Novel *in situ* synthesis of MWNT-hydroxyapatite composites, *Carbon*, **42**, 423–460.

308. Bai, Y., Neupane, M. P., Park, I. S., Lee, M. H., Bae, T. S., Watari, F., and Uo, M. (2010) Electrophoretic deposition of carbon nanotubes-hydroxyapatite nanocomposites on titanium substrate, *Mater. Sci. Eng. C*, **30**, 1043–1049.

309. Rautaray, D., Mandal, S., and Sastry, M. (2005) Synthesis of hydroxyapatite crystals using amino acid-capped gold nanoparticles as a scaffold, *Langmuir*, **21**, 5185–5191.

310. Wang, X. J., Li, Y., Wei, J., and de Groot, K. (2002) Development of biomimetic nano-hydroxyapatite/poly(hexamethylene adipamide) composites, *Biomaterials*, **23**, 4787–4791.

311. Wei, J., and Li, Y. (2004) Tissue engineering scaffold material of nano-apatite crystals and polyamide composite, *Eur. Polym. J.*, **40**, 509–515.

312. Memoto, R., Nakamura, S., Isobe, T., and Senna, M. (2001) Direct synthesis of hydroxyapatite-silk fibroin nano-composite sol via a mechano-chemical route, *J. Sol Gel Sci. Technol.*, **21**, 7–12.

313. Fujiwara, M., Shiokawa, K., Morigaki, K., Tatsu, Y., and Nakahara, Y. (2008) Calcium phosphate composite materials including inorganic

powders, BSA or duplex DNA prepared by W/O/W interfacial reaction method, *Mater. Sci. Eng. C*, **28**, 280–288.

314. Nagata, F., Miyajima, T., and Yokogawa, Y. (2006) A method to fabricate hydroxyapatite/poly(lactic acid) microspheres intended for biomedical application, *J. Eur. Ceram. Soc.*, **26**, 533–535.

315. Russias, J., Saiz, E., Nalla, R. K., and Tomsia, A. P. (2006) Microspheres as building blocks for hydroxyapatite/polylactide biodegradable composites, *J. Mater. Sci.*, **41**, 5127–5133.

316. Khan, Y. M., Cushnie, E. K., Kelleher, J. K., and Laurencin, C. T. (2007) *In situ* synthesized ceramic-polymer composites for bone tissue engineering: bioactivity and degradation studies, *J. Mater. Sci.*, **42**, 4183–4190.

317. Kim, H. W., Knowles, J. C., and Kim, H. E. (2005) Hydroxyapatite and gelatin composite foams processed via novel freeze-drying and crosslinking for use as temporary hard tissue scaffolds, *J. Biomed. Mater. Res. A*, **72A**, 136–145.

318. Sinha, A., Das, G., Sharma, B. K., Roy, R. P., Pramanick, A. K., and Nayar, S. (2007) Poly(vinyl alcohol)-hydroxyapatite biomimetic scaffold for tissue regeneration, *Mater. Sci. Eng. C*, **27**, 70–74.

319. Sun, L., Berndt, C. C., and Gross, K. A. (2002) Hydroxyapatite/polymer composite flame-sprayed coatings for orthopedic applications, *J. Biomater. Sci. Polym. Edn.*, **13**, 977–990.

320. Sugawara, A., Yamane, S., and Akiyoshi, K. (2006) Nanogel-templated mineralization: polymer-calcium phosphate hybrid nanomaterials, *Macromol. Rapid Commun.*, **27**, 441–446.

321. Kickelbick, G. (2003) Concepts for the incorporation of inorganic building blocks into organic polymers on a nanoscale, *Progress Polym. Sci.*, **28**, 83–114.

322. Liu, Q., de Wijn, J. R., and van Blitterswijk, C. A. (1997) Nanoapatite/polymer composites: mechanical and physicochemical characteristics, *Biomaterials*, **18**, 1263–1270.

323. Uskokovic, P. S., Tang, C. Y., Tsui, C. P., Ignjatovic, N., and Uskokovic, D. P. (2007) Micromechanical properties of a hydroxyapatite/poly-L-lactide biocomposite using nanoindentation and modulus mapping, *J. Eur. Ceram. Soc.*, **27**, 1559–1564.

324. Todo, M., and Kagawa, T. (2008) Improvement of fracture energy of HA/PLLA biocomposite material due to press processing, *J. Mater. Sci.*, **43**, 799–801.

325. Woo, K. M., Seo, J., Zhang, R. Y., and Ma, P. X. (2007) Suppression of apoptosis by enhanced protein adsorption on polymer/hydroxyapatite composite scaffolds, *Biomaterials*, **28**, 2622–2630.

326. Ma, P. X., Zhang, R., Xiao, G., and Franceschi, R. (2001) Engineering new bone tissue *in vitro* on highly porous poly(α-hydroxyl acids)/hydroxyapatite composite scaffolds, *J. Biomed. Mater. Res.*, **54**, 284–293.

327. Wang, M., Chen, L. J., Ni, J., Weng, J., and Yue, C. Y. (2001) Manufacture and evaluation of bioactive and biodegradable materials and scaffolds for tissue engineering, *J. Mater. Sci. Mater. Med.*, **12**, 855–860.

328. Baji, A., Wong, S. C., Srivatsan, T. S., Njus, G. O., and Mathur, G. (2006) Processing methodologies for polycaprolactone-hydroxyapatite composites: a review, *Mater. Manuf. Process.*, **21**, 211–218.

329. Wei, G., and Ma, P. X. (2006) Macroporous and nanofibrous polymer scaffolds and polymer/bone-like apatite composite scaffolds generated by sugar spheres, *J. Biomed. Mater. Res. A*, **78A**, 306–315.

330. Guan, L., and Davies, J. E. (2004) Preparation and characterization of a highly macroporous biodegradable composite tissue engineering scaffold, *J. Biomed. Mater. Res. A*, **71A**, 480–487.

331. Teng, X. R., Ren, J., and Gu, S. Y. (2007) Preparation and characterization of porous PDLLA/HA composite foams by supercritical carbon dioxide technology, *J. Biomed. Mater. Res. B (Appl. Biomater.)*, **81B**, 185–193.

332. Ren, J., Zhao, P., Ren, T., Gu, S., and Pan, K. (2008) Poly (D,L-lactide)/nano-hydroxyapatite composite scaffolds for bone tissue engineering and biocompatibility evaluation, *J. Mater. Sci. Mater. Med.*, **19**, 1075–1082.

333. Wang, M., Yue, C. Y., and Chua, B. (2001) Production and evaluation of hydroxyapatite reinforced polysulfone for tissue replacement, *J. Mater. Sci. Mater. Med.*, **12**, 821–826.

334. Chlopek, J., Rosol, P., and Morawska-Chochol, A. (2006) Durability of polymer-ceramics composite implants determined in creep tests, *Compos. Sci. Technol.*, **66**, 1615–1622.

335. Szaraniec, B., Rosol, P., and Chlopek, J. (2005) Carbon composite material and polysulfone modified by nano-hydroxyapatite, *e-Polymers*, **030**, 1–7.

336. Nayar, S., and Sinha, A. (2004) Systematic evolution of a porous hydroxyapatite-poly(vinylalcohol)-gelatin composite, *Colloids Surf. B*, **35**, 29–32.

337. Chang, M. C., Ko, C. C., and Douglas, W. H. (2005) Modification of hydroxyapatite/gelatin composite by polyvinylalcohol, *J. Mater. Sci.*, **40**, 505–509.

338. Chang, M. C., Ko, C. C., and Douglas, W. H. (2005) Modification of hydroxyapatite/gelatin composite by polyvinylalcohol, *J. Mater. Sci.*, **40**, 2723–2727.

339. You, C., Miyazaki, T., Ishida, E., Ashizuka, M., Ohtsuki, C., and Tanihara, M. (2007) Fabrication of poly(vinyl alcohol)-apatite hybrids through biomimetic process, *J. Eur. Ceram. Soc.*, **27**, 1585–1588.

340. Xu, F., Li, Y., Yao, X., Liao, H., and Zhang, L. (2007) Preparation and *in vivo* investigation of artificial cornea made of nano-hydroxyapatite/poly (vinyl alcohol) hydrogel composite, *J. Mater. Sci. Mater. Med.*, **18**, 635–640.

341. Xu, F., Li, Y., Deng, Y., and Xiong, G. (2008) Porous nano-hydroxyapatite/poly(vinyl alcohol) composite hydrogel as artificial cornea fringe: characterization and evaluation *in vitro*, *J. Biomater. Sci. Polymer Edn.*, **19**, 431–439.

342. Nayar, S., Pramanick, A. K., Sharma, B. K., Das, G., Kumar, B. R., and Sinha, A. (2008) Biomimetically synthesized polymer-hydroxyapatite sheet like nano-composite, *J. Mater. Sci. Mater. Med.*, **19**, 301–304.

343. Poursamar, S. A., Orang, F., Bonakdar, S., and Savar, M. K. (2010) Preparation and characterisation of poly vinyl alcohol/hydroxyapatite nanocomposite via *in situ* synthesis: a potential material as bone tissue engineering scaffolds, *Int. J. Nanomanufacturing*, **5**, 330–334.

344. Guha, A., Nayar, S., and Thatoi, H. N. (2010) Microwave irradiation enhances kinetics of the biomimetic process of hydroxyapatite nanocomposites, *Bioinspiration and Biomimetics*, **5**, 024001 (5 pages).

345. Pramanik, N., Biswas, S. K., and Pramanik, P. (2008) Synthesis and characterization of hydroxyapatite/poly(vinyl alcohol phosphate) nanocomposite biomaterials, *Int. J. Appl. Ceram. Technol.*, **5**, 20–28.

346. Bigi, A., Boanini, E., Gazzano, M., and Rubini, K. (2005) Structural and morphological modifications of hydroxyapatite-polyaspartate composite crystals induced by heat treatment, *Cryst. Res. Technol.*, **40**, 1094–1098.

347. Bertoni, E., Bigi, A., Falini, G., Panzavolta, S., and Roveri, N. (1999) Hydroxyapatite polyacrylic acid nanocrystals, *J. Mater. Chem.*, **9**, 779–782.

348. Qiu, H. J., Yang, J., Kodali, P., Koh, J., and Ameer, G. A. (2006) A citric acid-based hydroxyapatite composite for orthopedic implants, *Biomaterials*, **27**, 5845–5854.

349. Greish, Y. E., and Brown, P. W. (2001) Chemically formed HAp-Ca poly(vinyl phosphonate) composites, *Biomaterials*, **22**, 807–816.

350. Greish, Y. E., and Brown, P. W. (2001) Preparation and characterization of calcium phosphate-poly(vinyl phosphonic acid) composites, *J. Mater. Sci. Mater. Med.*, **12**, 407–411.

351. Greish, Y. E., and Brown, P. W. (2002) Formation and properties of hydroxyapatite-calcium poly(vinyl phosphonate) composites, *J. Am. Ceram. Soc.*, **85**, 1738–1744.

352. Nakahira, A., Tamai, M., Miki, S., and Pezotti, G. (2002) Fracture behavior and biocompatibility evaluation of nylon-infiltrated porous hydroxyapatite, *J. Mater. Sci.*, **37**, 4425–4430.

353. Sailaja, G. S., Velayudhan, S., Sunny, M. C., Sreenivasan, K., Varma, H. K., and Ramesh, P. (2003) Hydroxyapatite filled chitosan-polyacrylic acid polyelectrolyte complexes, *J. Mater. Sci.*, **38**, 3653–3662.

354. Zhang, H., Xu, J. J., and Chen, H. Y. (2007) Electrochemically deposited 2D nanowalls of calcium phosphate-PDDA on a glassy carbon electrode and their applications in biosensing, *J. Phys. Chem. C*, **111**, 16564–16570.

355. Piticescu, R. M., Chitanu, G. C., Albulescu, M., Giurginca, M., Popescu, M. L., and Łojkowski, W. (2005) Hybrid HAp-maleic anhydride copolymer nanocomposites obtained by *in-situ* functionalisation, *Solid State Phenomena*, **106**, 47–56.

356. Enlow, D., Rawal, A., Kanapathipillai, M., Schmidt-Rohr, K., Mallapragada, S., Lo, C. T., Thiyagarajan, P., and Akin, M. (2007) Synthesis and characterization of self-assembled block copolymer templated calcium phosphate nanocomposite gels, *J. Mater. Chem.*, **17**, 1570–1578.

357. Kaito, T., Myoui, A., Takaoka, K., Saito, N., Nishikawa, M., Tamai, N., Ohgushi, H., and Yoshikawa, H. (2005) Potentiation of the activity of bone morphogenetic protein-2 in bone regeneration by a PLA-PEG/hydroxyapatite composite, *Biomaterials*, **26**, 73–79.

358. Song, J., Saiz, E., and Bertozzi, C. R. (2003) A new approach to mineralization of biocompatible hydrogel scaffolds: an efficient process toward 3-dimensional bonelike composites, *J. Am. Chem. Soc.*, **125**, 1236–1243.

359. Meenan, B. J., McClorey, C., and Akay, M. (2000) Thermal analysis studies of poly(etheretherketone)/hydroxyapatite biocomposite mixtures, *J. Mater. Sci. Mater. Med.*, **11**, 481–489.

360. Fan, J. P., Tsui, C. P., Tang, C. Y., and Chow, C. L. (2004) Modeling of the mechanical behavior of HA/PEEK biocomposite under quasi-static tensile load, *Biomaterials*, **25**, 5363–5373.

361. Abu Bakar, M. S., Cheng, M. H. W., Tang, S. M., Yu, S. C., Liao, K., Tan, C. T., Khor, K. A., and Cheang, P. (2003) Tensile properties, tension-tension fatigue and biological response of polyetheretherketone-hydroxyapatite composites for load-bearing orthopedic implants, *Biomaterials*, **24**, 2245–2250.

362. Abu Bakar, M. S., Cheang, P., and Khor, K. A. (2003) Mechanical properties of injection molded hydroxyapatite-polyetheretherketone biocomposites, *Compos. Sci. Technol.*, **63**, 421–425.

363. Abu Bakar, M. S., Cheang, P., and Khor, K. A. (2003) Tensile properties and microstructural analysis of spheroidized hydroxyapatite-poly(etheretherketone) biocomposites, *Mater. Sci. Eng. A*, **345**, 55–63.

364. Fan, J. P., Tsui, C. P., and Tang, C. Y. (2004) Modeling of the mechanical behavior of HA/PEEK biocomposite under quasi-static tensile load, *Mater. Sci. Eng. A*, **382**, 341–350.

365. Yu, S., Hariram, K. P., Kumar, R., Cheang, P., and Aik, K. K. (2005) *In vitro* apatite formation and its growth kinetics on hydroxyapatite/polyetheretherketone biocomposites, *Biomaterials*, **26**, 2343–2352.

366. Gong, X. H., Tang, C. Y., Hu, H. C., and Zhou, X. P. (2004) Improved mechanical properties of HIPS/hydroxyapatite composites by surface modification of hydroxyapatite via *in situ* polymerization of styrene, *J. Mater. Sci. Mater. Med.*, **15**, 1141–1146.

367. Wang, L., Weng, L., Song, S., and Sun, Q. (2010) Mechanical properties and microstructure of polyetheretherketone-hydroxyapatite nanocomposite materials, *Mater. Lett.*, **64**, 2201–2204.

368. Laurencin, C. T., Attawia, M. A., Elgendy, H. E., and Herbert, K. M. (1996) Tissue engineered bone-regeneration using degradable polymers: the formation of mineralized matrices, *Bone*, **91**, S93–S99.

369. Laurencin, C. T., Attawia, M. A., Lu, L. Q., Borden, M. D., Lu, H. H., Gorum, W. J., and Lieberman, J. R. (2001) Poly(lactide-*co*-glycolide)/hydroxyapatite delivery of BMP-2-producing cells: a regional gene therapy approach to bone regeneration, *Biomaterials*, **22**, 1271–1277.

370. Kim, S. S., Ahn, K. M., Park, M. S., Lee, J. H., Choi, C. Y., and Kim, B. S. (2007) A poly(lactide-*co*-glycolide)/hydroxyapatite composite scaffold with enhanced osteoconductivity, *J. Biomed. Mater. Res. A*, **80A**, 206–215.

371. Oliveira, J., Miyazaki, T., Lopes, M., Ohtsuki, C., and Santos, J. (2005) Bonelike®/PLGA hybrid materials for bone regeneration: preparation route and physicochemical characterization, *J. Mater. Sci. Mater. Med.*, **16**, 253–259.

372. Kim, S., Kim, S. S., Lee, S. H., Ahn, S. E., Gwak, S. J., Song, J. H., Kim, B. S., and Chung, H. M. (2008) *In vivo* bone formation from human embryonic stem cell-derived osteogenic cells in poly(D,L-lactic-*co*-glycolic acid)/hydroxyapatite composite scaffolds, *Biomaterials*, **29**, 1043–1053.

373. Petricca, S. E., Marra, K. G., and Kumta, P. N. (2006) Chemical synthesis of poly(lactic-*co*-glycolic acid)/hydroxyapatite composites for orthopaedic applications, *Acta Biomater.*, **2**, 277–286.

374. Sato, M., Slamovich, E. B., and Webster, T. J. (2005) Enhanced osteoblast adhesion on hydrothermally treated hydroxyapatite/titania/poly(lactide-*co*-glycolide) sol-gel titanium coatings, *Biomaterials*, **26**, 1349–1357.

375. Gu, S.Y., Zhan, H., Ren, J., and Zhou, X. Y. (2007) Sol-gel synthesis and characterisation of nano-sized hydroxyapatite powders and hydroxyapatite/poly(D,L-lactide-*co*-glycolide) composite scaffolds, *Polymers & Polymer Composites*, **15**, 137–144.

376. Aboudzadeh, N., Imani, M., Shokrgozar, M. A., Khavandi, A., Javadpour, J., Shafieyan, Y., and Farokhi, M. (2010) Fabrication and characterization of poly(D,L-lactide-*co*-glycolide)/ hydroxyapatite nanocomposite scaffolds for bone tissue regeneration, *J. Biomed. Mater. Res. A*, **94A**, 137–145.

377. Verheyen, C. C. P. M., de Wijin, J. R., van Blitterswijk, C. A., and de Groot, K. (1992) Evaluation of hydroxyapatite/poly(L-lactide) composites: mechanical behavior, *J. Biomed. Mater. Res.*, **26**, 1277–1296.

378. Balac, I., Uskokovic, P. S., Aleksic, R., and Uskokovic, D. (2002) Predictive modeling of the mechanical properties of particulate hydroxyapatite reinforced polymer composites, *J. Biomed. Mater. Res.*, **63**, 793–799.

379. Dawes, E., and Rushton, N. (1994) The effects of lactic acid on PGE2 production by macrophages and human synovial fibroblasts: a possible explanation for problems associated with the degradation of poly(lactide) implants? *Clin. Mater.*, **17**, 157–163.

380. Verheyen, C. C. P. M., Klein, C. P. A. T., de Blieck-Hogervorst, J. M. A., Wolke, J. G. C., de Wijin, J. R., van Blitterswijk, C. A., and de Groot, K. (1993) Evaluation of hydroxylapatite poly(L-lactide) composites-physicochemical properties, *J. Mater. Sci. Mater. Med.*, **4**, 58–65.

381. Li, H., and Chang, J. (2005) pH-compensation effect of bioactive inorganic fillers on the degradation of PLGA, *Compos. Sci. Technol.*, **65**, 2226–2232.

382. Agrawal, C. M., and Athanasiou, K. A. (1997) Technique to control pH in vicinity of biodegrading PLA-PGA implants, *J. Biomed. Mater. Res. Appl. Mater.*, **38**, 105–114.

383. Peter, S. J., Miller, S. T., Zhu, G., Yasko, A. W., and Mikos, A. G. (1998) *In vivo* degradation of a poly(propylene fumarate)/β-tricalcium phosphate injectable composite scaffold, *J. Biomed. Mater. Res.*, **41**, 1–7.

384. Ara, M., Watanabe, M., and Imai, Y. (2002) Effect of blending calcium compounds on hydrolitic degradation of poly(D,L-lactic acid-*co*-glycolic acid), *Biomaterials*, **23**, 2479–2483.

385. Linhart, W., Peters, F., Lehmann, W., Schwarz, K., Schilling, A., Amling, M., Rueger, J. M., and Epple, M. (2001) Biologically and chemically optimized composites of carbonated apatite and polyglycolide as bone substitution materials, *J. Biomed. Mater. Res.*, **54**, 162–171.

386. Schiller, C., and Epple, M. (2003) Carbonated apatites can be used as pH-stabilizing filler for biodegradable polyesters, *Biomaterials*, **24**, 2037–2043.

387. Schiller, C., Rasche, C., Wehmöller, M., Beckmann, F., Eufinger, H., Epple, M., and Weihe, S. (2004) Geometrically structured implants for cranial reconstruction made of biodegradable polyesters and calcium phosphate/calcium carbonate, *Biomaterials*, **25**, 1239–1247.

388. Shikinami, Y., and Okuno, M. (2001) Bioresorbable devices made of forged composites of hydroxyapatite (HA) particles and poly L-lactide (PLLA). Part II: practical properties of miniscrews and miniplates, *Biomaterials*, **22**, 3197–3211.

389. Russias, J., Saiz, E., Nalla, R. K., Gryn, K., Ritchie, R. O., and Tomsia, A. P. (2006) Fabrication and mechanical properties of PLA/HA composites: a study of *in vitro* degradation, *Mater. Sci. Eng. C*, **26**, 1289–1295.

390. Kim, H. W., Lee, H. H., and Knowles, J. C. (2006) Electrospinning biomedical nanocomposite fibers of hydroxyapaite/poly(lactic acid) for bone regeneration, *J. Biomed. Mater. Res. A*, **79A**, 643–649.

391. Gross, K. A., and Rodríguez-Lorenzo, L. M. (2004) Biodegradable composite scaffolds with an interconnected spherical network for bone tissue engineering, *Biomaterials*, **25**, 4955–4962.

392. Zhang, H., and Chen, Z. (2010) Fabrication and characterization of electrospun PLGA/MWNTs/ hydroxyapatite biocomposite scaffolds for bone tissue engineering, *J. Bioactive and Compatible Polymers*, **25**, 241–259.

393. Durucan, C., and Brown, P. W. (2000) Low temperature formation of calcium-deficient hydroxyapatite-PLA/PLGA composites, *J. Biomed. Mater. Res.*, **51**, 717–725.

394. Durucan, C., and Brown, P. W. (2000) Calcium-deficient hydroxyapatite-PLGA composites: mechanical and microstructural investigation, *J. Biomed. Mater. Res.*, **51**, 726–734.

395. Ignjatovic, N., Suljovrujic, E., Biudinski-Simendic, J., Krakovsky, I., and Uskokovic, D. (2004) Evaluation of hot-presses hydroxyapatite/poly-L-lactide composite biomaterial characteristics, *J. Biomed. Mater. Res. B (Appl. Biomater.)*, **71B**, 284–294.

396. Nazhat, S. N., Kellomäki, M., Törmälä, P., Tanner, K. E., and Bonfield, W. (2001) Dynamic mechanical characterization of biodegradable composites of hydroxyapatite and polylactides, *J. Biomed. Mater. Res.*, **58**, 335–343.

397. Hasegawa, S., Tamura, J., Neo, M., Goto, K., Shikinami, Y., Saito, M., Kita, M., and Nakamura, T. (2005) *In vivo* evaluation of a porous hydroxyapatite/poly-D,L-lactide composite for use as a bone substitute, *J. Biomed. Mater. Res. A*, **75A**, 567–579.

398. Hasegawa, S., Neo, M., Tamura, J., Fujibayashi, S., Takemoto, M., Shikinami, Y., Okazaki, K., and Nakamura, T. (2007) *In vivo* evaluation of a porous hydroxyapatite/poly-D,L-lactide composite for bone tissue engineering, *J. Biomed. Mater. Res. A*, **81A**, 930–938.

399. Higashi, S., Yamamuro, T., Nakamura, T., Ikada, Y., Hyon, S. H., and Jamshidi, K. (1986) Polymer-hydroxyapatite composites for biodegradable bone fillers, *Biomaterials*, **7**, 183–187.

400. Ylinen, P. (1994) Filling of bone defects with porous hydroxyapatite reinforced with polylactide or polyglycolide fibres, *J. Mater. Sci. Mater. Med.*, **5**, 522–528.

401. Reis, R. L., Cunha, A. M., and Bevis, M. J. (1998) Using nonconventional processing routes to develop anisotropic and biodegradable composites of starch-based thermoplastics reinforced with bone-like ceramics, *J. Appl. Med. Polym.*, **2**, 49–53.

402. Reis, R. L., and Cunha, A. M. (2000) New degradable load-bearing biomaterials composed of reinforced starch based blends, *J. Appl. Med. Polym.*, **4**, 1–5.

403. Sousa, R. A., Mano, J. F., Reis, R. L., Cunha, A. M., and Bevis, M. J. (2002) Mechanical performance of starch based bioactive composites moulded with preferred orientation for potential medical applications, *Polym. Eng. Sci.*, **42**, 1032–1045.

404. Marques, A. P., and Reis, R. L. (2005) Hydroxyapatite reinforcement of different starch-based polymers affects osteoblast-like cells adhesion/spreading and proliferation, *Mater. Sci. Eng. C*, **25**, 215–229.

405. Reis, R. L., Cunha, A. M., Allan, P. S., and Bevis, M. J. (1997) Structure development and control of injection-molded hydroxylapatite-reinforced starch/EVOH composites, *J. Polym. Adv. Tech.*, **16**, 263–277.

406. Vaz, C. M., Reis, R. L., and Cunha, A. M. (2002) Use of coupling agents to enhance the interfacial interactions in starch-EVOH/hydroxylapatite composites, *Biomaterials*, **23**, 629–635.

407. Leonor, I. B., Ito, A., Onuma, K., Kanzaki, N., and Reis, R. L. (2003) *In vitro* bioactivity of starch thermoplastic/hydroxyapatite composite biomaterials: an *in situ* study using atomic force microscopy, *Biomaterials*, **24**, 579–585.

408. Vaz, C. M., Reis, R. L., and Cunha, A. M. (2001) Degradation model of starch-EVOH+HA composites, *Mat. Res. Innovat.*, **4**, 375–380.

409. Boeree, N., Dove, J., Cooper, J. J., Knowles, J. C., and Hastings, G. W. (1993) Development of a degradable composite for orthopaedic use – mechanical evaluation of an hydroxyapatite-polyhydroxybutyrate composite materials, *Biomaterials*, **14**, 793–796.

410. Doyle, C., Tanner, E. T., and Bonfield, W. (1991) *In vitro* and *in vivo* evaluation of polyhydroxybutyrate and of polyhydroxybutyrate reinforced with hydroxyapatite, *Biomaterials*, **12**, 841–847.

411. Chen, L. J., and Wang, M. (2002) Production and evaluation of biodegradable composites based on PHB-PHV copolymer, *Biomaterials*, **23**, 2631–2639.

412. Ni, J., and Wang, M. (2002) *In vitro* evaluation of hydroxyapatite reinforced polyhydroxybutyrate composite, *Mater. Sci. Eng. C*, **20**, 101–109.

413. Knowles, J. C., Hastings, G. W., Ohta, H., Niwa, S., and Boeree, N. (1992) Development of a degradable composite for orthopedic use

– *in vivo* biomechanical and histological evaluation of two bioactive degradable composites based on the polyhydroxybutyrate polymer, *Biomaterials*, **13**, 491–496.

414. Luklinska, Z. B., and Bonfield, W. (1997) Morphology and ultrastructure of the interface between hydroxyapatite-polybutyrate composite implant and bone, *J. Mater. Sci. Mater. Med.*, **8**, 379–383.

415. Reis, E. C. C., Borges, A. P. B., Fonseca, C. C., Martinez, M. M. M., Eleotério, R. B., Morato, G. O., and Oliveira, P. M. (2010) Biocompatibility, osteointegration, osteoconduction, and biodegradation of a hydroxyapatite-polyhydroxybutyrate composite, *Brazilian Archives of Biology and Technology*, **53**, 817–826.

416. Chen, D. Z., Tang, C. Y., Chan, K. C., Tsui, C. P., Yu, P. H. F., Leung, M. C. P., and Uskokovic, P. S. (2007) Dynamic mechanical properties and *in vitro* bioactivity of PHBHV/HA nanocomposite, *Compos. Sci. Technol.*, **67**, 1617–1626.

417. Rai, B., Noohom, W., Kithva, P. H., Grøndahl, L., and Trau, M. (2008) Bionanohydroxyapatite/poly(3-hydroxybutyrate-*co*-3-hydroxyvalerate) composites with improved particle dispersion and superior mechanical properties, *Chem. Mater.*, **20**, 2802–2808.

418. Wang, Y. W., Wu, Q., Chen, J., and Chen, G. Q. (2005) Evaluation of three-dimensional scaffolds made of blends of hydroxyapatite and poly(3-hydroxybutyrate-*co*-3-hydroxyhexanoate) for bone reconstruction, *Biomaterials*, **26**, 899–904.

419. Linhart, W., Lehmann, W., Siedler, M., Peters, F., Schilling, A. F., Schwarz, K., Amling, M., Rueger, J. M., and Epple, M. (2006) Composites of amorphous calcium phosphate and poly(hydroxybutyrate) and poly(hydroxybutyrate-*co*-hydroxyvalerate) for bone substitution: assessment of the biocompatibility, *J. Mater. Sci.*, **41**, 4806–4813.

420. Azevedo, M., Reis, R. L., Claase, M., Grijpma, D., and Feijen, J. (2003) Development and properties of polycaprolactone/hydroxyapatite composite biomaterials, *J. Mater. Sci. Mater. Med.*, **14**, 103–107.

421. Choi, D., Marra, K. G., and Kumta, P. N. (2004) Chemical synthesis of hydroxyapatite/poly(ε-caprolactone) composites, *Mater. Res. Bull.*, **39**, 417–432.

422. Hao, J., Yuan, M., and Deng, X. (2003) Biodegradable and biocompatible nanocomposites of poly(ε-caprolactone) with hydroxyapatite nanocrystals: thermal and mechanical properties, *J. Appl. Polymer Sci.*, **86**, 676–683.

423. Walsh, D., Furuzono, T., and Tanaka, J. (2001) Preparation of porous composite implant materials by *in situ* polymerization of

porous apatite containing ε-caprolactone or methyl methacrylate, *Biomaterials*, **22**, 1205–1212.

424. Cerrai, P., Guerra, G. D., Tricoli, M., Krajewski, A., Guicciardi, S., Ravaglioli, A., Maltinti, S., and Masetti, G. (1999) New composites of hydroxyapatite and bioresorbable macromolecular material, *J. Mater. Sci. Mater. Med.*, **10**, 283–289.

425. Verma, D., Katti, K., and Katti, D. (2006) Bioactivity in *in situ* hydroxyapatite-polycaprolactone composites, *J. Biomed. Mater. Res. A*, **78A**, 772–780.

426. Kim, H. W. (2007) Biomedical nanocomposites of hydroxyapatite/polycaprolactone obtained by surfactant mediation, *J. Biomed. Mater. Res. A*, **83A**, 169–177.

427. Kim, H. W., Knowles, J. C., and Kim, H. E. (2004) Development of hydroxyapatite bone scaffold for controlled drug release via poly(ε-caprolactone) and hydroxyapatite hybrid coatings, *J. Biomed. Mater. Res. A*, **70A**, 240–249.

428. Guerra, G. D., Cerrai, P., Tricoli, M., Krajewski, A., Ravaglioli, A., Mazzocchi, M., and Barbani, N. (2006) Composites between hydroxyapatite and poly(ε-caprolactone) synthesized in open system at room temperature, *J. Mater. Sci. Mater. Med.*, **17**, 69–79.

429. Bianco, A., di Federico, E., Moscatelli, I., Camaioni, A., Armentano, I., Campagnolo, L., Dottori, M., Kenny, J. M., Siracusa, G., and Gusmano, G. (2009) Electrospun poly(ε-caprolactone)/Ca-deficient hydroxyapatite nanohybrids: microstructure, mechanical properties and cell response by murine embryonic stem cells, *Mater. Sci. Eng. C*, **29**, 2063–2071.

430. Chuenjitkuntaworn, B., Inrung, W., Damrongsri, D., Mekaapiruk, K., Supaphol, P., and Pavasant, P. (2010) Polycaprolactone/hydroxyapatite composite scaffolds: preparation, characterization, and *in vitro* and *in vivo* biological responses of human primary bone cells, *J. Biomed. Mater. Res. A*, **94A**, 241–251.

431. di Foggia, M., Corda, U., Plescia, E., Taddei, P., and Torreggiani, A. (2010) Effects of sterilisation by high-energy radiation on biomedical poly-(ε-caprolactone)/hydroxyapatite composites, *J. Mater. Sci. Mater. Med.*, **21**, 1789–1797.

432. Lebourg, M., Antón, J. S., and Ribelles, J. L. G. (2010) Hybrid structure in PCL-HAp scaffold resulting from biomimetic apatite growth, *J. Mater. Sci. Mater. Med.*, **21**, 33–44.

433. Makarov, C., Gotman, I., Jiang, X., Fuchs, S., Kirkpatrick, C. J., and Gutmanas, E. Y. (2010) *In situ* synthesis of calcium phosphate-

polycaprolactone nanocomposites with high ceramic volume fractions, *J. Mater. Sci. Mater. Med.*, **21**, 1771–1779.

434. Causa, F., Netti, P. A., Ambrosio, L., Ciapetti, G., Baldini, N., Pagani, S., Martini, D., and Giunti, A. (2006) Poly-ε-caprolactone/hydroxyapatite composites for bone regeneration: *in vitro* characterization and human osteoblast response, *J. Biomed. Mater. Res. A*, **76A**, 151–162.

435. Thomas, V., Jagani, S., Johnson, K., Jose, M. V., Dean, D. R., Vohra, Y. K., and Nyairo, E. (2006) Electrospun bioactive nanocomposite scaffolds of polycaprolactone and nanohydroxyapatite for bone tissue engineering, *J. Nanosci. Nanotechol.*, **6**, 487–493.

436. Dunn, A., Campbell, P., and Marra, K. G. (2001) The influence of polymer blend composition on the degradation of polymer/hydroxyapatite biomaterials, *J. Mater. Sci. Mater. Med.*, **12**, 673–677.

437. Calandrelli, L., Immirzi, B., Malinconico, M., Volpe, M., Oliva, A., and Ragione, F. (2000) Preparation and characterization of composites based on biodegradable polymers for *in vivo* application, *Polymer*, **41**, 8027–8033.

438. Chen, B., and Sun, K. (2005) Poly(ε-caprolactone)/hydroxyapatite composites: effects of particle size, molecular weight distribution and irradiation on interfacial interaction and properties, *Polymer Testing*, **24**, 64–70.

439. Ural, E., Kesenci, K., Fambri, L., Migliaresi, C., and Piskin, E. (2000) Poly(D,L-lactide/ε-caprolactone)/hydroxyapatite composites, *Biomaterials*, **21**, 2147–2154.

440. Takayama, T., and Todo, M. (2010) Improvement of fracture properties of hydroxyapatite particle filled poly(l-lactide)/poly(ε-caprolactone) biocomposites using lysine tri-isocyanate, *J. Mater. Sci.*, **45**, 6266–6270.

441. Shokrollahi, P., Mirzadeh, H., Scherman, O. A., and Huck, W. T. S. (2010) Biological and mechanical properties of novel composites based on supramolecular polycaprolactone and functionalized hydroxyapatite, *J. Biomed. Mater. Res. A*, **95A**, 209–221.

442. Yoon, B. H., Kim, H. W., Lee, S. H., Bae, C. J., Koh, Y. H., Kong, Y. M., and Kim, H. E. (2005) Stability and cellular responses to fluorapatite-collagen composites, *Biomaterials*, **26**, 2957–2963.

443. Kim, H. W., Lee, E. J., Kim, H. E., Salih, V., and Knowles, J. C. (2005) Effect of fluoridation of hydroxyapatite in hydroxyapatite/polycaprolactone composites on osteoblast activity, *Biomaterials*, **26**, 4395–4404.

444. Busch, S., Dolhaine, H., DuChesne, A., Heinz, S., Hochrein, O., Laeri, F., Podebrad, O., Vietze, U., Weiland, T., and Kniep, R. (1999) Biomimetic

morphogenesis of fluorapatite-gelatin composites: fractal growth, the question of intrinsic electric fields, core/shell assemblies, hollow spheres and reorganization of denatured collagen, *Eur. J. Inorg. Chem.*, 1643–1653.

445. Busch, S., Schwarz, U., and Kniep, R. (2003) Chemical and structural investigations of biomimetically grown fluorapatite-gelatin composite aggregates, *Adv. Funct. Mater.*, **13**, 189–198.

446. Simon, P., Carrillo-Cabrera, W., Formanek, P., Göbel, C., Geiger, D., Ramlau, R., Tlatlik, H., Buder, J., and Kniep, R. (2004) On the real-structure of biomimetically grown hexagonal pristamic seed of fluoroapatite-gelatin-composites: TEM investigations along [001], *J. Mater. Chem.*, **14**, 2218–2224.

447. Göbel, C., Simon, P., Buder, J., Tlatlik, H., and Kniep, R. (2004) Phase formation and morphology of calcium phosphate-gelatin-composites grown by double diffusion technique: the influence of fluoride, *J. Mater. Chem.*, **14**, 2225–2230.

448. Simon, P., Schwarz, U., and Kniep, R. (2005) Hierarchical architecture and real structure in a biomimetic nano-composite of fluorapatite with gelatine: a model system for steps in dentino- and osteogenesis? *J. Mater. Chem.*, **15**, 4992–4996.

449. Tlatlik, H., Simon, P., Kawska, A., Zahn, D., and Kniep, R. (2006) Biomimetic fluorapatite-gelatin nanocomposites: pre-structuring of gelatin matrices by ion impregnation and its effect on form development, *Angew. Chem. Int. Ed. Engl.*, **45**, 1905–1910.

450. Simon, P., Zahn, D., Lichte, H., and Kniep, R. (2006) Intrinsic electric dipole fields and the induction of hierarchical form developments in fluorapatite-gelatin nanocomposites: a general principle for morphogenesis of biominerals? *Angew. Chem. Int. Ed. Engl.*, **45**, 1911–1915.

451. Kniep, R., and Simon, P. (2007) Fluorapatite-gelatin-nanocomposites: self-organized morphogenesis., real structure and relations to natural hard materials, *Top. Curr. Chem. 270*, Springer, Berlin, pp. 73–125.

452. Kniep, R., and Simon, P. (2008) "Hidden" hierarchy of microfibrils within 3D-periodic fluorapatite-gelatin nanocomposites: development of complexity and form in a biomimetic system, *Angew. Chem. Int. Ed. Engl.*, **47**, 1405–1409.

453. Brickmann, J., Paparcone, R., Kokolakis, S., Zahn, D., Duchstein, P., Carrillocabrera, W., Simon, P., and Kniep, R. (2010) Fluorapatite-gelatine nanocomposite superstructures: new insights into a

biomimetic system of high complexity, *ChemPhysChem*, **11**, 1851–1853.

454. However, there are some reports about a lack of TCP biodegradation after implantation in calvarial defects [455].

455. Handschel, J., Wiesmann, H. P., Stratmann, U., Kleinheinz, J., Meyer, U., and Joos, U. (2002) TCP is hardly resorbed and not osteoconductive in a non-loading calvarial model, *Biomaterials*, **23**, 1689–1695.

456. Kikuchi, M., Tanaka, J., Koyama, Y., and Takakuda, K. (1999) Cell culture tests of TCP/CPLA composites, *J. Biomed. Mater. Res.*, **48**, 108–110.

457. Yaszemski, M. J., Payne, R. G., Hayes, W. C., Langer, R., and Mikos, A. G. (1996) *In vitro* degradation of a poly(propylene fumarate)-based composite material, *Biomaterials*, **17**, 2117–2130.

458. Wang, M., Wang, J., and Ni, J. (2000) Developing tricalcium phosphate/polyhydroxybutyrate composite as a new biodegradable material for clinical applications, *Biomechanics*, **192**, 741–744.

459. Kikuchi, M., Koyama, Y., Takakuda, K., Miyairi, H., Shirahama, N., and Tanaka, J. (2002) *In vitro* change n mechanical strength of β-tricalcium phosphate/copolymerized poly-L-lactide composites and their application for guided bone regeneration, *J. Biomed. Mater. Res.*, **62**, 265–272.

460. Ignatius, A. A., Augat, P., and Claes, L. E. (2001) Degradation behaviour of composite pins made of tricalcium phosphate and poly(L, D,L-lactide), *J. Biomater. Sci. Polym. Edn.*, **12**, 185–194.

461. Ignatius, A. A., Wolf, S., Augat, P., and Claes, L. E. (2001) Composites made of rapidly resorbable ceramics and poly(lactide) show adequate mechanical properties for use as bone substitute materials, *J. Biomed. Mater. Res.*, **57**, 126–131.

462. Kikuchi, M., and Tanaka, J. (2000) Chemical interaction in β-tricalcium phosphate/copolymerized poly-L-lactide composites, *J. Ceram. Soc. Jpn.*, **108**, 642–645.

463. Aunoble, S., Clement, D., Frayssinet, P., Harmand, M. F., and le Huec, J. C. (2006) Biological performance of a new β-TCP/PLLA composite material for applications in spine surgery: *in vitro* and *in vivo* studies, *J. Biomed. Mater. Res. A*, **78A**, 416–422.

464. Haaparanta, A. M., Haimi, S., Ellä, V., Hopper, N., Miettinen, S., Suuronen, R., and Kellomäki, M. (2010) Porous polylactide/β-tricalcium phosphate composite scaffolds for tissue engineering applications, *J. Tiss. Eng. Regen. Med.*, **4**, 366–373.

465. Kikuchi, M., Koyama, Y., Yamada, T., Imamura, Y., Okada, T., Shirahama, N., Akita, K., Takakuda, K., and Tanaka, J. (2004) Development of guided bone regeneration membrane composed of β-tricalcium phosphate and poly(L-lactide-*co*-glycolide-*co*-ε-caprolactone) composites, *Biomaterials*, **25**, 5979–5986.

466. Chen, T. M., Yao, C. H., Wang, H. J., Chou, G. H., Lee, T. W., and Lin, F. H. (1998) Evaluation of a novel malleable, biodegradable osteoconductive composite in a rabbit cranial defect model, *Mater. Chem. Phys.*, **55**, 44–50.

467. Dong, G. C., Chen, H. M., and Yao, C. H. (2008) A novel bone substitute composite composed of tricalcium phosphate, gelatin and drynaria fortunei herbal extract, *J. Biomed. Mater. Res. A*, **84A**, 167–177.

468. Ji, J., Yuan, X., Xia, Z., Liu, P., and Chen, J. (2010) Porous β-tricalcium phosphate composite scaffold reinforced by K_2HPO_4 and gelatin, *Key Eng. Mater.*, 434–435, 620–623.

469. Yao, C. H., Liu, B. S., Hsu, S. H., Chen, Y. S., and Tsai, C. C. (2004) Biocompatibility and biodegradation of a bone composite containing tricalcium phosphate and genipin crosslinked gelatin, *J. Biomed. Mater. Res. A*, **69A**, 709–717.

470. Lin, F. H., Yao, C. H., Sun, J. S., Liu, H. C., and Huang, C. W. (1998) Biological effects and cytotoxicity of the composite composed by tricalcium phosphate and glutaraldehyde cross-linked gelatin, *Biomaterials*, **19**, 905–917.

471. Eslaminejad, M. B., Mirzadeh, H., Mohamadi, Y., and Nickmahzar, A. (2007) Bone differentiation of marrow-derived mesenchymal stem cells using β-tricalcium phosphate-alginate-gelatin hybrid scaffolds, *J. Tissue Eng. Regen. Med.*, **1**, 417–424.

472. Takahashi, Y., Yamamoto, M., and Tabata, Y. (2005) Osteogenic differentiation of mesenchymal stem cells in biodegradable sponges composed of gelatin and β-tricalcium phosphate, *Biomaterials*, **26**, 3587–3596.

473. Bigi, A., Cantelli, I., Panzavolta, S., and Rubini, K. (2004) α-tricalcium phosphate-gelatin composite cements, *J. Appl. Biomat. Biomech.*, **2**, 81–87.

474. Yang, S. H., Hsu, C. K., Wang, K. C., Hou, S. M., and Lin, F. H. (2005) Tricalcium phosphate and glutaraldehyde crosslinked gelatin incorporating bone morphogenetic protein – a viable scaffold for bone tissue engineering, *J. Biomed. Mater. Res. B (Appl. Biomater.)*, **74B**, 468–475.

475. Kato, M., Namikawa, T., Terai, H., Hoshino, M., Miyamoto, S., and Takaoka, K. (2006) Ectopic bone formation in mice associated with a lactic acid/dioxanone/ethylene glycol copolymer-tricalcium phosphate composite with added recombinant human bone morphogenetic protein-2, *Biomaterials*, **27**, 3927–3933.

476. Muramatsu, K., Oba, K., Mukai, D., Hasegawa, K., Masuda, S., and Yoshihara, Y. (2007) Subacute systemic toxicity assessment of β-tricalcium phosphate/carboxymethyl-chitin composite implanted in rat femur, *J. Mater. Sci. Mater. Med.*, **18**, 513–522.

477. Panzavolta, S., Fini, M., Nicoletti, A., Bracci, B., Rubini, K., Giardino, R., and Bigi, A. (2009) Porous composite scaffolds based on gelatin and partially hydrolyzed α-tricalcium phosphate, *Acta Biomater.*, **5**, 636–643.

478. Uchino, T., Kamitakahara, M., Otsuka, M., and Ohtsuki, C. (2010) Hydroxyapatite-forming capability and mechanical properties of organic-inorganic hybrids and α-tricalcium phosphate porous bodies, *J. Ceram. Soc. Jpn.*, **118**, 57–61.

479. Boguń, M., and Rabiej, S. (2010) The influence of fiber formation conditions on the structure and properties of nanocomposite alginate fibers containing tricalcium phosphate or montmorillonite, *Polymer Composites*, **31**, 1321–1331.

480. Bleach, N. C., Tanner, K. E., Kellomäki, M., and Törmälä, P. (2001) Effect of filler type on the mechanical properties of self-reinforced polylactide-calcium phosphate composites, *J. Mater. Sci. Mater. Med.*, **12**, 911–915.

481. Liu, L., Xiong, Z., Yan, Y. N., Hu, Y. Y., Zhang, R. J., and Wang, S. G. (2007) Porous morphology, porosity, mechanical properties of poly(α-hydroxy acid)-tricalcium phosphate composite scaffolds fabricated by low-temperature deposition, *J. Biomed. Mater. Res. A*, **82A**, 618–629.

482. Zhang, Y., and Zhang, M. Q. (2001) Synthesis and characterization of macroporous chitosan/calcium phosphate composite scaffolds for tissue engineering, *J. Biomed. Mater. Res.*, **55**, 304–312.

483. Rai, B., Teoh, S. H., Hutmacher, D. W., Cao, T., and Ho, K. H. (2005) Novel PCL-based honeycomb scaffolds as drug delivery systems for rhBMP-2, *Biomaterials*, **26**, 3739–3748.

484. Rai, B., Teoh, S. H., Ho, K. H., Hutmacher, D. W., Cao, T., Chen, F., and Yacob, K. (2004) The effect of rhBMP-2 on canine osteoblasts seeded onto 3D bioactive polycaprolactone scaffolds, *Biomaterials*, **25**, 5499–5506.

485. Lei, Y., Rai, B., Ho, K. H., and Teoh, S. H. (2007) *In vitro* degradation of novel bioactive polycaprolactone – 20% tricalcium phosphate composite scaffolds for bone engineering, *Mater. Sci. Eng. C*, **27**, 293–298.

486. Miyai, T., Ito, A., Tamazawa, G., Matsuno, T., Sogo, Y., Nakamura, C., Yamazaki, A., and Satoh, T. (2008) Antibiotic-loaded poly-ε-caprolactone and porous β-tricalcium phosphate composite for treating osteomyelitis, *Biomaterials*, **29**, 350–358.

487. Takahashi, Y., Yamamoto, M., and Tabata, Y. (2005) Enhanced osteoinduction by controlled release of bone morphogenetic protein-2 from biodegradable sponge composed of gelatin and β-tricalcium phosphate, *Biomaterials*, **26**, 4856–4865.

488. Ignatius, A. A., Betz, O., Augat, P., and Claes, L. E. (2001) *In vivo* investigations on composites made of resorbable ceramics and poly(lactide) used as bone graft substitutes, *J. Biomed. Mater. Res. (Appl. Biomater.)*, **58**, 701–709.

489. Miao, X., Lim, W. K., Huang, X., and Chen, Y. (2005) Preparation and characterization of interpenetrating phased TCP/HA/PLGA composites, *Mater. Lett.*, **59**, 4000–4005.

490. Brodie, J. C., Goldie, E., Connel, G., Merry, J., and Grant, M. H. (2005) Osteoblast interactions with calcium phosphate ceramics modified by coating with type I collagen, *J. Biomed. Mater. Res. A*, **73A**, 409–421.

491. Zhang, L. F., Sun, R., Xu, L., Du, J., Xiong, Z. C., Chen, H. C., and Xiong, C. D. (2008) Hydrophilic poly (ethylene glycol) coating on PDLLA/BCP bone scaffold for drug delivery and cell culture, *Mater. Sci. Eng. C*, **28**, 141–149.

492. Ignjatovic, N., Ninkov, P., Ajdukovic, Z., Konstantinovic, V., and Uskokovic, D. (2005) Biphasic calcium phosphate/poly-(D,L-lactide-*co*-glycolide) biocomposite as filler and blocks for reparation of bone tissue, *Mater. Sci. Forum*, **494**, 519–524.

493. Ignjatovic, N., Ninkov, P., Ajdukovic, Z., Vasiljevic-Radovic, D., and Uskokovic, D. (2007) Biphasic calcium phosphate coated with poly-D,L-lactide-*co*-glycolide biomaterial as a bone substitute, *J. Eur. Ceram. Soc.*, **27**, 1589–1594.

494. Yang, W., Yin, G., Zhou, D., Gu, J., and Li, Y. (2010) *In vitro* characteristics of surface-modified biphasic calcium phosphate/poly(L-Lactide) biocomposite, *Adv. Eng. Mater.*, **12**, B128–B132.

495. Ignjatovic, N., Ninkov, P., Kojic, V., Bokurov, M., Srdic, V., Krnojelac, D., Selakovic, S., and Uskokovic, D. Cytotoxicity and fibroblast properties

during *in vitro* test of biphasic calcium phosphate/poly-D,L-lactide-*co*-glycolide biocomposites and different phosphate materials, *Microsc. Res. Techniq.*, **69**, 976–982.

496. Ajdukovic, Z., Ignjatovic, N., Petrovic, D., and Uskokovic, D. (2007) Substitution of osteoporotic alveolar bone by biphasic calcium phosphate/poly-D,L-lactide-*co*-glycolide biomaterials, *J. Biomater. Appl.*, **21**, 317–328.

497. Kim, H. W., Knowles, J. C., and Kim, H. E. (2004) Effect of biphasic calcium phosphates on drug release and biological and mechanical properties of poly(ε-caprolactone) composite membranes, *J. Biomed. Mater. Res. A*, **70A**, 467–479.

498. Bakhtiari, L., Rezai, H. R., Hosseinalipour, S. M., and Shokrgozar, M. A. (2010) Investigation of biphasic calcium phosphate/gelatin nanocomposite scaffolds as a bone tissue engineering, *Ceram. Int.*, **36**, 2421–2426.

499. Bakhtiari, L., Rezai, H. R., Hosseinalipour, S. M., and Shokrgozar, M. A. (2010) Preparation of porous Biphasic Calcium phosphate-gelatin nanocomposite for bone tissue engineering, *J. Nano Res.*, **11**, 67–72.

500. Matsuda, A., Ikoma, T., Kobayashi, H., and Tanaka, J. (2004) Preparation and mechanical property of core-shell type chitosan/calcium phosphate composite fiber, *Mater. Sci. Eng. C*, **24**, 723–728.

501. Rattanachan, S., Lorprayoon, C., and Boonphayak, P. (2008) Synthesis of chitosan/brushite powders for bone cement composites, *J. Ceram. Soc. Jpn.*, **116**, 36–41.

502. Ohsawa, H., Ito, A., Sogo, Y., Yamazaki, A., and Ohno, T. (2007) Synthesis of albumin/DCP nano-composite particles, *Key Eng. Mater.*, 330–332, 239–242.

503. Xu, H. H. K., Sun, L., Weir, M. D., Antonucci, J. M., Takagi, S., Chow, L. C., and Peltz, M. (2006) Nano DCPA-whisker composites with high strength and Ca and PO_4 release, *J. Dent. Res.*, **85**, 722–727.

504. Xu, H. H. K., Weir, M. D., Sun, L., Takagi, S., and Chow, L. C. (2007) Effects of calcium phosphate nanoparticles on Ca-PO_4 composite, *J. Dent. Res.*, **86**, 378–383.

505. Xu, H. H. K., Weir, M. D., and Sun, L. (2007) Nanocomposites with Ca and PO_4 release: effects of reinforcement, dicalcium phosphate particle size and silanization, *Dent. Mater.*, **23**, 1482–1491.

506. Tortet, L., Gavarri, J. R., Nihoul, G., and Dianoux, A. J. (1997) Proton mobilities in brushite and brushite/polymer composites, *Solid State Ionics*, **97**, 253–256.

507. Tortet, L., Gavarri, J. R., Musso, J., Nihoul, G., and Sarychev, A. K. (1998) Percolation and modeling of proton conduction in polymer/brushite composites, *J. Solid State Chem.*, **141**, 392–403.

508. Park, M. S., Eanes, E. D., Antonucci, J. M., and Skrtic, D. (1998) Mechanical properties of bioactive amorphous calcium phosphate/methacrylate composites, *Dent. Mater.*, **14**, 137–141.

509. Skrtic, D., Antonucci, J. M., Eanes, E. D., Eichmiller, F. C., and Schumacher, G. E. (2000) Physicochemical evaluation of bioactive polymeric composites based on hybrid amorphous calcium phosphates, *J. Biomed. Mater. Res.*, **53**, 381–391.

510. Skrtic, D., Antonucci, J. M., Eanes, E. D., and Eidelman, N. (2004) Dental composites based on hybrid and surface-modified amorphous calcium phosphates, *Biomaterials*, **25**, 1141–1150.

511. Skrtic, D., and Antonucci, J. M. (2003) Effect of bifunctional comonomers on mechanical strength and water sorption of amorphous calcium phosphate- and silanized glass-filled Bis-GMA-based composites, *Biomaterials*, **24**, 2881–2888.

512. Gutierrez, M. C., Jobbágy, M., Ferrer, M. L., and del Monte, F. (2008) Enzymatic synthesis of amorphous calcium phosphate-chitosan nanocomposites and their processing into hierarchical structures, *Chem. Mater.*, **20**, 11–13.

513. Hakimimehr, D., Liu, D. M., and Troczynski, T. (2005) *In-situ* preparation of poly(propylene fumarate) – hydroxyapatite composite, *Biomaterials*, **26**, 7297–7303.

514. Perkin, K. K., Turner, J. L., Wooley, K. L., and Mann, S. (2005) Fabrication of hybrid nanocapsules by calcium phosphate mineralization of shell cross-linked polymer micelles and nanocages, *Nano Lett.*, **5**, 1457–1461.

515. Wang, K. W., Zhu, Y. J., Chen, F., and Cao, S. W. (2010) Calcium phosphate/block copolymer hybrid porous nanospheres: preparation and application in drug delivery, *Mater. Lett.*, **64**, 2299–2301.

516. LeGeros, R. Z., Chohayeb, A., and Shulman, A. (1982) Apatitic calcium orthophosphates: possible dental restorative materials, *J. Dent. Res.*, **61**, Special issue, 343.

517. Brown, W. E., and Chow, L. C. (1983) A new calcium orthophosphate setting cement, *J. Dent. Res.*, **62**, Special issue, 672.

518. Brown, W. E., and Chow, L. C. Dental restorative cement pastes. US Patent No. 4518430. May 21, 1985.

519. Dorozhkin, S. V. (2009) Calcium orthophosphate cements and concretes, *Materials*, **2**, 221–291.

520. Barrelet, J. A., Hofmann, M., Grover, L. M., and Gbureck, U. (2003) High-strength apatitic cement by modification with α-hydroxyacid salts, *Adv. Mater.*, **15**, 2091–2094.

521. Bigi, A., Bracci, B., and Panzavolta, S. (2004) Effect of added gelatin on the properties of calcium phosphate cement, *Biomaterials*, **25**, 2893–2899.

522. Bigi, A., Panzavolta, S., Sturba, L., Torricelli, P., Fini, M., and Giardino, R. (2006) Normal and osteopenic bone-derived osteoblast response to a biomimetic gelatin-calcium phosphate bone cement, *J. Biomed. Mater. Res. A*, **78A**, 739–745.

523. Panzavolta, S., Torricelli, P., Sturba, L., Bracci, B., Giardino, R., and Bigi, A. (2008) Setting properties and *in vitro* bioactivity of strontium-enriched gelatin-calcium phosphate bone cements, *J. Biomed. Mater. Res. A*, **84A**, 965–972.

524. Rammelt, S., Neumann, M., Hanisch, U., Reinstorf, A., Pompe, W., Zwipp, H., and Biewener, A. (2005) Osteocalcin enhances bone remodeling around hydroxyapatite/collagen composites, *J. Biomed. Mater. Res. A*, **73A**, 284–294.

525. Wang, X., Ye, J., Wang, Y., and Chen, L. (2007) Reinforcement of calcium phosphate cement by bio-mineralized carbon nanotube, *J. Am. Ceram. Soc.*, **90**, 962–964.

526. According to Wikipedia, the free encyclopedia: "Concrete is a construction material that consists of a cement (commonly Portland cement), aggregates (generally gravel and sand) and water. It solidifies and hardens after mixing and placement due to a chemical process known as hydration. The water reacts with the cement, which bonds the other components together, eventually creating a stone-like material". http://en.wikipedia.org/wiki/Concrete (accessed in October 2010).

527. Kim, S. B., Kim, Y. J., Yoon, T. L., Park, S. A., Cho, I. H., Kim, E. J., Kim, I. A., and Shin, J. W. (2004) The characteristics of a hydroxyapatite-chitosan-PMMA bone cement, *Biomaterials*, **25**, 5715–5723.

528. Vallo, C. I., Montemartini, P. E., Fanovich, M. A., Lópes, J. M. P., and Cuadrado, T. R. (1999) Polymethylmethacrylate-based bone cement modified with hydroxyapatite, *J. Biomed. Mater. Res.* (*Appl. Biomater.*), **48**, 150–158.

529. Sogal, A., and Hulbert, S. F. (1992) Mechanical properties of a composite bone cement: polymethylmethacrylate and hydroxyapatite, *Bioceramics*, 213–224.

530. Harper, E. J., Behiri, J. C., and Bonfield, W. (1995) Flexural and fatigue properties of a bone cement based upon polymethylmethacrylate and hydroxyapatite, *J. Mater. Sci. Mater. Med.*, **6**, 799–803.

531. Harper, E. J., Braden, M., and Bonfield, W. (2001) Mechanical properties of hydroxyapatite reinforced poly(methylmethacrylate) bone cement after immersion in a physiological solution: influence of a silane coupling agent, *J. Mater. Sci. Mater. Med.*, **11**, 491–497.

532. Moursi, A. M., Winnard, A. V., Winnard, P. L., Lannutti, J. J., and Seghi, R. R. (2002) Enhanced osteoblast response to a PMMA – HA composite, *Biomaterials*, **23**, 133–144.

533. Dalby, M. J., di Silvio, L., Harper, E. J., and Bonfield, W. (2001) Initial interaction of osteoblasts with the surface of a hydroxyapatite-poly(mefhylmethacrylate) cement, *Biomaterials*, **22**, 1739–1747.

534. Dalby, M. J., di Silvio, L., Harper, E. J., and Bonfield, W. (2002) Increasing hydroxyapatite incorporation into poly(methylmethacrylate) cement increases osteoblast adhesion and response, *Biomaterials*, **23**, 569–576.

535. Itokawa, H., Hiraide, T., Moriya, M., Fujimoto, M., Nagashima, G., Suzuki, R., and Fujimoto, T. (2007) A 12 month *in vivo* study on the response of bone to a hydroxyapatite-polymethylmethacrylate cranioplasty composite, *Biomaterials*, **28**, 4922–4927.

536. Cheang, P., and Khor, K. A. (2003) Effect of particulate morphology on the tensile behaviour of polymer-hydroxyapatite composites, *Mater. Sci. Eng. A*, **345**, 47–54.

537. Dalby, M. J., di Silvio, L., Harper, E. J., and Bonfield, W. (1999) *In vitro* evaluation of a new polymethylmethacrylate cement reinforced with hydroxyapatite, *J. Mater. Sci. Mater. Med.*, **10**, 793–796.

538. Tham, W. L., Chow, W. S., and Ishak, Z. A. M. (2010) The effect of 3-(trimethoxysilyl) propyl methacrylate on the mechanical, thermal, and morphological properties of poly(methyl methacrylate)/hydroxyapatite composites, *J. Appl. Polym. Sci.*, **118**, 218–228.

539. Tham, W. L., Chow, W. S., and Mohd Ishak, Z. A. (2010) Flexural and morphological properties of Poly(Methyl Methacrylate)/hydroxyapatite composites: effects of planetary ball mill grinding time, *J. Reinforced Plastics and Composites*, **29**, 2065–2075.

540. Pattanayak, D. K., Rao, B. T., and Mohan, T. R. R. (2011) Calcium phosphate bioceramics and bioceramic composites, *J. Sol Gel Sci. Technol.*, **59**, 432–447.

541. Deb, S., Braden, M., and Bonfield, W. (1995) Water absorption characteristics of modified hydroxyapatite bone cements, *Biomaterials*, **16**, 1095–1100.

542. Borzacchiello, A., Ambrosio, L., Nicolais, L. Harper, E. J., Tanner, K. E., and Bonfield, W. (1998) Comparison between the polymerization behavior of a new bone cement and a commercial one: modeling and *in vitro* analysis, *J. Mater. Sci. Mater. Med.*, **9**, 835–838.

543. Ohgaki, M., and Yamashita, K. (2003) Preparation of polymethylmethacrylate-reinforced functionally graded hydroxyapatite composites, *J. Am. Ceram. Soc.*, **86**, 1440–1442.

544. del Real, R. P., Padilla, S., and Vallet-Regi, M. (2000) Gentamicin release from hydroxyapatite/poly(ethyl methacrylate)/poly(methyl methacrylate)composites, *J. Biomed. Mater. Res.*, **52**, 1–7.

545. Saito, M., Maruoka, A., Mori, T., Sugano, N., and Hino, K. (1994) Experimental studies on a new bioactive bone cement: hydroxyapatite composite resin, *Biomaterials*, **15**, 156–160.

546. Watson, K. E., Ten Huisen, K. S., and Brown, P. W. (1999) The formation of hydroxyapatite – calcium polyacrylate composites, *J. Mater. Sci. Mater. Med.*, **10**, 205–213.

547. Reed, C. S., Ten Huisen, K. S., Brown, P. W., and Allcock, H. R. (1996) Thermal stability and compressive strength of calcium-deficient hydroxyapatite – poly[bis(carboxylatophenoxy)phosphazene] composites, *Chem. Mater*, **8**, 440–447.

548. Peter, S. J., Kim, P., Yasko, A. W., Yaszemski, M. J., and Mikos, A. G. (1999) Crosslinking characteristics of an injectable poly(propylene fumarate)/β-tricalcium phosphate paste and mechanical properties of the crosslinked composite for use as a biodegradable bone cement, *J. Biomed. Mater. Res.*, **44**, 314–321.

549. He, S., Yaszemski, M. J., Yasko, A. W., Engel, P. S., and Mikos, A. G. (2000) Injectable biodegradable polymer composites based on poly (propylene fumarate) crosslinked with poly(ethylene glycol)-dimethacrylate, *Biomaterials*, **21**, 2389–2394.

550. Ignjatovic, N., Jovanovic, J., Suljovrujic, E., and Uskokovic, D. (2003) Injectable polydimethylsiloxane/hydroxyapatite composite cement, *Biomed. Mater. Eng.*, **13**, 401–410.

551. Fei, Z., Hu, Y., Wu, D., Wu, H., Lu, R., Bai, J., and Song, H. (2008) Preparation and property of a novel bone graft composite consisting of rhBMP-2 loaded PLGA microspheres and calcium phosphate cement, *J. Mater. Sci. Mater. Med.*, **19**, 1109–1116.

552. Link, D. P., van den Dolder, J., van den Beucken, J. J. J. P., Cuijpers, V. M., Wolke, J. G. C., Mikos, A. G., and Jansen, J. A. (2008) Evaluation of the biocompatibility of calcium phosphate cement/PLGA microparticle composites, *J. Biomed. Mater. Res. A*, **87A**, 760–769.

553. Chiang, T. Y., Ho, C. C., Chen, D. C. H., Lai, M. H., and Ding, S. J. (2010) Physicochemical properties and biocompatibility of chitosan oligosaccharide/gelatin/calcium phosphate hybrid cements, *Mater. Chem. Phys.*, **120**, 282–288.

554. Fujishiro, Y., Takahashi, K., and Sato, T. (2001) Preparation and compressive strength of α-tricalcium phosphate/gelatin gel composite cement, *J. Biomed. Mater. Res.*, **54**, 525–230.

555. Miyazaki, K., Horibe, T., Antonucci, J. M., Takagi, S., and Chow, L. C. (1993) Polymeric calcium phosphate cements: analysis of reaction products and properties, *Dental Mater.*, **9**, 41–45.

556. Miyazaki, K., Horibe, T., Antonucci, J. M., Takagi, S., and Chow, L. C. (1993) Polymeric calcium phosphate cements: setting reaction modifiers, *Dental Mater*, **9**, 46–50.

557. dos Santos, L. A., de Oliveira, L. C., Rigo, E. C. S., Carrodeguas, R. G., Boschi, A. O., and de Arruda, A. C. F. (1999) Influence of polymeric additives on the mechanical properties of α-tricalcium phosphate cement, *Bone*, **25**, 99S–102S.

558. Greish, Y. E., Brown, P. W., Bender, J. D., Allcockm, H. R., Lakshmim, S., and Laurencin, C. T. (2007) Hydroxyapatite-polyphosphazane composites prepared at low temperatures, *J. Am. Ceram. Soc.*, **90**, 2728–2734.

559. Greish, Y. E., Bender, J. D., Lakshmi, S., Brown, P. W., Allcock, H. R., and Laurencin, C. T. (2006) Formation of hydroxyapatite-polyphosphazene polymer composites at physiologic temperature, *J. Biomed. Mater. Res. A*, **77A**, 416–425.

560. Greish, Y. E., Bender, J. D., Lakshmi, S., Brown, P. W., Allcock, H. R., and Laurencin, C. T. (2005) Low temperature formation of hydroxyapatite-poly(alkyl oxybenzoate) phosphazene composites for biomedical applications, *Biomaterials*, **26**, 1–9.

561. Mickiewicz, R. A., Mayes, A. M., and Knaack, D. (2002) Polymer – calcium phosphate cement composites for bone substitutes, *J. Biomed. Mater. Res.*, **61**, 581–592.

562. Carey, L. E., Xu, H. H. K., Simon, C. G., Takagi, S., and Chow, L. C. (2005) Premixed rapid-setting calcium phosphate composites for bone repair, *Biomaterials*, **26**, 5002–5014.

563. Miao, X., Tan, L. P., Tan, L. S., and Huang, X. (2007) Porous calcium phosphate ceramics modified with PLGA – bioactive glass, *Mater. Sci. Eng. C*, **27**, 274–279.

564. Lickorish, D., Guan, L., and Davies, J. E. (2007) A three-phase, fully resorbable, polyester/calcium phosphate scaffold for bone tissue engineering: evolution of scaffold design, *Biomaterials*, **28**, 1495–1502.

565. Xu, H. H. K., and Simon, C. G. (2005) Fast setting calcium phosphate-chitosan scaffold: mechanical properties and biocompatibility, *Biomaterials*, **26**, 1337–1348.

566. Zhang, L., Li, Y., Zhou, G., Lu, G. Y., and Zuo, Y. (2006) Setting mechanism of nano-hydroxyapatite/chitosan bone cement, *J. Inorg. Mater.*, **21**, 1197–1202.

567. Ruhe, P. Q., Hedberg, E. L., Padron, N. T., Spauwen, P. H. M., Jansen, J. A., and Mikos, A. G. (2005) Biocompatibility and degradation of poly(D,L-lactic-*co*-glycolic acid)/calcium phosphate cement composites, *J. Biomed. Mater. Res. A*, **74A**, 533–544.

568. Guo, D. G., Sun, H. L., Xu, K. W., and Han, Y. (2007) Long-term variations in mechanical properties and *in vivo* degradability of CPC/PLGA composite, *J. Biomed. Mater. Res. B* (*Appl. Biomater.*), **82B**, 533–544.

569. Habraken, W. J. E. M., Wolke, J. G. C., Mikos, A. G., and Jansen, J. A. (2006) Injectable PLGA microsphere/calcium phosphate cements: physical properties and degradation characteristics, *J. Biomater. Sci. Polym. Edn.*, **17**, 1057–1074.

570. Ruhe, P. Q., Hedberg-Dirk, E. L., Padron, N. T., Spauwen, P. H. M., Jansen, J. A., and Mikos, A. G. (2006) Porous poly (D,L-lactic-*co*-glycolic acid)/calcium phosphate cement composite for reconstruction of bone defects, *Tissue Eng.*, **12**, 789–800.

571. Ruhe, P. Q., Hedberg, E. L., Padron, N. T., Spauwen, P. H. M., Jansen, J. A., and Mikos, A. G. (2003) rhBMP-2 release from injectable poly (D,L-lactic-*co*-glycolic acid)/calcium phosphate composites, *J. Bone Joint Surg.* (*Am.*), **85A**(3), 75–81.

572. Ruhe, P. Q., Boerman, O. C., Russel, F. G. M., Spauwen, P. H. M., Mikos, A. G., and Jansen, J. A. (2005) Controlled release of rhBMP-2 loaded poly (D,L-lactic-*co*-glycolic acid)/calcium phosphate cement composites *in vivo*, *J. Control. Release*, **106**, 162–171.

573. Plachokova, A., Link, D., van den Dolder, J., van den Beucken, J., and Jansen, J. A. (2007) Bone regenerative properties of injectable PGLA-CaP composite with TGF-β1 in a rat augmentation model, *J. Tissue. Eng. Regen. Med.*, **1**, 457–464.

574. Webster, T. J., Siegel, R. W., and Bizios, R. (1999) Osteoblast adhesion on nanophase ceramics, *Biomaterials*, **20**, 1221–1227.

575. Webster, T. J., Ergun, C., Doremus, R. H., Siegel, R. W., and Bizios, R. (2000) Specific proteins mediate enhanced osteoblast adhesion on nanophase ceramics, *J. Biomat. Med. Res.*, **51**, 475–483.

576. Webster, T. J., Ergun, C., Doremus, R. H., Siegel, R. W., and Bizios, R. (2000) Enhanced functions of osteoblasts on nanophase ceramics, *Biomaterials*, **21**, 1803–1810.

577. Li, G., Huang, J., Li, Y., Zhang, R., Deng, B., Zhang, J., and Aoki, H. (2007) *In vitro* study on influence of a discrete nano-hydroxyapatite on leukemia P388 cell behavior, *Biomed. Mater. Eng.*, **17**, 321–327.

578. Tadic, D., Peters, F., and Epple, M. (2002) Continuous synthesis of amorphous carbonated apatites, *Biomaterials*, **23**, 2553–2559.

579. Xu, H. H. K., Sun, L., Weir, M. D., Takagi, S., Chow, L. C., and Hockey, B. (2007) Effects of incorporating nanosized calcium phosphate particles on properties of whisker-reinforced dental composites, *J. Biomed. Mater. Res. B* (*Appl. Biomater.*), **81B**, 116–125.

580. Deng, X. M., Hao, J. Y., and Wang, C. S. (2001) Preparation and mechanical properties of nanocomposites of poly(D,L-lactide) with Ca-deficient hydroxyapatite nanocrystals, *Biomaterials*, **22**, 2867–2873.

581. Hong, Z. K., Zhang, P. B., He, C. L., Qiu, X. Y., Liu, A. X., Chen, L., Chen, X., and Jing, X. (2005) Nanocomposite of poly(L-lactide) and surface grafted hydroxyapatite: mechanical properties and biocompatibility, *Biomaterials*, **26**, 6296–6304.

582. Deng, C., Weng, J., Cheng, Q. Y., Zhou, S. B., Lu, X., Wan, J. X., Qu, S. X., Feng, B., and Li, X. H. (2007) Choice of dispersants for the nano-apatite filler of polylactide-matrix composite biomaterial, *Curr. Appl. Phys.*, **7**, 679–682.

583. Deng, C., Weng, J., Lu, X., Zhou, S. B., Wan, J. X., Qu, S. X., Feng, B., Li, X. H., and Cheng, Q. Y. (2008) Mechanism of ultrahigh elongation rate of poly(D,L-lactide)-matrix composite biomaterial containing nano-apatite fillers, *Mater. Lett.*, **62**, 607–610.

584. Kothapalli, C. R., Shaw, M. T., and Wei, M. (2005) Biodegradable HA-PLA 3-D porous scaffolds: effect of nano-sized filler content on scaffold properties, *Acta Biomater.*, **1**, 653–662.

585. Hong, Z., Qiu, X., Sun, J., Deng, M., Chen, X., and Jing, X. (2004) Grafting polymerization of L-lactide on the surface of hydroxyapatite nanocrystals, *Polymer*, **45**, 6699–6706.

586. Xiao, Y., Li, D., Fan, H., Li, X., Gu, Z., and Zhang, X. (2007) Preparation of nano-HA/PLA composite by modified-PLA for controlling the growth of HA crystals, *Mater. Lett.*, **61**, 59–62.

587. Qiu, X., Han, Y., Zhuang, X., Chen, X., Li, Y., and Jing, X. (2007) Preparation of nano-hydroxyapatite/poly(L-lactide) biocomposite microspheres, *J. Nanoparticle Res.*, **9**, 901–908.

588. Nejati, E., Firouzdor, V., Eslaminejad, M. B., and Bagheri, F. (2009) Needle-like nano hydroxyapatite/poly(l-lactide acid) composite scaffold for bone tissue engineering application, *Mater. Sci. Eng. C*, **29**, 942–949.

589. Deplaine, H., Ribelles, J. L. G., and Ferrer, G. G. (2010) Effect of the content of hydroxyapatite nanoparticles on the properties and bioactivity of poly(l-lactide) – hybrid membranes, *Composites Sci. Technol.*, **70**, 1805–1812.

590. Kim, S. S., Park, M. S., Jeon, Q., Choi, C. Y., and Kim, B. S. (2006) Poly(lactide-*co*-glycolide)/hydroxyapatite composite scaffolds for bone tissue engineering, *Biomaterials*, **27**, 1399–1409.

591. Hong, Z., Zhang, P., Liu, A., Chen, L., Chen, X., and Jing, X. (2007) Composites of poly(lactide-*co*-glycolide) and the surface modified carbonated hydroxyapatite nanoparticles, *J. Biomed. Mater. Res. A*, **81A**, 515–522.

592. Huang, Y. X., Ren, J., Chen, C., Ren, T. B., and Zhou, X. Y. (2008) Preparation and properties of poly(lactide-*co*-glycolide) (PLGA)/nano-hydroxyapatite (NHA) scaffolds by thermally induced phase separation and rabbit MSCs culture on scaffolds, *J. Biomaterials Applications*, **22**, 409–432.

593. Xue, D., Zheng, Q., Zong, C., Li, Q., Li, H., Qian, S., Zhang, B., Yu, L., and Pan, Z. (2010) Osteochondral repair using porous poly(lactide-*co*-glycolide)/nano-hydroxyapatite hybrid scaffolds with undifferentiated mesenchymal stem cells in a rat model, *J. Biomed. Mater. Res. A*, **94A**, 259–270.

594. Du, C., Cui, F. Z., Zhu, X. D., and de Groot, K. (1999) Three-dimensional nano-HAp/collagen matrix loading with osteogenic cells in organ culture, *J. Biomed. Mater. Res.*, **44**, 407–415.

595. Wang, R. Z., Cui, F. Z., Lu, H. B., Wen, H. B., Ma, C. L., and Li, H. D. (1995) Synthesis of nanophase hydroxyapatite/collagen composite, *J. Mater. Sci. Lett.*, **14**, 490–492.

596. Du, C., Cui, F. Z., Feng, Q. L., Zhu, X. D., and de Groot, K. (1998) Tissue response to nano-hydroxyapatite/collagen composite implants in marrow cavity, *J. Biomed. Mater. Res.*, **42**, 540–548.

597. Kikuchi, M., Itoh, S., Ichinose, S., Shinomiya, K., and Tanaka, J. (2001) Self-organization mechanism in a bone-like hydroxyapatite/collagen nanocomposite synthesized *in vitro* and its biological reaction *in vivo*, *Biomaterials*, **22**, 1705–1711.

598. Kikuchi, M., Matsumoto, H. N., Yamada, T., Koyama, Y., Takakuda, K., and Tanaka, J. (2004) Glutaraldehyde cross-linked hydroxyapatite/collagen self-organized nanocomposites, *Biomaterials*, **25**, 63–69.

599. Lynn, A. K., Nakamura, T., Patel, N., Porter, A. E., Renouf, A. C., Laity, P. R., Best, S. M., Cameron, R. E., Shimizu, Y., and Bonfield, W. (2005) Composition-controlled nanocomposites of apatite and collagen incorporating silicon as osseopromotive agent, *J. Biomed. Mater. Res. A*, **74A**, 447–453.

600. Chang, M. C., and Tanaka, J. (2002) FTIR study for hydroxyapatite/collagen nanocomposite cross-linked by glutaraldehyde, *Biomaterials*, **23**, 4811–4818.

601. Chang, M. C., and Tanaka, J. (2002) XPS study for the microstructure development of hydroxyapatite-collagen nanocomposites cross-linked using glutaraldehyde, *Biomaterials*, **23**, 3879–3885.

602. Murugan, R., and Ramakrishna, S. (2006) *In situ* formation of recombinant humanlike collagen-hydroxyapatite nanohybrid through bionic approach, *Appl. Phys. Lett.*, **88**, 193124.

603. Wang, Y., Yang, C., Chen, X., and Zhao, N. (2006) Biomimetic formation of hydroxyapatite/collagen matrix composite, *Adv. Eng. Mater.*, **8**, 97–100.

604. Thomas, V., Dean, D. R., Jose, M. V., Mathew, B., Chowdhury, S., and Vohra, Y. K. (2007) Nanostructured biocomposite scaffolds based on collagen coelectrospun with nanohydroxyapatite, *Biomacromolecules*, **8**, 631–637.

605. Fukui, N., Sato, T., Kuboki, Y., and Aoki, H. (2008) Bone tissue reaction of nano-hydroxyapatite/collagen composite at the early stage of implantation, *Biomed. Mater. Eng.*, **18**, 25–33.

606. Kim, T. G., Park, S. H., Chung, H. J., Yang, D. Y., and Park, T. G. (2010) Microstructured scaffold coated with hydroxyapatite/collagen nanocomposite multilayer for enhanced osteogenic induction of human mesenchymal stem cells, *J. Mater. Chem.*, **20**, 8927–8933.

607. Liao, S. S., Tamura, K., Zhu, Y., Wang, W., Uo, M., Akasaka, T., Cui, F. Z., and Watari, F. (2006) Human neutrophils reaction to the biodegraded

nano-hydroxyapatite/collagen and nano-hydroxyapatite/collagen/poly(L-lactic acid) composites, *J. Biomed. Mater. Res. A*, **76A**, 820–825.

608. Liao, S. S., Cui, F. Z., and Zhu, Y. (2004) Osteoblasts adherence and migration through three-dimensional porous mineralized collagen based composite: nHAC/PLA, *J. Bioact. Compat. Polym.*, **19**, 117–130.

609. Liao, S. S., Cui, F. Z., Zhang, W., and Feng, Q. L. Hierarchically biomimetic bone scaffold materials: nano-HA/collagen/PLA composite, *J. Biomed. Mater. Res. B (Appl. Biomater.)*, **69B**, 158–165.

610. Liao, S. S., and Cui, F. Z. (2004) *In vitro* and *in vivo* degradation of the mineralized collagen based composite scaffold: nanohydroxyapatite/collagen/poly(L-lactide), *Tissue Eng.*, **10**, 73–80.

611. Liao, S. S., Wang, W., Uo, M., Ohkawa, S., Akasaka, T., Tamura, K., Cui, F. Z., and Watari, F. (2005) A three-layered nano-carbonated hydroxyapatite/collagen/PLGA composite membrane for guided tissue regeneration, *Biomaterials*, **26**, 7564–7571.

612. Li, X., Feng, Q., and Cui, F. Z. (2006) *In vitro* degradation of porous nano-hydroxyapatite/collagen/PLLA scaffold reinforced by chitin fibres, *Mater. Sci. Eng. C*, **26**, 716–720.

613. Zhou, D. S., Zhao, K. B., Li, Y., Cui, F. Z., and Lee, I. S. (2006) Repair of segmental defects with nano-hydroxyapatite/collagen/PLA composite combined with mesenchymal stem cells, *J. Bioactive and Compatible Polymers*, **21**, 373–384.

614. Zhang, C., Hu, Y. Y., Cui, F. Z., Zhang, S. M., and Ruan, D. K. (2006) A study on a tissue-engineered bone using rhBMP-2 induced periosteal cells with a porous nano-hydroxyapatite/collagen/poly(L-lactic acid) scaffold, *Biomed. Mater.*, **1**, 56–62.

615. Liao, S., Watari, F., Zhu, Y., Uo, M., Akasaka, T., Wang, W., Xu, G., and Cui, F. Z. (2007) The degradation of the three layered nano-carbonated hydroxyapatite/collagen/PLGA composite membrane *in vitro*, *Dental Materials*, **23**, 1120–1128.

616. Degirmenbasi, N., Kalyon, D. M., and Birinci, E. (2006) Biocomposites of nanohydroxyapatite with collagen and poly(vinyl alcohol), *Colloids Surf. B*, **48**, 42–49.

617. Zhang, S. M., Cui, F. Z., Liao, S. S., Zhu, Y., and Han, L. (2003) Synthesis and biocompatibility of porous nanohydroxyapatite/collagen/alginate composite, *J. Mater. Sci. Mater. Med.*, **14**, 641–645.

618. Sotome, S., Uemura, T., Kikuchi, M., Chen, J., Itoh, S., Tanaka, J., Tateishi, T., and Shinomiya, K. (2004) Synthesis and *in vivo* evaluation of a

novel hydroxyapatite/collagen-alginate as a bone filler and a drug delivery carrier of bone morphogenetic protein, *Mater. Sci. Eng. C*, **24**, 341–347.

619. Chang, M. C., Ko, C. C., and Douglas, W. H. (2003) Conformational change of hydroxyapatite/gelatin nanocomposite by glutaraldehyde, *Biomaterials*, **24**, 3087–3094.

620. Kim, H. W., Kim, H. E., and Vehid, S. (2005) Stimulation of osteoblast responses to biomimetic nanocomposites of gelatin-hydroxyapatite for tissue engineering scaffolds, *Biomaterials*, **26**, 5221–5230.

621. Chang, M. C., Ikoma, T., and Tanaka, J. (2004) Cross-linkage of hydroxyapatite/gelatin nanocomposite using EGDE, *J. Mater. Sci.*, **39**, 5547–5550.

622. Teng, S., Shi, J., Peng, B., and Chen, L. (2006) The effect of alginate addition on the structure and morphology of hydroxyapatite/gelatin nanocomposites, *Compos. Sci. Technol.*, **66**, 1532–1538.

623. Chang, M. C., Ko, C. C., and Douglas, W. H. (2003) Preparation of hydroxyapatite-gelatin nanocomposite, *Biomaterials*, **24**, 2853–2862.

624. Mobini, S., Javadpour, J., Hosseinalipour, M., Ghazi-Khansari, M., Khavandi, A., and Rezaie, H. R. (2008) Synthesis and characterisation of gelatin-nano hydroxyapatite composite scaffolds for bone tissue engineering, *Adv. Appl. Ceram.*, **107**, 4–8.

625. Lewandrowski, K. U., Bondre, S. P., Wise, D. L., and Trantolo, D. J. (2003) Enhanced bioactivity of a poly(propylene fumarate) bone graft substitute by augmentation with nano-hydroxyapatite, *Biomed. Mater. Eng.*, **13**, 115–124.

626. Jayabalan, M., Shalumon, K. T., Mitha, M. K., Ganesan, K., and Epple, M. (2010) Effect of hydroxyapatite on the biodegradation and biomechanical stability of polyester nanocomposites for orthopaedic applications, *Acta Biomater.*, **6**, 763–775.

627. Jayabalan, M., Shalumon, K. T., Mitha, M. K., Ganesan, K., and Epple, M. (2010) The effect of radiation processing and filler morphology on the biomechanical stability of a thermoset polyester composite, *Biomed. Mater.*, **5**, 025009.

628. Wei, J., Li, Y., and He, Y. (2005) Processing properties of nano apatite-polyamide biocomposites, *J. Mater. Sci.*, **40**, 793–796.

629. Wei, J., Li, Y., Chen, W., and Zuo, Y. (2003) A study on nano-composite of hydroxyapatite and polyamide, *J. Mater. Sci.*, **38**, 3303–3306.

630. Wang, H., Li, Y., Zuo, Y., Li, J., Ma, S., and Cheng, L. (2007) Biocompatibility and osteogenesis of biomimetic nano-hydroxyapatite/polyamide composite scaffolds for bone tissue engineering, *Biomaterials*, **28**, 3338–3348.

631. Sender, C., Dantras, E., Dantras-Laffont, L., Lacoste, M. H., Dandurand, J., Mauzac, M., Lacout, J. L., Lavergne, C., Demont, P., Bernes, A., and Lacabanne, C. (2007) Dynamic mechanical properties of a biomimetic hydroxyapatite/polyamide 6,9 nanocomposite, *J. Biomed. Mater. Res. B (Appl. Biomater.)*, **83B**, 628–635.

632. Yang, K., Wei, J., Wang, C. Y., and Li, Y. (2007) A study on *in vitro* and *in vivo* bioactivity of nano hydroxyapatite/polymer biocomposite, *Chinese Sci. Bull.*, **52**, 267–271.

633. Zhang, X., Li, Y., Lv, G. Y., Zuo, Y., and Mu, Y. H. (2006) Thermal and crystallization studies of nano-hydroxyapatite reinforced polyamide 66 biocomposites, *Polym. Degrad. Stabil.*, **91**, 1202–1207.

634. Huang, M., Feng, J., Wang, J., Zhang, X., Li, Y., and Yan, Y. (2003) Synthesis and characterization of nano-HA/PA66 composite, *J. Mater. Sci. Mater. Med.*, **14**, 655–660.

635. Zhang, X., Li, Y., Zuo, Y., Lv, G. Y., Mu, Y. H., and Li, H. (2007) Morphology, hydrogen-bonding and crystallinity of nano-hydroxyapatite/polyamide 66 biocomposites, *Composites A*, **38**, 843–848.

636. Lan, W., Li, Y., Yi, Z., Li, Z., Mu, Y. H., and Jimei, H. (2006) Study on the biomimetic properties of bone substitute material: nano-hydroxyapatite/polyamide 66 composite, *Mater. Sci. Forum*, 510–511, 938–941.

637. Zhang, L., Li, Y., Wang, X., Wei, J., and Peng, X. (2005) Studies on the porous scaffold made of the nano-HA/PA66 composite, *J. Mater. Sci.*, **40**, 107–110.

638. Zhang, X., Li, Y., Lv, G. Y., Zuo, Y., Mu, Y. H., and Lan, W. (2005) The study on interaction mechanism between n-HA and PA66 in n-HA/PA66 biocomposites, *Funct. Mater.*, **36**, 896–899.

639. Menon, D., Anoop Anand, K., Anitha, V. C., and Nair, S. (2010) Hydroxyapatite-reinforced polyamide 6,6 nanocomposites through melt compounding, *Int. J. Polym. Mater.*, **59**, 498–509.

640. Yusong, P., Dangsheng, X., and Xiaolin, C. (2007) Mechanical properties of nanohydroxyapatite reinforced poly(vinyl alcohol) gel composites as biomaterial, *J. Mater. Sci.*, **42**, 5129–5134.

641. Xu, F., Li, Y., Wang, X., Wei, J., and Yang, A. (2004) Preparation and characterization of nano-hydroxyapatite/poly(vinyl alcohol) hydrogel biocomposite, *J. Mater. Sci.*, **39**, 5669–5672.

642. Wang, H. S., Wang, G. X., and Pan, Q. X. (2005) Electrochemical study of the interactions of DNA with redox-active molecules based on the immobilization of dsDNA on the sol-gel derived nano porous hydroxyapatite-polyvinyl alcohol hybrid material coating, *Electroanalysis*, **17**, 1854–1860.

643. Pramanik, N., Mohapatra, S., Pramanik, P., and Bhargava, P. (2007) Processing and properties of nano-hydroxyapatite (n-HAp)/poly(ethylene-*co*-acrylic acid) (EAA) composite using a phosphonic acid coupling agent for orthopedic applications, *J. Am. Ceram. Soc.*, **90**, 369–375.

644. Pramanik, N., Bhargava, P., Alam, S., and Pramanik, P. (2006) Processing and properties of nano- and macro-hydroxyapatite/poly(ethylene-*co*-acrylic acid) composites, *Polym. Compos.*, **27**, 633–641.

645. Zhang, L., Li, Y., Yang, A., Peng, X., Wang, X., and Zhang, X. (2005) Preparation and *in vitro* investigation of chitosan/nano-hydroxyapatite composite used as bone substitute materials, *J. Mater. Sci. Mater. Med.*, **16**, 213–219.

646. Zhang, Y. F., Cheng, X. R., Chen, Y., Shi, B., Chen, X. H., Xu, D. X., and Ke, J. (2007) Three-dimensional nanohydroxyapatite/chitosan scaffolds as potential tissue engineered periodontal tissue, *J. Biomaterials Applications*, **21**, 333–349.

647. Kong, L., Gao, Y., Lu, G., Gong, Y., Zhao, N., and Zhang, X. (2006) A study on the bioactivity of chitosan/nano-hydroxyapatite composite scaffolds for bone tissue engineering, *Eur. Polymer J.*, **42**, 3171–3179.

648. Lu, X. Y., Wang, X. H., Qu, S. X., and Weng, J. (2008) Preparation of nano-hydroxyapatite/chitosan hybrids, *J. Inorg. Mater.*, **23**, 332–336.

649. Chen, J. D., Wang, Y., and Chen, X. (2009) *In situ* fabrication of nano-hydroxyapatite in a macroporous chitosan scaffold for tissue engineering, *J. Biomater. Sci. Polymer Edn.*, **20**, 1555–1565.

650. Thein-Han, W. W., and Misra, R. D. K. (2009) Biomimetic chitosan-nanohydroxyapatite composite scaffolds for bone tissue engineering, *Acta Biomater.*, **5**, 1182–1197.

651. Chen, J., Nan, K., Yin, S., Wang, Y., Wu, T., and Zhang, Q. (2010) Characterization and biocompatibility of nanohybrid scaffold prepared via *in situ* crystallization of hydroxyapatite in chitosan matrix, *Colloids Surf. B*, **81**, 640–647.

652. Lu, Y., Zhu, A., Wang, W., and Shi, H. (2010) New bioactive hybrid material of nano-hydroxyapatite based on N-carboxyethylchitosan for bone tissue engineering, *Appl. Surf. Sci.*, **256**, 7228–7233.

653. Zhou, G., Li, Y., Zhang, L., Li, H., Wang, M., Cheng, L., Wang, Y., Wang, H., and Shi, P. (2007) The study of tri-phasic interactions in nano-hydroxyapatite/konjac glucomannan/chitosan composite, *J. Mater. Sci.*, **42**, 2591–2597.

654. Huang, J., Lin, Y. W., Fu, X. W., Best, S. M., Brooks, R. A., Rushton, N., and Bonfield, W. (2007) Development of nano-sized hydroxyapatite reinforced composites for tissue engineering scaffolds, *J. Mater. Sci. Mater. Med.*, **18**, 2151–2157.

655. Lee, H. J., Choi, H. W., Kim, K. J., and Lee, S. C. (2006) Modification of hydroxyapatite nanosurfaces for enhanced colloidal stability and improved interfacial adhesion in nanocomposites, *Chem. Mater.*, **18**, 5111–5118.

656. Lee, H. J., Kim, S. E., Choi, H. W., Kim, C. W., Kim, K. J., and Lee, S. C. (2007) The effect of surface-modified nano-hydroxyapatite on biocompatibility of poly(ε-caprolactone) / hydroxyapatite nanocomposites, *Eur. Polymer J.*, **43**, 1602–1608.

657. Grande, C. J., Torres, F. G., Gomez, C. M., and Bañó, M. C. (2009) Nanocomposites of bacterial cellulose/hydroxyapatite for biomedical applications, *Acta Biomater.*, **5**, 1605–1615.

658. Zadegan, S., Hossainalipour, M., Ghassai, H., Rezaie, H. R., and Naimi-Jamal, M. R. (2010) Synthesis of cellulose-nanohydroxyapatite composite in 1-*n*-butyl-3-methylimidazolium chloride, *Ceram. Int.*, **36**, 2375–2381.

659. Jia, N., Li, S. M., Zhu, J. F., Ma, M. G., Xu, F., Wang, B., and Sun, R. C. (2010) Microwave-assisted synthesis and characterization of cellulose-carbonated hydroxyapatite nanocomposites in NaOH-urea aqueous solution, *Mater. Lett.*, **64**, 2223–2225.

660. Pang, P., Li, W., and Liu, Y. (2007) Effect of ball milling process on the microstructure of titanium-nanohydroxyapatite composite powder, *Rare Metals*, **26**, 118–123.

661. Li, W., Pang, P., and Liu, Y. (2007) Microstructure and phase composition of Ti-based biocomposites with different contents of nano-hydroxyapatite, *Transactions of Nonferrous Metals Society of China*, **17**, special issue, S1148–S1151.

662. Niespodziana, K., Jurczyk, K., Jakubowicz, J., and Jurczyk, M. (2010) Fabrication and properties of titanium-hydroxyapatite nanocomposites, *Mater. Chem. Phys.*, **123**, 160–165.

663. Hao, J. Y., Liu, Y., Zhou, S., Li, Z., and Deng, X. (2003) Investigation of nanocomposites based on semi-interpenetrating network of [L-poly (ε-caprolactone)]/[net-poly (ε-caprolactone)] and hydroxyapatite nanocrystals, *Biomaterials*, **24**, 1531–1539.

664. Yan, Y., Li, Y., Zheng, Y., Yi, Z., Wei, J., Xia, C., and Chen, Y. (2003) Synthesis and properties of a copolymer of poly(1,4-phenylene sulfide)-poly(2,4-phenylene sulfide acid) and its nano-apatite reinforced composite, *Eur. Polym. J.*, **39**, 411–416.

665. Bhattacharyya, S., Nair, L. S., Singh, A., Krogman, N. R., Bender, J., Greish, Y. E., Brown, P. W., Allcock, H. R., and Laurencin, C. T. (2005) Development of biodegradable polyphosphazene-nanohydroxyapatite composite nanofibers via electrospinning, *MRS Symp. Proc.*, **845**, 91–96.

666. Sinha, A., Nayar, S., Agrawal, A., Bhattacharyya, D., and Ramachandrarao, P. (2003) Synthesis of nanosized and microporous precipitated hydroxyapatite in synthetic polymers and biopolymers, *J. Am. Ceram. Soc.*, **86**, 357–359.

667. Zuo, Y., Li, Y., Wei, J., Han, J., and Xu, F. (2004) The preparation and characterization of n-HA/PA series biomedical composite, *Funct. Mater.*, **35**, 513–516.

668. Zhou, G., Li, Y., Zhang, L., Zuo, Y., and Jansen, J. A. (2007) Preparation and characterization of nano-hydroxyapatite/chitosan/konjac glucomannan composite, *J. Biomed. Mater. Res. A*, **83A**, 931–939.

669. Daniel-da-Silva, A. L., Lopes, A. B., Gil, A. M., and Correia, R. N. (2007) Synthesis and characterization of porous κ-carrageenan/calcium phosphate nanocomposite scaffolds, *J. Mater. Sci.*, **42**, 8581–8591.

670. Furuzono, T., Kishida, A., and Tanaka, J. (2004) Nano-scaled hydroxyapatite/polymer composite I. Coating of sintered hydroxyapatite particles on poly(γ-methacryloxypropyl trimethoxysilane)-grafted silk fibroin fibers through chemical bonding, *J. Mater. Sci. Mater. Med.*, **15**, 19–23.

671. Korematsu, A., Furuzono, T., Yasuda, S., Tanaka, J., and Kishida, A. (2004) Nano-scaled hydroxyapatite/polymer composite II. Coating of sintered hydroxyapatite particles on poly(2-(o-[1'-methylpropylideneamino] carboxyamino) ethyl methacrylate)-grafted silk fibroin fibers through covalent linkage, *J. Mater. Sci.*, **39**, 3221–3225.

672. Korematsu, A., Furuzono, T., Yasuda, S., Tanaka, J., and Kishida, A. (2005) Nano-scaled hydroxyapatite/polymer composite III. Coating of sintered hydroxyapatite particles on poly(4-methacryloyloxyethyl

trimellitate anhydride)-grafted silk fibroin fibers, *J. Mater. Sci. Mater. Med.*, **16**, 67–71.

673. Yang, K., Wang, C., and Wei, J. (2007) A study on biocomposite of nano apatite/poly (1,4-phenylene-sulfide)-poly (2,4-phenylene sulfide acid), *Composites B*, **38**, 306–310.

674. Jiang, L., Li, Y., Zhang, L., and Wang, X. J. (2008) Study on nano-hydroxyapatite/chitosan-carboxymethyl cellulose composite scaffold, *J. Inorg. Mater.*, **23**, 135–140.

675. Jiang, L., Li, Y., Zhang, L., and Liao, J. (2008) Preparation and properties of a novel bone repair composite: nano-hydroxyapatite/chitosan/carboxymethyl cellulose, *J. Mater. Sci. Mater. Med.*, **19**, 981–987.

676. Liou, S. C., Chen, S. Y., and Liu, D. M. (2004) Phase development and structural characterization of calcium phosphate ceramics-polyacrylic acid nanocomposites at room temperature in water-methanol mixtures, *J. Mater. Sci. Mater. Med.*, **15**, 1261–1266.

677. Liu, L., Liu, J., Wang, M., Min, S., Cai, Y., Zhu, L., and Yao, J. (2008) Preparation and characterization of nano-hydroxyapatite/silk fibroin porous scaffolds, *J. Biomater. Sci. Polymer Edn.*, **19**, 325–338.

678. Ren, Y. J., Sun, X. D., Cui, F. Z., Wei, Y. T., Cheng, Z. J., and Kong, X. D. (2007) Preparation and characterization of *antheraea pernyi* silk fibroin based nanohydroxyapatite composites, *J Bioactive and Compatible Polymers*, **22**, 465–474.

679. Mikołajczyk, T., Rabiej, S., and Bogun, M. (2006) Analysis of the structural parameters of polyacrylonitrile fibers containing nanohydroxyapatite, *J. Appl. Polym. Sci.*, **101**, 760–765.

680. Wei, J., Li, Y., and Lau, K. T. (2007) Preparation and characterization of a nano apatite/polyamide$_6$ bioactive composite, *Composites B*, **38**, 301–305.

681. Sundaraseelan, J., and Sastry, T. P. (2007) Fabrication of a biomimetic compound containing nano hydroxyapatite – demineralised bone matrix, *J. Biomed. Nanotechnol.*, **3**, 401–405.

682. Leeuwenburgh, S. C. G., Jansen, J. A., and Mikos, A. G. (2007) Functionalization of oligo(poly(ethylene glycol)fumarate) hydrogels with finely dispersed calcium phosphate nanocrystals for bone-substituting purposes, *J. Biomater. Sci. Polymer Edn.*, **18**, 1547–1564.

683. Hesaraki, S., Ebadzadeh, T., and Ahmadzadeh-Asl, S. (2010) Nanosilicon carbide/hydroxyapatite nanocomposites: structural, mechanical and *in vitro* cellular properties, *J. Mater. Sci. Mater. Med.*, **21**, 2141–2149.

684. Zandi, M., Mirzadeh, H., Mayer, C., Urch, H., Eslaminejad, M. B., Bagheri, F., and Mivehchi, H. (2010) Biocompatibility evaluation of nano-rod hydroxyapatite/gelatin coated with nano-HAp as a novel scaffold using mesenchymal stem cells, *J. Biomed. Mater. Res. A*, **92A**, 1244–1255.

685. Sun, T. S., Guan, K., Shi, S. S., Zhu, B., Zheng, Y. J., Cui, F. Z., Zhang, W., and Liao, S. S. (2004) Effect of nano-hydroxyapatite/collagen composite and bone morphogenetic protein-2 on lumbar intertransverse fusion in rabbits, *Chin. J. Traumatol.*, **7**, 18–24.

686. Itoh, S., Kikuehi, M., Koyama, Y., Takakuda, K., Shinomiya, K., and Tanaka, J. (2004) Development of a hydroxyapatite/collagen nanocomposite as a medical device, *Cell Transp.*, **13**, 451–461.

687. Kester, M., Heakal, Y., Fox, T., Sharma, A., Robertson, G. P., Morgan, T. T., Altinoğlu, E. I., Tabaković, A., Parette, M. R., Rouse, S. M., Ruiz-Velasco, V., and Adair, J. H. (2008) Calcium phosphate nanocomposite particles for in vitro imaging and encapsulated chemotherapeutic drug delivery to cancer cells, *Nano Lett.*, **8**, 4116–4121.

688. Wang, K. W., Zhou, L. Z., Sun, Y., Wu, G. J., Gu, H. C., Duan, Y. R., Chen, F., and Zhu, Y. J. (2010) Calcium phosphate/PLGA-mPEG hybrid porous nanospheres: a promising vector with ultrahigh gene loading and transfection efficiency, *J. Mater. Chem.*, **20**, 1161–1166.

689. Gelinsky, M., Welzel, P. B., Simon, P., Bernhardt, A., and König, U. (2008) Porous three-dimensional scaffolds made of mineralized collagen: preparation and properties of a biomimetic nanocomposite material for tissue engineering of bone, *Chemical Eng. J.*, **137**, 84–96.

690. Hu, Q., Li, B. Q., Wang, M., and Shen, J. C. (2004) Preparation and characterization of biodegradable chitosan/hydroxyapatite nanocomposite rods via *in situ* hybridization: a potential material as internal fixation of bone fracture, *Biomaterials*, **25**, 779–785.

691. Wei, G., and Ma, P. X. (2004) Structure and properties of nano-hydroxyapatite/polymer composite scaffolds for bone tissue engineering, *Biomaterials*, **25**, 4749–4757.

692. Liou, S. C., Chen, S. Y., and Liu, D. M. (2003) Synthesis and characterization of needlelike apatitic nanocomposite with controlled aspect ratios, *Biomaterials*, **24**, 3981–3988.

693. Liou, S. C., Chen, S. Y., and Liu, D. M. (2005) Manipulation of nanoneedle and nanosphere apatite/poly(acrylic acid) nanocomposites, *J. Biomed. Mater. Res. B* (*Appl. Biomater.*), **73B**, 117–122.

694. Huang, J., Best, S. M., Bonfield, W., Brooks, R. A., Rushton, N., Jayasinghe, S. N., and Edirisinghe, M. J. (2004) *In vitro* assessment

of the biological response to nano-size hydroxyapatite, *J. Mater. Sci. Mater. Med.*, **15**, 441–415.

695. Kong, L., Gao, Y., Cao, W., Gong, Y., Zhao, N., and Zhang, X. (2005) Preparation and characterization of nano-hydroxyapatite/chitosan composite scaffolds, *J. Biomed. Mater. Res. A*, **75A**, 275–282.

696. Christenson, E. M., Anseth, K. S., van den Beucken, J. J. J. P., Chan, C. K., Ercan, B., Jansen, J. A., Laurencin, C. T., Li, W. J., Murugan, R., Nair, L. S., Ramakrishna, S., Tuan, R. S., Webster, T. J., and Mikos, A. G. (2007) Nanobiomaterial applications in orthopedics, *J. Orthop. Res.*, **25**, 11–22.

697. Prockop, D. J., and Kivirikko, K. I. (1995) Collagens: molecular biology, diseases, and potentials for therapy, *Annual Rev. Biochem.*, **64**, 403–434.

698. Fratzl, P. (2010) *Collagen: Structure and Mechanics*, Springer, USA, p. 528.

699. Olmo, N., Turnay, J., Herrera, J. I., Gavilanes, J. G., and Lizarbe, M. A. (1996) Kinetics of *in vivo* degradation of sepiolite-collagen complexes: effect of glutaraldehyde treatment, *J. Biomed. Mater. Res.*, **30**, 77–84.

700. Xie, J., Baumann, M. J., and McCabe, L. R. (2004) Osteoblasts respond to hydroxyapatite surfaces with immediate changes in gene expression, *J. Biomed. Mater. Res. A*, **71A**, 108–117.

701. Tcacencu, I., and Wendel, M. (2008) Collagen-hydroxyapatite composite enhances regeneration of calvaria bone defects in young rats but postpones the regeneration of calvaria bone in aged rats, *J. Mater. Sci. Mater. Med.*, **19**, 2015–2021.

702. Yamauchi, K., Goda, T., Takeuchi, N., Einaga, H., and Tanabe, T. (2004) Preparation of collagen/calcium phosphate multilayer sheet using enzymatic mineralization, *Biomaterials*, **25**, 5481–5489.

703. Du, C., Cui, F. Z., Zhang, W., Feng, Q. L., Zhu, X. D., and de Groot, K. (2000) Formation of calcium phosphate/collagen composites through mineralization of collagen matrix, *J. Biomed. Mater. Res.*, **50**, 518–527.

704. Hellmich, C., and Ulm, F. J. (2002) Are mineralized tissues open crystal foams reinforced by crosslinked collagen? – Some energy arguments, *J. Biomech.*, **35**, 1199–1212.

705. Boskey, A. L. (1989) Hydroxyapatite formation in a dynamic collagen gel system – effects of type I collagen, lipids and proteoglycans, *J. Phys. Chem.*, **93**, 1628–1633.

706. Mathers, N. J., amd Czernuszka, J. T. (1991) Growth of hydroxyapatite on type I collagen, *J. Mater. Sci. Lett.*, **10**, 992–993.

707. Sukhodub, L. F., Moseke, C., Sukhodub, L. B., Sulkio-Cleff, B., Maleev, V. Y., Semenov, M. A., Bereznyak, E. G., and Bolbukh, T. V. (2004) Collagen-hydroxyapatite-water interactions investigated by XRD, piezogravimetry, infrared and Raman spectroscopy, *J. Molecular Structure*, **704**, 53–58.

708. Roveri, N., Falini, G., Sidoti, M. C., Tampieri, A., Landi, E., Sandri, M., and Parma, B. (2003) Biologically inspired growth of hydroxyapatite nanocrystals inside self-assembled collagen fibers, *Mater. Sci. Eng. C*, **23**, 441–446.

709. Tampieri, A., Celotti, G., and Landi, E. (2005) From biomimetic apatites to biologically inspired composites, *Anal. Bioanal. Chem.*, **381**, 568–576.

710. Tampieri, A., Celotti, G., Landi, E., Sandri, M., Roveri, N., and Falini, G. (2003) Biologically inspired synthesis of bone-like composite: self-assembled collagen fibers/hydroxyapatite nanocrystals, *J. Biomed. Mater. Res. A*, **67A**, 618–625.

711. Mehlisch, D. R., Taylor, T. D., Leibold, D. G., Hiatt, R., Waite, D. E., Waite, P. D., Laskin, D. M., Smith, S. T., and Koretz, M. M. (1987) Evaluation of collagen/hydroxyapatite for augmenting deficient alveolar ridges., a preliminary report, *J. Oral Maxillofac. Surg.*, **45**, 408–413.

712. Okazaki, M., Ohmae, H., Takahashi, J., Kimura, H., and Sakuda, M. (1990) Insolubilized properties of UV-irradiated CO_3 apatite-collagen composites, *Biomaterials*, **11**, 568–572.

713. Ten Huisen, K. S., Martin, R. I., Klimkiewicz, M., and Brown, P. W. (1995) Formation and properties of a synthetic bone composite: hydroxyapatite-collagen, *J. Biomed. Mater. Res.*, **29**, 803–810.

714. Marouf, H. A., Quayle, A. A., and Sloan, P. (1990) *In vitro* and *in vivo* studies with collagen/hydroxyapatite implants, *Int. J. Oral Maxillofac. Implants*, **5**, 148–154.

715. Zerwekh, J. E., Kourosh, S., Scheinberg, R., Kitano, T., Edwards, M. L., Shin, D., and Selby, D. K. (1992) Fibrillar collagen-biphasic calcium phosphate composite as a bone graft substitute for spinal fusion, *J. Orthop. Res.*, **10**, 565–572.

716. Clarke, K. I., Graves, S. E., Wong, A. T. C., Triffit, J. T., Francis, M. J. O., and Czernuszka, J. T. (1993) Investigation into the formation and mechanical properties of a bioactive material based on collagen and calcium phosphate, *J. Mater. Sci. Mater. Med.*, **4**, 107–110.

717. Rovira, A., Bareille, R., Lopez, L., Rouasis, F., Bordenave, L., Rey, C., and Rabaud, M. (1993) Preliminary report on a new composite material made of calcium phosphate, elastin peptides and collagens, *J. Mater. Sci. Mater. Med*, **4**, 372–380.

718. Zhang, Q. Q., Ren, L., Wang, C., Liu, L. R., Wen, X. J., Liu, Y. H., and Zhang, X. D. (1996) Porous hydroxyapatite reinforced with collagen protein, *Artif. Cells Blood Substit. Immobiliz. Biotechnol.*, **24**, 693–702.

719. Bakoš, D., Soldán, M., and Hernández-Fuentes, I. (1999) Hydroxyapatite-collagen-hyaluronic acid composite, *Biomaterials*, **20**, 191–195.

720. John, A., Hong, L., Ikada, Y., and Tabata, Y. (2001) A trial to prepare biodegradable collagen-hydroxyapatite composites for bone repair, *J. Biomater. Sci. Polym. Edn.*, **12**, 689–705.

721. Itoh, S., Kikuchi, M., Takakuda, K., Koyama, Y., Matsumoto, H. N., Ichinose, S., Tanaka, J., Kawauchi, T., and Shinomiya, K. (2001) The biocompatibility and osteoconductive activity of a novel hydroxyapatite/collagen composite biomaterial and its function as a carrier of rhBMP-2, *J. Biomed. Mater. Res.*, **54**, 445–453.

722. Shinomiya, K., Itoh, S., Kawauchi, T., Kikuchi, M., and Tanaka, J. (2001) Development of a novel hydroxyapatite/collagen composite biomaterial, *Tissue Eng. Therap. Use*, **5**, 165–177.

723. Uskokovic, V., Ignjatovic, N., and Petranovic, N. (2002) Synthesis and characterization of hydroxyapatite-collagen biocomposite materials, *Mater. Sci. Forum*, **413**, 269–274.

724. Wahl, D. A., and Czernuszka, J. T. (2006) Collagen-hydroxyapatite composites for hard tissue repair, *Eur. Cell Mater.*, **11**, 43–56.

725. Ishikawa, H., Koshino, T., Takeuchi, R., and Saito, T. (2001) Effects of collagen gel mixed with hydroxyapatite power on interface between newly formed bone and grafted Achilles tendon in rabbit femoral bone tunnel, *Biomaterials*, **22**, 1689–1694.

726. Sachlos, E., Gotora, D., and Czernuszka, J. T. (2006) Collagen scaffolds reinforced with biomimetic composite nano-sized carbonate-substituted hydroxyapatite crystals and shaped by rapid prototyping to contain internal microchannels, *Tissue Eng.*, **12**, 2479–2487.

727. Venugopal, J., Ramakrishna, S., Low, S., Choon, A. T., and Kumar, T. S. S. (2008) Mineralization of osteoblasts with electrospun collagen/hydroxyapatite nanofibers, *J. Mater. Sci. Mater. Med.*, **19**, 2039–2046.

728. Teng, S. H., Lee, E. J., Park, C. S., Choi, W. Y., Shin, D. S., and Kim, H. E. (2008) Bioactive nanocomposite coatings of collagen/hydroxyapatite on titanium substrates, *J. Mater. Sci. Mater. Med.*, **19**, 2453–2461.

729. Song, J. H., Kim, H. E., and Kim, H. W. (2007) Collagen-apatite nanocomposite membranes for guided bone regeneration, *J. Biomed. Mater. Res. B* (*Appl. Biomater.*), **83B**, 248–257.

730. Pek, Y. S., Gao, S., Arshad, M. S. M., Leck, K. J., and Ying, J. Y. (2008) Porous collagen-apatite nanocomposite foams as bone regeneration scaffolds, *Biomaterials*, **29**, 4300–4305.

731. Mittelmeier, H., and Nizzard, M. (1982) Knochenregeneration mit industriell gefertigtem Collagen Apatit Implantat ("Collapat"), in *Osteogenese und Knochenwachstum* (ed. Hackenbroch, M. H., Refior, H. J., and Jäger, M. G.), Thieme, Stuttgart, pp. 194–197.

732. Serre, C. M., Papillard, M., Chavassieux, P., and Boivin, G. (1993) *In vitro* induction of a calcifying matrix by biomaterials constituted of collagen and/or hydroxyapatite: an ultrastructural comparison of three types of biomaterials, *Biomaterials*, **14**, 97–106.

733. Scabbia, A., and Trombelli, L. (2004) A comparative study on the use of a HA/collagen/chondroitin sulphate biomaterial (Biostite®) and a bovine-derived HA xenograft (Bio-Oss®) in the treatment of deep intraosseous defects, *J. Clin. Periodontol.*, **31**, 348–355.

734. Yamasaki, Y., Yoshida, Y., Okazaki, M., Shimazu, A., Kubo, T., Akagawa, Y., and Uchida, T. (2003) Action of FGMgCO_3Ap-collagen composite in promoting bone formation, *Biomaterials*, **24**, pp. 4913–4920.

735. Wang, X., Grogan, S. P., Rieser, F., Winkelmann, V., Maquet, V., Berge, M. L. and Mainil-Varlet, P. (2004) Tissue engineering of biphasic cartilage constructs using various biodegradable scaffolds: an *in vitro* study, *Biomaterials*, **25**, 3681–3688.

736. Chang, M. C., Ikonama, T., Kikuchi, M., and Tanaka, J. (2002) The cross-linkage effect of hydroxyapatite/collagen nanocomposites on a self-organization phenomenon, *J. Mater. Sci. Mater. Med.*, **13**, 993–997.

737. Iijima, M., Moriwaki, Y., and Kuboki, Y. (1996) Oriented growth of octacalcium phosphate on and inside the collagenous matrix *in vitro*, *Connect. Tissue Res.*, **32**, 519–524.

738. Miyamoto, Y., Ishikawa, K., Takechi, M., Toh, T., Yuasa, T., Nagayama, M., and Suzuki, K. (1998) Basic properties of calcium phosphate cement containing atelocollagen in its liquid or powder phases, *Biomaterials*, **19**, 707–715.

739. Iijima, M., Moriwaki, Y., and Kuboki, Y. (1994) *In vitro* crystal growth of octacalcium phosphate on type I collagen fiber, *J. Cryst. Growth*, **137**, 553–560.

740. Iijima, M., Iijima, K., Moriwaki, Y., and Kuboki, Y. (1994) Oriented growth of octacalcium phosphate crystals on type 1 collagen fibrils under physiological conditions, *J. Cryst. Growth*, **140**, 91–99.

741. Lawson, A. C., and Czernuszka, J. T. (1998) Collagen – calcium phosphate composites, *Proc. Instn. Mech. Engrs. Part H*, **212**, 413–425.

742. Kamakura, S., Sasaki, K., Honda, Y., Anada, T., and Suzuki, O. (2006) Octacalcium phosphate combined with collagen orthotopically enhances bone regeneration, *J. Biomed. Mater. Res. B (Appl. Biomater.)*, **79B**, 210–217.

743. Kawai, T., Anada, T., Honda, Y., Kamakura, S., Matsui, K., Matsui, A., Sasaki, K., Morimoto, S., Echigo, S, and Suzuki, O. (2009) Synthetic octacalcium phosphate augments bone regeneration correlated with its content in collagen scaffold, *Tiss. Eng. A*, **15**, 23–32.

744. Masuda, T., Kawai, T., Anada, T., Kamakura, S., and Suzuki, O. (2010) Quality of regenerated bone enhanced by implantation of octacalcium phosphate-collagen composite, *Tiss. Eng. C*, **16**, 471–478.

745. Rhee, S. H., and Tanaka, J. (1998) Hydroxyapatite coating on a collagen membrane by a biomimetic method, *J. Am. Ceram. Soc.*, **81**, 3029–3031.

746. Itoh, S., Kikuchi, M., Takakuda, K., Nagaoka, K., Koyama, Y., Tanaka, J., and Shinomiya, K. (2002) Implantation study of a novel hydroxyapatite/collagen (HAp/col) composite into weight-bearing sites of dogs, *J. Biomed. Mater. Res.*, **63**, 507–515.

747. Kikuchi, M., Ikoma, T., Itoh, S., Matsumoto, H. N., Koyama, Y., Takakuda, K., Shinomiya, K., and Tanaka, J. (2004) Biomimetic synthesis of bone-like nanocomposites using the self-organization mechanism of hydroxyapatite and collagen, *Compos. Sci. Technol.*, **64**, 819–825.

748. Yang, X. B., Bhataagar, R. S., Li, S., and Oreffo, R. O. (2004) Biomimetic collagen scaffolds for human bone cell growth and differentiation, *Tissue Eng.*, **14**, 1148–1159.

749. Doi, Y., Horiguchi, T., Moriwaki, Y., Kitago, H., Kajimoto, T., and Iwayama, Y. (1996) Formation of apatite – collagen complexes, *J. Biomed. Mater. Res.*, **31**, 43–49.

750. Bradt, J. H., Mertig, M., Teresiak, A., and Pompe, W. (1999) Biomimetic mineralization of collagen by combined fibril assembly and calcium phosphate formation, *Chem. Mat.*, **11**, 2694–2701.

751. Scharnweber, D., Born, R., Flade, K., Roessler, S., Stoelzel, M., and Worch, H. (2004) Mineralization behaviour of collagen type I immobilized on different substrates, *Biomaterials*, **25**, 2371–2380.

752. Yunoki, S., Ikoma, T., Monkawal, A., Ohtal, K., and Tanaka, J. (2007) Preparation and characterization of hydroxyapatite / collagen nanocomposite gel, *J. Nanosci. Nanotechnol.*, **7**, 818–821.

753. Li, X., and Chang, J. (2008) Preparation of bone-like apatite-collagen nanocomposites by a biomimetic process with phosphorylated collagen, *J. Biomed. Mater. Res. A*, **85A**, 293–300.

754. Sun, T., and Wang, M. (2010) Electrochemical deposition of apatite/ collagen composite coating on NiTi shape memory alloy and coating properties, *Mater. Res. Soc. Symp. Proc.*, **1239**, 141–146.

755. Zhao, H., Huang, C., Jin, H., and Cai, J. (2010) A novel route for collagen/hydroxyapatite preparation by enzymatic decomposition of urea, *J. Composite Mater.*, **44**, 2127–2133.

756. Ficai, A., Andronescu, E., Voicu, G., Ghitulica, C., Vasile, B. S., Ficai, D., and Trandafir, V. (2010) Self-assembled collagen/hydroxyapatite composite materials, *Chem. Eng. J.*, **160**, 794–800.

757. Jee, S. S., Thula, T. T., and Gower, L. B. (2010) Development of bone-like composites via the polymer-induced liquid-precursor (PILP) process, Part 1: Influence of polymer molecular weight, *Acta Biomater.*, **6**, 3676–3686.

758. Andronescu, E., Ficai, M., Voicu, G., Ficai, D., Maganu, M., and Ficai, A. (2010) Synthesis and characterization of collagen/hydroxyapatite: magnetite composite material for bone cancer treatment, *J. Mater. Sci. Mater. Med.*, **21**, 2237–2242.

759. Ficai, M., Andronescu, E., Ficai, D., Voicu, G., and Ficai, A. (2010) Synthesis and characterization of COLL-PVA/HA hybrid materials with stratified morphology, *Colloids Surf. B*, **81**, 614–619.

760. Tamimi, F., Kumarasami, B., Doillon, C., Gbureck, U., Nihouannen, D. L., Cabarcos, E. L., and Barralet, J. E. (2008) Brushite-collagen composites for bone regeneration, *Acta Biomater.*, **4**, 1315–1321.

761. Mai, R., Reinstorf, A., Pilling, E., Hlawitschka, M., Jung, R., Gelinsky, M., Schneider, M., Loukota, R., Pompe, W., Eckelt, U., and Stadlinger, B. (2008) Histologic study of incorporation and resorption of a bone cement-collagen composite: an *in vivo* study in the minipig, *Oral Surg. Oral Med. Oral Pathol. Oral Radiol. Endod.*, **105**, e9–e14.

762. Moreau, J. L., Weir, M. D., and Xu, H. H. K. (2009) Self-setting collagen-calcium phosphate bone cement: mechanical and cellular properties, *J. Biomed. Mater. Res. A*, **91A**, 605–613.

763. Liu, X., Wang, X. M., Chen, Z., Cui, F. Z., Liu, H. Y., Mao, K., and Wang, Y. (2010) Injectable bone cement based on mineralized collagen, *J. Biomed. Mater. Res. B (Appl. Biomater.)*, **94**, 72–79.

764. Otsuka, M., Nakagawa, H., Ito, A., and Higuchi, W. I. (2010) Effect of geometrical structure on drug release rate of a three-dimensionally perforated porous apatite/collagen composite cement, *J. Pharmaceutical Sci.*, **99**, 286–292.

765. Young, S. W., Andrews, W. A., Muller, H., and Constantz, B. (1991) Induction of fracture healing using fibrous calcium phosphate composite spherulites, *Investigative Radiol.*, **2**, 470–473.

766. Rovira, A., Amedee, J., Bareille, R., and Radaud, M. (1996) Colonization of a calcium phosphate/elastin-solubilized peptide – collagen composite material by human osteoblasts, *Biomaterials*, **17**, 1535–1540.

767. Kazim, M., Katowitz, J. A., Fallon, M., and Piest, K. L. (1992) Evaluation of a collagen/hydroxylapatite implant for cortical bone orbital reconstructive surgery, *Ophthalmic Plastic and Reconstructive Surg.*, **8**, 94–108.

768. Hirota, K., Nishihara, K., and Tanaka, H. (1993) Pressure sintering of apatite-collagen composite, *Biomed. Mater. Eng.*, **3**, 147–151.

769. Zahn, D., Hochrein, O., Kawska, A., Brickmann, J., and Kniep, R. (2007) Towards an atomistic understanding of apatite-collagen biomaterials: linking molecular simulation studies of complex-, crystal- and composite-formation to experimental findings, *J. Mater. Sci.*, **42**, 8966–8973.

770. Silva, C. C., Pinheiro, A. G., Figueiro, S. D., Goes, J. C., Sasaki, J. M., Miranda, M. A. R., and Sombra, A. S. B. (2002) Piezoelectric properties of collagen-nanocrystalline hydroxyapatite composites, *J. Mater. Sci.*, **37**, 2061–2070.

771. Yunoki, S., Ikoma, T., Tsuchiya, A., Monkawa, A., Ohta, K., Sotome, S., Shinomiya, K., and Tanaka, J. (2007) Fabrication and mechanical and tissue ingrowth properties of unidirectionally porous hydroxyapatite/collagen composite, *J. Biomed. Mater. Res. B (Appl. Biomater.)*, **80B**, 166–173.

772. Keeney, M., Collin, E., and Pandit, A. (2009) Multi-channelled collagen-calcium phosphate scaffolds: their physical properties and human cell response, *Tiss. Eng. C*, **15**, 265–273.

773. Chapman, M. W., Bucholz, R., and Cornell, C. (1997) Treatment of acute fractures with a collagen-calcium phosphate graft material: a randomized clinical trial, *J. Bone Joint Surg. (Am.)*, **79A**, 495–502.

774. Rodrigues, C. V. M., Serricella, P., Linhares, A. B. R., Guerdes, R. M., Borojevic, R., Rossi, M. A., Duarte, M. E. L., and Farina, M. (2003)

Characterization of a bovine collagen-hydroxyapatite composite scaffold for bone tissue engineering, *Biomaterials*, **24**, 4987–4997.

775. Lickorish, D., Ramshaw, J. A. M., Werkmeister, J. A., Glattauer, V., Howlett, C. R. (2004) Development of a collagen-hydroxyapatite composite biomaterial via biomimetic process, *J. Biomed. Mater. Res. A*, **68A**, 19–27.

776. Sionkowska, A., and Kozłowska, J. (2010) Characterization of collagen/hydroxyapatite composite sponges as a potential bone substitute, *Int. J. Biol. Macromol.*, **47**, 483–487.

777. Hsu, F. Y., Chueh, S. C., and Wang, J. Y. (1999) Microspheres of hydroxyapatite/reconstituted collagen as supports for osteoblast cell growth, *Biomaterials*, **20**, 1931–1936.

778. Wu, T. J., Huang, H. H., Lan, C. W., Lin, C. H., Hsu, F. Y., and Wang, Y. J. (2004) Studies on the microspheres comprised of reconstituted collagen and hydroxyapatite, *Biomaterials*, **25**, 651–658.

779. Liao, S. S., Watari, F., Uo, M., Ohkawa, S., Tamura, K., Wang, W., and Cui, F. Z. (2005) The preparation and characteristics of a carbonated hydroxyapatite/collagen composite at room temperature, *J. Biomed. Mater. Res. B (Appl. Biomater.)*, **74B**, 817–821.

780. Yokoyama, A., Gelinsky, M., Kawasaki, T., Kohgo, T., König, U., Pompe, W., and Watari, F. (2005) Biomimetic porous scaffolds with high elasticity made from mineralized collagen – an animal study, *J. Biomed. Mater. Res. B (Appl. Biomater.)*, **75B**, 464–472.

781. Zou, C., Weng, W., Deng, X. J., Cheng, K., Liu, X., Du, P., Shen, G., and Han, G. (2005) Preparation and characterization of porous β-tricalcium phosphate/collagen composites with an integrated structure, *Biomaterials*, **26**, 5276–5284.

782. Martins, V. C. A., and Goissis, G. (2000) Nonstoichiometric hydroxyapatite-anionic collagen composite as a support for the double sustained release of gentamicin and norfloxacin/ciprofloxacin, *Artif. Organs*, **24**, 224–230.

783. Gotterbarm, T., Richter, W., Jung, M., Berardi-Vilei, S., Mainil-Varlet, P., Yamashita, T., and Breusch, S. J. (2006) An *in vivo* study of a growth-factor enhanced, cell free, two-layered collagen-tricalcium phosphate in deep osteochondral defects, *Biomaterials*, **27**, 3387–3395.

784. Martins, V. C., Goissis, G., Ribeiro, A. C., Marcantonio, Jr. E., and Bet, M. R. (1998) The controlled release of antibiotic by hydroxyapatite: anionic collagen composites, *Artif. Organs*, **22**, 215–221.

785. Jayaraman, M., and Subramanian, M. V. (2002) Preparation and characterization of two new composites: collagen-brushite and

collagen-octacalcium phosphate, *Medical Sci. Monitor*, **8**, BR481–BR487.

786. Matsui, K., Matsui, A., Handa, T., Kawai, T., Suzuki, O., Kamakura, S., and Echigo, S. (2010) *Bone regeneration by octacalcium phosphate collagen composites in a dog alveolar cleft model*, **39**, 1218–1225.

787. Masuda, T., Kawai, T., Anada, T., Kamakura, S., and Suzuki, O. (2010) Quality of regenerated bone enhanced by implantation of octacalcium phosphate-collagen composite, *Tiss. Eng. C*, **16**, 471–478.

788. Iibuchi, S., Matsui, K., Kawai, T., Sasaki, K., Suzuki, O., Kamakura, S., and Echigo, S. (2010) Octacalcium phosphate (OCP) collagen composites enhance bone healing in a dog tooth extraction socket model, *Int. J. Oral Maxillofac. Surg.*, **39**, 161–168.

789. Ikeda, H., Yamaza, T., Yoshinari, M., Ohsaki, Y., Ayukawa, Y., Kido, M. A., Inoue, T., Shimono, M., Koyano, K., and Tanaka, T. (2000) Ultrastructural and immunoelectron microscopic studies of the peri-implant epithelium-implant (Ti-6Al-4V) interface of rat maxilla, *J. Periodontol.*, **71**, 961–973.

790. Uchida, M., Oyane, A., Kim, H. M., Kokubo, T., and Ito, A. (2004) Biomimetic coating of laminin-apatite composite on titanium metal and its excellent cell-adhesive properties, *Adv. Mater.*, **16**, 1071–1074.

791. Oyane, A., Uchida, M., and Ito, A. (2005) Laminin-apatite composite coating to enhance cell adhesion to ethylene-vinyl alcohol copolymer, *J. Biomed. Mater. Res. A*, **72A**, 168–174.

792. Oyane, A., Uchida, M., Onuma, K., and Ito, A. (2006) Spontaneous growth of a laminin-apatite nano-composite in a metastable calcium phosphate solution, *Biomaterials*, **27**, 167–175.

793. Oyane, A., Tsurushima, H., and Ito, A. (2010) Highly efficient gene transfer system using a laminin-DNA-apatite composite layer, *J. Gene Med.*, **12**, 194–206.

794. Liu, W. B., Qu, S. X., Shen, R., Jiang, C. X., Li, X. H., Feng, B., and Weng, J. (2006) Influence of pH values on preparation of hydroxyapatite/gelatin composites, *J. Mater. Sci.*, **41**, 1851–1853.

795. Yaylaoglu, M. B., Korkusuz, P., Ors, U., Korkusuz, F., and Hasirci, V. (1999) Development of a calcium phosphate-gelatin composite as a bone substitute and its use in drug release, *Biomaterials*, **20**, 711–719.

796. Kim, H. W., Song, J. H., and Kim, H. E. (2005) Nanofiber generation of gelatin-hydroxyapatite biomimetics for guided tissue regeneration, *Adv. Funct. Mater.*, **15**, 1988–1994.

797. Sivakumar, M., and Rao, K. P. (2002) Preparation, characterization and *in vitro* release of gentamicin from coralline hydroxyapatite-gelatin composite microspheres, *Biomaterials*, **23**, 3175–3181.

798. Kim, H. W., Knowles, J. C., and Kim, H. E. (2005) Porous scaffolds of gelatin-hydroxyapatite nanocomposites obtained by biomimetic approach: characterization and antibiotic drug release, *J. Biomed. Mater. Res. B (Appl. Biomater.)*, **74B**, 686–698.

799. Yin, Y. J., Zhao, F., Song, X. F., Yao, K. D., Lu, W. W., and Leong, J. C. (2000) Preparation and characterization of hydroxyapatite/chitosan-gelatin network composite, *J. Appl. Polym. Sci.*, **77**, 2929–2938.

800. Kim, H. W., Yoon, B. H., and Kim, H. E. (2005) Microsphere of apatite-gelatin nanocomposite as bone regenerative filler, *J. Mater. Sci. Mater. Med.*, **16**, 1105–1109.

801. Hillig, W. B., Choi, Y., Murtha, S., Natravali, N., and Ajayan, P. (2008) An open-pored gelatin/hydroxyapatite composite as a potential bone substitute, *J. Mater. Sci. Mater. Med.*, **19**, 11–17.

802. Chang, M. C., Douglas, W. H., and Tanaka, J. (2006) Organic-inorganic interaction and the growth mechanism of hydroxyapatite crystals in gelatin matrices between 37 and 80 °C, *J. Mater. Sci. Mater. Med.*, **17**, 387–396.

803. Chang, M. C., and Douglas, W. H. (2007) Cross-linkage of hydroxyapatite/gelatin nanocomposite using imide-based zero-length cross-linker, *J. Mater. Sci. Mater. Med.*, **18**, 2045–2051.

804. Teng, S., Chen, L., Guo, Y., and Shi, J. (2007) Formation of nano-hydroxyapatite in gelatin droplets and the resulting porous composite microspheres, *J. Inorg. Biochem.*, **101**, 686–691.

805. Liu, X., Smith, L. A., Hu, J., and Ma, P. X. (2009) Biomimetic nanofibrous gelatin/apatite composite scaffolds for bone tissue engineering, *Biomaterials*, **30**, 2252–2258.

806. Zhou, J., Tang, J., Quan-Li, L., Zhang, J. X., and Dou, X. C. (2010) Rabbit tibial defects repaired by nano-gelatin-apatite-minocycline bionic composite, *J. Clin. Rehabil. Tiss. Eng. Res.*, **14**, 4563–4567.

807. Zhao, F., Grayson, W. L., Ma, T., Bunnell, B., and Lu., W. W. (2006) Effects of hydroxyapatite in 3-D chitosan-gelatin polymer network on human mesenchymal stem cell construct development, *Biomaterials*, **27**, 1859–1867.

808. Lin, H. R., and Yeh, Y. J. (2004) Porous alginate/hydroxyapatite composite scaffolds for bone tissue engineering: preparation., characterization and *in vitro* studies, *J. Biomed. Mater. Res. B (Appl. Biomater.)*, **71B**, 52–65.

809. Turco, G., Marsich, E., Bellomo, F., Semeraro, S., Donati, I., Brun, F., Grandolfo, M., Accardo, A., and Paoletti, S. (2009) Alginate/hydroxyapatite biocomposite for bone ingrowth: a trabecular structure with high and isotropic connectivity, *Biomacromolecules*, **10**, 1575–1583.

810. Gelinsky, M., Eckert, M., and Despang, F. (2007) Biphasic, but mololithic scaffolds for the therapy of osteochondral defects, *Int. J. Mater. Res. (formerly Z. Metallkd.)*, **98**, 749–755.

811. Yamaguchi, I., Tokuchi, K., Fukuzaki, H., Koyama, Y., Takakuda, K., Monma, H., and Tanaka, J. (2001) Preparation and microstructure analysis of chitosan/hydroxyapatite nanocomposites, *J. Biomed. Mater. Res.*, **55**, 20–27.

812. Zhao, F., Yin, Y. J., Lu, W. W., Leong, J. C., Zhang, W. J., Zhang, J. Y., Zhang, M. F., and Yao, K. D. (2002) Preparation and histological evaluation of biomimetic three-dimensional hydroxyapatite/chitosan-gelatin network composite scaffolds, *Biomaterials*, **23**, 3227–3234.

813. Shen, X., Tong, H., Jiang, T., Zhu, Z., Wan, P., and Hu, J. (2007) Homogeneous chitosan/carbonate apatite/citric acid nanocomposites prepared through a novel in situ precipitation method, *Compos. Sci. Technol.*, **67**, 2238–2245.

814. Murugan, R., and Ramakrishna, S. (2004) Bioresorbable composite bone paste using polysaccharide based nano-hydroxyapatite, *Biomaterials*, **25**, 3829–3835.

815. Yoshida, A., Miyazaki, T., Ishida, E., and Ashizuka, M. (2004) Preparation of bioactive chitosan-hydroxyapatite nanocomposites for bone repair through mechanochemical reaction, *Mater. Trans.*, **45**, 994–998.

816. Zhang, Y., Ni, M., Zhang, M. Q., and Ratner, B. (2003) Calcium phosphate – chitosan composite scaffolds for bone tissue engineering, *Tissue Eng.*, **9**, 337–345.

817. Zhang, Y., and Zhang, M. Q. (2001) Microstructural and mechanical characterization of chitosan scaffolds reinforced by calcium phosphates, *J. Non-Cryst. Solids*, **282**, 159–164.

818. Zhang, Y., and Zhang, M. Q. (2002) Calcium phosphate/chitosan composite scaffolds for controlled *in vitro* antibiotic drug release, *J. Biomed. Mater. Res.*, **62**, 378–386.

819. Tachaboonyakiat, W., Ogomi, D., Serizawa, T., and Akashi, M. (2006) Evaluation of cell adhesion and proliferation on a novel tissue engineering scaffold containing chitosan and hydroxyapatite, *J. Bioactive and Compatible Polymers*, **21**, 579–589.

820. Sreedhar, B., Aparna, Y., Sairam, M., and Hebalkar, N. (2007) Preparation and characterization of HAP/carboxymethyl chitosan nanocomposites, *J. Appl. Polym. Sci.*, **105**, 928–934.

821. Rusu, V. M., Ng, C. H., Wilke, M., Tiersch, B., Fratzl, P., and Peter, M. G. (2005) Size-controlled hydroxyapatite nanoparticles as self-organized organic-inorganic composite materials, *Biomaterials*, **26**, 5414–5426.

822. Pinheiro, A. G., Pereira, F. F. M., Santos, M. R. P., Freire, F. N. A., Góes, J. C., and Sombra, A. S. B. (2007) Chitosan-hydroxyapatite-BIT composite films: preparation and characterization, *Polym. Compos.*, **28**, 582–587.

823. Tang, X. J., Gui, L., and Lü, X. Y. (2008) Hard tissue compatibility of natural hydroxyapatite/chitosan composite, *Biomed. Mater.*, **3**, 044115.

824. Zhang, Y., Venugopal, J. R., El-Turki, A., Ramakrishna, S., Su, B., and Lim, C. T. (2008) Electrospun biomimetic nanocomposite nanofibers of hydroxyapatite/chitosan for bone tissue engineering, *Biomaterials*, **29**, 4314–4322.

825. Madhumathi, K., Shalumon, K. T., Rani, V. V. D., Tamura, H., Furuike, T., Selvamurugan, N., Nair, S. V., and Jayakumar, R. (2009) Wet chemical synthesis of chitosan hydrogel-hydroxyapatite composite membranes for tissue engineering applications, *Int. J. Biological Macromolecules*, **45**, 12–15.

826. Cai, X., Tong, H., Shen, X., Chen, W., Yan, J., and Hu, J. (2010) Preparation and characterization of homogeneous chitosan-polylactic acid/hydroxyapatite nanocomposite for bone tissue engineering and evaluation of its mechanical properties, *Acta Biomater.*, **5**, 2693–2703.

827. Ge, H., Zhao, B., Lai, Y., Hu, X., Zhang, D., and Hu, K. (2010) From crabshell to chitosan-hydroxyapatite composite material via a biomorphic mineralization synthesis method, *J. Mater. Sci. Mater. Med.*, **21**, 1781–1787.

828. Swetha, M., Sahithi, K., Moorthi, A., Srinivasan, N., Ramasamy, K., and Selvamurugan, N. (2010) Biocomposites containing natural polymers and hydroxyapatite for bone tissue engineering, *Int. J. Biol. Macromol.*, **47**, 1–4.

829. Wan, A. C. A., Khor, E., and Hastings, G. W. (1998) Preparation of a chitin-apatite composite by *in situ* precipitation onto porous chitin scaffolds, *J. Biomed. Mater. Res.*, **41**, 541–548.

830. Wan, A. C. A., Khor, E., and Hastings, G. W. (1997) Hydroxyapatite modified chitin as potential hard tissue substitute material, *J. Biomed. Mater. Res.*, **38**, 235–241.

831. Geçer, A., Yldz, N., Erol, M., and Çalml, A. (2008) Synthesis of chitin calcium phosphate composite in different growth media, *Polymer Composites*, **29**, 84–91.

832. Dong, H., Ye, J. D., Wang, X. P., and Yang, J. J. (2007) Preparation of calcium phosphate cement tissue engineering scaffold reinforced with chitin fiber, *J. Inorg. Mater.*, **22**, 1007–1010.

833. Wang, X., Ma, J., Feng, Q., and Cui, F. Z. (2002) Skeletal repair in rabbits with calcium phosphate cements incorporated phosphorylated chitin, *Biomaterials*, **23**, 4591–4600.

834. Wang, J., Sun, Q. Z., Gao, J., Liu, D. M., Meng, X. C., and Li, M. Q. (2010) Preparation and properties on silk fibers reinforced hydroxyapatite/chitosan composites, *Adv. Mater. Res.*, pp. 105–106, 557–560.

835. Zhang, Y., Reddy, V. J., Wong, S. Y., Li, X., Su, B., Ramakrishna, S., and Lim, C. T. (2010) Enhanced biomineralization in osteoblasts on a novel electrospun biocomposite nanofibrous substrate of hydroxyapatite/collagen/chitosan, *Tiss. Eng. A*, **16**, 1949–1960.

836. Kousalya, G. N., Gandhi, R. M., Sundaram, S. C., and Meenakshi, S. (2010) Synthesis of nano-hydroxyapatite chitin/chitosan hybrid biocomposites for the removal of Fe(III), *Carbohydrate Polymers*, **82**, 594–599.

837. Sundaram, C. S., Viswanathan, N., and Meenakshi, S. (2008) Uptake of fluoride by nano-hydroxyapatite/chitosan, a bioinorganic composite, *Bioresource Technol.*, **99**, 8226–8230.

838. Sundaram, C. S., Viswanathan, N., and Meenakshi, S. (2009) Fluoride sorption by nano-hydroxyapatite/chitin composite, *J. Hazardous Mater.*, **172**, 147–151.

839. Wen, H. B., de Wijn, J. R., van Blitterswijk, C. A., and de Groot, K. (1999) Incorporation of bovine serum albumin in calcium phosphate coating on titanium, *J. Biomed. Mater. Res.*, **46**, 245–252.

840. Liu, T. Y., Chen, S. Y., Liu, D. M., and Liou, S. C. (2005) On the study of BSA-loaded calcium-deficient hydroxyapatite nano-carriers for controlled drug delivery, *J. Controlled Release*, **107**, 112–121.

841. Liu, Y., Hunziker, E., Randall, N., de Groot, K., and Layrolle, P. (2003) Proteins incorporated into biomimetically prepared calcium phosphate coatings modulate their mechanical strength and dissolution rate, *Biomaterials*, **24**, 65–70.

842. Dorozhkin, S. V., and Dorozhkina, E. I. (2003) The influence of bovine serum albumin on the crystallization of calcium phosphates from a revised simulated body fluid, *Colloids Surf. A*, **215**, 191–199.

843. Fu, H. H., Hu, Y. H., McNelis, T., and Hollinger, J. O. (2005) A calcium phosphate-based gene delivery system, *J. Biomed. Mater. Res. A*, **74A**, 40–48.

844. Bisht, S., Bhakta, G., Mitra, S., and Maitra, A. (2005) pDNA loaded calcium phosphate nanoparticles: highly efficient non-viral vector for gene delivery, *Int. J. Pharm.*, **288**, 157–168.

845. Kakizawa, Y., Miyata, K., Furukawa, S., and Kataoka, K. (2004) Size-controlled formation of a calcium phosphate-based organic-inorganic hybrid vector for gene delivery using poly(ethylene glycol)-block-poly(aspartic acid), *Adv. Mater.*, **16**, 699–702.

846. Singh, R., Saxena, A., and Mozumdar, S. (2008) Calcium phosphate – DNA nanocomposites: morphological studies and their bile duct infusion for liver-directed gene therapy, *Int. J. Appl. Ceram. Technol.*, **5**, 1–10.

847. Sporysh, I., Shynkaruk, E., Lysko, O., Shynkaruk, A., Dubok, V., Buzaneva, E., Ritter, U., and Scharff, P. (2010) Biomimetic hydroxyapatite nanocrystals in composites with C_{60} and Au-DNA nanoparticles: IR-spectral study, *Mater. Sci. Eng. B*, **169**, 128–133.

848. Taguchi, T., Kishida, A., and Akashi, M. (1998) Hydroxyapatite formation on/in poly(vinyl alcohol) hydrogel matrices using a novel alternate soaking process, *Chem. Lett.*, **8**, 711–712.

849. Tachaboonyakiat, W., Serizawa, T., and Akashi, M. (2001) Hydroxyapatite formation on/in iodegradable chitosan hydrogels by an alternate soaking process, *Polym. J.*, **33**, 177–181.

850. Schnepp, Z. A. C., Gonzalez-McQuire, R., and Mann, S. (2006) Hybrid biocomposites based on calcium phosphate mineralization of self-assembled supramolecular hydrogels, *Adv. Mater.*, **18**, 1869–1872.

851. Patel, M., Patel, K. J., Caccamese, J. F., Coletti, D. P., Sauk, J. J., and Fisher, J. P. (2010) Characterization of cyclic acetal hydroxyapatite nanocomposites for craniofacial tissue engineering, *J. Biomed. Mater. Res. A*, **94A**, 408–418.

852. Bigi, A., Boanini, E., Gazzano, M., Kojdecki, M. A., and Rubini, K. (2004) Microstructural investigation of hydroxyapatite-polyelectrolyte composites, *J. Mater. Chem.*, **14**, 274–279.

853. Bigi, A., Boanini, E., Gazzano, M., Rubini, K., and Torricelli, P. (2004) Nanocrystalline hydroxyapatite-polyaspartate composites, *Biomed. Mater. Eng.*, **14**, 573–579.

854. Boanini, E., Fini, M., Gazzano, M., and Bigi, A. (2006) Hydroxyapatite nanocrystals modified with acidic amino acids, *Eur. J. Inorg. Chem.*, pp. 4821–4826.

855. Boanini, E., Torricelli, P., Gazzano, M., Giardino, R., and Bigi, A. (2006) Nanocomposites of hydroxyapatite with aspartic acid and glutamic acid and their interaction with osteoblast-like cells, *Biomaterials*, **27**, 4428–4433.

856. Ikawa, N., Kimura, T., Oumi, Y., and Sano, T. (2009) Amino acid containing amorphous calcium phosphates and the rapid transformation into apatite, *J. Mater. Chem.*, **19**, 4906–4913.

857. Sánchez-Salcedo, S., Nieto, A., and Vallet-Regi, M. (2005) Hydroxyapatite/β-tricalcium phosphate/agarose macroporous scaffolds for bone tissue engineering, *Chem. Eng. J.*, **137**, 62–71.

858. Román, J., Cabañas, M. V., Peña, J., Doadrio, J. C., and Vallet-Regi, M. (2008) An optimized β-tricalcium phosphate and agarose scaffold fabrication technique, *J. Biomed. Mater. Res. A*, **84A**, 99–107.

859. Alcaide, M., Serrano, M. C., Pagani, R., Sánchez-Salcedo, S., Nieto, A., Vallet-Regí, M., and Portolés, M. T. (2009) L929 fibroblast and SAOS-2 osteoblast response to hydroxyapatite-βTCP/ agarose biomaterial, *J. Biomed. Mater. Res. A*, **89A**, 539–549.

860. Abiraman, S., Varma, H., Umashankar, P., and John, A. (2002) Fibrin sealant as an osteoinductive protein in a mouse model, *Biomaterials*, **23**, 3023–3031.

861. d'Arc, M. B., and Daculsi, G. (2003) Micro macroporous biphasic ceramics and fibrin sealant as a mouldable material for bone reconstruction in chronic otitis media surgery: a 15 years experience, *J. Mater. Sci. Mater. Med.*, **14**, 229–233.

862. Bonucci, E., Marini, E., Valdinucci, F., and Fortunato, G. (1997) Osteogenic response to hydroxyapatite-fibrin implants in maxillofacial bone defects, *Eur. J. Oral Sci.*, **105**, 557–561.

863. Fortunato, G., Marini, E., Valdinucci, F., and Bonucci, E. (1997) Long-term results of hydroxyapatite-fibrin sealant implantation in plastic and reconstructive craniofacial surgery, *J. Cranio Maxillofac. Surg.*, **25**, 124–135.

864. Jegoux, F., Goyenvalle, E., d'Arc, M. B., Aguado, E., and Daculsi, G. (2005) *In vivo* biological performance of composites combining micro-macroporous biphasic calcium phosphate granules and fibrin sealant, *Arch. Orthop. Trauma Surg.*, **125**, 153–159.

865. le Guehennec, L., Goyenvalle, E., Aguado, E., Pilet, P., d'Arc, M. B., and Daculsi, G. (2005) MBCP® biphasic calcium phosphate granules and

Tissucol® fibrin sealant in rabbit femoral defects: the effect of fibrin on bone ingrowth, *J. Mater. Sci. Mater. Med.*, **16**, 29–35.

866. Wittkampf, A. (1989) Fibrin sealant as sealant for hydroxyapatite granules, *J. Cranio Maxillofac. Surg.*, **17**, 179–181.

867. le Nihouannen, D., Guehennec, L. L., Rouillon, T., Pilet, P., Bilban, M., Layrolle, P., and Daculsi, G. (2006) Micro-architecture of calcium phosphate granules and fibrin glue composites for bone tissue engineering, *Biomaterials*, **27**, 2716–2722.

868. Le Nihouannen D., Saffarzadeh A., Aguado E., Goyenvalle E., Gauthier O., Moreau F., Pilet P., Spaethe R., Daculsi, G., and Layrolle, P. (2007) Osteogenic properties of calcium phosphate ceramics and fibrin glue based composites, *J. Mater. Sci. Mater. Med.*, **18**, 225–235.

869. le Guehennec, L., Goyenvalle, E., Aguado, E., Pilet, P., Spaethe, R., and Daculsi, G. (2007) Influence of calcium chloride and aprotinin in the *in vivo* biological performance of a composite combining biphasic calcium phosphate granules and fibrin sealant, *J. Mater. Sci. Mater. Med.*, **18**, 1489–1495.

870. le Nihouannen, D., Goyenvalle, E., Aguado, E., Pilet, P., Bilban, M., Daculsi, G., and Layrolle, P. (2007) Hybrid composites of calcium phosphate granules, fibrin glue, and bone marrow for skeletal repair, *J. Biomed. Mater. Res. A*, **81A**, 399–408.

871. le Nihouannen, D., Saffarzadeh, A., Gauthier, O., Moreau, F., Pilet, P., Spaethe, R., Layrolle, P., and Daculsi, G. (2008) Bone tissue formation in sheep muscles induced by a biphasic calcium phosphate ceramic and fibrin glue composite, *J. Mater. Sci. Mater. Med.*, **19**, 667–675.

872. Yoh, R., Matsumoto, T., Sasaki, J. I., and Sohmura, T. (2008) Biomimetic fabrication of fibrin/apatite composite material, *J. Biomed. Mater. Res. A*, **87A**, 222–228.

873. Cui, G., Li, J., Lei, W., Bi, L., Tang, P., Liang, Y., Tao, S., and Wang, Y. (2010) The mechanical and biological properties of an injectable calcium phosphate cement-fibrin glue composite for bone regeneration, *J. Biomed. Mater. Res. B (Appl. Biomater.)*, **92**, 377–385.

874. Boanini, E., Torricelli, P., Gazzano, M., Giardino, R., and Bigi, A. (2008) Alendronate-hydroxyapatite nanocomposites and their interaction with osteoclasts and osteoblast-like cells, *Biomaterials*, **29**, 790–796.

875. Li, L., Wei, K. M., Lin, F., Kong, X. D., and Yao, J. M. (2008) Effect of silicon on the formation of silk fibroin/calcium phosphate composite, *J. Mater. Sci. Mater. Med.*, **19**, 577–582.

876. Wang, L., Nemoto, R., and Senna, M. (2004) Effects of alkali pretreatment of silk fibroin on microstructure and properties of hydroxyapatite-silk fibroin nanocomposite, *J. Mater. Sci. Mater. Med.*, **15**, 261–265.

877. Wang, L., Nemoto, R., and Senna, M. (2002) Microstructure and chemical states of hydroxyapatite/silk fibroin nanocomposites synthesized via a wet-mechanochemical route, *J. Nanoparticle Res.*, **4**, 535–540.

878. Wang, L., Nemoto, R., and Senna, M. (2004) Three-dimensional porous network structure developed in hydroxyapatite-based nanocomposites containing enzyme pretreated silk fibroin, *J. Nanoparticle Res.*, **6**, 91–98.

879. Nemoto, R., Wang, L., Ikoma, T., Tanaka, J., and Senna, M. (2004) Preferential alignment of hydroxyapatite crystallites in nanocomposites with chemically disintegrated silk fibroin, *J. Nanoparticle Res.*, **6**, 259–265.

880. Wang, L., Li, C. Z., and Senna, M. (2007) High-affinity integration of hydroxyapatite nanoparticles with chemically modified silk fibroin, *J. Nanoparticle Res.*, **9**, 919–929.

881. Fan, C., Li, J., Xu, G., He, H., Ye, X., Chen, Y., Sheng, X., Fu, J., and He, D. (2010) Facile fabrication of nano-hydroxyapatite/silk fibroin composite via a simplified coprecipitation route, *J. Mater. Sci.*, **45**, 5814–5819.

882. Wang, L., and Li, C. Z. (2007) Preparation and physicochemical properties of a novel hydroxyapatite/chitosan-silk fibroin composite, *Carbohydr. Polym.*, **68**, 740–745.

883. Sogo, Y., Ito, A., Matsuno, T., Oyane, A., Tamazawa, G., Satoh, T., Yamazaki, A., Uchimura, E., and Ohno, T. (2007) Fibronectin-calcium phosphate composite layer on hydroxyapatite to enhance adhesion., cell spread and osteogenic differentiation of human mesenchymal stem cells *in vitro*, *Biomed. Mater.*, **2**, 116–123.

884. Rhee, S. H., Suetsugu, Y., and Tanaka, J. (2001) Biomimetic configurational arrays of hydroxyapatite nanocrystals on bio-organics, *Biomaterials*, **22**, 2843–2847.

885. Cross, K. J., Huq, N. L., Palamara, J. E., Perich, J. W., and Reynolds, E. C. (2005) Physicochemical characterization of casein phosphopeptide-amorphous calcium phosphate nanocomplexes, *J. Biol. Chem.*, **280**, 15362–15369.

886. Jung, J. Y., Hong, Y. J., Choi, Y. S., Jeong, S., and Lee, W. K. (2010) A new method for the preparation of bioactive calcium phosphate films

hybridized with 1α,25-dihydroxyvitamin D_3, *J. Mater. Sci. Mater. Med.*, **20**, 2441–2453.

887. Shchukin, D. G., Sukhorukov, G. B., and Möhwald, H. (2003) Biomimetic fabrication of nanoengineered hydroxyapatite/polyelectrolyte composite shell, *Chem. Mater.*, **15**, 3947–3950.

888. Jacoveila, P. F. (2008) Use of calcium hydroxylapatite (Radiesse®) for facial augmentation, *Clinical Interventions in Aging*, **3**, 161–174.

889. Lizzul, P. F., and Narurkar, V. A. (2010) The role of calcium hydroxylapatite (Radiesse®) in nonsurgical aesthetic rejuvenation, *J. of Drugs in Dermatology*, **9**, 446–450.

890. Klesing, J., Chernousova, S., Kovtun, A., Neumann, S., Ruiz, L., Gonzalez-Calbet, J. M., Vallet-Regi, M., Heumann, R., and Epple, M. (2010) An injectable paste of calcium phosphate nanorods, functionalized with nucleic acids, for cell transfection and gene silencing, *J. Mater. Chem.*, **20**, 6144–6148.

891. Thai, V. V., and Lee, B. T. (2010) Fabrication of calcium phosphate-calcium sulfate injectable bone substitute using hydroxy-propyl-methyl-cellulose and citric acid, *J. Mater. Sci. Mater. Med.*, **21**, 1867–1874.

892. Low, K. L., Tan, S. H., Zein, S. H. S., Roether, J. A., Mouriño, V., and Boccaccini, A. R. (2010) Calcium phosphate-based composites as injectable bone substitute materials, *J. Biomed. Mater. Res. B (Appl. Biomater.)*, **94B**, 273–286.

893. Weiss, P., Gauthier, O., Bouler, J. M., Grimandi, G., and Daculsi, G. (1999) Injectable bone substitute using a hydrophilic polymer, *Bone*, **25(2)**, 67S–70S.

894. Daculsi, G., Weiss, P., Bouler, J. M., Gauthier, O., Millot, F., and Aguado, E. (1999) Biphasic calcium phosphate/hydrosoluble polymer composites: a new concept for bone and dental substitution biomaterials, *Bone*, **25(2)**, 59S–61S.

895. Turczyn, R., Weiss, P., Lapkowski, M., and Daculsi, G. (2000) *In situ* self-hardening bioactive composite for bone and dental surgery, *J. Biomater. Sci. Polymer Edn.*, **11**, 217–223.

896. Bennett, S., Connolly, K., Lee, D. R., Jiang, Y., Buck, D., Hollinger, J. O., and Gruskin, E. A. (1996) Initial biocompatibility studies of a novel degradable polymeric bone substitute that hardens *in situ*, *Bone*, **19**, 101S–107S.

897. Liu, X., Okada, M., Maeda, H., Fujii, S., and Furuzono, T. (2011) Hydroxyapatite/biodegradable poly(L-lactide-*co*-ε-caprolactone)

composite microparticles as injectable scaffolds by a Pickering emulsion route, *Acta Biomater.*, **7**, 821–828.

898. Daculsi, G., Rohanizadeh, R., Weiss, P., and Bouler, J. M. (2000) Crystal polymer interaction with new injectable bone substitute; SEM and HrTEM study, *J. Biomed. Mater. Res.*, **50**, 1–7.

899. Grimande, G., Weiss, P., Millot, F., and Daculsi, G. (1998) *In vitro* evaluation of a new injectable calcium phosphate material, *J. Biomed. Mater. Res.*, **39**, 660–666.

900. Weiss, P., Lapkowski, M., LeGeros, R. Z., Bouler, J. M., Jean, A., and Daculsi, G. (1997) FTIR spectroscopic study of an organic/mineral composite for bone and dental substitute materials, *J. Mater. Sci. Mater. Med.*, **8**, 621–629.

901. Weiss, P., Bohic, S., Lapkowski, M., and Daculsi, G. (1998) Application of FTIR microspectroscopy to the study of an injectable composite for bone and dental surgery, *J. Biomed. Mater. Res.*, **41**, 167–170.

902. Schmitt, M., Weiss, P., Bourges, X., del Valle, G. A., and Daculsi, G. (2002) Crystallization at the polymer/calcium-phosphate interface in a sterilized injectable bone substitute IBS, *Biomaterials*, **23**, 2789–2794.

903. Gauthier, O., Müller, R., von Stechow, D., Lamy, B., Weiss, P., Bouler, J. M., Aguado, E., and Daculsi, G. (2005) *In vivo* bone regeneration with injectable calcium phosphate biomaterial: a three-dimensional micro-computed tomographic, biomechanical and SEM study, *Biomaterials*, **26**, 5444–5453.

904. Weiss, P., Layrolle, P., Clergeau, L. P., Enckel, B., Pilet, P., Amouriq, Y., Daculsi, G., and Giumelli, B. (2007) The safety and efficacy of an injectable bone substitute in dental sockets demonstrated in a human clinical trial, *Biomaterials*, **28**, 3295–3305.

905. Fatimi, A., Tassin, J. F., Axelos, M. A. V., and Weiss, P. (2010) The stability mechanisms of an injectable calcium phosphate ceramic suspension, *J. Mater. Sci. Mater. Med.*, **21**, 1799–1809.

906. Trojani, C., Boukhechba, F., Scimeca, J. C., Vandenbos, F., Michiels, J. F., Daculsi, G., Boileau, P., Weiss, P., Carle, G. F., and Rochet, N. (2006) Ectopic bone formation using an injectable biphasic calcium phosphate/Si-HPMC hydrogel composite loaded with undifferentiated bone marrow stromal cells, *Biomaterials*, **27**, 3256–3264.

907. Zhang, S. M., and Lü, G. (2009) Clinical application of compound injectable bone substitutes in bone injury repair, *J. Clin. Rehabil. Tiss. Eng. Res.*, **13**, 10117–10120.

908. Daculsi, G., Uzel, P. A., Bourgeois, N., le François, T., Rouvillain, J. L., Bourges, X., and Baroth, S. (2009) New injectable bone substitute using reversible thermosensitive hydrogel and BCP granules: *in vivo* rabbit experiments, *Key Eng. Mater.*, pp. 396–398, 457–460.

909. Iooss, P., le Ray, A. M., Grimandi, G., Daculsi, G., and Merle, C. (2001) A new injectable bone substitute combining poly(ε-caprolactone) microparticles with biphasic calcium phosphate granules, *Biomaterials*, **22**, 2785–2794.

910. Bohner, M. (2010) Design of ceramic-based cements and putties for bone graft substitution, *Eur. Cell Mater.*, **20**, 1–12.

911. Evis, Z., Ergun, C., and Doremus, R. H. (2005) Hydroxylapatite-zirconia composites: thermal stability of phases and sinterability as related to the CaO-ZrO_2 phase diagram, *J. Mater. Sci.*, **40**, 1127–1134.

912. Rao, R. R., and Kannan, T. S. (2002) Synthesis and sintering of hydroxyapatite-zirconia composites, *Mater. Sci. Eng. C*, **20**, 187–193.

913. Mansur, C., Pope, M., Pascucci, M. R., and Shivkumar, S. (1998) Zirconia-calcium phosphate composites for bone replacement, *Ceram. Int.*, **24**, 77–79.

914. Kim, H. W., Kim, H. E., Salih, V., and Knowles, J. C. (2004) Dissolution control and cellular responses of calcium phosphate coatings on zirconia porous scaffold, *J. Biomed. Mater. Res. A*, **68A**, 522–530.

915. Milella, E., Cosentino, F., Licciulli, A., and Massaro, C. (2001) Preparation and characterisation of titania/hydroxyapatite composite coatings obtained by sol-gel process, *Biomaterials*, **22**, 1425–1431.

916. Goller, G., Demirkiran, H., Oktar, F. N., and Demirkesen, E. (2003) Processing and characterization of Bioglass reinforced hydroxyapatite composites, *Ceram. Int.*, **29**, 721–724.

917. Tancred, D. C., Carr, A. J., and McCormack, B. A. (2001) The sintering and mechanical behaviour of hydroxyapatite with bioglass additions, *J. Mater. Sci. Mater. Med.*, **12**, 81–93.

918. Lopes, M. A., Silva, R. F., Monteiro, F. J., and Santos, J. D. (2000) Microstructural dependence of Young's moduli of P_2O_5 glass reinforced hydroxyapatite for biomedical applications, *Biomaterials*, **21**, 749–754.

919. Juang, H. Y., and Hon, M. H. (1994) Fabrication and mechanical properties of hydroxyapatite-alumina composites, *Mater. Sci. Eng. C*, **2**, 77–81.

920. Li, J., Forbreg, S., and Hermansson, L. (1991) Evaluation of the mechanical properties of hot isotatically pressed titania and titania-calcium phosphate composites, *Biomaterials*, **12**, 438–440.

921. Noma, T., Shoji, N., Wada, S., and Suzuki, T. (1993) Preparation of spherical Al_2O_3 particle dispersed hydroxyapatite ceramics, *J. Ceram. Soc. Jpn.*, **101**, 923–927.

922. Gautier, S., Champion, E., and Bernache-Assollant, D. (1999) Toughening characterization in alumina platelet-hydroxyapatite matrix composites, *J. Mater. Sci. Mater. Med.*, **10**, 533–540.

923. Fang, Y., Roy, D. M., Cheng, J., Roy. R., and Agrawal, D. K. (1993) Microwave sintering of hydroxyapatite-based composites, *Ceram. Trans.*, **36**, 397–407.

924. Park, K., and Vasilosa, T. (1997) Microstructure and mechanical properties of silicon carbide whisker/calcium phosphate composites produced by hot pressing, *Mater. Lett.*, **32**, 229–233.

925. Jin, H. B., Oktar, F. N., Dorozhkin, S., and Agathopoulos, S. (2011) Sintering behavior and properties of reinforced hydroxyapatite/TCP biphasic bioceramics with ZnO-whiskers, *J. Composite Mater.*, **45**, 1435–1445.

926. de With, G., and Corbijn, A. T. (1989) Metal fibre reinforced hydroxyapatite ceramics, *J. Mater. Sci.*, **24**, 3411–3415.

927. Ruys, A. J., Simpson, S. A., and Sorrell, C. C. (1994) Thixotropic casting of fibre-reinforced ceramic matrix composites, *J. Mater. Sci. Lett.*, **13**, 1323–1325.

928. Miao, X., Ruys, A. J., and Milthorpe, B. K. (2001) Hydroxyapatite-316L fibre composites prepared by vibration assisted slip casting, *J. Mater. Sci.*, **36**, 3323–3332.

929. Kim, H. M., Chae, W. P., Chang, K. W., Chun, S., Kim, S., Jeong, Y., and Kang, I. K. (2010) Composite nanofiber mats consisting of hydroxyapatite and titania for biomedical applications, *J. Biomed. Mater. Res. B (Appl. Biomater.)*, **94B**, 380–387.

930. Hirakura, S., Kobayashi, T., Ono, S., Oaki, Y., and Imai, H. (2010) Fibrous nanocrystals of hydroxyapatite loaded with TiO_2 nanoparticles for the capture and photocatalytic decomposition of specific proteins, *Colloids Surf. B*, **79**, 131–135.

931. Wu, J. M., and Yeh, T. S. (1988) Sintering of hydroxylapatite-zirconia composite materials, *J. Mater. Sci.*, **23**, 3771–3777.

932. Li, J., Liao, H., and Hermansson, L. (1996) Sintering of partially-stabilized zirconia and partially-stabilized zirconia-hydroxyapatite

composites by hot isostatic pressing and pressureless sintering, *Biomaterials*, **17**, 1787–1790.

933. Takagi, M., Mochida, M., Uchida, N., Saito, K., and Uematsu, K. (1992) Filter cake forming and hot isostatic pressing for TZP-dispersed hydroxyapatite composite, *J. Mater. Sci. Mater. Med.*, **3**, 199–203.

934. Silva, V. V., and Domingues, R. Z. (1997) Hydroxyapatite-zirconia composites prepared by precipitation method, *J. Mater. Sci. Mater. Med.*, **8**, 907–910.

935. Silva, V. V., Lameiras, F. S., and Domingues, R. Z. (2001) Synthesis and characterization of calcia partially stabilized zirconia-hydroxyapatite powders prtepared by co-precipitation method, *Ceram. Int.*, **27**, 615–620.

936. Rapacz-Kmita, A., Slosarczyk, A., Paszkiewicz, Z., and Paluch, D. (2004) Evaluation of HAp-ZrO_2 composites and monophase HAp bioceramics, *In vitro* study, *J. Mater. Sci.*, **39**, 5865–5867.

937. Rapacz-Kmita, A., Slosarczyk, A., and Paszkiewicz, Z. (2005) HAp ZrO_2 composite coatings prepared by plasma spraying for biomedical applications, *Ceram. Int.*, **31**, 567–571.

938. Sung, Y. M., and Kim, D. H. (2003) Crystallization characteristics of yttria-stabilized zirconia/hydroxyapatite composite nanopowder, *J. Crystal Growth*, **254**, 411–417.

939. Silva, V. V., and Lameiras, F. S. (2000) Synthesis and characterization of composite powders of partially stabilized zirconia and hydroxyapatite, *Mater. Character.*, **45**, 51–59.

940. Shen, Z., Adolfsson, E., Nygren, M., Gao, L., Kawaoka, H., and Niihara, K. (2001) Dense hydroxyapatite-zirconia ceramic composites with high strength for biological applications, *Adv. Mater.*, **13**, 214–216.

941. Adolfsson, E., Alberiushenning, P., and Hermansson, L. (2000) Phase-analysis and thermal-stability of hot isostatically pressed zirconia-hydroxyapatite composites, *J. Am. Ceram. Soc.*, **83**, 2798–2802.

942. Kim, H. W., Noh, Y. J., Koh, Y. H., Kim, H. E., and Kim, H. M. (2002) Effect of CaF_2 on densification and properties of hydroxyapatite-zirconia composites for biomedical applications, *Biomaterials*, **23**, 4113–4121.

943. Li, W., and Gao, L. (2003) Fabrication of HAp-ZrO_2 (3Y) nano-composite by SPS, *Biomaterials*, **24**, 937–940.

944. Kim, H. W., Knowles, J. C., Li, L. H., and Kim, H. E. (2005) Mechanical performance and osteoblast-like cell responses of fluorine-

substituted hydroxyapatite and zirconia dense composite, *J. Biomed. Mater. Res. A*, **72A**, 258–268.

945. Xiao, X. F., Liu, R. F., and Zheng, Y. Z. (2006) Hydrothermal-electrochemical codeposited hydoxyapatite/yttria-stabilized zirconia composite coating, *J. Mater. Sci.*, **41**, 3417–3424.

946. Kumar, B. R., Prakash, K. H., Cheang, P., and Khor, K. A. (2005) Microstructure and mechanical properties of spark plasma sintered zirconia-hydroxyapatite nano-composite powders, *Acta Materialia*, **53**, 2327–2335.

947. Ahn, E. S., Gleason, N. J., Nakahira, A., and Ying, J. Y. (2001) Nanostructure processing of hydroxyapatite-based bioceramics, *Nano Lett.*, **1**, 149–153.

948. Silva, V. V., Lameiras, F. S., and Domingues, R. Z. (2001) Microstructural and mechanical study of zirconia-hydroxyapatite (ZH) composite ceramics for biomedical applications, *Compos. Sci. Technol.*, **61**, 301–310.

949. Khor, K. A., Fu, L., Lim, J. P., and Cheang, P. (2001) The effects of ZrO_2 on the phase compositions of plasma sprayed HA/YSZ composite coatings, *Mater. Sci. Eng. A*, **316**, 160–166.

950. Fu, L., Khor, K. A., and Lim, J. P. (2002) Effects of yttria-stabilized zirconia on plasma-sprayed hydroxyapatite/yttria-stabilized zirconia composite coatings, *J. Am. Ceram. Soc.*, **85**, 800–806.

951. Chou, B. Y., Chang, E., Yao, S. Y., and Chen, J. M. (2002) Phase transformation during plasma spraying of hydroxyapatite-10-wt%-zirconia composite coating, *J. Am. Ceram. Soc.*, **85**, 661–669.

952. Wang, Q., Ge, S., and Zhang, D. (2005) Nano-mechanical properties and biotribological behaviors of nanosized HA/partially-stabilized zirconia composites, *Wear*, **259**, 952–957.

953. Murugan, R., and Ramakrishna, S. (2003) Effect of zirconia on the formation of calcium phosphate bioceramics under microwave irradiation, *Mater. Lett.*, **58**, 230–234.

954. Fu, L., Khor, K. A., and Lim, J. P. (2000) Processing., microstructure and mechanical properties of yttria stabilized zirconia reinforced hydroxyapatite coatings, *Mater. Sci. Eng. A*, **276**, 46–51.

955. Nagarajan, V. S., and Rao, K. J. (1993) Structural., mechanical and biocompatibility studies of hydroxyapatite-derived composites toughened by zirconia addition, *J. Mat. Chem.*, **3**, 43–51.

956. Tamari, N., Kondo, I., Mouri, M., and Kinoshita, M. (1988) Effect of calcium fluoride addition on densification and mechanical properties

of hydroxyapatite-zirconia composite ceramics, *J. Ceram. Soc. Jpn.*, **96**, 1200–1202.

957. Evis, Z., and Doremus, R. H. (2007) Hot-pressed hydroxylapatite/monoclinic zirconia composites with improved mechanical properties, *J. Mater. Sci.*, **42**, 2426–2431.

958. Evis, Z., and Doremus, R. H. (2007) Effect of MgF_2 on hot pressed hydroxylapatite and monoclinic zirconia composites, *J. Mater. Sci.*, **42**, 3739–3744.

959. Ahn, E. S., Gleason, N. J., and Ying, J. Y. (2005) The effect of zirconia reinforcing agents on the microstructure and mechanical properties of hydroxyapatite-based nanocomposites, *J. Am. Ceram. Soc.*, **88**, 3374–3379.

960. Erkmen, Z. E., Genç, Y., and Oktar, F. N. (2007) Microstructural and mechanical properties of hydroxyapatite-zirconia composites, *J. Am. Ceram. Soc.*, **90**, 2885–2892.

961. Rapacz-Kmita, A., Slosarczyk, A., and Paszkiewicz, Z. (2006) Mechanical properties of HAp-ZrO_2 composites, *J. Eur. Ceram. Soc.*, **26**, 1481–1488.

962. Sung, Y. M., Shin, Y. K., and Ryu, J. J. (2007) Preparation of hydroxyapatite/zirconia bioceramic nanocomposites for orthopaedic and dental prosthesis applications, *Nanotechnology*, **18**, 065602 (6 pp.).

963. Quan, R., Yang, D., Wu, X., Wang, H., Miao, X., and Li, W. (2008) *In vitro* and *in vivo* biocompatibility of graded hydroxyapatite-zirconia composite bioceramic, *J. Mater. Sci. Mater. Med.*, **19**, 183–187.

964. Khalil, K. A., Kim, S. W., and Kim, H. Y. (2007) Consolidation and mechanical properties of nanostructured hydroxyapatite-(ZrO_2 + 3 mol% Y_2O_3) bioceramics by high-frequency induction heat sintering, *Mater. Sci. Eng. A*, **456**, 368–372.

965. Kong, Y. M., Kim, S., Kim, H. E., and Lee, I. S. (1999) Reinforcement of hydroxyapatite bioceramic by addition of ZrO_2 coated with Al_2O_3, *J. Am. Ceram. Soc.*, **82**, 2963–2968.

966. Wang, L. L., Wang, X. F., Jiang, H. T., and Yu, C. L. (2010) Preparation of porous hydroxyapatite-zirconia composite scaffolds by combination of gel-casting and polymer sponge methods, *Adv. Mater. Res.*, pp. 105–106, 616–619.

967. Choi, J. W., Kong, Y. M., Kim, H. E., and Lee, I. S. (1998) Reinforcement of hydroxyapatite bioceramic by addition of Ni_3Al and Al_2O_3, *J. Am. Ceram. Soc.*, **81**, 1743–1748.

968. Adolfsson, E., and Hermansson, L. (2000) Phase stability aspects of various apatite-aluminium oxide composites, *J. Mater. Sci.*, **35**, 5719–5723.

969. Li, J., Fartash, B., and Hermansson, L. (1995) Hydroxyapatite-alumina composites and bone-bonding, *Biomaterials*, **16**, 417–422.

970. Kim, S., Kong, Y. M., Kim, H. E., and Lee, I. S. (2002) Effect of calcinations of starting powder on mechanical properties of hydroxyapatite-alumina bioceramic composite, *J. Mater. Sci. Mater. Med.*, **13**, 307–310.

971. Chiba, A., Kimura, S., Raghukandan, K., and Morizono, Y. (2003) Effect of alumina addition on hydroxyapatite biocomposites fabricated by underwater-shock compaction, *Mater. Sci. Eng. A*, **350**, 179–183.

972. Pang, Y. X., Bao, X., and Weng, L. (2004) Preparation of tricalcium phosphate/alumina composite nanoparticles and self-reinforcing composites by simultaneous precipitation, *J. Mater. Sci.*, **39**, 6311–6323.

973. Jun, Y. K., Kim, W. H., Kweon, O. K., and Hong, S. H. (2003) The fabrication and biochemical evaluation of alumina reinforced calcium phosphate porous implants, *Biomaterials*, **24**, 3731–3739.

974. Epure, L. M., Dimitrievska, S., Merhi, Y., and Yahia, L. H. (2007) The effect of varying Al_2O_3 percentage in hydroxyapatite/Al_2O_3 composite materials: morphological., chemical and cytotoxic evaluation, *J. Biomed. Mater. Res. A*, **83A**, 1009–1023.

975. Evis, Z., and Doremus, R. H. (2007) A study of phase stability and mechanical properties of hydroxylapatite-nanosize α-alumina composites, *Mater. Sci. Eng. C*, **27**, 421–425.

976. Viswanath, B., and Ravishankar, N. (2006) Interfacial reactions in hydroxyapatite/alumina nanocomposites, *Scripta Mater.*, **55**, 863–866.

977. Lu, Y. P., Li, M. S., Li, S. T., Wang, Z. G., and Zhu, R. F. (2004) Plasma-sprayed hydroxyapatite+titania composite bond coat for hydroxyapatite coating on titanium substrate, *Biomaterials*, **25**, 4393–4403.

978. Li, H., Khor, K. A., and Cheang, P. (2003) Impact formation and microstructure characterization of thermal sprayed hydroxyapatite/titania composite coatings, *Biomaterials*, **24**, 949–957.

979. Zheng, X. B., and Ding, C. X. (2000) Characterization of plasma-sprayed hydroxyapatite/TiO_2 composite coatings, *J. Therm. Spray Technol.*, **9**, 520–525.

980. Lee, S. H., Kim, H. W., Lee, E. J., Li, L. H., and Kim, H. E. (2006) Hydroxyapatite-TiO_2 hybrid coating on Ti implants, *J. Biomater. Appl.*, **20**, 195–208.

981. Lin, C., and Yen, S. (2004) Characterization and bond strength of electrolytic HA/TiO_2 double layers for orthopedic applications, *J. Mater. Sci. Mater. Med.*, **15**, 1237–1246.

982. Balamurugan, A., Balossier, G., Kannan, S., Michel, J., and Rajeswari, S. (2007) *In vitro* biological., chemical and electrochemical evaluation of titania reinforced hydroxyapatite sol-gel coatings on surgical grade 316L SS, *Mater. Sci. Eng. C*, **27**, 162–171.

983. Gaona, M., Limab, R. S., and Marple, B. R. (2007) Nanostructured titania/hydroxyapatite composite coatings deposited by high velocity oxy-fuel (HVOF) spraying, *Mater. Sci. Eng. A*, **458**, 141–149.

984. Boyd, A. R., Duffy, H., McCann, R., and Meenan, B. J. (2008) Sputter deposition of calcium phosphate/titanium dioxide hybrid thin films, *Mater. Sci. Eng. C*, **28**, 228–236.

985. Fidancevska, E., Ruseska, G., Bossert, J., Linc, Y. M., and Boccaccini, A. R. (2007) Fabrication and characterization of porous bioceramic composites based on hydroxyapatite and titania, *Mater. Chem. Phys.*, **103**, 95–100.

986. Harle, J., Kim, H. W., Mordan, N., Knowles, J. C., and Salih, V. (2006) Initial responses of human osteoblasts to sol-gel modified titanium with hydroxyapatite and titania composition, *Acta Biomater.*, **2**, 547–556.

987. Kim, H. W., Kim, H. E., Salih, V., and Knowles, J. C. (2005) Hydroxyapatite and titania sol-gel composite coatings on titanium for hard tissue implants; mechanical and *in vitro* biological performance, *J. Biomed. Mater. Res. B (Appl. Biomater.)*, **72B**, 1–8.

988. Pushpakanth, S., Srinivasan, B., Sreedhar, B., and Sastry, T. P. (2008) An *in situ* approach to prepare nanorods of titania-hydroxyapatite (TiO_2-HAp) nanocomposite by microwave hydrothermal technique, *Mater. Chem. Phys.*, **107**, 492–498.

989. Sato, M., Aslani, A., Sambito, M. A., Kalkhoran, N. M., Slamovich, E. B., and Webster, T. J. (2008) Nanocrystalline hydroxyapatite/titania coatings on titanium improves osteoblast adhesion, *J. Biomed. Mater. Res.*, **84A**, 265–272.

990. Ramires, P. A., Romito, A., Cosentino, F., and Milella, E. (2001) The influence of titania/hydroxyapatite composite coatings on *in vitro* osteoblasts behaviour, *Biomaterials*, **22**, 1467–1474.

991. Que, W., Khor, K. A., Xu, J. L., and Yu, L. G. (2008) Hydroxyapatite/titania nanocomposites derived by combining high-energy ball milling with spark plasma sintering processes, *J. Eur. Ceram. Soc.*, **28**, 3083–3090.

992. Nath, S., Tripathi, R., and Basu, B. (2009) Understanding phase stability, microstructure development and biocompatibility in calcium phosphate-titania composites, synthesized from hydroxyapatite and titanium powder mix, *Mater. Sci. Eng. C*, **29**, 97–107.

993. Lee, J. H., Kim, H. E., and Koh, Y. H. (2009) Highly porous titanium (Ti) scaffolds with bioactive microporous hydroxyapatite/TiO_2 hybrid coating layer, *Mater. Lett.*, **63**, 1995–1998.

994. Ün, S., and Durucan, C. (2009) Preparation of hydroxyapatite-titania hybrid coatings on titanium alloy, *J. Biomed. Mater. Res. B (Appl. Biomater.)*, **90**, 574–583.

995. Nathanael, A. J., Mangalaraj, D., and Ponpandian, N. (2010) Controlled growth and investigations on the morphology and mechanical properties of hydroxyapatite/titania nanocomposite thin films, *Composites Sci. Technol.*, **70**, 1645–1651.

996. Ebrahimi-Kahrizsangi, R., Nasiri-Tabrizi, B., and Chami, A. (2010) Synthesis and characterization of fluorapatite-titania (FAp-TiO_2) nanocomposite via mechanochemical process, *Solid State Sci.*, **12**, 1645–1651.

997. Sun, R., Li, M., Lu, Y., and An, X. (2006) Effect of titanium and titania on chemical characteristics of hydroxyapatite plasma-sprayed into water, *Mater. Sci. Eng. C*, **26**, 28–33.

998. Lee, B. T., Lee, C. W., Gain, A. K., and Song, H. Y. (2007) Microstructures and material properties of fibrous Ap/Al_2O_3-ZrO_2 composites fabricated by multi-pass extrusion process, *J. Eur. Ceram. Soc.*, **27**, 157–163.

999. Kong, Y. M., Bae, C. J., Lee, S. H., Kim, H. W., and Kim, H. E. (2005) Improvement in biocompatibility of ZrO_2-Al_2O_3 nano-composite by addition of HA, *Biomaterials*, **26**, 509–517.

1000. Oktar, F. N., Agathopoulos, S., Ozyegin, L. S., Gunduz, O., Demirkol. N., Bozkurt, Y., and Salman, S. (2007) Mechanical properties of bovine hydroxyapatite (BHA) composites doped with SiO_2, MgO, Al_2O_3 and ZrO_2, *J. Mater. Sci. Mater. Med.*, **18**, 2137–2143.

1001. Gunduz, O., Erkan, E. M., Daglilar, S., Salman, S., Agathopoulos, S., and Oktar, F. N. (2008) Composites of bovine hydroxyapatite (BHA) and ZnO, *J. Mater. Sci.*, **43**, 2536–2540.

1002. Ajeesh, M., Francis, B. F., Annie, J., and Varma, P. R. H. (2010) Nano iron oxide-hydroxyapatite composite ceramics with enhanced radiopacity, *J. Mater. Sci. Mater. Med.*, **21**, 1427–1434.

1003. Yang, Z. P., Gong, X. Y., and Zhang, C. J. (2010) Recyclable Fe_3O_4/hydroxyapatite composite nanoparticles for photocatalytic applications, *Chem. Eng. J.*, **165**, 117–121.

1004. Liu, Y., Zhong, H., Li, L., and Zhang, C. (2010) Temperature dependence of magnetic property and photocatalytic activity of Fe_3O_4/hydroxyapatite nanoparticles, *Mater. Res. Bull.*, **45**, 2036–2039.

1005. Li, X. W., Yasuda, H. Y., and Umakoshi, Y. (2006) Bioactive ceramic composites sintered from hydroxyapatite and silica at 1200ºC: preparation., microstructures and *in vitro* bone-like layer growth, *J. Mater. Sci. Mater. Med.*, **17**, 573–581.

1006. Ragel, C. V., Vallet-Regi, M., and Rodríguez-Lorenzo, L. M. (2002) Preparation and *in vitro* bioactivity of hydroxyapatite/sol-gel glass biphasic material, *Biomaterials*, **23**, 1865–1872.

1007. Padilla. S., Sánchez-Salcedo. S., and Vallet-Regi, M. (2005) Bioactive and biocompatible pieces of HA/sol-gel glass mixtures obtained by the gel-casting method, *J. Biomed. Mater. Res. A*, **75A**, 63–72.

1008. Lopes, M. A., Monterio, F. J., and Santos, J. D. (1999) Glass-reinforced hydroxyapatite composites: fracture toughness and hardness dependence on microstructural characteristics, *Biomaterials*, **20**, 2085–2090.

1009. Fu, Q., Zhou, N., Huang, W., Wang, D., Zhang, L., and Li, H. (2004) Preparation and characterization of a novel bioactive bone cement: glass based nanoscale hydroxyapatite bone cement, *J. Mater. Sci. Mater. Med.*, **15**, 1333–1338.

1010. Fu, Q., Zhou, N., Huang, W., Wang, D., Zhang, L., and Li, H. (2005) Effects of nano HAP on biological and structural properties of glass bone cement, *J. Biomed. Mater. Res. A*, **74A**, 156–163.

1011. Oktar, F. N., and Goller, G. (2002) Sintering effects on mechanical properties of glass-reinforced hydroxyapatite composites, *Ceram. Int.*, **28**, 617–621.

1012. Padilla, S., Román, J., Sánchez-Salcedo, S., and Vallet-Regi, M. (2006) Hydroxyapatite/SiO_2-CaO-P_2O_5 glass materials: *in vitro* bioactivity and biocompatibility, *Acta Biomater.*, **2**, 331–342.

1013. Sych, E. E., Pinchuk, N. D., and Ivanchenko, L. A. (2010) Effect of sintering temperature on the properties of biogenic hydroxyapatite-glass composites, *Powder Metallurgy and Metal Ceramics*, **49**, 153–158.

1014. Liu, X., and Ei-Ghannam, A. (2010) Effect of processing parameters on the microstructure and mechanical behavior of silica-calcium phosphate nanocomposite, *J. Mater. Sci. Mater. Med.*, **21**, 2087–2094.

1015. Kokubo, T., Shigematsu, M., Nagashima, Y., Tashiro, M., Nakamura, T., Yamamuro, T., and Higashi, S. (1982) Apatite- and wollastonite-containing glass ceramics for prosthetic applications, *Bull. Inst. Chem. Res. 60. Kyoto University*, pp. 260–268.

1016. Kitsugi, T., Yamamuro, T., Nakamura, T., Higashi, S., Kakutani, Y., Hyakuna, K., Ito, S., Kokubo, T., Takagi, M., and Shibuya, T. (1986) Bone bonding behavior of three kinds of apatite containing glass ceramics, *J. Biomed. Mater. Res.*, **20**, 1295–1307.

1017. Kokubo, T., Ito, S., Shigematsu, M., Sakka, S., and Yamamuro, T. (1987) Fatigue and life-time of bioactive glass-ceramic A-W containing apatite and wollastonite, *J. Mater. Sci.*, **22**, 4067–4070.

1018. Kokubo, T., Ito, S., Shigematsu, M., Sakka, S., and Yamamuro, T. (1985) Mechanical properties of a new type of apatite-containing glass-ceramic for prosthetic application, *J. Mater. Sci.*, **20**, 2001–2004.

1019. Kokubo, T. (1991) Bioactive glass ceramics: properties and applications, *Biomaterials*, **12**, 155–163.

1020. Kokubo, T., Ito, S., Huang, Z. T., Hayashi, T., Sakka, S., Kitsugi, T., and Yamamuro, T. (1990) CaP-rich layer formed on high strength bioactive glass ceramic A-W, *J. Biomed. Mater. Res.*, **24**, 331–343.

1021. Nishio, K., Neo, M., Akiyama, H., Okada, Y., Kokubo, T., and Nakamura, T. (2001) Effects of apatite and wollastonite containing glass-ceramic powder and two types of alumina powder in composites on osteoblastic differentiation of bone marrow cells, *J. Biomed. Mater. Res.*, **55**, 164–176.

1022. Zhang, D., Chang, J., and Zeng, Y. (2008) Fabrication of fibrous poly(butylene succinate)/wollastonite/apatite composite scaffolds by electrospinning and biomimetic process, *J. Mater. Sci. Mater. Med.*, **19**, 443–449.

1023. Acharya, N. K., Mahajan, C. V., Kumar, R. J., Varma, H. K., and Menon, V. K. (2010) Can iliac crest reconstruction reduce donor site morbidity?: A study using degradable hydroxyapatite-bioactive glass ceramic composite, *J. Spinal Disorders and Techniques*, **23**, 266–271.

1024. Encinas-Romero, M. A., Aguayo-Salinas, S., Valenzuela-García, J. L., Payán, S. R., and Castillón-Barraza, F. F. (2010) Mechanical and bioactive behavior of hydroxyapatite-wollastonite sintered composites, *Int. J. Appl. Ceram. Technol.*, **7**, 164–177.

1025. Ha, N. R., Yang, Z. X., Hwang, K. H., Kim, T. S., and Lee, J. K. (2010) Improvement of the stability of hydroxyapatite through glass ceramic reinforcement, *J. Nanosci. Nanotechnol.*, **10**, 3459–3462.

1026. Nath, S., Biswas, K., Wang, K., Bordia, R. K., and Basu, B. (2010) Sintering, phase stability, and properties of calcium phosphate-mullite composites, *J. Am. Ceram. Soc.*, **93**, 1639–1649.

1027. Nath, S., Ummethala, R., and Basu, B. (2010) Fretting wear behavior of calcium phosphate-mullite composites in dry and albumin-containing simulated body fluid conditions, *J. Mater. Sci. Mater. Med.*, **21**, 1151–1161.

1028. Chaki, T. K., and Wang, P. E. (1994) Densification and strengthening of silver-reinforced hydroxyapatite-matrix composite prepared by sintering, *J. Mater. Sci. Mater. Med.*, **5**, 533–542.

1029. Zhang, X., Gubbels, G. H. M., Terpstra, R. A., and Metselaar, R. (1997) Toughening of calcium hydroxyapatite with silver particles, *J. Mater. Sci.*, **32**, 235–243.

1030. Chu, C., Lin, P., Dong, Y., Xue, X., Zhu, J., and Yin, Z. (2002) Fabrication and characterization of hydroxyapatite reinforced with 20 vol. % Ti particles for use as hard tissue replacement, *J. Mater. Sci. Mater. Med.*, **13**, 985–992.

1031. Shi, W., Kamiya, A., Zhu, J., and Watazu, A. (2002) Properties of titanium biomaterial fabricated by sinter-bonding of titanium/hydroxyapatite composite surface-coated layer to pure bulk titanium, *Mater. Sci. Eng. A*, **337**, 104–109.

1032. Ning, C. Q., and Zhou, Y. (2002) *In vitro* bioactivity of a biocomposite fabricated from HA and Ti powders by powder metallurgy method, *Biomaterials*, **23**, 2909–2915.

1033. Karanjai, M., Sundaresan, R., Rao, G. V. N., Mohan, R. T. R., and Kashyap, B. P. (2007) Development of titanium based biocomposite by powder metallurgy processing with in situ forming of Ca-P phases, *Mater. Sci. Eng. A*, **447**, 19–26.

1034. Karanjai, M., Kumarb, M. B. V., Sundaresan, R., Basu, B., Mohan, R. T. R., and Kashyap, B. P. (2008) Fretting wear study on Ti-Ca-P biocomposite in dry and simulated body fluid, *Mater. Sci. Eng. A*, **475**, 299–307.

1035. Ning, C. Q., and Zhou, Y. (2004) On the microstructure of biocomposites sintered from Ti., HA and bioactive glass, *Biomaterials*, **25**, 3379–3387.

1036. Chu, C., Xue, X., Zhu, J., and Yin, Z. (2004) Mechanical and biological properties of hydroxyapatite reinforced with 40 vol. % titanium particles for use as hard tissue replacement, *J. Mater. Sci. Mater. Med.*, **15**, 665–670.

1037. Smirnov, V. V., Egorov, A. A., Barinov, S. M., and Shvorneva, L. I. (2007) Composite calcium phosphate bone cements reinforced by particulate titanium, *Doklady Chem.*, **413**, 82–85.

1038. Chu, C., Xue, X., Zhu, J., and Yin, Z. (2006) Fabrication and characterization of titanium-matrix composite with 20 vol. % hydroxyapatite for use as heavy load-bearing hard tissue replacement, *J. Mater. Sci. Mater. Med.*, **17**, 245–251.

1039. Li, J., Habibovic, P., Yuan, H., van den Doel, M., Wilson, C. E., de Wijn, J. R., van Blitterswijk, C. A., and de Groot, K. (2007) Biological performance in goats of a porous titanium alloy-biphasic calcium phosphate composite, *Biomaterials*, **28**, 4209–4218.

1040. Pattanayak, D. K., Mathur, V., Rao, B. T., and Mohan, T. R. R. (2003) Synthesis and characterization of titanium – calcium phosphate composites for bio applications, *Trends Biomater. Artif. Organs*, **17**, 8–12.

1041. Ding, Y., Liu, J., Wang, H., Shen, G., and Yu, R. (2007) A piezoelectric immunosensor for the detection of α-fetoprotein using an interface of gold/hydroxyapatite hybrid nanomaterial, *Biomaterials*, **28**, 2147–2154.

1042. Miranda, M., Fernández, A., Díaz, M., Esteban-Tejeda, L., López-Esteban, S., Malpartida, F., Torrecillas, R., and Moya, J. S. (2010) Silver-hydroxyapatite nanocomposites as bactericidal and fungicidal materials, *Int. J. Mater. Res.*, **101**, 122–127.

1043. Younesi, M., and Bahrololoom, M. E. (2010) Optimizations of wear resistance and toughness of hydroxyapatite nickel free stainless steel new bio-composites for using in total joint replacement, *Mater. Des.*, **31**, 234–243.

1044. Murakoshi, Y., Kikuchi, K., Katoh, M., and Matsuzaki, K. (2010) Fabrication and property of degradable magnesium-calcium alloy composites with hydroxyapatite, *IFMBE Proceedings*, **31**, 1226–1229.

1045. Razavi, M., Fathi, M. H., and Meratian, M. (2010) Fabrication and characterization of magnesium-fluorapatite nanocomposite for biomedical applications, *Mater. Characterization*, **61**, 1363–1370.

1046. Damien, C. J., Parsons, J. R., Benedict, J. J., and Weisman, D. S. (1990) Investigation of a hydroxyapatite and calcium sulfate composite supplemented with an osteoinductive factor, *J. Biomed. Mater. Res.*, **24**, 639–654.

1047. Rauschmann, M. A., Wichelhaus, T. A., Stirnal, V., Dingeldein, E., Zichner, L., Schnettler, R., and Alt, V. (2005) Nanocrystalline hydroxyapatite and calcium sulphate as biodegradable composite carrier material for local delivery of antibiotics in bone infections, *Biomaterials*, **26**, 2677–2684.

1048. Urban, R. M., Turner, T. M., Hall, D. J., Inoue, N., and Gitelis, S. (2007) Increased bone formation using calcium sulfate-calcium phosphate composite graft, *Clin. Orthop. Relat. Res.*, **459**, 110–117.

1049. Rauschmann, M., Vogl, T., Verheyden, A., Pflugmacher, R., Werba, T., Schmidt, S., and Hierholzer, J. (2010) Bioceramic vertebral augmentation with a calcium sulphate/hydroxyapatite composite (Cerament™ SpineSupport) in vertebral compression fractures due to osteoporosis, *Eur. Spine J.*, **19**, 887–892.

1050. Kumar, G. S., Girija, E. K., Thamizhavel, A., Yokogawa, Y., and Kalkura, S. N. (2010) Synthesis and characterization of bioactive hydroxyapatite-calcite nanocomposite for biomedical applications, *J. Coll. Interf. Sci.*, **349**, 56–62.

1051. Smirnov, V. V., Goldberg, M. A., Shvorneva, L. I., Fadeeva, I. V., Shibaeva, T. V., and Barinov, S. M. (2010) Synthesis of composite biomaterials in the hydroxyapatite-calcite system, *Doklady Chem.*, **432**, 151–154.

1052. Gittings, J. P., Bowena, C. R., Turner, I. G., Baxter, F., and Chaudhuri, J. (2007) Characterisation of ferroelectric-calcium phosphate composites and ceramics, *J. Eur. Ceram. Soc.*, **27**, 4187–4190.

1053. Watanabe, Y., Ikoma, T., Suetsugu, Y., Yamada, H., Tamura, K., Komatsu, Y., Tanaka, J., and Moriyoshi, Y. (2006) The densification of zeolite/apatite composites using a pulse electric current sintering method: a long-term assurance material for the disposal of radioactive waste, *J. Eur. Ceram. Soc.*, **26**, 481–486.

1054. Lahiri, D., Singh, V., Benaduce, A. P., Seal, S., Kos, L., and Agarwal, A. (2011) Boron nitride nanotube reinforced hydroxyapatite composite: mechanical and tribological performance and *in-vitro* biocompatibility to osteoblasts, *J. Mech. Behavior Biomed. Mater.*, **4**, 44–56.

1055. Agathopoulos, S., Tulyaganov, D. U., Marques, P. A. A. P., Ferro, M. C., Fernandes, M. H. V., and Correia, R. N. (2003) The fluorapatite-anorthite system in biomedicine, *Biomaterials*, **24**, 1317–1331.

1056. Khor, K. A., Gu, Y. W., Pan, D., and Cheang, P. (2004) Microstructure and mechanical properties of plasma sprayed HA/YSZ/Ti-6Al-4V composite coatings, *Biomaterials*, **25**, 4009–4017.

1057. Gu, Y. W., Khor, K. A., Pan, D., and Cheang, P. (2004) Activity of plasma sprayed yttria stabilized zirconia reinforced hydroxyapatite/Ti-6Al-4V composite coatings in simulated body fluid, *Biomaterials*, **25**, 3177–3185.

1058. Kalmodia, S., Goenka, S., Laha, T., Lahiri, D., Basu, B., and Balani, K. (2010) Microstructure, mechanical properties, and *in vitro* biocompatibility of spark plasma sintered hydroxyapatite-aluminum oxide-carbon nanotube composite, *Mater. Sci. Eng. C*, **30**, 1162–1169.

1059. Best, S. M., Porter, A. E., Thian, E. S., and Huang, J. (2008) Bioceramics: past, present and for the future, *J. Eur. Ceram. Soc.*, **28**, 1319–1327.

1060. de Aza, P. N., Guitián, F., and de Aza, S. (1997) Bioeutectic: a new ceramic material for human bone replacement, *Biomaterials*, **18**, 1285–1291.

1061. Huang, X., Jiang, D., and Tan, S. (2004) Apatite formation on the surface of wollastonite/tricalcium phosphate composite immersed in simulated body fluid, *J. Biomed. Mater. Res. B (Appl. Biomater.)*, **69B**, 70–72.

1062. Zhang, F., Chang, J., Lin, K., and Lu, J. (2008) Preparation, mechanical properties and *in vitro* degradability of wollastonite/tricalcium phosphate macroporous scaffolds from nanocomposite powders, *J. Mater. Sci. Mater. Med.*, **19**, 167–173.

1063. Juhasz, J. A., Best, S. M., Kawashita, M., Miyata, N., Kokubo, T., Nakamura, T., and Bonfield, W. (2003) Bonding strength of the apatite layer formed on glass-ceramic apatite-wollastonite-polyethylene composites, *J. Biomed. Mater. Res. A*, **67A**, 952–959.

1064. Juhasz, J. A., Best, S. M., Bonfield, W., Kawashita, M., Miyata, N., Kokubo, T., and Nakamura, T. (2003) Apatite-forming ability of glass-ceramic apatite-wollastonite – polyethylene composites: effect of filler content, *J. Mater. Sci. Mater. Med.*, **14**, 489–495.

1065. Juhasz, J. A., Best, S. M., Brooks, R., Kawashita, M., Miyata, N., Kokubo, T., Nakamura, T., and Bonfield, W. (2004) Mechanical properties of glass-ceramic A-W-polyethylene composites: effect of filler content and particle size, *Biomaterials*, **25**, 949–955.

1066. Rea, S. M., Brooks, R. A., Best, S. M., Kokubo, T., and Bonfield, W. (2004) Proliferation and differentiation of osteoblast-like cells on apatite-wollastonite/polyethylene composites, *Biomaterials*, **25**, 4503–4512.

1067. Greish, Y. E., and Brown, P. W. (2001) Characterization of wollastonite-reinforced HAp-Ca polycarboxylate composites, *J. Biomed. Mater. Res.*, **55**, 618–628, *Erratum in: J. Biomed. Mater. Res.*, **56**, 459.

1068. Greish, Y. E., and Brown, P. W. (2000) Characterization of bioactive glass-reinforced HAP-polymer composites, *J. Biomed. Mater. Res.*, , **52**, 687–694.

1069. Radev, L., Hristov, V., Samuneva, B., and Ivanova, D. (2009) Organic/inorganic bioactive materials. Part II: *in vitro* bioactivity of collagen-calcium phosphate silicate/wollastonite hybrids, *Central Eur. J. Chem.*, **7**, 711–720.

1070. Radev, L., Hristov, V., Fernandes, M. H. V., and Salvado, I. M. M. (2010) Organic/inorganic bioactive materials part IV: *in vitro* assessment of bioactivity of gelatin-calcium phosphate silicate/wollastonite hybrids, *Central Eur. J. Chem.*, **8**, 278–284.

1071. Kangasniemi, I., de Groot, K., Wolke, J., Andersson, O., Luklinska, Z., Becht, J. G. M., Lakkisto, M., and Yli-Urpo, A. (1991) The stability of hydroxyapatite in an optimized bioactive glass matrix at sintering temperatures, *J. Mater. Sci. Mater. Med.*, **2**, 133–137.

1072. Kangasniemi, I. M. O., de Groot, K., Becht, J. G. M., and Yli-Urpo, A. (1992) Preparation of dense hydroxylapatite or rhenanite containing bioactive glass composites, *J. Biomed. Mater. Res.*, **26**, 663–674.

1073. Kangasniemi, I. M. O., Vedel, E., de Blick-Hogerworst, J., Yli-Urpo. A., and de Groot, K. (1993) Dissolution and scanning electron microscopic studies of Ca, P particle-containing bioactive glasses, *J. Biomed. Mater. Res.*, **27**, 1225–1233.

1074. Maruno, S., Ban, S., Wang, Y. F., Iwata, H., and Itoh, H. (1992) Properties of functionally gradient composite consisting of hydroxyapatite containing glass coated titanium and characters for bioactive implant, *J. Ceram. Soc. Jpn.*, **100**, 362–367.

1075. Kobayashi, S., and Kawai, W. (2007) Development of carbon nanofiber reinforced hydroxyapatite with enhanced mechanical properties, *Composites A*, **38**, 114–123.

1076. Balani, K., Anderson, R., Laha, T., Andara, M., Tercero, J., Crumpler, E., and Agarwal, A. (2007) Plasma-sprayed carbon nanotube reinforced hydroxyapatite coatings and their interaction with human osteoblasts *in vitro*, *Biomaterials*, **28**, 618–624.

1077. Chen, Y., Gan, C. H., Zhang, T. H., Yu, G., Bai, P., and Kaplan, A. (2005) Laser-surface-alloyed carbon nanotubes reinforced hydroxyapatite composite coatings, *Appl. Phys. Lett.*, **86**, 251905 (3 pages).

1078. Chen, Y., Zhang, T. H., Gan, C. H., and Yu, G. (2007) Wear studies of hydroxyapatite composite coating reinforced by carbon nanotubes, *Carbon*, **45**, 998–1004.

1079. Chen, Y., Zhang, Y. Q., Zhang, T. H., Gan, C. H., Zheng, C. Y., and Yu, G. (2006) Carbon nanotube reinforced hydroxyapatite composite coatings produced through laser surface alloying, *Carbon*, **44**, 37–45.

1080. Xu, J. L., Khor, K. A., Sui, J. J., and Chen, W. N. (2009) Preparation and characterization of a novel hydroxyapatite/carbon nanotubes composite and its interaction with osteoblast-like cells, *Mater. Sci. Eng. C*, **29**, 44–49.

1081. Lee, H. H., Shin, U. S., Won, J. E., and Kim, H. W. (2010) Preparation of hydroxyapatite-carbon nanotube composite nanopowders, *Mater. Lett.*, **65**, 208–211.

1082. White, A. A., Kinloch, I. A., Windle, A. H., and Best, S. M. (2010) Optimization of the sintering atmosphere for high-density hydroxyapatite-carbon nanotube composites, *J. Royal Soc. Interf.*, **7(5)**, S529–S539.

1083. Ding, Y., Liu, J., Jin, X., Lu, H., Shen, G., and Yu, R. (2008) Poly-L-lysine/hydroxyapatite/carbon nanotube hybrid nanocomposite applied for piezoelectric immunoassay of carbohydrate antigen 19-9, *Analyst*, **133**, 184–190.

1084. Zhao, H. Y., Xu, X. X., Zhang, J. X., Zheng, W., and Zheng, Y. F. (2010) Carbon nanotube-hydroxyapatite-hemoglobin nanocomposites with high bioelectrocatalytic activity, *Bioelectrochemistry*, **78**, 124–129.

1085. Slosarcyk, A., Klisch, M., Blazewicz, M., Piekarczyk, J., Stobierski, L., and Rapacz-Kmita, A. (2000) Hot pressed hydroxyapatite-carbon fibre composites, *J. Eur. Ceram. Soc.*, **20**, 1397–1402.

1086. Dorner-Reisel, A., Berroth, K., Neubauer, R., Nestler, K., Marx, G., Scislo, M., Müller, E., and Slosarcyk, A. (2004) Unreinforced and carbon fibre reinforced hydroxyapatite: resistance against microabrasion, *J. Eur. Ceram. Soc.*, **24**, 2131–2139.

1087. Fu, T., Zhao, J. L., Wei, J. H., Han, Y., and Xu, K. W. (2004) Preparation of carbon fiber fabric reinforced hydroxyapatite/epoxy composite by RTM processing, *J. Mater. Sci.*, **39**, 1411–1413.

1088. Pecheva, E., Pramatarova, L., Hikov, T., Fingarova, D., Tanaka, Y., Sakamoto, H., Doi, H., Tsutsumi, Y., and Hanawa, T. (2010) Apatitenanodiamond composite as a functional coating of stainless steel, *Surf. Interf. Analysis*, **42**, 475–480.

1089. Yoshimura, M. (1988) Phase stability of zirconia, *Am. Ceram. Soc. Bull.*, **67**, 1950–1955.

1090. Thompson, I., and Rawlings, R. D. (1990) Mechanical behaviour of zirconia and zirconia-toughened alumina in a simulated body environment, *Biomaterials*, **11**, 505–508.

1091. Monma, H. (1987) Tricalcium phosphate ceramics complexed with hydroxyapatite, *J. Ceram. Soc. Jpn.*, **96**, 60–64.

1092. Farzadi, A., Solati-Hashjin, M., Bakhshi, F., and Aminian, A. (2011) Synthesis and characterization of hydroxyapatite/β-tricalcium phosphate nanocomposites using microwave irradiation, *Ceram. Int.*, **37**, 65–71.

1093. Wu, C. C., Huang, S. T., Tseng, T. W., Rao, Q. L., and Lin, H. C. (2010) FT-IR and XRD investigations on sintered fluoridated hydroxyapatite composites, *J. Molecular Structure*, **979**, 72–76.

1094. Suchanek, W., Yashima, M., Kakihana, M., and Yoshimura, M. (1996) Processing and mechanical properties of hydroxyapatite reinforced with hydroxyapatite whiskers, *Biomaterials*, **17**, 1715–1723.

1095. Suchanek, W., Yashima, M., Kakihana, M., and Yoshimura, M. (1997) Hydroxyapatite/hydroxyapatite-whisker composites without sintering additives: mechanical properties and microstructural evolution, *J. Am. Ceram. Soc.*, **80**, 2805–2813.

1096. Kaito, T., Mukai, Y., Nishikawa, M., Ando, W., Yoshikawa, H., and Myoui, A. (2006) Dual hydroxyapatite composite with porous and solid parts: Experimental study using canine lumbar interbody fusion model, *J. Biomed. Mater. Res. B (Appl. Biomater.)*, **78B**, 378–384.

1097. Ramay, H. R., and Zhang, M. (2004) Biphasic calcium phosphate nanocomposite porous scaffolds for load-bearing bone tissue engineering, *Biomaterials*, **21**, 5171–5180.

1098. Matsumoto, K., Tsuru, K., Kawachi, G., Maruta, M., Matsuya, S., Takahashi, I., and Ishikawa, K. (2010) Reinforcement of carbonate apatite bone substitutes with carbonate apatite by Ca salt introduction, *J. Ceram. Soc. Jpn.*, **118**, 521–524.

1099. Tampieri, A., Celotti, G., Sprio, S., Delcogliano, A., and Franzese, S. (2001) Porosity-graded hydroxyapatite ceramics to replace natural bone, *Biomaterials*, **22**, 1365–1370.

1100. Werner, J., Linner-Krcmar, B., Friess, W., and Greil, P. (2002) Mechanical properties and *in vitro* cell compatibility of hydroxyapatite ceramics with graded pore structure, *Biomaterials*, **23**, 4285–4294.

1101. Hsu, Y. H., Turner, I. G., and Miles, A. W. (2007) Fabrication of porous bioceramics with porosity gradients similar to the bimodal structure of cortical and cancellous bone, *J. Mater. Sci. Mater. Med.*, **18**, 2251–2256.

1102. Macchetta, A., Turner, I. G., and Bowen, C. R. (2009) Fabrication of HA/TCP scaffolds with a graded and porous structure using a camphene-based freeze-casting method, *Acta Biomater.*, **5**, 1319–1327.

1103. Watari, F., Yokoyama, A., Saso, F., Uo, M., and Kawasaki, T. (1997) Fabrication and properties of functionally graded dental implant, *Composites B*, **28B**, 5–11.

1104. Watari, F., Yokoyama, A., Omori, M., Hirai, T. Kondo, H., Uo, M., and Kawasaki, T. (2004) Biocompatibility of materials and development to functionally graded implant for bio-medical application, *Compos. Sci. Technol.*, **64**, 893–908.

1105. Chu C., Zhu J., Yin Z., and Wang S. (1999) Hydroxyapatite-Ti functionally graded biomaterial fabricated by powder metallurgy, *Mater. Sci. Eng. A*, **271**, 95–100.

1106. Chu, C., Zhu, J., Yin, Z., and Lin, P. (2001) Structure optimization and properties of hydroxyapatite-Ti symmetrical functionally graded biomaterial, *Mater. Sci. Eng. A*, **316**, 205–210.

1107. Chu, C., Zhu, J., Yin, Z., and Lin, P. (2003) Optimal design and fabrication of hydroxyapatite-Ti asymmetrical functionally graded biomaterial, *Mater. Sci. Eng. A*, **348**, 244–250.

1108. Inagaki, M., Yokogawa, Y., and Kameyama, T. (2006) Effects of plasma gas composition on bond strength of hydroxyapatite/titanium composite coatings prepared by rf-plasma spraying, *J. Eur. Ceram. Soc.*, **26**, 495–499.

1109. Nonami, T., Kamiya, A., Naganuma, K., and Kameyana, T. (1998) Implantation of hydroxyapatite granules into superplastic titanium alloy for biomaterials, *Mater. Sci. Eng. C*, **6**, 281–284.

1110. Bai, X., More, K., Rouleau, C. M., and Rabiei, A. (2010) Functionally graded hydroxyapatite coatings doped with antibacterial components, *Acta Biomater.*, **6**, 2264–2273.

1111. Liu, C., Han, Z., and Czernuszka, J. T. (2009) Gradient collagen/nanohydroxyapatite composite scaffold: development and characterization, *Acta Biomater.*, **5**, 661–669.

1112. Erisken, C., Kalyon, D. M., and Wang, H. (2008) Functionally graded electrospun polycaprolactone and β-tricalcium phosphate nanocomposites for tissue engineering applications, *Biomaterials*, **29**, 4065–4073.

1113. Nindhia, T. G. T., Koyoshi, Y., Kaneko, A., Sawada, H., Ohta, M., Hirai, S., and Uo, M. (2008) Hydroxyapatite-silk functionally graded material by pulse electric current sintering, *Trends in Biomaterials and Artificial Organs*, **22**, 25–29.

1114. Ban, S., Hasegawa, J., and Maruno, S. (1999) Fabrication and properties of functionally gradient bioactive composites comprising hydroxyapatite containing glass coated titanium, *Mater. Sci. Forum*, pp. 308–311, 350–355.

1115. Stojanovic, D., Jokic, B., Veljovic, Dj., Petrovic, R., Uskokovic, P. S., and Janackovic, Dj. (2007) Bioactive glass-apatite composite coating for titanium implant synthesized by electrophoretic deposition, *J. Eur. Ceram. Soc.*, **27**, 1595–1599.

1116. Wong, L. H., Tio, B., and Miao, X. (2002) Functionally graded tricalcium phosphate/fluoroapatite composites, *Mater. Sci. Eng. C*, **20**, 111 115.

1117. Okazaki, M., Miake, Y., Tohda, H., Yanagisawa, T., Matsumoto, T., and Takahashi, J. (1999) Functionally graded fluoridated apatites, *Biomaterials*, **20**, 1421–1426.

1118. Okazaki, M., and Takahashi, J. (1999) Synthesis of functionally graded CO_3 apatite as surface biodegradable crystals, *Biomaterials*, **20**, 1073–1078.

1119. Peltola, T., Patsi, M., Rahiala, H., Kangasniemi, I., and Yli-Urpo, A. (1998) Calcium phosphate induction by sol-gel-derived titania coatings on titanium substrates *in vitro*, *J. Biomed. Mater. Res.*, **41**, 504–510.

1120. Heilmann, F., Standard, O. C., Müller, F. A., and Hoffman, M. (2007) Development of graded hydroxyapatite/$CaCO_3$ composite structures for bone ingrowth, *J. Mater. Sci. Mater. Med.*, **18**, 1817–1824.

1121. Cavalcanti, A., Shirinzadeh, B., Zhang, M., and Kretly, L. C. (2008) Nanorobot hardware architecture for medical defense, *Sensors*, **8**, 2932–2958.

1122. Ding, Y., Liu, J., Jin, X., Shen, G., and Yu, R. (2008) A novel piezoelectric immunosensor for CA125 using a hydroxyapatite/chitosan nanocomposite-based biomolecular immobilization method, *Australian J. Chem.*, **61**, 500–505.

1123. López, M. S. P., Tamimi, F., López-Cabarcos, E., and López-Ruiz, B. (2009) Highly sensitive amperometric biosensor based on a biocompatible calcium phosphate cement, *Biosensors and Bioelectronics*, **24**, 2574–2579.

1124. Wang, B., Zhang, J. J., Pan, Z. Y., Tao, X. Q., and Wang, H. S. (2009) A novel hydrogen peroxide sensor based on the direct electron transfer of horseradish peroxidase immobilized on silica-hydroxyapatite hybrid film, *Biosensors and Bioelectronics*, **24**, 1141–1145.

1125. Lu, L., Zhang, L., Zhang, X., Huan, S., Shen, G., and Yu, R. (2010) A novel tyrosinase biosensor based on hydroxyapatite-chitosan nanocomposite for the detection of phenolic compounds, *Analytica Chimica Acta*, **665**, 146–151.

1126. Wang, S., Lei, Y., Zhang, Y., Tang, J., Shen, G., and Yu, R. (2010) Hydroxyapatite nanoarray-based cyanide biosensor, *Analytical Biochem.*, **398**, 191–197.

1127. Tagaya, M., Ikoma, T., Migita, S., Okuda, M., Takemura, T., Hanagata, N., Yoshioka, T., and Tanaka, J. (2010) Fetal bovine serum adsorption onto hydroxyapatite sensor monitoring by quartz crystal microbalance with dissipation technique, *Mater. Sci. Eng. B*, **173**, 176–181.

1128. Wypych, G. (1999) *Handbook of fillers. 2nd Ed.*, ChemTec Publishing, New York.

1129. Almora-Barrios, N., and de Leeuw, N. H. (2010) A density functional theory study of the interaction of collagen peptides with hydroxyapatite surfaces, *Langmuir*, **26**, 14535–14542.

1130. Rhee, S. H., Lee, J. D., and Tanaka, J. (2000) Nucleation of hydroxyapatite crystal through chemical interaction with collagen, *J. Am. Ceram. Soc.*, **83**, 2890–2892.

1131. Lin, X., Li, X., Fan, H., Wen, X., Lu, J., and Zhang, X. (2004) *In situ* synthesis of bone-like apatite/collagen nano-composite at low temperature, *Mater. Lett.*, **58**, 3569–3572.

1132. Zhang, W., Liao, S. S., and Cui, F. Z. (2003) Hierarchical self-assembly of nanofibrils in mineralized collagen, *Chem. Mater.*, **15**, 3221–3226.

1133. Liu, Q., de Wijn, J. R., and van Blitterswijk, C. A. (1998) Covalent bonding of PMMA, PBMA and poly(HEMA) to hydroxyapatite particles, *J. Biomed. Mater. Res.*, **40**, 257–263.

1134. Li, J., Chen, Y. P., Yin, Y., Yao, F., and Yao, K. (2007) Modulation of nano-hydroxyapatite size via formation on chitosan-gelatin network film *in situ*, *Biomaterials*, **28**, 781–790.

1135. Ficai, A., Andronescu, E., Ghitulica, C., Voicu, G., Trandafir, V., Manzu, D., Ficai, M., and Pall, S. (2009) Colagen / hydroxyapatite interactions in composite biomaterials, *Materiale Plastice*, **46**, 11–15.

1136. Danilchenko, S. N., Kalinkevich, O. V., Kuznetsov, V. N., Kalinkevich, A. N., Kalinichenko, T. G., Poddubny, I. N., Starikov, V. V., Sklyar, A. M., and Sukhodub, L. F. (2010) Thermal transformations of the mineral component of composite biomaterials based on chitosan and apatite, *Crystal Res. Technol.*, **45**, 685–691.

1137. Boanini, E., Gazzano, M., Rubini, K., and Bigi, A. (2007) Composite nanocrystals provide new insight on alendronate interaction with hydroxyapatite structure, *Adv. Mater.*, **19**, 2499–2502.

1138. Tjandra, W., Yao, J., Ravi, P., Tam, K. C., and Alamsjah, A. (2005) Nanotemplating of calcium phosphate using a double-hydrophilic block copolymer, *Chem. Mater.*, **17**, 4865–4872.

1139. Tsuchiya, K., Yoshioka, T., Ikoma, T., and Tanaka, J. (2010) Chemical interaction between hydroxyapatite and organic molecules in biomaterials, *Ceram. Transactions*, **210**, 531–535.

1140. Grossman, R. F. (1996) Coupling agents, in *Plastics Additives and Modifiers Handbook*, 2nd edn., (ed. Edenbaum, J.), Chapman & Hall, New York.

1141. Chang, M. C., Ikoma, T., Kikuchi, M., and Tanaka, J. (2001) Preparation of a porous hydroxyapatite/collagen nanocomposite using glutataldehyde as a crosslinkage agent, *J. Mater. Sci. Lett.*, **20**, 1199–1201.

1142. Sousa, R. A., Reis, R. L., Cunha, A. M., and Bevis, M. J. (2003) Coupling of HDPE/hydroxyapatite composites by silane-based methodologies, *J. Mater. Sci. Mater. Med.*, **14**, 475–487.

1143. Dupraz, A. M. P., de Wijn, J. R., van der Meer, S. A. T., and Goedemoed, J. H. (1996) Biocompatibility screening of silane-treated hydroxyapatite powders., for use as filler in resorbable composites, *J. Mater. Sci. Mater. Med.*, **7**, 731–738.

1144. Dupraz, A. M. P., de Wijn, J. R., van der Meer, S. A. T., and de Groot, K. (1996) Characterization of silane-treated hydroxyapatite powders reinforced for use as filler in biodegradable composites, *J. Biomed. Mater. Res.*, **30**, 231–238.

1145. Liao, J. G., Wang, X. J., Zuo, Y., Zhang, L., Wen, J. Q., and Li, Y. (2008) Surface modification of nano-hydroxyapatite with silane agent, *J. Inorg. Mater.*, **23**, 145–149.

1146. Sousa, R. A., Reis, R. L., Cunha, A. M., and Bevis, M. J. (2002) Structure development and interfacial interactions in HDPE/HA composites moulded with preferred orientation, *J. Appl. Polym. Sci.*, **86**, 2866–2872.

1147. Misra, D. N. (1985) Adsorption of zirconyl salts and their acids on hydroxyapatite: use of salts as coupling agents to dental polymer composites, *J. Dent. Res.*, **12**, 1405–1408.

1148. Liu, Q., de Wijn, J. R., and van Blitterswijk, C. A. (1998) A study on the grafting reaction of isocyanates with hydroxyapatite particles, *J. Biomed. Mater. Res.*, **40**, 358–364.

1149. Tanaka, H., Yasukawa, A., Kandori, K., and Ishikawa, T. (1997) Surface modification of calcium hydroxyapatite with hexyl and decyl phosphates, *Colloids Surf. A*, **125**, 53–62.

1150. Tanaka, H., Watanabe, T., Chikazawa, M., Kandori, K., and Ishikawa, T. (1999) TPD, FTIR, and molecular adsorption studies of calcium hydroxyapatite surface modified with hexanoic and decanoic acids, *J. Coll. Interf. Sci.*, **214**, 31–37.

1151. Borum-Nicholas, L., and Wilson Jr., O. C. (2003) Surface modification of hydroxyapatite, *Part I. Dodecyl alcohol. Biomaterials*, **24**, 3671–3679.

1152. Borum, L., and Wilson Jr., O. C. (2003) Surface modification of hydroxyapatite, *Part II. Silica. Biomaterials*, **24**, 3681–3688.

1153. Li, Y., and Weng, W. (2008) Surface modification of hydroxyapatite by stearic acid: characterization and *in vitro* behaviors, *J. Mater. Sci. Mater. Med.*, **19**, 19–25.

1154. Lee, S. C., Choi, H. W., Lee, H. J., Kim, K. J., Chang, J. H., Kim, S. Y., Choi, J., Oh, K. S., and Jeong, Y. K. (2007) *In-situ* synthesis of reactive hydroxyapatite nano-crystals for a novel approach of surface grafting polymerization, *J. Mater. Chem.*, **17**, 174–180.

1155. Morita, S., Furuya, K., Ishihara, K., and Nakabayashi, N. (1998) Performance of adhesive bone cement containing hydroxyapatite particles, *Biomaterials*, **19**, 1601–1606.

1156. Shinzato, S., Nakamura, T., Tamura, J., Kokubo, T., and Kitamura, Y. (2001) Bioactive bone cement: effects of phosphoric ester monomer on mechanical properties and osteoconductivity, *J. Biomed. Mater. Res.*, **56**, 571–577.

1157. Dorozhkin, S. V. (2001) Is there a chemical interaction between calcium phosphates and hydroxypropylmethylcellulose (HPMC) in organic/inorganic composites? *J. Biomed. Mater. Res.*, **54**, 247–255.

1158. Omori, M., Okubo, A., Otsubo, M., Hashida, T., and Tohji, K. (2004) Consolidation of multi-walled carbon nanotube and hydroxyapatite coating by the spark plasma system (SPS), *Key Eng. Mater.*, pp. 254–256, 395–398.

1159. Zhao, B., Hu, H., Mandal, S. K., and Haddon, R. C. (2005) A bone mimic based on the self-assembly of hydroxyapatite on chemically functionalized single-walled carbon nanotubes, *Chem. Mater.*, **17**, 3235–3241.

1160. Kasuga, T., Yoshida, M., Ikushima, A. J., Tuchiya, M., and Kusakari, H. (1992) Bioactivity of zirconia-toughened glass-ceramics, *J. Am. Ceram. Soc.*, **75**, 1884–1888.

1161. Dorozhkin, S. V. (1999) Inorganic chemistry of the dissolution phenomenon: the dissolution mechanism of calcium apatites at the atomic (ionic) level, *Comments Inorg. Chem.*, **20**, 285–299.

1162. Dorozhkin, S. V. (2002) A review on the dissolution models of calcium apatites, *Prog. Crystal Growth Charact.*, **44**, 45–61.

1163. Furukawa, T., Matsusue, Y., Yasunaga, T., Shikinami, Y., Okuno, M., and Nakamura, T. (2000) Biodegradation behavior of ultra-high strength hydroxyapatite/poly(L-lactide) composite rods for internal fixation of bone fractures, *Biomaterials*, **21**, 889–898.

1164. Furukawa, T., Matsusue, Y., Yasunaga, T., Nakagawa, Y., Okada, Y., Shikinami, Y., Okuno, M., and Nakamura, T. (2000) Histomorphometric study on high-strength hydroxyapatite/poly(L-lactide) composite rods for internal fixation of bone fractures, *J. Biomed. Mater. Res.*, **50**, 410–419.

1165. Yasunaga, T., Matsusue, Y., Furukawa, T., Shikinami, Y., Okuno, M., and Nakamura, T. (1999) Bonding behaviour of ultrahigh strength unsintered hydroxyapatite particles/poly(L-lactide) composites to surface of tibial cortex in rabbits, *J. Biomed. Mater. Res.*, **47**, 412–419.

1166. Marques, A. P., Reis, R. L., and Hunt, J. A. (2002) *In vitro* evaluation of the biocompatibility of novel starch based polymeric and composite material, *Biomaterials*, **6**, 1471–1478.

1167. Mendes, S. C., Bovell, Y. P., Reis, R. L., Cunha, A. M., de Bruijn, J. D., and van Blitterswijk, C. A. (2001) Biocompatibility testing of novel starch-based materials with potential application in orthopaedic surgery, *Biomaterials*, **22**, 2057–2064.

1168. Habraken, W. J. E. M., Liao, H. B., Zhang. Z., Wolke, J. G. C., Grijpma, D. W., Mikos, A. G., Feijen, J., and Jansen, J. A. (2010) *In vivo* degradation of calcium phosphate cement incorporated into biodegradable microspheres, *Acta Biomater.*, **6**, 2200–2211.

1169. Ngiam, M., Liao, S., Patil, A. J., Cheng, Z., Chan, C. K., and Ramakrishna, S. (2009) The fabrication of nano-hydroxyapatite on PLGA and PLGA/collagen nanofibrous composite scaffolds and their effects in osteoblastic behavior for bone tissue engineering, *Bone*, **45**, 4–16.

1170. Hasegawa, S., Ishii, S., Tamura, J., Furukawa, T., Neo, M., Matsusue, Y., Shikinami, Y., Okuno, M., and Nakamura, T. (2006) A 5-7 year *in vivo* study of high-strength hydroxyapatite/poly(L-lactide) composite rods for the internal fixation of bone fractures, *Biomaterials*, **27**, 1327–1332.

1171. Bongio, M., van den Beucken, J. J. J. P., Leeuwenburgh, S. C. G., and Jansen, J. A. (2010) Development of bone substitute materials: from 'biocompatible' to 'instructive'. J, *Mater. Chem.*, **20**, 8747–8759.

1172. Meyers, M. A., Lin, A. Y. M., Seki, Y., Chen, P. Y., Kad, B. K., and Bodde, S. (2006) Structural biological composites: an overview, *JOM*, **58**, 36–43.

Chapter 7

The Dissolution Mechanism of Calcium Apatites in Acids

7.1 Introduction

Calcium apatites (HA and FA — see Table 1.1) appear to be of a special interest to the human being, because natural apatites are the main source of inorganic phosphorus in nature [1], while calcified tissues (bones and teeth) of mammals consist of ion-substituted calcium apatites of biological origin [2–4]. That is why such different processes like fertilizer production, dental caries, osteoporosis, as well as both *in vitro* and *in vivo* biodegradation of artificial bone grafts might be simulated by apatite dissolution in acids. The latter resulted in a great number of research papers published on apatites and calcium orthophosphates (extensively reviewed in both Refs. [5–7] and the previous chapters of this book) and their dissolution (see further in this chapter).

In chemistry, a reaction mechanism is the step-by-step sequence of elementary reactions by which the overall chemical change occurs [8]. Regarding the dissolution process of calcium apatites in acids, eight models have been already proposed to explain the processes involved. These models already provided the important information about the factors associated with solution (pH, composition, saturation, and hydrodynamics), bulk solid (chemical composition, solubility, particle sizes), and surface (defects, adsorbed ions, "history", phase transformation) of the apatite crystals. These

Calcium Orthophosphates: Applications in Nature, Biology, and Medicine
Sergey Dorozhkin

ISBN 978-981-4316-62-0 (Hardcover), 978-981-4364-17-1 (eBook)
www.panstanford.com

findings cannot be underestimated because they have already made a significant impact on understanding of the medical aspects of both dental caries and osteoporosis. Nevertheless, there are still a lot of elements missing in the overall explanation for the processes involved. To clarify possible directions of further investigations on apatite dissolution, a critical analysis of currently available models has been made in this chapter followed by an attempt to combine the data available into a general dissolution mechanism.

7.2 Critical Analysis of the Dissolution Models of Calcium Apatites

7.2.1 Diffusion (or Transport) and Kinetically (or Surface) Controlled Models

A diffusion (or transport) and a kinetically (or surface) controlled [9–26] dissolution models, as well as the two-site model [15, 27–29], are devoted to comparison of transport rates of chemical reagents (H^+ and anions of acid) from bulk solution to the surface of apatite crystals and those of products (Ca^{2+}, F^- and orthophosphate anions) obtained backwards with the kinetics of chemical interactions on the crystal surface. According to the diffusion-controlled and two-site models, a transport of either the chemical reagents and/or the products of chemical reactions mainly determine the overall dissolution rate, whereas, the kinetically controlled model emphasizes chemical transformations on the surface as the limiting factor (Fig. 7.1) [30]. An important consequence of the two types of rate-controlling mechanism is that the solution immediately next to the crystal surface is undersaturated with respect to apatite in the surface controlled process and this solution is saturated in the diffusion-controlled process [31]. Both models usually operate with a so-called driving force which means either a concentration gradient within the Nernst diffusion layer (the diffusion controlled model) or a gradient of ionic chemical potentials between the apatite crystal surface and bulk solution (the kinetically controlled model).

To investigate whether a dissolution process follows either of these models, one should study the influence of hydrodynamics, concentration of reagents, ionic strength, solution undersaturation, pH, temperature, crystal dimensions, and the presence of chemical modifiers on the dissolution kinetics. A rotary disk [10, 13, 14], constant composition [9, 16, 18, 19, 20, 32–34] or dual constant composition [22, 34, 35] techniques are used for experimental investigations. The results obtained are usually plotted as an uptake of H^+ ions (e.g., as a titrant volume added) and/or a release of calcium, orthophosphate, and fluoride ions versus the dissolution time. Afterwards, calculations of the numeric values for activation energy, rate constants, effective reaction order, diffusion layer thickness, characteristic adsorption impedance, diffusive jump distance, etc., are performed [9–29, 32–36]. The conclusion on whether a dissolution process is kinetically or diffusion controlled is made based on the numeric values calculated. For example, when analysis of the rate data gave a numeric value of the effective reaction order n = 6, a surface controlled dissolution mechanism was suggested [22].

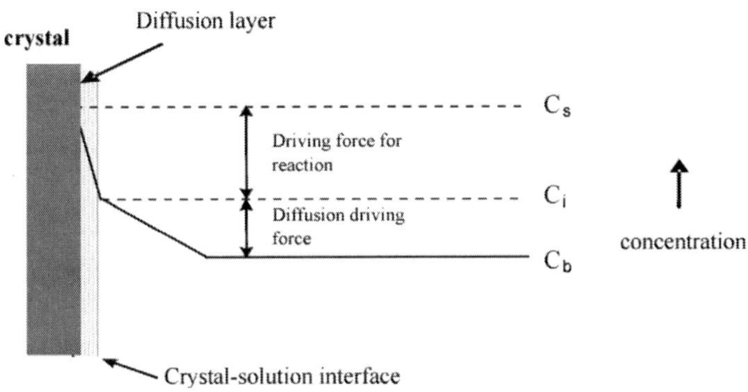

Figure 7.1 A dissolution process according to the diffusion and kinetically controlled models. Here: C_s — solute concentration on the surface, C_i — solute concentration on the interface, C_b — solute concentration in bulk. Reprinted from Ref. [30] with permission.

Basic thermodynamic principles predict that dissolution rates should increase with increasing driving force or chemical potential; however, the experimental studies show that this dependence is complex. Namely, dissolution of apatite in some cases was found to be diffusion controlled [19, 20, 23, 25], in some other cases—kinetically controlled [21, 22] and even intermediate (i.e. both kinetically and diffusion controlled) in still other cases [14]. Furthermore, the dissolution process might be composed of two stages: the first one was found to be both surface and diffusion controlled, while the second one was an exactly diffusion-controlled dissolution [24]. Therefore, the results obtained for these models are valid within the experimental conditions studied only; no extrapolation can be made outside the ranges studied. For example, after, say, an agitation decreasing or temperature increasing, an initially kinetically controlled dissolution might be controlled by diffusion [26].

Furthermore, in some cases, dissolution curves (a titrant volume against time) reach plateaus prior to complete dissolution of HA, thus indicating a creation of metastable states in which the reaction is effectively terminated even though apatite crystals remain in contact with the undersaturated solutions [37–48]. There is an opinion [49] that this effect resulted from a false solubility product value for HA. Thus, a high sensitivity to the experimental conditions used appears to be the main drawback of these models.

7.2.2 Mono- and Polynuclear Models

The surface dissolution may be described by three different models shown schematically in Fig. 7.2. As may be seen in Fig. 7.2A it is possible to start dissolution from one center from which originates the units detachment leading to step disintegration. This is a mononuclear dissolution. It is also possible that detachment takes place simultaneously from multiple centers (Figs. 7.2B and 7.2C) which may be developed in one or more steps on the crystal surface. This is a polynuclear dissolution. Alternatively, it has been suggested that the crystal units' detachment takes place along spirals originating from an active site on the crystal surface following an Archimedean spiral (Fig. 7.2D). The unit detachment along the spirals takes place at constant velocity [30].

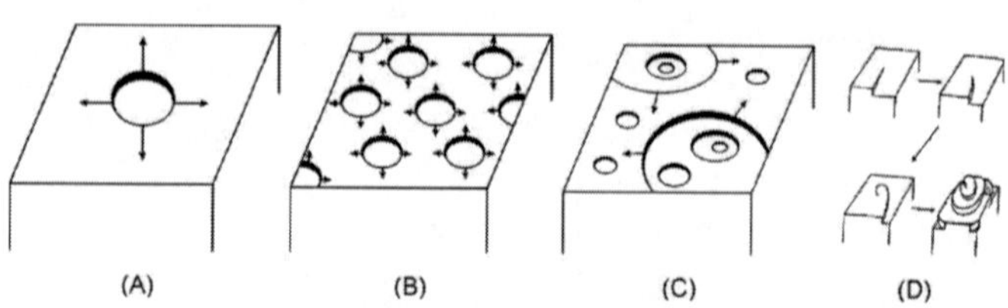

Figure 7.2 (A) Mononuclear model. (B) Polynuclear model in one crystal step. (C) Polynuclear model in multiple steps: birth and spread. (D) Spiral model. Reprinted from Ref. [30] with permission.

A polynuclear model [50–59] was created from the study on apatite dissolution and growth kinetics under the constant composition conditions. Polydispersed samples of apatite were put into a stirred undersaturated (for dissolution experiments) or supersaturated (for those on crystal growth) solutions and the solution pH and amount of chemicals added (an acid for the dissolution experiments and a base for those on crystal growth) were permanently recorded as functions of time. The results obtained were plotted versus either undersaturation or supersaturation values: straight lines were obtained in the specific logarithmic coordinates typical for this model [60]. According to this model, dissolution nuclei, i.e. collections of vacant sites for Ca^{2+}, PO_4^{3-} and OH^- ions, are formed on the crystal surface of apatite and spread over the surface with a definite lateral rate [51, p. 30].

In polynuclear model, the nucleation rate is assumed as a function of the mean ion activity. A lateral growth rate of the nuclei is assumed proportional to the difference between the total concentration of calcium ions in the saturated solution and in a solution, while the rate constant is related to the frequency for calcium ions to make a diffusion jump into a kink and, simultaneously, partly dehydrate [57, 58]. Investigations on apatite dissolution and crystal growth using atomic force microscopy revealed that the rate-determining step was not the diffusion but two-dimensional surface nucleation [61, 62]. These observations provided a valuable support for the polynuclear model.

The experimental data on dissolution and growth rates versus undersaturation ($S < 1$) and supersaturation ($S > 1$) values, respectively, were obtained. Numeric values of surface tension (40 mJ/m^2 for apatite dissolution and 100 mJ/m^2 for apatite growth [57, 58]), ionic frequencies to enter or leave a kink (1.6×10^5/s [56]), the mean distance between kinks in a surface nucleus (approximately 3.07 Å [55]) and the critical nuclei for both HA [50–54, 57, 58], and FA [55, 56] were calculated. Furthermore, within $5.0 < pH < 7.2$ the dissolution kinetics was found to follow the polynuclear model, while that of growth appeared to follow the polynuclear model within $5.5 < pH < 6.5$ and the combined mono-polynuclear model at $pH = 5.0$ [55, 56].

A new idea about hydroxide ions formation at the hydroxide sites of HA from water molecules trapped under the crystal surface appeared in the most recent papers of this series [57, 58]. Probably, this might be a reason for the above changing in the growth model found for HA at pH = 5.0. To conclude, the polynuclear model seems to be better elaborated among others because it is valid for apatite dissolution within solution pH of 5.0–7.2, saturation degrees of $0.1 < S < 7$, and Ca/P molar ratios of 0.1–20 [58]. However, nothing has been published on a validity of this model in more acidic solutions.

7.2.3 Self-Inhibition (Calcium-rich Layer Formation) Model

A model with a self-inhibition layer formation describes a formation of a calcium-rich layer on the surface of apatite during dissolution [19, 44–48]. This model was created from the results on dissolution kinetics of apatite powders in acidic buffers at solution pH within 3.7–6.9 (the constant composition technique was used). Following concentration of calcium and solution pH, the authors found that during the initial period (2–5 min) of dissolution release of calcium into the bulk solution was always less when compared with consumption of H^+. When dissolution progressed, rates of calcium release permanently increased and hydrogen uptake decreased until the equivalence (consumption of two hydrogen resulted in release of one calcium) was reached, while the over-all rate of apatite dissolution permanently decreased [19, 44–48]. Similar decreasing of apatite dissolution rates were obtained by other research groups [37–43, 63].

According to this model, apatite is dissolved by ionic detachment of calcium and orthophosphate ions from the surface to a solution. When an initial portion of apatite has been dissolved, some amount of calcium cations (presumably, in connection with anionic counter ions) is returned from the solution and adsorbed back onto the surface of apatite. The latter results in formation of a semi-permeable ionic membrane consisting of a positively charged layer containing strongly adsorbed calcium ions, i.e. a calcium-rich layer is formed [19, 44–48]. An indirect confirmation of the presence of a calcium-rich layer was obtained in the most recent paper of this series: for initial strong calcium undersaturation in solutions, attainment of a pseudo-steady state was very long [48]. The areas per adsorbed single PO_4^{3-} and Ca^{2+} ions on HA surface at the point of zero charge (pH = 8.5) were estimated to be 33.1 ± 2.7 and 23.0 ± 2.1 Å^2, respectively [64].

The authors of calcium-rich layer formation model suggested that the rate control of apatite dissolution was diffusion of calcium and/or orthophosphate ions through an interfacial surface layer formed during the initial portion of the dissolution process. This interface would be composed of the Nernst layer that would be adjacent to the calcium-rich layer. This last one would present low permeability making the dissolution process slower [19, 44–48]. It is important to notice that these conclusions were made from the experiments on following calcium concentration and solution pH only; no measurements of the surface composition of apatites were performed.

As said by this model, ionic diffusion in solution results in a situation when ionic concentration near the surface layer of apatite is always higher than that in the bulk solution. Calcium concentration in the solution is also inhomogeneous: it decreases when the distance from the liquid/solid interfaces increases [19, 44–48]. Therefore, the calcium-rich layer appears to be undoubtedly rich in calcium when compared with the bulk solution only. However, when compared with the bulk of solid apatite (Ca/P = 1.67) this layer appears to be calcium-rich (Ca/P > 1.67) when the counter anions A^{n-} (it is assumed, that apatite is dissolved in acid H_nA, where A^{n-} is an anion) adsorbed from the solution are not taken into account. If the counter anions are calculated, a numeric value for Ca/(P+A) molar ratio on the surface will be around 1.67 in the case of zero charge on the surface.

On the other hand, ions of calcium might be adsorbed without the counter anions. This results in an electrical double layer formation on the solid/liquid interface with positive charge on apatite. Indeed, apatite is charged positively in acidic solutions. This effect was explained by adsorption of either protons only [65] or protons and calcium cations from the solution [66, 67]. Certainly, the presence of a positive charge on apatite surface is in favor for the calcium-rich layer model but one should not forget that this charge might be caused by adsorption of protons only [65]. The dissolution rate reduction found by the authors and explained by the state of the interface (adsorption of calcium [48]) might also be explained by adsorption of some impurities those always present in the solution. Other explanations are also available [37–43]. Thus, unless the direct measurements of the calcium content on the surface are made, this model cannot be considered as experimentally proven.

A contradiction between the calcium-rich layer formation and polynuclear models can be outlined: sorption of calcium is likely to happen on the dissolution nuclei. In this case, some amount of the nuclei should disappear, as a result. No suggestions how to overcome this contradiction were found in the references.

To conclude this model, one should notice that due to the formation of a calcium-rich layer, a non-stoichiometric (incongruent) dissolution (see the next model) is suggested by default.

7.2.4 Stoichiometric/Non-Stoichiometric (congruent/incongruent) Dissolution

A congruent/incongruent (or stoichiometric/non-stoichiometric) dissolution [11, 17, 19, 21, 36, 45, 63, 68–84] is based on the direct measurements of either the ionic concentrations in a solution [19, 21, 36, 63, 69–71, 75, 78, 79, 82, 83] or the surface composition of apatites during dissolution [17, 72, 76]. According to the definitions, ions of a solid might be dissolved either simultaneously with the dissolution rates proportional to their molar concentrations (a congruent dissolution) or non-simultaneously with different values of the dissolution rates for each ion (an incongruent dissolution). The latter case always results in a situation when a surface layer with a chemical composition different from that of the bulk of solid is formed.

For acidic dissolution of FA, ions of F^- were found to dissolve faster (or previously) when compared to ions of calcium and orthophosphate [17, 63, 72, 78, 83, 84]. A similar phenomenon of previous (or faster) dissolution of calcium when compared to that of orthophosphate was also found [36, 70, 71, 74–77]. In the latter case, a surface layer of acidic calcium orthophosphates, presence of HPO_4^{2-} ions and/or Ca-depleting of the surface were either suggested [47, 56, 80] or found [76, 81, 83, 84]. Thus, if incongruent dissolution occurs from a stoichiometric HA surface, the surface Ca/P ratio will become lower than for the bulk and a CDHA (see Table 1.1) must then be present, even if there is no phase change [49]. However, there are contradictory results. For example, when FA was dissolved, the solution concentrations of calcium and fluoride ions were non-stoichiometrically lower with respect to that of orthophosphate ions due to a possibility that some amount of sparingly soluble calcium fluoride was precipitated [36, 81, 83–85]. Furthermore, one should mention on a study, in which release of calcium into the bulk solution was always less, when compared with release of orthophosphate [63]. On the other hand, having followed the release kinetics of calcium and orthophosphate ions into solution, apatites were found to dissolve stoichiometricaly (congruently) [21, 68, 86].

Interestingly, but apatites might be dissolved stoichiometrically or non-stoichiometrically depending on the experimental conditions. For example, CDHA was found to dissolve congruently at the solution pH < 4.5, while an incongruent dissolution with Ca/P ratio in solution attaining values as high as ~25 was detected in solutions within 4.5 < pH < 8.2. Furthermore, almost negligible dissolution with another type of a non-stoichiometry with Ca/P ratio close to zero in solutions was found at solution pH > 8.2 [82]. For 4.5 < pH < 8.2 conditions, the authors explained the incongruent dissolution by re-adsorption of already dissolved orthophosphate anions from the solution onto the positively charged CDHA surface. While in more basic solutions, the surface of CDHA was charged negatively and, thus, the authors suggested a possibility of both re-adsorption of already dissolved calcium cations onto CDHA surface and desorption of orthophosphate ions from the CDHA surface. However, it has to be realized that at solution pH > 8.2 the uncertainties in the chemical analysis of Ca (in particular) and P were relatively high, making any quantitative interpretation of this re-adsorption/desorption difficult [82]. The same effect was discovered in another study [84]. Furthermore, dissolution of both FA and natural sedimentary carbonated FA was found to be incongruent in the early stages but tended toward congruency as the dissolution reaction proceeded to steady state [36, 63, 87]. According to the authors, this initial incongruency was likely a result of the exchange of solution H^+ ion for mineral bound Ca [36]. Besides, a presence of simple inorganic salts (Na_2SO_4, CH_3COONa, or NaCl) in solution appeared to have an influence on the equilibrium concentrations of orthophosphate and calcium ions released from apatite: the order of salts increasing orthophosphate in solution was Na_2SO_4 > CH_3COONa > NaCl, while that increasing calcium was opposite [79]. This effect was explained by a specific affinity of some ions (e.g., sulfate) for apatite and might be a reason for transition of congruent to incongruent dissolution in some cases [88]. Similarly, presence of other compounds might influence the dissolution congruence [89]. Other reasonable explanations might be due to either experimental difficulties in precise following of Ca/P ratio both in the solution and on the crystal surface or differences in chemical and structural compositions of apatites studied by different researchers (both FA and HA are known to be often non-stoichiometric and calcium deficient, especially on the surface [2–4, 82–84, 90]).

The idea on a great influence of a surface layer is supported by the data on dissolution of stoichiometric HA (Ca/P = 1.67) in aqueous solutions within pH 4.90–9.94 [71]. The authors found that after removal of an irreversible surface, which generated excessive calcium and orthophosphate solution concentrations, further behavior of HA became reproducible. A surface of one atom-layer thickness or less was believed to have formed during the preparation of the solid [71, p. 697]. On the other hand, dissolution of HA in aqueous medium appeared to be always non-stoichiometric at the beginning, but when the solid was successively equilibrated at any given pH, the solution Ca/P ratio approached a limiting value of 1.67. Once this value was reached, the solid only maintained this solution ratio by dissolving stoichiometricaly [71, p. 700]. The authors explained these phenomena by a previous history of the samples studied. A similar conclusion was also made in another paper [69]. Thus, the experimental techniques of apatite preparation might have a great influence to stoichiometry/non-stoichiometry of dissolution. This is a serious drawback.

Finally, the results on theoretical analysis of surface layer formation on HA should be reported [80]. The main point of that paper is an idea that both the stoichiometric HA (Ca/P = 1.67) and any non-stoichiometric CDHA (Ca/P ratio within 1.5–1.67) might be described as the same substance. More precisely, there is a continuous amount of intermediate compounds within $1.5 \leq$ Ca/P ≤ 1.67 having the same crystal structure which is drawn as a line on the solubility diagram $CaO–P_2O_5–H_2O$ [80]. Furthermore, according to the authors, at the steady state, both HA and CDHA are dissolved congruently only when both the solution and solid have the same Ca/P ratio. Therefore, any sample of HA and CDHA is dissolved incongruently except the only one solution composition having the similar Ca/P ratio with that particular sample. In all other cases, a surface layer having a Ca/P ratio less than 1.67, which is often used as the indication of incongruent dissolution of apatite, will form as a result (this, however, depends on the accuracy with which the analyses can be made) [80]. Thus, whether apatites are dissolved congruently or incongruently might also depend on both their initial chemical composition and the solution properties.

To conclude, one should notice that all the aforementioned cases of incongruent dissolution of both FA and HA describe the formation of a surface layer of a CDHA and/or acidic calcium orthophosphates, which is calcium-depleted when compared to the bulk apatite. Thus, there is a disagreement between the incongruent dissolution and calcium-rich layer formation model. Perhaps, this might be due to the differences in experimental conditions.

7.2.5 Chemical Model

A chemical dissolution model was proposed as well [91–93]. This model was developed from a self-evident supposition that it would be highly unlikely if apatite were dissolved by detachment of "single molecules" equal to the unit cells and consisting of 18 ions. Moreover, in the crystal lattice all ions are shared with neighboring unit cells and often cannot be attributed to the given "single molecule". Therefore, based on the experimental results obtained on the one hand and an analysis of the data found in references on the other hand, a sequence of four successive chemical reactions:

$$Ca_5(PO_4)_3(F, OH) + H_2O + H^+ = Ca_5(PO_4)_3(H_2O)^+ + HF, H_2O \quad (7.1)$$

$$2Ca_5(PO_4)_3(H_2O)^+ = 3Ca_3(PO_4)_2 + Ca^{2+} + 2H_2O \quad (7.2)$$

$$Ca_3(PO_4)_2 + 2H^+ = Ca^{2+} + 2CaHPO_4 \quad (7.3)$$

$$CaHPO_4 + H^+ = Ca^{2+} + H_2PO_4^- \quad (7.4)$$

was proposed to describe the chemical processes of apatite dissolution [91–93]. According to this model, Eqs. (7.1)–(7.4) should be used instead of the well-known net-reactions (7.5) and (7.6):

$$Ca_5(PO_4)_3(F, OH) + 7H^+ = 5Ca^{2+} + 3H_2PO_4^- + HF, H_2O \quad (7.5)$$

$$Ca_5(PO_4)_3(F, OH) \rightarrow 5Ca^{2+} + 3PO_4^{3-} + F^-, OH^- \quad (7.6)$$

Chemical model (7.1)–(7.4) was initially created for description of apatite dissolution in strong acidic media [92] and appeared to be a further development of some ideas proposed earlier. For example, already in 1960s reactions (7.5) and (7.6) had been suggested "not to give the mechanism but merely express net-reactions only" [94, p. 333]. Later, Pearce, [95] and Chow [31] discussed which of net-reactions (7.5) or (7.6) described the dissolution process better.

For acidic dissolution of apatites, net-reaction (7.5) represents the direct chemical interaction between the surface of solid apatite and an acidic solution. However, Pearce suggested that, in principle, the dissolution process might also happen according to net-reaction (7.6) followed by a chemical interaction in the solution among the ions of apatite and acid near the crystal surface [95]:

$$5Ca^{2+} + 3PO_4^{3-} + F^-, OH^- + 7H^+ = 5Ca^{2+} + 3H_2PO_4^- + HF, H_2O \qquad (7.7)$$

In spite of a lack of the experimental techniques, able to follow the detachment of single ions, a reasonable ionic description of apatite dissolution as a system of chemical equations (7.1)–(7.4) was created from the indirect measurements [91–93]. According to this model, chemical dissolution of apatite occurs step by step on the surface via intermediate formation of a mono-atomic layer of some virtual compounds, like $Ca_3(PO_4)_2$ and $CaHPO_4$ [91–93]. Both latter substances are known as precursors of apatite formation during chemical crystallization [3, 96]. Moreover, based on the sequence of ionic detachment (first fluorine for FA or hydroxide for HA, followed by calcium and afterwards by orthophosphate) it is suggested that dissolution of apatites is always incongruent (non-stoichiometric) at the atomic (ionic) level. However, a lack of the precise experimental verification transforms the chemical model into a reasonable hypothesis only.

To conclude this model, one may assume that some new relations between the kinetically controlled and chemical models might be established in future. Before the chemical model was created, chemistry of apatite dissolution had been always described by equations (7.5) or (7.6) + (7.7). For this reason, the kinetically controlled model was also adapted for these equations. For example, the overall rate constants for net-reactions (7.5), (7.6), and (7.7) are still used for calculations, whereas they are probably equal to either the lowest of the rate constants of (7.1)–(7.4) (the rate-determining step) or some combination.

7.2.6 Etch Pit Formation

A process of etch pit formation describes acidic dissolution of apatite crystals containing structural defects (dislocations and inclusions) [37–43, 87, 97–106]. A radius of dislocations outlets (hollow cores) on apatite was calculated to be within 8.3–20 Å for screw dislocations and 18–43 Å for edge ones [100]. The presence of dislocations accelerates dissolution, because dislocations give rise to continuous steps on the surface (Fig. 7.2D) and the strain energy they cause in crystals favors etch pit formation. Thus, the dissolution process of solids is induced by formation of pits (Fig. 7.3) and continues with spreading of their stepwaves [107, 108]. As a result, the surface of apatite becomes rough and the total edge length and edge free energy increases [37–43]. These pits provide dissolution sites and the entire reaction proceeds via nucleation and growth of the pits accompanying step flow. The pits appear at the dislocation outlets, usually they are 0.1–10 μm in size (their dimensions depend on dissolution kinetics and dissolution time: they increase when the dissolution progresses; furthermore, the crystals must be sufficiently big to provide enough room for large pits to form). For dissolution of apatites, the pits have a hexagonal shape (Fig. 7.3) according to the crystal symmetry $P6_3/m$ of pure HA and FA [97–102, 109–112]. Furthermore, their walls and bottom consist of crystal faces possessing the highest dissolution rates.

However, in the case of nearly physiological conditions and sufficiently low solution undersaturation, a free energy barrier becomes too high for vacancy nucleation to occur on a time scale that is competitive with other processes. Thus, the dissolution process of apatites becomes spontaneous only when etch pits of critical sizes (determined by the Gibbs-Thomson effect, a well-known thermodynamic principle) are reached [37–43]. At this critical size, the free energy change goes through a maximum that defines an energy barrier to a pit formation. Therefore, only relatively large pits (of sizes greater than a critical value) appear to be active, with stepwaves contributing to dissolution, while the spreading velocities are also dependent on the pit sizes, decreasing with pit size decreasing. Moreover, during dissolution, the crystals become smaller and the average lengths of dissolution steps decrease (which leads to a decrease in dissolution rates), and approach the critical value. When dimensions of dissolving crystals is sufficiently reduced, in some cases dissolution is thought to be dynamically stabilized (dissolution suppression) due to a lack of space for active pits/defects formation on the very small crystallite surfaces [37–43]. Although, the authors claim this effect as "a new dissolution model incorporating particle size considerations" [37–43], in fact, they just have introduced some boundary conditions (the minimal pit sizes and crystal dimensions) to the etch pit formation process. One should remind, that a similar effect of the dissolution self-inhibition was detected by the authors of the calcium-rich layer formation model [19, 44–48], although the interpretation of this effect was quite different (see above).

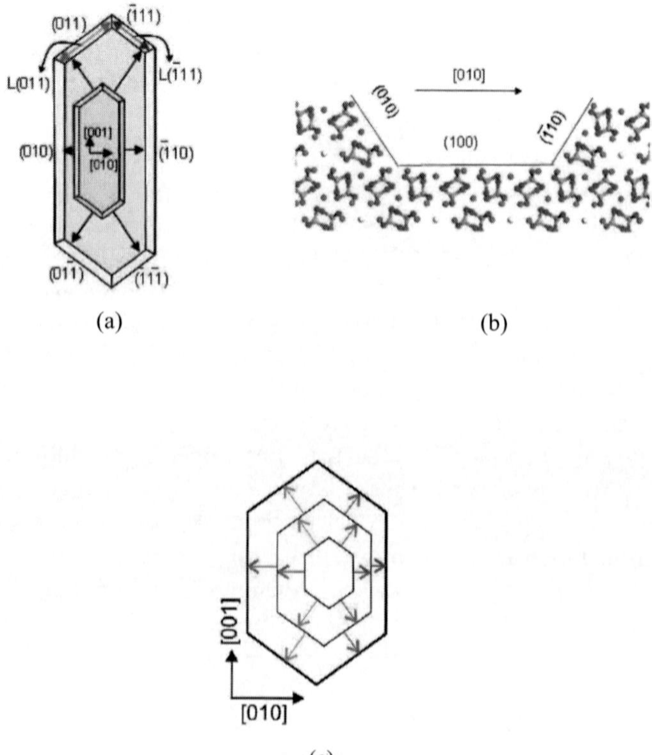

Figure 7.3 A typical etch pit on the surface of calcium apatites. (a) Top view and a model of its evolution during dissolution. Arrows represent relative step velocities displaying different facets. (b) View through the [001] direction in the (100) surface. The exposed step faces can be seen to be un-equivalent owing to the opposite orientations of the orthophosphate groups. Reprinted from Ref. [109] with permission. (c) Schematic of the evolution of hexagonal etch pits. Four differently colored arrows represent relative step velocities. Reprinted from Ref. [111] with permission.

Different acids were used in experimental investigations but, unless the conditions of surface coating creation [113], no specific influence of the acids to etch pit formation and growth was found

[97–106]. A theoretical basis for the pit formation and growth at the surface outlets of dislocations includes detachment of single ions from the kink sites. The bulk rate associated with dissolution stepwaves arises quite naturally from the equations describing the spreading of step trains from structure defects and densities [39]. However, until recently, the experimental investigations on pits formation and growth on apatites were made when their dimensions became at least 0.3–0.5 μm and even larger. Smaller dimensions of pits were out of the range of old microscopes. That is why, the initial stages of etch pits formation are not quite clear yet.

The investigations on apatite dissolution using atomic force microscopy provided new data on this point [37–43, 61, 62, 109–112]. For example, the growth process of a single pit on apatite was followed since its dimensions were 50 nm [61, 62]. Therefore, one may expect new results on even smaller dimensions of pits to appear soon.

7.2.7 Ion Exchange Model

An ion exchange dissolution model is based on a supposition about adsorption of protons [36] and anions of acid (e.g., citrate anions [114–116]) from a solution onto the surface of apatite and removing ions of calcium and orthophosphate into the solution instead. An uptake of citric acid [114] and citrate ions [115] from dilute (4–12.5 mM for citric acid [114] and 0.05–1 mM for sodium citrate [115]) aqueous solutions by a synthetic HA was found to be time independent and follow Langmuir isotherms rather good. However, it was not the case when solutions of higher concentrations were used. In more concentrated solutions (13–100 mM of citrate), this interaction was "reactive" and not "adsorptive" [114, p. 1421].

Adsorption was found to take place by ionic exchange of orthophosphate by citrate ions at the solid-solution interface, caused by a higher affinity of citrate than orthophosphate species for the Ca-sites on the surface. Citrate ions were proposed to interact in two ways: $C_6H_8O_7^{3-}$ interacted weakly in a bi-dentate manner (1 citrate per 2 calcium sites), whereas $C_6H_9O_7^{2-}$ interacted strongly in a mono-dentate manner (1 citrate per 1 calcium site) [115].

Later this model was updated by investigation of apatite dissolution in other acids and an adhesion-decalcification concept was introduced [117, 118] According to this concept, the mechanism of acidic interaction with HA was found to involve two phases. In the first phase, an acid bonds to surface calcium of HA with formation of calcium-acid complexes. Depending on both a solubility and a diffusion rate of the calcium-acid complexes from the HA surface to a solution, the acid will in the second phase either remain attached to the HA surface with only limited decalcification involved or the calcium-acid complexes will detach, resulting in a substantial decalcification effect. Rather similar data, in which carboxylic groups of polyalkenoic acid were adsorbed on the surface of an HA substrate, replaced orthophosphate ions and made ionic bonds with calcium ions of HA, were obtained in another study [119].

Recently, *ab initio* total energy methods were used to study the atomic structure and surface chemistry of HA [120]. All surfaces of HA were found to react strongly with water and a loss of Ca from the surface in exchange for two H appeared to be very favorable. Since these calculations were performed for the case of water adsorption on HA, it remained unclear whether they can be applied to the acidic solutions. Nevertheless, other researchers found this effect experimentally [81, 83]. Furthermore, another type of the exchange was proposed for HA, in which one calcium and one hydroxyl were replaced by one proton [82].

To conclude this model, the time independent ionic exchange of orthophosphate by citrate was detected in dilute solutions only. In more concentrated solutions it transferred into chemical reaction (7.5) followed by time dependent precipitation of calcium citrate when its solubility was exceeded [114, 116]. Application of this model to HA dissolution in other acids revealed its dependency on both the solubility and the diffusion rate of the calcium-acid complexes [117, 118]. Thus, new experimental data are necessary, to elaborate the ion exchange model further.

7.2.8 Hydrogen Catalytic Model

A hydrogen catalytic model is based on a reasonable suggestion about adsorption of protons onto negatively charged oxygen ions of the surface orthophosphate groups of apatites [121]. Sorption of protons results in partial transformation of surface PO_4^{3-} groups into HPO_4^{2-} and catalyses the dissolution process [9, 10, 15, 18, 20, 48–56, 84]. A schematic representation of this process at different solution pH is shown in Fig. 7.4 [110]. Besides, recent computer simulations revealed that both OH^- (for HA) and F^- (for FA) located on the surface are protonated as well and, of them, fluoride ions exhibited much lower proton affinities (pK_B = 10.8, a weak base) compared to hydroxide (pK_B = 0, a strong base) [122]. Furthermore, the surface protonation of apatites is intimately related to hydrogen bonding. The details on the latter process are available in literature [123].

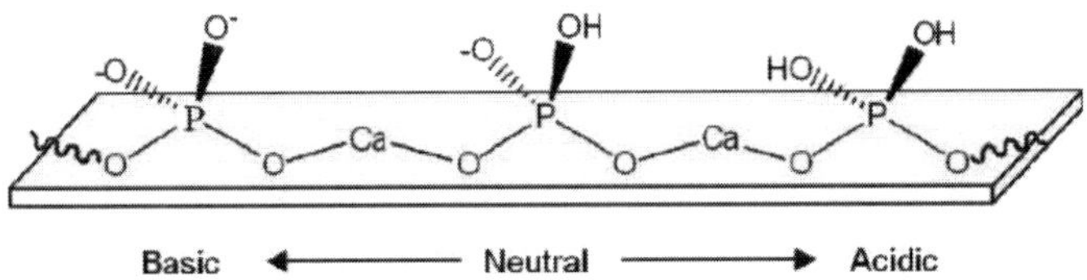

Figure 7.4 A schematic representation of surface protonation of apatites at different solution pH. Reprinted from Ref. [110] with permission.

One can notice, that the authors of different dissolution models (the diffusion and kinetically controlled [9, 10, 15, 18, 20] and polynuclear [51–56] models) have suggested a similar possibility of a catalytic effect of the adsorbed protons. At present, indirect data only are available to support this hypothesis. Namely, at neutral pH, dissociation of ions is penalized by more than 150 kJ/mol giving rise to very stable apatite-water interfaces [122]. This picture changes drastically with decreasing pH, as the protonation of orthophosphate and hydroxide ions lowers the free energy of calcium ions dissociation. An obvious trend in apatite dissolution is that the dissolution rate is increased as the solution pH decreases, which is a general phenomenon observed in dissolution of other minerals. This indicates that dissolution proceeds by surface protonation, which promotes breaking of Ca–O bonds.

In the course of proton uptake, the charge of the surface anions is reduced and, the Coulomb attraction to the crystal surface is lowered considerably, which might be considered as "catalysis". Currently, no precise information about a catalytic activity of adsorbed protons on the dissolution of apatites is available. However, it does not mean that this effect does not exist. Up to now, nobody has ever succeeded to reveal a catalytic effect unambiguously due to a lack of an experimental technique able to provide the experimental data necessary. Obviously, the general idea on a catalytic activity of the adsorbed protons seems reasonable and, perhaps, this effect will be better elucidated in future.

7.3 Summary on the Dissolution Models

There is an old tale on six blind men, who wanted to determine what an elephant looked like. An elephant was delivered and the men were allowed to feel different parts of its body. The blind man who felt a leg said the elephant was like a pillar; the one who felt the tail said the elephant was like a rope; the one who felt the trunk said the elephant was like a tree branch; the one who felt an ear said the elephant was like a hand fan; the one who felt the belly said the elephant was like a wall; and the one who felt the tusk said the elephant was like a solid pipe. Thus, six blind men appeared to be in complete disagreement. Then, a wise man explained to them: "All of you are right. The reason every one of you is telling it differently is because each one of you touched the different part of the elephant. So, actually the elephant has all the features you mentioned." The story is widely used to indicate that the reality may be viewed differently depending upon one's perspective, suggesting that what seems the absolute truth may be relative due to a deceptive nature of half-truths.

As the complete physical picture that explains widely observed variations in dissolution behavior is still lacking, a similar approach must be adapted to the aforementioned dissolution models of calcium apatites. Undoubtedly, each from the aforementioned models appears to have limitations and drawbacks of its own and none of them is able to describe the apatite dissolution in general. Moreover, most of the models were elaborated for apatite dissolution in either slightly acidic or nearly neutral ($4 < pH < 8$) aqueous solutions, relatively small values of solution undersaturation and temperatures of 25 and 37°C only. Nothing is known about the validity of these models for apatite dissolution in strong inorganic acids (HCl, HNO_3, H_2SO_4, H_3PO_4) for solution $pH < 2$ and temperatures above 70°C. Moreover, in such "tough" experimental conditions, the numerical value of solution undersaturation by HA and FA becomes uncertain due to impossibility to determine their solubility [1–4]. The latter cases serious problems for application of current versions of calcium-rich layer formation and polynuclear models to apatite dissolution in strong acids, because both models operate with the solution supersaturation/undersaturation phenomena. The abovementioned "tough" dissolution conditions are common in industry of fertilizer production [124] and have to be taken into account by the final version of any dissolution model.

However, there are no doubts that each of the models discussed is correct for the specific experimental conditions studied. The different models are just devoted to diverse aspects of the same process of apatite dissolution and are likely to complement each other. Undoubtedly, the dissolution process of calcium apatites has all the features elaborated within eight dissolution models. Unfortunately, it is still impossible to obtain the complete understanding for all the processes involved because experimental techniques able to provide direct following after diffusion, sorption, detachment, and transformation of the single atoms, ions, and molecules currently are lacking. Therefore, researchers are still forced to make conclusions about the events happening at the atomic (ionic) level based on indirect measurements and/or calculations only. That is why, combining the experimental and theoretical findings, obtained by different investigators (often under hardly comparable experimental conditions), currently, seems to be the only way to create a general depiction of the dissolution mechanism of calcium apatites in acids.

7.4 A Reasonable Classification of the Dissolution Models

At the fundamental level reactions between solids and liquids involve a coupled sequence of mass transport, adsorption/desorption phenomena, heterogeneous reaction, chemical transformations of intermediates, etc., whose identification, separation, and kinetic quantification are all necessary if the mechanism of the process is to be fully understood and described [125]. It is also generally agreed that during dissolution lattice ions undergo the following processes: (1) detachment from a kink site, (2) surface diffusion away from the crystal steps, (3) desorption from the surface and (4) diffusion into the bulk solution [126]. Therefore, the steady-state conditions of apatite dissolution in aqueous acidic media include the following simultaneous processes or steps:

1. diffusion of chemical reagents (H^+ and anions of acid A^{n-}) from bulk solution to the solid/liquid interface;
2. adsorption of the chemicals onto the surface of apatite;

3. chemical transformations on the surface;
4. desorption of products (ions of fluoride, calcium, and orthophosphate) from the crystal surface;
5. their diffusion into the bulk solution.

All the steps mentioned above are likely to be much more complicated. For example, processes 1 and 5 include chemical transformations happen with the ions during diffusion, because solution pH is known to depend on the distance from the solid/liquid interface (in acidic solutions it is higher near the surface of apatite and decreases when the distance increases) [127, 128]. Processes 2 and 4 include ionic diffusion along the surface both to (step 2a) and away from (step 4a) the crystal steps, respectively, as well as a diffusive jump (steps 2b and 4b) [126]. Finally, process 3 consists of several successive chemical transformations (step 3a — chemical equations (7.1)–(7.4)) [91–93] and includes ionic detachment from the kink sites (step 3b) [126]. Moreover, for adsorption (step 2) to take place, the following intermediate steps are necessary: (i) dehydration of a surface site, (ii) partial dehydration of ions and (iii) a diffusive jump toward the surface [126, p. 1692]. One can also expect three similar intermediate steps (diffusion jump from the surface, hydration both the ions and the surface site) for desorption (step 4).

As shown above, different models describe just diverse aspects of the same phenomenon of apatite dissolution. Commonly, the main distinction is made between the transport control and surface-reaction control. For example, in the diffusion controlled, two-site and polynuclear models steps 1 and 5 limit a reaction rate. On the contrary, in the kinetically controlled, chemical and hydrogen catalytic models a reaction rate is limited by step 3. In the cases of the ion exchange and self-inhibition (or calcium-rich layer formation) models, a reaction rate is likely to be controlled by both ionic diffusion (steps 1 and 5) and kinetics of the adsorption/desorption phenomena (steps 2 and 4). Processes of etch pit formation and dissolution stoichiometry—are the experimental facts, not models. Thus, they may be controlled by factors pertaining to any step from 1–5 and even all of them.

7.5 Brief Information on Apatite Structure

The crystallographic data on HA and FA are mentioned in Table 1.3, while the crystal structure of HA is shown in Fig. 2.11. Briefly, crystals of HA and FA consist of individual orthophosphate tetrahedra linked by calcium cations. The orthophosphate groups forms the skeleton having two types of calcium sites. The first (Ca(1) site) is occupied by four calcium atoms that have a nine-fold coordination, while the second (Ca(2) site) is occupied by six calcium atoms having a seven-fold coordination. Fluoride (for FA) or hydroxide (for HA) anions are located within the same channels as the Ca(2) sites and have a high mobility [1–4, 129, 130]. Thus, the apatite structure contains Ca–F (for FA), O–H (for HA), Ca–O, and P–O bonds; consideration of the relative electrostatic strength of these bonds suggests that their relative destruction rate is consistent with: Ca–F ≈ Ca–OH > Ca(1)–O > Ca(2)–O > P–O. Considering its structure, once all Ca–F and Ca–O bonds are broken, apatite is destroyed; while breaking of P–O bonds is unnecessary for apatite dissolution [87].

7.6 Necessary Assumptions and Limitations

When apatite comes into a contact with an acidic aqueous solution, dissolution steps 1–5 (see section 7.4) begin. Before giving the detailed description of the processes involved, the following assumptions and limitations must be made:

1. The stoichiometric apatite (Ca/P = 1.67) is dissolved and neither non-stoichiometric layers nor other ions except of calcium, orthophosphate, hydroxide (for HA) or fluoride (for FA) are initially present in the crystal lattice, whereas volume and surface defects (dislocations and dislocation outlets respectively) might be present and, if present, they are distributed randomly.
2. Except of the presence of dislocation outlets, the initial surface of apatite is perfect (molecularly sooth). Otherwise, each imperfection might act as a possible dissolution nucleus.
3. In spite of the limitations and drawbacks discussed above, all the afore-discussed dissolution models are correct and complementary to each other.
4. In some cases, anions of acid might have an influence on apatite dissolution due to a specific affinity (e.g., citrate) or by means of formation of insoluble compounds (e.g., sulfate, fluoride). That is why, anions are specified neither here nor

below. It is just assumed that dissolution of apatite occurs in acid H_nA, where A^{n-} is an anion.

5. Only soluble compounds are formed during the apatite dissolution in acids. No precipitates are obtained.
6. Hydration effects for all the ions and molecules involved as well as those for crystal surface of apatite [90, 131–134] are omitted for simplicity.
7. All crystal faces of a solid apatite are equal and have a similar ionic arrangement. Although, it was found that HA {001} facets could be more resistant against acidic dissolution than {100} ones [135], no specific influence of different crystal faces on the dissolution process is considered.
8. Since the afore-described dissolution models have been elaborated for micron-sized apatite crystals, while the nanodimensional calcium orthophosphates possess the special properties (chapter 2), to eliminate the size effects, dissolution of large crystals is described. No specific influence of crystal dimensions on the dissolution mechanism is considered.

Based on these assumptions and limitations, an attempt to create the general description of the apatite dissolution mechanism at the atomic (ionic) level is given below.

7.7 Creation of the General Dissolution Mechanism

7.7.1 Atomic (Ionic) Description for the Perfect Crystals

Immediately after being placed to acidic aqueous solutions, the surface of apatite crystals becomes in contact with water molecules and ions of the acid. Transport of the chemicals (H^+ and A^{n-}) to a solid/liquid interface always happens by ionic diffusion through the Nernst diffusion layer (Fig. 7.1). A thickness of this layer depends on both solution hydrodynamics and dimensions of the crystals dissolving: it decreases when agitation intensity increases and crystal sizes decrease. However, in no cases the thickness becomes equal to the zero [136]. Thus, a concentration gradient is always present near the solid/liquid interface, the diffusion laws always remain valid, and all findings of the diffusion-controlled model (e.g., an adsorption resistance and a diffusion jump) remain applicable for any case of apatite dissolution. A similar approach is correct for the surface phenomena: when dissolution occurs, adsorption and desorption of ions, as well as chemical transformations (7.1)–(7.4) always take place on the surface regardless the experimental conditions chosen.

After being delivered by diffusion to the solid/liquid interface, ions of H^+ and A^{n-} are adsorbed onto the surface of apatite. A great number of various surface complexes might be formed as a result [81–84, 137–139]. According to the diffusion-controlled model, there is an adsorption resistance for ions to be adsorbed onto the surface and in order to overcome the resistance the ions make a diffusive jump toward the surface to a distance corresponding to their size [126, 140]. Being charged positively, protons are adsorbed onto oxygen ions of orthophosphate groups [53] as well as onto ions of fluoride (in the case of FA) and hydroxide (in the case of HA). After the model by Wu et al., the surface protonation of apatites proceeds via formation of $\equiv$ POH surface groups ("$\equiv$" stands for the surface) at $5 < pH < 7$, while apatite surfaces become fully protonated at $pH < 5$ [137]. One may expect to find some differences in the adsorption kinetics of protons onto fluoride, hydroxide and orthophosphate ions. Namely, due to a higher electro negativity of fluoride (when compared to oxygen ions of orthophosphate groups) and a higher basicity of hydroxide (when compared to orthophosphate), adsorption of protons might happen faster (or previously) onto these ions when compared with orthophosphate. However, recent results of computer simulations indicated a possibility of OH^- protonation by the nearest HPO_4^{2-} ion [122], indicating to the fact, that orthophosphate ions might be protonated faster (or previously) if compared to that for hydroxide ions. Obviously, this topic needs to be clarified in future.

Anyway, negatively charged anions A^{n-} may be adsorbed onto calcium cations only. According to the ion exchange model, the exchange process shows an adsorption of about one anion per unit-cell of apatite [114–116]. On the other hand, the surface of apatites is charged positively in aqueous acidic media and negatively in basic solutions (the point of zero charge is at solution pH within 6.8–8.5) with an electric double layer formation at the solid/liquid interface [65–67, 73, 81–83, 137–139, 141]. The latter points out to a non-equivalent ionic adsorption of H^+ and A^{n-} onto apatite: in acidic solutions, adsorption of protons always exceeds that of A^{n-}, while in basic solutions the situation is opposite. Therefore, in acidic solutions there is at least one extra-adsorbed proton per several neighboring unit-cells of apatite when compared to the number of adsorbed anions. The fact of a non-equivalent ionic adsorption might be explained by differences in sizes and mobility between H^+ and A^{n-} [142]. Great difference in the surface sites available is another reason for the non-equivalent ionic adsorption. Namely, the concentration of the lattice ions being on the surface of HA were found to be equal to 2.57, 3.02, and 4.50 ions/nm^2 for OH^-, PO_4^{3-}, and Ca^{2+}, respectively [64]. Furthermore, according to drawings of the apatite structure, the most part of its surface (up to 80–90%) is covered by oxygen ions of orthophosphate groups [143].

Ions of the crystal lattice of apatites are known to have different crystallographic positions: so-called X-ions (F^-, OH^-, Cl^- and substitutes for them) are located in Ca(2) channels parallel to *c*-axis [1–4, 129, 130]. This position results in a relatively high mobility of X-ions: for example, they might be mutually substituted rather easily [83, 84, 144–146] with formation of intermediate compounds, like fluorhydroxyapatites (FHA) or hydroxyfluorapatites (HFA) [17, 78, 147] — see section 1.3.11. for further details. Based on these findings, dissolution of apatite in acids was suggested to start with detachment of X-ions (more likely, as HX) from the surface [91–93] followed by their diffusion away from the crystal steps and further into the bulk solution [126, 140]. Recent computer simulations revealed some differences in the dissolution kinetics between F^- and OH^- anions in apatites: surface fluoride is dissolved not as readily as hydroxide is [130]. Initial leaving of X-ions was also detected in the incongruent dissolution model [17, 63, 72, 78, 83, 84]. Moreover, numerical values for the surface tension found for dissolution of HA appeared to be clearly lower than those found for HA growth (the polynuclear model). The latter data were explained by partial protonation of orthophosphate groups and missing of some hydroxide ions on the crystal surface [55, pp. 299–300].

After detachment of X-ions from the surface, there are calcium, orthophosphate, H^+ (as HPO_4, and/or H_2PO_4 groups) and A^{n-} ions left on it (hydration effect is omitted for simplicity). From now on, any differences between FA and HA disappear. In any cases, adsorption of protons onto calcium is impossible due to charge repulsion. Therefore, an electrical double layer with a positive charge on apatite cannot be continuous at the atomic (ionic) scale. There should be local uncharged places in it, say, at the positions of calcium cations. According to the chemical model, an interaction between calcium cations and anions of acid adsorbed from a solution is believed to happen in such places [91–93]. This interaction results in breaking of surface ≡O–Ca bonds ("≡" stands for the surface) and detachment of some calcium from the kink sites followed by their diffusion away from crystal steps and further into the bulk solution (the diffusion-controlled model [126, 140]). Detachment of calcium might occur as calcium-acid complexes (the ion exchange model). Recent computer simulations revealed that a local excess charge of +3 and +4 must be created to cause exothermic Ca^{2+} displacement from Ca(1) and Ca(2) cites, respectively [122]. If so, under otherwise equal conditions, detachment of Ca^{2+} ions from Ca(1) cites of the apatite surface should happen faster and/or easily than that from Ca(2) cites.

Detachment of calcium ions results in formation of dissolution nuclei. These nuclei are defined as collections of vacant sites for Ca^{2+}, PO_4^{3-}, and X^- ions on the crystal surface of apatite [51, p. 30]. According to the polynuclear model, removal even of one ion might result in further dissolution because critical nuclei (x^*) were calculated to consist of 1–26 ions for dissolution of HA and 0.3–34 ions for dissolution of FA. (According to the authors, $x^* < 1$ means that there is no nucleation barrier to be overcome [56, p. 309]. Since one unit cell of apatite contains 18 ions (Table 1.3), $x^* > 18$ means that more than one unit cell should be dissolved for a nucleus to form). The numeric values for x^* were found to depend on solution pH and solution undersaturation [55, 56]. Furthermore, due to charge repulsion, adsorbed protons (as HPO_4 and/or H_2PO_4 groups) might catalyze detachment of calcium ions from the kink sites (the hydrogen catalytic model).

After being detached, calcium cations (possibly, as calcium-acid complexes (the ion exchange model)) diffuse along the surface away from the dissolution steps before entering the solution [126]. For dissolution of OCP (see Table 1.1) in slightly acidic (pH = 5.66) solutions numeric values for the mean surface diffusion distance of the lattice ions detached from steps were calculated. These values were found to depend on the edge free energy and be within $(17 \pm 4)\alpha - (41 \pm 10)\alpha$, where $\alpha \approx 3.7$ Å is the size of a growth unit (or mean ionic diameter) [140]. Numeric values of mean ionic diameter for apatites were also calculated. They appeared to be less than that for OCP: 3.09 Å for HA and 3.07 Å for FA [57]. Taking into the consideration great similarities found between the chemical composition and crystal structures of OCP and HA [148], one might assume that the numerical values of surface diffusion distance for HA should be close to the values for OCP.

Ionic detachment of some calcium results in removing of a local positive charge from apatite, which is immediately compensated by adsorption of other protons from the acidic solution. This conclusion is based on the surface charge measurements: no information on influence of dissolution time to values of the surface charge on apatite has been found in literature [65–67, 73, 81–83, 137–141]. So, one may assume, that if a chemical composition of the solution is kept constant the surface charge on apatite will also remain constant during dissolution. Replacement of calcium by protons results in formation of mono-atomic layer acidic calcium orthophosphates on the crystal surface. Chemical phenomena of these transformations are described by the chemical model [91–93] and the kinetics (it strongly depends on the experimental conditions chosen)—by the surface controlled model.

After being desorbed from the surface, calcium cations (presumably, coupled with anions A^{n-}) diffuse through the Nernst layer into the bulk solution (the diffusion controlled model) [126, 140]. However, according to the calcium-rich layer formation model, a part of the already dissolved calcium ions might be adsorbed back onto apatite. Adsorption of calcium was suggested to happen at the very beginning of dissolution only and resulted in formation of a perm-selective layer of calcium cations with or without counter ions (A^{n-} and/or acidic orthophosphate) [19, 44–48, 63]. In this case, there should be a competition between protons and calcium cations for the exposed ions of oxygen on the surface to be adsorbed onto. Experimental data on a shift of the point of surface zero charge into more basic solution pH in presence of calcium in a solution [141] are in favor of such competition.

An idea on formation of some layer on the surface of apatite was raised in several dissolution models. For example, there is either a supposition (the polynuclear, diffusion, and kinetically controlled models) or a description (the ion exchange, hydrogen catalytic and chemical models) of a very thin virtual layer of acidic calcium orthophosphates (more exactly presence of HPO_4^{2-} ions) on the surface of apatite. Besides, according to the calcium-rich layer formation model, dissolution of apatite happens by ionic diffusion through a layer containing calcium with or without counter ions [19, 44–48, 63]. In part, this idea can be found within the ion exchange model where a monolayer of calcium citrate [114–116] and/or calcium-acid complexes [117, 118] are assumed to form onto apatite during dissolution. However, no other model requires such suggestion. According to the rest of them, formation of any calcium containing compounds (different from acidic orthophosphates) on apatite is not discussed at all.

According to the chemical model, the initial stages of apatite dissolution consist of calcium detachment from the surface and incorporation of protons instead. Orthophosphate groups are assumed to keep their positions without any relocation [91–93]. Due to the fact, that calcium occupies definite lattice positions, whereas protons are bound to oxygen ions of orthophosphate groups, removing of each calcium results in decreasing of attraction forces between the nearest (to this calcium) orthophosphate group, and rest part of the crystals. When all (or almost all) neighboring cations of calcium have been removed, orthophosphate groups (as $H_2PO_4^-$, $CaH_2PO_4^+$ or H_3PO_4 — it is not clear yet) also detach from the surface. Then they diffuse along the surface away from the dissolution steps before entering the solution as described above for calcium [126, 140].

In the case of crystal faces with perfectly smooth surface (dissolution steps are absent), detachment of one or several ions results in formation of dissolution nuclei, which the polynuclear model is based on [50–58]. According to this model, all sites in nuclei edges are the kink positions [57] and, after appearance, the nuclei grow and spread over the surface with a definite lateral rate [51], giving rise to formation of dissolution steps. If dissolution steps already existed on the crystal faces of apatite, detachment of one calcium or one orthophosphate would result in dissolution step movement jump-wise over a distance equal to the sizes of these ions (approximately for 1 and 3 Å, respectively).

The latter is also valid for dissolution of crystal edges and corners. Since the classical paper by Stranski [149], it is generally considered, that ions are less strongly bound along edges of a crystal than in the middle of faces and still less strongly at the corners [150]. No information about the ionic arrangements on edges and corners of apatite crystals has been found in literature. Therefore, two boundary conditions are possible: the edges and corners mainly consist of ions of either calcium or orthophosphate. However, the edges parallel to the *c*-axis, in principle, may consist of X-ions (F^-, OH^-, Cl^-). The last case is the simplest one. According to the chemical model [91–93], dissolution of apatite starts with protonation and detachment of X-ions, followed by removing of calcium (one should remind here that "walls" of the channels where X-ions are located consist of Ca(2) ions [87]) and afterwards by protonation and removing of orthophosphate. In the case, when the edges and corners of apatite mainly consist of calcium, again, according to the chemical model, detachment of weakly bounded calcium precedes protonation and removing of orthophosphate. Finally, if the edges and corners mainly consist of orthophosphate, the dissolution will start by chemical interaction of protons with these weakly bounded orthophosphate groups. The latter decreases electrostatic attraction forces between orthophosphate anions and neighboring calcium cations (attraction forces between Ca^{2+} and PO_4^{3-} are evidently stronger when compared with those between Ca^{2+} and HPO_4^{2-}). These forces become still weaker after adsorption of the second proton onto the given orthophosphate anion, causing its transformation to $H_2PO_4^-$ followed by detachment.

7.7.2 The Influence of Dislocations and Surface Defects

Before now, no data on crystal defects have been used. For this reason, the above description is valid for dissolution of the perfect single crystals without defects and with the molecularly smooth surface. However, like other solids, crystals of apatite always contain both surface irregularities (e.g., steps, missing ions, and dislocation outlets) and structural defects inside the bulk of crystals (e.g., dislocations and inclusions). The schematic depiction of a crystal surface is shown in Fig. 7.5. This model is known as the "Kossel model" [151]. In principle, any surface irregularities may act as dissolution nuclei because sometimes even one missing ion might become the critical nuclei (the polynuclear model) [55, 56].

Each dislocation outlet on apatite was found to be a hollow core with radius within 8.3–20 Å for the screw dislocations and even 18–43 Å for the edge ones [100]. Dimensions of the hexagonal unit-cell of apatite are: $a \approx 9.43$ Å, $c \approx 6.88$ Å (Table 1.3). If the numeric values for the dislocation outlets were calculated correctly, their dimensions (diameter of the hollow cores) would be 1.8–4.2 times a or 2.4–5.8 times c (for the screw dislocations), and even 3.8–9.1 times a or 5.2–12.5 times c (for the edge dislocations). Thus, each dislocation outlet on the crystal surface looks like a large dissolution nucleus equal to detachment of 3–19 unit cells, (if this is taken to be the product ac) for the screw dislocations and even of 15–90 unit cells for the edge ones.

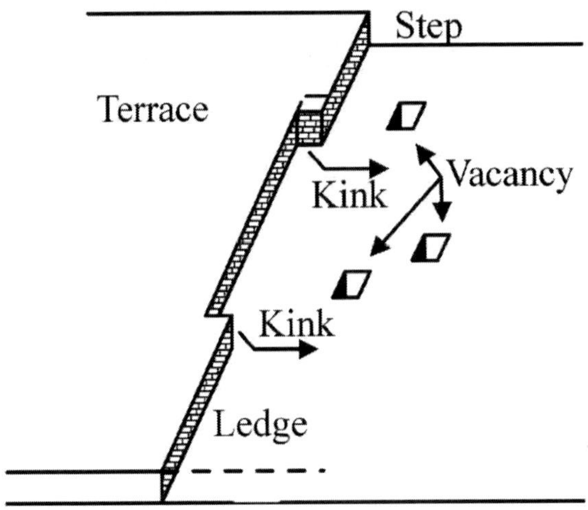

Figure 7.5 Surface of a crystal according to Kossel [151]. The various types of imperfections are shown. Reprinted from Ref. [30] with permission.

The process of etch pits formation and growth on apatites was described in details elsewhere [37–43, 61, 62, 97–112]. In the case of single crystals of pure HA, an acid attack was found to start formation of one etch pit in the basal plane which was the preferred dissolution spot or active site. Immediately afterwards the acid formed a longitudinal hole parallel to the c-axis [101]. On the contrary, dissolution of natural FA was found to happen with formation of a great number of etch pits (about 5×10^5 cm^{-2}) on the surface (Fig. 7.6) [104–106]. The latter just points out to the great differences in amount of crystal defects and impurities between the chemically pure [37–43, 61, 62, 97–103, 109–112] and natural [104–106] apatites.

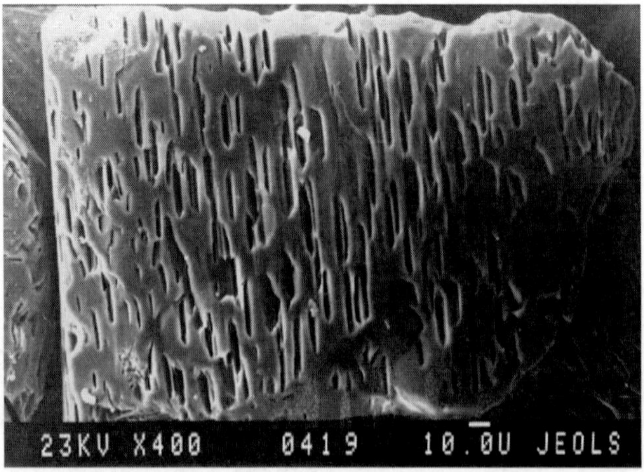

Figure 7.6 An example of etched crystal surface of natural FA with typical hexagonal etch pits. Bar 10 μm.

In any case, after being formed, the pits were found to grow in all directions (length, width, and depth—see Fig. 7.3) with definite rates until they either covered all crystal faces of apatite (dissolution of natural crystals [104–106]) or made longitudinal holes parallel to the *c*-axis (dissolution of the single crystals of pure HA [97–102]). However, according to "a new dissolution model incorporating particle size considerations", the dissolution process becomes spontaneous only when the pits are reached the critical sizes [37–43]. The unit-step movement of etch pit growth was suggested to consist of removing of a single ion, followed by surface diffusion away from the place of ionic detachment, desorption from the surface and diffusion into the bulk solution. All these processes evidently occur according to the basic findings of the ion exchange, chemical, calcium-rich layer formation, hydrogen catalytic, and diffusion controlled dissolution models.

To conclude this part, except the dissolution rates increasing and surface etching, no other influences of dislocations on the dissolution mechanism were found.

7.8 Conclusions

Thus, based on eight dissolution models drawn from the previously published papers, an attempt to create the general description of apatite dissolution process at the atomic (ionic) level has been made. This dissolution mechanism is based on current level of knowledge and will be updated, corrected, and modified simultaneously with appearance of new experimental data. Now some missing and unclear points to be investigated in future can be outlined:

1. Chemical transformations happening with all ions during diffusion through the Nernst layer;
2. Chemical composition of the apatite surface during dissolution;
3. Processes of ionic sorption and detachment;
4. Processes of dissolution nuclei formation and growth;
5. Initial stages of etch pits formation at the dislocation outlets;
6. Specific influence of anions to the ionic sorption, surface diffusion, and detachment.

The above topics should be investigated in order to have better understanding of all the processes involved. For example, atomic force microscopy and phase shift interferometry might be useful tools in investigations on dissolution nuclei and etch pits formation [61, 62]. Micron-sized (better nano-sized) electrodes might be helpful in studying ionic transformations during diffusion through the Nernst layer [127]. Modern techniques of the surface state analysis might be helpful in understanding the surface structure and its chemical composition.

References and Notes

1. McConnell, D. (1973) Apatite: its crystal chemistry, mineralogy, utilization, and geologic and biologic occurrences, *Applied Mineralogy,* vol. 5., Springer-Verlag, New York, p. 111.
2. Aoki, H. (1991) *Science and Medical Applications of Hydroxyapatite*, JAAS, Tokyo, p. 245.
3. Elliott, J. C. (1994) Structure and chemistry of the apatites and other calcium orthophosphates, *Studies in Inorganic Chemistry*, vol. 18, Elsevier, Amsterdam, p. 389.
4. LeGeros, R. Z. (1991) Calcium phosphates in oral biology and medicine, *Monographs in Oral Science*, vol. 15, Karger, Basel, p. 201.
5. Dorozhkin, S. V. (2009) Calcium orthophosphate cements and concretes, *Materials*, **2**, 221–291.

6. Dorozhkin, S. V. (2009) Calcium orthophosphates in nature, biology and medicine, *Materials*, **2**, 399–498.
7. Dorozhkin, S. V. (2009) Calcium orthophosphate-based biocomposites and hybrid biomaterials, *J. Mater. Sci.*, **44**, 2343–2387.
8. March, J. (1974) *March's Advanced Organic Chemistry: Reactions, Mechanisms, and Structure*, 5th edn, Wiley, New York, p. 2083.
9. Nancollas, G. H. (1974) Physicochemistry of demineralization and remineralization, *J. Dent. Res.*, **53**, 297–302.
10. Wu, M. S., Higuchi, W. I., Fox, J. L., and Friedman, M. (1976) Kinetics and mechanism of hydroxyapatite crystal dissolution in weak acid buffers using the rotating disk method, *J. Dent. Res.*, **55**, 496–505.
11. Smith, A. N., Posner, A. M., and Quirk, J. P. (1977) A model describing the kinetics of dissolution of hydroxyapatite, *J. Coll. Interf. Sci.*, **62**, 475–494.
12. Okazaki, M., Moriwaki, Y., Aoba, T., Doi, Y., and Takahashi, J. (1981) Dissolution rate behavior of fluoridated apatite pellets, *J. Dent. Res.*, **60**, 1907–1911.
13. Crommelin, D. J., Higuchi, W. I., and Fox, J. L. (1983) Dissolution rate behavior of hydroxyapatite-fluorapatite mixtures, *Caries Res.*, **17**, 289–296.
14. Nelson, D. G. A., Featherstone, J. D. B., Duncan, J. F., and Cutress, T. W. (1983) Effect of carbonate and fluoride on the dissolution behaviour of synthetic apatites, *Caries Res.*, **17**, 200–211.
15. Higuchi, W. I., Cesar, E. Y., Cho, P. W., and Fox, J. L. (1984) Powder suspension method for critically re-examining the two-site model for hydroxyapatite dissolution kinetics, *J. Pharm. Sci.*, **73**, 146–153.
16. Budz, J. A., LoRe, M., and Nancollas, G. H. (1987) Hydroxyapatite and carbonated apatite as models for the dissolution behavior of human dental enamel, *Adv. Dent. Res.*, **1**, 314–321.
17. Wong, L., Cutress, T. W., and Duncan, J. F. (1987) The influence of incorporated and adsorbed fluoride on the dissolution of powdered and pelletized hydroxyapatite in fluoridated and non-fluoridated acid buffers, *J. Dent. Res.*, **66**, 1735–1741.
18. Budz, J. A., and Nancollas, G. H. (1988) The mechanism of dissolution of hydroxyapatite and carbonated apatite in acidic solutions, *J. Cryst. Growth*, **91**, 490–496.
19. Thomann, J. M., Voegel, J. C., and Gramain, P. (1990) Kinetics of dissolution of calcium hydroxyapatite powder. III: pH and sample conditioning effects, *Calcif. Tiss. Int.*, **46**, 121–129.

20. Chin, K. O. A., and Nancollas, G. H. (1991) Dissolution of fluorapatite. A constant-composition kinetics study, *Langmuir*, **7**, 2175–2179.

21. Margolis, H. C., and Moreno, E. C. (1992) Kinetics of hydroxyapatite dissolution in acetic, lactic, and phosphoric acid solutions, *Calcif. Tiss. Int.*, **50**, 137–143.

22. Paschalis, E. P., Wikiel, K., and Nancollas, G. H. (1994) Dual constant composition kinetics characterization of apatitic surfaces, *J. Biomed. Mater. Res.*, **28**, 1411–1418.

23. Calmanovici, C. E., Gilot, B., and Laguerie, C. (1997) Mechanism and kinetics for the dissolution of apatitic materials in acid solutions, *Brazilian J. Chem. Eng.*, **14**, 95–102.

24. Mohammadi, Z., Ziaei-Moayyed, A. A., and Sheikh-Mehdi, M. A. (2008) *In vitro* dissolution of plasma-sprayed hydroxyapatite coatings with different characteristics: experimental study and modeling, *Biomed. Mater.*, **3**, 015006 (7 pages).

25. van der Sluis, S., Meszaros, Y., Marchee, W. G. J., Wesselingh, H. A., and van Rosmalen, G. M. (1987) The digestion of phosphate ore in phosphoric acid, *Ind. Eng. Chem. Res.*, **26**, 2501–2505.

26. Brahim, K., Antar, K., Khattech, I., and Jemal, M. (2008) Effect of temperature on the attack of fluorapatite by a phosphoric acid solution, *Scientific Research and Essays,* **3**, 035–039.

27. Fawzi, M. B., Fox, J. L., Dedhiya, M. G., Higuchi, W. I., and Hefferren, J. J. (1978) A possible second site for hydroxyapatite dissolution in acidic media, *J. Coll. Interf. Sci.*, **67**, 304–311.

28. Fox, J. L., Higuchi, W. I., Fawzi, M. B., and Wu, M. S. (1978) A new two-site model for hydroxyapatite dissolution in acidic media, *J. Coll. Interf. Sci.*, **67**, 312–330.

29. Griffith, E. N., Katdare, A., Fox, J. L., and Higuchi, W. I. (1978) Transmission electron microscopic confirmation of the morphological predictions of the two-site model for hydroxyapatite dissolution, *J. Coll. Interf. Sci.*, **67**, 331–335.

30. Ehrlich, H., Koutsoukos, P. G., Demadis, K. D., and Pokrovsky, O. S. (2009) Principles of demineralization: modern strategies for the isolation of organic frameworks, *Part II. Decalcification. Micron*, **40**, 169–193.

31. Chow, L. C. (1988) Response to Dr. E.I.F. Pearce's letter to the Editor, *J. Dent. Res.*, **67**, 1058–1059.

32. Tomson, M. B., and Nancollas, G. H. (1978) Mineralization kinetics: a constant composition approach, *Science*, **200**, 1059–1060.

33. Tang, R., Henneman, Z. J., and Nancollas, G. H. (2003) Constant composition kinetics study of carbonated apatite dissolution, *J. Cryst. Growth*, **249**, 614–624.

34. Tang, R., Hass, M., Wu, W., Gulde, S., and Nancollas, G. H. (2003) Constant composition dissolution of mixed phases. II. Selective dissolution of calcium phosphates, *J. Coll. Interf. Sci.*, **260**, 379–384.

35. Ebrahimpour, A., Zhang, J., and Nancollas, G. H. (1991) Dual constant composition method and its application to studies of phase transformation and crystallization of mixed phases, *J. Cryst. Growth*, **113**, 83–91.

36. Guidry, M. W., and Mackenzie, F. T. (2003) Experimental study of igneous and sedimentary apatite dissolution: control of pH, distance from equilibrium, and temperature on dissolution rates, *Geochim. Cosmochim. Acta*, **67**, 2949–2963.

37. Tang, R., Nancollas, G. H., and Orme, C. A. (2001) Mechanism of dissolution of sparingly soluble electrolytes, *J. Am. Chem. Soc.*, **123**, 5437–5443.

38. Tang, R., Wang, L., Orme, C. A., Bonstein, T., Bush, P. J., and Nancollas, G. H. (2004) Dissolution at the nanoscale: self-preservation of biominerals, *Angew. Chem. Int. Ed.*, **43**, 2697–2701.

39. Tang, R., Wang, L., and Nancollas, G. H. (2004) Size-effects in the dissolution of hydroxyapatite: an understanding of biological demineralization, *J. Mater. Chem.*, **14**, 2341–2346.

40. Tang, R., Orme, C. A., and Nancollas, G. H. (2004) Dissolution of crystallites: surface energetic control and size effects, *ChemPhysChem*, **5**, 688–696.

41. Wang, L. J., Tang, R., Bonstein, T., Bush, P., and Nancollas, G. H. (2006) Enamel demineralization in primary and permanent teeth, *J. Dent. Res.*, **85**, 359–363.

42. Wang, L. J., and Nancollas, G. H. (2008) Calcium orthophosphates: crystallization and dissolution, *Chem. Rev.*, **108**, 4628–4669.

43. Wang, L. J., and Nancollas, G. H. (2009) Pathways to biomineralization and biodemineralization of calcium phosphates: the thermodynamic and kinetic controls, *Dalton Trans.*, 2665–2672.

44. Mafé, S., Manzanares, J. A., Reiss, H., Thomann, J. M., and Gramain, P. (1992) Model for the dissolution of calcium hydroxyapatite powder, *J. Phys. Chem.*, **96**, 861–866.

45. Gasser, P., Voegel, J. C., and Gramain, P. (1993) Surface reactions on hydroxyapatite in the presence of fluoride ions. 1. Saturated and congruent conditions, *Colloids Surf. A*, **74**, 275–286.

46. Thomann, J. M., Voegel, J. C., and Gramain, P. (1993) Quantitative model for the dissolution of calcium hydroxyapatite with a permselective ionic interface, *J. Coll. Interf. Sci.*, **157**, 369–374.

47. Gasser, P., Haikel, Y., Voegel, J. C., and Gramain, P. (1994) Surface reactions of hydroxyapatite in the presence of fluoride ions. 2. Effects of calcium and phosphate in saturated solutions, *Colloids Surf. A*, **88**, 157–168.

48. Schaad, P., Poumier, F., Voegel, J. C., and Gramain, P. (1997) Analysis of calcium hydroxyapatite dissolution in non-stoichiometric solutions, *Colloids Surf. A*, **121**, 217–228.

49. Pan, H. B., and Darvell, B. W. (2009) Calcium phosphate solubility: the need for re-evaluation, *Cryst. Growth Des.*, **9**, 639–645.

50. Christoffersen, J., Christoffersen, M. R., and Kjaergaard, N. (1978) The kinetics of dissolution of calcium hydroxyapatite in water at constant pH, *J. Cryst. Growth*, **43**, 501–511.

51. Christoffersen, J. (1980) Kinetics of dissolution of calcium hydroxypatite. III. Nucleation-controlled dissolution of a polydisperse sample of crystals, *J. Cryst. Growth*, **49**, 29–44.

52. Christoffersen, J. (1981) Dissolution of calcium hydroxyapatite, *Calcif. Tiss. Int.*, **33**, 557–560.

53. Christoffersen, J., and Christoffersen, M. R. (1982) Kinetics of dissolution of calcium hydroxyapatite. V. The acidity constant for the hydrogen phosphate surface complex, *J. Cryst. Growth*, **57**, 21–26.

54. Christoffersen, M. R., and Christoffersen, J. (1992) Possible mechanisms for the growth of the biomaterial, calcium hydroxyapatite microcrystals, *J. Cryst. Growth*, **121**, 617–630.

55. Christoffersen, J., Christoffersen, M. R., and Johansen, T. (1996) Kinetics of growth and dissolution of fluorapatite, *J. Cryst. Growth*, **163**, 295–303.

56. Christoffersen, J., Christoffersen, M. R., and Johansen, T. (1996) Some new aspects of surface nucleation applied to the growth and dissolution of fluorapatite and hydroxyapatite, *J. Cryst. Growth*, **163**, 304–310.

57. Christoffersen, J., Dohrup, J., and Christoffersen, M. R. (1998) The importance of formation of hydroxyl ions by dissociation of trapped water molecules for growth of calcium hydroxyapatite crystals, *J. Cryst. Growth*, **186**, 275–282.

58. Christoffersen, M. R., Dohrup, J., and Christoffersen, J. (1998) Kinetics of growth and dissolution of calcium hydroxyapatite in suspensions with variable calcium to phosphate ratio, *J. Cryst. Growth*, **186**, 283–290.

59. Zhang, H., Li, S., and Yan, Y. (2001) Dissolution behavior of hydroxyapatite powder in hydrothermal solution, *Ceram. Int.*, **27**, 451–454.

60. Gilmer, G. H., and Bennema, P. (1972) Simulation of crystal growth with surface diffusion, *J. Appl. Physics*, **43**, 1347–1360.

61. Onuma, K, Ito, A, Tateishi, T, and Kameyama, T. (1995) Growth kinetics of hydroxyapatite crystal revealed by atomic force microscopy, *J. Cryst. Growth*, **154**, 118–125.

62. Onuma, K., Ito, A., and Tateishi, T. (1996) Investigation of a growth unit of hydroxyapatite crystal from the measurements of step kinetics, *J. Cryst. Growth*, **167**, 773–776.

63. Zhu, Y., Zhang, X., Chen, Y., Xie, Q., Lan, J., Qian, M., and He, N. (2009) A comparative study on the dissolution and solubility of hydroxylapatite and fluorapatite at 25°C and 45°C, *Chemical Geology*, **268**, 89–96.

64. Kukura, M., Bell, L. C., Posner, A. M., and Quirk, J. P. (1972) Radioisotope determination of the surface concentrations of calcium and phosphorus on hydroxyapatite in aqueous solution, *J. Phys. Chem.*, **76**, 900–904.

65. Somasundaran, P. (1972) Zeta potential of apatite in aqueous solutions and its change during equilibration, *J. Coll. Interf. Sci.*, **27**, 659–666.

66. Bell, L. C., Posner, A. M., and Quirk, J. P. (1972) Surface charge characteristics of hydroxyapatite and fluorapatite, *Nature*, **239**, 515–517.

67. Bell, L. C., Posner, A. M., and Quirk, J. P. (1973) The point of zero charge of hydroxyapatite and fluorapatite in aqueous solutions, *J. Coll. Interf. Sci.*, **42**, 250–261.

68. Moreno, E. C., Gregory, T. M., and Brown, W. E. (1968) Preparation and solubility of hydroxyapatite, *J. Res. Natl. Bur. Stand.*, **72A**, 773–782.

69. Chuong, R. (1973) Experimental study of surface and lattice effects on the solubility of hydroxyapatite, *J. Dent. Res.*, **52**, 911–914.

70. Smith, A. N., Posner, A. M., and Quirk, J. P. (1974) Incongruent dissolution and surface complexes of hydroxyapatite, *J. Coll. Interf. Sci.*, **48**, 442–449.

71. Mika, H., Bell, L. C., and Kruger, B. J. (1976) The role of surface reactions in the dissolution of stoichiometric hydroxyapatite, *Arch. Oral Biol.*, **21**, 697–701.

72. Moreno, E. C., Kresak, M., and Zahradnik, R. T. (1977) Physicochemical aspects of fluoride-apatite systems relevant to the study of dental caries, *Caries Res.*, **11**(1), 142–171.

73. Bell, L. C., Mika, H., and Kruger, B. J. (1978) Synthetic hydroxyapatite-solubility product and stoichiometry of dissolution, *Arch. Oral. Biol.*, **23**, 329–336.

74. Ingram, G. S. (1999) Chemical events during tooth dissolution, *J. Dent. Res.*, **69**, Spec. Iss., 581–586, discussion 634–636.

75. Kaufman, H. W., and Kleinberg, I. (1979) Studies on the incongruent solubility of hydroxyapatite, *Calcif. Tiss. Int.*, **27**, 143–151.

76. Amrah-Bouali, S., Rey, C., Lebugle, A., and Bernache, D. (1994) Surface modifications of hydroxyapatite ceramics in aqueous media, *Biomaterials*, **15**, 269–272.

77. Nordstrom, E. G., Hara, T., and Hero, H. (1996) Solubility of hydroxyapatite/mica composites, *Biomed. Mater. Eng.*, **6**, 73–78.

78. Pearce, E. I. F, Guha-Chowdhury, N., Iwami, Y., and Cutress, T. W. (1995) Stoichiometry of fluoride release from fluorhydroxyapatite during acid dissolution, *Caries Res.*, **29**, 130–136.

79. Shimabayashi, S., and Matsumoto, M. (1993) Non-stoichiometric dissolution of hydroxyapatite in the presence of simple salts, *Nippon Kagaku Kaishi*, **10**, 1118–1122.

80. Brown, P. W., and Martin, R. I. (1999) An analysis of hydroxyapatite surface layer formation, *J. Phys. Chem. B*, **103**, 1671–1675.

81. Chaïrat, C., Oelkers, E. H., Schott, J., and Lartigue, J. E. (2007) Fluorapatite surface composition in aqueous solution deduced from potentiometric, electrokinetic, and solubility measurements, and spectroscopic observations, *Geochim. Cosmochim. Acta*, **71**, 5888–5900.

82. Bengtsson, Å., Shchukarev, A., Persson, P., and Sjöberg, S. (2009) A solubility and surface complexation study of a non-stoichiometric hydroxyapatite, *Geochim. Cosmochim. Acta*, **73**, 257–267.

83. Bengtsson, Å., Shchukarev, A., Persson, P., and Sjöberg, S. (2009) Phase transformations, ion-exchange, adsorption, and dissolution processes in aquatic fluorapatite systems, *Langmuir*, **25**, 2355–2362.

84. Bengtsson, Å., and Sjöberg, S. (2009) Surface complexation and proton-promoted dissolution in aqueous apatite systems, *Pure Appl. Chem.*, **81**, 1569–1584.

85. Misra, D. N. (1999) Interaction of citric or hydrochloric acid with calcium fluorapatite: precipitation of calcium fluoride, *J. Coll. Interf. Sci.*, **220**, 387–391.

86. Harouiya, N., Chaïrat, C., Köhler, S. J., Gout, R., and Oelkers, E. H. (2007) The dissolution kinetics and apparent solubility of natural apatite in

closed reactors at temperatures from 5 to 50 °C and pH from 1 to 6, *Chemical Geology*, **244**, 554–568.

87. Chaïrat, C., Schott, J., Oelkers, E. H., Lartigue, J. E., and Harouiya, N. (2007) Kinetics and mechanism of natural fluorapatite dissolution at 25 °C and pH from 3 to 12, *Geochim. Cosmochim. Acta*, **71**, 5901–5912.
88. Melikhov, I. V., Dorozhkin, S. V., Nikolaev, A. L., and Voronina, N. Y. (1992) Transition from congruent to incongruent dissolution, *Russ. J. Phys. Chem.*, **66**, 1108–1111.
89. Tanaka, H, Miyajima, K, Nakagaki, M, and Shimabayashi, S. (1991) Incongruent dissolution of hydroxyapatite in the presence of phosphoserine, *Coll. Polymer Sci.*, **269**, 161–165.
90. Park, C., Fenter, P., Zhang, Z., Cheng, L., and Sturchio, N. C. (2004) Structure of the fluorapatite (100)-water interface by high-resolution X-ray reflectivity, *Am. Mineral.*, **89**, 1647–1654.
91. Dorozhkin, S. V. (1997) Surface reactions of apatite dissolution, *J. Coll. Interf. Sci.*, **191**, 489–497.
92. Dorozhkin, S. V. (1997) Acidic dissolution mechanism of natural fluorapatite. II. Nanolevel of investigations, *J. Cryst. Growth*, **182**, 133–140.
93. Dorozhkin, S. V. (1999) Inorganic chemistry of the dissolution phenomenon: the dissolution mechanism of calcium apatites at the atomic (ionic) level, *Comments Inorg. Chem.*, **20**, 285–299.
94. Higuchi, W. I., Gray, J. A., Hefferren, J. J., and Patel, P. R. (1965) Mechanisms of enamel dissolution in acid buffers, *J. Dent. Res.*, **44**, 330–341.
95. Pearce, E. I. F. (1988) On the dissolution of hydroxyapatite in acid solutions, *J. Dent. Res.*, **67**, 1056–1058.
96. Eanes, E. D., and Meyer, J. L. (1977) The maturation of crystalline calcium phosphates in aqueous suspensions at physiologic pH, *Calcif. Tiss. Res.*, **23**, 259–269.
97. Lowell, L. C. (1958) Dislocation etch pits in apatite, *Acta Metall.*, **6**, 775–778.
98. Jongebloed, W. L., Molenaar, I., and Arends, J. (1973) Orientation-dependent etchpit penetration and dissolution of fluoroapatite, *Caries Res.*, **7**, 154–165.
99. Jongebloed, W. L., van den Berg, P. J., and Arends, J. (1974) The dissolution of single crystals of hydroxyapatite in citric and lactic acids, *Calcif. Tiss. Res.*, **15**, 1–9.
100. Arends, J., and Jongebloed, W. L. (1977) Dislocations and dissolution in apatites: theoretical considerations, *Caries Res.*, **11**, 186–188.

101. Arends, J., and Jongebloed, W. L. (1979) Ultrastructural studies of synthetic apatite crystals, *J. Dent. Res.*, **58**, Spec. Iss. B, pp. 837–843.

102. Daculsi, G., Kerebel, B., Kerebel, L.M. Mechanisms of acid dissolution of biological and synthetic apatite crystals at the lattice pattern level. Caries Res. 1979, 13, 277-289.

103. Daculsi, G., LeGeros, R. Z., and Mitre, D. (1989) Crystal dissolution of biological and ceramic apatites, *Calcif. Tiss. Int.*, **49**, 95–103.

104. Melikhov, I. V., Dorozhkin, S. V., Nikolaev, A. L., Kozlovskaya, E. D., and Rudin, V. N. (1990) Dislocations and the rate of dissolution of solids, *Russ. J. Phys. Chem.*, **64**, 1746–1750.

105. Dorozhkin, S. V. (1995) Chemical etching of natural fluorapatite crystals in acid solutions studied with the scanning electron microscope, *Scanning*, **17**, 355–360.

106. Dorozhkin, S. V. (1997) Acidic dissolution mechanism of natural fluorapatite, I. Milli- and microlevels of investigations, *J. Cryst. Growth*, **182**, 125–132.

107. Lasaga, A. C., and Luttge, A. (2001) Variation of crystal dissolution rate based on a dissolution stepwave model, *Science*, **291**, 2400–2404.

108. Dove, P. M., Han, N., and de Yoreo, J. J. (2005) Mechanisms of classical crystal growth theory explain quartz and silicate dissolution behavior, *Proc. Natl. Acad. Sci.*, **102**, 15357–15362.

109. Kwon, K. Y., Wang, E., Chung, A., Chang, N., Saiz, E., Choe, U. J., Koobatian, M., and Lee, S. W. (2008) Defect induced asymmetric pit formation on hydroxyapatite, *Langmuir*, **24**, 11063–11066.

110. Kwon, K. Y., Wang, E., Chung, A., Chang, N., and Lee, S. W. (2009) Effect of salinity on hydroxyapatite dissolution studied by atomic force microscopy, *J. Phys. Chem. C*, **113**, 3369–3372.

111. Kwon, K. Y., Wang, E., Chang, N., and Lee, S. W. (2009) Characterization of the dominant molecular step orientations on hydroxyapatite (100) surfaces, *Langmuir*, **25**, 7205–7208.

112. Dong, S. S., Hwan, K., and Jong, K. L. (2007) AFM study on the surface dissolution of hydroxyapatite, *Key Eng. Mater.*, 336–338, 1553–1555.

113. Dorozhkin, S. V. (1996) Process of epitaxial crystal growth for $CaSO_4 \cdot 0.5H_2O$ on a surface of dissolving fluorapatite crystals studied by scanning electron microscopy, *Scanning*, **18**, 119–124.

114. Misra, D. N. (1996) Interaction of citric acid with hydroxyapatite: surface exchange of ions and precipitation of calcium citrate, *J. Dent. Res.*, **75**, 1418–1425.

115. López-Macipe, A., Gómez-Morales, J., Rodríguez-Clemente, R. (1998) The role of pH in the adsorption of citrate ions on hydroxyapatite, *J. Coll. Interf. Sci.*, **200**, 114–120.

116. Misra, D. N. (1999) Adsorption from solutions on synthetic hydroxyapatite: nonaqueous vs. aqueous solvents, *J. Biomed. Mater. Res. (Appl. Biomater.)*, **48**, 848–855.

117. Yoshida, Y., van Meerbeek, B., Nakayama, Y., Yoshioka, M., Snauwaert, J., Abe, Y., Lambrechts, P., Vanherle, G., and Okazaki, M. (2001) Adhesion to and decalcification of hydroxyapatite by carboxylic acids, *J. Dent. Res.*, **80**, 1565–1569.

118. Yoshioka, M., Yoshida, Y., Inoue, S., Lambrechts, P., Vanherle, G., Nomura, Y., Okazaki, M., Shintani, H., and van Meerbeek, B. (2002) Adhesion/decalcification mechanisms of acid interaction with human hard tissues, *J. Biomed. Mater. Res.*, **59**, 56–62.

119. Yoshida, Y., van Meerbeek, B., Nakayama, Y., Snauwaert, J., Hellemans, L., Lambrechts, P., Vannerle, G., and Wakasa, K. (2000) Evidence of chemical bonding at biomaterial-hard tissue interfaces, *J. Dent. Res.*, **79**, 709–714.

120. Astala, R., and Stott, M. J. (2008) First-principles study of hydroxyapatite surfaces and water adsorption, *Phys. Rev. B*, **78**, pp. 075427 (11 pages).

121. Posner, A. S. (1985) The structure of bone apatite surfaces, *J. Biomed. Mater. Res.*, **19**, 241–250.

122. Hochrein, O., and Zahn, D. (2010) On the molecular mechanisms of the acid-induced dissociation of hydroxy-apatite in water, *J. Mol. Model.*, **17**, 1525–1528.

123. Jarlbring, M., Gunneriusson, L., and Forsling, W. Characterisation of the protolytic properties of synthetic carbonate free fluorapatite, *J. Coll. Interf. Sci.*, **285**, 206–211.

124. Becker, P. (1989) *Phosphates and Phosphoric Acid: Raw Materials Technology and Economics of the Wet Process*, 2nd edn., *Fertilizer Science and Technology Series*, Marcel Dekker, New York, pp. 760.

125. Compton, R. G., Harding, M. S., Pluck, M. R., Atherton, J. H., and Brennan, C. M. (1993) Mechanism of solid/liquid interfacial reactions. The dissolution of benzoic acid in aqueous solution, *J. Phys. Chem.*, **97**, 10416–10420.

126. Zhang, J., and Nancollas, G. H. (1994) Unexpected pH dependence of dissolution kinetics of dicalcium phosphate dihydrate, *J. Phys. Chem.*, **98**, 1689–1694.

127. de Aza, P. N., Guitian, F., Merlos, A., Lora-Tamayo, E., and de Aza, S. (1996) Bioceramics — simulated body fluid interfaces: pH and its influence of hydroxyapatite formation, *J. Mater. Sci. Mater. Med.*, **7**, 399–402.

128. Bohner, M., Lemaître, J., and Ring, T. A. (1997) Kinetics of dissolution of β-tricalcium phosphate, *J. Coll. Interf. Sci.*, **190**, 37–48.

129. Young, R. A. (1974) Implications of atomic substitutions and other structural details in apatites, *J. Dent. Res.*, **53**, 193–203.

130. de Leeuw, N. H. (2010) Computer simulations of structures and properties of the biomaterial hydroxyapatite, *J. Mater. Chem.*, **20**, 5376–5389.

131. Pareek, A., Torrelles, X., Angermund, K., Rius, J., Magdans, U., and Gies, H. (2008) Structure of interfacial water on fluorapatite (100) surface, *Langmuir*, **24**, 2459–2464.

132. Pan, H., Tao, J., Wu, T., and Tang, R. (2007) Molecular simulation of water behaviors on crystal faces of hydroxyapatite, *Frontiers of Chemistry in China*, **2**, 156–163.

133. Zahn, D., and Hochrein, O. (2003) Computational study of interfaces between hydroxyapatite and water, *Phys. Chem. Chem. Phys.*, **5**, 4004–4007.

134. Corno, M., Busco, C., Bolis, V., Tosoni, S., and Ugliengo, P. (2009) Water adsorption on the stoichiometric (001) and (010) surfaces of hydroxyapatite: a periodic B3LYP study, *Langmuir*, **25**, 2188–2198.

135. Pan, H., Tao, J., Yu, X., Fu, L., Zhang, J., Zeng, X., Xu, G., and Tang, R. (2008) Anisotropic demineralization and oriented assembly of hydroxyapatite crystals in enamel: smart structures of biominerals, *J. Phys. Chem. B*, **112**, 7162–7165.

136. Levich, V. G. (1962) *Physicochemical Hydrodynamics*, Prentice Hall, New York.

137. Wu, L., Forsling, W., and Schindler, P. W. (1991) Surface complexation of calcium minerals in aqueous solution. 1. Surface protonation at fluorapatite-water interfaces, *J. Coll. Interf. Sci.*, **147**, 178–185.

138. Skartsila, K., and Spanos, N. (2007) Surface characterization of hydroxyapatite: potentiometric titrations coupled with solubility measurements, *J. Coll. Interf. Sci.*, **308**, 405–412.

139. Vučinić, D. R., Radulović, D. S., and Deušić, S. D. (2010) Electrokinetic properties of hydroxyapatite under flotation conditions, *J. Coll. Interf. Sci.*, **343**, 239–245.

140. Zhang, J., and Nancollas, G. H. (1992) Kinetics and mechanisms of octacalcium phosphate dissolution at 37 °C, *J. Phys. Chem.*, **96**, 5478–5483.

141. Doss, S. K. (1976) Surface properties of hydroxyapatite: I. The effect of various inorganic ions on the electrophoretic behavior, *J. Dent. Res.*, **55**, 1067–1075.

142. Nielsen, A. E. (1980) Transport control in crystal growth from solution, *Croat. Chem. Acta*, **53**, 255–279.

143. Okazaki, M., Sato, M., Takahashi, J. (1995) Space-cutting model of hydroxyapatite, *Biomaterials*, **16**, 45–49.

144. Iijima, M., Nelson, D. G. A., Pan, Y., Kreinbrink, A. T., Adachi, M., Goto, T., and Moriwaki, Y. (1996) Fluoride analysis of apatite crystals with a central planar OCP inclusion: Concerning the role of F^- ions on apatite/OCP/apatite structure formation, *Calcif. Tiss. Int.*, **59**, 377–384.

145. Aoba, T. (1997) The effect of fluoride on apatite structure and growth, *Crit. Rev. Oral Biol. Med.*, **8**, 136–153.

146. de Leeuw, N. H. (2004) Resisting the onset of hydroxyapatite dissolution through the incorporation of fluoride, *J. Phys. Chem. B*, **108**, 1809–1811.

147. Liu, Y., Sethuraman, G., Wu, W., Nancollas, G. H., and Grynpas, M. (1997) The crystallization of fluorapatite in the presence of hydroxyapatite seeds and of hydroxyapatite in the presence of fluorapatite seeds, *J. Coll. Interf. Sci.*, **186**, 102–109.

148. Dickens, B., and Schroeder, L. W. (1980) Investigation of epitaxy relationships between $Ca_5(PO_4)_3OH$ and other calcium orthophosphates, *J. Res. Natl. Bur. Stand.*, **85**, 347–362.

149. Stranski, I. N. (1928) Zur Theorie des Kristallwachstums, *Z. Phys. Chem.*, **136**, 259–278.

150. Nangia, S., Garrison, B. J. (2009) *Ab initio* study of dissolution and precipitation reactions from the edge, kink, and terrace sites of quartz as a function of pH, *Molecular Phys.*, **107**, 831–843.

151. Kossel, W. (1934) Zur Energetik von Oberflächenvorgangen, *Annalen der Physik* ,**21**, 457–480.

Chapter 8

The History of Calcium Orthophosphates from 1770s till 1950

8.1 Introduction

By virtue of abundance in the nature and presence in the living organisms, calcium apatites [1] and other calcium orthophosphates (Table 1.1) appear to be the chemical compounds of a special interest in many fields of science, including geology, chemistry, biology and medicine [2, 3]. As follows from the designation, calcium orthophosphates contain both calcium (Ca, atomic number 20) and phosphorus (P, atomic number 15) as the major constituencies. Concerning the history of both chemical elements, according to Wikipedia, the free encyclopedia, calcium (from Latin *calx*, genitive *calcis*, meaning "lime") compounds were known as early as the first century, when the ancient Romans prepared lime as calcium oxide. However, calcium sulfate (also known as plaster of Paris or lime plaster) had been known much earlier: three statues were discovered in a buried pit at 'Ain Ghazal in Jordan those were sculpted with lime plaster over armatures of reeds and twine. They were made in the pre-pottery Neolithic period, around 7200 BC. However, calcium was not isolated until 1808, when Sir Humphry Davy (1778–1829) electrolyzed a mixture of lime and mercuric oxide [4, 5]. Phosphorus is a bit younger. The discovery of this chemical element (its name given from Greek mythology, Φωσφόρος meaning "light-bearer" (Latin Lucifer), referring to the "Morning Star", the planet Venus)

Calcium Orthophosphates: Applications in Nature, Biology, and Medicine
Sergey Dorozhkin

ISBN 978-981-4316-62-0 (Hardcover), 978-981-4364-17-1 (eBook)
www.panstanford.com

is credited to a German merchant and alchemist Hennig Brand (ca. 1630–ca. 1710) in 1669, although other alchemists might have discovered phosphorus around the same time. Brand experimented with urine, which contains considerable quantities of dissolved phosphates from normal metabolism. However, it was Antoine Laurent Lavoisier (1743–1794), who recognized phosphorus as a chemical element in 1777. Interestingly, but phosphorus appears to be the first element discovered since antiquity. To conclude these introductive exercises, the earliest research paper, I have been able to find, containing the word "phosphorus" in the title was written by Robert Boyle (1627–1691) and published in 1693 after his death [6].

8.2 Knowledge on Calcium Orthophosphates in the 18th Century

Due to the big problems with accessing to the scientific literature published in the first half of the 19th century and even earlier, a deep invasion into the history of the subject still remains to be both fragmental and incomplete. That time, the scientific concepts were rather different from the modern ones and chemical formulae of the substances had not been introduced yet. Furthermore, scientific journals were rare; luckily, many scientific books published before the 20th century have been scanned by Google as a part of its project to make the world's books discoverable online. This timely project by Google combined with the power of the modern electronic databases of scientific publications allows reconstructing the major historical milestones on calcium orthophosphates, which was often impossible for earlier writers. For example, a paper of 1994 by Driskell entitled: "Early history of calcium phosphate materials and coatings" [7] starts from the classical publication of 1920 by Albee assisted by Morrison [8]. In 1999, Shackelford published a paper: "Bioceramics — an historical perspective" [9], in which the same publication by Albee and Morrison [8] was mentioned as the earliest reference. The same is valid for the historical papers by Hulbert, *et al.* [10, 11] and Shepperd [12]. Thus, it seemed that calcium phosphates had been unknown before 1920. Certainly, this is not the case; nevertheless, the precise sequence of the scientific events happened in the 18th century still remain poorly restorable, while the historical time scale of even earlier scientific events remain almost irrecoverable. This is mainly due to a lack of the citation practice existed in the scientific literature published in the 19th century and before.

To the best of my findings, according to Shepperd [12], as early as in the end of the 18th century, a German chemist Martin Heinrich Klaproth (1743–1817) and a French chemist Joseph-Louis Proust (1754–1826) proposed that calcium phosphates were the major inorganic component of bones. Unfortunately, Shepperd has not provided any references to the publications by those great chemists. However, according to Roscoe and Schorlemmer [13], other researches had discovered this fact a bit earlier: "Gahn [14], in 1769, discovered the existence of calcium phosphate in bones, but it was not until this fact was published by Scheele [15] in 1771 that phosphorus was obtained from bone-ash, which has from that time invariably served for its preparation." (p. 458). Furthermore, let me cite a publication of 1777 [16] (please, note the old-fashioned using a long, medial or descending letter "ſ", which is a form of the minuscule letter "s" formerly used where "s" occurred in the middle or at the beginning of a word): "I have only been informed of this diſcovery, by the *Gazette, Salutaire de Bouillon*, October, 1775. It is there ſaid, that Mr. *Henry Gahn*, a phyſician at Stockholm, has communicated a proceſs for extracting from bones the ſaline matter in question; and that Mr. *Scheele* had aſcertained, that the earth of animals was compoſed of a calcareous ſubſtance united with the phoſphoric acid. This diſcovery, continues the author of the article of the Gazette, belongs to Mr. *Gahn*, and has been confirmed by later experiments." (p. 383). Presumably, this citation might be considered as one of the earliest mentioning on calcium phosphates.

Furthermore, as written in a book by Lavoisier [17], the production process of orthophosphoric acid by decomposition of calcined bones in sulfuric acid has been known since, at least, 1790 (again, the old-fashioned replacement of letter "s" by "ſ"): "The bones of adult animals being calcined to whiteneſs, are pounded, and paſſed through a fine ſilk ſiewe; pour upon the fine powder a quantity of dilute ſulphuric acid, leſs than is ſufficient for diſſolving the whole. This acid unites with the calcareous earth of the bones into a ſulphat of lime, and the phoſphoric acid remains free in the liquor." (p. 205). Further, the production process of white phosphorus has been described: "The liquid is decanted off, and the reſiduum waſhed with boiling water; this water which has been uſed to waſh out the adhering acid is joined with what was before decanted off, and the whole is gradually evaporated; the diffolved fulphat of lime criftallizes in form of filky threads, which are removed, and by continuing the evaporation we procure the phofphoric acid under the appearance of a white pellucid glafs. When this is powdered, and mixed with one third its weight of charcoal, we procure very pure phofphorus by fublimation." (p. 206).

8.3 Chemical Investigations on Calcium Orthophosphates in the 19th Century and the First Half of the 20th Century

In the 19th century, various investigations of calcium apatites and other calcium orthophosphates were performed by Jöns Jacob Berzelius (1779–1848) [18, 19], M. Baruel [20], C. Morfit [21], Robert Warington (there were 2 chemists with this name, presumably, a father (1807–1867) and a son (1838–1907)), who performed the earliest well documented systematic studies of the outstanding quality [22–25], and R. Fresenius [26]. Furthermore, a German chemist Eilhard Mitscherlich (1794–1863) also worked in this area (unfortunately, I failed to find any reference to his publications). To provide some proofs of this statement, let me cite a paragraph from Ref. [23]: "Mitscherlich tells us, that when chloride of calcium is added to ordinary disodic phosphate, the latter being maintained in excess, the precipitate formed is tricalcic phosphate, while the solution becomes acid from the production of monosodic phosphate. Berzelius, on the contrary, states, that the precipitate formed under these conditions is not tricalcic phosphate, but the octocalcic triphosphate, which lie has elsewhere described. All experimenters agree, that when the operation is reversed, and discc1.c. phosphate is poured into an excess of chloride of calcium, the precipitate is neither tricalcic nor octocalcic, but dicalcic phosphate." (pp. 296–297). Thus, TCP, DCP and OCP have been known since, at least, 1866, while in fact a bit earlier. Furthermore, the standard preparation procedure of the sparingly soluble calcium orthophosphates has been known since the same time.

Among the available publications written by two Robert Waringtons [22–25], Ref. [23] by Robert Warington Jr. deserves both a special attention and extensive citations. For example, to prove, that OCP indeed was known in 1866, let me make another citation from Ref. [23]: "Octocalcic phosphate can only be produced by the simultaneous formation of monosodic phosphate:

$8CaCl_2 + 5Na_4H_2P_2O_8 = Ca_8H_2P_6O_{24} + 16NaCl + 2Na_2H_4P_2O_8$." (p. 300).

One can see a balanced chemical equation, fully identical to the modern ones. It is hard to believe, that it was published in 1866! More to the point, the hydrated forms of calcium orthophosphates were known in 1866: "8·73 grs. of the vacuum-dried salt, lost on ignition 12·30 grains, or 26·35 per cent.; the formula $Ca_2H_2P_2O_8{\cdot}4H_2O$, demands 26·16 per cent. of water." (p. 299). Needless to explain, that "$Ca_2H_2P_2O_8{\cdot}4H_2O$" represents 2 molecules of DCPD (see Table 1.1). Furthermore: "It is interesting to observe that while disodic phosphate is of an alkaline nature, dicalcic phosphate possesses faint acid properties." (p. 300). The form and shape of DCPD crystals were described as well: "The crystalline form of the dicalcic tetrahydrated phosphate has been examined by Professor Church. He describes the crystals as thin rhomboïdal plates, of which the diagonally opposite acute angles are sometimes truncated, hexagonal forms being thus produced. This truncation seems to be occasionally hemihedral, and then may proceed up to the diagonal between obtuse angles; from this change triangular forms arise. Other modifications are also met with." (pp. 300–301). Another interesting conclusion might be found here: "We may then safely affirm that whenever dicalcic phosphate, octocalcic triphosphate, or any phosphate of intermediate composition, is precipitated from solution by ammonia, the salt obtained will be the octocalcic triphosphate; a tricalcic phosphate cannot be obtained in this manner. The following is probably a type of the reaction:

$4Ca_2H_2P_2O_8 + 6NH_3 = Ca_8H_2P_6O_{24} + 6NH_4.P_2O_8$." (pp. 301–302).

This seems to be the earliest mentioning on the fact that TCP cannot be precipitated from the aqueous solutions (currently we know that ACP or CDHA are precipitated instead). However, the following citation from the same publication: "It is quite possible that precipitated tricalcic phosphate may possess somewhat different solubilities, when prepared by different methods; this difference can, however, scarcely be great." (p. 304) means that this fact was not quite clear in 1866.

The latest available publication by Warington [25] was devoted to the hydrolysis of a freshly precipitated TCP (*i.e.*, either ACP or

CDHA) to the stoichiometric HA under continuous (up to 50 hours) boiling in distilled water. From the results of numerous chemical analyses the author concluded that during boiling an aqueous suspension of the TCP was slowly transformed to a suspension of $3Ca_3P_2O_8.CaOH_2O$ (*i.e.*, HA–see Table 1.1) and soluble acidic calcium orthophosphates. The following conclusion was made: "Since it appears that all phosphates of calcium less basic than apatite are unstable under the continued action of pure water, it seems probable that a more exact examination of natural phosphates would show that many phosphates now regarded as tricalcic are in fact of a more basic nature." (p. 989). Thus, the apatitic nature of the majority of natural calcium orthophosphates has been predicted in 1873. The next available study on the TCP hydrolysis was published in 1929 only [27].

To give the appropriate tribute to other researchers, one should mention that existence of acidic calcium orthophosphates, currently known as MCP and DCP (it is impossible to discover, whether they were hydrated or anhydrous), has been known, since, at least 1806: "In this way were distinguished among the salts two combinations, one neutral, and one with an excess of acid; and these were supposed to be determinate, as in sulphate and super-sulphate of potassa, or the phosphate and super-phosphate of lime." [28, p. 38]. Furthermore, in 1849, it was written that "M. J. L. Lassaigne, at the meeting of the Academy of Sciences of Paris, of the 15th January, presented a memoir upon this subject, showing by experiments that the phosphate and carbonate of lime are introduced into plants in solution in water containing carbonic acid, which had before been shown as to the phosphate by M. Dumas, and has long been known as to the carbonate." [29]. Thus, a higher solubility of both calcium orthophosphates and calcium carbonate in weak acids was already known in 1849. The first accessible paper on detection and preservation techniques of various deposits, including calcium orthophosphates, was published shortly afterwards [30]. Presence of calcium orthophosphates in teakwood was established in 1862 [31]. Various analytical topics on calcium orthophosphate chemistry has been studied since, at least, 1863 [32] and remained to be a subject of active research approx. until the 1910s [33–42]. A popular fertilizer superphosphate, which represents "a mixture of the last-mentioned compound and sulphate of lime" [43] has been known since, at least, 1868 [44]. More than 20 research papers on superphosphate were published by the beginning of the 20th century, which indicated to the importance of this chemical for the human being. To give the appropriate tribute to those early investigators and restore the historical perspective, it is worth citing them all [45–66]. One can see that almost all of these studies were performed and published by single researchers (see individual comments to Refs. [45, 52]), while the majority of the investigations were devoted to analytical chemistry.

In *Chemistry* by Wilson, published in 1850 [67], one can read the following: "797. *Phosphates of Lime.* — There are many compounds of lime and phosphoric acid, owing to the peculiarity of that acid in relation to the number of equivalents of base it combines with at once. The most interesting phosphate of lime is that which occurs in bones, and is distinguished as the bone-earth phosphate, $3CaO,PO_5$." (p. 219). Thus, various calcium orthophosphates were already known in 1850. However, the preparation technique sounds unusual to the modern readers: "The phosphorus combines in part with the oxygen of the lime, CaO, to form phosphoric acid, and this with undecomposed lime, to form phosphate of lime, CaO,PO_5. At the same time another portion of the phosphorus combines with the calcium of the lime, forming phosphuret of calcium, CaP." (p. 164). In *Chemistry* by Brande and Taylor, published in 1863 [68], one can find the following statements: "COMMON PHOSPHATE OF LIME; TRIBASIC PHOSPHATE OF LIME; BONE PHOSPHATE; $(3(CaO),PO_5)$. — This salt occurs abundantly in bone-ash, and is found as a mineral product." (p. 331). Furthermore, "*Native phosphate of lime* (bone phosphate) occurs in *apatite*, *moroxite*, *phosphorite*, and *asparagus stone*; its primitive form is a six-sided prism: it also occurs in some volcanic products." (p. 332). Thus, a similarity between the inorganic phase of bones and calcium orthophosphate rocks of natural origin (apatite and phosphorites) was already known in 1863. "When a solution of bone-earth in hydrochloric or nitric acid is boiled to expel all carbonic acid, and decomposed by caustic ammonia, the bone-phosphate separates in the form of a bulky precipitate, which, when perfectly dried, is white and amorphous." (p. 331). This statement is really astonishing because it might be considered as the first mention on ACP (Chapter 2), 32 years before Wilhelm Conrad Röntgen (1845–1923) discovered X-rays in 1895! Furthermore, the presence of carbonates in bones was already known. Next citation: "The substance known under the name of *coprolites*, and which appear to be the excrements of fossil reptiles, also abound in phosphate of lime." (p. 332) means that already in 1863 researchers were aware on this fact.

Furthermore, let me make a citation from a publication by Wells of 1906 [69]: "The apparent constancy of the proportion of carbonate and phosphate of calcium in bones made an impression on Hoppe-Seyler in I862, and we find him speculating on the possibility of the components of the two salts being joined together to form one giant molecule: 3 (Ca_3 (Po_4 $)_2$)-$CaCo_3$, which he imagined might be united in some such way" (p. 522) as shown in Fig. 8.1. Further, Wells mentions: "This formula is interesting chiefly from the historical standpoint, but it serves to emphasize the tendency of these salts to appear in nearly constant proportions in the animal body, a fact possibly of some importance." (p. 523). Obviously, the atomic arrangement shown in Fig. 8.1 represents the earliest structural drawing of a single molecule of carbonateapatite. An attentive reader will notice two different types of calcium (currently known as Ca(1) and Ca(2)) in that structure.

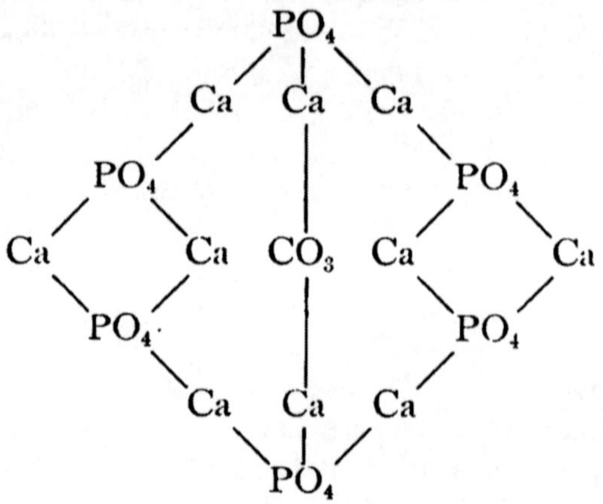

Figure 8.1 The first available structure of a bone mineral (carbonateapatite). Reprinted from Ref. [69].

Vitrification properties of some forms of calcium orthophosphates at heating have been known since, at least, 1877 [70]. The modern chemical formula of perfectly transparent crystals of natural FA as $Ca_5(PO_4)_3F$ has been known since, at least, 1873 [71], while the major crystallographic faces of a natural calcium apatite were described in

1883 [72]. Chemical formulae of DCP (as "Mono-Hydrogen Calcium Orthophosphate, $HCaPO_4$") and MCP (as "Tetra-Hydrogen Calcium Phosphate, $H_4Ca(PO_4)_2$") have been known since, at least, 1879 [43, pp. 205–206]. Interestingly that in a publication of 1871, the chemical formulae of calcium orthophosphates were written in different ways: $3CaO\ PO_5$ for apatite and $CaO\ 2HO\ PO_5$ for "some acid phosphate of lime" [46].

Neutral phosphates of lime have been known since, at least, 1872 [73]. Besides, in the 19th century, calcium apatites were considered as combined compounds, which results from this citation: "Calcium phosphate, combined with calcium chloride or calcium fluoride, occurs in the well-known minerals, apatite and osteolite." [43, p. 188]. One might guess that, in the 19th century, the atomic arrangement of single molecule of carbonateapatite (Fig. 8.1) could have inspired researchers to compose similar drawings for the single molecules of FA, HA and/or chlorapatite; however, I have not succeeded to find anything on this matter.

Chemical equations, describing various interactions between calcium phosphates and other chemicals have been known since, at least, 1863. For example, the afore-cited production processes of both orthophosphoric acid and white phosphorus from the Lavoisier book [17], in 1863 were written using chemical equations [68]: "When bone-phosphate is digested in dilute sulphuric acid, it is resolved into sulphate of lime and (if a sufficiency of sulphuric acid be used) phosphoric acid: $3(CaO),PO_5 + 3SO_3 = 3[CaO,SO_3] + PO_5$." (p. 331). Furthermore, various types of phosphates (namely, metaphosphate, acid phosphate) and differences in their solubility have been already known. Let me cite: "In order to prepare phosphorus, the bone-ash is first mixed with so much dilute sulphuric acid as to form the acid phosphate:

$$Ca_3\ (PO_4)_2 + 2H_2SO_4 = CaH_4\ (PO_4)_2 + 2CaSO_4.$$

The solution of this soluble acid phosphate is next poured off from the precipitated gypsum, and evaporated to dryness, after which, the solid residue being heated to redness, water is given off and calcium metaphosphate formed:

$$CaH_4\ (PO_4)_2 = Ca\ (PO_3)_2 + 2H_2O.$$

This salt is then carefully mixed with charcoal, and heated to bright redness in earthenware retorts shown in Fig. 147, when the following change takes place:

$Ca (PO_3)_2 + 10C = P_4 + Ca_3 (PO_4)_2 + 10CO$." [13, p. 460].

The quantitative analysis of a calcium orthophosphate was performed in 1884 [74], followed by remarks by C. Glaser in 1885 [75]. In the 1880s, occurrence of a calcium apatite [76] and TTCP [77–79] in metallurgical slags was discovered. A chemical interaction of TCP with carbon dioxide and iron hydroxide was investigated in 1891 [80], while, at the beginning of the 20th century, the systematic studies of calcium orthophosphates were performed by F. K. Cameron with co-workers [66, 81–87] and H. Bassett [88–91]. Namely, the first version of the solubility diagram $CaO–P_2O_5–H_2O$ was composed by Bassett [90]. As follows from the available literature [13, 43, 92], by that time, the researchers already operated with individual calcium orthophosphates. Binary salts of calcium orthophosphates with orthophosphates of other chemical elements have been known since, at least, 1911 [93]. Currently well-known chemical route of CDHA preparation by mixing of aqueous solutions of calcium nitrate and K_2HPO_4 (now $(NH_4)_2HPO_4$ is used instead) was investigated conductometrically in 1915 [94], while a neutralization reaction of H_3PO_4 by $Ca(OH)_2$ was investigated in 1923 [95]. The first paper on *in vitro* mineralization using aqueous solutions containing ions of calcium and orthophosphate was published in 1926 [96], followed by a publication of 1933 [97]. The crystal structure of FA was established in 1931 [98], followed by a great study on the structural characteristics of apatite-like calcium orthophosphates of various origin in 1933 [99]. Such terms as hydroxyfluorapatite, $Ca_{10}(F,OH)_2(PO_4)_6$, OA and carbonateapatite were already known. Besides, currently unknown substance "a hydrate of tricalcium phosphate, $Ca_9(H_2O)_2(PO_4)_6$" with the molecular weight 966.4 and apatite-like diffraction pattern (obviously, it was CDHA with $x = 1$, see Table 1.1) was mentioned as well [99]. Determination of the amounts of calcium orthophosphates in spinal fluid [100] and serum [101, 102] were performed in the 1920s, while that in saliva — in the beginning of the 1930s [103]. Furthermore, the solubility data of several calcium orthophosphates were updated in 1931 [104, 105]. Very popular at the turn of the millennium silicon- (or silica-) contained calcium orthophosphates, in fact, appear to be known since, at least, 1937 [106], while the earliest available paper on application of calcium orthophosphates in chromatography was published in 1942 [107].

In 1940, the available level of knowledge on basic calcium orthophosphates (TCP, TTCP and apatites) was summarized into a big review [108], which might be considered as the first comprehensive review on the subject. Interestingly, that the authors of this review have written about HA that "Existence as a unique stoichiometric compound doubtful" [108, p. 259, Table). Furthermore, according to this review, α- and β-modifications of TCP (see Table 1.1) have been differentiated in 1931–1933 by Schneiderhöhn and Bredig *et al.*, who also created the initial versions of the high-temperature diagram for the binary system $CaO–P_2O_5$. That time, the existence of OA was uncertain [108]. Besides, in 1940, a fundamental study on the equilibrium in the system $CaO–P_2O_5–H_2O$ was published [109].

To finalize this chemical section, one should mention that by 1928 it was clearly known that TCP could not be prepared by wet-precipitation. Let me cite: "*We have been unable to find any evidence of the existence of a molecular species with the formula* $Ca_3(PO_4)_2$. Precipitates of this "substance" rarely have the correct empirical composition and they cannot be purified by recrystallization. The evidence is such as to lead us to suspect that there may be no such chemical entity as $Ca_3(PO_4)_2$. No one has succeeded in preparing it by precipitation (12, 35–48), bearing out the theoretical objections to such a reaction on the grounds that fifth order reactions do not occur (49). We do not take the position that there can be no compound with this formula. It may be found possible to synthesize it by other methods, but so far no one has succeeded in preparing it by precipitation." [110, p. 128]. Similar conclusions were made by other researchers [111]. Nevertheless, this knowledge was not common yet, since, in 1935, a report was published that "Tricalcium phosphate monohydrate was prepared by the slow addition of calcium chloride solution to a constantly agitated alkaline solution of disodium phosphate, maintained at 65° to 70°C." [112]. This controversy has been explained in Ref. [108] by a matter of definitions: "The terms "tricalcium phosphate" and "hydroxyapatite" are very widely and very loosely used. For example, some authors use the former for any precipitate more basic than dicalcium phosphate, although such precipitates have been frequently shown to have an apatite lattice or to be mixtures of dicalcium phosphate and an apatite. Others confine the use of the term to those precipitates with P_2O_5:CaO ratios approaching that of $Ca_3(PO_4)_2$." (p. 265).

More recent (after 1940) publications on chemistry of calcium orthophosphates are not considered, since they are well known.

8.4 Early Publications on Calcium Orthophosphates of Geological Origin

The earliest research paper, I have been able to find, devoted to natural apatites was published in 1873 [71], while that dedicated to deposits of natural calcium orthophosphates was published in 1889 [113]. Presumably, the latter publication might be considered as the first review paper on calcium orthophosphates, in which the author described various types of natural phosphate deposits, their formation, as well as possible utilization, mainly as fertilizers. Next available scientific publications on natural calcium orthophosphates were published in 1895 [114], 1901 [115, 116], 1902 [117–119] and 1904 [120]. An important study on the presence of orthophosphates of various chemical elements, including calcium, in soil was published in 1906 [121]; however, this subject is beyond the scope of this Chapter.

8.5 Early Studies on Calcium Orthophosphates in Living Organisms

As mentioned above, the presence of calcium orthophosphates in bones has been known since the 1770s [13, 16]. Further investigations in this field were performed in the 19th century. For example, as seen from the book by Humphry Davy [122], the organic/inorganic composite nature of bones and horns has been known since, at least, 1814: "Bones in particular produce it very plentifully, as they consist of phosphat of lime combined or cemented by gelatine. Horns which are a species of bone, will yield abundance of gelatine." (p. 294). Furthermore, the presence of calcium orthophosphates in milk has been known since, at least, the same time: "But milk contains, besides, a considerable portion of phosphat of lime, the purposes of which, I have already pointed out." (p. 331). Besides, the general principles of bone and teeth formation (biomineralization) were already known: "During the earliest period of animal life they consist almost entirely of a gelatinous membrane of the form of the bones, but of a loose spongy texture, the cells or cavities of which are destined to be filled with phosphat of lime; it is the gradual acquisition of this salt which gives to the bones their subsequent hardness and durability. Infants first receive it from their mother's milk, and afterwards derive it from all animal and from most vegetable food, especially farinaceous substances, such as wheat floor, which contain it in sensible quantities. A portion of the phosphat after the bones of the infant have been sufficiently expanded and solidified, is deposited in the teeth, which consist at first of only a gelatinous membrane or case, fitted for the reception of this salt; and which, after acquiring hardness within the gum, gradually protrude from it." (p. 302). What is more, such disease as rickets (rachitis) was already known to be caused by deficiency of calcium phosphates: "Pray, Mrs. B. does not the disease of the bones of children, called the rickets, proceed from a deficiency of phosphat of lime?" and response: "... The other cause of this disorder is an imperfect digestion and assimilation of the food, attended with an excess of acid, which counteracts the formation of phosphat of lime." (p. 303). The next publication, I have been able to find, devoted to the importance of calcium orthophosphates to bone health was published in 1842 [123], followed by studies published in 1895 [124], 1906 [125] and 1910 [126], while those on calcium phosphates of bones were published by Aeby in 1872 [127] and 1873 [128, 129], followed by papers written by other researchers [130, 131]. The presence of calcium orthophosphates in urine has been known, since, at least, 1861 [132, 133], while the presence of calcium orthophosphates in the auditory system of mammals has been known since, at least, 1879 [134]. Papers on the importance of calcium orthophosphates for lactation were published in 1914 [135], 1919 [136] and 1922 [137], while the fact that the optimum Ca:P ratio in a diet is the same as that found in human bones or in mother's milk has been known since, at least, 1930 [138].

Back on topic of bones, one should note that the differences between the chemical compositions of bones belonged to young and old individuals were investigated in the 19th century. Since I have failed to extract the original references, let me cite a summary on this topic, published in 1928 [139]: "In 1855 Frémy (4) concluded that in humans there is, with increasing age, a relative increase of $CaCO_3$ with respect to calcium phosphate. Von Recklinghausen (5) in 1858 concluded from his own analyses and from the work of Bibra (1844), of Stark (1845), of Heintz (1849), and of Frémy, that there was no observable difference in composition between the bones of young and those of old individuals. In 1860 Milne Edwards (6) reported that his own analyses agreed with those of Bibra and of Frémy in showing that the proportion of carbonate in the bones of young animals was smaller than in adult bone. Zalesky (7) in 1866 reviewed the literature and concluded that the bones of children of various ages have the same composition as those of adults, not only as a whole, but also as regards the proportion of individual constituents. He pointed out that the methods of determining CO_2 employed by previous investigators were defective. In 1872 Wildt (8) analyzed the ashed bones of rabbits of different ages and found an increase in the proportion of CO_2 with increasing age. Weiske (10) in 1889 analyzed the ashed bones of birds and found an increase in CaO and CO_2 with advancing age, while the P_2O_5 remained constant. Graffenberger (11) in 1891 analyzed the ashed bones of rabbits and also found an increase of the ratio calcium carbonate: calcium phosphate with increasing age." (pp. 155–156). The citation shows that this was a hot topic of research in the 19th century.

The earliest research paper, I have been able to find, on undesirable *in vivo* formed calcifications of biologically formed calcium orthophosphates was published in 1856 [140], followed by a book [141] written by a famous German doctor, anthropologist, pathologist, prehistorian, biologist and politician Rudolf Ludwig Karl Virchow (1821–1902), in which he introduced a term "metastatic calcification". To the best of my findings, next studies on undesirable *in vivo* formed calcifications containing calcium orthophosphates were published in 1882 [142] and 1883 [143], followed by publications of 1901 [144], 1906 [69] and 1913 [145]. An important conclusion was made in 1906 [69]: "The composition of the inorganic salts in calcified areas in the body seems to be practically the same, if not identical, whether the salts are laid down under normal conditions (ossification) or under pathological conditions." (p. 520). Thus, the similarity between the calcium orthophosphates of bones and those found in metastatic calcification has been established, at least, in 1906.

The earliest study, I have been able to find, on a behavior of an undisclosed calcium orthophosphate in organisms of carnivores was published in 1883 [146]. Next available studies on a similar topic were in print in 1912 [147] and 1921 [148]. A paper on the importance of calcium and phosphorus for growing children was published in 1913 [149], while that on the necessity of normal calcium and phosphorus metabolism was published in 1930 [150]. To conclude this topic, one should mention on a huge review published in 1935 [151] and another one on bone composition published in 1937 [152] summarizing the available by that time knowledge on calcium orthophosphates in living organisms.

8.6 Early Attempts to Treat Various Diseases by Calcium Orthophosphates

Concerning the attempts to treat various diseases by calcium orthophosphates, the earliest publications, I have been able to find, were in print as early as in 1851 [153, 154]. The author performed a thorough study on the biomedical influence of phosphate of lime that was "taken with the breakfast, dinner, and supper, in order to have it mixed as well as possible with the food taken" on treatment of various diseases. According to the author: "... the phosphate of lime proved most beneficial in children who suffered from scrofula, diarrhoea, ulcerations and excoriations of the skin and the bowels, general waste of the cellular tissue, loss of power, &c. In these cases, the mere use of six to ten grains per diem was often quite sufficient to effect the cure, and I have met with such striking and satisfactory results, as to leave no doubt that the want of the phosphate of lime was the real cause of the symptoms alluded to." [153, p. 434, left column]. Furthermore, the author states: "As to other affections, in which I have tried the phosphate of lime, I have to mention rickets, caries, inflammations, and consecutive abundant suppurations of the cellular tissue, and also fractures of the bones. In all these cases, the administration proved most beneficial, and I would strongly advocate its further experimental use. With respect to fractures of the bones, I have to state in particular, that the consolidation of the callus took place in a much shorter period than is generally the case; however, too large doses of phosphate of lime must be shunned in these cases, as I have observed an abundant callus, causing a deformity of the bones, produced by the daily administration of twenty-four grains of the phosphate for a fortnight." [153, p. 434, right column]. In the second paper, the same author mentioned an interesting hypothesis that "phosphate of lime contained in a certain quantity of blood" (thus, the presence of calcium orthophosphates in blood was already known in 1851) might be partly "originated from the wear and tear of the bones and the muscular tissue". Let me cite: "Supposing now the wear and tear of bones to be abnormally increased, an abnormal quantity of phosphates will consequently re-enter into the composition of the blood, and by analysing the latter we shall find an increased quantity of phosphates in it. The general waste of bones is therefore leading to an increase of the phosphates in the blood, and nothing really would be more erroneous, than to suppose the general quantity of phosphates in the system to be increased, whenever we find their quantity increased in the blood." [154, p. 668, right column]. The readers are strongly advised to get through the both papers to rediscover the medical ideas and approaches published 160 years ago!

The next available medical study involving phosphate of lime was published in 1874 [155]. An interesting idea to cure cancer "by embedding a piece of bone material or calcium phosphate in a cancer and exposing it to a pencil of X rays ..." to create a secondary radiation in calcium orthophosphates (either natural or implanted ones) was announced in 1913 [156]. Strange enough, but seeking in the scientific databases on farther development of this idea has resulted in the zero publications (researchers introduced radioactive isotopes of other chemical elements instead [157–160]). Nevertheless, this might be a worth reinvestigating approach.

To conclude this topic, the earliest paper, I have been able to find, on addition of calcium orthophosphates to forage of animals was published in 1953 [161].

8.7 Artificial Grafts in the 19th Century and Before

The artificial generation of tissues, organs or even more complex living organisms was throughout the history of mankind a matter of myth and dream. Unfortunately, due to the practice of cremation in many societies, little is known about the prehistoric materials used to replace bones lost to accident or disease. Nevertheless, according to the available literature, introduction of non-biological materials into the human body was noted far back in prehistory. For example, the remains of a human found near Kennewick, Washington, USA (often referred to as the "Kennewick Man") was dated to be up to

9000 years old. This individual, described by archeologists as a tall, healthy, active person, wandered through the region now known as southern Washington with a spear point embedded in his hip. It had apparently healed in and did not significantly impede his activity. This unintended implant illustrates the body's capacity to deal with implanted foreign materials. The spear point has little resemblance to modern biomaterials, but it was a tolerated foreign material implant, just the same [162].

Later, man's attempts to repair the human body with the use of implant materials were recorded in the early medical writings of the Hindu, Egyptian and Greek civilizations. The earliest successful implants were in the skeletal system. For example, the famous painting by Fra Angelico (ca. 1395–1455) "The Healing of Justinian by Saint Cosmas and Saint Damian" (Fig. 8.2) a visualization of the legend of twins Sts. Cosmas and Damien (died ca. 287 AD) depicting a transplantation of a homograft limb onto an injured soldier, is one early instance of the vision of a regenerative medicine. Historically, a selection of the materials was based on their availability and an ingenuity of the individual making and applying the prosthetic [163]. Archaeological findings exhibited in museums showed that materials used to replace missing human bones and teeth included animal or human (from corpses) bones and teeth, shells, corals, ivory (elephant tusk), wood, as well as some metals (gold or silver). For instance, the father of Western medicine Hippocrates (ca. 460 BC–ca. 370 BC) apparently used gold wire and linen thread for ligatures in the repair of bone fractures. Aulus Cornelius Celsus (ca. 25 BC–ca. 50 AD) recommended the filling of large cavities with lint, lead and other substances before attempting extraction to prevent the tooth from breaking under the pressure of the instrument. This may have been the beginning of filling materials for carious teeth. The Etruscans learned to substitute missing teeth with bridges made from artificial teeth carved from the bones of oxen, while in ancient Phoenicia loose teeth were bound together with gold wires for tying artificial ones to neighboring teeth. Popp states that ancient Egyptians also made artificial ears, noses and eyes [164]. The Chinese recorded the first use of dental amalgam to repair decayed teeth in the year 659 AD, while in Americas the pre-Columbian civilizations used gold sheets to heal cranial cavities following trepanation [165]. Besides, while excavating Mayan burial sites in Honduras in 1931, archaeologists found a fragment of mandible of Mayan origin, dating from about 600 AD. This mandible, which is considered to be that of a woman in her twenties, had three tooth-shaped pieces of shell placed into the sockets of three missing lower incisor teeth. In 1970, a Brazilian dental academic Prof. Amadeo Bobbio studied the mandible specimen and took a series of radiographs. He noted compact bone formation around two of the implants, which led him to conclude that the implants were placed during life [166]. This may be the first recorded use of dental implants. More to the point, an iron dental implant in a corpse dated ~ 200 AD was found in Europe. This implant, too, was described as properly bone integrated [167].

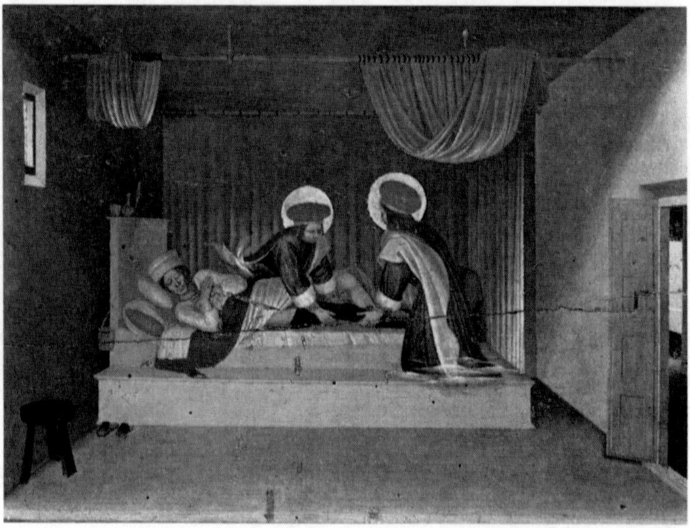

Figure 8.2 Fra Angelico (ca. 1395–1455) "The Healing of Justinian by Saint Cosmas and Saint Damian" (approx. 1439) is exhibited at Museo di San Marco, Florence, Italy.

In middle ages, one of the first scientific descriptions of congenital and acquired defects of the maxilla and their treatments was given by Ambroise Paré (1510–1590) in his *Dix livres de la chirurgie, avec le magasin des instrumens necessaires à icelle* in 1564. He specifically described defects of the palate with bone destruction caused by arquebus shots, stab wounds or syphilitic gumma, describing also the accompanying speech deficiency and giving general principles of treatment. He used a flat, vaulted, metallic plate in gold or silver with a sponge attached to it. The sponge was introduced into the defect, where it expanded with readily absorbed nasal and oral secretions, thus holding the obturator base in position. Paré is also credited with having prepared artificial teeth from bones and ivory. In the 17th century, a piece of dog skull was successfully transplanted into the damaged skull of a Dutch duke.

A common method in the 18th century to replace teeth was the homologous transplantation of teeth in humans. John Hunter (1728–1793) investigated in his pioneering work the effect of transplantation not only at a clinical level (he claimed, that homologous transplanted teeth lasted for years in the host) but also performed animal experimental work on the fate of transplants, thereby setting the basis for a scientific approach on transplantation medicine [168]. Besides, various restorative materials might be used for feeling defects, capping exposed pulps and dental cavities. Currently they include zinc orthophosphate, zinc polyacrylate (polycarboxylate), various amalgams, glass ionomer cement of complicated formulations, calcium hydroxide, etc.; however, this is another story. Within the scope of this review, calcium orthophosphate-based formulations will be considered only. According to the available literature, Dr. Junius E. Cravens (1844–1920) from USA proffered creative concepts in pulp capping in the 1870s. He had the opinion that dentin-like material would be the best to keep the pulp vital. Therefore, Cravens used a calcium orthophosphate powder, which was mixed with lactic acid to low viscosity. The result was a soluble calcium lactic orthophosphate, which was applied onto the exposed pulp tissue [169]. This pulp-capping agent was brought to the market by the S.S. White company with the trade name "Lacto-Phosphate of Lime" (Fig. 8.3). To the best of my findings, this study might be considered as the first mentioning on calcium orthophosphate-based biocomposites and hybrid biomaterials (Chapter 6). Thus, one might claim that the grafting history of calcium orthophosphates starts from the 1870s.

LACTO-PHOSPHATE OF LIME.

This article has been recommended by Dr. J. E. Cravens for the treatment of sensitive dentine, and to promote deposition of dentine in exposure of pulps.

Put up in half-ounce bottles, in the form of paste, ready for application.

Price.. 50 cents.

SAMUEL S. WHITE.

Figure 8.3 An advertisement of the S.S. White company for "Lacto-Phosphate of Lime" 1873. Reprinted from *Dent. Cosmos* 1873, 15, 683.

One must stress that in the past, many implantations failed due to various infections, which tended to be exacerbated in the presence of implants, since they provided a region inaccessible to the body's immunologically competent cells. Thus, the use of implantable biomaterials did not become practical until the advent of an aseptic surgical technique developed by a British surgeon Sir Joseph Lister (1827–1912) in the 1860s. Furthermore, there was a lack of knowledge about a toxicity of the selected materials. The exact chemical composition of the normal calcified tissues of mammals has been known since, at least, the 1870s, which might be concluded from both Table 8.1 [24] and the following citations: "The bones and teeth of animals contain large amounts of calcium phosphate, together with some carbonate and fluoride." [43, p. 188] and "Calcium phosphate is also the chief inorganic constituent of bones, forming about 80 per cent of burnt bones; the other constituents being magnesium phosphate, calcium carbonate, and calcium fluoride." [43, p. 205]. Those data were updated in 1894 [170]. However, the apatitic structure of the calcium orthophosphates of bones and teeth has been established by X-ray diffraction in 1926 [171] and confirmed in 1932 [172] and 1933 [173]. Since that time, bone mineral has been frequently identified as HA. In this frame, application of calcium orthophosphates as artificial grafts appears to be logical due to their similarity with the mineral phases of bones and teeth.

Table 8.1 The chemical composition of two bone samples taken from a publication of 1871 [24].

	Commercial bone-ash	Pure ox bone-ash
Moisture and volatile matter	6.70	1.86
Siliceous matter	9.69	0.51
Oxide of iron	0.58	0.17
Lime	43.37	52.46
Magnesia	1.14	1.02
Phosphoric acid	33.68	39.55
Carbonic acid, alkalies, and other substances undetermined	4.84	4.43
Total	100.00	100.00

To conclude this topic, one should stress that the performance of living tissues is the result of millions years of evolution, while the performance of acceptable artificial substitutions those humankind has designed to repair damaged tissues are only a few decades old. This explains the greatest differences between them. To get the historical perspective on the development of artificial grafts prepared from other materials, the interested readers are referred to the special literature [7, 9–12, 174–177].

8.8 Calcium Orthophosphates as Bone Graft Substitutes: A Historical Perspective

Historically, plaster of Paris (calcium sulfate) was the first widely tested artificial bioceramics. For example, according to Wikipedia, the free encyclopedia, literature dating back to 975 AD notes that calcium sulfate was useful for setting broken bones. However, those were *ex vivo* applications. According to the available literature, by the end of the 19th century, surgeons already used plaster of Paris as a bone-filling substitute [178]. Nevertheless, it was a famous German surgeon Themistocles Gluck (1853–1942), who, amongst his range of contributions, on 20 May 1890 performed the first well documented ivory (virtually, pure biological apatite) knee replacement bedded in a calcium sulfate based cement, which was followed by a total wrist replacement in another patient three weeks later [179]. Later in 1890, Gluck presented a further case of a total knee replacement to the Berlin Medical Society: at only 35 days after operation, the patient was pain free with active knee flexion and extension. All the joint arthroplasties performed by Gluck were remarkably successful in the short term; however, all ultimately failed because of chronic infections [180, 181]. After the abovementioned case of lacto-phosphate of lime (Fig. 8.3), this seems to be the second well-documented grafting application of calcium orthophosphates.

However, in the aforementioned cases, either the biomedical applications of biologically produced calcium orthophosphates (Gluck) or dental applications, not requiring any surgery (Cravens) have been described. According to both the electronic databases and previous reviews on the subject [7, 9–12, 174–177], the first attempt to implant a laboratory produced calcium orthophosphate (it was TCP) as an artificial material to repair surgically created defects in rabbit bones was performed in 1920 [8] by an US surgeon Fred Houdlette Albee (1876–1945), who invented bone grafting [182] and some other advances in orthopedic surgery. The researchers injected either 0.5 or 1 c.c. of 5 % slurries of TCP in distilled water (which was then sterilized for three successive days in the Arnold sterilizer, at 60º C) into surgically created radial bone gaps of rabbits, leaving the periosteum intact [8]. Radiographic analysis revealed that the TCP injected defect demonstrated more rapid bone growth and union than the control. The average length of time for bony union with TCP was 31 days, compared to 41 days for the controls. No appreciable bone growth was stimulated by injecting TCP beneath the periosteum in non-defective radii or into subcutaneous tissues. Although this seems to be the first scientific study on use of an artificially prepared calcium orthophosphate for *in vivo* repairing of bone defects, it remains unclear whether that TCP was a precipitated or a ceramic material and whether it was in a powder or a granular form. Unfortunately, the researchers published nothing further on this topic. In 1927, Hey Groves (1872–1944) described pure ivory hip hemi-arthroplasty for fracture [183]. In 1931, Murray also reported an acceleration of healing following implantation of calcium salts composed of 85% TCP and 15% $CaCO_3$ in canine long bone defects [184, 185].

At the beginning of the 1930s, the classic osteoinductive phenomenon was defined well by Huggins [186], who demonstrated that autoimplantation of transitional epithelium of the urinary bladder to abdominal wall muscle in dogs provoked ectopic bone formation. A bit later, Levander demonstrated that crude alcoholic extracts of bones induced a new bone formation when injected into muscle tissue [187, 188].

Simultaneously, in the 1930s, Haldeman and Moore [189], Stewart [190], Key [191] and Shands [192] discovered the fact that only certain types of calcium orthophosphates mentioned in Table 1.1 really influence the bone healing process. Namely, Haldeman and Moore implanted various calcium orthophosphates such as MCP and DCP (it remains unclear whether they were in hydrated or anhydrous forms), TCP, as well as calcium glycerophosphate as dry powdered salts into 0.5 to 1.0 cm defects in radii of 17 rabbits, while the opposite side served as control. Radiographic analysis demonstrated that in no case did the presence of MCP, DCP or calcium glycerophosphate had a favorable influence delayed healing compared to control, while the presence of TCP at the site of the fracture appeared to favor the union [189]. Furthermore, Key [191] suggested that "if a defect in bone could be filled by a non-irritating, slowly soluble mass, which was porous and which contained calcium phosphate and carbonate in a form in which they could be resorbed, it would be reasonable to expect osteoblasts to invade the mass, utilize the calcium, and build new bone which would replace the mass of calcium and cause the bone to be restored to its original form. The ideal material would appear to be rather dense cancellous bone from which a large percentage of the organic material had been removed." (p. 176). However, Key found that "Neither calcium phosphate and carbonate in the proportions in which they occur in bone, nor bone powder, made by removing the organic matter from bone, appear to stimulate osteogenesis of bone when implanted in a bone defect." (p. 184). Stewart [190] concluded that "1. Lime salts and boiled bone when placed into a bone defect with either traumatized muscle or fascia do not serve as a source of available calcium resulting in supersaturation of connective tissue and regeneration of missing bone. 2. Live bone chips placed in bone defects regenerate the missing bone." (p.871). Shands [192] also reported conflicting effects of several calcium salts (calcium glycerophosphate, a mixture of TCP (3 parts) and $CaCO_3$ (1 part), bone ash and calcium gluconate) on bone repair. Namely, in defects in the ulna of dogs, the investigated calcium salts appeared to stimulate bone formation, while in operations upon the spine, calcium glycerophosphate did not stimulate bone formation and appeared rather to exert an inhibiting influence. In 1948, Schram and Fosdick confirmed the fact that only certain types of calcium orthophosphates influence the bone healing process [193]. Similar conclusions were obtained in 1951 by Ray and Ward [194].

In 1950, the history of calcium orthophosphate cements was started [195]. The author of that important publication investigated mixtures of both oxides and hydroxides of various metals with aqueous solutions of orthophosphoric acid and discovered a number of cold-setting formulations. For example, he found that CaO, sintered at 1100°C, did not set in H_3PO_4, while that in liquid containing 9.6% CaO was found to set after ~ 12 h in presence of H_3PO_4 [195]. The latter mixture might be considered as the first prototype of self-setting calcium orthophosphate cements (Chapter 5); however, the real history of this subject started in 1982.

More recent historical events are reported very briefly. Namely, the modern history of ACP started in 1955 (Chapter 2), while more than twenty years afterwards the first dental application of a calcium orthophosphate (erroneously described as TCP) in surgically created periodontal defects [196] and the use of dense HA cylinders for immediate tooth root replacement [197] were reported. According to the available databases, the first paper with the term "bioceramics" in the abstract was published in 1971 [198], while with that in the title were published in 1972 [199, 200]. However, application of the ceramic materials as prostheses had been known before [201–204]. On April 26, 1988, the first international symposium on bioceramics was held in Kyoto, Japan.

The history of calcium orthophosphate based biocomposites and hybrid biomaterials (Chapter 6) started in 1981, while that of calcium orthophosphate coatings (reviewed in Refs. [205, 206]) started in 1980 [207]. An extensive commercialization of the dental and surgical applications of calcium orthophosphate (mainly, HA) bioceramics occurred in the 1980s, largely due to the pioneering efforts by Jarcho [208–211] in the USA, de Groot [212–214] in Europe and Aoki [215–218] in Japan. Shortly afterwards HA has become a bioceramic of reference in the field of calcium orthophosphates for biomedical applications. Preparation and biomedical applications of apatites derived from sea corals (coralline HA) [219–221] and bovine bone [222] have been reported at the same time [223], while the history of nanodimensional and nanocrystalline calcium orthophosphates (Chapter 3) started in 1995.

The interested readers are able to get further details on a recent history of calcium orthophosphates, bioceramics and biomaterials from other reviews on the subject [7, 9–12, 174–177].

8.9 Conclusions

Even though calcium orthophosphates have been investigated for more then two centuries, commercial implants containing calcium orthophosphates appeared only a few decades ago. However, the advantages and obvious benefits of these chemical compounds are still both an inspiration and a hope of many researchers and clinicians. The reported historical findings should encourage both scientists and clinicians to reinvestigate the already forgotten and poorly known facts and approaches in further biomedical applications of calcium orthophosphates.

References and Notes

1. As a mineral species, apatite was first recognized by the father of German geology Abraham Gottlob Werner (1750–1817) in 1786 and named by him from the ancient Greek απατάω (apatao) — "to mislead" or "to deceive", because it had previously been mistaken for other minerals, such as beryl, tourmaline, chrysolite, amethyst, fluorite, etc. Currently, apatite is the name for a group of minerals with the same crystallographic structure and does not indicate one chemical composition. That is why the term "calcium apatite" is used in this Chapter.
2. Dorozhkin, S. V. (2007) Calcium orthophosphates, *J. Mater. Sci.*, **42**, 1061–1095.
3. Dorozhkin, S. V. (2009) Calcium orthophosphates in nature, biology and medicine, *Materials*, **2**, 399–498.
4. Davy, H. (1808) The Bakerian lecture: On some new phenomena of chemical changes produced by electricity, particularly the decomposition of the fixed alkalies, and the exhibition of the new substances which constitute their bases; and on the general nature of alkaline bodies, *Phil. Trans. R. Soc. Lond.*, **98**, 1–44.
5. Davy, H. (1808) Electro-chemical researches on the decomposition of the earths; with observations on the metals obtained from the alkaline earths, and on the amalgam procured from ammonia, *Phil. Trans. R. Soc. Lond.*, **98**, 333–370.
6. Boyl, R. (1693) A paper of the honourable Robert Boyl's, deposited with the secretaries of the Royal Society, Octob. 14. 1680. and opened since his death; being an account of his making the phosphorus, etc., *Phil. Trans.*, **17**, 583–584.
7. Driskell, T. D. (1994) Early history of calcium phosphate materials and coatings. In: *Characterization and Performance of Calcium Phosphate Coatings for Implants*. ASTM STP 1196. Horowitz, E. Parr, J. E. (Eds.), American Society for Testing and Materials, Philadelphia, USA, pp. 1–9.
8. Albee, F. H., assisted by Morrison, H. F. (1920) Studies in bone growth. Triple calcium phosphate as a stimulus to osteogenesis, *Ann. Surg.*, **71**, 32–39.

9. Shackelford J. F. (1999) Bioceramics — an historical perspective, *Mater. Sci. Forum*, **293**, 1–4.
10. Hulbert, S. F., Hench, L. L., Forbers, D., and Bowman, L. S. (1982) History of bioceramics, *Ceram. Int.*, **8**, 131–140.
11. Hulbert, S. F., Hench, L. L., Forbers, D., and Bowman, L. S. (1983) History of bioceramics. In: *Ceramics in Surgery*. Vincenzini, P. (Ed.), Elsevier, Amsterdam, The Netherlands, pp. 3–29.
12. Shepperd, J. (2004) The early biological history of calcium phosphates. In: *Fifteen Years of Clinical Experience with Hydroxyapatite Coatings in Joint Arthroplasty*. Epinette, J. A., Manley, M. T. (Eds.), Springer, France, pp. 3–8.
13. Roscoe, H. E., and Schorlemmer, C. (1881) *A Treatise on Chemistry. Volume I: The Non-Metallic Elements*. Macmillan and Co., London, 751 pp.
14. Johan Gottlieb Gahn (1745–1818) was a Swedish chemist and metallurgist, who discovered manganese in 1774.
15. Carl Wilhelm Scheele (1742–1786) was a German-Swedish pharmaceutical chemist, who discovered oxygen (although Joseph Priestley published his findings first), as well as identified molybdenum, tungsten, barium and chlorine before Humphry Davy.
16. A dictionary of chemistry. Containing the theory and practice of that science: its application to natural philosophy, natural history, medicine, and animal economy: with full explanations of the qualities and modes of action of chemical remedies: and the fundamental principles of the arts, trades, and manufactures, dependent on chemistry. Translated from the French. With notes, additions, and plates. The second edition. To which is added, as an appendix, a treatise on the various kinds of permanently elastic fluids, or gases. Vol. III. Printed for T. Cadell, and P. Elmsly, in the Strand. London, 1777, 666 pp.
17. Lavoisier, Elements of chemistry, in a new systematic order, containing all the modern discoveries. Translated from the French, by Robert Kerr. Edinburgh: printed for William Creech, and sold in London by G. G. and J. J. Robinsons. 1790, 511 pp.
18. Berzelius, J. (1816) Untersuchungen über die Zusammensetzung der Phosphorsäure, der phosphorigen Säure und ihrer Salze, *Ann. Physik*, **53**, 393–446.
19. Berzelius, J. (1845) Ueber basische phosphorsaure Kalkerde, *Justus Liebigs Annalen der Chemie*, **53**, 286–288.
20. Baruel, M. (1838) Analysis of a double phosphate of lead and lime, *J. Franklin Inst.*, **25**, 343.

21. Morfit, C. (1855) On Colombian guano; and certain peculiarities in the chemical behavior of "bone phosphate of lime", *J. Franklin Inst.*, **30**, 325–329.

22. Warington, R. CXIX (1843) On a curious change in the composition of bones taken from the guano, *Mem. Proc. Chem. Soc.*, **2**, 223–226.

23. Warington, R., Jr. XXVII. (1866) Researches on the phosphates of calcium, and upon the solubility of tricalcic phosphate, *J. Chem. Soc.*, **19**, 296–318.

24. Warington, R. X. (1871) On the solubility of the phosphates of bone-ash in carbonic water, *J. Chem. Soc.*, **24**, 80–83.

25. Warington, R. XL (1873) On the decomposition of tricalcic phosphate by water, *J. Chem. Soc.*, **26**, 983–989.

26. Fresenius, R. (1867) Ueber die Bestimmung der Phosphorsäure im Phosphorit nebst Mittheilung der Analysen des Phosphorits und Staffelits aus dem Lahnthal, *Zeitschrift für Analytische Chemie*, **6**, 403–409.

27. Lorah, J. R., Tartar, H. V., and Wood, L. (1929) A basic phosphate of calcium and of strontium and the adsorption of calcium hydroxide by basic calcium phosphate and by tricalcium phosphate, *J. Am. Chem. Soc.*, **51**, 1097–1106.

28. Murray, J. (1806) *A System of Chemistry*. Vol. I. Edinburgh. Printed for Longman, Hurst, Rees & Orme, London; and William Creech, and A. Constable & Co., 592 pp.

29. F. (1849) On the method by which the phosphate and carbonate of lime is introduced into the organs of plants, *J. Franklin Inst*, **48**, 156.

30. Hassall, A. (1852) On the detection and preservation of crystalline deposits of uric acid, urate of ammonia, phosphate of lime, triple phosphate, oxalate of lime, and other salts, *Lancet*, **59**, 466–467.

31. Abel, F. A. XVIII (1862) On the occurrence of considerable deposits of crystallized phosphate of lime in teak-wood, *J. Chem. Soc.*, **15**, 91–93.

32. Stammer, C. (1863) Bestimmung kohlensauren Kalkes neben phosphorsaurem Kalk, *Zeitschrift für Analytische Chemie*, **2**, 96–97.

33. Roussin, Z. (1868) Prüfung des Wismuthsubnitrats auf eine Verfälschung mit Kalkphosphat, *Zeitschrift für Analytische Chemie*, **7**, 511.

34. Janovsky, J. V. (1872) Ueber die verschiedenen Methoden der Phosphorsäure-Bestimmung neben Eisenoxyd, Thonerde, Kalk und Magnesia, *Zeitschrift für Analytische Chemie*, **11**, 153–167.

35. Pellet, H. (1882) Die Zusammensetzung des Niederschlags, welcher durch Ammoniak aus sauren Lösungen von Phosphorsäure, Baryt, Kalk und Magnesia gefällt wird, *Zeitschrift für Analytische Chemie*, **21**, 261.
36. Stokvis, B. J., Salkowski, E., and Smith, W. G. (1884) Ueber die Löslichkeitsverhältnisse des phosphorsauren Kalks im Harn, *Zeitschrift für Analytische Chemie*, **23**, 273–274 (in reality, this publication represents a set of 3 short independent studies performed by 3 individual authors but combined under a general title).
37. Ott, A. (1886) Die Löslichkeitsverhältnisse des phosphorsauren Kalks im Harn, *Zeitschrift für Analytische Chemie*, **25**, 279–280.
38. Kennepohl, G. (1889) Zur Bestimmung von Eisenoxyd und Thonerde neben Kalk und Phosphorsäure, *Zeitschrift für Analytische Chemie*, **28**, 343.
39. Immendorff, H., and Reitmair, O. (1892) Zur Bestimmung des Kalks in Gegenwart von Phosphorsäure, Eisen, Thonerde und Mangan, *Zeitschrift für Analytische Chemie*, **31**, 313–316 (in reality, this publication represents a set of 2 short independent studies performed by 2 individual authors but combined under a general title).
40. Fingerling, G., and Grombach, A. (1907) Eine neue Modifikation der Bestimmung der zitratlöslichen Phosphorsäure in den Futterkalken nach Petermann, *Zeitschrift für Analytische Chemie*, **46**, 756–760.
41. Schulze, B. (1911) Untersuchung des phosphorsauren Futterkalkes, *Zeitschrift für Analytische Chemie*, **50**, 126–127.
42. Hinden, F. (1915) Anreicherungsmethode zur Bestimmung der Phosphorsäure in phosphorsäurearmen Kalksteinen, *Zeitschrift für Analytische Chemie*, **54**, 214–216.
43. Roscoe, H. E., and Schorlemmer, C. (1879) *A Treatise on Chemistry. Volume II: Metals*. Part 1. Macmillan and Co., London, UK, 504 pp.
44. Fresenius, R. (1868) Zur Analyse der Superphosphate, *Zeitschrift für Analytische Chemie*, **7**, 304–309.
45. Chesshire, J. A., Hughes, J., Sutton, F., and Sibson, A. (1870) Ueber die Bestimmung des Betrags an "reducirten" Phosphaten in Superphosphaten, *Zeitschrift für Analytische Chemie*, **9**, 524–527 (in reality, this publication represents a set of 4 short independent studies performed by 4 individual authors but combined under a general title).
46. Williams, C. P. (1871) On the solubility of some forms of phosphate of lime, *J. Franklin Inst.*, **92**, 419–423.

47. Rümpler, A. (1873) Ueber eisen- und thonerdehaltige Superphosphate und deren analytische Untersuchung, *Zeitschrift für Analytische Chemie*, **12**, 151–163.

48. Albert, H., and Siegfried, L. (1877) Beiträge zur Werthbestimmung der Superphosphate, *Zeitschrift für Analytische Chemie*, **16**, 182–188.

49. Albert, H., and Siegfried, L. (1879) Beiträge zur Werthbestimmung der Superphosphate, *Zeitschrift für Analytische Chemie*, **18**, 220–224.

50. Pavec, A. (1879) Zur maassanalytischen Bestimmung der Phosphorsäure im Superphosphat und Spodium mittelst Uranlösung, *Zeitschrift für Analytische Chemie*, **18**, 360–361.

51. Mohr, C. (1880) Ein maassanalytisches Bestimmungsverfahren der in Rohphosphaten und Superphosphaten enthaltenen Phosphorsäure mit Uran bei Gegenwart von Eisenoxyd, *Zeitschrift für Analytische Chemie*, **19**, 150–153.

52. Erlenmeyer, E., Wattenberg, H., Wein, E., Rösch, L., Lehmann, J., Johnson, S. W., and Jenkins, E. H. (1880) Zur Analyse der Superphosphate, *Zeitschrift für Analytische Chemie*, **19**, 243–246 (in reality, this publication represents a set of 7 short independent studies performed by 7 individual authors but combined under a general title).

53. Meyer, C. F. (1880) Weitere Mittheilungen über das Zurückgehen der eisen- und thonerdehaltigen Superphosphate–Berichtigung, *Zeitschrift für Analytische Chemie*, **19**, 309–311.

54. Drewsen, S. (1881) Zur Bestimmung der löslichen Phosphorsäure in Superphosphaten, *Zeitschrift für Analytische Chemie*, **20**, 54–57.

55. Mollenda, A. (1883) Eine neue Methode zur maassanalytischen Bestimmung der Phosphorsäure in den Superphosphaten, *Zeitschrift für Analytische Chemie*, **22**, 155–159.

56. Phillips, W. B. (1884) Rate of reversion in superphosphates prepared from red navassa rock, *J. Am. Chem. Soc.*, **6**, 224–228.

57. Wagner, P. (1886) Eine neue Methode zur Feststellung des Handelswerthes der Superphosphate, *Zeitschrift für Analytische Chemie*, **25**, 272–278.

58. Emmerling, A. (1887) Eine Methode zur Bestimmung der wasserlöslichen Phosphorsäure in Superphosphaten auf maassanalvtischem Wege, *Zeitschrift für Analytische Chemie*, **26**, 244–247.

59. Stoklasa, J. (1890) Bestimmung des Wassers in den Superphosphaten. I, *Zeitschrift für Analytische Chemie*, **29**, 390–397.

60. Crispo, D. (1891) Belgische Methode zur Bestimmung der in Wasser löslichen Phosphorsäure in den Superphosphaten, *Zeitschrift für Analytische Chemie*, **30**, 301–303.

61. Güssefeld, O. (1892) Eine Schüttelmaschine für die Analyse von Superphosphaten, *Zeitschrift für Analytische Chemie*, **31**, 556.

62. Keller, A. (1893) Eine Schüttelmaschine für Superphosphaten, *Zeitschrift für Analytische Chemie*, **32**, 590–591.

63. Kalmann, W., and Meissels, K. (1894) Eine Methode zur maassanalytischen Schätzung der wasserlöslichen Phosphorsäure in Superphosphaten, *Zeitschrift für Analytische Chemie*, **33**, 764–766.

64. Glaser, C. (1895) Zur maassanalytischen Bestimmung der wasserlöslichen Phosphorsäure in Superphosphaten, *Zeitschrift für Analytische Chemie*, **34**, 768–769.

65. Seib, O. (1905) Bestimmung der zitratlöslichen Phosphorsäure in Superphosphaten, *Zeitschrift für Analytische Chemie*, **44**, 397–398.

66. Cameron, F. K., and Bell, J. M. (1906) The phosphates of calcium, III: Superphosphate, *J. Am. Chem. Soc.*, **28**, 1222–1229.

67. Wilson, G. (1850) *Chemistry*. William and Robert Chambers, Edinburgh, UK, 316 p.

68. Brande, W. T., and Taylor, A. S. (1863) *Chemistry*. Blanchard and Lea, Philadelphia, USA, 696 p.

69. Wells, H. G. (1906) Pathological calcification, *J. Med. Res.*, **14**, 491–525.

70. C. (1877) Vitreous phosphate of lime, *J. Franklin Inst.*, **104**, 315.

71. Church, A. H. (1873) New analyses of certain mineral arseniates and phosphates. 1. Apatite; 2. Arseniosiderite; 3. Childrenite; 4. Ehlite; 5. Tyrolite; 6. Wavellite, *J. Chem. Soc.*, **26**, 101–111.

72. L., W. J. (1883) On a fine specimen of apatite from Tyrol, lately in the possession of Mr. Samuel Henson, *Nature*, **27**, 608–609.

73. Reichardt, E. (1872) Ueber neutralen phosphorsauren Kalk, Darstellung und Löslichkeit desselben, *Zeitschrift für Analytische Chemie*, **11**, 275–277.

74. Mohr, C. (1884) Ueber die quantitative Bestimmung der zurückgegangenen Phosphorsäure und der Phosphorsäure im Dicalciumphosphat, *Zeitschrift für Analytische Chemie*, **23**, 487–491.

75. Glaser, C. (1885) Bemerkungen zu der Abhandlung des Herrn Carl Mohr über die quantitative Bestimmung der zurückgegangenen Phosphorsäure und der Phosphorsäure im Dicalciumphosphat, *Zeitschrift für Analytische Chemie*, **24**, 180.

76. Hutchings, W. M. (1887) Occurrence of apatite in slag, *Nature*, **36**, 460.

77. Hilgenstock, G. (1883) Eine neue Verbindung von P_2O_5 und CaO, *Stahl und Eisen*, **3**, 498.

78. Hilgenstock, G. (1887) Das vierbasische Kalkphosphat und die Basicitätsstufe des Silicats in der Thomas-Schlacxke, *Stahl und Eisen*, **7**, 557–560.

79. Scheibler C. (1886) Ueber die Herstellung reicher Kalkphosphate in Verbindung mit einer Verbesserung des Thomasprocesses, *Ber. Dtsch. Chem. Ges.*, **19**, 1883–1893.

80. Georgievics, G. V. (1891) Über das Verhalten des Tricalciumphosphats gegen Kohlensäure und Eisenhydroxyd, *Monatshefte für Chemie*, **12**, 566–581.

81. Cameron, F. K., and Hurst, L. A. (1904) The action of water and saline solutions upon certain slightly soluble phosphates, *J. Am. Chem. Soc.*, **26**, 885–913.

82. Cameron, F. K., and Seidell, A. (1904) The action of water upon the phosphates of calcium, *J. Am. Chem. Soc.*, **26**, 1454–1463.

83. Cameron, F. K., and Seidell, A. (1905) The phosphates of calcium. I, *J. Am. Chem. Soc.*, **27**, 1503–1512.

84. Cameron, F. K., and Bell, J. M. (1905) The phosphates of calcium. II, *J. Am. Chem. Soc.*, **27**, 1512–1514.

85. Cameron, F. K., and Bell, J. M. (1910) The phosphates of calcium. IV, *J. Am. Chem. Soc.*, **32**, 869–873.

86. Cameron, F. K., and McCaughey, W. J. (1911) Apatite and spodiosite, *J. Phys. Chem.*, **15**, 463–470.

87. Mebane, W. M., Dobbins, J. T., and Cameron, F. K. (1929) The solubility of the phosphates of calcium in aqueous solutions of sulfur dioxide, *J. Phys. Chem.*, **33**, 961–969.

88. Bassett, H., Jr. (1907) Beiträge zum Studium der Calciumphosphate. I. Die Hydrate der Calcium-Hydroorthophosphate, *Z. Anorg. Chem.*, **53**, 34–48.

89. Bassett, H., Jr. (1907) Beiträge zum Studium der Calciumphosphate. II. Die Einwirkung von Ammoniakgas auf Calcium-Hydroorthophosphate, *Z. Anorg. Chem.*, **53**, 49–62.

90. Bassett, H., Jr. (1908) Beiträge zum Studium der Calciumphosphate. III. Das System $CaO–P_2O_5–H_2O$, *Z. Anorg. Chem.*, **59**, 1–55.

91. Bassett, H., Jr. (1917) The phosphates of calcium. Part IV. The basic phosphates, *J. Chem. Soc.*, **111**, 620–642.

92. Norton, T. H., and Newman, H. E. LV (1897) On a soluble compound of hydrastine with monocalcium phosphate, *J. Am. Chem. Soc.*, **19**, 838–840.

93. Rolfe, B. H. (1911) Autunite (hydrated uranium-calcium phosphate), *Lancet*, **177**, 766.

94. Withers, W. A., and Field, A. L. (1915) A conductivity study of the reaction between calcium nitrate and dipotassium phosphate in dilute solution, *J. Am. Chem. Soc.*, **37**, 1091–1105.

95. Wendt, G. L. (1923) An electrometric study of the neutralization of phosphoric acid by calcium hydroxide, *J. Am. Chem. Soc.*, **45**, 881–887.

96. Shipley, P. G., Kramer, B., and Howland, J. (1926) Studies upon calcification *in vitro*, *Biochem J.*, **20**, 379–387.

97. Benjamin, H. R. (1933) The forms of the calcium and inorganic phosphorus in human and animal sera II. The nature and significance of the filtrable, adsorbable calcium-phosphorus complex, *J. Biol. Chem.*, **100**, 57–78.

98. Náray-Szabó, S. (1931) The structure of apatite $(CaF)Ca_4(PO_4)_3$, *Zeitschrift für Kristallographie*, **75**, 387–398.

99. Hendricks, S. B., Hill, W. L., Jacob, K. D., and Jefferson, M. E. (1931) Structural characteristics of apatite-like substances and composition of phosphate rock and bone as determined from microscopical and X-ray diffraction, *Ind. Eng. Chem.*, **23**, 1413–1418.

100. Barrio, N. G. (1923) Comparative studies in the chemistry of blood and cerebrospinal fluid. II. Calcium, magnesium, and phosphorus, *J. Labor. Clin. Med.*, **9**, 54–56.

101. Wang, C. C., and Felsher, A. R. (1925) The effect of hemolysis on the calcium and inorganic phosphorus content of serum and plasma, *J. Labor. Clin. Med.*, **10**, 269–272.

102. Nitschke, A. (1929) Darstellung Einer den Calciumgehalt und Einer den Phosphatgehalt des Serum Senkenden Substanz — II. Mitteilung. Nachweis in Milz und Lymphknoten. Klinische Wochenschrift, **8**, 794.

103. Krasnow, F., Karshan, M., and Krejci, L. E. (1932) The determination of calcium and phosphorus in saliva, *J. Labor. Clin. Med.*, **17**, 1148–1152.

104. Clark, N. A. (1931) The system P_2O_5–CaO–H_2O and the recrystallization of monocalcium phosphate, *J. Phys. Chem.*, **35**, 1232–1238.

105. Lugg, J. W. H. (1931) A study of aqueous salt solutions in equilibrium with solid secondary calcium phosphate at 40°C, *Trans. Faraday Soc.*, **27**, 297–309.

106. Nagelschmidt, G. (1937) A new calcium silicophosphate, *J. Chem. Soc.* 865–867.

107. Moore, L. A. (1942) Activation of dicalcium phosphate for the chromatographic determination of carotene, *Ind. Eng. Chem.*, **14**, 707–708.

108. Elsenberger, S., Lehrman, A., and Turner, W. D. (1940) The basic calcium phosphates and related systems. Some theoretical and practical aspects, *Chem. Rev.*, **26**, 257–296.

109. Elmore, K. L., and Farr, T. D. (1940) Equilibrium in the system calcium oxide–phosphorus pentoxide–water, *Ind. Eng. Chem.*, **32**, 580–586.

110. Shear, M. J., and Kramer, B. (1928) Composition of bone. III. Physicochemical mechanism, *J. Biol. Chem.*, **79**, 125–145.

111. Trömel, G., and Möller, H. (1932) Die Bildung schwer löslicher Calciumphosphate aus wäßriger Lösung und die Beziehungen dieser Phosphate zur Apatitgruppe, *Zeitschrift für anorganische und allgemeine Chemie*, **206**, 227–240.

112. Larson, H. W. E. (1935) Preparation and properties of mono-, di-, and tricalcium phosphates, *Ind. Eng. Chem. Anal. Ed.*, **7**, 401–406.

113. Shaler, N. S. (1889) Nature and origin of phosphate of lime, *Science*, **13**, 144–148.

114. Graber, H. XV (1895) Diopsid und Apatit von Zöptau, *Tschermaks Mineralogische und Petrographische Mitteilungen*, **14**, 265–270.

115. Church, A. H. (1901) Apatite in Ceylon, *Nature*, **63**, 464.

116. Klein, C. (1901) Brushite from the island of Mona (between Haiti and Puerto Rico), *Sitzber. K. Preuss. Aka.* 720–725.

117. Wolff, J. E., and Palache, C. (1902) Apatite from Minot, Maine, *Proc. Am. Acad. Arts Sci.*, **37**, 517–528.

118. Wolff, J. E., and Palache, C. (1902) On apatite from Minot, Maine, *Zeitschrift für Krystallographie und Mineralogie*, **36**, 438–448.

119. de Mercey, N. (1902) Sur des gites de phosphate de chaux de la Craie à Bélemnites, formés avant le soulèvement du Bray, *Comptes Rendus Hebdomadaires des Seances de l'Academie des Sciences.*, **135**, 1137–1138.

120. Zimanyi, K. (1904) On the green apatite from Malmberget in Sweden, *Zeitschrift für Krystallographie und Mineralogie*, **39**, 505–519.

121. Fraps, G. S. (1906) Availability of phosphoric acid of the soil, *J. Am. Chem. Soc.*, **28**, 823–834.

122. Davy, H. (1814) Conversations on chemistry, in which the elements of that science are familiarly explained and illustrated by experiments and plates. To which are added, some late discoveries on the subject of the fixed alkalies. Didnep's Press, 383 p.

123. Nasse, H. (1842) Ueber die Bestandtheile der Knochen in einigen Krankheiten, *J. Prakt. Chem.*, **27**, 274–280.

124. Levy, M. (1895) Chemische Untersuchungen über osteomalacische Knochen, *Zeitschrift für physiologische Chemie*, **19**, 239–270.

125. Hotz, G. (1906) Phosphorsäure- und Kalkstoffwechsel bei Osteomalacie unter dem Einfluss der Phosphortherapie, *Zeitschrift für Experimentelle Pathologie und Therapie*, **3**, 605–632.

126. Gaßmann, T. (1910) Chemische Untersuchungen von gesunden und rachitischen Knochen, *Zeitschrift für physiologische Chemie*, **70**, 161–170.

127. Aeby, C. (1872) Ueber die Constitution des phosphorsauren Kalkes der Knochen, *J. Prakt. Chem.*, **5**, 308–311.

128. Aeby, C. (1873) Ueber die nähern Bestandtheile des Knochenphosphates, *J. Prakt. Chem.*, **6**, 169–171.

129. Aeby, C. (1873) Ueber die Metamorphose der Knochen, *J. Prakt. Chem.*, **7**, 37–44.

130. Wibel, F. (1874) Die Constitution des Knochenphosphates, insbesondere die Existenz und Bildung einer basischen Verbindung ($Ca_3P_2O_8$+xCaO), *J. Prakt. Chem.*, **9**, 113–132.

131. Weiske, H. (1883) Beitrag zur Knochenanalyse, *Zeitschrift für Physiologische Chemie*, **7**, 474–478.

132. Jones, H. B. II (1862) On the occurrence of deposits of crystallised phosphate of lime in human urine, *J. Chem. Soc.*, **15**, 8–15.

133. Hassal (1862) Phosphorsaurer Kalk im Urinsediment, *Zeitschrift für Analytische Chemie*, **1**, 125.

134. Moss (1879) Ueber das Vorkommen und die Bedeutung phosphorsaurer Kalkconcremente im Stamme des Gehörnerven, *Archiv für Psychiatrie und Nervenkrankheiten*, **9**, 122–128.

135. Zuckmayer, F. (1914) Über die Frauenmilch der ersten Laktationszeit und den Einfluss einer Kalk- und Phosphorsäurezulage auf ihre Zusammensetzung, *Pflüger's Archiv für die Gesammte Physiologie des Menschen und der Tiere*, **158**, 209–218.

136. Meigs, E. B., Blatherwick, N. R., and Cary, C. A., with the collaboration of Woodward, T. E. (1919) Further contributions to the physiology of

phosphorus and calcium metabolism of dairy cows, *J. Biol. Chem.*, **40**, 469–500.

137. de Wesselow, O. L. V. (1922) The calcium and inorganic phosphorus content of the maternal blood during pregnancy and lactation, *Lancet*, **200**, 227–228.

138. Schneider, J. Z. (1930) The calcium to phosphorus ratio as related to mineral metabolism, *Int. J. Orthodontia Oral Surg. Radiography*, **16**, 277–285.

139. Kramer, B., and Shear, M. J. (1928) Composition of bone. IV. Primary calcification, *J. Biol. Chem.*, **79**, 147–160.

140. Under the care of Mr. Canton (1856) Stone in the bladder of a boy for upwards of three years; lithotomy; extraction of a large oval calculus of phosphate of lime; recovery, *Lancet*, **68**, 650–651.

141. Virchow, R. (1857) Untersuchungen über die Entwicklung des Schädelgrundes im gesunden und krankhaften Zustande und über den Einfluss derselben auf Schädelform, Gesichtsbildung und Gehirnbau, *Druck und Verlag von Georg Reimer, Berlin*, 128 p.

142. Jones, S. (1882) St. Thomas's hospital.: Lateral lithotomy; removal of large stone composed of oxalate of lime covered with phosphates; recovery, *Lancet*, **119**, 781–782.

143. Hill, B. (1883) University college hospital.: Two cases of calculus of oxalate and phosphate of lime removed by lithotrity at a single sitting; singularly long duration of symptoms, with remarks upon an improved clover's evacuating bottle, *Lancet*, **121**, 59–60.

144. Fisher, T. (1901) A singular case of extensive deposit of phosphate of lime in the lungs, *Lancet*, **157**, 244–245.

145. Moody, D. W. K. (1913) Note on a case of a large concretion composed mainly of calcium phosphate found inside the fistulous channel in a case of fistula in ano, *Lancet*, **181**, 684.

146. Tereg, A. (1883) Das Verhalten der Calciumphosphate im Organismus der Fleischfresser. Pflüger, Archiv für die Gesammte Physiologie des Menschen und der Thiere, *Pflügers Archiv Eur. J. Physiology*, **32**, 122–170.

147. Zuckmayer, F. (1912) Beitrag zur Aufnahme und Verwertung von Kalk und Phosphorsäure durch den Darm, *Pflüger's Archiv für die Gesammte Physiologie des Menschen und der Tiere*, **148**, 225–256.

148. Blühdorn, K. (1921) Untersuchungen des Kalk- und Phorphorsäurestoffwechsels bei Verabfolgung großer Gaben von Kalk und Natriumphosphat, *Zeitschrift für Kinderheilkunde*, **29**, 43–55.

149. Herbst, O. (1913) Calcium und Phosphor beim Wachstum am Ende der Kindheit, *Zeitschrift für Kinderheilkunde*, **7**, 161–192.

150. Lowy, R. (1930) The necessity of normal calcium phosphorus metabolism, *Int. J. Orthodontia, Oral Surgery and Radiography*, **16**, 947–950.

151. Schmidt, C. L. A., and Greenberg, D. M. (1935) Occurrence, transport and regulation of calcium, magnesium and phosphorus in animal organism, *Physiol. Rev.*, **15**, 297–434.

152. Huggins, C. (1937) The composition of bone and the function of the bone cell, *Physiol. Rev.*, **17**, 119–143.

153. Beneke, W. (1851) On the physiology and pathology of the phosphate and oxalate of lime, and their relation to the formation of cells. Part I, *Lancet*, **57**, 431–434.

154. Beneke, W. (1851) On the physiology and pathology of the phosphate and oxalate of lime, and their relation to the formation of cells. Part II, *Lancet*, **57**, 668–670 and 699–700.

155. Blanc, H. (1874) On the treatment of phthisis by the phosphate of lime and the juice of raw meat, *Lancet*, **103**, 831–832.

156. Fessenden, R. A. (1913) Cancer and secondary rays from calcium phosphate, *Lancet*, **182**, 1804.

157. Thomas, S., Assi, P., Marycel, B., Correa, M., Liberato, W., and Brito, V. (2008) Yttrium 90-hydroxyapatite, a new radioisotope for chronic synovitis in hemophilia, *Haemophilia*, **14**, 77.

158. Chinol, M., Vallabhajosula, S., Goldsmith, S. J., Klein, M. J., Deutsch, K. F., Chinen, L. K., Brodack, J. W., Deutsch, E. A., Watson, B. A., and Tofe, A. J. (1993) Chemistry and biological behavior of samarium-153 and rhenium-186-labeled hydroxyapatite particles: potential radiopharmaceuticals for radiation synovectomy, *J. Nucl. Med.*, **34**, 1536–1542.

159. Argüelles, M. G., Berlanga, I. S. L., and Torres, E. A. (1999) Preparation of ^{153}Sm-particles for radiosynovectomy, *J. Radioanal. Nucl. Chem.*, **240**, 509–511.

160. O'Duffy, E., Clunie, G., Lui, D., Edwards, J., and Ell, P. (1999) Double blind glucocorticoid controlled trial of samarium-153 particulate hydroxyapatite radiation synovectomy for chronic knee synovitis, *Ann. Rheum. Dis.*, **58**, 554–558.

161. de Franciscis, G. (1953) Bicalcium phosphate as adjunct in the feeding of domestic animals. [Il fosfato bicalcico quale integrante

dell'alimentazione del bestiame.], *Bollettino della Società italiana di biologia sperimentale*, **29**, 1794–1796.

162. Ratner, B. D. (2004) A history of biomaterials. In: *Biomaterials Science: An Introduction to Materials in Medicine*. Ratner, B. D., Hoffman, A. S., Schoen, F. J., and Lemons, J. E. (Eds.), 2nd Ed. Academic Press, San Diego, CA, USA. pp. 10–19.

163. Huebsch N., and Mooney, D. J. (2009) Inspiration and application in the evolution of biomaterials, *Nature*, **462**, 426–432.

164. Popp. H. (1939) Zur Geschichte der Prosthesen, *Med. Welt*, **13**, 961–964.

165. Ring, M. E. (1992) *Dentistry: An Illustrated History*. Harry N. Abrams, New York, USA, 320 pp.

166. Bobbio, A. (1972) The first endosseous alloplastic implant in the history of man, *Bull. Hist. Dent.*, **20**, 1–6.

167. Crubezy, E., Murail, P., Girard, L., and Bernadou, J. P. (1998) False teeth of the Roman world, *Nature*, **391**, 29.

168. Hunter, J. (1778) *The Natural History of the Human Teeth*. Printed for J. Johnson, London, 256 pp.

169. Cravens, J. E. (1876) Lacto-phosphate of lime: pathology and treatment of exposed dental pulps and sensitive dentine, *Dent. Cosmos*, **18**, 463–469; discussion 469–476.

170. Gabriel, S. (1894) Chemische Untersuchungen über die Mineralstoffe der Knochen und Zähne, *Zeitschrift für physiologische Chemie*, **18**, 257–303.

171. de Jong, W. F. (1926) La substance minérale dans les os, *Recueil des Travaux Chimiques des Pays-Bas*, **45**, 445–448.

172. Henschen, C., Straumann, R., and Bucher, R. (1932) Ergebnisse röntgenspektrographischer Untersuchungen am Knochen — I. Mitteilung. Krystallitbau des anorganischen und des organischen Knochens, *Deutsche Zeitschrift für Chirurgie*, **236**, 485–514.

173. Bredig, M. A. (1933) Zur Apatitstruktur der anorganischen Knochen- und Zahnsubstanz. Hoppe-Seylers, *Zeitschrift für Physiologische Chemie*, **216**, 239–243.

174. Weinberger, B. W. (1948) *An Introduction to the History of Dentistry with Medical and Dental Chronology and Bibliographic Data*. Two volumes. The C. V. Mosby Company, St. Louis, USA, 992 pp.

175. Baden, E. (1955) Prosthetic therapy of congenital and acquired clefts on the palate: an historical essay, *J. Hist. Med. Alld. Sci.*, **10**, 290–301.

176. Ratner, B. D., and Bryant, S. J. (2004) Biomaterials: where we have been and where we are going, *Annu. Rev. Biomed. Eng.*, **6**, 41–75.

177. Dammaschke, T. (2008) The history of direct pulp capping, *J. Hist. Dentistry*, **56**, 9–23.

178. Dreesmann, H. (1892) Ueber Knochenplombierung, *Beitr. Klin. Chir.*, **9**, 804–810.

179. Gluck, T. (1891) Referat über die durch das moderne chirurgische, *Langenbecks Archiv für Klinische Chirurgie*, **41**, 187–239.

180. Muster, D. (1990) Themistocles Gluck, Berlin 1890: a pioneer of multidisciplinary applied research into biomaterials for endoprostheses, *Bull. Hist. Dent.*, **38**, 3–6.

181. Eynon-Lewis, N. J., Ferry, D., and Pearse, M. F. (1992) Themistocles Gluck: an unrecognised genius, *BMJ*, **305**, 1534–1536.

182. Albee, F. H. (1915) *Bone-Graft Surgery*. W. B. Saunders Company, Philadelphia and London, 417 pp.

183. Hey Groves, E. W. (1927) Some contributions to the reconstructive surgery of the hip, *Br. J. Surg.*, **14**, 486–517.

184. Murray, C. R. (1931) Delayed and non-union in fractures in the adult, *Ann. Surg.*, **93**, 961–967.

185. Murray, C. R. (1931) The modern conception of bone formation and its relation to surgery, *J. Dent. Res.*, **11**, 837–845.

186. Huggins, C. (1931) The formation of bone under the influence of epithelium of the urinary tract, *Arch. Surg.*, **22**, 377–408.

187. Levander, G. (1934) On the formation of new bone in bone transplantation, *Acta Chir. Scand.*, **74**, 425–426.

188. Levander, G. (1938) A study of bone regeneration, *Surg. Gynecol. Obstet.*, **67**, 705–714.

189. Haldeman, K. O., and Moore, J. M. (1934) Influence of a local excess of calcium and phosphorus on the healing of fractures. An experimental study, *Arch. Surg.*, **29**, 385–396.

190. Stewart, W. J. (1934) Experimental bone regeneration using lime salts and autogenous grafts as sources of available calcium, *Surg. Gynec. Obstet.*, **59**, 867–871.

191. Key, J. A. (1934) The effect of a local calcium depot on osteogenesis and healing of fractures, *J. Bone Joint Surg.*, **16**, 176–184.

192. Shands, A. R., Jr. (1937) Studies in bone formation: the effect of the local presence of calcium salts on osteogenesis, *J. Bone Joint Surg.*, **19**, 1065–1076.

193. Schram, W. R., and Fosdick, L. S. (1948) Stimulation of healing in long bones by use of artificial material, *J. Oral Surg.*, **6**, 209–217.

194. Ray, R. D., and Ward, A. A., Jr. (1951) A preliminary report on studies of basic calcium phosphate in bone replacement, *Surg. Form.*, **3**, 429–434.

195. Kingery, W. D. II (1950) Cold-setting properties, *J. Am. Ceram. Soc.*, **33**, 242–246.

196. Nery, E. B., Lynch, K. L., Hirthe W. M., and Mueller, K. H. (1975) Bioceramic implants in surgically produced infrabony defects, *J. Periodontol.*, **46**, 328–347.

197. Denissen, H. W., and de Groot, K. (1979) Immediate dental root implants from synthetic dense calcium hydroxylapatite, *J. Prosthet. Dent.*, **42**, 551–556.

198. Blakeslee, K. C., and Condrate, R. A., Sr. (1971) Vibrational spectra of hydrothermally prepared hydroxyapatites, *J. Am. Ceram. Soc.*, **54**, 559–563.

199. Garrington, G. E., and Lightbody, P. M. (1972) Bioceramics and dentistry, *J. Biomed. Mater. Res.*, **6**, 333–343.

200. Cini, L., Sandrolini, S., Paltrinieri, M., Pizzoferrato, A., and Trentani, C. (1972) Materiali bioceramici in funzione sostitutiva. Nota preventiva. (Bioceramic materials for replacement purposes. Preliminary note.), *La Chirurgia Degli Organi Di Movimento*, **60**, 423–430.

201. Rivault, M. A. (1966) Evolution, conception et technologie des travaux de prothèse fixe, réalisés en céramo-métallique. (Evolution, conception and technology of fixed prosthesis made of ceramic and metal), *Revue Francaise d'Odonto-Stomatologie*, **13**, 1367–1402.

202. Dumont, A., Appel, M., and Favard, E. (1968) Prothèses plurales en céramique sur métal. Soudage et artifices de jonction. (Multiple prostheses made of ceramics on metal. Soldering and artifacts of the junction), *Annales Odonto-Stomatologiques*, **25**, 231–240.

203. Hulbert, S. F., Young, F. A., Mathews, R. S., Klawitter, J. J., Talbert, C. D., and Stelling, F. H. (1970) Potential of ceramic materials as permanently implantable skeletal prostheses, *J. Biomed. Mater. Res.*, **4**, 433–456.

204. Hench, L. L., Splinter, R. J., Allen, W. C., and Greenlee, T. K. (1971) Bonding mechanisms at the interface of ceramic prosthetic materials, *J. Biomed. Mater. Res.*, **2**, 117–141.

205. Paital, S. R., and Dahotre, N. B. (2009) Calcium phosphate coatings for bio-implant applications: materials, performance factors, and methodologies, *Mater. Sci. Eng. R*, **66**, 1–70.

206. León, B., and Jansen, J. A. (Eds.) (2009) *Thin Calcium Phosphate Coatings for Medical Implants*. Springer, New York, USA, 326 pp.

207. Ducheyne, P., Hench, L. L., Kagan II, Martens, A., Bursens, A., and Mulier, J. C. (1980) Effect of hydroxyapatite impregnation on skeletal bonding of porous coated implants, *J. Biomed. Mater. Res.*, **14**, 225–237.

208. Jarcho, M., Bolen, C. H., Thomas, M. B., Bobick, J., Kay, J. F., and Doremus, R. H. (1976) Hydroxylapatite synthesis and characterization in dense polycrystalline form, *J. Mater. Sci.*, **11**, 2027–2035.

209. Jarcho, M., O'Connor, J. R., and Paris, D. A. (1977) Ceramic hydroxylapatite as a plaque growth and drug screening substrate, *J. Dental Res.*, **56**, 151–156.

210. Jarcho, M., Salsbury, R. L., Thomas, M. B., and Doremus, R. H. (1979) Synthesis and fabrication of β-tricalcium phosphate ceramics for potential prosthetic applications, *J. Mater. Sci.*, **14**, 142–150.

211. Jarcho, M. (1981) Calcium phosphate ceramics as hard tissue prosthetics, *Clin. Orthop. Rel. Res.*, **157**, 259–278.

212. Rejda, B. V., Peelen, J. G. J., and de Groot, K. (1977) Tricalcium phosphate as a bone substitute, *J. Bioeng.*, **1**, 93–97.

213. de Groot, K. (1980) Bioceramics consisting of calcium phosphate salts, *Biomaterials*, **1**, 47–50.

214. de Groot, K. (Ed.) (1983) *Bioceramics of Calcium Phosphate*. CRC Press, Boca Raton, Fl, 146 pp.

215. Aoki, H., Kato, K. M., Ogiso, M., and Tabata, T. (1977) Studies on the application of apatite to dental materials, *J. Dent. Eng.*, **18**, 86–89.

216. Kato, K., Aoki, H., Tabata, T., and Ogiso, M. (1979) Biocompatibility of apatite ceramics in mandibles, *Biomater. Med. Dev. Artif. Organs*, **7**, 291–297.

217. Akao, M., Aoki, H., and Kato, K. (1981) Mechanical properties of sintered hydroxyapatite for prosthetic applications, *J. Mater. Sci.*, **16**, 809–812.

218. Akao, M., Aoki, H., Kato, K., and Sato, A. (1982) Dense polycrystalline β-tricalcium phosphate for prosthetic applications, *J. Mater. Sci.*, **17**, 343–346.

219. Roy, D. M., and Linnehan, S. K. (1974) Hydroxyapatite formed from coral skeletal carbonate by hydrothermal exchange, *Nature*, **247**, 220–222.

220. Holmes, R. E. (1979) Bone regeneration within a coralline hydroxyapatite implant, *Plast. Reconstr. Surg.*, **63**, 626–633.

221. Elsinger, E. C., and Leal, L. (1996) Coralline hydroxyapatite bone graft substitutes, *J. Foot Ankle Surg.*, **35**,396–399.

222. Shipman, P., Foster, G., and Schoeninger, M. (1984) Burnt bones and teeth: an experimental study of color, morphology, crystal structure and shrinkage, *J. Archaeological Sci.*, **11**, 307–325.

223. LeGeros, R. Z., and LeGeros, J. P. (2003) Calcium phosphate bioceramics: past, present, future, *Key Eng. Mater.*, **240–242**, 3–10.

Index